ORIGIN AND EVOLUTION
OF VIRUSES

ORIGIN AND EVOLUTION OF VIRUSES

Edited by

ESTEBAN DOMINGO
Centro de Biología Molecular
"Severo Ochoa"
Universidad Autónoma de Madrid
28049 Madrid
Spain

ROBERT WEBSTER
St Jude's Children's Research Hospital
Memphis, TN 38105-2794
USA

JOHN HOLLAND
University of California
San Diego, CA 92093-0116
USA

ACADEMIC PRESS

San Diego   London   Boston   New York   Sydney   Tokyo   Toronto

This book is printed on acid-free paper.

Copyright © 1999 by ACADEMIC PRESS

All Rights Reserved.
No part of this publication may be reproduced or transmitted in any form or by any
means, electronic or mechanical, including photocopying, recording, or any
information storage and retrieval system, without permission in writing from the
publisher.

Academic Press
24–28 Oval Road, London NW1 7DX, UK
http://www.hbuk.co.uk/ap/

Academic Press
a division of Harcourt Brace & Company
525 B Street, Suite 1900, San Diego, California 92101-4495, USA
http://www.apnet.com

ISBN 0-12-220360-7

A catalogue for this book is available from the British Library

Library of Congress Catalog Card Number: 99-62165

Typeset by Phoenix Photosetting, Chatham, Kent
Printed in Great Britain by The Bath Press, Bath

99 00 01 02 03 04 BP 9 8 7 6 5 4 3 2 1

Contents

Contributors

Christof K. Biebricher
*Max-Planck-Institut für Biophysikalische Chemie,
Karl-Friedrich-Bonhoeffer-Institut, Am Fassberg 11,
D-37077, Göttingen, Germany*

William F. Carman
*Institute of Virology, University of Glasgow, Glasgow
G11 5GR, UK*

John M. Coffin
*Department of Molecular Biology and Microbiology,
Tufts University School of Medicine, 136 Harrison
Avenue, Boston, MA 02111, USA*

Andrew J. Davison
*MRC Virology Unit, Church Street, Glasgow G11 5JR,
UK*

Esteban Domingo
*Centro de Biología Molecular 'Severo Ochoa', Consejo
Superior de Investigaciones Científicas, Universidad
Autónoma de Madrid, 28049 Madrid, Spain*

N. Duran-Vila
*Istituto Valenciano de Investigaciones Agrarias,
Moncada (Valencia), Spain*

Cristina Escarmís
*Centro de Biología Molecular 'Severo Ochoa', Consejo
Superior de Investigaciones Científicas, Universidad
Autónoma de Madrid, 28049 Madrid, Spain*

Juan I. Esteban
*Area d'Investigació Basica, Hospital General Vall
d'Hebron, Passeig Vall d'Hebron, 119–129, 08035
Barcelona, Spain*

F. García-Arenal
*Departamento de Biotecnología, E.T.S.I. Agrónomos,
Universidad Politécnica de Madrid, 28040 Madrid,
Spain*

Adrian J. Gibbs
*Research School of Biological Sciences, Australian
National University, PO Box 475, Canberra,
ACT 2601, Australia*

M. J. Gibbs
*Research School of Biological Sciences, Australian
National University, PO Box 475, Canberra,
ACT 2601, Australia*

Jordi Gómez
*Area d'Investigació Basica, Hospital General Vall
d'Hebron, Passeig Vall d'Hebron, 119–129, 08035
Barcelona, Spain*

Alexander E. Gorbalenya
*Advanced Biomedical Computing Center, 430 Miller
Drive, Room 235, SAIC/NCI-FCRDC, PO Box B,
Frederick, MD 21702-1201, USA*

Matthias Gromeier
*Department of Molecular Genetics and Microbiology,
School of Medicine, State University of New York at
Stony Brook, Stony Brook, NY 11794-5222, USA*

John J. Holland
*Department of Biology and Center for Molecular
Genetics, University of California, San Diego, La Jolla,
CA 92093-0116, USA*

P. L. Keese
*CAMBIA, PO Box 3200 Canberra, ACT 2601,
Australia*

Marcella A. McClure
Department of Biological Sciences, University of Nevada, 4505 Maryland Parkway, Box 454004 Las Vegas, NV 89145-4004, USA

Duncan J. McGeoch
MRC Virology Unit, Church Street, Glasgow G11 5JR, UK

Maria Martell
Area d'Investigació Basica, Hospital General Vall d'Hebron, Passeig Vall d'Hebron, 119–129, 08035 Barcelona, Spain

Luis Menéndez-Arias
Centro de Biología Molecular 'Severo Ochoa', Consejo Superior de Investigaciones Científicas, Universidad Autónoma de Madrid, 28049 Madrid, Spain

Andreas Meyerhans
Abteilung Virologie, Institut für Medizinische Mikrobiologie und Hygiene, Klinikum Homburg, Universität des Saarlandes, 66421 Homburg/Saar, Germany

Olivia D. Neel
Department of Biochemistry, Weill Medical College of Cornell University, 1300 York Avenue, New York, NY 10021, USA

Martin A. Nowak
Institute for Advanced Study, Olden Lane, Princeton, NJ 08540, USA

Colin R. Parrish
James A. Baker Institute, College of Veterinary Medicine, Cornell University, Ithaca, NY 14853, USA

Hugh D. Robertson
Department of Biochemistry, Weill Medical College of Cornell University, 1300 York Avenue, New York, NY 10021, USA

Igor M. Rouzine
Department of Molecular Biology and Microbiology, Tufts University, 136 Harrison Avenue, Boston MA 02111, USA

Monica Sala
Unité de Rétrovirologie Moléculaire, Institut Pasteur, 28 rue du Dr Roux, 75724, Paris cedex 15, France

José Salas
Centro de Biología Molecular 'Severo Ochoa', Consejo Superior de Investigaciones Científicas, Universidad Autónoma de Madrid, 28049 Madrid, Spain

María L. Salas
Centro de Biología Molecular 'Severo Ochoa', Consejo Superior de Investigaciones Científicas, Universidad Autónoma de Madrid, 28049 Madrid, Spain

Peter Schuster
Institut für Theoretische Chemie und Molekulare Strukturbiologie, Universität Wien, Währingerstraße 17, A-1090 Vienna, Austria and Santa Fé Institute, Santa Fé, NM 87501, USA

J. S. Semancik
Department of Plant Pathology, University of California, Riverside, CA 92521-0122, USA

Peter F. Stadler
Institut für Theoretische Chemie und Molekulare Strukturbiologie, Universität Wien, Währingerstraße 17, A-1090 Vienna, Austria and Santa Fé Institute, Santa Fé, NM 87501, USA

Uwe Truyen
Institut für Medizinische Mikrobiologie, Ludwig-Maximiliens-Universität, Veterinärstr. 13, 80539 Munich, Germany

Jean-Pierre Vartanian
Unité de Rétrovirologie Moléculaire, Institut Pasteur, 28 rue du Dr Roux, 75725 Paris cedex 15, France

Luis P. Villarreal
Irvine Research Unit on Animal Viruses, Department of Molecular Biology and Biochemistry, 3232 Biological Science 2, University of California, Irvine, CA 92697, USA

Eladio Viñuela
Centro de Biología Molecular 'Severo Ochoa', Consejo Superior de Investigaciones Científicas, Universidad Autónoma de Madrid, 28049 Madrid, Spain

Simon Wain-Hobson
Unité de Rétrovirologie Moléculaire, Institut Pasteur, 28 rue du Dr Roux, 75724 Paris cedex 15, France

Robert G. Webster
Department of Virology and Molecular Biology, St Jude Children's Research Hospital, PO Box 318, 332 North Lauderdale, Memphis, TN 38105-2794, USA

Eckard Wimmer
Department of Molecular Genetics and Microbiology, School of Medicine, State University of New York at Stony Brook, 280 Life Sciences Building, Stony Brook, NY 11794-5222, USA

Dominik Wodarz
Institute for Advanced Study, Olden Lane, Princeton, NJ 08540, USA

Preface

Viruses differ greatly in their molecular strategies of adaptation to the organisms they infect. RNA viruses utilize continuous genetic change as they explore sequence space to improve their fitness, and thereby to adapt to the changing environments of their hosts. Variation is intimately linked to their disease-causing potential. Paramount to the understanding of RNA viruses is the concept of quasispecies, first developed to describe the early replicons thought to be components of a primitive RNA world devoid of DNA or proteins. The first chapters of the book deal with theoretical concepts of self-organization, RNA-mediated catalysis and the adaptive exploration of sequence space by RNA replicons. Likely descendants of the RNA world that we can study today are the plant-infecting viroids, and the δ agent (hepatitis D), a unique RNA genome associated with some cases of hepatitis B infection. δ provides an example of a simple, bifunctional molecule that contains a viroid-like replication domain, and a minimal protein-coding domain. It may be a relic of the type of recombinant molecules that may have participated in the transition to the DNA world from the RNA world. The impact of genetic variability of pathogenic RNA viruses is addressed in several chapters that cover specific viruses of animals and plants.

Retroid agents probably had an essential role in early evolution. Not only are they widely distributed and capable of copying RNA into DNA, but they may also have provided regulatory elements, and promoted genetic modifications for adaptation of DNA genomes. Among the retroelements, retroviruses are transmitted as RNA-containing particles, prior to intracellular copying of their RNA genomes into DNA, which can be stably maintained as an insert into the DNA of their hosts. The book discusses retroid agents and retroviruses, with emphasis on human immunodeficiency virus, the most thoroughly scrutinized retrovirus of all. Experiments and modeling meet to try to understand how variation and adaptation of this dreaded pathogen lead to a collapse of the human immune system.

DNA viruses are likely to have coevolved with their hosts while the DNA world was developing. The last chapters of the book deal with the interplay between host evolution and DNA virus evolution, including chapters on the simplest and the most complex of the DNA viral genomes known. This broad coverage of topics would not have been possible without the contributions of many experts. We express our most sincere gratitude to all of these authors for having joined in the effort. The strong interdisciplinary flavor of the book is due to their different points of view. We expect the book to take the reader on a long journey (in time and in concepts) from the primitive and basic to the modern and complex.

While this book was in press, Professor Eladio Viñuela passed away on March 9, 1999. Eladio was an outstanding scientist, a pioneer of Virology in Spain, and a friend. The editors dedicate this volume to his memory.

E. Domingo, R.G. Webster, J.J. Holland

1

Nature and Evolution of Early Replicons

Peter Schuster and Peter F. Stadler

SIMPLE REPLICONS AND THE ORIGIN OF REPLICATION

A large number of successful experimental studies that tried to work out plausible chemical scenarios for the origin of early *replicons*, being molecules capable of replication, have been conducted in the past (Mason, 1991). A sketch of such a possible sequence of events in prebiotic evolution is shown in Figure 1.1. Most of the building blocks of present-day biomolecules are available from different prebiotic sources, from extraterrestrial origins as well as from processes taking place in the primordial atmosphere or near hot vents in deep oceans. Condensation reactions and polymerization reactions formed non-instructed polymers, for example random oligopeptides of the protenoid type (Fox and Dose, 1977).

Template catalysis opens the door to molecular copying and self-replication. Several small templates were designed by Julius Rebek and co-workers: these molecules indeed show complementarity and undergo self-replication (see, for example, Tjivikua *et al.*, 1990; Nowick *et al.*, 1991). Like nucleic acids they consist of a backbone whose role is to bring "molecular digits" in sterically appropriate positions, so that they can be read by their complements. Complementarity is also based on essentially the same principle as in nucleic acids: specific patterns of hydrogen bonds allow recognition of complementary digits and discrimination between "letters" of an alphabet. The hydrogen bonding pattern in these model replicons may be assisted by opposite electric charges carried by the complements. We shall encounter the same principle later in the discussion of Ghadiri's replicons based on stable coiled coils of oligopeptide α-helices (Lee *et al.*, 1996). Autocatalysis in small model systems is certainly interesting because it reveals some mechanistic details of molecular recognition. These systems are, however, highly unlikely to be the basis of biologically significant replicons because they cannot be extended to large polymers in a simple manner and hence they are unsuitable for storing a sizeable amount of (sequence) information. Ligation of small pieces to larger units, on the other hand, is a source of combinatorial complexity providing sufficient capacity for information storage and, hence, evolution. Heteropolymer formation thus seems inevitable and we shall therefore focus on replicons that have this property: nucleic acids and proteins.

A first major transition leads from a world of simple chemical reaction networks to autocatalytic processes that are able to form self-organized systems, which are capable of replication and mutation as required for darwinian evolution. This transition can be seen as the interface between chemistry and biology since an early darwinian scenario is tantamount to the onset of biological evolution. Two suggestions were made in this context: (1) autocatalysis arose in a network of reactions catalyzed by oligopeptides (Kauffman, 1993); and (2) the first autocatalyst was a representative of a class of molecules with

Origin and Evolution of Viruses
ISBN 0–12–220360-7

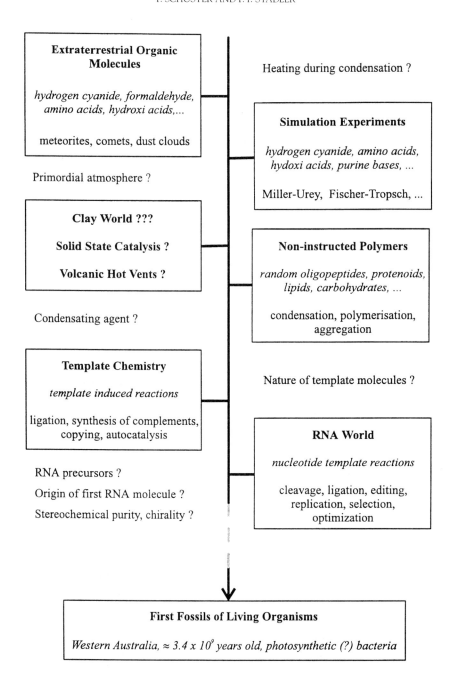

FIGURE 1.1 The RNA world. The concept of a precursor world preceding present-day genetics based on DNA, RNA and protein is based on the idea that RNA can act as both a means of storage of genetic information and a specific catalyst for biochemical reactions. An RNA world is the first scenario on the route from prebiotic chemistry to present-day organisms that allows for darwinian selection and evolution. Problems and open questions are indicated by question marks. Little is known about further steps (not shown here explicitly) from early replicons to the first cells (Eigen and Schuster, 1982; Maynard Smith and Szathmáry, 1995).

"obligatory" template function (Eigen, 1971; Orgel, 1987). The first suggestion works with molecules that are easily available under prebiotic conditions but lacks plausibility because the desired properties, conservation and propagation of mutants, are unlikely to occur with oligopeptides. The second concept suffers from opposite reasons: it is very hard to obtain the first nucleic-acid-like molecules but they would fulfill all functional requirements.

Until the 1980s, biochemists had an empirically well-established but nevertheless prejudiced view on the natural and artificial functions of proteins and nucleic acids. Proteins were thought to be nature's unbeatable universal catalysts, highly efficient as well as ultimately specific and, as in the case of immunoglobulins, even tunable to recognize previously unseen molecules. After Watson and Crick's famous discovery of the double helix, DNA was considered to be the molecule of inheritance, capable of encoding genetic information and sufficiently stable to allow for essential conservation of nucleotide sequences over many replication rounds. RNA's role in the molecular concert of nature was reduced to the transfer of sequence information from DNA to protein, be it as mRNA or as tRNA. Ribosomal RNA and some rare RNA molecules did not fit well into this picture: some sort of scaffolding functions were attributed to them, such as holding supramolecular complexes together or bringing protein molecules into the correct spatial positions required for their functions.

This conventional picture was based on the idea of a complete "division of labor". Nucleic acids, DNA as well as RNA, were the templates, ready for replication and read-out of genetic information, but not to do catalysis. Proteins were the catalysts and thus not capable of template function. In both cases these rather dogmatic views turned out to be wrong. Tom Cech and Sidney Altman discovered RNA molecules with catalytic functions (Guerrier-Takada *et al.*, 1983; Cech 1983, 1986, 1990). The name *ribozyme* was created for this new class of biocatalysts because they combine properties of ribonucleotides and enzymes (see next section). Their examples were dealing with RNA cleavage reactions catalyzed by RNA: without the help of a protein catalyst a non-coding region of an RNA transcript, a group I intron, cuts itself out during mRNA maturation. The second example concerns the enzymatic reaction of RNase P, which catalyzes tRNA formation from the precursor poly-tRNA. For a long time biochemists had known that this enzyme consists of a protein and an RNA moiety. It was tacitly assumed that the protein was the catalyst while the RNA component had only a backbone function. The converse, however, is true: the RNA acts as catalyst and the protein is merely a scaffold required to enhance efficiency.

The second prejudice was disproved only about 2 years ago by the demonstration that oligopeptides can act as templates for their own synthesis and thus show autocatalysis (Lee *et al.*, 1996, 1997; Severin *et al.*, 1997). In this very elegant work, Reza Ghadiri and his co-workers have demonstrated that template action does not necessarily require hydrogen bond formation. Two smaller oligopeptides of chain lengths 17 (E) and 15 (N) are aligned on the template (T) by means of the hydrophobic interaction in a coiled coil of the leucine zipper type and the 32-mer is produced by spontaneous peptide bond formation between the activated carboxygroup and the free amino residue (Figure 1.2). The hydrophobic cores of template and ligands consist of alternating valine and leucine residues and show a kind of knobs-into-holes packing in the complex. The capability for template action of proteins is a consequence of the three-dimensional structure of the protein α-helix, which allows the formation of coiled coils. It requires that the residues making the contacts between the helices fulfill the condition of space filling and thus stable packing. Modification of the oligopeptide sequences allows alteration of the interaction in the complex and thereby modifies the specificity and efficiency of catalysis. A highly relevant feature of oligopeptide self-replication concerns easy formation of higher replication complexes: coiled-coil formation is not restricted to two interacting helices; triple helices and higher complexes are known to be very stable too. Autocatalytic oligopeptide formation may thus involve not only a template and two substrates but, for example, a template and a catalyst that form a triple helix together

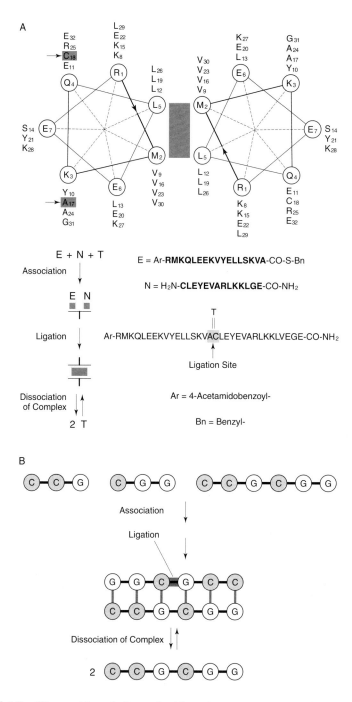

FIGURE 1.2 Oligopeptide and oligonucleotide replicons. **A.** An autocatalytic oligopeptide that makes use of the leucine zipper for template action. The upper part illustrates the stereochemistry of oligopeptide template–substrate interaction by means of the helix-wheel. The ligation site is indicated by arrows. The lower part shows the mechanism (Lee *et al.*, 1996; Severin *et al.*, 1997). **B.** Template-induced self-replication of oligonucleotides (von Kiedrowski, 1986) follows essentially the same reaction mechanism. The critical step is the dissociation of the dimer after bond formation, which commonly prevents these systems from exponential growth and darwinian behavior (see below).

with the substrates (Severin *et al.*, 1997). Only a very small fraction of all possible peptide sequences fold into three-dimensional structures that are suitable for leucine zipper formation and hence a given autocatalytic oligopeptide is very unlikely to retain the capability of template action upon mutation. Peptides thus are *occasional* templates and replicons based upon peptides are rare.

In contrast to the volume filling principle of protein packing, specificity of catalytic RNAs is provided by base pairing and to a lesser extent by tertiary interactions. Both are the results of hydrogen bond specificity. Metal ions, in particular Mg^{2+}, are often involved in RNA structure formation and catalysis too. Catalytic action of RNA on RNA is exercised in the cofolded complexes of ribozyme and substrate. Since the formation of a ribozyme's catalytic center, which operates on another RNA molecule, requires sequence complementarity in parts of the substrate, ribozyme specificity is thus predominantly reflected by the sequence and not by the three-dimensional structure of the isolated substrate. Template action of nucleic acid molecules, being the basis for replication, results directly from the structure of the double helix. It requires an appropriate backbone provided by the antiparallel ribose–phosphate or 2'-deoxyribose–phosphate chains and a suitable geometry of the complementary purine–pyrimidine pairs. All RNA (and DNA) molecules, however, share these features, which, accordingly, are independent of sequence. Every RNA molecule has a uniquely defined complement. Nucleic acid molecules, in contrast to proteins, are therefore *obligatory* templates. This implies that mutations are conserved and readily propagated into future generations.

Enzyme-free template-induced synthesis of longer RNA molecules from monomers, however, has not been successfully achieved so far (see, for example, Orgel, 1986). A major problem, among others, is the dissociation of double-stranded molecules at the temperature of efficient replication. If monomers bind with sufficiently high binding constants to the template in order to guarantee the desired accuracy of replication, then the new molecules are too sticky to dissociate after the synthesis has been completed. Autocatalytic template-induced synthesis of oligonucleotides from smaller oligonucleotide precursors was nevertheless successful: a hexanucleotide through ligation of two trideoxynucleotide precursors was carried out by Günter von Kiedrowski (1986). His system is the oligonucleotide analog of the autocatalytic template-induced ligation of oligopeptides discussed above (Figure 1.2). In contrast to the latter system the oligonucleotides do not form triple-helical complexes. Isothermal autocatalytic template-induced synthesis, however, cannot be used to prepare longer oligonucleotides because of the same duplex dissociation problem as mentioned for the template-induced polymerization of monomers (see also Parabolic and exponential growth, below).

RNA CATALYSIS AND THE RNA WORLD

The natural ribozymes discovered early were all RNA cleaving molecules: the RNA moiety of RNase P (Guerrier-Takada *et al.*, 1983), the class I introns (Cech, 1983) as well as the first small ribozyme called "hammerhead" (Figure 1.3) because of its characteristic secondary structure shape (Uhlenbeck, 1987). Three-dimensional structures are now available for three classes of RNA-cleaving ribozymes (Pley *et al.*, 1994; Scott *et al.*, 1995; Cate *et al.*, 1996; Ferré-D'Amaré *et al.*, 1998) and these data revealed the mechanism of RNA-catalyzed cleavage reactions in full molecular detail. Additional catalytic RNA molecules were obtained through selection from random or partially random RNA libraries and subsequent evolutionary optimization (see Evolution of phenotypes, below). RNA catalysis in nonnatural ribozymes is not restricted only to RNA cleavage: some ribozymes show ligase activity (Bartel and Szostak, 1993; Ekland *et al.*, 1995) and many efforts were undertaken to prepare a ribozyme with full RNA replicase activity. The attempt that comes closest to the goal yielded a ribozyme that catalyzes RNA polymerization in short stretches (Ekland and Bartel, 1996). RNA catalysis is not restricted to operating on RNA, nor do nucleic acid catalysts require the ribose

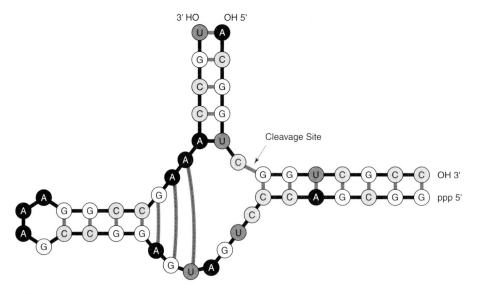

FIGURE 1.3   The hammerhead ribozyme. The substrate is a tridecanucleotide forming two double-helical stacks together with the ribozyme ($n = 34$) in the co-folded complex (Pley *et al.*, 1994). Some tertiary interactions indicated by broken lines in the drawing determine the detailed structure of the hammerhead ribozyme complex and are important for the enzymatic reaction cleaving one of the two linkages between the two stacks. Substrate specificity of ribozyme catalysis is caused by the secondary structure in the co-folded complex between substrate and catalyst.

backbone: ribozymes were trained by evolutionary techniques to process DNA rather than their natural RNA substrate (Beaudry and Joyce, 1992), and catalytically active DNA molecules were evolved as well (Breaker and Joyce, 1994; Cuenoud and Szostak, 1995). Polynucleotide kinase activity has been reported (Lorsch and Szostak, 1994, 1995) as well as self-alkylation of RNA on base nitrogens (Wilson and Szostak, 1995).

Systematic studies also revealed examples of RNA catalysis on non-nucleic acid substrates. RNA catalyzes ester, amino acid and peptidyl transferase reactions (Lohse and Szostak, 1996; Zhang and Cech, 1997; Jenne and Famulok, 1998). The latter examples are particularly interesting because they revealed close similarities between the RNA catalysis of peptide bond formation and ribosomal peptidyl transfer (Zhang and Cech, 1998). A spectacular finding in this respect was that oligopeptide bond cleavage and formation is catalyzed by ribosomal RNA and not by protein: more than 90% of the protein fraction can be removed from ribosomes without losing the catalytic effect on peptide

bond formation (Noller *et al.*, 1992; Green and Noller, 1997). In addition, ribozymes were prepared that catalyze alkylation on sulfur atoms (Wecker *et al.*, 1996) and, finally, RNA molecules were designed that are catalysts for typical reactions of organic chemistry, for example an isomerization of biphenyl derivatives (Prudent *et al.*, 1994).

For two obvious reasons RNA was chosen as candidate for the leading molecule in a simple scenario at the interface between chemistry and biology: (1) RNA is thought to be capable of storing retrievable information because it is an obligatory template; and (2) it has catalytic properties. Although the catalytic properties of RNA are less universal than those of proteins, they are apparently sufficient for processing RNA. RNA molecules operating on RNA molecules form a self-organizing system that can develop a form of molecular organization with emerging properties and functions. This scenario has been termed the *RNA world* (see, for example, Gilbert, 1986, Joyce, 1991, as well as the collective volume by Gesteland and Atkins, 1993). The idea of an RNA world turned out to be fruitful in a different aspect too: it initiated the search for molecular templates and created an entirely new field, which may be characterized as *template chemistry* (Orgel, 1992). Series of systematic studies were performed, for example, on the properties of nucleic acids with modified sugar moieties (Eschenmoser, 1993). These studies revealed the special role of ribose and provided explanations why this molecule is basic to all life processes.

Chemists working on the origin of life see a number of difficulties for an RNA world being a plausible direct successor of the functionally unorganized prebiotic chemistry (see Figure 1.1 and the reviews Orgel, 1987, 1992, Joyce, 1991, Schwartz, 1997): (1) no convincing prebiotic synthesis has been demonstrated for all RNA building blocks; (2) materials for successful RNA synthesis require a high degree of purity that can hardly be achieved under prebiotic conditions; (3) RNA is a highly complex molecule whose stereochemically correct synthesis (3′–5′ linkage) requires an elaborate chemical machinery; and (4) enzyme-free template-induced synthesis of RNA molecules from monomers has not been achieved so far. In particular, the dissociation of duplexes into single strands and the optical asymmetry problem are of major concern. Template-induced synthesis of RNA molecules requires pure optical antipodes. Enantiomeric monomers (containing L-ribose instead of the natural D-ribose) are "poisons" for the polycondensation reaction on the template since their incorporation causes termination of the polymerization process. Several suggestions postulating more "intermediate worlds" between chemistry and biology were made. Most of the intermediate information carriers were thought to be more primitive and easier to synthesize than RNA but nevertheless still having the capability of template action (Schwartz, 1997). Glycerol, for example, was suggested as a substitute for ribose because it is structurally simpler and it lacks chirality. However, no successful attempts to use such less sophisticated backbone molecules together with the natural purine and pyrimidine bases for template reactions have been reported so far.

Starting from a world of replicating molecules, it took a series of many not yet well-understood steps (Eigen and Schuster, 1982) to arrive at the first organisms that formed the earliest identified fossils (Warrawoona, Western Australia, 3.4×10^9 years old; Schopf, 1993) and possibly the even older kerogen found in the Isua formation (Greenland, 3.8×10^9 years old; Pflug and Jaeschke-Boyer, 1979; Schidlowski, 1988; Figure 1.1). It has been speculated that functionally correlated RNA molecules have developed a primitive translation machinery based on an early genetic code. After such a relation between RNA and proteins had been established the stage was set for concerted evolution of proteins and RNA. Proteins may induce vesicle formation into lipid-like materials and eventually lead to the formation of compartments. After a number of steps such an ensemble might have developed a primitive metabolism and thus led to the first protocells (Eigen and Schuster, 1982). DNA, being now the backup copy of genetic information, is seen as a latecomer in prebiotic evolution.

A successful experimental approach to self-reproduction of micelles and vesicles is highlighting one of the many steps enumerated above: prebiotic formation of vesicle structures (Bachmann *et al.*, 1992). The basic reaction leading to autocatalytic production of amphiphilic materials is the hydrolysis of ethyl caprilate. The combination of vesicle formation with RNA replication represents a particularly important step towards the construction of a kind of minimal synthetic cell (Luisi *et al.*, 1994). Despite these elegant experimental studies and the attempts to build comprehensive models, satisfactory answers to the problems of compartment formation and cell division are not at hand yet.

PARABOLIC AND EXPONENTIAL GROWTH

It is relatively easy to derive a kinetic rate equation displaying the elementary behavior of replicons if one assumes that catalysis proceeds through the complementary binding of reactant(s) to free template and that autocatalysis is limited by the tendency of the template to bind to itself as an inactive "product inhibited" dimer (Von Kiedrowski, 1993). However, in order to achieve an understanding of what is likely to happen in systems where there is a diverse mixture of reactants and catalytic templates, it is desirable to develop a comprehensive kinetic description of as many individual steps in the reaction mechanism of template synthesis as is feasible and tractable from the mathematical point of view.

Szathmáry and Gladkih (1989) oversimplified the resulting dynamics to a simple parabolic growth law $\dot{x}_k \propto x_k^p$, $0<p<1$ for the concentrations of the interacting template species. Their model suffers from a conceptual and a technical problem: (1) under no circumstances does one observe extinction of a species in any parabolic growth model; and (2) the vector fields are not Lipschitz-continuous on the boundary of the concentration simplex, indicating that we cannot expect a physically reasonable behavior in this area.

In a recent paper (Wills et al., 1998) we have derived the kinetic equations of a system of coupled template-instructed ligation reactions of the form

$$
A_i + B_j + C_{kl} \xrightleftharpoons[\bar{a}_{ijkl}]{a_{ijkl}} A_i B_j C_{kl} \xrightarrow{b_{ijkl}} C_{ij} C_{kl} \xrightleftharpoons[\bar{d}_{ijkl}]{d_{ijkl}} C_{ij} + C_{kl}
$$

(1)

Here A and B denote the two substrate molecules which are ligated on the template C, for example, the electrophilic, E, and the nucleophilic, N, oligopeptide in peptide template reactions or the two different trinucleotides, GGC and GCC, in the autocatalytic hexanucleotide formation (Figure 1.2). This scheme thus encapsulates the experimental results on both peptide and nucleic acid replicons (Von Kiedrowski, 1986; Lee et al., 1996).

The following assumptions are straightforward and allow for a detailed mathematical analysis:

1. The concentrations of the intermediates are stationary in agreement with the "quasi-steady-state" approximation (Segel and Slemrod, 1989).
2. The total concentration c_0 of all replicating

species is constant in the sense of constant organization (Eigen, 1971).
3. The formation of heteroduplices of the form $C_{ij} C_{kl}$, $ij \neq kl$ is neglected.
4. Only reaction complexes of the form $A_k B_l C_{kl}$ lead to ligation.

Assumptions 3 and 4 are closely related. They make immediate sense for hypothetical macromolecules for which the template instruction is direct instead of complementary. It has been shown, however, that the dynamics of complementary replicating polymers is very similar to direct replication dynamics if one considers the two complementary strands as "single species" by simply adding their concentrations (Stadler, 1991).

Assumptions 3 and 4 suggest a simplified notation of the reaction scheme:

$$
A_k + B_k + C_k \xrightleftharpoons[\bar{a}_k]{a_k} A_k B_k C_k \xrightarrow{b_k} C_k C_k \xrightleftharpoons[\bar{d}_k]{d_k} 2C_k
$$

(2)

It can be shown that equation (2), together with the assumptions 1 and 2, leads to the following system of differential equations for the frequencies or relative total concentrations x_k, i.e. $\sum_k^M x_k = 1$ of the template molecules C_k in the system (note that x_k accounts not only for the free template molecules but also for those bound in the complexes $C_k C_k$ and $A_k B_k C_k$):

$$
\dot{x}_k = x_k \left(\alpha_k \varphi(\beta_k x_k) - \sum_j^M \alpha_j x_j \varphi(\beta_j x_j) \right), \quad k = 1, \ldots, M,
$$

(3)

where

$$
\varphi(z) = \frac{1}{z} \left(\sqrt{z+1} - 1 \right), \quad \varphi(0) = \frac{1}{2}
$$

(3')

and the effective kinetic constants α_k and β_k can be expressed in terms of the physical parameters a_k, $\bar{a}_k$, etc. It will turn out that survival of replicon species is determined by the constants α_k, which we characterize therefore as darwinian fitness parameters.

Equation 3 is a special form of a replicator equation with the non-linear response functions $f_k(x) := \alpha_k \varphi(\beta_k x_k)$. Its behavior depends strongly on the values of β_k: for large values of z we have $\varphi(z) \sim 1/\sqrt{z}$. Hence equation (3) approaches Szathmáry's expression (Szathmáry and Gladkih, 1989):

$$\dot{x}_k = h_l \sqrt{x_k} - x_k \sum_j^M h_j \sqrt{x_j}$$

with suitable constants h_k. This equation exhibits a very simple dynamics: the mean fitness $\Phi(x) = \sum_j^M h_j \sqrt{x_j}$ is a Ljapunov function, i.e. it increases along all trajectories, and the system approaches a globally stable equilibrium at which all species are present (Varga and Szathmáry, 1997; Wills et al., 1998). Szathmáry's parabolic growth model thus does not lead to selection.

On the other hand, if z remains small, that is, if β_k is small, then $\varphi(\beta_k x_k)$ is almost constant at ½ (since the relative concentration x_k is of course a number between 0 and 1). Thus we obtain:

$$\dot{x}_k = \frac{1}{2} x_k \left(\alpha_k - \sum_j^M \alpha_j x_j \right) \qquad (4)$$

which is the "no-mutation" limit of Eigen's kinetic equation for replication (Eigen, 1971). (If condition (4) above is relaxed, we in fact arrive at Eigen's model with a mutation term.) Equation (4) leads to survival of the fittest: the species with the largest value of α_k will eventually be the only survivor in the system. It is worth noting that the mean fitness also increases along all orbits of equation (4) in agreement with the no-mutation case (Schuster and Swetina, 1988).

The constants β_k that determine whether the system shows darwinian selection or unconditional coexistence are proportional to the total concentration c_0 of the templates. For small total concentration we obtain equation (4), while for large concentrations, when the formation of the dimers $C_k C_k$ becomes dominant, we enter the regime of parabolic growth.

Equation (3) is a special case of a class of replicator equations studied in Hofbauer et al. (1981). Restating the previously given result yields the following. All orbits or trajectories starting from physically meaningful points (these are points in the interior of the simplex S_M with $x_j > 0$ for all $j = 1, 2, \ldots, M$) converge to a unique equilibrium point $\bar{x} = (\bar{x}_1, \bar{x}_2, \ldots, \bar{x}_M)$ with $\bar{x}_i \geq 0$, which is called the ω-limit of the orbits. This means that species may go extinct in the limit $t \to \infty$. If $\bar{x}$ lies on the surface of S_M (which is tantamount to saying that at least one component $\bar{x}_i = 0$) then it is also the ω-limit for all orbits on this surface. If we label the replicon species according to decreasing values of the darwinian fitness parameters, $\alpha_1 \geq \alpha_2 \geq \ldots \geq \alpha_M$, then there is an index $l \geq 1$ such that $\bar{x}$ is of the form $x_i > 0$ if $i \leq l$ and $\bar{x}_i = 0$ for $i > l$. In other words, l replicon species survive and the $M - l$ least efficient replicators die out. This behavior is in complete analogy to the reversible exponential competition case (Schuster and Sigmund, 1985) where the darwinian fitness parameters α_k are simply the rate constants a_k. If the smallest concentration-dependent value $\beta_s(c_0) = \min \{\beta_j(c_0)\}$ is sufficiently large, we find $l = M$ and no replicon goes extinct ($\bar{x}$ is an interior equilibrium point).

The condition for survival of species k is explicitly given by:

$$\alpha_k > 2\Phi(\bar{x}).$$

It is interesting to note that the darwinian fitness parameters α_k determine the order in which species go extinct whereas the concentration-dependent values $\beta_k(c_0)$ collectively influence the flux term and hence set the "extinction threshold". In contrast to Szathmáry's model equation the extended replicon kinetics leads to both competitive selection and coexistence of replicons depending on total concentration and kinetic constants.

MOLECULAR EVOLUTION EXPERIMENTS

In the first half of this century it was apparently out of the question to do conclusive and interpretable experiments on evolving populations because of two severe problems: (1) Time scales of evolutionary processes are prohibitive for laboratory investigations; and (2) the numbers of possible genotypes are outrageously large and thus only a negligibly small fraction of all possible sequences can be realized and evaluated by selection. If generation times could be reduced to a minute or less, thousands of generations, numbers sufficient for the observation of optimization and adaptation, could be recorded in the laboratory. Experiments with RNA molecules in the test-tube indeed fulfill this time-scale criterion for observability. With respect to the "combinatorial explosion" of the numbers of possible genotypes the situation is less clear. Population sizes of nucleic acid molecules of 10^{15}–10^{16} individuals can be produced by random synthesis in conventional automata. These numbers cover roughly all sequences up to chain lengths of $n = 27$ nucleotides. These are only short RNA molecules but their length is already sufficient for specific binding to predefined target molecules, for example antibiotics (Jiang *et al.*, 1997). In addition, sequence-to-structure-to-function mappings of RNA are highly redundant and thus only a small fraction of all sequences has to be searched in order to find solutions to given evolutionary optimization problems (Fontana *et al.*, 1993; Schuster *et al.*, 1994).

The first successful attempts to study RNA evolution *in vitro* were carried out in the late 1960s by Sol Spiegelman and his group (Mills *et al.*, 1967; Spiegelman, 1971). They created a "protein-assisted RNA replication medium" by adding an RNA replicase isolated from *Escherichia coli* cells infected by the RNA bacteriophage Qβ to a medium for replication that also contains the four ribonucleoside triphosphates (GTP, ATP, CTP and UTP) in a suitable buffer solution. Qβ RNA and some of its smaller variants start instantaneously to replicate when transferred into this medium. Evolution experiments were carried out by means of the serial transfer technique: materials consumed in RNA replication are replenished by transfer of small samples of the current solution into fresh stock medium. The transfers were made after equal time steps. In series of up to 100 transfers the rate of RNA synthesis increased by orders of magnitude. The increase in the replication rate

occurs in steps and not continuously as one might have expected. Analysis of the molecular weights of the replicating species showed a drastic reduction of the RNA chain lengths during the series of transfers: the initially applied Qβ RNA was 4220 nucleotides long and the finally isolated species contained little more than 200 bases. What happened during the serial transfer experiments was a kind of degradation due to suspended constraints on the RNA molecule. In addition, to perform well in replication the viral RNA has to code for four different proteins in the host cell and needs also a proper structure in order to enable packing into the virion. In test-tube evolution these constraints are released and the only remaining requirement is recognition of the RNA by Qβ replicase and fast replication.

Evidence for a non-trivial evolutionary process came a few years later when the Spiegelman group published the results of another serial transfer experiment that gave evidence for adaptation of an RNA population to environmental change. The replication of an optimized RNA population was challenged by the addition of ethidium bromide to the replication medium (Kramer *et al.*, 1974). This dye intercalates into DNA and RNA double helices and thus reduces replication rates. Further serial transfers in the presence of the intercalating substance led to an increase in the replication rate until an optimum was reached. A mutant was isolated from the optimized population that differed from the original variant by three point mutations. Extensive studies on the reaction kinetics of RNA replication in the Qβ replication assay were performed by Christof Biebricher in Göttingen (Biebricher and Eigen, 1988). These studies revealed consistency of the kinetic data with a many-step reaction mechanism. Depending on concentration, the growth of template molecules allows one to distinguish three phases of the replication process.

1. At low concentration all free template molecules are instantaneously bound by the replicase, which is present in excess, and therefore the template concentration grows exponentially.

2. Excess of template molecules leads to saturation of enzyme molecules, then the rate of RNA synthesis becomes constant and the concentration of the template grows linearly.
3. Very high template concentrations impede dissociation of the complexes between template and replicase, and the template concentration approaches a constant in the sense of product inhibition.

We neglect plus–minus complementarity in replication by assuming constancy in relative concentrations of plus and minus strands (Eigen, 1971) and consider the plus–minus ensemble as a single species. Then, RNA replication may be described by the overall mechanism:

$$A + I_i + E \underset{k_i}{\overset{k_i}{\rightleftharpoons}} A + I_i \cdot E \overset{a_i}{\longleftarrow} I_i \cdot E \cdot I_i \underset{k_i'}{\overset{k_i'}{\rightleftharpoons}} I_i \cdot E + I_i.$$

$$(5)$$

Here E represents the replicase and A stands for the low-molecular-weight material consumed in the replication process. This simplified reaction scheme reproduces all three characteristic phases of the detailed mechanism (Figure 1.4) and can be readily extended to replication and mutation.

Despite the apparent complexity of RNA

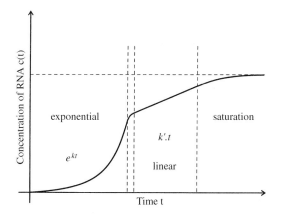

FIGURE 1.4 Replication kinetics of RNA with Qβ replicase. In essence, three different phases of growth are distinguished: (1) exponential growth under conditions with excess of replicase; (2) linear growth when all enzyme molecules are loaded with RNA; and (3) a saturation phase that is caused by product inhibition.

replication kinetics the mechanism at the same time fulfills an even simpler overall rate law provided the activated monomers, ATP, UTP, GTP, and CTP, as well as Qβ replicase are present in excess. In that case, the rate of increase for the concentration x_i of RNA species I_i follows the simple relation $\dot{x}_i \propto x_i$, which in the absence of constraints ($\Phi = 0$) leads to exponential growth. This growth law is identical to that found for asexually reproducing organisms and hence replication of molecules in the test-tube leads to the same principal phenomena that are found with evolution proper. RNA replication in the Qβ system requires specific recognition by the enzyme, which implies sequence and structure restrictions. Accordingly only RNA sequences that fulfill these criteria can be replicated. In order to be able to amplify RNA free of such constraints many-step replication assays have been developed. The discovery of the DNA polymerase chain reaction (PCR; Mullis, 1990) was a milestone towards sequence-independent amplification of DNA sequences. It has one limitation: double helix separation requires higher temperatures and conventional PCR therefore works with a temperature program. PCR is combined with reverse transcription and transcription by means of bacteriophage T7 RNA polymerase in order to yield a sequence-independent amplification procedure for RNA. This assay contains two possible amplification steps: PCR and transcription. Another frequently used assay makes use of the isothermal self-sustained sequence replication reaction of RNA (3SR; Fahy *et al.*, 1991). In this system the RNA–DNA hybrid obtained through reverse transcription is converted into single-stranded DNA by RNAse digestion of the RNA strand, instead of melting the double strand. DNA double-strand synthesis and transcription complete the cycle. Here, transcription by T7 polymerase represents the amplification step. Artificially enhanced error rates needed for the creation of sequence diversity in populations can be achieved readily with PCR. Reverse transcription and transcription are also susceptible to increase in mutation rates. These two and other new techniques for RNA amplification provided universal and efficient tools for the study of molecular evolution under laboratory conditions and made the usage of viral replicases with their undesirable sequence specificities obsolete.

ERROR PROPAGATION AND QUASISPECIES

Evolution of molecules based on replication and mutation exposed to selection at constant population size has been formulated and analyzed in terms of chemical reaction kinetics (Eigen, 1971; Eigen and Schuster, 1977; Eigen *et al.*, 1989). Error-free replication and mutation are parallel chemical reactions:

$$A + I_i \xrightarrow{a_i Q_{ij}} I_j + I_i, \tag{6}$$

and form a network that in principle allows formation of every RNA genotype as a mutant of any other genotype. The materials required for, or consumed by, RNA synthesis, again denoted by A, are replenished by continuous flow in a reactor resemblimg a chemostat for bacterial cultures (Figure 1.5). The object of interest is now the distribution of genotypes in the population and its time-dependence. We present here a short account of the most relevant features of such replication–mutation assays, in particular the existence of thresholds in error propagation.

Selection in populations is described by ordinary differential equations. It has been shown for systems of type (6) that the outcome of selection is independent of the selection constraint applied. In particular, the flow reactor and constant organization yield essentially the same results (Schuster and Sigmund, 1985; Happel and Stadler, 1999) and thus we used the latter simpler condition without losing generality. Variables are again the frequencies of individual genotypes, x_i measuring that of genotype or RNA sequence I_i. The frequencies are nomalized, $\Sigma_{i=1}^{M} x_i = 1$ (due to constant organization), the population size is denoted by N and the number of different genotypes by M. The time-dependence of the sequence distribution is described by the kinetic equation:

$$\dot{x} = x_i \left(a_i Q_{ii} - \bar{E}(t) \right) + \sum_{j=1, j\neq i}^{M} a_j Q_{ji} x_j, \quad i = 1, \ldots, M. \tag{7}$$

The rate constants for replication of the molecular species are a_i. Once a reaction has been initiated it can lead to a correct copy, $I_i \to I_i$, or to a mutant, $I_i \to I_j$. The frequencies of the individual reaction channels are contained in the mutation matrix $Q \doteq \{Q_{ij}; i, j = 1, \ldots, M\}$, in particular the fraction of error copies of genotype I_i falling into

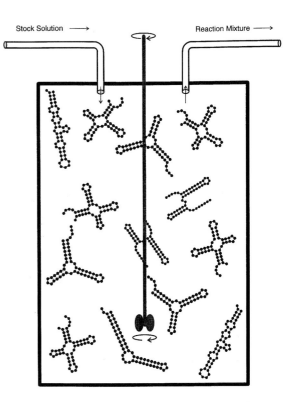

FIGURE 1.5 A flow reactor for the evolution of RNA molecules. A stock solution containing all materials for RNA replication including an RNA polymerase flows continuously into a well-stirred tank reactor and an equal volume containing a fraction of the reaction mixture leaves the reactor. The population in the reactor fluctuates around a mean value, $N \pm \sqrt{N}$. RNA molecules replicate and mutate in the reactor, and the fastest replicators are selected. The RNA flow reactor has been used also as an appropriate setup for computer simulations (Fontana and Schuster, 1987, 1998; Huynen *et al.*, 1996). There, other criteria than fast replication can be used for selection. For example, fitness functions are defined that measure the distance to a predefined target structure and fitness increases during the approach towards the target (Huynen *et al.*, 1996; Fontana and Schuster, 1998).

species I_j is given by Q_{ij} and thus we have $\Sigma Q_{ij} = 1$. The diagonal elements of Q are the replication accuracies, i.e. the fractions of correct replicas produced on the corresponding templates. The time-dependent excess productivity which is compensated by the flow in the reactor is the mean value $\bar{E}(t) = \Sigma a_i x_i(t)$. The quantities determining then the outcome of selection are the products of replication rate constants and mutation frequencies subsumed in the value matrix: $W \doteq \{w_{ij} = a_i Q_{ij}; i, j = 1, \ldots, M\}$; its diagonal elements, w_{ii}, were called the selective values of the individual genotypes.

The selective value of a genotype is tantamount to its fitness in the case of vanishing mutational backflow and hence the genotype with maximal selective value, I_m:

$$w_m = \max\{w_{ii} \mid i = 1,\ldots, M\}, \qquad (8)$$

dominates a population after it has reached the selection equilibrium and hence it is called the *master* sequence. The notion *quasispecies* was introduced for the stationary genotype distribution in order to point at its role as the genetic reservoir of the population.

A simple expression for the stationary frequency can be found, if the master sequence is derived from the single-peak model landscape that assigns a higher replication rate to the master and identical values to all others, for example $a_m = \sigma_m \cdot a$ and $a_i = a$ for all $i \neq m$ (Swetina and Schuster, 1982; Tarazona, 1992; Alves and Fontanari, 1996). The (dimensionless) factor σ_m is called the superiority of the master sequence. The assumption of a single-peak landscape is tantamount to lumping all mutants together into a mutant cloud with average fitness. The probability of being in the cloud is simply $x_c = \Sigma_{j=1, j\neq m}^{M} x_j = 1 - x_m$ and the replication–mutation problem boils down to an exercise in a single variable, x_m, the frequency of the master. The single-peak model can be interpreted as a kind of mean field approximation since the mutant cloud is characterizable by "mean-except-the-master" properties, for example by the mean-except-the-master replication rate constant $\bar{a} = \Sigma_{j \neq m} a_j x_j / (1 - x_m)$. The superiority then reads: $\sigma_m = a_m / \bar{a}$. Neglecting mutational backflow we can readily compute the stationary frequency of the master sequence:

$$\bar{x}_m = \frac{a_m Q_{mm} - \bar{a}}{a_m - \bar{a}} = \frac{\sigma_m Q_{mm} - 1}{\sigma_m - 1}. \qquad (9)$$

In this expression the master sequence vanishes at some finite replication accuracy, $Q_{mm} \mid \bar{x}_{m=0} = Q_{min} = \sigma_m^{-1}$. Non-zero frequency of the master thus requires $Q_{mm} > Q_{min}$. We introduce the uniform error rate model, which assumes that the mutation rate is p per site and replication event independently of the nature of the nucleotide to be copied and the position in the sequence (Eigen and Schuster, 1977). Then, the single digit accuracy $q = 1 - p$ is the mean fraction of correctly incorporated nucleotides and the elements of the mutation matrix for a polynucleotide of chain length n are of the form:

$$Q_{ij} = q^n \left(\frac{1-q}{q} \right)^{d_{ij}},$$

with d_{ij} being the Hamming distance between two sequences I_i and I_j. The critical condition occurs at the minimum accuracy:

$$q_{min} = 1 - p_{max} = \sqrt[n]{Q_{min}} = \sigma_m^{-1/n}, \qquad (10)$$

which was called the *error threshold*. Above threshold no stationary distribution of sequences is formed. Instead, the population drifts randomly through sequence space. This implies that all genotypes have only finite lifetimes, inheritance breaks down and evolution becomes impossible.

Figure 1.6 shows the stationary frequency of the master sequence as a function of the error rate. Variations in the accuracy of *in-vitro* replication can indeed be easily achieved because error rates can be tuned over many orders of magnitude (Leung *et al.*, 1989; Martinez *et al.*, 1994) The range of replication accuracies that are suitable for evolution is limited by the maximum accuracy that can be achieved by the replication machinery and the minimum accuracy determined by the error threshold. Populations in constant environments have an

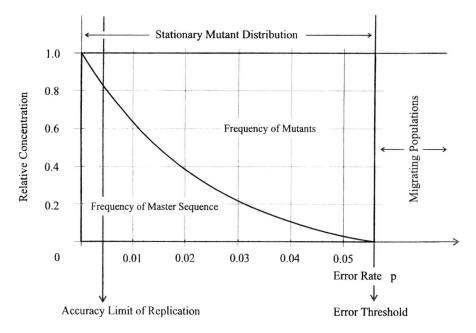

FIGURE 1.6 The genotypic error threshold. The fraction of mutants in stationary populations increases with the error rate p. Stable stationary mutant distributions called quasispecies require sufficient accuracy of replication: the single-digit accuracy has to exceed a minimal value known as error threshold, $1 - p = q > q_{min}$. Above threshold populations migrate through sequence space in random walk-like manner (Huynen *et al.*, 1996). There is also a lower limit to replication accuracy, which is given by the maximum accuracy of the replication machinery.

advantage when they operate near the maximum accuracy because then they lose as few copies as possible through mutation. In highly variable environments the opposite is true: it pays to produce as many mutants as possible because then the chance is largest to cope successfully with change.

In order to be able to study stochastic features of population dynamics around the error threshold, the replication–mutation system was modeled by a multitype branching process (Demetrius *et al.*, 1985) The main result of this study is the derivation of an expression for the probability of survival to infinite time for the master sequence and its mutants. In the regime of sufficiently accurate replication the survival probability is non-zero and decreases with increasing error rate. At the critical accuracy q_{min} this probability becomes zero. This implies that all molecular species that are currently in the populations, master and mutants, will die out in finite times and new variants will appear. This scenario is tantamount to migration of the pop-ulation through sequence space. The critical accuracy q_{min}, commonly seen as an error threshold for replication, can also be understood as the localization threshold of the population in sequence space (McCaskill, 1984). Later investigations aimed directly at a derivation of the error threshold in finite populations (Nowak and Schuster 1989; Alves and Fontanari, 1998).

In order to check the relevance of the error threshold for the replication of RNA viruses the minimum accuracy of replication can be transformed into a maximum chain length n_{max} for a given error rate p. The condition for stationarity of the quasispecies then reads:

$$n < n_{max} = -\frac{\ln \sigma}{\ln q} \approx \frac{\ln \sigma}{1-q}. \qquad (10a)$$

The populations of most RNA viruses were shown to live indeed near the above-mentioned critical value of replication accuracy (Domingo, 1996; Domingo and Holland, 1997). In particular, the chain length n was found to be roughly the inverse mutation rate per site and replication (Drake, 1993). According to previously mentioned expectations these viruses should live in very variable environments in agreement with the highly active defense mechanisms of the host cells.

EVOLUTION OF PHENOTYPES

If several molecular species have the same maximal fitness we are dealing with a case of neutrality (Kimura, 1983). The superiority of the master sequence becomes $\sigma_m = 1$ in this case, and the localization threshold of the quasispecies converges to the limit of absolute replication accuracy, $q_{min} = 1$. Accordingly, the deterministic model fails, and we have to modify the kinetic equations. Genotypes are ordered with respect to non-increasing selective values. The first k_1 different genotypes have maximal selective value: $w_1 = w_2 = \ldots = w_{k_1} = w_{max} = \tilde{w}_1$ (where ˜ indicates properties of groups of neutral phenotypes). The second group of neutral genotypes has highest-but-one selective value: $w_{k_1+1} = w_{k_1+2} = \ldots = w_{k_1+k_2} = \tilde{w}_2 < \tilde{w}_1$, etc. Replication rate constants are assigned in the same way: $a_1 = a_2 = \ldots = a_{k_1} = \tilde{a}_1$, etc. In addition, we define new variables, y_j ($j = 1, \ldots, L$), that lump together all genotypes folding into the same phenotype:

$$y_j = \sum_{i=k_{j-1}+1}^{k_j} x_i \quad \text{with} \quad \sum_{j=1}^{L} y_j = \sum_{i=1}^{M} x_i = 1. \tag{11}$$

Without loss of generality we denote the phenotype with maximal fitness, the *master phenotype*, by m. Since we are heading again for a kind of zeroth-order solution, we consider only the master phenotype and put $k_1 = k$. With $y_m = \sum_{i=1}^{k} x_i$ we obtain the following kinetic differential equation for the set of sequences forming the neutral network of the master phenotype:

$$\dot{y}_m = \sum_{i=1}^{k} \dot{x}_i = y_m(\tilde{a}_m Q_{mm} - \overline{E}) + \sum_{i=1}^{k}\sum_{j\neq i} a_j Q_{ji} x_j. \tag{12}$$

The mean excess productivity of the population is, of course, independent of the choice of variables:

$$\overline{E} = \sum_{j=1}^{L} \tilde{a}_j y_j = \sum_{i=1}^{M} a_i x_i.$$

In order to derive a suitable expression for a phenotypic error threshold we split the mutational backflow into two contributions: (1) mutational backflow on the neutral network and (2) mutational backflow from genotypes not on the network:

$$\sum_{i=1}^{k}\sum_{j\neq i} a_j Q_{ji} x_j = \left\{ \tilde{a}_m \sum_{i=1}^{k} \sum_{j=1, j\neq i}^{k} Q_{ji} x_j \right\} + \left\{ \sum_{i=1}^{k} \sum_{j=k+1}^{M} a_j Q_{ji} x_j \right\}.$$

We approximate by assuming a constant fraction of selectively neutral neighbors of the master phenotype (λ_m) and equal mutation rates ($Q_{ji} = \tilde{Q}$; $i, j = 1, \ldots, k$; $i \neq j$) on the master network and find:

$$\sum_{i=1}^{k}\sum_{j=1, j\neq i}^{k} Q_{ji} x_j \approx \frac{\lambda_m(1 - Q_{mm})}{k-1} \sum_{i=1}^{k}\sum_{j=1, j\neq i}^{k} x_j$$
$$= \frac{\lambda_m(1 - Q_{mm})}{k-1} \sum_{i=1, j\neq i}^{k}\sum_{i=1}^{k} x_j = \lambda_m(1 - Q_{mm})y_m.$$

Mutational backflow from other networks (y_j, $j \neq m$) need not be evaluated explicitly since it has also been neglected in the derivation of the genotypic error threshold. The kinetic equation for the master phenotype can now be rewritten:

$$\dot{y}_m = (\tilde{a}_m \tilde{Q}_{mm} - \overline{E})y_m + \text{Mutational Backflow}.$$

They are identical with those in the variables expressing genotype concentrations except the use of an effective replication accuracy of:

$$Q_{mm} = Q_{mm} + \lambda_m(1 - Q_{mm}) = q^n(\phi(q)\lambda_m + 1),$$

$$\text{with } \phi(q) = \left(\frac{1}{q^n - 1}\right).$$

The last part of the equation has the advantage that the overall accuracy can be factorized into contributions from classes of nucleotides corresponding to positions on the sequence with different degrees of neutrality, $\lambda^{(k)}$:

$$\tilde{Q}_{mm} = q^n \prod_k \left(\Phi(q)\lambda_m^{(k)} + 1 \right), \quad \text{with } \Phi(q) = \left(\frac{1}{q^{n_k} - 1} \right).$$

The numbers of nucleotides in class k is denoted by n_k; clearly we have $\Sigma_k \, n_k = n$. Recently, it has been shown that a four-class approximation of the distribution of λ-values yields excellent results for tRNAs (Reidys $et\ al.$, 1999).

Neglecting mutational backflow from non-master phenotypes we finally find complete analogy with the derivation of the genotypic error threshold:

$$\tilde{Q}_{\min} = Q_{mm} + \lambda_m(1 - Q_{mm}) = \sigma_m^{-1},$$

where σ_m is the superiority of the "master phenotype". Introducing the uniform error rate model we obtain by neglecting mutational backflow for the stationary frequency of master phenotypes:

$$\bar{y}_m(p) = \frac{\tilde{Q}_{mm}(p)\sigma_m - 1}{\sigma_m - 1} = \frac{(1-p)^n \sigma_m (1 - \lambda_m) + \sigma_m \lambda_m - 1}{\sigma_m - 1}.$$

Eventually we find the phenotypic error threshold by applying the "zeroth-order approximation" ($\bar{y}_m = 0$):

$$q_{\min} = (1 - p_{\max}) = \left(\frac{1 - \lambda_m \sigma_m}{(1 - \lambda_m)\sigma_m} \right)^{1/n}.$$

The function $q = q_{\min}\,(n, \lambda_m, \sigma_m)$ is illustrated in Figure 1.7. The limits are easily visualized: (1) the phenotypic error threshold converges to the genotypic value $q_{\min} = \sigma_m^{-1/n}$ in the limit $\lambda_m \to 0$, and (2) the minimal replication accuracy $q_{\min}$ approaches zero in the limit $\lambda_m \to \sigma_m^{-1}$. The second case implies that single-digit accuracy plays no role when the degree of neutrality is larger than the reciprocal value of the superiority.

Recapitulating the results on stationary distributions of phenotypes derived in this section we state that selective neutrality allows tolerance for more replication errors than in the non-neutral case. We are dealing with a distribution of changing genotypes corresponding to a population that drifts randomly (Huynen $et\ al.$, 1996) on the neutral network of the fittest or master phenotype. In this drift the master phenotype is conserved as long as the replication accuracy is above a critical minimal value, $q_{\min}$. When the accuracy falls also below this critical value the population drifts through sequence space and through shape space and no more stasis, neither with genotypes nor with phenotypes, is observed. It is particularly interesting to note

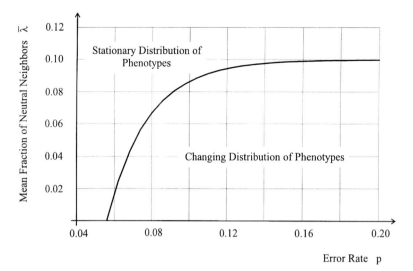

FIGURE 1.7 The phenotypic error threshold. The error threshold is shown as a function of the error rate p and the mean degree of neutrality $\bar{\lambda}$. The line separates the domains of stationary quasispecies and migrating populations. More replication errors can be tolerated at higher degrees of neutrality.

that there is a degree of neutrality related to the superiority of the master phenotype ($\lambda = \sigma^{-1}$) above which the error rate does not matter. In other words, the master phenotype will never be lost when the degree of neutrality exceeds a limit which is the inverse of the superiority.

So far, phenotypes have only been considered in terms of parameters contained in the kinetic equations. Mutation acts on genotypes whereas selection deals with phenotypes, since fitness is a property of the phenotype. The relations between genotypes and phenotypes are thus an intrinsic part of evolution and no theory can be complete without considering them. A comprehensive theory of evolution that deals explicitly with phenotypes was introduced a few years ago (Schuster, 1995; 1997a,b). The model is shown in Figure 1.8. The complex process of

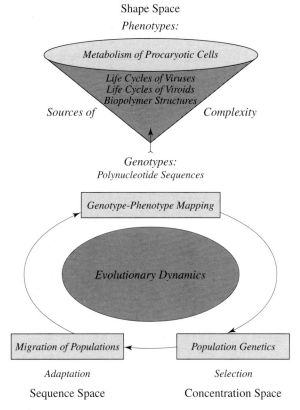

FIGURE 1.8 A comprehensive model of molecular evolution. The highly complex process of biological evolution is partitioned into three simpler dynamical phenomena: (1) population genetics; (2) migration of populations; and (3) genotype–phenotype mapping. Population genetics describes how optimal genotypes with optimal genes are chosen from a given reservoir by natural (or artificial) selection. The basis of population genetics is replication, mutation and recombination modeled by differential equations as derived from chemical reaction kinetics. In essence, population genetics is concerned with selection and other evolutionary phenomena occurring on short time-scales. Population support dynamics describes how the genetic reservoirs change when populations migrate in the huge space of all possible genotypes. Issues are the internal structure of populations and the mechanisms by which the regions of high fitness are found in sequence or genotype space. Support dynamics deals with the long-term phenomena of evolution, for example, with optimization and adaptation to changes in the environment. Genotype–phenotype mapping represents a core problem of evolutionary thinking since the dichotomy between genotypes and phenotypes is the basis of Darwin's principle of variation and selection: all genetically relevant variation takes place on the genotypes whereas the phenotypes are subjected to selection. Variations and their results are quantitatively uncorrelated in the sense that a mutation yielding a fitter phenotype does not occur more frequently because of the increase in fitness. The problem is the enormous complexity of the unfolding of genotypes that involves sophisticated processes from the formation of biopolymer structures to cellular metabolism and higher up to the almost open-ended increase in complexity with the development of multicellular organisms.

evolution is partitioned into three simpler phenomena: (1) population genetics; (2) migration of populations; and (3) genotype–phenotype mapping. Conventional population genetics is extended by two more aspects: population support dynamics, describing the migration of populations through sequence space, and genotype–phenotype mapping, providing the source of the parameters for population genetics. In general, phenotypes and their formation from genotypes are so complex that they cannot be handled appropriately. In test-tube evolution of RNA, however, the phenotypes are molecular structures. Then, genotype and phenotype are two features of the same molecule. In this simplest known case the relations between genotypes and phenotypes are reduced to the mapping of RNA sequences onto structures. Folding RNA sequences into structures is an essential part of the RNA optimization process and can be considered explicitly provided a coarse-grained version of structure, the secondary structure, is used. The model is self-contained in the sense that it is based on the rules of RNA secondary structure formation, the kinetics of replication and mutation as well as the structure of sequence space, and it needs no further inputs. The three processes shown in Figure 1.8 are indeed connected by a cyclic mutual dependence in which each process is driven by the previous one in the cycle and provides the input for the next one.

1. Folding sequences into structures yields the input for population genetics.
2. Population genetics describes the arrival of new genotypes through mutation and the dying of old ones through selection, and determines thereby how and where the population migrates.
3. Migration of the population in sequence space finally defines the new genotypes that are to be mapped into phenotypes and thus completes the cycle.

The model of evolutionary dynamics has been applied to interpret the experimental data on molecular evolution and it was implemented for computer simulations of neutral evolution and RNA optimization in the flow reactor (Fontana and Schuster, 1998). The computer simulations

allow one to follow the optimization process in full detail on the molecular level. Individual runs are monitored as time series of structures that eventually lead to the optimized molecule. The simulations helped to clarify the role of neutral variants in evolution. Recording of evolution experiments (Elena et al., 1996) as well as computer simulations (Huynen, 1996; Huynen et al., 1996; Fontana and Schuster, 1998) have shown first that optimization does not occur continuously. Instead, stepwise increases of fitness are observed. The periods of increase are interrupted by long phases of almost constant fitness. Inspection of populations during the quasi-static phases revealed that constancy is restricted to the level of phenotypes or their properties, respectively. The genotypes are changing all the time and the apparent stasis is a result of selective neutrality or, in other words, populations drift randomly through sequence space but stay on neutral networks.

Selective neutrality plays an active role in optimization. On a rugged landscape in a constant environment without neutrality, populations are regularly caught in evolutionary traps: whenever a population reaches a local optimum in sequence space, i.e. a point that has no neighbors with higher fitness values, optimization comes to an end. If we are dealing with a sufficiently high degree of neutrality, however, the landscape consists of extended neutral networks for all common phenotypes (Reidys et al., 1997). Almost all points having no further advantageous neighbors belong to one of the extended neutral networks. When a population reaches such a point at the end of an adaptive phase, it starts drifting randomly on the network until it comes to an area that contains also points of higher fitness. There, the next adaptive period starts and the population continues the hill-climbing process. The role of neutral variants is to enable populations to leave local fitness optima and to proceed towards areas of higher fitness in sequence space. Optimization on realistic landscapes is a process on two time scales: fast adaptive phases with substantial increase in fitness are interrupted by periods of random drift during which fitness is essentially constant. The combination of adaptation and drift allows escape from evolutionary traps and, depending on the degree of neutrality, eventually leads to the global optimum of the landscape.

RNA PERSPECTIVES

Molecular evolution experiments with RNA molecules and the accompanying theoretical descriptions made three important contributions to evolutionary biology:

1. The role of replicative units in the evolutionary process has been clarified, the conditions for the occurrence of error thresholds have been laid down and the role of neutrality has been elucidated.
2. The darwinian principle of (natural) selection has shown to be no privilege of cellular life since it is valid also in serial transfer experiments, flow-reactors and other laboratory assays such as SELEX.
3. Evolution in molecular systems is faster than organismic evolution by many orders of magnitude and thus allows observation of optimization and adaptation on easily accessible time-scales, i.e. within days or weeks.

The third issue made selection and adaptation subjects of laboratory investigations. In all these systems the coupling between different replicons is weak: in the simplest case there is merely competition for common resources, for example the raw materials for replication. With more realistic chemical reaction mechanisms a sometimes substantial fraction of the replicons is unavailable as long as templates are contained in complexes. None of these systems, however, comes close to the strong interactions and interdependencies characteristic of ecosystems.

In contrast to the weakly coupled networks of replicons considered in this contribution, *hypercycles* (Eigen, 1971; Eigen and Schuster, 1979) involve specific catalysis beyond mere template instruction (Figure 1.9). In the simplest case, where we consider catalyzed replication reactions explicitly, the reaction equations are of the form:

$$(A) + I_k + I_l \rightarrow 2I_k + I_l. \tag{13}$$

Here a copy of I_k is produced using another macromolecular species I_l as a specific catalyst for the replication reaction. A more realistic ver-

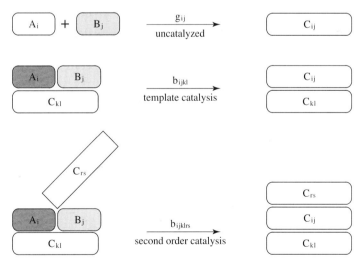

FIGURE 1.9 Modes of template formation. In complex systems of mixed template and depending on the underlying mechanism of template synthesis, different modes of dynamic behavior are possible. Uncatalyzed synthesis generally corresponds to linear growth. Template-instructed synthesis gives parabolic or exponential growth. The coupling of systems involving second order autocatalysis can also give rise to hyperbolic growth, as has been predicted for hypercycles (Eigen and Schuster, 1979).

sion of (13) that might be experimentally feasible is

$$A_i + B_j + C_{kl} + C_{rs} \xrightleftharpoons[\overline{a}_{ijkl}]{a_{ijkl}} A_i B_j C_{kl} + C_{rs} \xrightleftharpoons[\overline{e}_{ijklrs}]{e_{ijklrs}} A_i B_j C_{kl} C_{rs} \xrightarrow{b_{ijklrs}}$$

$$C_{ij} C_{kl} C_{rs} \xrightarrow{f_{ijklrs}} C_{ij} + C_{kl} C_{rs} \xrightleftharpoons[\overline{d}_{ijkl}]{d_{ijkl}} C_{ij} + C_{kl} + C_{rs}$$

Here the template C_{rs} plays the role of a ligase for the template-directed replication step.

The kinetic differential equation

$$\dot{x}_k = x_k \left(\sum_l a_{kl} x_l - \Phi(x) \right),$$

corresponding to the mechanism (13), has been termed *second-order replicator equation* (Schuster and Sigmund, 1983). These systems can display enormous diversity of dynamic behavior (Hofbauer and Sigmund, 1998) depending on the structure of the matrix (a_{kl}) of coupling constants which describes the catalytic activity of one species (I_l) on the replication of another one (I_k). Second-order replicator equations are mathemically equivalent to Lotka–Volterra equations used in mathematical ecology (Hofbauer, 1981). Indeed, recent research in the group of John McCaskill in Jena (McCaskill, 1997; Wlotzka and McCaskill, 1997) deals with *molecular ecologies* of strongly interacting replicons.

The work with RNA replicons has had a pioneering character. Both the experimental approach to evolution in the laboratory and the development of a theory of evolution are much simpler for RNA than for proteins or viruses. On the other hand, genotype and phenotype are more closely linked in RNA than in any other system. The next logical step in theory (Eigen and Schuster, 1979; Happel *et al.*, 1996) and experiment (Eigen *et al.*, 1991) consists of the development of a coupled RNA–protein system that makes use of both replication and translation. This achieves the effective decoupling of genotype and phenotype that is characteristic of all living organisms: RNA is the genotype, protein the usual phenotype and thus genotype and phenotype are no longer housed in the same molecule. The development of a theory of evolution in the "RNA–protein world" requires little more

than an understanding of the sequence–structure relations in proteins. There, a huge body of theoretical and empirical knowledge is already available and the daily growing sequence and structure databanks provide a substantial amount of not yet exploited information.

Virus life-cycles represent the next logical step in increasing complexity of genotype–phenotype interactions. RNA viruses are the simplest candidates and indeed the development of a phage in a bacterial cell has already been modeled in a pioneering paper by Charles Weissmann (1974). Complete viral RNA-genomes are now accessible to computational investigations searching for functional substructures (Hofacker *et al.*, 1998) and we can expect progress in understanding viral phenotypes in the not-too-distant future.

ACKNOWLEDGEMENTS

The work reported here was supported financially by the Austrian Fonds zur Förderung der Wissenschaftlichen Forschung, Projects No. 11065-CHE, 12591-INF, and 13093-GEN, by the European Commission, Project No. PL970189, and by the Santa Fe Institute.

REFERENCES

Alves, D. and Fontanari, J.F. (1996) Population genetics approach to the quasispecies model. *Phys. Rev. E*, **54**, 4048–4053.

Alves, D. and Fontanari, J.F. (1998) Error threshold in finite populations. *Phys. Rev. E*, **57**, 7008–7013.

Bachmann, P.A., Luisi, P.L. and Lang, J. (1992) Autocatalytic self-replicating micelles as models for prebiotic structures. *Nature*, **357**, 57–59.

Bartel, D.P. and Szostak, J.W. (1993) Isolation of new ribozymes from a large pool of random sequences. *Science*, **261**, 1411–1418.

Beaudry, A.A. and Joyce, G.F. (1992) Directed evolution of an RNA enzyme. *Science*, **257**, 635–641.

Biebricher, C.K. and Eigen, M. (1998) Kinetics of RNA replication by Qβ replicase. In: *RNA Genetics, vol. I: RNA Directed Virus Replication* (eds Domingo, E., Holland, J.J. and Ahlquist, P.), pp. 1–21. CRC Press, Boca Raton, FL.

Breaker, R.R. and Joyce, G.F. (1994) Emergence of a replicating species from an *in vitro* RNA evolution reaction. *Proc. Natl Acad. Sci. USA*, **91**, 6093–6097.

Cate, J.H., Gooding, A.R., Podell, E. *et al.* (1996) Crystal structure of a group I ribozyme domain: principles of RNA packing. *Science*, **273**, 1678–1685.

Cech, T.R. (1983) RNA splicing: three themes with variations. *Cell*, **34**, 713–716.

Cech, T.R. (1986) RNA as an enzyme. *Sci. Am.*, **255**(5), 76–84.

Cech, T.R. (1990) Self-splicing of group I introns. *Ann. Rev. Biochem.*, **59**, 543–568.

Cuenoud, B. and Szostak, J.W. (1995) A DNA metalloenzyme with DNA ligase activity. *Nature*, **375**, 611–614.

Demetrius, L., Schuster, P. and Sigmund, K. (1985) Polynucleotide evolution and branching processes. *Bull. Math. Biol.*, **47**, 239–262.

Domingo, E. (1996) Biological significance of viral quasispecies. *Viral Hepatitis Rev*, **2**, 247–261.

Domingo, E. and Holland, J.J. (1997) RNA virus mutations and fitness for survival. *Ann. Rev. Microbiol.*, **51**, 151–178.

Drake, J.W. (1993) Rates of spontaneous mutation among RNA viruses. *Proc. Natl Acad. Sci. USA*, **90**, 4171–4175.

Eigen, M. (1971) Selforganization of matter and the evolution of macromolecules. *Naturwissenschaften*, **58**, 465–523.

Eigen, M. and Schuster, P. (1977) The hypercycle. A principle of natural self-organization. Part A: Emergence of the hypercycle. *Naturwissenschaften*, **64**, 541–565.

Eigen, M. and Schuster, P. (1979) *The Hypercycle – A Principle of Natural Self-Organization*. Springer-Verlag, Berlin.

Eigen, M. and Schuster, P. (1982) Stages of emerging life – five principles of early organization. *J. Mol. Evol.*, **19**, 47–61.

Eigen, M., McCaskill, J. and Schuster, P. (1989) The molecular quasispecies. *Adv. Chem. Phys.*, **75**, 149–263.

Eigen, M., Biebricher, C.K., Gebinoga, M. and Gardiner Jr, W.C. (1991) The hypercycle. Coupling of RNA and protein biosynthesis in the infection cycle of an RNA bacteriophage. *Biochemistry*, **30**, 11005–11018.

Ekland, E.H. and Bartel, D.P. (1996) RNA-catalysed RNA polymerization using nucleoside triphosphates. *Nature*, **382**, 373–376.

Ekland, E.H., Szostak, J.W. and Bartel, D.P. (1995) Structurally complex and highly active RNA ligases derived from random RNA sequences. *Science*, **269**, 364–370.

Elena, S.F., Cooper, V.S. and Lenski, R.E. (1996) Punctuated evolution caused by selection of rare beneficial mutants. *Science*, **272**, 1802–1804.

Eschenmoser, A. (1993) Hexose nucleic acids. *Pure Appl. Chem.*, **65**, 1179–1188.

Fahy, E., Kwoh, D.Y. and Gingeras, T.R. (1991) Self-sustained sequence replication (3SR): An isothermal transcription-based amplification system alternative to PCR. *PCR Methods Appl.*, **1**, 25–33.

Ferré-D'Amaré, A.R., Zhou, K. and Doudna, J.A. (1998) Crystal structure of a hepatitis delta virus ribozyme. *Nature*, **395**, 567–574.

Fontana, W. and Schuster, P. (1987) A computer model of evolutionary optimization. *Biophys. Chem.*, **26**, 123–147.

Fontana, W. and Schuster, P. (1998) Continuity in evolution. On the nature of transitions. *Science*, **280**, 1451–1455.

Fontana, W., Konings, D.A.M., Stadler, P.F. and Schuster, P. (1993) Statistics of RNA secondary structures. *Biopolymers*, **33**, 1389–1404.

Fox, S.W. and Dose, H. (1977) *Molecular Evolution and the Origin of Life*. Academic Press, New York.

Gesteland, R.F. and Atkins, J.F. (eds) (1993) *The RNA World*. Cold Spring Harbor Laboratory Press, Plainview, NY.

Gilbert, W. (1986) The RNA world. *Nature*, **319**, 618.

Green, R. and Noller, H.F. (1997) Ribosomes and translation. *Ann. Rev. Biochem.*, **66**, 679–716.

Guerrier-Takada, C., Gardiner, K., Marsh, T., Pace, N. and Altman, S. (1983) The RNA moiety of ribonuclease P is the catalytic subunit of the enzyme. *Cell*, **35**, 849–857.

Happel, R. and Stadler, P.F. (1999) Autocatalytic replication in a cstr and constant organization. *J. Math. Biol.*, in press. SFI preprint 95–07–062.

Happel, R., Hecht, R. and Stadler, P.F. (1996) Autocatalytic networks with translation. *Bull. Math. Biol.*, **58**, 877–905.

Hofacker, I.L., Fekete, M., Flamm, C. *et al.* (1998) Automatic detection of conserved RNA structure elements in complete RNA virus genomes. *Nucl. Acids Res.*, **26**, 3825–3836.

Hofbauer, J. (1981) On the occurrence of limit cycles in the Volterra–Lotka differential equation. *Nonlin. Anal.*, **5**, 1003–1007.

Hofbauer, J. and Sigmund, K. (1998) *Dynamical Systems and the Theory of Evolution*. Cambridge University Press, Cambridge.

Hofbauer, J., Schuster, P. and Sigmund, K. (1981) Competition and cooperation in catalytic selfreplication. *J. Math. Biol.*, **11**, 155–168.

Huynen, M.A. (1996) Exploring phenotype space through neutral evolution. *J. Mol. Evol.*, **43**, 165–169.

Huynen, M.A., Stadler, P.F. and Fontana, W. (1996) Smoothness within ruggedness. The role of neutrality in adaptation. *Proc. Natl Acad. Sci. USA*, **93**, 397–401.

Jenne, A. and Famulok, M. (1998) A novel ribozyme with ester transferase activity. *Chem. Biol.*, **5**, 23–34.

Jiang, L., Suri, A.K., Fiala, R. and Patel, D.J. (1997) Saccharide-RNA recognition in an aminoglycoside antibiotic-RNA aptamer complex. *Chem. Biol.*, **4**, 35–50.

Joyce, G.F. (1991) The rise and fall of the RNA world. *New Biol.*, **3**, 399–407.

Kauffman, S.A. (1993) *The Origins of Order. Self-Organization and Selection in Evolution*. Oxford University Press, New York.

Kimura, M. (1983) *The Neutral Theory of Molecular Evolution*. Cambridge University Press, Cambridge.

Kramer, F.R., Mills, D.R., Cole, P.E., Nishihara, T. and Spiegelman, S. (1974) Evolution *in vitro*: sequence and phenotype of a mutant RNA resitant to ethidium bromide. *J. Mol. Biol.*, **89**, 719–736.

Lee, D.H., Granja, J.R., Martinez, J.A., Severin, K. and Ghadiri, M.R. (1996) A self-replicating peptide. *Nature*, **382**, 525–528.

Lee, D.H., Severin, K., Yokobayashi, Y. and Ghadiri, M.R. (1997) Emergence of symbiosis in peptide self-replication through a hypercyclic network. *Nature*, **390**, 591–594.

Leung, D.W., Chen, E. and Goeddel, D.V. (1989) A method for random mutagenesis of a defined DNA segment using a modified polymerase chain reaction. *Technique*, **1**, 11–15.

Lohse, P.A. and Szostak, J.W. (1996) Ribozyme-catalyzed amino-acid transfer reactions. *Nature*, **381**, 442–444.

Lorsch, J.R. and Szostak, J.W. (1994) *In vitro* evolution of new ribozymes with polynucleotide kinase activity. *Nature*, **371**, 31–36.

Lorsch, J.R. and Szostak, J.W. (1995) Kinetic and thermodynamic characterization of the reaction catalyzed by a polynucleotide kinase ribozyme. *Biochemistry*, **33**, 15315–15327.

Luisi, P.L., Walde, P. and Oberholzer, T. (1994) Enzymatic RNA synthesis in self-reproducing vesicles: An approach to the construction of a minimal synthetic cell. *Ber. Bunsenges. Phys. Chem.*, **98**, 1160–1165.

Martinez, M.A., Vartanian, J.P. and Wain-Hobson, S. (1994) Hypermutagenesis of RNA using human immunodeficiency virus type 1 reverse transcriptase and biased dNTP concentrations. *Proc. Natl Acad. Sci. USA*, **91**, 11787–11791.

Mason, S.F. (1991) *Chemical Evolution. Origin of the Elements, Molecules, and Living Systems*. Clarendon Press, Oxford.

Maynard Smith, J. and Szathmáry, E. (1995) *The Major Transitions in Evolution*. W.H. Freeman, Oxford.

McCaskill, J. (1984) A localization threshold for macromolecular quasispecies from continuously distributed replication rates. *J. Chem. Phys.*, **80**, 5194–5202.

McCaskill, J.S. (1997) Spatially resolved *in vitro* molecular ecology. *Biophys. Chem*, **66**, 145–158.

Mills, D.R., Peterson, R.L. and Spiegelman, S. (1967) An extracellular Darwinian experiment with a self-duplicating nucleic acid molecule. *Proc. Natl Acad. Sci. USA*, **58**, 217–224.

Mullis, K.B. (1990) The unusual origin of the polymerase chain reaction. *Sci. Am.*, **262**(4), 36–43.

Noller, H.F., Hoffarth, V. and Zimniak, L. (1992) Unusual resistance of peptidyl transferase to protein extraction procedures. *Science*, **256**, 1416–1419.

Nowak, M. and Schuster, P. (1989) Error thresholds of replication in finite populations. Mutation frequencies and the onset of Muller's ratchet. *J. Theor. Biol.*, **137**, 375–395.

Nowick, J.S., Feng, Q., Ballester, T. and Rebek Jr, J. (1991) Kinetic studies and modeling of a self-replicating system. *J. Am. Chem. Soc.*, **113**, 8831–8839.

Orgel, L.E. (1986) RNA catalysis and the origin of life. *J. Theor. Biol.*, **123**, 127–149.

Orgel, L.E. (1987) Evolution of the genetic apparatus. A review. *Cold Spring Harbor Symp. Quant. Biol.*, **52**, 9–16.

Orgel, L.E. (1992) Molecular replication. *Nature*, **358**, 203–209.

Pflug, H.D. and Jaeschke-Boyer, H. (1979) Combined structural and chemical analysis of 3.800-Myr-old microfossils. *Nature*, **280**, 483–486.

Pley, H., Flaherty, K. and McKay, D. (1994) Three-dimensional structures of a hammerhead ribozyme. *Nature*, **372**, 68–74.

Prudent, J.R., Uno, T. and Schultz, P.G. (1994) Expanding the scope of RNA catalysis. *Science*, **264**, 1924–1927.

Reidys, C.M., Stadler, P.F. and Schuster, P. (1997) Generic properties of combinatory maps: natural networks of RNA secondary structures. *Bull. Math. Biol.*, **59**, 339–397.

Reidys, C., Forst, C. and Schuster, P. (1999) Replication and mutation on neutral networks. *Bull. Math. Biol.*, submitted. Also published as: Preprint No. 98–04–036, Santa Fe Institute, Santa Fe, NM 1998.

Schidlowski, M. (1988) A 3.800-million-year isotope record of life from carbon in sedimentary rocks. *Nature*, **333**, 313–318.

Schopf, J.W. (1993) Microfossils of the early archean apex chert: new evidence of the antiquity of life. *Science*, **260**, 640–646.

Schuster, P. (1995) Artificial life and molecular evolutionary biology. In: *Advances in Artificial Life. Proceedings of the Third European Conference on Artificial Life, Ganada, 1995, vol. 929 of Lecture Notes in Artificial Intelligence* (eds Morán, F., Moreno, A., Merelo, J.J. and Chacón, P.), pp. 3–19. Springer-Verlag, Berlin.

Schuster, P. (1997a) Genotypes with phenotypes: adventures in an RNA toy world. *Biophys. Chem.*, **66**, 75–110.

Schuster, P. (1997b) Landscapes and molecular evolution. *Physica D*, **107**, 351–365.

Schuster, P. and Sigmund, K. (1983) Replicator dynamics. *J. Theor. Biol.*, **100**, 533–538.

Schuster, P. and Sigmund, K. (1985) Dynamics of evolutionary optimization. *Ber. Bunsenges. Phys. Chem.*, **89**, 668–682.

Schuster, P. and Swetina, J. (1988) Stationary mutant distribution and evolutionary optimization. *Bull. Math. Biol.*, **50**, 635–660.

Schuster, P., Fontana, W., Stadler, P.F. and Hofacker, I.L. (1994) From sequences to shapes and back: a case study in RNA secondary structures. *Proc. R. Soc. Lond. B*, **255**, 279–284.

Schwartz, A.W. (1997) Speculation on the RNA precursor problem. *J. Theor. Biol.*, **187**, 523–527.

Scott, W.G., Finch, J.T. and Klug, A. (1995) The crystal structure of an all-RNA A proposed mechanism for RNA catalytic cleavage. *Cell*, **81**, 991–1002.

Segel, L.A. and Slemrod, M. (1989) The quasi-steady state assumption: a case study in perturbation. *SIAM Rev.*, **31**, 446–477.

Severin, K., Lee, D.H., Granja, J.R., Martinez, J.A. and Ghadiri, M.R. (1997) Peptide self-replication via template directed ligation. *Chemistry*, **3**, 1017–1024.

Spiegelman, S. (1971) An approach to the experimental analysis of precellular evolution. *Rev. Biophys.*, **4**, 213–253.

Stadler, P.F. (1991) Complementary replication. *Math. Biosci.*, **107**, 83–109.

Swetina, J. and Schuster, P. (1982) Self-replication with errors – a model for polynucleotide replication. *Biophys. Chem.*, **16**, 329–345.

Szathmáry, E. and Gladkih, I. (1989) Sub-exponential growth and coexistence of non-enzymatically replicating templates. *J. Theor. Biol.*, **138**, 55–58.

Tarazona, P. (1992) Error-thresholds for molecular quasi-species as phase transitions: from simple landscapes to spinglass models. *Phys. Rev. A[15]*, **45**, 6038–6050.

Tjivikua, T., Ballester, P. and Rebek Jr, J. (1990) A self-replicating system. *J. Am. Chem. Soc.*, **112**, 1249–1250.

Uhlenbeck, O.C. (1987) A small catalytic oligori-bonucleotide. *Nature*, **328**, 596–600.

Varga, S. and Szathmáry, E. (1997) An extremum principle for parabolic competition. *Bull. Math. Biol.*, **59**, 1145–1154.

Von Kiedrowski, G. (1986) A self-replicating hexadeoxynucleotide. *Angew. Chem. Int. Ed. Engl.*, **25**, 932–935.

Von Kiedrowski, G. (1993) Minimal replicator theory I: Parabolic versus exponential growth. In: *Bioorganic Chemistry Frontiers*, vol. 3, pp. 115–146. Springer-Verlag, Berlin.

Wecker, M., Smith, D. and Gold, L. (1996) *In vitro* selection of a novel catalytic RNA: character-ization of a sulfur alkylation reaction and interaction with a small peptide. *RNA*, **2**, 982–994.

Weissmann, C. (1974) The making of a phage. *FEBS Lett. (Suppl.)*, **40**, S10–S12.

Wills, P.R., Kauffman, S.A., Stadler, B.M. and Stadler, P.F. (1998) Selection dynamics in autocatalytic systems: templates replicating through binary ligation. *Bull. Math. Biol.*, in press, Santa Fe Institute Preprint 97–07–065.

Wilson, C. and Szostak, J.W. (1995) *In vitro* evo-lution of a self-alkylating ribozyme. *Nature*, **374**, 777–782.

Wlotzka, B. and McCaskill, J.S. (1997) A molecu-lar predator and its prey: coupled isothermal amplification of nucleic acids. *Chem. Biol.*, **4**, 25–33.

Zhang, B. and Cech, T.R. (1997) Peptide bond formation by *in vitro* selected ribozymes. *Nature*, **390**, 96–100.

Zhang, B. and Cech, T.R. (1998) Peptidyl-trans-ferase ribozymes: *trans* reactions, structural characterization and ribosomal RNA-like fea-tures. *Chem. Biol.*, **5**, 539–553.

Virus Origins: Conjoined RNA Genomes as Precursors to DNA Genomes

Hugh D. Robertson and Olivia D. Neel

INTRODUCTION

The rapid and unexpected progress in RNA research during the past two decades has led to many theoretical and practical advances. RNA's unprecedented ability to act both as a template for information storage and as an enzymatic molecule has led to the proposal that primitive living systems were based on RNA, with protein synthesis and DNA templates for information storage added later. If this "RNA world" hypothesis is to be taken seriously, it is necessary to explain a number of developments during evolution at the RNA level, including not only coding and self-replication but also the ability of genetic information to rearrange, recombine and expand itself, creating ever more complex living systems. It is the purpose of this chapter to interpret certain RNA-level events that must have taken place in simple viral or pre-viral systems in this light.

In what follows, we will seek, in what is known today about simple self-replicating RNAs, enlightenment regarding how they acquired their singular nature. We will focus on a process that has been called "RNA conjunction" (Branch *et al.*, 1989) or "RNA capture" (Diener, 1989), in which two independent, functional RNA molecules become associated in such a way that each retains function and contributes properties to the resulting "conjoined RNA" or "RNA mosaic". As we shall see, RNA conjunction differs from both random RNA rearrangements and recombination between the RNAs of closely related RNA viruses, in that two independent activities, each embodied in a separately evolved RNA, are required to survive the conjunction or capture process that joins them together. This is not to suggest that the mechanisms that drive random or homologous RNA recombination are not used in RNA conjunction but rather that conjoined RNAs comprise a highly selected subset of successful multifunctional RNA mosaics.

Assuming that a prebiotic system of chemical evolution somehow produced RNA in the first place, there has been intense speculation about how early, small RNAs might have replicated. And, assuming that a genetic code leading from nucleic acids to proteins also evolved, there has been equally thorough scrutiny as to how the replicating and coding RNAs in such a hypothetical primitive time might have combined and expanded leading to viral and, ultimately, cellular RNAs. Once converted to the more stable DNA storage system, these molecules may have formed the basis for modern DNA viruses. Until recently, studies on the creation and properties of conjoined RNAs have been theoretical in nature, emphasizing computer modeling and mutational probabilities rather than experiments or molecular prototypes. Recent work to be reviewed below shows that there is one class of primitive life forms – the viroid-like

Origin and Evolution of Viruses
ISBN 0–12–220360-7

pathogens – whose properties today could help us to understand how primitive RNA-based self-replication may have been compatible with expansion to produce more complex RNAs. In summary, the causative agent for human hepatitis delta contains two specialized domains, one concerned with replication and the other encoding a single protein (Branch *et al.*, 1989; Purcell and Gerin, 1996; Taylor, 1996). From both theoretical and practical considerations, it now seems likely that the two RNA domains that embody these two functions arose separately and were subsequently joined together. The existence of a prototype conjoined viral RNA that is functional in the modern world provides a singular opportunity for testing some of the above ideas, and has already caused a redoubling of efforts to find other examples as well as to understand the one we have.

In this chapter, we will first review briefly current knowledge about RNA rearrangement and recombination, principally in viruses. We will cite evidence for various mechanisms catalyzing these events. We will also review some recent evidence that, at least in one system, RNA recombination can occur in what appears to be a spontaneous manner, as if it were an inherent property of the RNA. Such a potential, even at low frequency, would expand opportunities for RNA conjunction. Second, we will outline the significance of work on viroid-like pathogens, circular RNA replication and their potential relation to early RNA. We will then put delta agent RNA in context, discussing the relative significance of its dual RNA nature to the RNA recombination systems already cited. Finally, we will relate the early emergence of RNA mosaics to developments leading to today's DNA-based systems of viral gene expression.

RNA REARRANGEMENT: MECHANISMS OF VIRAL RNA RECOMBINATION

While most studies on genetic recombination have been carried out on DNA-based organisms, there are several examples from the field of RNA virology that clearly demonstrate that RNA recombination is a reality. Animal, plant and bacterial virus systems have all been identified in which RNA recombination takes place (reviewed in Lai 1992a,b, 1995, Nagy and Simon, 1997). In most of these studies, emphasis is placed on the types of RNA molecules that are joined (homologous versus non-homologous) and the mechanism by which two separate RNAs become recombined. The majority of RNA recombination appears to involve closely related, or homologous, sequences, in which mutant markers are reassorted in an orderly fashion. Most picornavirus and coronavirus RNA recombination takes place in this way (Lai, 1992b; Zhang and Lai, 1994; Pilipenko *et al.*, 1995; Duggal *et al.*, 1997), although there are exceptions. The same is true of most plant virus RNA recombination events (Gibbs and Cooper, 1995; Le Gall *et al.*, 1995; Figlerowicz *et al.*, 1997; Fraile *et al.*, 1997; Nagy and Bujarski, 1997, 1998), including those involving the well-studied brome mosaic virus (BMV) system. However, there are exceptions, as exemplified by the turnip crinkle virus (TCV)/satellite system (Carpenter *et al.*, 1995; Carpenter and Simon, 1996a,b; Nagy and Simon, 1997).

The favored recombination mechanism for viral RNA molecules is one involving template switching (analogous to the copy/choice mechanism of DNA recombination), in which the RNA-dependent RNA polymerase of the virus ceases the copying of a particular strand in midreaction, moves to a second strand with nascent RNA still attached and resumes synthesis at a point in the second strand near the place where copying ended in the first one (Lai, 1992a, b). Other possibilities include a cleavage/ligation reaction resembling trans RNA splicing (Maroney *et al.*, 1996) and a recently identified transesterification process (Chetverin *et al.*, 1997). The majority of well-studied examples of both plant and animal viruses have been assigned to the template switching category of RNA recombination, and much effort has gone into the identification of regions of sequence or secondary structure which would promote the template switching event.

An orthodox view of viral RNA recombination would thus include the involvement of known components – the viral RNA-dependent

RNA polymerase and viral RNA strands for template switching; previously known ribozymes or conventional RNA processing enzymes for the break/rejoin reactions analogous to trans RNA splicing. The ability to harness such mechanisms to reassort viral RNA genomes and promote new, and perhaps more fit, combinations is viewed as a significant contributing factor to the evolution of RNA viruses.

Included in the catalog of RNA recombination studies are a few cases in which viruses have picked up host-cell RNA sequences. These examples will be important when we consider the nature of the reactions that produced delta agent RNA. In plant viruses, Mayo and Jolley (1991) have shown the occasional uptake of RNA sequences encoded in host chloroplast DNA. Because the acquired sequence is part of an open reading frame, and the recombination site is within 7 bases of an exon–intron boundary, the authors speculate that the recombination occurred by a trans splicing (or break/rejoin) mechanism. That plant viruses can recombine with cellular mRNAs to promote new sequence combinations was proved unmistakably when Greene and Allison (1994) demonstrated that, in transgenic plants containing a viral RNA sequence now expressed in the cell as a conventional mRNA, recombination with exogenous virus could take place as its RNA replicated in the cells of the transgenic plant. Presumably such events can occur with more conventional cellular mRNAs as well, although none were reported in this system.

In animal virus systems, there are several examples in which host cell RNAs are incorporated into viral RNA. In influenza viral RNA, for example (Khatchkian et al., 1989) a segment of host 28S rRNA is incorporated into the hemagglutinin gene by a mechanism involving nonhomologous recombination. It is reported that the viral species containing the host RNAs have increased viral pathogenicity, although it is not known whether this trait conferred a selective advantage upon the recombinant influenza virus population. In Sindbis virus (Monroe and Schlesinger, 1983), tRNA sequences are sometimes incorporated into the 5' termini of defective RNAs; while in the TCV satellite system, the incorporation of nonviral sequences has been reported (Carpenter et al., 1995; Nagy and Simon, 1997).

Perhaps the most striking example of RNA recombination between a virus and host gene sequences is the bovine viral diarrhea virus, BVDV, a pestivirus with a single-stranded RNA genome 12.5 kb in length that encodes a single polyprotein (Collett et al., 1989; Meyers et al., 1991). In the course of an investigation of changes in cytopathogenicity among different BVDV strains, many were found to have acquired cellular RNA sequences into a domain encoding a non-structural protein. The most frequently observed inserts consisted of sequences from the host ubiquitin gene (Meyers et al., 1991). It is not known whether expression of the acquired sequences took place, but the virus was clearly able to survive this acquisition into its polyprotein, and in some cases to acquire a selective advantage. This phenomenon could be reproduced, and the investigators concluded that some unknown features of BVDV RNA not only facilitated the recombination process but also conferred some selective advantage to the recombinants.

Thus today's viral RNA recombination mechanisms can occasionally lead to the acquisition of host sequences apparently unrelated to the virus. Before concluding that such events must always be mediated by one of only two mechanisms – template switching by the viral RNA polymerase or specific cleavage and ligation of RNA resembling trans splicing – it is as well to consider some recent findings from a phage system, which suggest that RNA recombination may take place by a more general, chemical mechanism. In the Qβ phage system, Chetverin et al. (1997) have discovered non-homologous RNA recombination, which takes place in a cell-free system at a variety of sequence locations. The non-homologous recombinations observed are entirely dependent on the 3' hydroxyl group of the 5' fragment in the joining reaction. Chetverin et al. (1997) believe that the mechanism by which these recombinants are generated is "entirely different from copy choice".

Nagy and Simon (1997), in reviewing the above work from a perspective favoring template switching, concede that the data of Chetverin et al. can all be explained by an RNA-mediated transesterification mechanism, but that a template-switching mechanism is not excluded. While further controls need to be done, uncoupling the recombination events from the Qβ replicase-dependent amplification needed to detect the results, it seems probable that an RNA-mediated breakage and ligation accounts for at least a fraction of Qβ RNA recombinants. And, while Nagy and Simon (1997) correctly point out that "it is difficult to estimate how widespread [such a system] might be in natural virus systems", the prospect that RNA molecules have a certain probability for spontaneous rearrangement provides additional scope for the evolution of viral RNAs.

VIROID-LIKE AGENTS, CIRCULAR RNA REPLICATION AND EARLY RNA GENOMES

Early reports of viroid-like RNA pathogens centered on plant viroids and their relatives (Gross *et al.*, 1978; Diener, 1979; Semancik, 1987; Branch *et al.*, 1990). More recently, the causative agent for delta hepatitis in humans was confirmed to be a circular viroid-like RNA (Kos *et al.*, 1986; Wang *et al.*, 1986; Makino *et al.*, 1987; Taylor *et al.*, 1987). Delta RNA is about four times the size of plant viroids. The principal effort which led to the working out of the replication cycle for these agents took place between 1981 and 1987, at a time when the role of RNA in the evolution of primitive, self-replicating systems was just coming into focus. For example, two proposals based on both the template and enzymatic qualities of RNA (Sharp, 1985; Gilbert, 1986) appeared during that time. The potential for RNA circles to simplify the tasks required for replication in a primitive environment is considerable, and includes at least four elements. First, as with circular DNA genomes (Reanney and Ralph, 1968), there are advantages involving the ability to tolerate gene duplication and subsequent variation while preserving the initial sequence; second, as pointed out previously (Robertson, 1992), the synthesis of multimeric copies on a circular complementary template leads automatically to the unwinding of each

copy from duplex structure with its template as it is displaced by the next copy; third, as also pointed out by Diener (1989), the need for a specific initiation point at one end of a linear genome is eliminated; and fourth, circular RNAs with no free ends – especially if they also contain extensive secondary structure as do the RNAs of viroid-like pathogens – are less susceptible to breakdown by ribonucleases than normal RNA molecules.

Many advocates of the "RNA world" hypothesis (Gesteland and Atkins, 1993) have proposed a set of common assumptions. One is that RNA molecules evolved self-replication first, then the property of protein coding and finally an information storage system using DNA copies. This idea leads to the prediction that genetic systems of today will contain features reflecting such a history. In the context of viroid-like RNAs, one way to test these assumptions is to consider the way today's viroid-like RNAs, including that of the delta agent, are thought to replicate. We proposed the rolling circle pathway as a general mechanism for viroid-like RNA replication (Branch *et al.*, 1981; Branch and Robertson, 1984), in which multimeric copies of RNA strands are synthesized and then processed to yield monomeric progeny molecules (Figure 2.1). This pathway has been demonstrated for a number of viroid-like RNAs, including the delta agent (Chen *et al.*, 1986). Host enzymes are required for the RNA synthetic steps of this pathway, and are the only proteins absolutely required for replication (since examples of RNA-catalyzed cleavage of multimers and ligation to form circles have been documented in several systems).

Studies by Cech and others (reviewed in Cech, 1989) have begun to demonstrate how RNA may have first begun to copy itself. These proposals reveal several potential problems, e.g. how to copy accurately (and protect from exconuclease cleavage) the ends of such molecules; how to unwind the newly synthesized RNA strand from a stable duplex with its template so that subsequent rounds of copying can proceed; and how to initiate synthesis without a pre-existing set of initiation factors in a way that guarantees accurate inheritance of every base by the progeny RNA. As mentioned above, a circu-

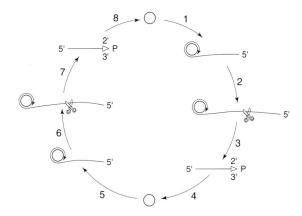

FIGURE 2.1 The rolling circle replication pathway for viroid-like RNAs. In this depiction, the circular genomic ("+") strands are copied into multimeric antigenomic ("−") strands (steps 1 and 2), cleaved to unit length (step 3) and ligated to give minus strand circles (step 4). These antigenomic monomeric circles serve as templates for multimeric genomic strands (steps 5 and 6) which are cleaved to unit length (step 7) and circularized to produce progeny genomic RNAs (step 8).

lar template simplifies all of these difficulties. Since it has no ends, its structure is stabilized in a fashion impossible for linear molecules. In addition, a circular template undergoing copying by an RNA polymerase (whether a primitive one composed of RNA or a modern host protein) will lead to production of greater than unit length multimeric copies, so that the first copy will be displaced from its duplex association with the template as the second copy is synthesized, overcoming the unwinding dilemma. Furthermore, initiation at any point on a circular template leads to a complete copy with no risk of losing ends or other domains. We conclude that the case for primitive circular self-replicating RNAs is a persuasive one.

Another line of thought concerning viroid-like RNAs in evolution emerged shortly after the discovery of eukaryotic RNA splicing, in which a number of investigators speculated on the potential relationship between viroids and intervening sequences, or introns. Shortly after split genes and the need for mRNA splicing were first announced, Roberts (1978) speculated about a connection between viroid-like RNAs and introns, guessing (correctly) that RNA splic-

ing mechanisms might turn out to be reciprocal, not only joining exons but producing circular introns as well. Crick (1979) observed that introns might be excised as circles, while both Diener (1981) and Dickson (1981) pointed out sequence homologies between the plant viroid PSTV and the small nuclear RNAs involved in mRNA splicing. Gilbert (1987) focused on the possibility that early ribozyme-containing introns in the "RNA world" might somehow have served as insertion sequences.

The idea that emerges, then, is that introns originally arose as circular self-replicating RNAs, with a replication pattern that presaged both RNA capture and modern mRNA splicing. The earlier speculations cited above (Roberts, 1978; Crick, 1979; Diener, 1981; Dickson, 1981; Gilbert, 1987) did not focus on the way in which the rolling circle mode of replication used by viroid-like RNAs (Branch and Robertson, 1984) combines stable RNA circles with ribozymes that cleave and ligate RNA (although Crick (1979) does point out that, if introns were excised as circles, "There is little difficulty in thinking of interesting functions which such a single-stranded circular RNA might perform," and gives a reference to viroids). Subsequent publications (Robertson and Branch, 1987; Diener, 1989; Branch *et al.*, 1989; Robertson, 1992) recognized these and other advantages of circular RNA, leading to the idea that viroid-like RNA circles could have developed into introns over evolutionary time. Indeed, if the rolling circle pathway was in fact employed in the RNA world (Gesteland and Atkins, 1993), events taking place during each cycle of RNA synthesis – in which linear monomers built into multimers by repeated copying of a circular RNA template are cut apart and then circularized by ribozyme action – could foreshadow the development of RNA introns. The existence of self-replicating RNA circles equipped with the ribozyme machinery to cleave newly synthesized chains and then join the newly formed ends would lead naturally to events in which cleavage could be followed by the joining of two different molecules – perhaps rarely at first, and then more often. Alternatively, template switching during rolling circle replication could also lead to the joining of a viroid-like RNA and coding segment. In either case, as the population of circular self-replicating RNAs increased, the likelihood of their acquiring newly arisen coding sequences would go up in parallel.

DELTA AGENT RNA

The genome of the hepatitis delta agent or virus is a circular, highly structured single-stranded RNA. As shown in Figure 2.2, delta RNA has two domains, each with features of two quite different RNA types. Shown on the left is the viroid-like region, comprising about one-quarter of the bases. This domain is named after the 350–400 base viroid RNAs that infect plants, and contains most of the sites important for the rolling circle replication pathway, including the ribozyme activities, whose cleavage sites are indicated. The three-quarters of the delta RNA map depicted on the right of Figure 2.2 is very different: as shown, it encodes mRNA for the delta antigen, a minor structural protein of delta particles. The protein-coding region is much more divergent in sequence than the viroid-like region. The differences between these two domains led to the postulate that the delta RNA

genome arose from two domains and the suggestion that this RNA in its present form is the product of RNA conjunction (Branch *et al.*, 1989); and the broadening of this hypothesis (Robertson, 1992).

In common with plant viroids, many of which have a core "central conserved region" involving about 50 bases, which are nearly invariant (Keese and Symons, 1985), part of the delta viroid-like region comprising bases 613–771 and 842–957 shows similar evidence of conservation (Krushkal and Li, 1995). In contrast, the protein coding region can have up to 20% variability between two given strains whose viroid-like sequences are completely conserved (Imazeki *et al.*, 1990). Thus the viroid-like and protein-coding regions have many properties to suggest they originated separately and were joined by a process which – if better understood – could help us to understand not only RNA evolution but also its present-day function. These conclusions would remain valid regardless of whether delta RNA itself turned out to be a primitive molecule – a "missing link" in molecular evolution – or whether delta represents a recent recurrence of processes that have a constant potential to occur among informational RNA molecules which are active in present-day cells.

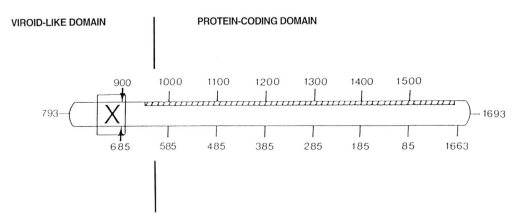

FIGURE 2.2 The genomic RNA of the delta hepatitis agent. As shown, the circular single-stranded RNA molecule comprises two separate domains, the viroid-like region (left) and the protein-coding region (right). Residues 613 and 980 form the boundary between the two regions. The hatched area corresponds to the residues that encode the mRNA for the delta antigen protein, whose coding sequence is complementary to the region shown here. Residues 685 and 900 (arrows) depict the sites of ribozyme cleavage in the genomic and antigenomic strands respectively. The box to their left marked with an "X" is a region of local tertiary structure thought to be involved in interaction with host proteins.

The possibility that some of the above models might be subject to experimental test arose when Brazas and Ganem (1996) detected an mRNA molecule in uninfected liver cells encoding a protein that can strongly influence the replication of hepatitis delta RNA, and which appears to be related in amino-acid sequence and composition to the delta antigen protein, HDAg, for which it has a binding affinity. This host protein, named Dip-A (for Delta-interacting protein A), has been isolated and its gene cloned. The 202-amino-acid Dip-A protein is similar in size to HDAg and 56% of its amino acids are identical or very similar to the corresponding delta antigen residues. The question of whether this sequence similarity is statistically significant has been the subject of two additional publications (Brazas and Ganem, 1997; Long *et al.*, 1997). The predominant conclusion, reached using the latest protein sequence comparison programs devised by Dr S. Altschul of NIH (cited in Brazas and Ganem, 1997), is that the similarity between Dip-A and HDAg remains striking enough for the possibility that they have diverged from a common ancestor to be vigorously pursued.

If such events have taken place, we need to propose testable mechanisms for the creation of viable RNA mosaics. The rolling circle mechanism illustrated in Figure 2.1 and the map of delta RNA in Figure 2.2 allow us to visualize one process of RNA conjunction that might accomplish this end. If a primitive viroid-like RNA 368 bases long occupies residues 613–980 (reading 5′ to 3′ (clockwise) in Figure 2.2) of the delta map, then there must have been a way either during or after RNA synthesis to join residue 980 to the 5′ end of a copy of the cellular RNA to be captured. Brazas and Ganem (1996) have suggested a copy-choice mechanism, in which the enzyme system copying the viroid negative strand template switches, presumably at low frequency, to copying a cellular RNA such as Dip-A mRNA. As outlined in the section on RNA rearrangent above, a number of RNA viruses that undergo RNA rearrangement are thought to use such a template-switching mechanism, in which the virally encoded RNA polymerase moves from one RNA to another while continuing synthesis (Kierkegaard and Baltimore, 1986; Lai, 1992a,b; Carpenter *et al.*, 1995; Nagy and Bujarski, 1996). In contrast, delta agent RNA encodes no RNA polymerase, using instead a host activity.

An alternate mechanism for RNA capture by delta would involve the sites of RNA cleavage and ligation shown in Figure 2.2 (whereby the multimeric RNA precursors produced by rolling circle replication yield monomers that are then circularized): the termini of linear monomers could occasionally be misaligned, joining a linear negative viroid RNA strand to a cellular mRNA species. Such a trans splicing mechanism was proposed by Mayo and Jolly (1991) to account for the occasional acquisition by potato leafroll virus RNA of unexpressed host RNA fragments. The formation of RNA mosaics by either mechanism would be expected to preserve the ability of the delta viroid-like domain to replicate, and the resulting RNA would encode an advantageous protein.

While other cases in which RNA viruses acquire altered RNAs from each other or even from host sources by RNA recombination have been cited above, the delta agent is the only known case where a coding sequence thought to be cellular in origin has entered a functional association with a replicating virus-like element at the level of RNA only, with no DNA involvement, so that the translation-level expression of the protein has survived *and* the replicating element remains functional. If this is indeed the result of RNA capture, leading to a molecule that fits the definition of a conjoined RNA given in the introduction to this chapter, we would assume that such successful events would be rare, and when they happened it is likely that they would usually prove fatal to the capturing organism. If an RNA mosaic were to survive, we would assume both that its ability to replicate has been preserved and that the encoded protein must contribute somehow to the welfare of the mosaic. For example, Brazas and Ganem (1996) show that Dip-A has sequence elements in common with nucleic acid binding proteins and transcription factors. RNA encoding a protein like this, which might help a highly structured RNA molecule like delta to simplify replication, might confer increased survival value to the replicating RNA.

More recent work from the Ganem laboratory on the nature and function of the Dip-A protein and the origin of its mRNA has focused on two issues: the likelihood that delta RNA replication indeed involves host RNA polymerase II and the nature and cellular function of the Dip-A protein itself. The former consideration is important for any speculation about delta RNA recombination mediated by a template-switching mechanism. If host RNA polymerase II can copy delta RNA, then its occasional switch – during delta genomic strand synthesis – to a host nuclear mRNA template would not be entirely far-fetched. The resulting conjoined RNA would contain the delta genomic strand attached to the template for the host mRNA.

Additional mapping experiments carried out by Brazas and Ganem (personal communication) using the mouse Dip-A locus have strengthened the idea that this gene may encode a protein related to transcription. The mouse Dip-A locus (which turns out to be 98% homologous at the amino-acid level to the human Dip-A previously reported (Brazas and Ganem, 1997) maps only 600 bases away from a mouse homolog of the Fra (fos-related antigen) transcription factor family. Since previous work (Brazas and Ganem, 1996) had establishd that Dip-A protein is homologous to Fra-1 over a limited range, this mapping result supports the notion that Dip-A is a member of the Fra family. It is important to recall that – in addition to the finding that Dip-A is a host cellular protein with extensive homology to the delta antigen protein – Dip-A also has an affinity for the delta antigen which may play a role in delta replication. To test this, knockout mice are being developed by Ganem and co-workers, and it should soon be possible to evaluate whether the host Dip-A protein plays a part in the delta RNA synthetic cycle (R. Brazas and D. Ganem, personal communication).

RELATION TO DNA SYSTEMS

It is evident that if RNA conjunction leading to delta-like RNA mosaics with both replicating and functionally translatable protein-coding domains has taken place, we need to consider the consequences both for the evolution of primitive RNA systems, yielding today's DNA-based cellular information system, and for present-day RNA-level events. We suggest that what have now become modern mRNA introns arose because their ancestors carried activities essential for RNA replication in pre-DNA times. The rolling circle yields progeny multimers to be cleaved out and ligated to give progeny circles. This primitive RNA replication pathway could have led first to sequence capture and the formation of RNA mosaics and then to RNA splicing, both using the same pre-existing ribozyme activities. RNA mosaics would build up in this way, interspersing viroid-like replicating elements with newly evolved coding segments (some of which could produce larger proteins by developing the ability to be periodically spliced together using ribozyme action). It is clear that template switching by the primitive RNA-synthesizing machinery could also contribute to this process.

In this context, the ongoing possibility of RNA variation would have been accompanied by a constant tendency to combine, re-assort and test new RNA combinations. The resulting RNA mosaics would eventually be copied into DNA with the arrival of reverse transcriptase activities. The RNAs transcribed from such DNAs, since they in turn would need to be spliced for proper expression of their encoded proteins, would be expected to preserve the built-in splicing capability inherited from their RNA-coded ancestors. Such reactions would be further enhanced by the presence of RNA segments encoding RNA-binding proteins, some of which would evolve into transcription factors for the DNA-directed enzymes now in place. However, the role in RNA replication previously associated with the viroid-like RNA domains would be lost.

What we see, then, is a system in which DNA encoding RNA mosaics emerges with a ready-made RNA splicing system to liberate the coding segments for mRNA function. This early emergence of RNA splicing may explain why it has been retained so far by eukaryotic genomes despite its many problems. Furthermore, the potential for intervening sequences to retain at least those functions required for cleavage and ligation steps in RNA splicing suggests that they may be less inert than previously suspected. The idea considered here – that intervening sequences began as viroid-like RNAs – carries with it the possibility that self-replicating RNAs may re-emerge from transcripts of DNA, combine with other RNA segments and give rise to new combinations of genetic material. In this regard, it is possible that directed versions of such a process could lead to a new class of RNA-level vectors capable of changing the genetic expression of the cells containing them. It is also possible that naturally occurring RNA mosaics with pathogenic potential continue to arise with some regularity *in vivo*. Whether the delta agent RNA arose in this way, or whether it instead represents an ancestral intermediate in molecular evolution, remains to be determined.

Whatever their role may have been in the evolution of RNA viruses, it is clear that the mechanisms and processes for RNA-level genetic assortment and reassortment are alive and well today. Whether they play a role in the continued development of DNA-based viral or retroviral systems needs further evaluation. Now that intron-containing RNAs of both viral and cellular origin are being transcribed from DNA templates, and their rates of sequence variation studied, it is well to remember that such rates would be profoundly influenced if it should turn out that what was thought to be a DNA gene had spent even a short time in RNA form. Such an "escaped gene" would temporarily be free of the usual DNA-based constraints on mutation and recombination, and could thus rapidly turn into something quite different. This "RNA uncertainty principle" applies, for example, in the case of the Dip-A versus delta antigen mRNA sequence comparison, where the divergence of at least one of this pair of sequences descended from a hypothetical common ancestor has taken place at the RNA, not the DNA, level. Further studies on the Dip-A gene sequences and their RNA-based counterparts should allow the extent of "RNA uncertainty" to be discerned and then used to enhance our understanding of DNA *and* RNA virus evolution.

ACKNOWLEDGEMENT

We thank Drs D. Ganem and R. Brazas for helpful discussions and sharing of results prior to publication. Work in our laboratory has been supported by the US National Institutes of Health and by the New York State Science and Technology Foundation.

REFERENCES

Branch, A.D. and Robertson, H.D. (1984) A replication cycle for viroids and other small infectious RNAs. *Science*, **223**, 450–454.

Branch, A.D., Robertson, H.D. and Dickson, E. (1981) Longer-than-unit-length viroid minus strands are present in RNA from infected plants. *Proc. Natl Acad. Sci. USA*, **78**, 6381–6385.

Branch, A.D., Benenfeld, B.J., Baroudy, B.M., Wells, F.V., Gerin, J.L. and Robertson, H.D. (1989) An ultraviolet-sensitive RNA structural element in a viroid-like domain of the hepatitis delta virus. *Science*, **243**, 659–652.

Branch, A.D., Levine, B.J. and Robertson, H.D. (1990) The brotherhood of circular RNA pathogens: viroids, circular satellites, and the delta agent. *Sem. Virol.*, **1**, 143–152.

Brazas R. and Ganem, D. (1996) A cellular homolog of hepatitis delta antigen: implications for viral replication and evolution. *Science*, **274**, 90–94.

Brazas R. and Ganem, D. (1997) Response to [Delta-interacting protein A and the origin of hepatitis delta antigen]. *Science*, **276**, 825.

Carpenter, C.D. and Simon, A.E. (1996a) Changes in locations of crossover sites over time in de novo generated RNA recombinants. *Virology*, **223**, 165–173.

Carpenter, C.D. and Simon, A.E. (1996b) In vivo restoration of biologically active 3' ends of virus-associated RNAs by nonhomologous RNA recombination and replacement of a terminal motif. *J. Virol.*, **70**, 478–486.

Carpenter, C.D., Oh, J. W., Zhang, C. and Simon, A.E. (1995) Involvement of a stem-loop structure in the location of junction sites in viral RNA recombination. *J. Mol. Biol.*, **245**, 608–622.

Cech, T.R. (1989). RNA enzymes. *Adv. Enyzmol*, **62**, 1–36.

Chen, P.J., Kalpana, G., Goldberg, J. *et al.* (1986) Structure and replication of the genome of the hepatitis delta virus. *Proc. Natl Acad. Sci. USA*, **83**, 8774–8778.

Chetverin, A.B., Chetverina, H.V., Demidenko, A.A. and Ugarov, V.I. (1997) Nonhomologous RNA recombination in a cell-free system: evidence for a transesterification mechanism guided by secondary structure. *Cell*, **88**, 503–513.

Collett, M.S., Moening, V. and Horzinek, M.C. (1989) Recent advances in pestivirus research. *J. Gen. Virol.*, **70**, 253–266.

Crick, F. (1979) Split genes and RNA splicing. *Science*, **204**, 264–271.

Dickson, E. (1981) A model for the involvement of viroids in RNA splicing. *Virology*, **115**, 216–221.

Diener, T.O. (1979) *Viroids and Viroid Diseases*, John Wiley, New York.

Diener, T.O. (1981) Are viroids escaped introns? *Proc. Natl Acad. Sci. USA*, **78**, 5104–5105.

Diener, T.O. (1989) Circular RNAs: relics of precellular evolution? *Proc. Natl Acad. Sci. USA*, **86**, 9370–9374.

Duggal, R., Cuconati, A., Gromeier, M. and Wimmer, E. (1997) Genetic recombination of poliovirus in a cell-free system. *Proc. Natl Acad. Sci. USA*, **94**, 13786–13791.

Figlerowicz, M., Nagy, P.D. and Bujarksi, J.J. (1997) A mutation in the putative RNA polymerase gene inhibits nonhomologous, but not homologous, genetic recombination in an RNA virus. *Proc. Natl Acad. Sci. USA*, **94**, 2073–2078.

Fraile, A., Alonso-Prados, J.L., Aranda, M.A., Bernal, J.J., Malpica, J.M. and Garcia-Arenal, F. (1997) Genetic exchange by recombination or reassortment is infrequent in natural populations of a tripartite RNA plant virus. *J. Virol.*, **71**, 934–940.

Gesteland, R.F. and Atkins, J.F. (1993) *The RNA World*. Cold Spring Harbor Laboratory Press, Cold Spring Harbor, NY.

Gibbs, M.J. and Cooper, J.I. (1995) A recombina-tional event in the history of luteoviruses probably induced by base-pairing between the genomes of two distinct viruses. *Virology*, **206**, 1129–1132.

Gilbert, W. (1986) The RNA world. *Nature*, **319**, 618.

Gilbert, W. (1987) The exon theory of genes. *Cold Spring Harbor Symp. Quantit. Biol.*, **52**, 901–905.

Golding, B.G., Tsao, N. and Pearlman, R.E. (1994) Evidence for intron capture: an unusual path for the evolution of proteins. *Proc. Natl Acad. Sci. USA*, **91**, 7506–7509.

Greene, A.E. and Allison, R.F. (1994) Recombination between viral RNA and transgenic plant transcripts. *Science*, **263**, 1423–1425.

Gross, H.J., Domdey, H., Lossow, C. *et al.* (1978) Nucleotide sequence and secondary structure of potato spindle tuber viroid. *Nature*, **273**, 203–211.

Imazeki, F., Omata, M. and Ohto, M. (1990) Heterogeneity and evolution rates of delta virus RNA sequences. *J.Virol.*, **64**, 5594–5599.

Keese, P. and Symons, R.H. (1985) Domains in viroids: evidence of intermolecular RNA rearrangements and their contribution to viroid evolution. *Proc. Natl Acad. Sci. USA*, **82**, 4582–4586.

Khatchkian, D., Orlich, M. and Rott, R. (1989) Increased viral pathogenicity after insertion of a 28S ribosomal RNA sequence into the hemagglutinin gene of an influenza virus. *Nature*, **340**, 156–157.

Kierkegaard, K. and Baltimore, D. (1986) The mechanism of RNA recombination in poliovirus. *Cell*, **47**, 433–443.

Kos, A., Kijema, R., Arnberg, A. C., van der Meide, P.H. and Schellekens, H. (1986) The hepatitis delta virus possesses a circular RNA. *Nature*, **323**, 558–560.

Krushkal, J. and Li, W.H. (1995) Substitution rates in hepatitis delta virus. *J. Mol. Evol.*, **41**, 721–726.

Lai, M.M.C. (1992a) RNA recombination in animal and plant viruses. *Microbiol. Rev.*, **56**, 61–79.

Lai, M.M.C. (1992b) Genetic recombination in RNA viruses. *Curr. Topics Microbiol. Immunol.*, **176**, 21–32.

Lai, M.M.C. (1995) Recombination and its evolutionary effect on viruses with RNA genomes. In: *Molecular Basis of Virus Evolution* (eds Gibbs, A.J., Calisher, C.H., and Garcia-Arenal, F.), pp.119–132. Cambridge University Press, Cambridge

Le Gall, O., Lanneau, M., Candresse, T. and Dumez, J. (1995) The nucleotide sequence of the RNA-2 of an isolate of the English serotype of tomato black ring virus: RNA recombination in the history of nepoviruses. *J. Gen. Virol.*, **76**, 1279–1283.

Long, M., deSouza, S.J. and Gilbert, W. (1997) Delta-interacting protein A and the origin of hepatitis delta antigen. *Science*, **276**, 824–825.

Makino, S., Chang, M.F., Sheih C.K. *et al.* (1987) Molecular cloning and sequencing of a human hepatitis delta virus RNA. *Nature*, **329**, 343–346.

Maroney, P.A., Yu, Y.T., Jankowska, M. and Nilsen, T.W. (1996) Direct analysis of nematode *cis-* and *trans*-spliceosomes: a functional role for U5 snRNA in *spliced* leader addition *trans-splicing* and the identification of novel Sm snRNPs. *RNA*, **2(8)**, 735–745.

Mayo, M.A. and Jolly, C.A. (1991) The 5′-terminal sequence of potato leafroll virus RNA: evidence of recombination between virus and host RNA. *J. Gen. Virol.*, **72**, 2591–2595.

Meyers, G., Tautz, N., Dubovi, E.J. and Thiel., H.-J. (1991) Viral cytopathogenicity correlated with integration of ubiquitin-coding sequences. *Virology*, **180**, 602–616.

Monroe, S.S. and Schlesinger, S. (1983) RNAs from two independently isolated defective interfering particles of Sindbis virus contain a cellular tRNA sequence at their 5′-ends. *Proc. Natl Acad. Sci. USA*, **80**, 3279–3283.

Nagy, P.D. and Bujarksi, J.J. (1996) Homologous RNA recombination in brome mosaic virus: AU-rich sequences decrease the accuracy of crossovers. *J. Virol.*, **70**, 415–426.

Nagy, P.D. and Bujarksi, J.J. (1997) Engineering of homologous recombination hotspots with AU-rich sequences in Brome mosaic virus. *J. Virol.*, **71**, 3799–3810.

Nagy, P.D. and Bujarksi, J.J. (1998) Silencing homologous RNA recombination hot spots with GC-rich sequences in Brome mosaic virus. *J. Virol.*, **72**, 1122–1130.

Nagy, P.D. and Simon, A.E. (1997) New insights into the mechanisms of RNA recombination. *Virology*, **234**, 1–9.

Pilipenko, E.V., Gmyl, A.P. and Agol, V.I. (1995) A model for rearrangements in RNA genomes. *Nucl. Acids Res.*, **23**, 1870–1875.

Purcell, R.H. and Gerin, J.L. (1996) Hepatitis delta virus. In: *Fields Virology*, vol 2 (eds Fields, B. N., Knipe, D.M. and Howley, P.M.), pp. 2819–2829. Lippincott-Raven, Philadelphia, PA.

Reanney, D.C. and Ralph, R.K. (1968) Genetic circularity and evolution. *J. Theoret. Biol.*, **21**, 217–228.

Roberts, R.J. (1978) Intervening sequences excised *in vitro*. *Nature*, **274**, 530.

Robertson, H.D. (1992) Replication and evolution of viroid-like pathogens. *Curr. Topics Microbiol. Immunol.*, **176**, 214–219.

Robertson, H.D. and Branch, A.D. (1987) The viroid replication process. In: *Viroids and Viroid-like Pathogens* (ed. Semancik, J. S.), pp. 49–70. CRC Press, Boca Raton, FL.

Semancik, J.S. (ed.) (1987) Viroids and Viroid-like Pathogens. CRC Press, Boca Raton, FL.

Sharp, P. A. (1985) On the origin of RNA splicing and introns. *Cell*, **42**, 397–400.

Taylor, J.M. (1996) Hepatitis delta virus and its replication. In: *Fields Virology*, vol. 2 (eds Fields, B.N., Knipe, D.M. and Howley, P.M.), pp. 2809–2818. Lippincott-Raven, Philadelphia, PA.

Taylor, J., Mason, W., Summers, J. *et al.* (1987) Replication of human hepatitis delta virus in primary cultures of woodchuck hepatocytes. *J. Virol.*, **61**, 2891–2895.

Wang, K.S., Choo, Q.L., Weiner, A. J. *et al.* (1986) Structure, sequence, and expression of the hepatitis delta viral genome. *Nature*, **323**, 508–514.

Zhang, X. and Lai, M.M.C. (1994) Unusual heterogeneity of leader-mRNA fusion in a murine coronavirus: implications for the mechanism of RNA transcription and recombination. *J. Virol.*, **68**, 6626–6633.

3

Viroids in Plants: Shadows and Footprints of a Primitive RNA

J. S. Semancik and N. Duran-Vila

INTRODUCTION

In retrospect, viroid research from the outset was characterized by the disguise of a set of fundamentally unique properties identified principally by biological activity within the range of conventional plant viruses. In the initial description of the causal agents of the potato spindle tuber and citrus exocortis diseases, the attempt was made to frame these plant pathogens of economic importance within the context of existing scientific limits. Thus, the natural tendency to organize the unknown into a more understandable framework was challenged by the need to accommodate another form of plant pathogen, a molecular phenomenon unprecedented in structure. From this introduction as a subviral curiosity has emerged a genuine class of unique molecules restricted in the ability to self-replicate only within the plant kingdom. With the advent of surprises contributed by experimental research did the need arise to adapt a new nomenclature for the identification of the "viroids".

With this perspective, we have attempted to acknowledge the status of the generally accepted mechanisms of viroid structure and function, while at the same time testing the broad application of these processes within a central theme of the origin and evolution of viroids. By nature, this topic encourages both a presumptive and speculative license. It is not our intention to be comprehensive but simply to offer yet another, hopefully constructive, perspective. To add to the musing of Sol Spiegelman (1971), "one might well wonder why DNA invented man?" and with the benefit of time, we might now extend this query to include, Why did RNA invent viroids? And just as the viroids emerged as an unusual form of self-replicating molecule restricted to the plant kingdom, the natural inclination to position viroids may still fail to appreciate the unique offerings of this molecular species.

PERSPECTIVES OF VIROID ORIGIN AND ANCESTRY

The subject of origin and evolution of viroids as the smallest self-replicating biological entities, which were also pathogenic, presented an obvious attraction to the attention of the scientific community. When viroids were still considered a curiosity or a biological accident, hypotheses focused on the assumption that the origin of viroids was a recent event. Following the first descriptions of viroids as disease-causing agents of plants (Diener, 1971; Semancik and Weathers, 1972), plant virologists regarded these unusual RNAs as related to viral RNAs. However, with increasing information on the unprecedented properties of the viroid molecule when compared with the virus genome, this relationship was not sustained.

Early hybridization studies using crude preparations from viroid-infected plants suggested the existence of homologies between viroids and the host DNA from infected (Semancik and Geelen, 1975) and even healthy plants (Hadidi *et al.*, 1976). These were further postulated as evidence for a cellular origin of viroids (Diener and Hadidi, 1977). However, with refinement of the hybridization techniques, these early observations were rendered unreliable and a general acceptance of viroid synthesis through an RNA intermediate (Grill and Semancik, 1978; Grill *et al.*, 1980) in a rolling circle mechanism (Branch *et al.*, 1981; Branch and Robertson, 1984) has come about.

With the advent of computer-assisted searches for sequence homology, specific nucleic acid sequences showing homology with viroids were exposed. On the basis of these similarities some mechanisms were advanced by which viroids might interfere with the normal functions of the host cell. These included: (1) interference with the processing of mRNA (Dickson, 1981) or rRNA (Jakab *et al.*, 1986); (2) specific interactions with host rRNA (Hass *et al.*, 1988; Meduski and Velten, 1990) or with those nuclear events in which snRNAs are involved (Kiss *et al.*, 1983). Some of the observed homologies between viroids and other RNAs were also taken into consideration to sustain theories about a recent origin of viroids (Table 3.1).

The sequence homologies and structural similarities identified between viroids and snRNAs were considered as evidence for phylogenetic and functional relationships between viroids and mammalian snRNAs (Kiss and Solymosy, 1982) as probably the consequence of structural convergence (Kiss *et al.*, 1983). An evolutionary mechanism at the transcriptional or post-transcriptional level based on the circularization of a molecule similar to U3 snRNA and fusion with some other small circular RNA molecule was proposed.

Viroids and viroid-like satellites also share similarities with split genes and introns as discussed by Dinter-Gottlieb (1986).

- Viroids contain in the lower portion of the CCR the 16-nucleotide Group I consensus sequence.
- Viroids can acquire suboptimal tertiary structure elements similar to those resulting from pairing of conserved sequence elements of Group I introns required for splicing (Michel and Dujon, 1983).
- Negative strand of PSTV presents a striking homology with the 5′ terminus of snRNA U1 also involved in the splicing role of spliceosomes (Dickson, 1981).
- Spliced introns of Tetrahymena and yeast may ligate into circular RNA molecules.

These observations led to the proposal that viroid-like elements had originated as a result of circularization of spliced introns (Diener, 1981). However, this hypothesis has not been sustained by thermodynamic parameters, since the intron-like active secondary/tertiary structure that viroids may acquire would not be stable.

TABLE 3.1 Genomic relationships and evolutionary mechanisms between viroids and other nucleic acid sequences

Nucleic acid sequence	References	Evolutionary mechanism
snRNA U1	Dickson, 1981	Circularization and fusion with other small circularized RNAs
snRNA U3	Kiss *et al.*, 1983	or snRNAs
snRNA U5	Kiss and Solymosy, 1982	
Group I introns	Diener, 1981 Dinter-Gottlieb, 1986	Circularization of spliced introns
Transposable elements	Kieffer *et al.*, 1983	Deletion of interior portions
Delta agent RNA	Robertson, 1992	Viroids as precursors of introns

Additionally, there is the functional considera-tion that self-splicing has not been shown to occur in the greater number of typical viroids, whereas self-cleaving viroids involve the forma-tion of "hammerhead" structures with no resemblance to group I introns (Symons, 1991).

The similarities between viroids and split genes, coupled with the observation that the genomic RNA of the hepatitis virus delta agent contains a non-coding viroid-like region, led to the proposition that introns retaining cleavage and ligation functions had actually evolved from viroid-like replicating RNAs combined with DNA transcripts or other RNA segments, giving rise to new combinations of genetic material (Robertson, 1992).

Viroids also display partial sequence similari-ties with transposable genetic elements (Roberts, 1978; Crick, 1979; Kiefer *et al.*, 1983) with an imperfect direct repeat of the strictly conserved 18-base in the center of the viroid central conserved region. Similarities were also observed in the flanking inverted repeats in the CA dinucleotide at the 3' end and a UG at the 5' end, mimicking the TG ... CA dinucleotides flanking the terminal repeats of transposons and retroviroid proviruses. These observations led to the proposal that viroids had originated from transposable elements or retroviral proviruses by deletion of interior portions of the viral (or element) DNA.

With the discovery that RNAs may possess enzymatic properties involved in the processing of rRNAs, tRNAs, mRNAs (Kruger *et al.*, 1982; Guerrier-Takada *et al.*, 1983; Altman, 1987; Michel *et al.*, 1989; Cech, 1990), the idea of a pre-cellular world consisting of RNAs in which reactions would have been catalyzed mostly by ribozymes gained support. Similar ribozyme activities have also been identified in the pro-cessing of replication intermediates of a subset of viroids, ASBVd, PLMVd and CChMVd (Hutchins *et al.*, 1986; Symons, 1992; Hernández and Flores, 1992; Lafontaine *et al.*, 1995; Navarro and Flores, 1997). In addition, viroid-like satel-lite RNAs (Bruening, 1989) and the viroid-like domain of hepatitis delta virus (Taylor, 1990) also display this activity.

Although the similarities between viroids and other nucleic acid sequences may suggest a pos-sible evolutionary relationship, a direct lineage should be considered as well as the possibility of convergent evolution of molecular species with common sequence-dependent constraints. The acceptance of the viroid as a conformation-al pathogen may further subordinate any sequence homology to the ultimate structural configuration required for self-replication and pathogenesis. Therefore, in the absence of func-tional similarities, the observed homologies in sequence may be fortuitous or reflect the foot-print of convergent evolutionary relationships possibly inherent in early RNAs.

A CHANGING VIEW OF THE VIROID WORLD

Structure Diversity and Population Complexity

Over the course of the 30 years of viroid research, generally accepted conventions have emerged to accommodate a relatively simple viroid model. Recently, however, a transition to an enlarged scope has gained importance. Thus, the tightly base-paired rod-like model including the transitional structures proposed during denaturation to a single-stranded circular RNA was accepted (Gross *et al.*, 1978; Riesner *et al.*, 1979). This *in vitro* structure predicted by sequence analysis has been enlarged by com-puter modeling (Zuker, 1989) to include a col-lection of branched structures as alternative suboptimal or metastable folding patterns (Bussiere *et al.*, 1996), which may be favored within a biological context over structures of greater thermodynamic stability (Gultyaev *et al.*, 1998).

These provide a wider range of molecular structures for potential interaction *in vivo* for replication and host interaction. In this manner, the highly limited viroid genotype (sequence) has the potential of a greatly enlarged number of phenotypes providing the complexity critical for evolutionary selection and diversity. Although it is difficult to attribute functional significance to any or all of these structures within the microenvironment of cell compartments, model systems have attempted to define specific conformations involved in viroid replication (Baumstark and Riesner, 1995).

The importance of the T1 domain (Gast *et al.*, 1996) and the metastable hairpin (HPm) in maintaining the balance between metastable and stable structural forms for replication efficiency of PSTVd is common to the larger viroids (Gultyaev *et al.*, 1998). The structure reinforces the previous reports of HPII hairpin structures of the T1 domain influencing PSTVd infectivity (Loss *et al.*, 1991; Qu *et al.*, 1993). A metastable form of RNA provided the preferred template to a stable RNA for the model system of Qβ replicase (Biebricher and Luce, 1992). Thus, the rod structure of the viroid may be the non-functional equivalent of viral nucleic acid packaged away within a protein coat – in essence, an accumulation quiescent structure characterized principally by genomic stability and not biological activity.

To these well-accepted descriptions of viroid structure can be added the simple but empirical observations of the ability of non-denatured viroids to hybridize to specific probes (Francis *et al.*, 1995; Romero-Durban *et al.*, 1995). This evidence supports the existence *in situ* of molecular forms other than the highly base-paired rod-like structure. The stem-loop configurations proposed by Francis *et al.* (1995) as acting as a template for the specific probes should not be available in the minimum free energy structure predicted by thermodynamic modeling. This conformational complexity is especially significant to the survival of viroids lacking a translated sequence and thus dependent upon structural interactions with host components.

A Blueprint for Replication Events

It is generally accepted that viroid replication occurs by a rolling circle process (Branch and Robertson, 1984) for synthesis of single-stranded RNA via an RNA intermediate (Grill and Semancik, 1978) catalyzed by DNA-dependent RNA polymerase II (Muhlbach and Sanger, 1979). The necessity of an RNA ligase as opposed to an autocatalytic reaction to accomplish the closure of the circular product remains

a topic of discussion (Symons, 1992). The fundamental importance of the ribozyme structure as an essential component of all viroid replication schemes has been suggested by recent studies that extend the activity to include CCCVd (Liu and Symons, 1998). The question persists, however, as to whether ribozyme activity, characteristic of only a few viroids, reflects a common origin for all viroid development or is simply a curiosity of a small subset of atypical viroids, ASBVd (Hutchins *et al.*, 1986), PLMVd (Hernandez and Flores, 1992) and CChMVd (Navarro and Flores, 1997), which followed an independent evolutionary track.

From an evolutionary perspective, a more extreme proposition that might be entertained could attribute the origin of the ribozyme to the molecular enlargement stage of RNA evolution preceding autonomous replication. Even more simply, it could be regarded as the self-perpetuation consequence of random nucleotide sequence rearrangement during the process of sequence diversity. Regardless of the stage of appearance, an obvious evolutionary advantage is recouped from ribozyme activity for the replication and/or survival of RNA species.

The question of the replicating enzyme has centered on the DNA-directed RNA polymerase II (RNAPII). This consensus acceptance, as most recently cautioned by Flores *et al.* (1997), relies almost exclusively on the sensitivity of the synthesis reaction to alpha-amanitin. In the absence of further corroborating evidence to sustain this data, such as an indication of essential binding proteins and specific transcription factors (Hernandez, 1993; Kim *et al.*, 1997) to validate the functioning of RNAPII, it seems prudent to employ the term *RNAPII-like* in describing the most probable catalyst for replication of non-self-cleaving viroids.

Further characterization of the putative viroid-associated ribonuclearprotein particle (RNP) structures detected in subnuclear preparations (Rivera-Bustamante and Semancik, 1989; Pace *et al.*, 1992) and the tomato signal recognition particle (SRP) RNA with the potential for base pairing with several viroids replicating in tomato (Hass *et al.*, 1988) offer experimental approaches. In this regard, it has already been established that an enzyme other than RNAPII-like functions in the replication of ASBVd, since alpha-amanitin does not inhibit this chloroplast-concentrated viroid (Bonfiglioli *et al.*, 1994).

With consideration of the distinction observed in organelle association between ASBVd and the main body of viroids, we might conjecture whether the differences and synthetic enzyme activity could provide a clue to viroid evolutionary development. From an ASBVd-like ancestor introduced into the plant cell by a pre-organelle organism, such as a cyanobacterium (Chela-Flores, 1994) possessing bacterial-like RNAP activity exploited by the viroid for self-replication, did a progression to a nuclear locus ensue, as in the main body of viroids, utilizing a nuclear RNAPII-like enzyme? The absence of viroids in the animal kingdom might be employed to sustain this proposition. The relationship among the various enzyme activities associated with viroid replication, including ribozyme activity probably characteristic of the more primitive viroids, the two RNAP activities centered in the chloroplast and nucleus as well as a possible host RNA ligase function, provides a subject for further speculation about evolutionary relationships.

The acceptance of DNA-RNAPII as the primary replicating enzyme for the majority of viroids requires the ability of the viroid-RNA to simulate the structure and conformation of a DNA molecule. From this assumption, a more comprehensive proposition can be queried. Is the viroid limited to this single activity that mimics DNA or are there additional DNA functions in which the viroid either interacts or interferes? One such possibility can be developed from the demonstration that endoribonuclease T1 promotes the cleavage and ligation of viroid RNA (Tabler *et al.*, 1992). The correlation made between DNA endoribonuclease with cell division (Grafi and Larkins, 1995) presents an interesting relationship between the potential ability of viroids to mimic DNA and the affinity of viroids for mitotically active cells (Marton *et al.*, 1982; Lin and Semancik, 1985). This admittedly tenuous example is presented simply to provoke a broader view for the potential interactions of viroids with host components in the guise of a DNA species.

Viroid–Host Interactions

While characterized as a self-replicating RNA species, the total dependence of the viroid on the host for synthesis and survival establishes the precedent for a more intimate evolutionary linkage than for any other autonomously replicating molecule. It also presents an excellent experimental system for investigating pathogenic responses or abnormal host metabolism at a minimal level. Existing information of viroid–host interaction is dominated by data from experimentally malleable systems, usually with herbaceous hosts and especially on tomato cultivars. Exploitation of different viroid–host systems may be informative or even critical to the further understanding of viroid structure and function relationships.

As an example, although CEVd displays a relatively broad host range, including long-term stable infection in vegetatively propagated citrus, when transmitted to an alternate host, tomato, unusual CEVd-related species with terminal repeats and modifications were detected. These structures were found exclusively in an experimental tomato host, a first-generation hybrid between *Lycopersicon esculentum* and *L. peruvianum* maintained by vegetative propagation (Semancik *et al.*, 1994). Clearly, the unique contribution of this specific hybrid was probably essential for the generation and/or survival of these viroid-related molecules.

Viroid Lineage

It has been speculated that the low titer of the small circular RNA (scRNA) associated with rice yellow mottle virus (RYMV) may be reflective of a recent evolutionary association (Collins *et al.*, 1998). Subscribing to this proposition, the extremely high titer of ASBVd in avocado (Semancik and Szychowski, 1994) would suggest an ancient acquisition. The narrow host range of ASBVd restricted almost exclusively to *Persea americanum* as well as the extremely low transmission rate of a non-symptomatic variant, further promote the case for a long-term accommodation of the viroid. In addition, ASBVd exhibits an unusual highly conserved structural periodicity for viroids (Juhasz *et al.*, 1988).

In contrast, within the citrus viroids (Duran-Vila *et al.*, 1988) are contained viroids with both high and low titers in citron (*Citrus medica*), postulated as the oldest citrus cultivar in the Mediterranean region (Bar-Joseph, 1996). This would favor the accumulation of a mixture of viroids of diverse lineage. Surveys of citrus throughout the world support the widespread occurrence of CEVd, CVd-II, and CVd-III, with a lesser occurrence of CVd-I and especially CVd-IV (Duran-Vila *et al.*, 1988; Gillings *et al.*, 1991). With the common movement of citrus germplasm, the most recently emerging viroid associated with citrus would probably be assigned to CVd-IV. Similarly, the least common viroid found in grapevines, characterized as a historically vegetatively propagated plant species, is Australian grapevine viroid (AGVd), considered to be a recombinant viroid species (Rezaian, 1990).

The suggestion was made that the Mexican pepita viroid (MPVd) from *Solanum cardiphyllum* is the progenitor of tomato planta macho viroid (TPMVd), derived from wild solanum species in Mexico by chance transmission to cultivated tomato, as well as the closely-related PSTVd (Martinez-Soriano *et al.*, 1996). Mexico, as the site for the domestication of tomato, may have acquired a number of transmissible agents perpetuated in the wild tomato. The observation that an identical nucleotide sequence to TPMVd or PSTVd were not found in six isolates of *S. cardiophyllum* does not negate the possibility that both viroids were derived from less frequent quasi-species contained in the viroid population.

Although the search for PSTVd in relatives of potato species proved negative (Owens *et al.*, 1992), the common ancestry of tomato and potato coupled with the domestication of tomato in Mexico suggest that the search for the putative ancestor of PSTVd might also include primitive tomato species. The proposed progenitor of TPMVd, PSTVd, and possibly even MPVd, might be traced to a primitive tomato source.

POPULATION DIVERSITY – VARIANTS AND QUASI-SPECIES

Mutation Frequencies and Viroids as Quasi-species

RNA viruses form complex populations which evolve very rapidly due to high mutation frequencies (Holland *et al.*, 1992). The term "quasi-species" was introduced to describe the rapidly evolving RNA populations postulated as the earlier self-replicating entities, and later adapted to virus (Eigen, 1993). Specific sequences have been assigned to type viroid isolates with differences in biological properties, which are usually the symptoms induced in the original host plants (Gross *et al.*, 1981; Ohno *et al.*, 1983; Visvader and Symons, 1985; Herold *et al.*, 1992; Laksman and Tavantzis, 1993; Gora *et al.*, 1994; Ben-Shaul *et al.*, 1995). The proposition that a viroid variant is composed of a master nucleotide sequence along with a collection of lower-titer but still competent additional sequences is compatible with the quasi-species concept.

This convention did not take into consideration the possibility that the symptoms expressed result from the interaction of a population of variants rather than the specific sequence assigned to a given isolate. Depending upon the host environment, a variation in the viroid population may elicit a distinct host response reflecting differences in the viroid–host relationship. The unique properties of these processes is stylized in Figure 3.1. That viroids exist in host plants as populations of heterogeneous sequences has been well illustrated for CEVd (Visvader and Symons, 1985), ASBVd (Rakowski and Symons, 1989), GYSVd (Koltunow and Rezaian, 1988; Rigden and Rezaian, 1993; Szychowski *et al.*, 1998) and PSTVd (Góra *et al.*, 1994). Base error frequencies due to the absence of proofreading and repair mechanisms have been estimated to range from about 10^{-3} to less than 10^{-5} per site in RNA virus genomes (Holland *et al.*, 1992). This estimate would easily account for an enormous variation potential in viroids and the resulting viroid heterogeneity as found in a single plant.

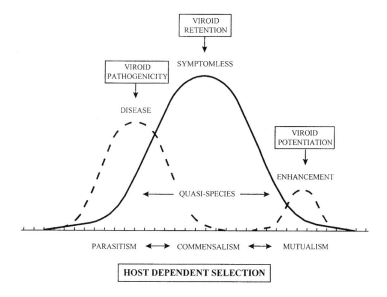

FIGURE 3.1   Viroid population diversity as related to host selection and biological interaction.

For the purpose of the material presented here, we wish to use the term "variant" to describe a population of genomic or sequence quasi-species (Góra-Sochacka *et al.*, 1997), which can be discriminated from the type variant by biological properties. Thus, CVd-II is a variant of HSVd. The predominant or master sequence should be reflected in the sequence of a randomized RT-PCR product, yet a number of closely related yet distinct quasi-species can be derived by sequence analysis of individual clones. Inoculation of these clones to the same host environment may have the potential of reconstructing the population of the original variant.

Although the number of possible variants of a viroid is very large, in a range of 4^{30} for a viroid of 300 nucleotides, site-mutagenesis (Owens *et al.*, 1986) has identified structural constraints that affect infectivity and pathogenesis. Conservation of structure appears to be essential for the viability of viroids (Owens *et al.*, 1991; Riesner, 1991 (see Flores 1997)). A number of structural alterations have been identified as critical to maintaining viroid infectivity (Hammond and Owens, 1987): (1) alterations in the terminal loops of the proposed minimal free energy structure; (2) single transitions disrupting base pairing in the P domain; (3) changes that may destabilize the proposed formation of hairpin I even when the changes are partially compensatory; and (4) changes affecting the region that forms the premelting loop 2. RNA folding models also identify a conserved metastable hairpin structure (HPm) in the left terminal domain of larger viroids, which regulates functional folding patterns while permitting sequence variation (Gultyaev *et al.*, 1998).

Infectivity studies using viroid cDNA clones illustrate the role of mutations on viroid evolution. A "fitness landscape" exists, with a high frequency of variants representing peaks of the most fit individuals and the progressive accumulation of new molecular variants (Góra-Sochacka *et al.*, 1997). When the inoculated cDNA clone is a sequence that is well fitted to the experimental host, the distribution of the progeny falls around the parental sequence; however, when the inoculated clone is not fit, new molecular variants may outcompete the original genotype, resulting in a new population of variants and a corresponding change in the "fitness landscape".

Population Diversity and Host Selection

Evolving heterogeneous viroid populations are also subjected to selection pressures acting upon viroid quasi-species. Any selection pressure will act through the host cell, with a major impact provided by the specific cell genotype as well as the effects of environmental factors on the plant. From that point of view, typical environmental parameters such as temperature, radiation and water availability may act as selection pressures through their effects upon the host plant. Cultural practices imposed by modern agriculture also play a selective role by providing: (1) the long-term contact between the viroids and their hosts through vegetative propagation (rooting of cuttings and scion/rootstock grafting); (2) the worldwide distribution of certain viroids through the international exchange of plant materials; (3) the mechanical dissemination of viroids across species by the intensive use of cutting devices; and (4) the eradication of viroids inducing diseases of economic importance.

With a few exceptions, most of the viroids known today were identified as disease-causing agents in agricultural crops and therefore our knowledge about their origin, variability and evolution is limited because of the constraints and specific selection pressures imposed by agricultural practices. In vegetatively propagated woody species such as citrus, fruit trees and grapevines the development of the commercial cultivars grown today is mainly the result of clonal selection and further propagation of high-performing individuals. Therefore the identification of viroids in those selected cultivars indicates that viroid infection is not necessarily linked to disease induction but may be to a desirable performance.

The widespread occurrence of several viroids in citrus and grapevines illustrates a long-term subtle adaptation of the viroids to their host that only became apparent under specific conditions triggering undesirable traits. In citrus, infection by CEVd was recognized only when CEVd-sensitive rootstocks were employed. Yet the same CEVd infection was correlated with tolerance to the diseases induced by fungi (Rossetti et al., 1980; Solel et al., 1995). In grapevines, infection with GYSVd was only perceptible under specific environmental conditions (yellow speckle disease in Australia) or as a result of interactions with other disease-causing agents, as in the synergistic reaction between GYSVd and grapevine fanleaf virus in the induction of vein banding disease (Szychowski et al., 1995): In both of these cases the host response was also markedly influenced by environmental stress conditions.

Viroids as a class of pathogenic RNA owe their recognition as an unusual molecular species to the expression of biological activity in the form of plant disease symptoms. In addition, most new viroids are described as a result of some disease expression in crop plants. These undesirable plant responses are, however, coupled with the existence of symptomless carrier plant species for numerous viroids. Thus, within the host range of any one viroid species there exists a selection process for parasitism or commensalism in the accommodation of specific quasi-species.

This selection can undoubtedly be influenced by other factors, such as environment and host genome variations. Thus, replication and retention of viroids can occur in the absence of any obvious pathogenic response. This relationship among quasi-species and host interactions are represented in the distribution of quasi-species (Figure 3.1). The distinction in the sequences of viroids inducing disease symptoms or simply replication and retention can be in the range of a few to a single nucleotide as for CEVd (Semancik et al, 1993; Fagoaga et al., 1995; Fagoaga and Duran-Vila, 1996) and PSTVd (Wassenegger et al., 1996) respectively.

Less frequently have the positive effects of viroids on the performance and survival of the host been emphasized. The association with disease as well as the virus-like nature implied by the term "viroid" has fostered the principle of production of viroid-tested plant materials in plant protection programs throughout the world. The implicit guiding philosophy fostered by these agencies considers viroid-free plants to be a means of providing disease control and superior agricultural return. Although valid for most cases involving viroids inducing diseases, as a result of the certification programs aiming at the control of plant diseases, growers may miss the beneficial effects of certain viroid–host interactions. Several diverse observations suggest a mutualistic role for viroids in specific host interactions ranging from experimental systems as cells and *in-vitro* cultured tissues or field performance of crops.

A "mutualistic" viroid–cell relationship can be found further extended to some viroid–plant systems and it is illustrated by a number of active defense responses that are induced as a result of viroid infection (Conejero *et al.*, 1990; Vera *et al.*, 1993; Domingo *et al.*, 1994; Tornero *et al.*, 1994; Gadea *et al.*, 1996). Several observations based on *in-vitro* cell cultures are illustrative of the interactions between viroids and host cells lacking the necessary organization and specialization implicit in tissues and whole plants.

Viroid-infected *in-vitro* cultures have been found to exhibit: (1) changes in the composition of the cell wall (Wang *et al.*, 1986) accounting for its tolerance to high osmotic pressure and degradation (Marton *et al.*, 1982); (2) general tolerance to chemicals affecting the growth of viroid-free cells (Duran-Vila and Semancik, 1982; Duran-Vila *et al.*, 1995); (3) enhanced chlorophyll content and biomass accumulation (Stöcker *et al.*, 1993; Duran-Vila *et al.*, 1995); and (4) extended longevity (Duran-Vila *et al.*, 1995). These indirect observations support the premise that in specific instances viroids might provide survival advantages to their host cells, which can be considered as a "mutualistic" relationship.

Commercial citrus infected with CEVd in Brazil (Rossetti *et al.*, 1980) and Israel (Solel *et al.*, 1995) have been observed to be more resistant to damage by *Phytophthora*. It is feasible that this response might also be due, in part, to a thickening of the cell wall inhibiting penetration of the fungal pathogen.

Interference of cachexia-disease-inducing viroid, CVd-IIb, by the closely related non-pathogenic CVd-IIa (Semancik *et al.*, 1992) represents another positive effect associated with transmission of specific viroids. Dwarfing of plant size is considered a recurrent symptom of "disease". However, when reduction in vegetative growth is accompanied by an increase in fruit yield and size coupled with no adverse affect on fruit quality, an economic benefit may ensue. This response has been documented for citrus production in both Australia (Broadbent *et al.*, 1986) and California (Semancik *et al.*, 1997a).

CHIMERIC VIROIDS: RECOMBINATION AND/OR RELATIVES?

Sequence Homology and Recombination

In molecular evolution, as in all living organisms, the driving forces are variability and selection. Sequence homologies among different viroids suggest that intermolecular recombination and genome reassortment may have played a role in the generation of new viroids. In addition, the high mutation frequencies common to RNA synthesis and resulting from replication in the absence of any proofreading and repair mechanisms as with DNA accounts for a continuing generation of viroid variants. An alternative or complementing scenario might propose a core sequence similarity required for the survival of all viroid-like molecules, this accounting for the homology attributed to recombination events. Further variations to this core are reflected by a consequence of mutation, errors in replication and core enlargement.

Non-homologous recombination is well documented for at least some virus families (Domingo and Holland, 1997) and has been offered as a likely mechanism involved in the generation of some viroids. With the identification of a number of viroids with sequences sharing well-defined segments of their primary structure with other viroids, the term "chimeric viroids" has been used to describe this type of partial homology, for example, TASVd and TPMVd with T1, C, and T2 domains of PSTVd and CEVd (Keese and Symons, 1985). Although the homology values are not as convincing for CSVd, a similar pattern of parentage to TASVd might be proposed, suggesting common origins from similar evolutionary events. Since CSVd has only been found naturally occurring in chrysanthemum, the lower sequence homology between CSVd and its putative parental species may be the result of further adaptation of the new emerging viroid in a different host environment.

The hypothesis that these chimeric viroids were the result of recombination and rearrangement among viroids co-infecting a single plant has been accepted and greatly enlarged. However, the lack of experimental evidence showing the emergence of a new chimeric viroid after co-inoculation of several viroids in a single plant supports the proposition that, if recombination does occur, it is an infrequent and even an unreproducible event (Biebricher and Luce, 1992). In addition, computer-generated random sequences of the same size and base composition as viroids gave sequence homolo-

gies of 29.5 ± 4.2% (Keese et al., 1988).

Comparison of the most conserved domains (T1, C, T2) was applied for the testing homology using significant segments of the viroid genome rather than a limited run of nucleotides (Table 3.2). In all cases a lower degree of sequence homology is observed in the P and V domains (not shown) and may be a consequence of further variation impairing the replication ability of new emerging viroid populations. Compatibility with a set of structural constraints resulting from the accumulation of mutations may exercise a control over variation. In this regard, the most common domains indicated as parental sequences are also the most frequently predicted sites for stem-loop configurations in the suboptimal structures from MFold modeling developed here. Perhaps both analy-

TABLE 3.2   Nucleotide sequence similarity between recombinant and punitive parental viroids

Genus	Recombinant	Parental(s)	Sequence homology (%)			References
			T1	C	T2	
Pospiviroids	TASVd	CEVd	90	80	56	Keese and Symons, 1985
		PSTVd	69	74	88	
	TPMV	PSTVd	70	89	92	Keese and Symons, 1985
		CEVd	77	76	55	
	CSVd	CEVd	75	89	52	Haseloff and Symons, 1981
		PSTVd	75	80	77	
	CLVd	PSTVd	90	60	89	Hammond et al., 1989a
		TASVd	77	59	96	
		HSVd	37	64	34	
Cocadviroids	CCCVd	PSTVd	20	73	36	Keese and Symons, 1985
		HSVd	37	54	57	
	CVd-IV	CEVd	34	56	88	Puchta et al., 1991
		HSVd	63	52	40	
Apscaviroids	GYSVd-2	GYSVd-1	82	87	57	Koltunow and Rezaian, 1989
		TPMVd	63	46	43	
	AGVd	GYSVd-1	56	49	43	Rezaian, 1990
		ASSVd	55	47	39	
		CEVd	61	45	54	
	ADFVd	CEVd	56	43	34	DiSerio et al., 1996
		HSVd	36	44	43	
		CVd-III	82	73	46	
Coleoviroids	CBVd-2	CBVd-1	42	89	100	Sänger and Spieker, 1997
		CBVd-3	92	38	40	

Sequence domain comparisons were made using the ALIGN program on the BCM Search Launcher http://kiwi.imgen.bcm.tmc.edu:8088/search-launcher/launcher.html modified from the algorithm described by Myers and Miller, 1988.

ses provide evidence for the vestigial viroid-like RNA "core" persisting in today's viroids.

Among the list of putative recombinants, the identification in *Coleus blumei* of CbVd 2 composed of two virtually unchanged parental sequences derived from the right portion of CbVd 1 and the left portion of CbVd 3 with a sharply defined boundary is unusual. The fact that CbVd 2 and the two hypothetical parentals have been found coinfecting a single plant provides the best available biological evidence (Sänger and Speaker, 1997) for a true recombination event. In addition, the vertical transmissibility of these viroids through seed (Singh and Boucher, 1991; Ramachandran *et al.*, 1992) promotes the coexistence of the presumptive parental viroids for extended periods allowing even an infrequent event to occur and to be retained. In contrast, it has also been proposed that three viroids were putative parents of CLVd (Hammond *et al.*, 1989a). However, the contribution of HSVd to the C domain (64% homology) is essentially comparable (59–60%) among the other two proposed contributors to the recombinant (Table 3.2).

The hypothesis that pospiviroids generated as a result of homologous recombination points to PSTVd and CEVd as the major contributors to the sequences of other viroids. It is even more surprising to find that these same two viroids are also proposed as major contributors to recombinants reported in the cocadviroids and apscaviroids. The observation that cocadviroids and apscaviroids present some level of sequence homology with pospiviroids and to a lesser degree with hostuviroids has also been considered as evidence for speculating about their origin as chimeric viroids with parentals from separate genera. If viewed as chimeric viroids, the main contributors to the sequence of cocadviroids would be the pospiviroids and HSVd, whereas apscaviroids seem to have incorporated only partial sequences or small oligonucleotide stretches of specific pospiviroids into other apscaviroids or a putative ancestor.

Sequence homology analysis has indicated the probability of recombination among viroid genomes existing as natural viroid chimeras. Although postulated to be rare and unreproducible events in RNA by *in-vitro* studies with the Qb replicase system (Biebricher and Luce, 1992), perpetuation of viroid RNA in vegetatively propagated plant materials provides the environment for observing this activity. Examples can be found for such genome rearrangement within both citrus, CVd-IV (Puchta *et al.*, 1991), and grapevines, AGVd (Rezaian, 1990), known to harbor five viroids. AGVd shares a length segment sequence of 112 nucleotides with largely the T1 domain of CEVd, also known to infect grapevines (Rezaian *et al.*, 1988; García-Arenal *et al.*, 1987).

However, this proposition must compete with the idea that coevolution of domains of a single species may result in "functional interdependence and coadaptation" (Góra *et al.*, 1996) as well as the forces of convergent evolution on small molecules with similar structural constraints.

However, until recombination is demonstrated under appropriate experimental systems, the possibility that the observed homologies were simply the result of primitive oligonucleotide growth and rearrangement followed by structural constraints remains a viable alternative. As an alternative consideration, the sequence homologies among small segments of different viroid molecules can be viewed as the vestige of convergent evolution (Branch *et al.*, 1993), perhaps driven by common structural requirements. As discussed by Hammond *et al.* (1989a), convergent evolution in terms of function is common but convergence in terms of structure is not (Lewin, 1985). Viroids, however, present the most restrictive relationship between structure and function and therefore a case for convergent evolution should not be so readily dismissed.

Origins and Host Associations

The hypothesis that the original source of PSTVd must be searched for in wild solanaceous plants growing in Mexico from where it was probably introduced to potatoes through breeding programs has been put forward (Diener, 1996). Vertical transmission of PSTVd through seed and pollen renders this hypothesis the most likely but it still remains untested. Efforts to identify PSTVd in wild plants in potato-growing areas as well as in the collections of wild solanaceous available in the International Potato Center in Peru (Owens *et al.*, 1992) were fruitless.

The finding of a new pospiviroid in wild *Solanum cardiophyllum* as well as other uncharacterized viroids in wild solanaceous in Mexico (Martinez-Soriano *et al.*, 1996) indicates that solanaceous species remaining in their center of origin and diversification may be reservoirs of unidentified viroids as well as putative ancestors of the pospiviroids. Although different pospiviroids have not been found coinfecting a single plant, they share the same host range and suggest the implication of solanaceous plants as the natural hosts that probably sustained these recombination events. Exceptions to this relationship between pospiviroids and solanaceous species as their primary hosts include CSVd and CLVd, which have only been found naturally occurring in chrysanthemum and *Columnea erytrophae* respectively. However, since CSVd, CLVd and their putative parentals all share solanaceous species in their host range, the possibility that CSVd and CLVd may have been originated from other pospiviroids can still be entertained.

The emergence of CEVd is even more obscure, since it has been found worldwide in commercial citrus spread through the exchange of propagative budwood. However, the fact that CEVd is not seed-borne and that old cultivars still growing in China are viroid-free indicates that infection of citrus with CEVd probably occurred as a chance transfer from unknown sources and was perpetuated by vegetative propagation. Where and how citrus budwood became infected with CEVd is not known but the fact that CEVd occurs naturally in hosts other than citrus as symptomless carriers suggests that, like PSTVd, CEVd may show some level of vertical transmissibility and that it may be more widespread in alternate hosts than was initially anticipated (Fagoaga *et al.*, 1995; Fagoaga and Duran-Vila, 1996).

HSVd, a viroid with probably the widest host range, is the single member of the hostuviroids (Flores *et al.*, 1998), with numerous sequence variants that can be clustered into five phylogenetic groups (Kofalvi *et al.*, 1997). Statistical evaluation of the probability of recombination events between HSVd variants indicates that a number of HSVd isolates probably derived from intraspecific recombination. These observations suggest that hostuviroids acquired an early independent evolutionary lineage as shown in the consensus phylogenetic tree. In addition, the phylogenetic positioning between pospiviroids and cocadviroids is compatible with the partial homology found between HSVd and CLVd (Hammond *et al.*, 1989a), CCCVd (Keese and Symons, 1985) and CVd-IV (Puchta *et al.*, 1991).

The hypothesis that homologous recombination may have played a role in the generation of new viroids in the pospiviroids and coleoviroids as well as in originating sequence variants of HSVd is supported by sequence homology, compatibility in host range, and the successful laboratory synthesis of infectious interspecific chimeric viroids (Hammond *et al.*, 1989b; Sano *et al.*, 1992).

Since viroids are most commonly identified as disease-causing agents, the information available about viroid sources in symptomless wild plants is still limited. It may be anticipated that the identification of viroids independent of agricultural practices will probably offer a more revealing view of viroid evolution.

SUBOPTIMAL STRUCTURES AND PHENOTYPIC VARIATION

Phylogeny and Suboptimal Structures

Phylogenetic schemes have been proposed based on sequence homology (Elena *et al.*, 1991) as well as common clusters in discrete locations as the central conserved region (CCR), the terminal conserved region (TCR) and the terminal conserved hairpin (TCH; Flores *et al.*, 1997). An ordered relationship (Figure 3.2) has been drawn between the main body of self-replicating viroids (PSTVd, HSVd, CCCVd, ASSVd) and virus-dependent self-cleaving virusoids and satellites bridged by ASBVd with self-cleaving activity.

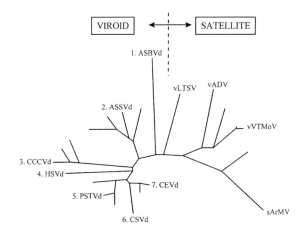

FIGURE 3.2 Selected viroids (1–7) from major phylogenetic groups derived from sequence homology (adapted from Elena *et al.*, 1991).

A survey from the major viroid clusters proposed by Elena *et al.* (1991) of the conformational structures predicted by MFold in the viroid structures from the more primitive satellite-like ASBVd to the larger pospiviroid provides a range of suboptimal species from a single structure for CCCVd to eight for CEVd (Table 3.3) within the same parameters.

With the exception of the branched structures predicted for the minimum free energy forms of ASBVd and ASSVd, the others conformed to the rod-like structure typical of viroid representations (Table 3.2). When the temperature for MFold analysis was increased to 37°C, branched structures become more common, with only CCCVd, HSVd, and PSTVd retaining the rod-like form. It is difficult to predict which of these modeling parameters more accurately represent the struc-

ture of the viroid molecule *in planta* while interacting with host components within the confines of a particular cellular compartment.

With the pre-eminence of conformation for viroid replication and survival, an analysis of the compatibility of phylogenetic schemes with predicted suboptimal folding patterns might be instructive in the search for common structure–function alternatives. Since viroids of sequence homology greater than 90% have the ability to express distinct differences in the number and forms of suboptimal structures (Reanwarakorn and Semancik, 1998), a single nucleotide change may significantly alter biological activity in the absence of major changes in the overall sequence homology, resulting in a modification or even loss of infectivity (Wassenegger *et al.*, 1996). The fact that the driving force in these relationships may be the convergent evolution of molecular species with either limited size or constraints imposed by a common conformation is at times subordinated to the overall sequence homology.

Stem-loop Structures Predicted from MFold

Most suboptimal forms contained a number of stem-loop configurations that might be viewed as structural loci for potential sites of biological activity. These should not be equated with the hairpins (HP-I, -II, or -III) identified as transient structures in the melting profile of the viroid molecule (Riesner *et al.*, 1992).

When the location of the stem-loop structures

TABLE 3.3   Properties of suboptimal structures predicted by MFold modeling for a phylogenetic progression of selected viroids

Viroid	nt	No. MFold @ 24°C	Minimum $-\Delta G$ range	$-\Delta G$ structure
1. ASBVd	247	3	78.8–76.2	T1 branched
2. ASSVd	330	3	137–131.9	T1 branched
3. CCCVd	246	1	120.9	Rod
4. HSVd	297	4	113.4–108.2	Rod
5. PSTVd	359	4	161.2–156.9	Rod
6. CSVd	356	7	149.7–142.8	Rod
7. CEVd	371	8	175.6–167.8	Rod

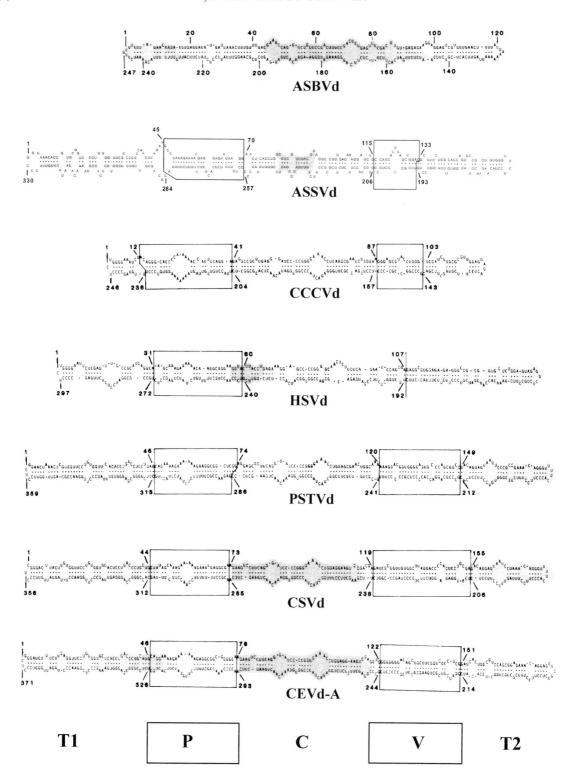

FIGURE 3.3 Regions (shaded) of selected (see Figure 3.1) viroid genomes in which stem-loop structures are predicted from MFold modeling. Pathogenic (P) and variable (V) domains boxed (adapted from Keese and Symons, 1987).

in all possible suboptimals was mapped according to domain, a non-random distribution in the terminal (T1, T2) domains and the central (C) was observed (Figure 3.3). With the exception of minor portions of the HSVd and CSVd sequences, no stem-loops were predicted for the pathogenic (P) and variable (V) domains.

The frequency of the stem-loop structures correlates directly with the domains in which the lowest numbers of nucleotide changes have been reported, indicating a conservation of genome in these regions. Whether this conservation is reflective of essential structural integrity and/or biological activity can only be conjectured. The absence of any totally uniform pattern argues for an equivalence of function by these structures and not sequence or locus within the viroid genome.

A CASE FOR TRANSIENT VIROID PROGENITORS

Viroid Genome Enlargement by Terminal Repeat

Sequence duplications in CCCVd (Haseloff *et al.*, 1982) offers a replication advantage by increasing the capacity of the viroid genome to interact with host components. The prediction made by Symons (1991) that a similar sequence duplication could have occurred in other members of the PSTVd group has been sustained by the report of an increase in the CEVd genome by 92 nt (CEVd-D92; Semancik *et al.*, 1994) or 104 nt (CEVd-D104) as a repeated sequence from the V-T2 region. The distinction between the two enlarged forms resides in the apparent "stutter" or repeat 12 nucleotides including a cluster of 5 base pairs at the juncture of the main repeated sequence (Figure 3.4).

This phenomenon has thus been extended beyond the scope of a viroid with a highly restricted host range, as for CCCVd. These demonstrations of enlargement in viroid genomes by jumps in molecular weight (Juhasz *et al.*, 1988) may reflect a genome enlargement phase necessary for rearrangement of viroid genome variants or even the

origin of distinct viroid species. This consideration of an enlargement of a genome may be viewed as a prelude not only to genome rearrangement but also to modification by recombination and mutation.

An extension of the proposal that enlarged structures provide some replication advantage is the proposition that new viroid species might be generated from sequence-duplicated molecules. These genome-redundant structures, which are independently transmissible and display significantly different symptom expressions, might provide a "progenitor platform" from which either the homologous or complementary sequence might contribute to the appearance of a new viroid species as in the *in vitro* terminal elongation of RNA by Qβ replicase (Biebricher and Luce, 1992).

Although CEVd displays a much broader host range than CCCVd, the only plant cultivar found competent to generate the terminal enlargement of CEVd is a hybrid of the domesticated *Lycopersicon esculentum* and the wild species *L. peruvianum* following continuous infection for the extended period of 9–12 months. The wild tomato-relative, *L. peruvianum*, is characterized by more indeterminate and vinous growth, simulating almost a perennial plant capable of extended growth. This feature may provide the extended exposure interval and numbers of replication events required for the generation and survival of low-titer intermediates involved in the evolution of viroid structure. In addition, the wild *L. peruvianum* is the most genetically variable of all tomato species and is strictly self-incompatible (Esquinas-Alcazar, 1981). Thus, by virtue of its growth characteristics and gene pool, it provides perhaps a more suitable or unique host background for the support of viroid replication.

Although capable of independent replication in tomato with no change in symptom expression, the genome enlargement resulted in a marked reduction in symptom expression in *Gynura aurantiaca* by CEVd-D92 as compared with the parental CEVd. This marked distinction in host response in the selective CEVd host, Gynura, is not reflected in hybrid tomato, the host generating the enlarged forms.

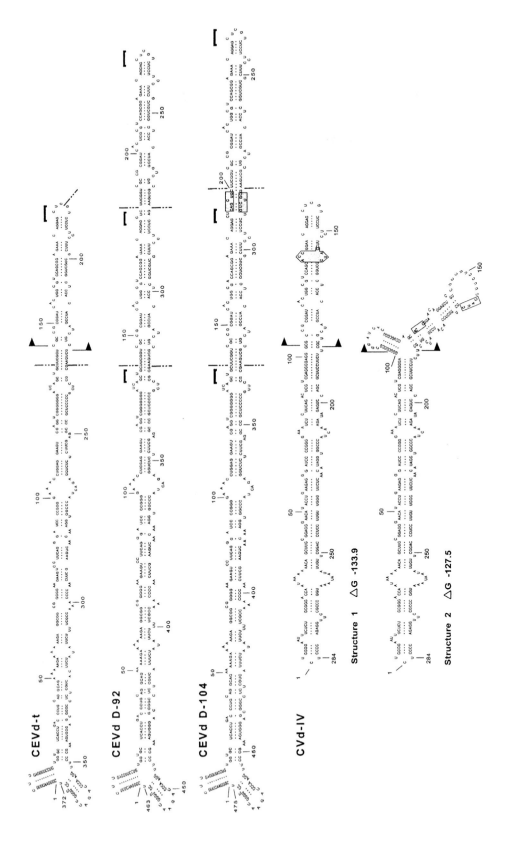

FIGURE 3.4 Nucleotide sequence and MFold modeling of CEVd variant from tomato (CEVd-t) and related species CEVd D-92 and CEVd D-104 containing a terminal repeated sequence (vertical divisions) with the 12 nucleotide distinction between the enlarged species shaded. The minimum free-energy rod-like form of CVd-IV is indicated as "structure 1", with the less stable suboptimal model as "structure 2". Area of sequence similarity (flags) of CVd-IV with CEVd-t with seven-nucleotide difference in sequence is shaded.

The repeated sequences of CCCVd in coconut have been attributed to "jumping polymerase" activity. With the extension of this property to CEVd, might the appearance and/or the accumulation of these enlarged structures be extremely host-dependent but still mimic the general enlargement phase of RNA proposed to precede rearrangement and modification? As further support for this possible scenario, citrons inoculated with CEVd-D92 were found to contain a variant of CVd-IV in the absence of replication of the inoculating viroid (Fig. 3.4).

CEVd-D92 as a Transient Progenitor of CVd-IV

The 92 nt terminal repeated portion of the CEVd sequence found in CEVd-D92 is virtually identical to the 81 nt CEVd-like sequences found in CVd-IV (Puchta *et al.*, 1991). Adding to this suggestion of transient progenitor forms is the absence of replication of the inoculum CEVd-D92 in citron, but the appearance of a CVd-IV variant (Semancik *et al.*, 1997b). It is important to note that CEVd-D92 is independently replicated to high titers in the hybrid tomato source but only with difficulty in *Gynura aurantiaca*, a selective host supporting high CEVd titers. This proposition supports a critical role for the host system in not only the survival and ultimate detection of viroid-like molecules potentially involved in the origin of new viroids but also in providing conditions which permit such events to occur.

A relationship must exist between the terminal repeated sequence in CEVd-D92 and the virtually identical 81 nucleotides found in the T2 domain of CVd-IV. Included in the suboptimal forms predicted for CVd-IV by MFold can be found a configuration in which the CEVd-related sequence does not conform to the rod-like viroid model (Figure 3.4, structure 1) but is represented by a double stem-loop branched structure at nucleotide 103–184 (Figure 3.4, structure 2), containing virtually the entire sequence of CEVd similarity. Since this structure does not represent a

repeated sequence in CVd-IV, it may represent a configuration that persists *in vivo* as a modular segment of the CEVd genome rather than the product of a "jumping polymerase" mechanism as suggested for the T2 repeated region.

Although CVd-IV has been considered a recombinant of CEVd and HSVd (Puchta *et al.*, 1991) as well as a member of the cocadviroids (Flores *et al.*, 1997), the model of a viroid with strikingly similar properties (Table 3.4) to CVd-IV can be achieved by selective deletion of portions of the CEVd genome (Figure 3.5). Employing the computer program ALIGN to identify regions of least homology, the 372 nt. sequence of CEVd was reduced to a hypothetical molecule of 280 nt, similar to the 284 nt size of CVd-IV. This derived form reflected remarkably similar properties with CVd-IV centered in homologies with HSVd and CCCVd, G+C content and a rod-like model as the minimum free energy model.

Although both CEVd-D92 and CVd-IV may be viewed as derivatives of CEVd, with recent findings a consideration of CEVd-D92 as a transient progenitor of both CEVd and CVd-IV also seems valid. In the absence of the fortuitous exploitation of the special relationship between CEVd and the hybrid tomato, these possible demonstrations of viroid genome enlargement and rearrangement (Figure 3.6) might not have become so apparent.

TABLE 3.4 Comparative properties of CVd-IV and CVd-IVδ derived from CEVd

	CVd-IV	CVd-IVδ
1. Nucleotides	284	280
2. Homology (%):		
CVd-IV	100	65
CEVd	57	75
HSVd (CVd-IIa)	57	51
CCCVd	48	44
Central domain:		
CEVd	55	100
HSVd	49	54
CCCVd	48	31
3. G+C (%)	56	58
4. MFold modeling		
• suboptimal structures	3	4
• −ΔG range	133.9–127.5	117.5–112.5
• minimum ΔG form	Rod	Rod

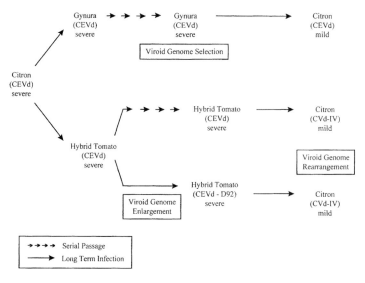

FIGURE 3.5   Host impact on the selection of CEVd variants and the origin of CEVd-related species.

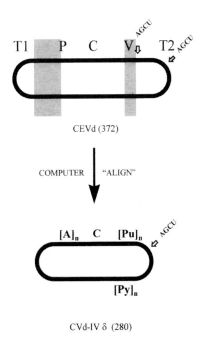

FIGURE 3.6   CVd-IV-like structure derived from CEVd sequence by deletion of selective sequences of main dissimilarity (shaded) in the pathogenic (P) and variable (V) domains as directed by the program ALIGN. Principal regions of the terminal (T1 and T2) and the central domains (C) are conserved, along with the poly A tract ([A]$_n$) and the purine ([Pu]$_n$) and pyrimidine ([Py]$_n$) rich clusters.

CONSIDERATIONS FOR VIROID GENEALOGY

A phylogenetic scheme relating classes of small RNAs (Figure 3.2) has been proposed (Elena *et al.*, 1991). Since ribozyme-catalyzed reactions are now viewed as remnants of an ancestral RNA world in which proteins carrying enzymatic activities were not available, the identification of ribozyme activities in some viroid and viroid-like satellite RNAs has rendered the hypothesis of an extremely early origin of viroids the most plausible, and expressions such as "relics of precellular evolution" (Diener, 1989) or "molecular fossils of the RNA world" (Chela-Flores, 1994) have been offered to illustrate the early origin of these molecules.

An attempt to adapt processes of nucleic acid metabolism to a scheme for the origin and evolution of viroids is offered in Figure 3.7. A primitive phase of a non-enzymatic origin and enlargement can be separated from the enzymatic growth and diversity phase by the crucial barrier imposed by the property of self-replication. The period of molecule building en route to the ability for replication is characterized by enlargement. This may be reflected in the periodicity noted by Juhasz *et al.* (1988) as evidence

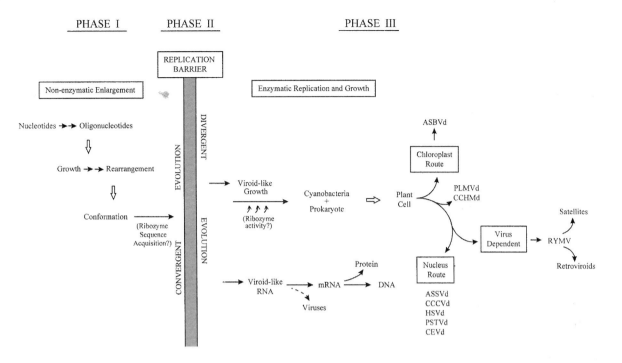

FIGURE 3.7   Considerations for the evolution of viroids and viroid-like molecules as related to viruses and other nucleic acid species.

of growth by jumps with rearrangement following to achieve the nucleotide sequence required for survival.

If some critical sequence is postulated as being necessary to maintain a survival conformation, the force of convergent evolution to that sequence would be pre-eminent. Footprints or remnants of this phase might still persist within the sequence of existing self-replicating viroid-like molecules (Branch *et al.*, 1993). In addition, some of the similarities noted between viroids and host nucleic acid sequences such as introns and transposons may also reflect vestiges of a distant but common past.

As a consequence of this growth and rearrangement process the sequence and configuration required for ribozyme activity could have been acquired. This event may signal a stage either as a prelude to bridging the stylized "replication barrier" or as the introduction of the expansive phase of divergent evolution and the generation of different self-replicating molecules.

The classical evolutionary experiments of Spiegelman (1971) employing Qβ replicase *in vitro* can be used to support the concept of the "replication barrier". In this system, in which the product's ability to serve as a template was challenged by rapid passage, the 4000 nt Qβ genome could be reduced by 83% while still retaining the property of self-replication (Mills *et al.*, 1967). The MDV-1 RNA species, reflecting the viroid-like properties not only of small size (221 nt) but also of extensive regions of GC-rich anti-parallel base pairing in stem-loop configurations, was successfully replicated in this system (Nishihara *et al.*, 1983).

This system can be offered as a hypothetical test system for the existence of the "replication barrier" as well as evidence for selection of RNA phenotypes for replication in the absence of any translation. With the introduction of this evidence for the generation of molecules with viroid-like properties from a virus, one must logically entertain the possibility that viroids may disguise a vestigial lineage from viruses.

Although this proposition is not favored by the current consensus view, the viability of both pathways may not be mutually exclusive.

The length of primitive oligonucleotides with non-enzymatic self-replication was probably limited but they had to assume some kind of growth before acquiring the characteristic conformation necessary for survival and a more efficient replication (replication barrier). Non-enzymatic growth probably occurred through a series of oligonucleotide duplications and rearrangements, as illustrated by the periodical repeat units of 12, 60 or 80 nucleotide residues (Juhasz et al., 1988) and the partial homologies found in chimeric viroids. In addition, the enlargement of oligonucleotides beyond a critical size either non-enzymatically or by primitive ribozymes was probably enhanced after circularization (Robertson, 1992). This observation suggests that the circularization of RNA precursors of viroids and viroid-like satellites probably occurred through convergent evolution and before passing over the replication "barrier". Outlined as essential (Chela-Flores, 1994), in addition to the ribozyme activity of some viroid and viroid-like satellites, were short nucleotide length or minimal target size, built-in UV defense mechanism with structural complexity and the absence of mRNA activity.

Viroids have only been found associated with angiosperm plants of which the earliest fossils date from the end of the Mesozoic Era. As pointed out by Chela-Flores (1994), the hypothesis of the emergence of viroid-like molecules from an early RNA world must be compatible with their survival until the appearance of their host plants. How did viroid-like molecules develop and survive as free molecules until 1000 million years ago when abiotic synthesis of organic compounds probably stopped? Which were the early precursors of eukaryotic cells that became associated with viroid-like RNAs?.

With this perspective, there are many unknown gaps and comparatively few hypothetical clues:

- The idea of an early association with free-living cyanobacteria has been entertained by Chela-Flores (1994) and is compatible with the detection of at least one viroid (ASBVd) in the chloroplast. However, evidence of a previous association of a viroid-like molecule with a cyanobacterium before their symbiotic association with prokaryotic cells resulting in eukaryotic plant cells is still lacking.

- The observation that replication of ASBVd is insensitive to high levels of alpha-amanitin (Marcos and Flores, 1992) is also compatible with an enzymatic replication of RNA mediated by chloroplast or even bacterial RNA polymerases.

- The involvement of ribozyme activities via the hammerhead structures found in each polarity strand of ASBVd (Hutchins et al., 1986) can be viewed as a characteristic of early viroid-like molecules as well as a requirement for the processing of replication oligomeric intermediaries of both polarities before they became totally dependent on the host cell for cleavage and ligation mechanisms.

- Replication through a symmetric pathway with two rolling cycles represents a strategy in which circular replication intermediaries are more stable to degradation and less dependent on specific initiation sites (Robertson, 1992).

Viroid-like circular RNA with ribozyme activity replicating by a rolling circle pathway has been postulated by Robertson (1992) as the predecessor of RNA with the property of protein coding ultimately stored in DNA copies. The unusual properties of ASBVd provide some clues, which are compatible with the hypothesis of an early origin of viroids from viroid-like ancestors. These observations, which suggest the existence of similarities between ASBVd and the putative precursors of viroids and viroid-like RNAs, are also compatible with the phylogenetic study which assigned to this viroid a central location in the consensus phylogenetic tree as well as an evolutionary link between typical viroids and satellite RNAs (Elena et al., 1991).

Two other viroids, PLMVd (Hernández and Flores, 1992) and CChMVd (Navarro and Flores, 1997) share with ASBVd the formation of hammerhead structures in both polarity strands with self-cleaving properties. These are also consistent with the proposed monophyletic origin of viroids and provide an additional and even closer phylogenetic link between PLMVd and viroid-like satellite RNAs (Hernández and Flores, 1992). Although PLMVd and CChMVd share a ribozyme activity mediated by the formation of hammerhead structures with ASBVd, a distinct lineage group must be represented by a GC-rich nucleotide sequence with an unusual branched secondary structure as compared to the AU-rich ASBVd. Further characterization of the location for accumulation of these two viroids, as well as their replication sites, is required to provide a more complete picture of possible evolutionary relationships.

Viroids and viroid-like satellites share structural properties and replication strategies, but connecting evolutionary links are difficult to verify. It has been suggested that viroids evolved from satellite RNAs while still free-living molecules, and that both acquired a dependence on their host (viroids) or helper virus (satellite RNAs) only after becoming intracellular pathogens (Diener, 1987). However, the observation that PSTVd can be encapsidated into particles of potato leafroll virus coinfecting the same plant (Francki et al., 1986; Querci et al., 1997) can be considered as indirect evidence supporting the hypothesis that viroid-like satellites may have arisen as a result of fortuitous encapsidation into a virus particle followed by further evolution resulting in the viral replication dependence observed today.

On the basis of the hypothesis that viroids may be direct descendants of early RNAs, regardless of whether or not primitive viroid-like molecules were first associated with prokaryotic or eukaryotic cells, it is tempting to speculate about the type of interactions that probably existed between viroids and their early hosts. Although viroids are basically considered as plant pathogens, it can be argued that a deleterious effect upon their hosts would have endangered their long-term persistence and survival.

Once eukaryotic cell precursors became associated with viroid-like molecules, their persistence was probably sustained by cell division, as has been observed in in vitro cell cultures (Marton et al., 1982; Stöcker et al., 1993), but the same mechanism does not apply to whole plants. Our present understanding of viroid epidemiology in crop plants indicates that the spread and persistence of viroids is mainly associated with vegetative propagation and agricultural practices that enhance mechanical transmission. Since these mechanisms were unlikely to act efficiently in natural environments or in primitive agricultural practices, natural transmission through seed and pollen has probably been more critical in the past.

Therefore, viroids with connections with the most primitive viroid-like forms, if still present in existing biological systems, must be sought among those viroid–host combinations that exhibit vertical transmission. From that point of view, this principle reinforces the central evolutionary role postulated for ASBVd as a seed-borne viroid with virtually 100% transmission through a symptomless carrier host (Wallace and Drake, 1962). However, although different levels of vertical transmission have been demonstrated for PSTVd, CSVd, HSVd in tomato (Kryczynski et al., 1988), PSTVd in potato (Fernow et al., 1970; Singh et al., 1992), CbVd (Singh and Boucher, 1991) and CCCVd (Hanold and Randles, 1991), most viroids known to be pathogens of agricultural crops are mostly or only spread horizontally. The identification in ASBVd of sequence variants unevenly distributed or even segregated by the host functions in symptomatic and symptomless tissues (Semancik and Szychowski, 1994) illustrates how propagation by grafting may have acted by selecting specific variants that may not be efficiently transmissible or even lead to the loss of vertical transmission through seed and pollen.

THE UNWRITTEN LAST CHAPTER

Just as the field of viroid research was conceived in the realm of the unknown, from the perspective of considerable creative research we are compelled to ask:

- Have all the surprises offered by viroids been revealed?
- Is the existing fund of knowledge of nucleic acid metabolism and self-replicating

molecules adequate to accommodate the processes of viroid structure and replication, much less the more challenging process of pathogenesis?

● Have all of the pertinent questions already even been posed?

ACKNOWLEDGEMENTS

The authors wish to acknowledge the assistance of J.A. Szychowski and E. De Jager-O'Shea in this effort.

REFERENCES

Altman, S. (1987) Ribonuclease P: an enzyme with a catalytic RNA subunit. *Adv. Enzymol.,* **62**, 1–36.

Bar-Joseph, M. (1996) A contribution to the natural history of viroids. In *Proceedings of the 13th IOCV Conference* (eds da Graca, J.V, Moreno, P and Yokomi, R. K.), pp. 226–229. IOCV, Riverside.

Baumstark, T. and Riesner, D. (1995) Only one of four possible secondary structures of the central conserved region of potato spindle tuber viroid is a substrate for processing in a potato nuclear extract. *Nucleic Acids Res.,* **23**, 4246–4254.

Ben-Shaul, A., Guang, Y., Mogilner, N. (1995) Genomic diversity among populations of two citrus viroids from different graft-transmissible dwarfing complexes in Israel. *Phytopathology,* **85**, 359–369.

Biebricher, C.K. and Luce, R. (1992) *In vitro* recombination and terminal elongation of RNA by Qβ replicase. *EMBO J.,* **11**, 5129–5135.

Bonfiglioli, R. G., McFadden, G.I. and Symons, R. H. (1994) *In situ* hybridisation localises avocado sunblotch viroid on chloroplast thylakoid membranes and coconut cadang cadang viroid in the nucleus. *Plant J.,* **6**, 99–103.

Branch, A.D. and Robertson, H.D. (1984) A repli-

cation cycle for viroids and other small infectious RNAs. *Science,* **223**, 450–454.

Branch A.D., Robertson H.D. and Dickson, E. (1981). Longer than unit length viroid minus strands are present in RNA from infected plants. *Proc. Natl Acad. Sci. USA,* **78**, 6381–6385.

Branch, A.D., Lee, S.E., Neel, O.D. and Robertson, H.D. (1993) Prominent polypurine and polypyrimidine tracts in plant viroids and in RNA of the human hepatitis delta-agent. *Nucleic Acids Res.,* **21**, 3529–3535.

Broadbent, P., Forsyth, J.B., Bevington, K.B. and Hutton, R.J. (1986) Citrus tree size control with dwarfing agents. *Calif. Citrogr.,* **71**, 8–10.

Bruening G. (1989) Compilation of self-cleaving sequences from plant virus satellite RNAs and other sources. *Methods Enzymol.,* **180**, 546–558.

Bussiere, F., Lafontaine, D. and Perreault, J.P. (1996) Compilation and analysis of viroid and viroid-like RNA sequences. *Nucleic Acids Res.,* **24**, 1793–1798.

Cech, T.R. (1990) Self-splicing of group-I introns. *Annu. Rev. Biochem.,* **59**, 543–568.

Chela-Flores, J. (1994) Are viroids molecular fossils of the RNA world? *J. Theor. Biol.,* **166**, 163–166.

Collins, R.F., Gellatly, D.L., Sehgal, O.P. and Abouhaidar, M.G. (1998) Self-cleaving circular RNA associated with rice yellow mottle virus is the smallest viroid-like RNA. *Virology,* **241**, 269–275.

Conejero, V., Bellés, J.M., García-Breijo, F. (1990) Signaling in viroid pathogenesis. In: *Recognition and Response in Plant–Virus Interactions* (ed. Frase, R.S.S.), pp. 233–261. NATO Springer-Verlag, Berlin.

Crick, F.H.C. (1979) Split genes and RNA splicing. *Science,* **204**, 264–271.

Dickson, E. (1981) A model for the involvement of viroids in RNA splicing. *Virology,* **115**, 216–221.

Diener, T. (1971) Potato spindle tuber "virus" IV. A replicating low-molecular weight RNA. *Virology,* **45**, 411–428.

Diener, T.O. (1981) Are viroids escaped introns? *Proc. Natl Acad. Sci. USA,* **78**, 5014–5015.

Diener, T.O. (1987) The frontier of life: the

viroids and viroid-like satellite RNAs. In: *The Viroids* (ed. Diener T.O.)), pp. 1–20. Plenum Press, New York.

Diener, T.O. (1989) Circular RNAs: relics of pre-cellular evolution? *Proc. Natl Acad. Sci. USA*, **86**, 9370–9374.

Diener, T.O. (1996) Origin and evolution of viroids and viroid-like satellite RNAs. *Virus Genes*, **11**, 119–131.

Diener, T.O. and Hadidi, A. (1977) Viroids. In: *Comprehensive Virology* (eds Fraenkel-Conrat, H. and Wagner R.R.), pp. 285–337. Plenum Press, New York.

Dinter-Gottlieb, G. (1986) Viroids and virusoids are related to group 1 introns. *Proc. Natl Acad. Sci. USA*, **83**, 6520–6534.

DiSerio, F., Aparico, F., Alioto, D., Ragozzino, A. and Flores, R. (1996) Identification and molecular properties of a 306 nucleotide viroid associated with apple dimple fruit disease. *J. Gen. Virol.*, **77**, 2833–2837.

Domingo, E. and Holland, J.J. (1997) RNA virus mutations and fitness for survival. *Annu. Rev. Microbiol.*, **51**, 151–178.

Domingo, C., Gómez, M.D., Cañas, L., Hernández-Yago, J., Conejero, V. and Vera, P. (1994) A novel extracellular matrix protein from tomato associated with lignified secondary walls. *Plant Cell*, **6**, 1035–1047.

Duran-Vila, N. and Semancik, J.S. (1982) Effects of exogenous auxins on tomato tissues infected with the citrus exocortis viroid. *Phytopathology*, **72**, 777–781.

Duran-Vila, N., Roistacher, C.N., Rivera-Bustamante, R. and Semancik, J.S. (1988) A definition of citrus viroid groups and their relationship to the exocortis disease. *J. Gen. Virol.*, **69**, 3069–3080.

Duran-Vila, N., Carbonell, E.A., Pérez Boada, S. and Semancik, J.S. (1995) Growth of healthy and viroid-infected tomato cells in vitro. *Plant Sci.*, **105**, 111–120.

Eigen, M. (1993) The origin of genetic information: viruses as models. *Gene*, **135**, 37–47.

Elena, S.F., Dopazo, J., Flores, R., Diener, T.O. and Moya, A. (1991) Phylogeny of viroids, viroidlike satellite RNAs, and viroidlike domain of hepatitis δ virus RNA. *Proc. Natl Acad. Sci. USA*, **88**, 5631–5634.

Esquinas-Alcazar, J.T. (1981) Genetic resources of tomatoes and wild relatives. AGP:IBPGR/80/103 Rome.

Fagoaga, C. and Duran-Vila, N. (1996) Natural occurrence of variants of the citrus exocortis viroid in vegetable crop species. *Plant Pathol.*, **45**, 45–53.

Fagoaga, C., Semancik, J.S. and Duran-Vila. (1995) A citrus exocortis viroid variant from broad bean (*Vicia faba* L.): infectivity and pathogenesis. *J. Gen. Virol.*, **76**, 2271–2277.

Fernow, K.H., Peterson, L.C. and Plaisted, R.L. (1970) Spindle tuber virus in seeds and pollen of infected potato plants. *Am. Potato J.*, **46**, 75–80.

Flores, R., DiSerio, F. and Hernandez, C. (1997) Viroids: the noncoding genomes. *Sem. Virol.*, **8**, 65–73.

Flores, R., Randles, J.W., Bar-Joseph, M. and Diener, T.O. (1998) A proposed scheme for viroid classification and nomenclature. *Arch. Virol.*, **143**, 623–629.

Francis, M.I., Szychowski, J.A. and Semancik, J.S. (1995) Structural sites specific to citrus viroid groups. *J. Gen. Virol.*, **76**, 1081–1089.

Francki, R.I.B., Zaitlin, M. and Palukaitis, P. (1986) *In vivo* encapsidation of potato spindle tuber viroid by velvet tobacco mottle virus particles. *Virology*, **155**, 469–473.

Gadea, J., Mayda, M.E., Conejero, V. and Vera, P. (1996) Characterization of defense-related genes ectopically expressed in viroid-infected tomato plants. *Mol. Plant–Microbe Interact.*, **9**, 409–415.

García-Arenal, F., Pallás, V. and Flores, R. (1987) The sequence of a viroid from grapevine closely related to severe isolates of citrus exocortis viroid. *Nucleic Acids Res.*, **15**, 4203–4210.

Gast, F.-U., Kempe, D., Spieker, R.L. and Sanger, H.L. (1996) Secondary structure probing of potato spindle tuber viroid (PSTVd) and sequence comparison with other small pathogenic RNA replicons provides evidence for central non-canonical base-pairs, large A-rich lops, and a terminal branch. *J. Mol. Biol.*, **262**, 652–670.

Gillings, M.R., Broadbent, P. and Gollnow, B.I. (1991) Viroids in Australian citrus: relationship to exocortis, cachexia and citrus dwarfing. *Aust. J. Plant Physiol.*, **18**, 559–570.

Góra, A., Candresse, T. and Zagorski, W. (1994) Analysis of the population structure of three phenotypically different PSTVd isolates. *Arch. Virol.*, **138**, 233–245.

Góra, A., Candresse, T. and Zagorski, W. (1996) Use of intramolecular chimeras to map molecular determinants of symptom severity of potato spindle tuber viroid (PSTVd). *Arch. Virol.*, **141**, 2045–2055.

Góra-Sochacka, A., Kierzek, A., Candresse, T. and Zagorski, W. (1997) The genetic stability of potato spindle tuber viroid (PSTVd) molecular variants. *RNA*, **3**, 68–74.

Grafi, G and Larkins, B.A. (1995) Activity of single-stranded DNA endonucleases in mung bean is associated with cell division. *Plant Mol. Biol.*, **29**, 703–710.

Grill, L.K. and Semancik, J.S. (1978) RNA sequences complementary to citrus exocortis viroid in nucleic acid preparations from infected *Gynura aurantiaca. Proc. Natl Acad. Sci. USA.*, **75**, 896–900.

Grill, L.K., Negruk, V.I. and Semancik, J.S. (1980) Properties of the complementary RNA sequences associated with infection by the citrus exocortis viroid. *Virology*, **107**, 24–33.

Gross, H.J., Domdey, H., Lossow, C. (1978) Nucleotide sequence and secondary structure of potato spindle tuber viroid. *Nature*, **273**, 203–208.

Gross, H.J., Liebel, U., Alberty, H. (1981) A severe and mild potato spindle tuber viroid isolate differ in three nucleotide exchanges only. *Bioscience*, **1**, 235–241.

Guerrier-Takada, C., Gardiner, K., Marsh, T., Pace, N. and Altman, S. (1983) The RNA moiety of ribonuclease P is the catalytic subunit of the enzyme. *Cell*, **35**, 849–857.

Gultyaev, A.P., van Batenburg, F.H.D. and Pleij, C.W.A. (1998) Dynamic competition between alternative structures in viroid RNAs simulated by an RNA folding algorithm. *J. Mol. Biol.*, **276**, 43–55.

Hadidi, A., Jones, D.M., Gillespie, D.H., Wong-Staal, F. and Diener, T.O. (1976) Hybridization of potato spindle tuber viroid to cellular DNA of normal plants. *Proc. Natl Acad. Sci. USA*, **73**, 2453–2457.

Hammond, R.W. and Owens, R.A. (1987) Mutational analysis of potato spindle tuber viroid reveals complex relationships between structure and infectivity. *Proc. Natl Acad. Sci. USA*, **84**, 3967–3971.

Hammond, R.W., Smith, D.R. and Diener, T.O. (1989a) Nucleotide sequence and proposed secondary structure of *Columnea* latent viroid: a natural mosaic of viroid sequences. *Nucleic Acids Res.*, **17**, 10083–10094.

Hammond, R.W., Diener, T.O. and Owens, R.A. (1989b) Infectivity of chimeric viroid transcripts reveals the presence of alternative processing sites in potato spindle tuber viroid. *Virology*, **170**, 486–495.

Hanold, D. and Randles, J.W. (1991) Coconut cadang-cadang disease and its viroid agent. *Plant Dis.*, **75**, 330–335.

Haseloff, J. and Symons, R.H. (1981) Chaysanthemum stunt viroid: primary sequence and secondary structure. *Nucleic Acids Res.*, **9**, 2741–2752.

Haseloff, J., Mohammed, N.A. and Symons, R.H. (1982) Viroid RNAs of cadang-cadang disease of coconut. *Nature*, **299**, 316–321.

Hass, B., Klanner, A., Ramm, K. and Sänger, H.L. (1988) The 7s RNA from tomato leaf tissue resembles a signal recognition particle RNA and exhibits a remarkable sequence complementarity to viroids. *EMBO J.*, **7**, 4063–4074.

Hernández, N. (1993) TBP, a universal eukaryotic transcription factor? *Genes Dev.*, **7**, 1291–1308.

Hernández, C. and Flores, R. (1992) Plus and minus RNAs of peach latent mosaic self-cleave *in vitro* via hammerhead structures. *Proc. Natl Acad. Sci. USA*, **89**, 3711–3715.

Herold, T., Haas, B., Singh, R.P., Boucher, A. and Sanger, H.L. (1992) Sequence analysis of five new field isolates demonstrates that the chain length of potato spindle tuber viroid (PSTVd) is not strictly conserved but variable as in other viroids. *Plant Mol. Biol.*, **19**, 329–333.

Holland J.J., De La Torre, J.C. and Steinhauer, D.A. (1992) RNA virus populations as quasi-species. *Curr. Top. Microbiol. Immunol.*, **176**, 1–20.

Hutchins, C.J., Rathjen, P.D., Forster, A.C. and Symons, R.H. (1986) Self-cleavage of plus and minus RNA transcripts of avocado sunblotch viroid. *Nucleic Acids Res.*, **14**, 3627–3640.

Jakab, G., Kiss, T. and Solymosy, F. (1986) Viroid pathogenicity and pre-rRNA processing: a model amenable to experimental testing. *Biochim. Biophys. Acta,* **868,** 190–197.

Juhasz, A., Hegyi, H. and Solymosy, F. (1988) A novel aspect of the information content of viroids. *Biochim. Biophys. Acta,* **950,** 455–458.

Keese, P. and Symons, R.H. (1985) Domains in viroids: evidence of intermolecular RNA rearrangements and their contribution to viroid evolution. *Proc. Natl Acad. Sci. USA,* **82,** 4582–4586.

Keese, P. and Symons, R.H. (1987) Physical–chemical properties: Molecular structure (primary and, secondary). In: *The Viroids,* (ed. Diener, T.O.), pp. 37–62, Plenum Press, New York.

Keese, P., Visvader, J.F. and Symons, R.H. (1988) Sequence variability in plant viroid RNAs. In: *RNA genetics, Variability of RNA Genomes,* vol. III (eds Domingo, E., Holland, J.J. and Alquist P.), pp. 71–98. CRC Press, Boca Raton, Florida.

Kiefer, M.C., Owens, R.A. and Diener, T.O. (1983) Structural similarities between viroids and transposable genetic elements. *Proc. Natl Acad. Sci. USA,* **80,** 6234–6238.

Kim, T.K., Lagrange, T., Wang, Y.H., Griffith, J.D., Reinberg, D. and Ebright, R.H. (1997) Trajectory of DNA in the RNA polymerase II transcription preinitiation complex. *Proc. Natl Acad. Sci. USA,* **94,** 12268–12273.

Kiss, T. and Solymosy F. (1982) Sequence homologies between a viroid and a small nuclear RNA (snRNA) species of mammalian origin. *FEBS Lett.,* **144,** 318–320.

Kiss, T., Pósfai, J. and Solymosy, F. (1983) Sequence homology between potato spindle tuber viroid and U3B snRNA. *FEBS Lett.,* **163,** 217–220.

Kofalvi, S.A., Marcos, J.F., Cañizares, M.C., Pallás, V. and Candresse, T. (1997) Hop stunt viroid (HSVd) sequence variants from *Prunus* species: evidence for recombination between HSVd isolates. *J. Gen. Virol.,* **78,** 3177–3186.

Koltunow, A.M. and Rezaian, M.A. (1988) Grapevine yellow speckle viroid: structural features of a new viroid group. *Nucleic Acids Res.,* **16,** 849–864.

Koltunow, A.M. and Rezaian, M.A. (1989) Grapevine viroid 1B, a new member of the apple scar skin viroid group contains the left terminal region of tomato planta macho viroid. *Virology,* **170,** 575–578.

Kruger, K., Grabowski, P.J., Zaug, A.J., Sands, J., Gottschling, D.E. and Cech, T.R. (1982) Self-splicing RNA: autoexcision and auto-cyclization of the ribosomal RNA intervening sequence of *Tetrahymena. Cell,* **31,** 147–157.

Kryczynski, S., Paduch-Cichal, E. and Skrzeczkowski, L.J. (1988) Transmission of three viroids by seed and pollen of tomato plants. *Phytopathology,* **121,** 51–57.

Lafontaine, D., Beaudry, D., Maequis, P. and Perrault, J.P. (1995) Intra- and intermolecular nonenzymatic ligations occur within transcripts derived from the peach latent mosaic viroid. *Virology,* **212,** 705–709.

Lakshman, D.K. and Tavantzis, S.M. (1993) Primary and secondary structure of a 360-nucleotide isolate of potato spindle tuber viroid. *Arch. Virol.,* **128,** 319–331.

Lewin, R. (1985) Fish to bacterium gene transfer. *Science,* **227,** 1020.

Lin, J.J. and Semancik, J.S. (1985) Coordination between host nucleic acid metabolism and citrus exocortis viroid turnover. *Virus Res.,* **3,** 213–230.

Lin, Y.-H. and Symons, R.H. (1998) Specific RNA self-cleavage in coconut cadang-cadang viroid: potential for a role in rolling circle replication. *RNA,* **4,** 418–429.

Loss, P., Schmitz, M., Steger, G. and Riesner, D. (1991) Formation of a thermodynamically metastable structure containing hairpin II is critical for infectivity of potato spindle tuber viroid RNA. *EMBO J.,* **10,** 719–727.

Marcos, J. and Flores, R. (1992) Characterization of RNAs specific to avocado sunblotch viroid synthesized *in vitro* by a cell-free system from infected avocado leaves. *Virology,* **186,** 482–488.

Martinez-Soriano, J.P., Galindo-Alonso, J., Maroon, C.J.M., Yucel, I., Smith, D.R. and Diener, T.O. (1996) Mexican papita viroid: putative ancestor of crop viroids. *Proc. Natl Acad. Sci. USA,* **93,** 9397–9401.

Marton, L., Duran-Vila, N., Lin, J.J. and Semancik, J.S. (1982) Properties of cell cultures containing the citrus exocortis viroid. *Virology,* **122,** 229–238.

Meduski, C.J. and Velten, J. (1990) PSTV sequence similarity to large rRNA. *Plant Mol. Biol.*, **14**, 625–627.

Michel, F. and Dujon, B. (1983) Conservation of RNA secondary structures in two intron families including mitochondrial, choloroplast, and nuclear coded members. *EMBO J.*, **2**, 33–38.

Michel, F., Umesono, K. and Ozeki. H. (1989) Comparative and functional anatomy of group II catalytic introns – a review. *Gene*, **52**, 5–20.

Mills, D., Pace, N. and Spiegelman, S. (1967) The *in vitro*, synthesis of a non-infectious complex containing biologically active viral RNA. *Proc. Natl Acad. Sci. USA*, **56**, 1778–1785.

Muhlbach, H.P. and Sanger, H.L. (1979) Viroid replication is inhibited by alpha-amanitin. *Nature*, **278**, 185–188.

Myers, E. and Miller, W. (1988) Optical alignments in linear space. *CABIOS*, **4**, 11–17.

Navarro, B. and Flores, R. (1997) Chrysanthemum chlorotic mottle viroid: unusual structural properties of a sub-group of viroids with hammerhead ribozymes. *Proc. Natl Acad. Sci. USA*, **94**, 11262–11267.

Nishihara, T., Mills, D.R. and Kramer, F.R. (1983) Localization of the Qβ replicase recognition site in MDV-1 RNA. *J. Biochem.*, **93**, 669–674.

Ohno, T., Takamatsu, N., Meshi, T. and Okada, Y. (1983) Hop stunt viroid: molecular cloning and nucleotide sequence of the complete cDNA copy. *Nucleic Acids Res.*, **11**, 6185–61196.

Owens, R.A., Hammond, R.W., Gardner, R.C., Kiefer, M.C., Thompson, S.M. and Cress, D.E. (1986) Site-specific mutagenesis of potato spindle tuber viroid cDNA. *Plant Mol. Biol.*, **6**, 179–192.

Owens, R.A., Thompson, S.M. and Steger, G. (1991) Effects of random mutagenesis upon potato spindle tuber viroid replication and symptom expression. *Virology*, **185**, 18–31.

Owens, R.A., Khurana, S.M.P., Smith D.R., Singh, M.N. and Garg, L.D. (1992) A new mild strain of potato spindle tuber viroid isolated from wild *Solanum* spp. in India. *Plant Dis.*, **76**, 527–529.

Pace, U., Branch, A.D. and Robertson, H.D. (1992) Generation of viroid conformational isomers that are stable to incubation with magnesium ions and in a nuclear extract from tomato plants. *Nucleic Acids Res.*, **20**, 6681–6686.

Puchta H., Ramm K., Luckinger R., Hadas R., Bar-Joseph M. and Sänger H.L. (1991) Primary and secondary structure of citrus viroid IV (CVd IV), a new chimeric viroid present in dwarfed grapefruit in Israel. *Nucleic Acids Res.*, **19**, 6640.

Qu, F., Heinrich, C., Loss, P., Steger, G., Tien, P. and Riesner, D. (1993) Multiple pathways of reversion in viroids for conservation of structural elements. *EMBO J.*, **12**, 2129–2139.

Querci, M., Owens, R.A. Bartolini, I., Lazarte, V. and Salazar, L.F. (1997) Evidence for heterologous encapsidation of potato spindle tuber viroid in particles of potato leafroll virus. *J. Gen. Virol.*, **78**, 1207–1211.

Rakowski, A.G. and Symons, R.H. (1989) Comparative sequence studies of variants of avocado sunblotch viroid. *Virology*, **73**, 352–356.

Ramachandran, P., Kumar, D., Varma, A., Pandex, P.K. and Singh, R.P. (1992) Coleus viroid in India. *Curr. Sci.*, **62**, 271–272.

Reanwarakorn, K. and Semancik, J.S. (1998) Regulation of pathogenicity in HSDd-related group II citrus viroids. *J. Gen. Virol.*, **79**, 3163–3171.

Rezaian, M.A. (1990) Australian grapevine viroid – evidence for extensive recombination between viroids. *Nucleic Acids Res.*, **18**, 1813–1818.

Rezaian, M.A., Koltunow, A.M. and Krake, L.R. (1998) Isolation of three viroids and a circular RNA from grapevines. *J. Gen. Virol.*, **60**, 413–422.

Riesner, D. (1991) Viroids: from thermodynamics to cellular structure and function. *Mol. Plant – Microbe Interact.*, **4**, 122–131.

Riesner, D., Henco, K., Rokohl, U. (1979) Structure and structure formation of viroids. *J. Mol. Biol.*, **133**, 85–115.

Riesner, D., Baumstark, T., Qu, F. (1992) Physical basis and biological examples of metastable RNA structures. In: *Structural Tools for the Analysis of Protein – Nucleic Acid Complexes* (eds Lilley, D.M.J., Heumann, H. and Suck D.), pp. 401–435. Birkhauser, Basel.

Rigden, J.E. and Rezaian, M.A. (1993) Analysis of sequence variation in grapevine yellow speckle viroid 1 reveals two distinct alternative structures for the pathogenic domain. *Virology*, **193**, 474–477.

Rivera-Bustamante, R. and Semancik, J.S. (1989) Properties of a viroid replicating complex solubilized from nuclei. *J. Gen. Virol.*, **70**, 2707–2716.

Roberts, R.J. (1978) Intervening sequences excised *in vitro*. *Nature*, **274**, 530.

Robertson, H.D. (1992) Replication and evolution of viroid-like pathogens. *Curr. Topics Microbiol. Immunol.*, **176**, 213–219.

Romero-Durban, J., Cambra, M. and Duran-Vila, N. (1995) A simple imprint hybridization method for detection of viroids. *J. Virol. Methods*, **55**, 37–47.

Rossetti, V., Pompeu, J., Rodriguez, O. (1980) Reaction of exocortis-infected and healthy trees to experimental Phytophthora inoculations. In: *Proceedings of the 8th IOCV Conference*, pp. 209–214. IOCV, Riverside.

Sänger, H.L. and Spieker, R.L. (1997) RNA recombination between viroids. In: *Plant Viroids and Viroid-like Satellite RNAs from Plants, Animals and Fungi*, p. 13. Instituto Juan March de Estudios e Investigaciones, Madrid.

Sano, T., Candresse, T., Hammond, R.W., Diener, T.O. and Owens, R.A. (1992) Identification of multiple structural domains regulating viroid pathogenicity. *Proc. Natl Acad. Sci. USA*, **89**, 10104–10108.

Semancik, J.S. and Geleen, J.L.M.C. (1975) Detection of DNA complementary to pathogenic viroid RNA in exocortis disease. *Nature*, **256**, 753–756.

Semancik, J.S. and Szychowski, J.A. (1994) Avocado sunblotch disease: a persistent viroid infection in which variants are associated with differential symptoms. *J. Gen. Virol.*, **75**, 1543–1549.

Semancik, J.S. and Weathers, L.G. (1972) Exocortis virus: evidence for a new species of "infectious" low molecular weight RNA in plants. *Nature*, **237**, 242–244.

Semancik, J.S., Gumpf, D.J. and Bash, J.A. (1992) Interference between viroids inducing exocortis and cachexia disease in citrus. *Ann. Appl. Biol.*, **121**, 577–583.

Semancik, J.S., Szychowski, J.A., Rakowski, A.G. and Symons, R.H. (1993) Isolates of citrus exocortis viroid recovered by host and tissue selection. *J. Gen. Virol.*, **74**, 2427–2436.

Semancik, J.S., Szychowski, J.A., Rakowski, A.G. and Symons, R.H. (1994) A stable 463 nucleotide variant of citrus exocortis viroid produced by terminal repeats. *J. Gen. Virol.*, **75**, 727–732.

Semancik, J.S., Rakowski, A.G., Bash, J.A. and Gumpf, D.J. (1997a) Application of elected viroids for dwarfing and enhancement of production of "Valencia" orange. *J. Hortic. Sci.*, **72**, 563–570.

Semancik, J.S., Reanwarakorn, K., de Jager, E. and Szychowski, J.A. (1997b) Alteration in biological activity by host selection and engineered rearrangements of viroid genomes. In: *Plant Viroids and Viroid-like Satellite RNAs from Plants, Animals and Fungi*, pp. 15–16. Instituto Juan March de Estudios e Investigaciones, Madrid.

Singh, R.P. and Boucher, A. (1991) High incidence of transmission and occurrence of a viroid in commercial seeds of *Coleus* in Canada. *Plant Dis.*, **75**, 184–187.

Singh, R.P., Boucher, A. and Somerville, T.H. (1992) Detection of potato spindle tuber viroid in the pollen and various parts of potato pollinated with viroid-infected pollen. *Plant Dis.*, **76**, 951–953.

Solel, Z., Mogilner, N., Gafny, R. and Bar-Joseph, M. (1995) Induced tolerance to mal secco disease in Etrog citron and Rangpur lime by infection with the citrus exocortis viroid. *Plant Dis.*, **79**, 60–62.

Spiegelman, S. (1971) An approach to the experimental analysis of precellular evolution. *Q. Rev. Biophys.*, **4**, 213–253.

Stöcker, S., Guitton, M.C., Barth, A. and Mülbach, J.P. (1993) Photosynthetically active suspension cultures of potato spindle tuber viroid infected tomato cells as tools for studying viroid-host cell interaction. *Plant Cell Rep.*, **12**, 597–602.

Symons, R.H. (1991) The intriguing viroids and virusoids: what is their information content and how did they evolve? *Mol. Plant–Microbe Interact.*, **4**, 111–121.

Symons, R.H. (1992) Small catalytic RNAs. *Annu. Rev. Biochem.*, **61**, 641–671.

Szychowski, J.A., McKenry, M.V., Walker, M.A., Wolpert, J.A., Credi, R. and Semancik, J.S. (1995) The vein-banding disease syndrome: a synergistic reaction between grapevine viroids and fanleaf virus. *Vitis*, **34**, 229–232.

Szychowski, J.A., Credi, R., Reanwarakorn, K. and Semancik, J.S. (1998) Population diversity in grapevine yellow speckle viroid-1 and the relationship to disease expression. *Virology*, **248**, 432–444.

Tabler, M., Tzortzakaki, S. and Tsagris, M. (1992) Processing of linear longer-than-unit-length potato spindle tuber viroid RNAs into infectious monomeric circular molecules by a G-specific endoribonuclease. *Virology*, **190**, 746–753.

Taylor, J. (1990) Structure and replication of hepatitis delta virus. *Sem. Virol.*, **1**, 135–141.

Tornero, P., Conejero, V. and Vera, P. (1994) A gene encoding a novel isoform of the PR-1 protein family from tomato in induced upon viroid infection. *Mol. Gen. Genet.*, **243**, 46–53.

Vera, P., Tornero, P. and Conejero, V. (1993) Cloning and expression analysis of a viroid-induced peroxidase from tomato plants. *Mol. Plant–Microbe Interact.*, **6**, 790–794.

Visvader, J.E. and Symons, R.H. (1985) Eleven new sequence variants of citrus exocortis viroid and the correlation of sequence with pathogenicity. *Nucleic Acids Res.*, **13**, 2907–2920.

Wallace, J.M. and Drake, R.J. (1962) A high rate of seed transmission of avocado sun-blotch virus from symptomless trees and the origin of such trees. *Phytopathology*, **52**, 237–241.

Wang, M.C., Lin, J.J., Duran-Vila, N. and Semancik, J.S. (1986) Alteration in cell wall composition and structure in viroid infected cells. *Physiol. Plant Pathol.*, **28**, 107–124.

Wassenegger, M., Spieker, R.L., Thalmeir, S., Gast, F.-U., Riedel, L. and Sanger, H.L. (1996) A single nucleotide substitution converts potato spindle tuber viroid (PSTVd) from a noninfectious to an infectious RNA for *Nicotiana tabacum*. *Virology*, **226**, 191–197.

Zuker, M. (1989) On finding all suboptimal foldings of an RNA molecule. *Science*, **244**, 48–52.

4

Mutation, Competition and Selection as Measured with Small RNA Molecules

Christof K. Biebricher

INTRODUCTION

Darwin's theory of natural selection is one of the greatest milestones in science. It provides answers to deep questions that are otherwise unanswerable. As Dobzhansky put it: "Nothing makes sense in Biology except in light of evolution" (Dobzhansky *et al.*, 1977).

Yet even a century after Darwin we still cannot understand organismic evolution in detail, let alone make quantitative predictions about its course. To be sure, neo-darwinistic theory does provide insights into evolution processes in quantitative terms, but they essentially comprise only quantitative descriptions of reproduction, in particular in mendelian populations, not of evolution itself. For the fundamental processes operating in evolution – mutation and selection – its parameters have to be adjusted to fit, more or less, the evolutionary outcome. They cannot be derived from measurable properties of the organisms themselves or of their genes.

In contrast to our lack of quantitative descriptions of evolution, its molecular basis is very well understood: the information needed for morphogenesis and function is encoded into the *genotype* as the sequence of nucleotides in each organism's genome (a few exceptions notwithstanding). The key step in reproduction is copying that nucleotide sequence. While replication error rates leading to accepted mutations vary over the genome, and also depend to some degree upon environmental conditions, these errors are more or less random. No teleology directing mutations to an advantageous result has ever been observed.

Information informs only if it is understood; thus genetic information has to be decoded by cellular machinery in order to operate a life-sustaining program of biochemical reactions. The program depends on the environment and leads to the properties of the organism that we observe as its *phenotype*. Evolutionary success depends on two components of the phenotype: those that determine survival in the prevailing environment and those that establish the rate of producing viable offspring. The combination of these two determines the population trend that we call *fitness*.

Environments are typically complex and variable. Species interact by competition for resources, predation or symbiosis; individuals of the same species influence one another socially, and the individuals themselves may be composed of large numbers of specialized cells, all containing the same genetic information, that have to cooperate with one another for the organism they compose to survive and reproduce. While the way that gene expression produces the metabolic apparatus is now felt to be well understood, we know much less about how cells acquire information about the environment and how this information triggers appropriate genetic responses. Biologists still debate the identity of the target of selection: is it an ecosystem, a species, a variant, a subpopulation, an individual, a gene or merely a "replicator unit" (Dawkins, 1982)? There is no ultimate answer to this question: selection takes place at all of these levels. Which of the selection levels dominates depends on the environment.

Evolution is a dynamical self-organization process in which causal correlations between the performance of the process as a whole and its component subprocesses are not identifiable (Biebricher et al., 1995). We can correlate fitness to the function of a single gene only if this gene happens to be absolutely required for survival under prevailing conditions. Further, translation of a genotype into a phenotype is far too complicated a process to be evaluated. Fitness values have to be determined a posteriori, i.e. so as to describe the observed changes in the composition of the population under study. Prediction of an evolutionary outcome from fitness values obtained from process-independent parameters is generally impossible. Is the darwinian concept of evolution then merely a tautomerism describing the survival of the survivors, as some have criticised? No, it is not. The studies with RNA viruses described in this volume witness that in many cases the molecular basis of fitness can be clearly identified.

Nevertheless, even the simplest RNA viruses are too complicated to allow quantitative descriptions of their mutation, competition and selection, in particular because their complex interactions with host cells are inevitably involved in the evolutionary process. The 1961 discovery of RNA bacteriophages by Loeb and Zinder, more than 80 years after the discovery of plant RNA viruses, was instrumental in accelerating progress in understanding molecular processes in the infection cycles. Ten years after their discovery they were already by far the best understood viruses. In particular, because of their essential nature as parasitic messenger RNA, they became invaluable experimental tools in studying the expression of genetic information.

THE EXPERIMENTAL SYSTEM

Most RNA bacteriophages belong to the plus strand virus family leviviridae. Except for a few members of the reoviridae family, no other RNA virus families have been found to infect prokaryotes. Leviviridae are particularly simple, in all respects, and their genome sizes are the smallest among autonomously infecting viruses. Their properties, gene map and infection cycle are described in Figures 4.1 and 4.2.

Shortly after their discovery, several research groups succeeded in detecting a novel RNA-dependent RNA polymerase in levivirus-infected cells (August et al., 1963; Haruna et al., 1963; Weissmann et al., 1963). After being found to be highly specific in amplifying viral RNA it became known as a replicase. The replicase of the coliphage Qβ was found to be particularly stable and thus the most suitable one for in-vitro studies. Purification of the Qβ replicase (Kamen, 1970; Kondo et al., 1970) revealed four subunits, one coded by the R gene of the phage, the others provided by the host. Together with an additional host factor (Franze de Fernandez et al., 1968) they perform all steps necessary to amplify viral RNA.

The experimental procedure for replication experiments is simple. An RNA template is incubated with replicase purified from Qβ infected cells, appropriate amounts of the four nucleoside triphosphate precursors and an appropriate buffer. When using viral RNA as template, the progeny RNA synthesized in vitro was found to be infectious (Spiegelman et al., 1965). However, when Spiegelman and collaborators tried to dilute out parental RNA by serially diluting aliquots of growing RNA into fresh test tubes containing replicase and precursors, further production of infectious RNA stopped after the fifth serial tranfers, while incorporation of nucleoside triphosphates into RNA continued, even at steadily increasing rates. Spiegelman and collaborators recognized that they had performed an "extra-cellular Darwinian

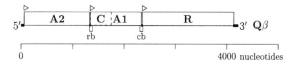

FIGURE 4.1   Gene map of the bacteriophage Qβ. ▷, ribosome binding sites; rb, replicase binding site; cb, coat protein binding site. Thick lines are untranslated regions, boxes the cistrons for adsorption protein A2, coat protein C and replicase R. Gene product A1 is produced by reading through a stop codon at the dashed line.

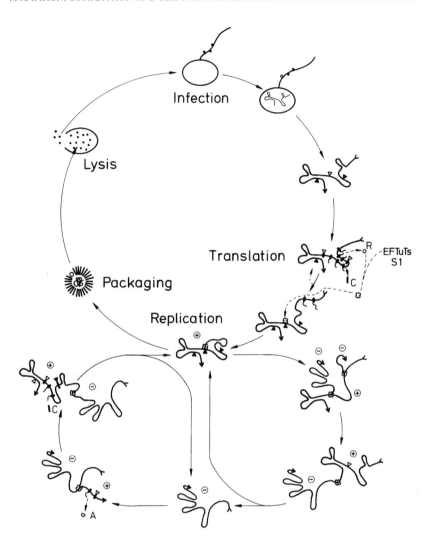

FIGURE 4.2 Infection cycle of phage Qβ. ●, ribosome; ⊠, replicase; ←, 5′ end of RNA; ——<, 3′ end of RNA; ▷, ribosome binding site available; ▶, ribosome binding site unavailable. Reproduced from Biebricher and Eigen, 1987, in: *RNA Genetics, vol. I: RNA-directed Virus Replication* (eds Domingo, E., Ahlquist, P. and Holland, J.J.), CRC Press, Boca Raton, FL, with permission.

At the beginning of the infection cycle, a phage particle adsorbs to an F-pilus of the host and injects its RNA into the host cell. Ribosomes bind to the binding site of the C-gene, which is the only available site. During translation of the C-gene the RNA structure is modified in such a way that the ribosome site of the R-gene also becomes available (Weissmann, 1974). Both pC and pR proteins are synthesized, the latter combining with EFTu·Ts and ribosomal S1 protein to form mature replicase. The replicase binds to a site (rb) directly at the ribosome site of the C-gene. Therefore, replicase and ribosome compete for the RNA. If sufficient replicase has been synthesized it binds to the RNA, but cannot start replication before completion of translation and clearance from the ribosome. The replicase produces minus strands that bind neither ribosomes nor coat protein but are excellent templates for the production of plus strands. With increasing amounts of coat protein, binding of coat protein occurs at a site (cb) next to the ribosome binding site of the R-gene. Binding of coat protein dimer shuts off R-gene translation and is the nucleus for the capsid formation. A2-gene translation is only available when a nascent plus strand is formed in minus strand replication (Beekwilder *et al.*, 1996).

The rate of RNA synthesis is hardly measurable in the first 15 min, because replicase must first be synthesized, then it explosively increases at a hyperbolic (faster than exponential) rate until 20 min after infection (Eigen *et al.*, 1991). After all the replicase is bound (nearly exclusively to minus strand), linear growth of RNA at maximum rate is observed. Mature phages accumulate in the interior of the host cell and are liberated at an eclipse time of about 50 min. Liberation occurs by host cell lysis triggered by A2 protein (Karnik and Billeter, 1983). A2 is also responsible for binding of the phage particles to the F-pilus.

experiment with a self-duplicating nucleic acid molecule", the title of a paper that became seminal for *in-vitro* evolution studies (Mills *et al.*, 1967). After 74 serial tranfers, the RNA was analysed and found to have eliminated 83% of its chain length in the course of increasing its replication rate. Experiments under different replication conditions, e.g. reduced levels of one nucleoside triphosphate or in the presence of inhibitors (Levisohn and Spiegelman, 1969; Saffhill *et al.*, 1970) were performed; the RNA was shown to adapt to these conditions, "revealing an unexpected wealth of phenotypic differences which a replicating nucleic acid can exhibit".

The experiment had an immediate impact. While most scientists were enthusiastic about the new possibilities, some scoffed that the experiment represented "a search for the best carcass". Indeed, the evolution experiments did start with an infectious RNA able to perform many different roles – to serve as messenger, to replicate and to be packed in a protein coat – and end with a "variant RNA" that had lost most of the information and was only able to replicate. Evolution thus produced degeneration, as have many other experimental evolution trials.

Furthermore, the reaction and its products were declared to be "unphysiological". Indeed, they have to be, for otherwise the consequences for phage reproduction would be disastrous. This criticism applies also to the other evolution studies reported below; the templates and the reactions are all unphysiological. This does not detract from their value, however, in studying evolution. The experiments I describe abstract the essential features from the enormously complex net of physiological interactions at work in physiological phage infection cycles *in vivo*. While replicase and RNA synthesis precursors must be synthesized during infection cycles, they are environmental factors in the *in-vitro* experiments. The *in-vitro* experiments provide precisely controllable and reproducible conditions indispensable for quantitative studies.

Some of the key results of the classical *in-vitro* RNA evolution studies were difficult to understand. The RNA species selected in many different such evolution experiments were indistinguishable from one another in their physical properties. One would have expected that

RNA evolution would proceed by rare events leading to different and irreproducible products. Furthermore, in the early work there was the problem that the enzyme used was a partially purified crude extract not entirely free from RNA. Spiegelman and coworkers recognized these difficulties and searched for short-chained RNA molecules where genotypic changes could be tracked by sequence analysis (Mills *et al.*, 1973).

THE MECHANISM OF RNA REPLICATION

Qβ replicase has evolved features absolutely necessary for virus infection.

(1) It is highly specific in accepting its own RNA as template. It is vital for phage reproduction that only the viral RNA be amplified, while a huge excess of host RNA is ignored. Therefore, the RNA itself cannot be considered to be merely a substrate of the replicase: it shares the catalytic role. Its template activity, i.e. its efficiency in instructing the replicase to replicate it, is a phenotypic expression of the RNA species that is crucial for its evolutionary success.

(2) Only single-stranded RNA is accepted as template. The replica formed is complementary and antiparallel to the template, suggesting Watson–Crick base-pairing at the replication site. A few other templates, e.g. poly(C) and C-rich nucleotide copolymers that are accepted by Qβ replicase, result in a perfect double strand (Hori *et al.*, 1967; Mitsunari and Hori, 1973). Double strands are not replicated (Biebricher *et al.*, 1982) and thus synthesis using these templates stops after transcribing the RNA; neither template nor enzyme is recycled. Autocatalytic amplification of a template takes place only when template and replica strands separate during replication and are released individually in single-stranded form (Dobkin *et al.*, 1979; Biebricher, 1983). An RNA species thus always consists of both complementary sequences.

(3) The replicase is highly processive, i.e. it usually does not dissociate from the template before a round is completed, because there is no way to complete a released incomplete replica strand in a subsequent reaction. As a consequence, release of the template after a replica-

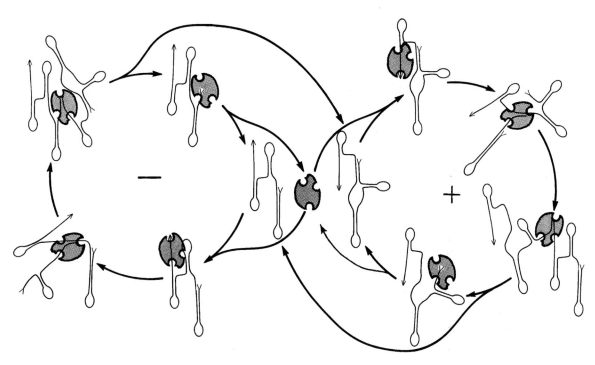

FIGURE 4.3 Diagram of the replication cycle. Enzyme is depicted here as having three sites: the catalytic site where triphosphates are bound and incorporated (replication fork) and two leaving sites for template and replica. Plus and minus strands have both a stem at their 5' termini and are unstructured at their 3' termini. While elongation proceeds, the replica is peeled off the template; secondary structure formation of both inhibits re-formation of the double strand. After completion of the replica strand, it is adenylated and released and is then immediately available as new template. The inactive template–replica complex where the enzyme has arrived at the 5' terminus has to be reactivated by slow decomposition. Plus and minus strand replication mechanisms are equivalent. Not depicted is the loss of single strands by double-strand formation.

tion round is the rate-limiting step in the whole cycle.

A large collection of artificial short-chained RNA species that are efficiently amplified by Qβ replicase have been selected, in most cases by the so-called template-free synthesis procedure described below under Creating biological information from scratch. They are replicated by the mechanism shown in Figure 4.3.

At the start of an amplification experiment, enzyme is typically present in large excess over RNA template. Newly synthesized replica as well as released template strands quickly bind to enzyme molecules, and exponential growth of the RNA concentration results. Once the enzyme is saturated with template, the RNA concentration increases linearly with time and the main products are free plus and minus strands. The overall replication rate in the linear growth phase is lower than in the exponential growth phase, because recycling of the enzyme becomes rate-limiting. Free complementary strands react to form double strands that are inactive as templates. Eventually, a steady state is reached where the concentration of single strands does not change, because the synthesis of new strands is balanced by loss of double-strand formation. At the final steady state, only the concentration of double-strand increases. (The double strands do bind to replicase, but they rapidly dissociate again. As the concentration of double strands increases, this slows down the synthesis rate, by lowering the concentration of available free enzyme.)

The essential chemical steps of the replication were identified in a series of experiments (Biebricher *et al.*, 1981, 1983, 1984, 1985), and rate coefficients for some of them were measured; for others, reasonable estimates could be introduced to enable kinetic modeling. A kinetic model set up this way is an oversimplification in the sense that the identified steps are not elementary chemical reactions. Binding of protein to RNA, for example, is not a simple bimolecular event. In reality, a cascade of chemical steps distorting bond angles, establishing short-range van der Waals interactions, hydrogen bonds and pushing out water molecules is involved for both macromolecular components. Fortunately, incorporating this level of detail in the model was not necessary to rationalize the experimental RNA concentration profiles. On the contrary, the kinetic model was able to describe the complicated experimental RNA growth profiles precisely, and is thus adequate for drawing conclusions about the roles of different parts of the replication process in determining the fitness of mutant RNA species. The quantitative determination of RNA replication is described in Box 4.1 and Figures 4.4 and 4.5.

SELECTION OF RNA SPECIES

When two or more RNA species are present in the starting template population, they compete with one another. If the sequences and the physical properties of the species differ sufficiently, the outcome of the selection process can be followed.

Box 4.1   Quantitative measurements of the replication rate

A simplified replication mechanism is shown in Figure 4.4. It is an adequate description for the replication time course when the replicase concentration exceeds 150 nmol/l and the concentrations of the triphosphates are higher than 300 μmol/l each. From the last equation in Figure 4.4, we can calculate the relative population increase $d[^iI]/([^iI_o]dt)$ of type i due to its fecundity $A_i = {}^iK_E[^iEI]/[^iI]$, which may be formulated in a more biological language as follows: the relative rate of population increase is equal to the proportion of the population in the reproducing age times its birth rate.

Experimentally, overall replication rates are determined by radioactive nucleotide incorporation measurements. The obtained profiles are shown in Figure 4.5. Two growth phases are clearly distinguished: an *exponential* one where enzyme is in excess and a *linear* one where enzyme is saturated with template. The overall growth rate in the linear growth phase is determined from the slope of the linear part. The overall growth rate in the exponential growth rate is calculated from the time displacement Δt of the profile caused by dilution by the factor F_{dil} according to $\kappa = \Delta t^{-1} \ln F_{dil}$. In the exponential growth phase, the intermediates and the total population show after an equilibration period coherent growth (Biebricher *et al.*, 1983):

$$d[I]/([I]dt) = d[EI]/([EI]dt) = d[IE]/([IE]dt) = d[I_o]/([I_o]dt) \equiv \kappa,$$

where κ is the overall (exponential) replication rate. It is easily determined from the displacement Δt on the time axis after dilution of the template by a factor F_{dil} by $\kappa = \Delta t^{-1} \ln F_{dil}$. In the linear growth phase, virtually all of the enzyme is bound to template, and a steady state is established where the intermediate concentrations do not change and the flux through each step is equal to the total flux v:

$$v = k_A[E][I] = k_E[EI] = k_D[IE] \equiv \rho[E_c].$$

In the linear growth phase of a single species, the enzyme is almost totally saturated with template and we obtain $\rho[E_c] \approx \rho[E_o]$, where $[E_o]$ is the total enzyme concentration, free and bound. The relative fecundities A_i are constant in the exponential growth phase ($A_i \approx {}^i\kappa$), but decrease in the linear growth phase with increasing $[^iI_o]$ ($A_i \approx \rho [E_o]/[^iI_o]$).

The population change of type i is of course also dependent on its mortality rate. Under the conditions of the described RNA replication experiments, the mortality is caused by the loss of template molecules by double-strand formation. Other contributions to mortality-like decomposition can be neglected. The loss rate can be described with the equation $-d[I]/dt = \frac{1}{2}d[II]/dt = k_{ds}[I]^2$. Hence, the relative mortality of type i is $D_i \approx k_{ds} [^iI]^2/[^iI_o]$. In the exponential growth phase, the concentrations of free strands are very small and the mortality is negligible.

The net population growth is the balance between fecundity and mortality: $E_i \equiv A_i - D_i$. In the exponential growth phase, we obtain $E_i = {}^i\kappa$. In the linear growth phase, A_i decreases and D_i increases until steady state is reached where $A_i = D_i$ and $E_i = 0$. The steady state concentrations can be calculated when the rates values have been determined (Biebricher *et al.*, 1984, 1991).

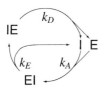

$$d[EI]/dt = k_A[E][I] - k_E[EI]$$
$$d[IE]/dt = k_E[EI] - k_D[IE]$$
$$d[E_c]/dt = k_A[E][I] - k_D[IE] = -d[E]/dt$$
$$d[I]/dt = k_E[EI] + k_D[IE] - k_A[E][I]$$
$$d[I_o]/dt = k_E[EI]$$

FIGURE 4.4 Simplified mechanism of RNA replication. Shown are the steps involving binding and releasing of complex components. All steps involved in replica synthesis are combined into a single step.

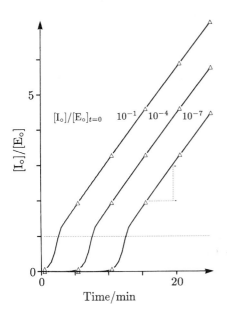

FIGURE 4.5 Incorporation profiles. MNV-11 was added at the indicated concentrations to a replication mix containing 200 nM replicase. The incorporation of labeled AMP was measured and calculated as number of strands synthesized per enzyme molecule. Two growth phases are clearly distinguished: an *exponential* one where enzyme is in excess and a *linear* one where enzyme is saturated with template. The overall growth rate in the linear growth phase is determined from the slope of the linear part. The overall growth rate in the exponential growth rate is calculated from the time displacement Δt of the profile caused by dilution by the factor F_{dil} according to $\kappa = \Delta t^{-1} \ln F_{dil}$. Experimental values under optimal growth conditions of MNV-11 were $\kappa = 1.92 \times 10^{-2} s^{-1}$ and $\rho = 6.37 \times 10^{-3} \ s^{-1}$ (Biebricher *et al.*, 1985).

The simplest case is exponential growth of small populations, which prevails when all resources required for amplification are present in excess. Under these conditions, each species grows as it would in the absence of the others. Each species grows independently with its own characteristic growth rate, but as this goes on the composition of the population changes: the population gradually becomes enriched in species with higher growth rates and relatively depleted in species that grow more slowly. A quantitative description of such experiments can be calculated (Box 4.2). The fitness of each species under exponential growth conditions is characterized by its fecundity, i.e. its replication rate, alone.

This strong selection makes working with an RNA replicase rather difficult. Assume that an RNA species with a replication rate one-tenth that of an optimized species has to be amplified by a factor of 10. While that happens, a single strand of the optimized species is amplified by a factor of 10^{10}, i.e. to macroscopic appearance! This illustrates that the techniques of amplification with replicase are technically not as easy as they might seem to be; severe precautions have to be made to avoid contamination of an RNA population with optimized species (Biebricher *et al.*, 1993). Synchronized amplification techniques like the polymerase chain reaction are much easier to handle. Purification of an RNA species by physico-chemical methods, e.g. by electrophoresis, must always be followed by a cloning procedure, because otherwise only the fastest species will be found. If we know that the separation method has reduced impurities to a level of say 1/1000, then it suffices to start an amplification experiment with fewer than 1000 template strands to get a pure species as product.

The outcomes of selection experiments carried out under linear growth conditions are at first glance surprising. Amplification of a mixed population by a factor of 10 can result in a change in the population composition by many orders of magnitude, and often a species with a lower replication rate is selected. As observed in organismic evolution, species with low fecundity can be quite successful if they are able to outcompete their competitors for limiting resources. In the linear growth phase of RNA replication, the limiting resource is the replicase itself, and the species that is fastest in binding to newly liberated replicase molecules will be selected whatever its replication rate may be. The quantitative description is somewhat more complicated. The kinetic model described previously is a good help. To describe competition in the linear growth phase, it is sufficient to set up the rate equations such that two species share the resources and the calculated profile again precisely matches the experimental

---

Box 4.2   Competition among species

For competition experiments two (or more) different RNA species share the resources enzyme and precursors; their concentrations and rates are distinguished by indices. Since the absolute concentrations vary, expecially when serial transfers are used, relative concentrations, i.e. the proportions of the species of the total population, x_i, are used to describe the population composition. Type conversion, i.e. reproduction producing a different type, is not possible in these experiments due to the species barrier.

Particularly easy is the calculation of the parameters important for selection in the exponential growth phase. The composition changes according to $[^1I_0]/[^2I_0]_t = [^1I_0]/[^2I_0]_{t=0}^{\exp(^1\kappa - ^2\kappa)t}$ (Kramer et al.; 1974; Biebricher et al., 1985). Instead of choosing the net growth rates E_i as selection values, it is more instructive to relate the net growth of each species to the total population change: we obtain the selection rate value $\zeta_i = E_i - \bar{E}$ where $\bar{E}$ is the weighted average of the net synthesis rates. A positive ζ value means that the species i becomes enriched in the population, a negative one means the population gets depleted in species i.

In the linear growth phase, calculation of selection values is more difficult. It can be shown that in the early linear growth phase the intrinsic selection rate value is proportional to ^ik_A [E]. At higher concentrations, the mortalities contribute also until finally an ecosystem is formed where each species occupies a constant fraction of the total population. The ratios of the free and bound types can be calculated (Biebricher *et al.*, 1985, 1991) as:

$$\left[\frac{^1I}{^2I}\right] = \frac{^1k_A\,^{22}k_{ds}}{^2k_A\,^{11}k_{ds}} \qquad \left[\frac{^1E_c}{^2E_c}\right] = \frac{(^1k_A)^2\,^{22}k_{ds}{}^2\rho}{(^2k_A)^2\,^1k_{ds}\,^1\rho} \ .$$

The complicated population dynamic can be illustrated by the growth profiles shown in Figure 4.6, obtained by computer simulation. The figure also shows the change of the parameters that are important for selection, A_i, D_i, E_i and $\bar{E}$. Experimental determinations were in full agreement with the calculated values.

---

results. Instead of fitness, we work better with selection (rate) values, the relative change of the relative population in time. The definitions are listed in Table 4.1. The selection values, which vary with time and the concentration of the competitors in the linear growth phase, can be precisely determined from the computed concentration profiles (Figure 4.6).

In the late linear growth phase, the loss terms caused by double-strand formation must also be taken into account in the model. For the case that the nucleotide sequences of two competing species are rather different, formation of heteroduplex strands can be neglected. Species with low concentrations of free single strands are favored by low loss rates through double strand formation and so the population can eventually reach a steady state, where its relative composition no longer changes: a stable ecosystem has been formed.

Even under the controlled external conditions of *in-vitro* evolution experiments, selection patterns can thus be quite complex, basically because the growing RNA species change their own environment. A typical example would be starting with two RNA species (MNV-11 and MDV-1), for which a computer simulation is shown in Figure 4.6. Initially both species are present in small equimolar amounts and exponential growth begins for both. When the enzyme is saturated, MNV-11 has conquered, because of its higher replication rate, most of the enzyme, and shortly afterwards it reaches the steady state of double-strand formation and its selection value vanishes. However, MDV-1 continues to grow, and eventually it displaces MNV-11 from the enzyme because of its higher enzyme binding rate. Eventually an ecosystem is formed where both species coexist; their selection values have both vanished.

MUTATION IN REPLICATING RNA

Any alteration of a genotype is called a mutation. Mutation may occur by chemical modification of a base, such as deamination of a cytidylate to a uridylate, but most mutations are produced by an erroneous replication, i.e. the

TABLE 4.1   Symbols and parameters used for the quantitative studies of *in vitro* evolution.

Concentrations

[^{i}I]	Concentration of free single-stranded RNA of type i	[mol/l]
[^{i}EI]	Concentration of active replication complex	[mol/l]
[^{i}IE]	Concentration of inactive replication complex	[mol/l]
[E]	Concentration of free enzyme	[mol/l]
[^{i}E$_c$]	Total concentration of template strands of type i complexed to enzyme	[mol/l]†
[E$_o$]	Total concentration of enzyme, bound or free	[mol/l]**
[^{i}I$_o$]	Total concentration of template strands of type i	[mol/l]†
Σ[^{i}I$_o$]	Total concentration of RNA	[mol/l]*
[^{ii}II]	Concentration of double strands (homoduplex) of type i	[mol/l]†
[^{ij}II]	Concentration of double strands between plus strand of type i and minus strand of type j (heteroduplex)	[mol/l]

Rate constants

$^i k_A$	Association rate constant for binding of replicase to RNA of type i	[l/mol s^{-1}]*
$^i k_E$	Rate constant for synthesizing and releasing a replica from a replication complex of type i	[s^{-1}]
$^i k_D$	Dissociation rate of inactive replication complex	[s^{-1}]*
$^{ij} k_{ds}$	Rate constant for double strand formation between plus strand of type i and minus strand of type j	[l/mol s^{-1}]*
$^i \kappa$	Overall replication rate constant of type i in the exponential growth phase	[s^{-1}]*
$^i \rho$	Experimentally measured relative rate of RNA synthesis per template strand of type i	[s^{-1}]*

Parameter definitions

x$_i$	Mutant frequency; Fraction of type i in the total population	†
A$_i$	Relative fecundity of type i	[s^{-1}]
D$_i$	Relative mortality of type i	[s^{-1}]
E$_i$	Relative net excess production rate of type i; E$_i$ = A$_i$ − D$_i$	[s^{-1}]†
$\bar{E}$	Relative net excess production rate for all types	[s^{-1}]*
Q$_{ij}$	Probability of producing type i per reproduction process of type j	
Q$_{ii}$	Probability of producing a correct copy per reproduction process of type i	
W$_{ii}$	Intrinsic selection rate value, W$_{ii}$ = Q$_{ii}$A$_i$ − D$_i$	[s^{-1}]
$^i \mu$	Mutational gain rate (synthesis by miscopying other templates)	[s^{-1}]
$^i \zeta$	Selection rate value	[s^{-1}]
$^i \xi$	Evolution rate value; relative rate for relative increase of type i, $^i \xi$ = $^i \zeta$ + $^i \mu$	[s^{-1}]†

* parameter can be readily measured
† parameter set at beginning of experiment
‡ parameter can be readily measured when types can be easily distinguished

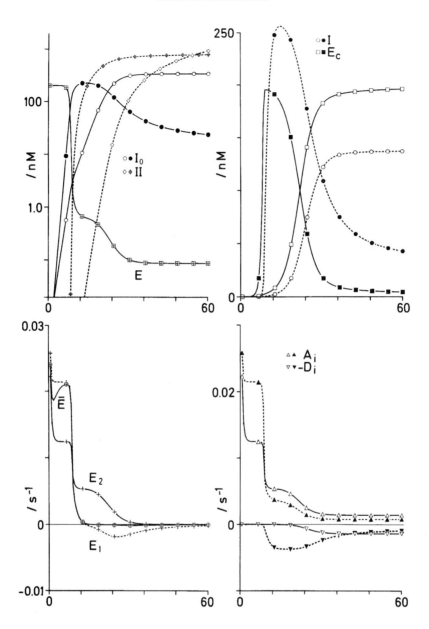

FIGURE 4.6 Competition among RNA species. Calculated growth profiles of two species, 1 (solid symbols) with standard values (MNV-11) and 2 (open symbols) having $^2k_A = 4 \times {}^1k_A$ and $^2k_D = 1/4 \, {}^1k_D$. Starting conditions were [E] = 200 nmol/l and [^{1}I] = [^{2}I] = 1 pmol/l. In the exponential growth phase (0–8 min), the smaller ^2k_D value is detrimental and species 1 grows more rapidly (see the semilogarithmic plot at upper left). It saturates the enzymes and enters the steady state, where its net growth (lower left) stops. Species 2 continues to grow exponentially at a lower rate (due to the smaller amount of free enzyme); it conquers most of the enzyme because of its higher binding rate (note the diminishing concentration of free enzyme). After 60 min the final steady state is reached where each species occupies a constant part of the enzyme. Calculations were done by numerical integration of the rate equations, using a much more detailed mechanism than shown in Figure 4.4. Approximate analytical solutions of the rate equations of the simplified mechanism can be found for certain cases. Reproduced from Biebricher *et al.*, 1985, in *Biochemistry*, **24**, 6550–6560, with permission.

progeny genotype differs at one or more positions from the parental genotype. Luria and Delbrück showed in a classic experiment in 1943 that the mutation event in bacteria is stochastic and that the mutated type may spread by error propagation. Different mutant types compete one with the other and selection occurs. Mutation can be studied quantitatively if selection is excluded by restricting amplification to a single replication round. Mutation rates can be measured, defined as the probability of incorporating a non-cognate base per incorporation event.

It is generally believed that DNA replaced RNA as genetic material during evolution because of its superior replication fidelity and chemical stability. Several energy-consuming error correction systems were invented to accomplish this fidelity. The systematic error caused by the tautomerization reaction of the pyrimidines is mainly corrected directly after phophodiester formation by proof-reading. A mismatch – particularly the one caused by a base that is returning from the wrong tautomeric structure to its favored one – is removed during the replication process itself. Extensive postreplicative repair systems detect and remove imperfections in newly formed DNA double helix. Neither fidelity-enhancing method is implemented in RNA synthesis. Because cellular RNA is normally produced in many copies per cell and is degraded after some time anyway, occasional errors do not cause permanent harm. Remarkably, no repair mechanism has yet been found among viruses with RNA genomes, where mutations can be lethal. Indeed, their mutation rates are so high that only a small fraction of the copies are identical to their parents. The high price for this – that most offspring of RNA viruses are defective – is apparently offset by the higher potential to adapt to changing environments, perhaps caused by host defenses, that is ultimately provided by error-prone replication.

Leviviruses were the first examples where the high mutant diversity of RNA viruses was detected. Watanabe collected a large number of leviviruses from all continents (Yonesaki et al., 1982). The leviviridae could be grouped into four classes. Their organization was nearly identical, yet the RNA fingerprints of species belonging to different classes did not indicate any sequence relationships. Today, while several such species have been sequenced, alignment of the sequences of species from different classes is difficult, because the information of the archetype founding the phylus has been almost totally diffused by mutations.

Within each virus species, however, RNA fingerprints were remarkably stable and well-defined, indicating a clearly defined wild-type sequence (Billeter et al., 1969; Fiers et al., 1976). The apparent simplicity of this result, however, was shattered when it was shown that clones derived from single phage plaques of a virus population showed differences in their fingerprint patterns (Domingo et al., 1978). It came as a shock to realize that viral populations are predominantly composed of an array of mutants, in which only a small fraction is what one would call a wild-type genome based on the dominant occupation of each nucleotide position in the sequence. Passaging the phage with a series of lysates restored the wild-type sequence. This result left only one explanation: the wild type sequence is nothing more than the average of all of the sequences present in the viral population.

Eigen and Schuster (1977) predicted this result with straightforward theoretical considerations. Error propagation causes a spread of mutations in the population, leading eventually to a stable population, the "quasispecies", where each mutant type maintains a constant share of the population, its mutant frequency, which depends on its production by mutation and its selective value. A high mutant frequency does not necessarily correlate with a particularly high mutation rate ("hot spot"); nearly-neutral multi-error mutants may have substantial mutant frequencies even though their rates of production by mutation are quite small. The theoretical background is covered in detail by Schuster and Stadler in Chapter 1, and Domingo et al., discuss the evolution of virus populations in vivo in Chapter 7.

From in-vitro studies (Batschelet et al., 1976) and in-vivo data (Drake, 1993) it was possible to estimate average RNA mutation rates per incorporated nucleotide; the two studies give values between 10^{-3} and 10^{-4}. On average, therefore, each phage RNA replica contains about one mutation. On the other hand, for the much shorter RNA sequences used in *in-vitro* experiments, the vast majorities of the copies should be correct.

A natural mutant spectrum of the replicating RNA species MNV-11 was investigated by Rohde *et al.* (1995). Each of their experiments began with a homogeneous RNA population created by cloning. A large number of serial transfers, under constant growth conditions, were then made to allow establishment of an equilibrium population. The same procedure was then repeated, starting with the same RNA clone, for different growth conditions, e.g. higher ionic strength or a different growth phase. In order to determine the sequence and other properties of the mutants in each equilibrated population, representative collections of the mutants had to be cloned. This cloning could not be accomplished by amplifying single RNA strands with Qβ replicase because of the high error rate and intrinsic bias introduced by the replicase. Lethal or seriously disadvantaged mutants, for example, would not show up at all, because they would undergo evolutionary optimization as the clones were amplified to levels where sequencing is possible. Cloning RNA first into DNA, and then amplifying the DNA, does not have these drawbacks, however, and was thus adopted for analyzing the equilibrium populations. The cloning procedure was designed in such a way that the same RNA sequence that provided each clone could be reconstructed from the DNA clone by DNA-directed RNA synthesis (Biebricher and Luce, 1993). It was shown that the RNA populations obtained by transcription were quite homogeneous. To be sure, the fidelity of transcription is no better than that of replication by viral replicase, but because transcription uses only the DNA and never the RNA copy as a template, error propagation is avoided. The sequences of some of the mutants found by Rohde *et al.* are shown in Figure 4.7.

The mutant spectra were found to be quite broad. When the linear growth phase was investigated, for example, "wild-type" RNA composed less than 40% of the quasispecies population. Mutations were not distributed randomly. At some positions mutations were frequent, while some regions were conserved, indicating parts of the RNA that are required for replication to occur. Single-error mutants were rare, and multi-error mutants appeared with up to 10% of the positions altered. Base transitions, transversions, deletions and insertions were observed, in one case even duplication of a 7-base segment.

It is clear that the mutant spectra observed by Rohde *et al.* are not simply correlated to mutation rates. Base transitions were not found more frequently than transversions, and multi-error mutants were strongly over-represented in comparison with what one would expect on the assumption that mutation rates governed the mutant spectra. Mutations themselves are essentially independent events, and if they generated the observed mutant spectra one would find a high frequency of one-error mutants, a much smaller frequency of two-error mutants, and multi-error mutants would be extremely rare. One has to conclude that the mutant spectra were governed instead by selection values. When this is true, frequently found mutants are expected to be neutral or nearly so. This was found to be true: the mutant replication rates were measured and found to be close to that of the wild type. The rate measurements also showed that the "wild type" found most frequently in equilibrated populations from the linear growth phase was not the mutant with the highest overall replication rate, but rather the best compromise among the rates of replication, replicase binding and double-strand formation.

The main reason for the high incidence of multi-error mutants must be that structural elements within the RNA are crucial for maintaining replication efficiency (Zamora *et al.*, 1995) and disturbance of such a structural element can be compensated by other mutations to restore replication efficiency.

Darwinian evolution of replicating RNA species offers more than an opportunity to do qualitative evolution experiments *in vitro*. It is also possible to predict evolutionary outcomes by deriving quantitative selection values from the physicochemical parameters of the competing RNA species, as outlined in Box 4.3. These parameters can readily be measured for individual RNA species. In addition, interactions among mutants such as formation of heteroduplex strands between single strands of different mutants must be taken into account. Since it has

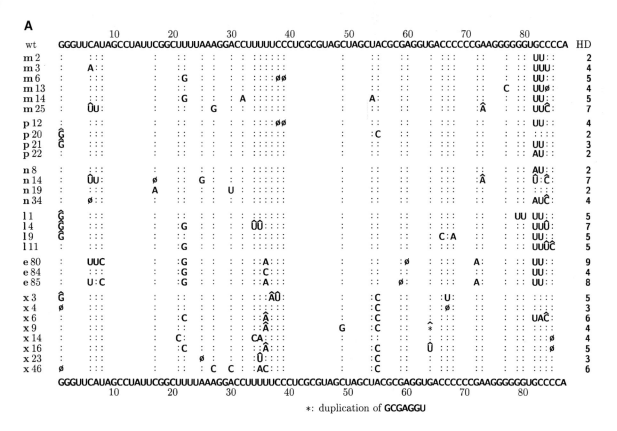

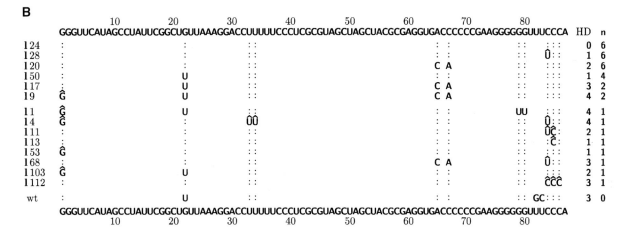

FIGURE 4.7 Mutant spectrum of MNV-11. **A.** Mutants within a population are indicated by number, different MNV-11 populations by letters: m, linear growth for 5 h; n,p, growth in the linear phase for 2 h followed by separation of plus (p) and minus (m) strands; 1l, growth in the exponential growth phase (100 replication rounds); e, exponential growth in the presence of 1 μM ethidium bromide; x, growth in the linear growth phase for 8 h in the presence of 50 mM (NH$_4$)$_2$ SO$_4$. HD is the Hamming distance, i.e. the number of base exchanges. Reproduced from Rohde *et al.*, 1995, in *J. Mol. Biol.*, **249**, 754–762, with permission. **B.** A more detailed analysis of a mutant spectrum of a quasispecies grown in the exponential growth phase. The last number in each column is the number of clones found in the population. Reproduced from Rohde *et al.*, 1995, in *J. Mol. Biol.*, **249**, 754–762, with permission.

Box 4.3   Mutation and selection

In Darwinian evolution, selection is complemented by mutation. The types i no longer represent different species but different mutants, and we must now take into account that type conversion occurs by mutation. In each replication round of type i there is a probability Q_{ji} to produce mutant j as copy. The relative fecundities A_i must therefore be corrected by the probabilities Q_i that the progeny is a correct copy of the template. With the short-chained model RNA templates this probability is close to unity, e.g. for RNA species MNV-11 the average Q_{ii} value is 0.97. For typical RNA viruses, however, the correction is rather dramatic, because the probability of producing correct offspring is much smaller than unity. The relative mortalities D_i are also influenced by the presence of other species. Under the experimental conditions used, loss is almost totally due to double-strand formation. With a few exceptions, double-strand formation does not discriminate between mutants, i.e. the loss is approximately proportional to the total RNA strand concentration of the opposite polarity. By combining the synthesis and loss terms, we can define the intrinsic selection value $W_{ii} = Q_{ii}A_i - D_i$. The relative population change $dx_i/(x_i dt)$ due to selection (fecundity and mortality) is the relative selection rate value $\zeta_i = W_{ii} - \bar{E}$. "Selection value" in the text means this value.

However, we must consider also that the population of mutants is affected not only by selection; strands of type i are also produced by erroneous replication of other mutants. The relative population phage $dx_i/(x_i dt_i)$ by the sum of all this contribution is called the mutational gain $\mu_i = \sum_{j \neq i} Q_{ij}A_j x_j / x_i$. It is always positive and usually dominated by the contributions of one or a few types that are sequentially closely related and highly populated.

The evolution rate $dx_i/(x_i dt) \equiv \zeta_i = \zeta_i + \mu_i$ is the total relative population change due to all contributions. Eventually, a steady state is formed where all evolution rates vanish; each mutant frequency has then reached a stable value, which does not change any more under constant conditions. The mutant distribution in the steady state is called a quasispecies. For a detailed quantitative description of the evolution of replicating RNA molecules, the reader should consult the literature (Biebricher et al., 1991).

been found that the rate constants for homoduplex and heteroduplex formation are essentially the same (Biebricher, unpublished measurements), these interactions can be quantified. Only when double-strand formation with other well-populated mutants is affected does a selective advantage result. Nevertheless, even in

this system with its minimal number of biochemical reactions, calculating selection values is a challenging exercise. The reason for this is that the experiments can be modeled using constant selective values only for the conditions of infinite dilution of the competing populations and unlimited resources. Under normal laboratory conditions, however, even in the constant environment of the test tube, the rates of production by mutation ("mutational gain") and the selection values change continuously as the population changes in composition and concentration (Eigen and Biebricher, 1987; Biebricher et al., 1991). Once again, computer simulations by numerical integration of the rate equations are of great help for getting insight, even though it is of course not possible, or reasonable, to try to account in computer simulations for all of the mutation possibilities that exist in the experiments.

Of particular interest is evolution in the pure exponential growth phase, because the selection values are then indeed constant and equal to the overall replication rate coefficient, which can be readily measured. The mutant distribution of such a quasispecies is shown in Figure 4.7B (Rohde et al., 1995). Among the 35 clones that were sequenced, the wild-type sequence was not found, because its replication rate is not the maximal one. There are fewer constraints on species evolving in the exponential phase, simply because competition and loss are excluded. A consequence of this is that the master sequence in the exponential growth phase is degenerate, the typical result being that several different mutants are nearly equally populated. These were not found in the mutant distributions in the linear growth phase, because their rate of binding replicase was reduced.

Adaptation to minor changes of growth conditions was found to be quite rapid. This is because the route of adaptation is different from what one might naively assume: when growth conditions change, there is no delay until appropriate mutations occur. Selection of the best adapted mutant already in the quasispecies is much faster than generation of new mutants. Its frequency rapidly increases and with it the (absolute) rate of producing mutants from it. The "center of gravity" within the existing quasispecies floats quickly through sequence space to a new position. Floating continues until a new evolutionary stable mutant spectrum emerges. What is thus observed is what has been described in organismic evolution as "punctuated equilibrium" (Gould and Eldredge, 1977).

The chance that a specific mutant is present depends strongly on the population size. When the population is small, more steps are required to reach a new equilibrium and adaptation takes longer. Furthermore, the route must then traverse a long staircase, on which each intermediate must have a selective advantage *per se* or it cannot be a part of the climb. We saw in analyzing the quasi-species, however, that this is seldom the case. Multi-error mutants were advantageous because the adverse effects of one mutation were compensated by subsequent ones. In a large population this is no problem, because some downward steps on the fitness landscape can be tolerated. The likelihood of generating a multi-error mutant depends on the number of steps necessary, the number of possible routes to reach it and the depth of the valleys that have to be crossed. Very deep canyons (i.e. where one of the intermediates represents a lethal mutation) must be crossed with a single jump, i.e. the two-error mutant be formed in one replication round.

Several adaptation experiments of short-chained RNA species have been reported. The first quantitative one was on replication of the species MDV-1 in the presence of a low concentration of ethidium bromide (Kramer *et al.*, 1974), which resulted in selection of a three-error mutant. Adaptation was achieved slowly, because each transfer began with a population of only 10^6 RNA strands. The first mutant was already present in the quasispecies population and the next mutations occurred in the seventh and 12th transfers, respectively. A disadvantage of the serial transfer technique is that small aliquots are used for inoculation of succeeding transfers. The probability of finding a newly formed mutant in an aliquot may be quite small, depending on its size and the time when the mutant emerged. Furthermore, each step in these experiments involved amplification in both the exponential and the linear growth phase. It was therefore not possible to calculate selection values.

Eigen and collaborators (Strunk and Ederhof, 1997) developed a machine that avoids these disadvantages. It always remains in the exponential growth phase, because the RNA concentration is measured in real time, triggering a serial transfer before the enzyme is saturated. The 1:10 aliquot used for the next transfer insures that mutant populations do not drop to low values. Using this machine a variant MNV-11 resistant to RNase A was selected after a rather large sequence change, including a deletion (Strunk and Ederhof, 1997). The evolution route taken in this process has not yet been reported.

Site-directed mutagenesis experiments with leviviridae genomes have shown that almost any mutation of the genome affects the fitness of the virus. Studies of the revertants and pseudo-revertants revealed an intricate influence of the RNA structure on replication, translation and regulation of the virus (Arora *et al.*, 1996; Klovins *et al.*, 1997; Poot *et al.*, 1997). These experiments brought many insights into the subtle control of the biochemical processes involving RNA and illustrated that the fitness of a viral type is a highly complex function that makes quantitative predictions extremely difficult.

RECOMBINATION AMONG RNA MOLECULES

In organisms with DNA genomes, the high replication fidelity makes large mutational jumps impossible as evolutionary routes. An alternative route is taken: DNA recombination. DNA from other organisms is occasionally inserted and sections of the native genome are occasionally deleted, duplicated, inverted or transposed to remote positions. Normal cells contain many enzymes involved in catalyzing DNA recombination, underscoring the importance of this process.

RNA recombination is far less frequent. In initial experiments using double infection of host with leviviruses containing defects in different cistrons no defect-free recombinants could be isolated. Recent experiments have shown that recombination does occur, but only at a very low rate (Palasingam and Shaklee, 1992). RNA recombination has been observed with many different viruses (King *et al.*, 1982; Lai *et al.*, 1985; Lai, 1992), but molecular devices that would catalyze it have not been identified. It seems that RNA recombination results from errors in the replication process itself. Several models have been proposed, the simplest and most plausible being "copy choice", i.e. a jump from one template to another (or to the same template, but on a different position) during replication (Lai, 1992).

An RNA species replicated by Qβ replicase has been isolated that is obviously a recombinant between part of the replicase gene of Qβ and host cell tRNA (Munishkin *et al.*, 1988). RNA recombination *in vitro* is a very rare event, but has also been reported (Biebricher and Luce, 1992). Even a very rare event, however, can quickly become evident in an evolution experiment if an advantageous mutant is created. Thus MNV-11 grown to equilibrium builds up a stable mutant spectrum (see above), but under conditions of high ionic strength and growth in the late linear growth phase a new RNA species with a higher chain length (135) than MNV-11 (86) eventually emerges and is rapidly selected. Repetition of the experiment under identical conditions showed that the eventual result is reproducible, indicating an instructed process, while the time lapse to emergence of the new species is not (Biebricher *et al.*, 1982).

RNA recombination events are more frequent at higher ionic strength. In the numerous cases we observed (Biebricher and Luce, 1992; Zamora *et al.*, 1995; Biebricher and Luce, unpublished work), a short repetitive sequence was usually found, indicating a copy choice mechanism. Chetverin *et al.* (1997) described examples where this is not the case. Since only the sequence changes that are genetically fixed can be observed, a clear decision between different models is not possible.

CREATING BIOLOGICAL INFORMATION FROM SCRATCH

So far I have described experiments that showed darwinian adaptation to the environment, i.e. optimization of a pre-existing biological function. Evolution, however, is able not only to adapt but also to create. Is it possible to generate a self-replicating RNA without offering a template? In the last years, many RNA species with novel functions have indeed been selected starting from completely random RNA sequences (Tuerk and Gold, 1990). In other words, new biological function has been formed without any ancestry at all. In these experiments human ingenuity (to set up the experiments in the first place), random chance and darwinian evolution are the driving forces that create information from nowhere. However, even a century after Darwin, doubts continue to be expressed that such information can be formed without human interference. The main conceptual difficulty that such doubters experience derives from the vanishingly low probability of creating a predefined sequence by chance. To find a specific sequence of chain length 50 would require a population of $4^{50} = 10^{30}$ strands. Fortunately, there is not just a single winner in the sequence lottery: the large number of total blanks partially is compensated by the large number of minor wins. Sequences with low fitness values, once any are created by chance, are optimized by adaptive evolution on a much quicker time scale.

Two basic strategies were found to create replicable RNA without any template being present in the starting mixture. (1) In the first it was intentionally extracted from a huge library of randomly assembled sequences (Biebricher and Orgel, 1973; Brown and Gold, 1995). (2) The second was an unexpected finding: incubation of an RNA replicase at high concentration in the presence of high nucleotide triphosphate concentrations produced replicable RNA after long incubation times despite the absence of detectable RNA in the starting material (Sumper and Luce, 1975). Different RNA species are selected in each experiment of this kind (Biebricher *et al.*, 1981; Biebricher, 1987). Evidence has been presented to show that in absence of template the replicase condenses nucleotides, at a rate five orders of magnitude less than that of template-instructed synthesis, to produce a random mixture of sequences (Biebricher *et al.*, 1986). Once any accepted template is produced, no matter how inefficient it may be, it is amplified and optimized. Indeed, replicability is a particularly sensitive function to select for, because the overwhelming majority of unaccepted RNA is ignored by the replicase.

Impurity RNA would have some genetic origin; however, the emerging species show no base homology to the genomes of the virus or of the host; moreover, they cannot be detected in infected or non-infected cells (Avota *et al.*, 1998). *In vitro*, template-free synthesis was even found to be suppressed by addition of non-replicable RNA or DNA. Aggregation of enzyme molecules increased the efficiency of template-free synthesis. Modification of the non-replicable RNA by instructed terminal elongation has been observed (Biebricher and Luce, 1992). *In vivo*, only weakly replicable RNA species were derived from host RNA, in particular from 16S ribosomal RNA (Avota *et al.*, 1998).

Sequence analysis and quantitative characterization of the properties of the early products of template-free synthesis were quite instructive (Biebricher and Luce, 1993). As mentioned earlier, the first feature noticed was that these short-chain RNA strands differ in chain length and sequence in each experiment. The low probability of assembling long replicable RNA strands favors small early products. Experimentally, strands with 25–40 nucleotides dominated. Their replication rates were low compared to those of optimized RNA. During subsequent serial transfers, early RNA products underwent rapid evolutionary optimization. During the optimization the molecular weight increased, in nearly all cases by recombination-like events such as duplications or insertion of sections of the complementary sequence. The optimization rate depended on experimental conditions. At high ionic strength optimization was fast. Otherwise it was so slow that a short inefficient template could be amplified at high amplification factors ($> 10^{20}$) during many serial transfers without changes of the average sequence.

The large number of short-chain replicable RNA species found offered a possibility to investigate the minimum sequence requirement for replication. Sequence comparison of the replicable species, however, did not reveal anything like a consensus sequence at all. Except for the invariant ends – pppGG[G] at the 5′ termini, CCA at the 3′ termini (a terminal A is attached without template instruction) – no homologies could be found. However, when the secondary structures were calculated, it appeared that the structures of all replicable RNA had a stem at the 5′ termini, while the 3′ termini were unpaired. The alternative folding, with 5′ and 3′ paired with each other, was energetically disfavored (Biebricher and Luce, 1993). The constraints for the more stable structure are more severe than it might seem at first. If base-pairing only involved the canonical base pairs, then a stem at the 5′ strand of one strand would correspond to a 3′ stem for the complementary sequence. Only non-canonical base pairs and outlooped bases at strategic positions makes the conserved replicable structure possible. Site-directed mutation replacing these positions with canonical base pairs destroyed the template activity of the RNA entirely (Zamora *et al.*, 1995).

What is the reason for this structure? It is not known yet, but there are arguments in favor of stems at the 5′ termini (Biebricher, 1994). As a replica is formed, the structure is transiently double-stranded. With progressing elongation the replica leaves the template. Rapid stem formation reduces the danger that replica and template reform a double strand. There are additional features common to many, but not all, replicable RNA species. A pyrimidine cluster in the interior of the sequence seems to be favorable for enzyme-RNA binding (Brown and Gold, 1993). However, binding strength to replicase is only poorly correlated with template activity. Some RNA sequences, notably 16S rRNA, bind quite well to Qβ replicase but have no template activity, while binding of some early products of template-free synthesis is only weak even though their replication rates are substantial.

The structural features described here appear to be necessary for RNA replication. To test whether they are sufficient for replication, the author's group designed and synthesized RNA strands with sequences predicted to give these structural features (Zamora *et al.*, 1995). Their template activities, however, were found to be barely measurable. Upon prolonged incubation with replicase, replicable RNA did grow out. The selected RNA species differed from each other, but were all clearly mutants of their respective initial templates. In some cases two or three base exchanges sufficed to make the species replicable, while in other cases recombination events were also involved. In all cases the above-described structural features were not only conserved but even enhanced. We conclude that there are unidentified additional requirements for adequate template activity. Clearly it is unlikely to strike a fitness peak when designing templates on the basis of the structural features we have been able to identify. During amplification, drifting of the designed sequences to nearby fitness peaks is thus inevitable.

CONCLUSIONS

Many quantitative insights into the nature of evolution have been gained from studying the model system provided by Qβ replicase. Are the often surprising results obtained from these experiments only a misleading caprice of nature, with no relevance to evolution in general? There are many reasons to think that this is not the case. The principal one is that studies using other enzymes, including some that are not viral, lead to similar results. If, as is generally believed, the origin of viruses is formation of intracellular parasites that eventually develop an apparatus for horizontal gene transfer among different hosts, then viral genes must derive from cellular ones. RNA replication with Qβ replicase requires catalytic participation of RNA; if template RNA is able to instruct Qβ replicase to replicate it, why should it be unlikely that cellular RNA can find ways to instruct a cellular DNA-dependent RNA polymerase to carry out RNA replication? Indeed, it has been shown that RNA templates exist that are accepted by the DNA-dependent RNA polymerases of the bacteriophages T7 and T3 (Konarska and Sharp, 1989, 1990; Biebricher and Luce, 1996) and *E. coli* (Biebricher and Luce, 1993; Wettich, 1995). Since for these enzymes no physiological RNA templates exist, replicable RNA species

were selected by the described methods from random nucleotide libraries (Biebricher and Orgel, 1973) or obtained by template-free synthesis (Biebricher and Luce, 1996; Wettich, 1995). The templates are specific for their cognate enzyme and not accepted by other RNA polymerases. (One exception that has been found is an RNA species replicated by T7 RNA polymerase as well as by T3 RNA polymerase.) The features like exponential and linear growth and strand separation during replication were nearly identical to what has been observed over the years with Qβ replicase.

Recently, newly developed RNA amplification methods with lower sequence specificity than Qβ replicase have been shown to be superior for artificial selection of functional RNA by evolutive biotechnology (Guatelli *et al.*, 1990; Breaker and Joyce, 1994). For quantitative studies of *natural* selection under controlled conditions, however, amplification of RNA by Qβ replicase is still unsurpassed.

ACKNOWLEDGEMENTS

I am indebted to Dr W.C. Gardiner for critical reading of the manuscript. Support from the Deutsche Forschungsgemeinschaft is gratefully acknowledged.

REFERENCES

Arora, R., Priano, C., Jacobson, A.B. and Mills, D.R. (1996) Cis-acting elements within an RNA coliphage genome – fold as you please, but fold you must. *J. Mol. Biol.*, **258**, 433–446.

August, J.T., Cooper, S., Shapiro, L. and Zinder, N.D. (1963) RNA phage-induced RNA polymerase. *Cold Spring Harbor Symp. Quant. Biol.*, **28**, 95–97.

Avota, E., Berzins, V., Grens, E., Vishnevsky, Y., Luce, R. and Biebricher, C.K. (1998) The natural 6S RNA found in Qβ-infected cells is derived from host and phage RNA. *J. Mol. Biol.*, **276**, 7–17.

Batschelet, E., Domingo, E. and Weissmann, C.

(1976) The proportion of revertant and mutant phage in a growing population, as a function of mutation and growth rate. *Gene*, **1**, 27–32.

Beekwilder, J., Nieuwenhuizen, R., Poot, R. and van Duin, J. (1996) Secondary structure model for the first 3 domains of Qβ RNA – control of A-protein synthesis. *J. Mol. Biol.*, **256**, 8–19.

Biebricher, C.K. (1983) Darwinian selection of self-replicating RNA. *Evol. Biol.*, **16**, 1–52.

Biebricher, C.K. (1987) Replication and evolution of short-chained RNA species replicated by Qβ replicase. *Cold Spring Harbor Symp. Quant. Biol.*, **52**, 299–306.

Biebricher, C.K. (1994) The role of RNA structure in RNA replication. *Ber. Bunsenges.*, **98**, 1122–1126.

Biebricher, C.K. and Eigen, M. (1987) Kinetics of RNA replication by Qβ replicase In: *RNA Genetics, vol. I: RNA-directed Virus Replication* (eds Domingo, E., Ahlquist, P. and Holland, J.J.), pp. 1–21. CRC Press, Boca Raton, FL.

Biebricher, C.K. and Luce, R. (1992) *In vitro* recombination and terminal elongation of RNA by Qβ replicase. *EMBO J.*, **11**, 5129–5135.

Biebricher, C.K. and Luce, R. (1993) Sequence analysis of RNA species synthesized by Qβ replicase without template. *Biochemistry*, **32**, 4848–4854.

Biebricher, C.K. and Luce, R. (1996) Template-free synthesis of RNA species replicating with T7 RNA polymerase. *EMBO J.*, **15**, 3458–3465.

Biebricher, C.K. and Orgel, L.E. (1973) An RNA that multiplies indefinitely with DNA-dependent RNA polymerase: selection from a random copolymer. *Proc. Natl Acad. Sci. USA*, **70**, 934–938.

Biebricher, C.K., Eigen, M. and Luce, R. (1981) Product analysis of RNA generated *de novo* by Qβ replicase. *J. Mol. Biol.*, **148**, 369–390.

Biebricher, C.K., Diekmann, S. and Luce, R. (1982) Structural analysis of self-replicating RNA synthesized by Qβ replicase. *J. Mol. Biol.*, **154**, 629–648.

Biebricher, C.K., Eigen, M. and Gardiner, W.C. (1983) Kinetics of RNA replication. *Biochemistry*, **22**, 2544–2559.

Biebricher, C.K., Eigen, M. and Gardiner, W. C. (1984) Kinetics of RNA replication: plus–minus asymmetry and double-strand formation. *Biochemistry*, **23**, 3186–3194.

Biebricher, C.K., Eigen, M. and Gardiner, W.C. (1985) Kinetics of RNA replication: competition and selection among self-replicating RNA species. *Biochemistry*, **24**, 6550–6560.

Biebricher, C.K., Eigen, M. and Luce, R. (1986) Template-free RNA synthesis by Qβ replicase. *Nature*, **321**, 89–91.

Biebricher, C.K., Eigen, M. and Gardiner, W.C. (1991) Quantitative analysis of selection and mutation in self-replicating RNA. In: *Biologically Inspired Physics,* (ed. Peliti, L.), NATO ASI Series B, vol. 263, pp. 317–337. Plenum Press, New York.

Biebricher, C.K., Eigen, M. and McCaskill, J.S. (1993) Template-directed and template-free RNA synthesis by Qβ replicase. *J. Mol. Biol.*, **231**, 175–179.

Biebricher, C.K., Nicolis, G. and Schuster, P. (1995) *Self-organization in the Physico-chemical and Life Sciences.* Office for Official Publications of the European Communities, Luxembourg.

Billeter, M.A., Dahlberg, J. E., Goodman, H.M., Hindley, J. and Weissmann, C. (1969) Sequence of the first 175 nucleotides from the 5′ terminus of Qβ RNA synthesized *in vitro*. *Nature, 224*, 1083–1086.

Breaker, R.R. and Joyce, G.F. (1994) Emergence of a replicating species from an in vitro RNA evolution reaction. *Proc. Natl Acad. Sci. USA*, **91**, 6093–6097.

Brown, D. and Gold, L. (1995) Selection and characterization of RNAs replicated by Qβ replicase. *Biochemistry*, **34**, 14775–14782.

Chetverin, A.B., Chetverina, H.V., Demidenko, A.A. and Ugarov, V.L. (1997) Non-homologous RNA recombination in a cell-free system: evidence for a transesterification mechanism guided by secondary structure. *Cell*, **88**, 503–513.

Dawkins, R. (1982) *The Extended Phenotype.* Freeman, San Francisco, CA.

Dobkin, C., Mills, D.R., Kramer, F.R. and Spiegelman, S. (1979) RNA replication: required intermediates and the dissociation of template, product, and Qβ replicase. *Biochemistry*, **18**, 2038–2044.

Dobzhansky, T., Ayala, F.J., Stebbins, G.L. and Valentine, J.W. (1977) *Evolution*. Freeman, San Francisco, CA.

Domingo, E., Sabo, D., Taniguchi, T. and Weissmann, C. (1978) Nucleotide sequence heterogeneity of an RNA phage population. *Cell*, **13**, 735–744.

Drake, J.W. (1993) Rates of spontaneous mutation among RNA viruses. *Proc. Natl Acad. Sci. USA*, **90**, 4171–4175.

Eigen, M. and Biebricher, C.K. (1987) Sequence space and quasispecies distribution. In: *RNA Genetics, vol. III: Variability of RNA Genomes* (eds Domingo, E., Ahlquist, P. and Holland, J.J.), pp. 211–245. CRC Press, Boca Raton, FL.

Eigen, M. and Schuster, P. (1977) The hypercycle – a principle of natural self-organization. Part A: Emergence of the hypercycle. *Naturwissenschaften*, **64**, 541–565.

Eigen, M., Biebricher, C.K., Gebinoga, M. and Gardiner, W.C. (1991) The hypercycle: coupling of RNA and protein biosynthesis in the infection cycle of an RNA bacteriophage. *Biochemistry*, **30**, 11005–11018.

Fiers, W., Contreras, R., Duerinck, F. *et al.* (1976) Complete nucleotide sequence of bacteriophage MS2 RNA: primary and secondary structure of the replicase gene. *Nature*, **260**, 500–507.

Franze de Fernandez, M.T., Eoyang, L. and August, J.T. (1968) Factor fraction required for the synthesis of bacteriophage Qβ RNA. *Nature*, **219**, 588–590.

Gould, S.J. and Eldredge, N. (1977) Punctuated equilibria: the tempo and mode of evolution reconsidered. *Palaeobiology*, **3**, 115–151.

Guatelli, J.C., Whitfield, K.M., Kwoh, D.Y., Barringer, K.E., Richman, D.D. and Gingeras, T.R. (1990) Isothermal in vitro amplification of nucleic acids by a multienzyme reaction modeled after retroviral replication. *Proc. Natl Acad. Sci. USA*, **87**, 1874–1878.

Haruna, I., Nozu, K., Ohtaka, Y. and Spiegelman, S. (1963) An RNA "replicase" induced by and selective for a viral RNA: isolation and properties. *Proc. Natl Acad. Sci. USA*, **50**, 905–911.

Hori, K.L., Eoyang, L., Banerjee, A.K. and August, J.T. (1967) Template activity of synthetic ribopolymers in the Qβ RNA polymerase reaction. *Proc. Natl Acad. Sci. USA*, **57**, 1790–1797.

Kamen, R. (1970) Characterization of the subunits of Qβ replicase. *Nature*, **228**, 527–533.

Karnik, S. and Billeter, M. (1983) The lysis function of RNA bacteriophage Qβ is mediated by the maturation (A2) protein. *EMBO J.*, **2**, 1521–1526.

King, A.M.Q., McCahon, D., Slade, W.R. and Newman, J.W.I. (1982) Recombination in RNA. *Cell*, **29**, 921–928.

Klovins, J., Tsareva, N.A., de Smith, M.H., Berzins, V. and van Duin, J. (1997) Rapid evolution of translational control mechanisms in RNA genomes. *J. Mol. Biol.*, **265**, 372–384.

Konarska, M.M. and Sharp, P.A. (1989) Replication of RNA by the DNA-dependent RNA polymerase of phage T7. *Cell* **57**, 423–431.

Kondo, M., Gallerani, R. and Weissmann, C. (1970) Subunit structure of Qβ replicase. *Nature*, **228**, 525–527.

Kramer, F.R., Mills, D.R., Cole, P.E., Nishihara, T. and Spiegelman, S. (1974) Evolution *in vitro*, sequence and phenotype of a mutant RNA resistant to ethidium bromide. *J. Mol. Biol.*, **89**, 719–736.

Lai, M.M.C. (1992) RNA recombination in animal and plant viruses. *Microbiol. Rev.*, **56**, 61–79.

Lai, M.M.C., Baric, R.S., Makino, S. (1985) Recombination between nonsegmental RNA genomes of muric coronaviruses. *J. Virol.*, **56**, 449–456.

Levisohn, R. and Spiegelman, S. (1969) Further extracellular Darwinian experiments periments with replicating RNA molecules: diverse variants isolated under different selective conditions. *Proc. Natl Acad. Sci. USA*, **63**, 807–811.

Loeb, T. and Zinder, N.D. (1961) A bacteriophage containing RNA. *Proc. Natl Acad. Sci. USA*, **47**, 282–289.

Luria, S.E. and Delbrück, M. (1943) Mutation of bacteria from virus sensitivity to virus resistance. *Genetics*, **28**, 491–511.

Mills, D.R., Peterson, R.L. and Spiegelman, S. (1967) An extracellular Darwinian experiment with a self-duplicating nucleic acid molecule. *Proc. Natl Acad. Sci. USA*, **58**, 217–224.

Mills, D.R., Kramer, F.R. and Spiegelman, S.

(1973) Complete nucleotide sequence of a replicating RNA molecule. *Science,* **180**, 916–927.

Mitsunari, Y. and Hori, K. (1973) Qβ replicase-associated, poly(C)-dependent poly(G) polymerase. *J. Biochem.,* **74**, 263–271.

Munishkin, A.V., Voronin, L.A. and Chetverin, A.B. (1988) An *in vivo* recombinant RNA capable of autocatalytic synthesis by Qβ replicase. *Nature,* **333**, 473–475.

Palasingam, K. and Shaklee, P.N. (1992) Reversion of Qβ RNA phage mutants by homologous RNA recombination. *J. Virol.,* **66**, 2435–2442.

Poot, R.A., Tsareva, N.V., Boni, I.V. and van Duin, J. (1997) RNA folding kinetics regulates translation of phage MS2 maturation gene. *Proc. Natl Acad Sci. USA,* **94**, 10110–10115.

Rohde, N., Daum, H. and Biebricher, C.K. (1995) The mutant distribution of an RNA species replicated by Qβ replicase. *J. Mol. Biol.,* **249**, 754–762.

Saffhill, R., Schneider-Bernloehr, H., Orgel, L.E. and Spiegelman, S. (1970) *In vitro* selection of bacteriophage Qβ RNA variants resistant to ethidium bromide. *J. Mol. Biol.,* **51**, 531–539.

Spiegelman, S., Haruna, I., Holland, I.B., Beaudreau, G. and Mills, D.R. (1965) The synthesis of a self-propagating and infectious nucleic acid with a purified enzyme. *Proc. Natl Acad. Sci. USA,* **54**, 919–927.

Strunk, G. and Ederhof, T. (1997) Machines for automated evolution experiments *in vitro* based on the serial-transfer concept. *Biophys. Chem.,* **66**, 193–202.

Sumper, M. and Luce, R. (1975) Evidence for *de novo* production of self-replicating and environmentally adapted RNA structures by bacteriophage Qβ replicase. *Proc. Natl Acad. Sci. USA,* **72**, 162–166.

Tuerk, C. and Gold, L. (1990) Systematic evolution of ligands by exponential enrichment. *Science,* **249**, 505–510.

Weissmann, C. (1974) The making of a phage. *FEBS Lett. (Suppl.),* **40**, S10–S18.

Weissmann, C., Simon, L., Borst, P. and Ochoa, S. (1963) Induction of RNA synthetase in *E. coli* after infection by the RNA phage MS2. *Cold Spring Harbor Symp. Quant. Biol.,* **28**, 99–104.

Wettich, A. (1995) RNA replication mechanisms of DNA-dependent RNA polymerases. Master's thesis, University of Göttingen.

Yonesaki, T., Furuse, K., Haruna, I. and Watanabe, I. (1982) Relationships among four groups of RNA coliphages based on the template specificity of GA replicase. *Virology,* **116**, 379–381.

Zamora, H., Luce, R. and Biebricher, C.K. (1995) Design of artificial short-chained RNA species that are replicated by Qβ replicase. *Biochemistry,* **34**, 1261–1266.

5

The Fidelity of Cellular and Viral Polymerases and its Manipulation for Hypermutagenesis

Andreas Meyerhans and Jean-Pierre Vartanian

INTRODUCTION

The copying of genetic information is a key process in all replicating biological entities. It involves enzyme catalysed synthesis of DNA or RNA strands complementary to a polynucleotide template. Mononucleoside triphosphates are selected on the basis of Watson–Crick base pairing and are added by step-wise 5'–3' phosphodiester bond formation into the growing nucleotide chain. Inevitably this complex biological process is error-prone, giving rise to variants of the original information. This provides the backdrop for evolution. The mutation rate, however, has an upper limit of approximately the inverse of the genome length above which the information becomes unstable.

The comparison of the spontaneous mutation rates per genome per replication revealed a remarkable similarity within broad groups of organisms (Drake, 1993; Drake *et al.*, 1998). This is suggestive of evolutionary forces balancing the costs of high fidelity replication, the deleterious effects of mutations and the necessity for mutations to adapt to changing environments. RNA viruses, for example, have genomic mutation rates around 1 per replication and thus replicate close to their error threshold (Eigen

and Biebricher, 1988; Eigen, 1993; Domingo and Holland, 1997). The rates for retroviruses and retrotransposons are one order of magnitude smaller. Replication of DNA-based microbes proceeds with an even higher fidelity, their rates varying about 2.5-fold at around 1/300 per genome per replication. Considering the approximate 6500-fold size difference in their genomes, the mutation rates per nucleotide per replication varies roughly 16 000-fold.

Genomic mutation rates are a composite of a number of processes, including enzymatic polymerization, proofreading and mismatch repair. In this review, we will concentrate on the discussion of two important aspects of DNA and RNA polymerization, the fidelity of the polymerases and the supply of the nucleotide triphosphate substrates by the intracellular microenvironment. After a brief introduction on the structure of polymerases and a description of the regulation of intracellular dNTP pools, polymerase fidelity and ways to measure it are described. Subsequently, we summarize the available data on viral hypermutation and discuss the underlying mechanisms, with its potential implication for genome evolution. Finally, the manipulation of fidelity for hypermutagenesis, which might have an impact on future antimicrobial therapies, is mentioned. Together

Origin and Evolution of Viruses
ISBN 0–12-220360-7

this highlights the intimate relationship between polymerase fidelity and the intracellular environment and thus may link genome evolution and spontaneous and hereditary disease.

TEMPLATE COPYING BY POLYMERASES

The superfamily of polymerases comprises a highly divergent set of nucleic acid copying enzymes. They are composed of modules with a particular function, linked together either non-covalently or by single polypeptide chains. Comparisons on the primary sequence level revealed short motifs that specify four polymerase families: the pol I, pol α, pol β and reverse transcriptase (RTase) families (Delarue *et al.*, 1990). Recently, availability of the three-dimensional structures from the Klenow fragment of DNA polymerase I (Beese *et al.*, 1993), T7 RNA polymerase (Doublié *et al.*, 1998), the heat-stable polymerase of *Thermus aquaticus* (Taq polymerase; Eom *et al.*, 1996), human DNA polymerase β (Pelletier *et al.*, 1994), polymerase α from bacteriophage RB69 (Wang *et al.*, 1997) and the RTase of the human immunodeficiency virus type 1 (HIV-1) has provided a new basis for comparisons (Kohlstaedt *et al.*, 1992; Jacobo-Molina *et al.*, 1993; Smerdon *et al.*, 1994; Ren *et al.*, 1995; Rodgers *et al.*, 1995). Together with biochemical studies, this has largely extended our knowledge on the individual steps of the poly-

merization reaction such as substrate binding and chain elongation, and has helped to understand some basic features of polymerases, like processivity and copying fidelity.

In brief, the overall structures of polymerases are similar and can be viewed as a right hand with "fingers", "palm" and "thumb" elements forming a groove that accomodates the nucleic acid substrate. While the fingers are involved in positioning of the template and nucleoside triphosphates, the palm harbors the conserved aspartate residues that bind the two catalytic metal ions. The thumb domain is substantial for enzyme processivity. Whereas the palm subdomains of the four DNA polymerase families are structurally similar, the finger and thumb subdomains are structurally highly divergent. However, they all function in a similar way and use corresponding secondary structural elements.

The polymerization reaction is outlined in Figure 5.1 (Donlin *et al.*, 1991; Kunkel, 1992; Kunz *et al.*, 1994). In the first step, the polymerase binds to the template/primer duplex. This reversible interaction is essentially non-specific, since the enzyme must bind to an "infinite" number of sequences during replication. The interaction with the nucleoside triphosphate having Watson–Crick complementarity to the next appropriate template residue stabilizes the enzyme/template/primer complex and shifts the equilibrium towards III. Topologically, the triphosphate is bound at the nucleotide binding site. By a conformational change into an

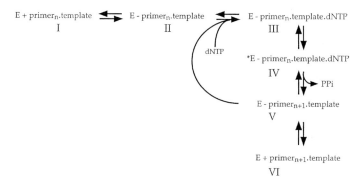

FIGURE 5.1 Reaction pathway of template copying by polymerases. The main components of the reaction are given. E, polymerase; PPi, pyrophosphate; * indicates an activated state. See text for details.

activated complex, the nucleoside triphosphate is brought next to the active site of the polymerase where the 3'-OH group of the primer is located (IV; Dalberg and Benkovic, 1991). The polymerase then catalyses the addition of the complexed triphosphate to the 3'-OH terminus of the primer and releases phyrophosphate. The nucleotide binding site is now free to capture another nucleoside triphosphate and polymerization can continue.

Maintenance of the polymerisation cycle is determined by the so-called "processivity" of the polymerase that is the number of nucleotides involved in sequential elongation steps due to a single enzyme/primer/template binding event. Low processive enzymes like the HIV-1 RTase tend to easily dissociate from such a complex and thus interrupt polymerization. In contrast, highly processive polymerases like the Klenow fragment (*Escherichia coli* polymerase I lacking the 5'-3' exonuclease activity) or Taq polymerase re-enter the polymerization cycle with high efficiency (Richetti and Buc, 1993).

In order to ensure incorporation of the correct complementary nucleotide into the growing primer chain, one can distinguish between two principal mechanisms used by polymerases with and without proofreading activity. According to a model for polymerases having proofreading activity, the 3' terminus of the DNA primer can shuttle within the enzyme/nucleic acid complex between the DNA polymerase active site and the 3'-5' exonuclease active site (Beese *et al.*, 1993). Here the terminal 5'-3' phosphodiester bond of a mismatched nucleotide can be hydrolysed and the complex can return to state II (Kohlstaedt *et al.*, 1992). Polymerases that lack proofreading activity, such as all RNA polymerases and RTases, seem to discriminate mismatched nucleotides already at stage IV. As documented for the HIV-1 RTase, the polymerase/template/primer/dNTP-complex can simply release dNTP before diester bond formation (Kunkel, 1992).

Nevertheless, in both cases, discrimination against incorrect nucleotides is not 100% and thus polymerization proceeds with a finite error rate. In addition to point mutations, the error spectrum of polymerases includes nucleotide insertions or deletions, complex rearrangements

and hypermutations. The frequency of such errors depends on the particular polymerase and is influenced by the type of mismatch formed and by all components involved in the polymerization reaction.

REGULATION OF INTRACELLULAR DEOXYNUCLEOSIDE TRIPHOSPHATE CONCENTRATIONS

A uniform supply of nucleoside triphosphates is essential to the fidelity of cellular replication processes. Determination of the dNTP and NTP levels from bulk cultures of mammalian cell lines revealed that the mean concentration of dNTPs is in the order of μM whereas those of NTPs is around two orders of magnitude higher (Meyerhans *et al.*, 1994; Reichard, 1988). Upon virus infection, the available cellular nucleoside triphosphates are also used as substrates for virus replication. Because many viruses carry enzymes that can influence intracellular dNTP levels, such as thymidine kinases, ribonucleotide reductases, thymidylate synthetases or dUTPases, one might expect that dNTP levels in particular would be important for efficient virus propagation (Honess, 1984; Honess *et al.*, 1986; McIntosh and Haynes, 1996).

Comparison of the individual dNTPs showed that their relative intracellular concentrations are usually in the order dTTP$\geq$dATP > dCTP$\geq$ dGTP (Mathews and Ji, 1992). Their magnitude and relative ratios can vary substantially, depending on the state of cellular activation (Bray and Brent, 1972). For example, intracellular dNTP concentrations are between 5 and 50 times greater in activated than in resting peripheral blood mononuclear cells (PBMCs), while the dTTP/dCTP ratio of around 1–2 in resting cells may increase four- to sixfold upon activation (Tattersall *et al.*, 1975; Tyrsted, 1982; Cohen *et al.*, 1983).

Only two of the seven relevant enzymes for *de novo* synthesis of dNTPs in mammalian cells, ribonucleotide reductase and dCMP deaminase, are subject to allosteric control and therefore targets of regulatory signals (Figure 5.2; Bray and

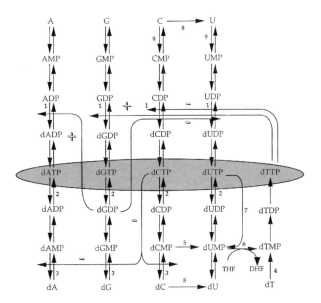

FIGURE 5.2 *De novo* pathway of deoxyribonucleoside triphosphate synthesis. The enzymes involved are indicated as follows. 1, ribonucleotide reductase; 2, nucleoside diphosphate kinase; 3, deoxycytidine kinase; 5, dCMP deaminase; 6, dTMP synthase; 7, dUTPase; 8, cytidine deaminase; 9, uridine/cytidine kinase; THF, tetrahydrofolate; DHF, dihydrofolate. Regulatory feedback loops are indicated by + and –.

Brent, 1972; Reichard, 1988, 1993). The ribonucleotide reductase is responsible for the reduction of all four ribonucleoside diphosphates (NDPs) to yield three of the four DNA precursors. The synthesis of the fourth DNA precursor, dTTP, derives from both scavenging of deoxythymidine (dThd) and metabolism of dUMP to dTMP. The dCMP deaminase activity controls the relative amounts of dCTP and dTTP, and provides most of the dUMP required for dTTP formation.

Intracellular steady-state dNTP concentrations are sufficient for around 2 min of chromosomal DNA synthesis and may be rapidly depleted by drug treatment. For example, hydroxyurea interferes with the ribonucleotide reductase Fe-Tyr free radical that is indispensable to NDP reduction (Reichard, 1988). Alternatively, dThd added to the culture medium may be scavenged by the cell and phosphorylated to yield the triphosphate. This not only increases the intracellular dTTP concentration but, by its allosteric effect on RR, specifically

decreases CDP reduction to dCDP, leading to a drop in intracellular dCTP (Dahbo and Eriksson, 1985; Reichard, 1988, 1993).

The replication of, for instance, retroviruses and pox viruses is susceptible to manipulation of the host cell dNTPs. Vaccinia virus and HIV are exquisitely sensitive to the effects of hydroxyurea while HIV replication could be abolished by the addition of thymidine (Meyerhans *et al.*, 1994; Slabaugh *et al.*, 1991). In both cases the effects could be reversed by the addition of antagonists. Furthermore HIV-1 replication could be enhanced by the addition of deoxycytidine (dCyd) supporting the notion that HIV-1 replication may be restricted by the concentration of intracellular dNTPs (Meyerhans *et al.*, 1994).

POLYMERASE ERRORS AND WAYS TO MEASURE INFIDELITY

Polymerases copy a nucleotide template only with a certain fidelity. The term "fidelity", however, is ambiguous. For a virologist it is used mainly for describing a certain mutation rate, i.e. the number of mutations per nucleotide per replication cycle. This rate is usually evaluated from errors occurring during *ex-vivo* virus passaging. In contrast, for an enzymologist, the fidelity f is strictly defined as a product of K_m and V_{max} values, measures of discrimination between the good (match) and the wrong (mismatch) nucleotide incorporated during polymerisation at a given template position (Fersht, 1985):

$$f = f_k^{-1} f_v,$$

where $f_k = K_m$ (mismatch)$/K_m$ (match), $f_v = V_{max}$ (mismatch)$/V_{max}$ (match), K_m is the Michaelis–Menten constant, and V_{max} is the maximum velocity.

In any case, the enzymatic fidelity is a function of many parameters, e.g. the type of the polymerase, the nature of the template (DNA or RNA), the concentration of the nucleotide triphosphates and the type and concentrations of cations and anions present. Thus, comparisons of particular values of polymerase fidelities should always bear in mind the specific experimental conditions under which the values were obtained.

In-vitro Assays of Polymerase Fidelity

Polymerase fidelities can be determined *in vitro*, *ex vivo* or *in vivo*. All *in-vitro* assays use purified enzymes to polymerize a primer on either a synthetic or a natural nucleotide template. The kinetic parameters V_{max} and K_m for matched and mismatched nucleotide incorporation can then be measured by the quantitation of the elongated radiolabelled primer after separation on acrylamide gels (Boosalis *et al.*, 1987; Randall *et al.*, 1987; Mendelman *et al.*, 1989). In these man-

ually carried-out experiments, one needs at least 30–60 s to stop the reaction after initiation of the polymerase reaction. In this time frame, the reaction is already in a steady-state condition and thus multiple enzyme recycling on the template/primer complex contributes to the values of K_m and V_{max}. In this case, by masking mechanistic details, K_m and V_{max} values may not entirely reflect events at the active site of the enzyme. Recently, exact K_m and V_{max} values for single-step polymerizations were determined by rapid quench experiments with reaction times ranging from milliseconds to several seconds (Kerr and Anderson, 1997; Suo and Johnson, 1997a,b). These measurements may give more precise information on the rate-limiting steps of nucleotide incorporation and misincorporation.

TABLE 5.1   Rate and type of mutation events for polymerization; error rates for polymerases are given (AMV, avian myoblastosis virus; HIV-1, human immunodeficiency virus type 1; MoLV, Moloney leukaemia virus)

Polymerase	Localisation/gene	Error rate	Reference
DNA			
Pol I	poly (dA-dT)	1.2×10^{-5}	Loeb and Kunkel, 1982
T4	poly (dA-dT)	2.3×10^{-5}	Loeb and Kunkel, 1982
α primase	M13mp2DNA	2×10^{-4}	Roberts and Kunkel, 1988
α	φ ×174 DNA	3.3×10^{-5}	Perrino and Loeb, 1989
α primase	φ ×174	2×10^{-6}	Reyland and Loeb, 1987
β	M13mp2	7×10^{-5}	Roberts *et al.*, 1988
	φ ×174	1.2×10^{-4}	Loeb and Kunkel, 1982
Thermostable			
Taq	M13mp2	1.1×10^{-4}	Tindall and Kunkel, 1988
			Eckert and Kunkel, 1990
	LacZα	$1.1–1.3 \times 10^{-4}$	Barnes, 1992
Vent exo (+)	M13mp2	$1–2 \times 10^{-5}$	Mattila *et al.*, 1991
	MutHLS-dependent cleavage	2.3×10^{-6}	Smith and Modrich, 1996
	P53	$1–7.3 \times 10^{-5}$	Flaman *et al.*, 1994
Vent exo (–)	M13mp2	$0.6–1.5 \times 10^{-4}$	Mattila *et al.*, 1991
Pfu	MutHLS-dependent cleavage	7.9×10^{-7}	Smith and Modrich, 1996
RT			
HIV-1	M13mp2	2×10^{-4}	Roberts *et al.*, 1988
	φ × 174	$1.3–2.5 \times 10^{-4}$	Bebenek *et al.*, 1989
			Weber and Grosse, 1989
AMV	M13mp2	6×10^{-5}	Preston *et al.*, 1988
	φ ×174	10^{-4}	Preston *et al.*, 1988
MoLV	M13mp2	3.3×10^{-5}	Roberts *et al.*, 1988

TABLE 5.2 Rate and type of mutation events for replication; error rates for various mutation events and organisms are given (BLV, bovine leukaemia virus; CAEV, caprine arthritis encephalitis virus; FMDV, foot-and-mouth-disease virus; HBV, hepatitis B virus; HIV-1, human immunodeficiency virus type 1; MoLV, Moloney leukaemia virus; RSV, Rous sarcoma virus; SNV, spleen necrosis virus; VSV, vesicular stomatitis virus)

Mutation type	Organism	Localization/gene	Error rate	Reference
Substitutions	**DNA**			
	Phage M13	LacZα	7×10^{-7}	Kunkel, 1985
				Drake, 1991
	1 phage	cI	8×10^{-8}	Sauer, 1978
				Drake, 1991
	Phage T2	rII	3×10^{-8}	Pribnow et al., 1981
				Drake, 1991
	Phage T4	rII	2×10^{-8}	Pribnow et al., 1981
				Drake, 1991
	E. coli	lac I	4×10^{-10}	Farabaugh, 1978
				Drake, 1991
	S. cerevisiae	URA3	3×10^{-10}	Drake, 1991
	S. cerevisiae	SUP4	8×10^{-9}	Kunz et al., 1990
				Drake, 1991
	Neurospora crassa	ad-3AB	4.5×10^{-11}	Orbach et al., 1988
				Drake, 1991
Substitutions	**RNA**			
	Qβ phage	Complete genome	3×10^{-4}	Domingo et al., 1976, 1978
		Position 39 in (–) strand		Batschelet et al., 1976
	Poliovirus	VP1	2×10^{-6}	Parvin et al., 1986
		Reversion amber mutation	2×10^{-6}	Sedivy et al., 1987
		2C	2×10^{-4}	De la Torre et al., 1990
		Different site of the viral genome	$3-5 \times 10^{-3}$	Ward and Flanegan, 1992
		3AB	2×10^{-5}	De la Torre et al., 1992
		VP1, VP2, VP3	$10^{-4}-10^{-5}$	Emini et al., 1982
				Minor et al., 1983, 1986
	VSV	L	10^{-4}	Schubert et al., 1984
		Different site of the viral genome	$10^{-3}-10^{-4}$	Steinhauer and Holland, 1986; Steinhauer et al., 1989
		L, G	$10^{-3}-10^{-4}$	White and McGeoch, 1987
		I1, I2, I3	10^{-4}	Holland et al., 1989
	FMDV	Complete genome	10^{-4}	Domingo et al., 1980
				Sobrino et al., 1983
		VP1	10^{-5}	Martinez et al., 1991
	Influenza A	NS	1.5×10^{-5}	Parvin et al., 1986
	Meningovirus	VP1, VP2, VP3	$3 \times 10^{-3}-5 \times 10^{-5}$	Boege et al., 1991
	Rhinovirus	VP1, VP2, VP3	$10^{-4}-10^{-5}$	Sherry et al., 1986
	Sindbis virus	E2	$<10^{-6}$	Durbin and Stollar, 1986
		E2	$10^{-2}-10^{-5}$	Hahn et al., 1989
		E1, E2	$10^{-3}-10^{-5}$	Stec et al., 1986

TABLE 5.2 (*continued*)

Mutation type	Organism	Localization/gene	Error rate	Reference
Substitutions	**Retrovirus**			
	RSV	env	10^{-4}	Coffin *et al.*, 1980
		gag-pol-env-src	10^{-4}	Leider *et al.*, 1988
	MoLV	LTR-gag-pol-env	2×10^{-5}	Monk *et al.*, 1992
		neo	2×10^{-6}	Varela-Echavarria *et al.*, 1992
	SNV	neo	2×10^{-5}	Leider *et al.*, 1988
		LacZα	6×10^{-6}	Pathak and Temin, 1990a,b, 1992
	BLV	LacZα	10^{-6}	Mansky and Temin, 1994
	HIV-1	LacZα	3.4×10^{-5}	Mansky and Temin, 1995
Frameshifts				
	SNV	LacZα	2×10^{-6}	Pathak and Temin, 1990a,b
		LTR	2×10^{-2}	Burns and Temin, 1994
	BLV	LacZα	10^{-6}	Mansky and Temin, 1994
	HIV-1	LacZα	7×10^{-6}	Mansky and Temin, 1995
Genetic rearrangement				
	SNV	LacZα	4×10^{-4}	Pathak and Temin, 1990a,b
		neo	10^{-7}	Dougherty and Temin, 1988
				Pulsinelli and Temin, 1991
	MoLV	gag-pol-env	3×10^{-5}	Shields *et al.*, 1978
		tk	2×10^{-5}	Varela-Echavarria *et al.*, 1992
				Parthasarathi *et al.*, 1995
	BLV	LacZα	8×10^{-7}	Mansky and Temin, 1994
	HIV-1	LacZα	2×10^{-6}	Mansky and Temin, 1995
Recombination				
Homologous				
	SNV	neo/hygro	3×10^{-5}	Hu and Temin, 1990, 1992
				Zhang and Temin, 1994
				Jones *et al.*, 1994
Non-homologous	SNV	neo/hygro	5×10^{-8}	Zhang and Temin, 1993
Hypermutation				
	SNV	LacZα	6×10^{-2}	Pathak and Temin, 1990a
	HIV-1	Complete genome	3×10^{-2}	Borman *et al.*, 1995
	CAEV	Complete genome	$2–4 \times 10^{-1}$	Wain-Hobson *et al.*, 1995
	HBV	Complete genome	$1.2–2.6 \times 10^{-1}$	Günter *et al.*, 1997

In other assays of polymerase fidelity, selectable marker sequences are copied *in vitro*, and mutant frequencies and/or mutation frequencies are determined. One such assay uses an amber mutant of the single-stranded circular DNA phage Φ×174 as a template (Kunkel *et al.*, 1982; Loeb and Kunkel, 1982; Kunkel and Alexander, 1986; Williams and Loeb, 1992). A single base substitution error that reverts the amber codon to a coding triplet can subsequently be visualized as a revertant plaque after transformation of bacteria. Another fidelity assay is based on the visual discrimination between a wild-type and a mutated bacteriophage M13mp2 (Kunkel *et al.*, 1982; Loeb and Kunkel, 1982; Kunkel and Alexander, 1986; Williams and Loeb, 1992). This single-stranded DNA phage contains the α-fragment of the *LacZ* gene. It can complement an α-fragment-lacking *E. coli* strain to produce functional β-galactosidase. The respective M13 plaques are visuable as dark blue using X-gal indicator plates while mutants that influence α-fragment translation or structure are white or light blue. The ratio of the numbers of altered plaques to wild-type directly gives the mutant frequency. Upon DNA sequencing of several α-fragment regions, the mutation frequencies can be determined. Such analyses display the existence of mutation hot spots and demonstrate that neighboring nucleotides significantly influence base substitution frequencies.

Misinsertion efficiencies for polymerases without proofreading activities, e.g. all RNA polymerases, RTases, DNA polymerase β and Taq polymerase, fall within a general range of 10^{-3}–10^{-6} (Tables 5.1 and 5.2; Preston *et al.*, 1988; Roberts *et al.*, 1988; Bebenek *et al.*, 1989; Mendelman *et al.*, 1989, 1990; Perrino *et al.*, 1989; Weber and Grosse, 1989; Richetti and Buc, 1990; Yu and Goodman, 1992). In the case of polymerases with proofreading activity, the copying fidelity is increased by two to three orders of magnitude. Concerning the type of mismatch formed during polymerisation, G–T mismatches are the most frequent while pyrimidine–pyrimidine and purine–purine mismatch frequencies can be 1000 times less. These mismatch preferences are correlated to mispairs' intrinsic thermodynamic stabilities (Galas and Branscomb, 1978; Clayton *et al.*, 1979; Goodman, 1988; Goodman and Fygenson, 1998).

RTases are particularly interesting polymerases because they can use either RNA or DNA as templates. In addition, the RTase of HIV is an important target for antiretroviral therapy and has therefore been analysed extensively. Kinetic investigations on base misincorporation (Preston *et al.*, 1988; Roberts *et al.*, 1988; Richetti and Buc, 1990; Boyer *et al.*, 1992; Yu and Goodman, 1992) and mispair extension (Perrino *et al.*, 1989; Richetti and Buc, 1990; Yu and

Goodman, 1992), and *in vitro* studies on primer–template misalignment that subsequently led to frameshifts (Bebenek *et al.*, 1989, 1993; Preston *et al.*, 1988) underlined the error-proneness of the HIV-1 RTase. As far as base substitutions are concerned, HIV-1 RTase fidelity is similar to that of other retroviral RTases (Preston *et al.*, 1988; Richetti and Buc, 1990; Yu and Goodman, 1992). However, by measuring the ability of several polymerases to extend an oligonucleotide primer beyond mismatches, a hierarchy of mispair extension in the order HIV-1 RTase > AMV RTase > MoMLV RTase was observed (Preston *et al.*, 1988; Roberts *et al.*, 1988; Martinez *et al.*, 1994, 1995; Vartanian *et al.*, 1994).

RTase inhibitors are the major drugs used to treat HIV-infected individuals. According to their mode of action, they can be grouped into two classes: nucleoside analogues and non-nucleoside inhibitors. The first group competes with natural nucleotides for the nucleotide binding site of the RTase, while non-nucleoside analogues act by allosteric inhibition (Condra and Emini, 1997). Nucleoside analogues lack the 3'-hydroxyl group. They can be incorporated into the growing chain during polymerization but cannot be further elongated and thus act as chain terminators of DNA synthesis (Drosopoulos and Prasad, 1998). Examples of therapeutically used nucleoside analogs include 2',3'-dideoxy-3'-azidothymidine (zidovudine); 2',3'-dideoxycytidine (ddC); 2',3'-dideoxyinosine (ddI); and 2',3'-dideoxy-3'-thiacytidine (3TC). Because of the rapid turnover of HIV *in vivo*, drug-resistant RTase mutants occur with variable efficiency. For some of these, e.g. mutant L74V resistant to zidovudine, mutant M184V resistant to 3TC, mutant Q89G resistant to ddGTP and ddITP, and the multiple resistance mutant E89G, a two- to 45-fold enhanced polymerase fidelity was observed (Pandey *et al.*, 1996; Wainberg *et al.*, 1996; Hsu *et al.*, 1997; Oude Essink *et al.*, 1997; Rubinek *et al.*, 1997; Drosopoulos and Prasad, 1998).

All these *in-vitro* assays have also illustrated the importance of template structure, salt and substrate concentrations for polymerase fidelity. A study of G_t:T mispair formation (t = template) of the RTase of HIV-1 on homologous DNA and RNA templates has shown striking fidelity differences: T misincorporation was not detected with the DNA templates but occurred at a high frequency with the homologous RNA templates ($f = 10^{-4}$–10^{-3}; Sala *et al.*, 1995; Kerr and Anderson, 1997). Further thermodynamic characterization of the primer–template duplexes by thermal melting measurements demonstrated that, although the DNA–DNA and RNA–DNA duplexes were of comparable stability in solution, the RNA–DNA duplexes presented more facile base-pair opening and a much higher conformational flexibility (Sala *et al.*, 1995). Thus, the release of helical constraint in the RNA–DNA duplex may facilitate base mispairing at the polymerase catalytic site. Furthermore, this study revealed that G_t:T mismatches located on the primer–template duplex upstream from the catalytic site can increase subsequent T misincorporation, suggesting long-range influences on polymerase fidelity. More recent pre-steady-state kinetic studies verified the increased nucleotide misincorporation of the HIV-1 RTase on DNA–DNA heteroduplexes and showed a 10–60-fold fidelity difference (Kerr and Anderson, 1997). In addition to the nature of the template, polymerase fidelity is significantly influenced by transition metal ions and nucleoside triphosphate imbalances. Both metals like cobalt (Co^{2+}) and manganese (Mn^{2+}) and an uneven supply of nucleoside triphosphates help mismatch formation and increase error rates (Filler and Lever, 1997; Richetti and Buc, 1996). Depending on the particular concentrations, polymerase fidelities can be manipulated in a directed fashion by several orders of magnitude (see Hypermutagenesis).

Ex-vivo Assays of Polymerase Fidelity

In polymerase fidelity assays *in vitro*, saturating concentrations of templates, primers and nucleotide triphosphates are used while the concentration of the enyme is limiting. In living cells, however, the conditions of template copying are markedly different. Taking reverse transcription of the two RNA copies of the HIV genome as an example, the natural reaction mix within the viral capsid contains approximately 50 molecules of the RTase, an imbalanced pool of dNTPs, each around μM concentration, and physiological salt concentrations and pH. Because these conditions may influence error rates, the fidelity of viral polymerases is most accurately measured by infection experiments. This is particularly important if one wants to determine error rates of virus replication. In addition to the viral polymerase, some viruses contain enzymes that may contribute to their overall replication fidelity, such as the dUTPase of several retroviruses and thymidine kinase and ribonucleotide reductase of herpes viruses (Honess, 1984; Honess *et al.*, 1986; McIntosh and Haynes, 1996).

The *ex-vivo* determination of virus replication errors was first analysed for the bacteriophage Qβ by serial passages on bacteria (Batschelet *et al.*, 1976), and subsequently for several mammalian viruses. Such measures had the intrinsic problem that the number of replication rounds involved was not precisely known and that, depending on the conditions of passaging, selection events or bottlenecking could influence the appearance of mutants. These problems were overcome most elegantly by Temin and co-workers. They have developed a single-cycle replication system to determine the error rates for the retroviruses spleen necrosis virus (SNV), bovine leukemia virus (BLV), murine leukaemia virus (MoMLV) and HIV (Dougherty and Temin, 1988; Hu and Temin, 1990; Pathak and Temin, 1990a,b, 1992; Pusinelli and Temin, 1991; Varela-Echavarria *et al.*, 1993; Burns and Temin, 1994; Mansky and Temin, 1994; 1995; Parthasarathi *et al.*, 1995). Non-essential selectable marker genes such as neomycin, *lacZ* or thymidine kinase were incorporated into respective retrovirus vectors. Infectious recombinant viruses were then generated after transfection and used to infect new target cells. The reverse-transcribed DNA was subsequently analysed for base substitutions exhibiting mutation rates of 2.5×10^{-5}, 6×10^{-6} and 10^{-6} for HIV-1, SNV and BLV respectively. This single-cycle replication error rate is a composite of two enzymatic steps: RNA transcription by the cellular RNA polymerase II and reverse transcription by the RTase. The values therefore imply that the error rate for the cellular RNA polymerase II might be lower than 10^{-6}. In addition to base substitutions, error rates for simple frame-shifts, genetic rearrangements and homologous recombinations were also determined. Their rates range from 2×10^{-5} to 5×10^{-8} (see Table 5.2).

Accumulation of Virus Mutations *In Vivo*

A large number of studies have analysed the sequential accumulation of virus mutants within the natural host. These include infections with foot-and-mouth disease virus, influenza virus, polio virus, measles virus, hepatitis viruses and retroviruses (Cattaneo and Billeter, 1992; Domingo *et al.*, 1992; Gorman *et al.*, 1992; Kinnunen *et al.*, 1992; Pelletier *et al.*, 1995; Preston and Dougherty, 1996; Plikat *et al.*, 1997; Mansky, 1998; Sala and Vartanian, 1998). The forces operative in these infections are extremely complex and processes like virus turnover rates, selection events, bottlenecking and clonal expansion all contribute to intrahost virus evolution. In addition, the complex nature of the virus population, the so-called quasispecies, can fluctuate both in space and time, implying that it is important to consider sampling (Meyerhans *et al.*, 1989; Delassus *et al.*, 1992; Cheynier *et al.*, 1994). All these parameters mask the evaluation of virus error rates *in vivo*. However, known error rates derived from *ex-vivo* studies, together with determined accumulation of mutations over time, can be used to estimate virus turnover rates within infected hosts (Ho *et al.*, 1995; Wei *et al.*, 1995). These estimates then may provide most interesting insights into the nature of a particular host–virus relationship (Wain-Hobson, 1993).

5. ERROR ESCALATION OR HYPERMUTATION OF RNA VIRUSES

Apart from the usual infidelity of template copying, a phenomenon of error escalation has been observed that might occur during the replication of several RNA viruses, including retroviruses. Upon sequencing individual virus genomes from blood, tissue specimen after *ex-vivo* passage, individual clones were found to have mutated extensively as compared to a reference virus sequence. To emphasize this extensive nature of the error process, it was termed hypermutation. Examples are given in Figure 5.3.

Three forms of hypermutations have been documented: (1) G→A hypermutation in retroviruses and hepadnaviruses (Pathak and Temin, 1990a; Johnson *et al.*, 1991; Vartanian *et al.*, 1991, 1994, 1997; Gao *et al.*, 1992; Perry *et al.*, 1992; Fitzgibbon *et al.*, 1993; Borman *et al.*, 1995; Günther *et al.*, 1997; Pelletier *et al.*, 1995; Wain-Hobson *et al.*, 1995); (2) A → G hypermutation in SNV, in measles and vesicular stomatitis viral genomes (Cattaneo *et al.*, 1988; Bass *et al.*, 1989; Cattaneo and Billeter, 1992; Felder *et al.*, 1994; Kim *et al.*, 1996); (3) insertional and complex hypermutations observed in SNV and in murine leukaemia virus (Pathak and Temin, 1992; Parthasarathi *et al.*, 1995). These error processes result in extraordinarily high substitution frequencies; for example, in the case of a hypermutated full-length HIV-1 genome, around 40% of all the guanosines were found to be substituted by adenosines. While the first two processes seemed to occur during and after replication respectively, the mechanism of complex hypermutation remains unknown.

G → A Hypermutation

This type of hypermutation has been described for the lentiviruses HIV-1 (Vartanian *et al.*, 1991, 1994; Fitzgibbon *et al.*, 1993; Borman *et al.*, 1995), HIV-2 (Gao *et al.*, 1992), simian immunodeficiency virus (SIV; Johnson *et al.*, 1991; Pelletier *et al.*, 1995), equine infectious anaemia virus (EIAV; Perry *et al.*, 1992) and caprine arthritis-encephalitis virus (CAEV; Wain-Hobson *et al.*,

Hypermutation associated with dNTP pool imbalances

HIV-1

```
M06 TGAAGAATGC CAATAGTAGT AGCGGGGGAA TGATGGAGAA AGGAGAAATG AAGAACTGCT
M04 .A........ ........A. .....A.A.. .A..A..A.. .AA.A.-..A ..A.......
M22 .A........ .......... .......A.. AA..AA.A.. .AA A....A ..A.......
```

HTLV-1

```
ATK CAGAACAGAC GAGGCCTTGA TCTCCTGTTC TGGGAGCAAG GAGGATTATG CAAAGCATTA
57  .......A.. ........A. .......... .......... ..AA...... ..........
ATK GCCGTTTTCC GAATATTACC AATTCCCATG TCTCAATACT ACAAGAAAGA CCCCCCCTTG
57  .......... .......... .......... .......... .......... .........A
```

SNV

```
cons CAATGATTTG ACCAGAATGT ACAAGAGCAG TGGGGAATGT GGGAGGGGCG GCCGCTGGGG
01   ....A..... .......... .......... ...AA..... .......... ..........
cons GGAATAGTGC TGGCTCGCTA ACTGCTATAT TAGCTTCTGT ACCCATGCTT GCTTGCCCTG
01   .......... .......... .......... .......... .......... ..........
cons GCCACTAACC GCCATATTAG CTTCTGTACA CATGCTTGCT TGCCTTAGCC GCCATTGTAC
01   .......... .......... .......... .......... .......... ..........
cons TTGATATGCC ATTTCTCGGA
01   ..A....... ........A.
```

HBV

```
cons AAGACTGTGT ATTTGCTGAG TGGGAGGAGT TGGGGGAGGA GATTAGGTTA AAGGTTTATG
02   ..A....... T...A..A.. ..AA..A... .AAAAA.AA. A....A.... ..........
cons TATTAGGAGG CTGTAGGCAT AAAATTGGTCT GTTCACCAGC
02   .....AA.A. .....AA... .....AA...  ..........
```

Hypermutation associated with unwinding activity

Measles

```
cons UCCGUGUUGUU UAUAUGAGC AUCACCCGUC UUUCGGAUAA CGGGUAUUAC ACCGUUCCUA
C.b  ....C.CC..C C.C...... .......... C...A..C.. ....C.CC.. ........C.
ZH   ....C..... ...C...... .......C. C......C.. ...A..CC.. ....C...C.
```

Human Parainfluenza Virus 3

```
cons UCUUCUUUGA ACAAACCUUU AUAUUUAAAU UUAAUUUUAA UUGAAUCCUA AUUUCUGUAA
P8   .....C.... ........C. .C..C...C C...C..... .......... .C.C......
P0   .........G G..G...... G.....G.G. ....C..C.. C....C...G .CCC....GG
cons CUGAUCUUCC AGUUCUUUUC CCUUGAGAUA UUAAAGUUUU
P8   .......... .......C.. ........C. .C....C...
P0   ...G...... ......C... .......GC. CCGG......
```

RAV

```
cons GATGGAATGT GCTGGTCTGG GGACGAGGTT ATGCAGCTGT GAAAAACAGG GACACTGATA
03   .......... .......... .......... G......... .......... .........G
04   .......... .......... .......GG .G........ .......... .....G..G.
cons AGGTTATTTG GGTACCCTCT CGGAAAGTTA AACCGGATGT CACCCAAAAG GATGAGGTAA
03   G....G.... ...G...... .....G...G .......... .....GGGG. ....G....
04   ....G..G.. .........G .......GG. .......G.G. .......... ..G....G..
cons CTAAGAAAGA TGAGGCGAGC CCTCTTTT
03   ...GG..... .......... ........
04   .G........ G......... .......G
```

FIGURE 5.3 Examples of viral hypermutation. Segments of viral hypermutated sequences are aligned to the appropriate reference sequence. Only sequence differences are given, a dot indicating sequence identity. A dash represents a frameshift mutation. HBV, hepatitis B virus; HIV-1, human immunodeficiency virus type 1; HTLV-1, human T-cell leukaemia/lymphoma virus type 1; RAV, Rous-associated virus; SNV, spleen necrosis virus. Sequences were taken from Vartanian *et al.*, 1991, 1997 (HIV-1, HTLV-1), Pathak and Temin, 1990a (SNV), Günter *et al.*, 1997 (HBV), Cattaneo and Billeter 1992 (measles), Murphy *et al.*, 1991 (parainfluenza), Felder *et al.*, 1994 (RAV).

1995), the oncoretrovirus SNV (Pathak and Temin, 1990a) and the hepatitis B virus (HBV; Günther *et al.*, 1997). The G → A transitions occur preferentially in the context of GpA and GpG; however, both the extent and the location of hypermutated regions vary between genomes thus underlining the erratic nature of the error process.

G → A hypermutation occurs during reverse transcription in a single viral replication cycle. Several alternative explanations can be excluded.

1. The observation of an HIV-1 provirus with a hypermutated 3′ U3-R region bearing a transcriptionally silent Tat-responsive element (TAR) rules out the involvement of polymerase II-mediated transcription (Vartanian *et al.*, 1994). Hypermutation during start of transcription at the 5′ TAR would only lead to an abberant transcript that could not be converted to a provirus. Hypermutation during transcription of the 5′ U3-R regions would after reverse transcription only give rise to a hypermutated 5′ U3 provirus but not to a hypermutated 3′ U3-R.

2. A hypermutated provirus was identified after single-cycle replication of a recombinant SNV vector, thus showing that hypermutation is not the result of an accumulation of consecutive rounds of mutations (Pathak and Temin, 1990a). This is also evident from the often observed defective genomes generated by G → A transitions. For example, tryptophan is solely coded by the triplet TGG, which may convert to a TGA, TAG or TAA stop codon. The respective proviruses would not be able to produce infectious progeny and continue error-prone replication. In addition, with the retrovirus error rate of 2.5×10^{-5} to 10^{-6}, at least four consecutive rounds of replication are needed to generate one substitution error in the roughly 10 kb virus genome. To generate, for instance, the observed 600 G → A monotonous transitions in the full-length clone of HIV-1 Vau (Borman *et al.*, 1995; Vartanian and Wain-Hobson, unpublished observation), one would need at least 2400 replication rounds or roughly 7 years of constant replication without the appearance of defective viruses, a possibility that simply excludes itself.

3. Hypermutation is not mediated by PCR (Cheynier *et al.*, 1997). The above-mentioned hypermutated recombinant SNV provirus was detected without the use of PCR-mediated amplification. Furthermore, the PCR reaction does not distinguish between the (+) and (−) DNA strands and thus would lead to a symmetrical appearance of the transitions, i.e. G → A plus C → T transitions, the (+) strand being the reference sequence.

4. Hypermutation is not a postreplicative process. The observed context dependence suggests that the nucleotide 3′ to the G → A substitution affects the error process. There is no evidence that this position could influence postreplication processes. In contrast, during reverse transcription, this position is seen by the polymerase just before the error occurs and thus could influence its appearance. This is consistent with *in-vitro* fidelity studies of purified RTases (Richetti and Buc, 1990; Martinez *et al.*, 1995).

5. Hypermutation is mediated by a wild-type RTase and does not depend on a special mutant RTase. The observed erratic distribution of hypermutated regions within a single virus genome argues against a mutant polymerase hypothesis, otherwise a smooth distribution of G → A transitions would be expected (Vartanian *et al.*, 1991). Formally, however, one cannot exclude the possibility that one or a few mutant RTases among the roughly 50 RTase molecules are present, for example, in an HIV capsid where reverse transcription takes place. Due to the low RTase processivity, a constant reshuffling of wild-type and mutant enzymes on the template could be imagined that would subsequently result in the patchy distribution of hypermutated regions as experimentally observed. This argument, however, does not hold for HBV because it seems that only a single polymerase molecule is incorporated into the virus capsid (Nassal and Schaller, 1993). Nonetheless, HBV hypermutants bear the same hallmarks as retrovirus hypermutants (Günther *et al.*, 1997). A final argument against the mutant polymerase hypothesis is the reproduction of G → A hypermutation *in vitro* using a cloned RTase, as well as *ex vivo* and *in vivo* by manipulation of the nucleoside triphosphate pools. As presented below, this demonstrates the importance of the microenvironment in this error escalation process.

G → A HYPERMUTATION AND IMBALANCES IN NUCLEOTIDE POOLS

Already in the first description of G → A hyper-mutation in HIV-1, it was hypothesized that imbalances in the local availability of dNTPs might be involved in the error process (Vartanian *et al.*, 1991, 1994, 1997): a depletion of dCTP during (–) cDNA synthesis would favor the formation of a rG-dT mismatch, the most stable of all mismatches. After degradation of the RNA in the heteroduplex by the RTase-associated RNaseH, a dA can be incorporated opposite to dT during (+) strand synthesis, thus leading to a G → A transition. In this scenario, the observed context dependence of hypermutation can be rationalized as follows (Figure 5.4A):

- Mutations occurring in the GpA context may be explained by a dislocation mechanism when the local concentration of dCTP is very low. The 3′ end of a primer that cannot incorporate dC opposite an rG might dislocate from the template. Incorporation of an additional dT opposite the rA might then give rise to a primer with a correct 3′ base pairing but an extrahelical dT. Upon realignment this would form the intermediate rG–dT mismatch. This hypothetical mechanism is supported by *in-vitro* experiments in which the dislocated intermediate was trapped (Vartanian *et al.*, 1994).
- Mutations in the context of GpG occur most frequently at the end of runs of Gs. This observation might be explained by progressive depletion of local dCTP levels due to the preceeding dCTP incorporation and the subsequent replacement by dTTP (Vartanian *et al.*, 1994, 1997).

More direct evidence that G → A hypermutation is mediated by dNTP pool imbalances came from the reproduction of this phenomenon *in vitro*, intravirion, *ex vivo* and *in vivo*. When an HIV-1 RNA template was reverse-transcribed by the purified HIV RTase *in vitro* with highly biased dCTP and dTTP concentrations, hypermutation was reiterated (Martinez *et al.*, 1994, 1995). Two main features are notable: first, the

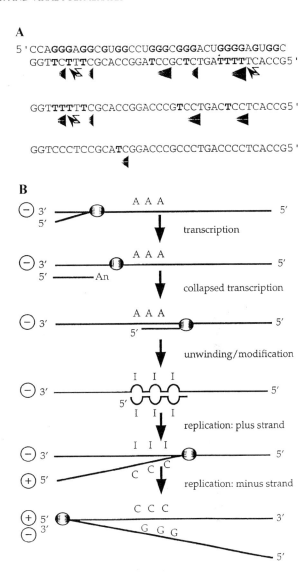

FIGURE 5.4 Proposed mechanisms of viral hypermutation. **A.** G → A transitions that occur in the context of GpG and GpA are represented. The plus-strand RNA reference sequence is given at the top and the hypermutated minus strand underneath. Arrowheads emphasize the decreasing gradient of intracellular dCTP as polymerization progresses opposite G residues. The zigzag arrows represent G → A transition within GpA dinucleotides by dislocation. Bold type indicates the G residues in the template and mismatched T residues (Vartanian *et al.*, 1994). **B.** Postreplicative hypermutation. Adenosines in double-stranded RNA regions can be modified to inosines by the unwinding/modifying activity of RNA-adenosine deaminase. Subsequent copying of the modified template leads to the incorporation of a C residue opposite inosine, resulting in a U → C or A → G substitution (Bass *et al.*, 1989).

proportion of G → A transitions was dose-dependent and inversely proportional to the concentration of dCTP; and second, by using low dATP and high dGTP concentrations, U → C hypermutants resulting from multiple rU–dG mismatches were generated. These data demonstrate that the generation of hypermutants and the direction of hypermutation is solely dependent on the available dNTP pool imbalance and not restricted by properties of the RTase.

To prove that hypermutation could also occur in the natural environment of a viral capsid, HIV-1 particles were purified by a sucrose gradient, permeabilized with the non-ionic detergent NP40 and incubated with biased dNTP concentrations (Vartanian et al., 1997). Both the direction of hypermutation and the misincorporation frequencies were comparable to the in-vitro experiments, suggesting that the natural replication complex does not enhance the mutation rate. However, unlike the in-vitro reaction, the known dinucleotide preference was reproduced when using physiological dCTP and high dTTP concentrations. In any case, the pool biases needed to obtain G → A hypermutation were far from anything physiological. Thus attempts were made to reproduce hypermutation ex vivo and in vivo. Yet, from the casual detection of hypermutants in virus isolates, a very low hypermutant frequency in the order of 1% is to be expected, rendering the detection of hypermutants cumbersome. To overcome these experimental limitations, a screening test based upon E. coli β-galactosidase complementation was established. G → A hypermutation effecting the unique tryptophan codon TGG generates stop codons. Thus amplification of a small segment of hypermutated DNA that previously encoded multiple tryptophan codons would, upon subsequent cloning in frame in the α-subunit segment of the β-galactosidase gene, allow identification of hypermutants, which show up as white plaques against the reference sequence, which is blue. Such a small trytophan-rich sequence may be found in the HIV-1 reverse transcriptase at the interface of the polymerization and RNaseH domains, and also in the gag region of gypsy, an active, replication competent endogenous retrovirus of Drosophila melanogaster. The respective regions were used

to screen for hypermutants after infection of thymidine-treated cell lines with HIV-1 or after feeding Drosophila with thymidine. In both cases, the G → A mutation rate was augmented about 10-fold, in agreement with the directed increase of the dTTP/dCTP ratio (Vartanian et al., 1997; Vartanian and Wain-Hobson, unpublished observation).

Why is hypermutation in HIV mediated by dNTP pool imbalances restricted to G → A transitions? As shown with in-vitro and intravirion assays, other than dTTP > dCTP pool changes may gave rise to novel hypermutants (Martinez et al., 1994, 1995; Vartanian et al., 1997). A dGTP > dATP bias produced U → C transitions, presumably from U–dG mispairing, while an excess of dTTP over dGTP resulted in multiple C → A transversions following elongation beyond rC–dT mismatches. That such forms have not been seen to date despite the huge size of the HIV sequence database, in excess of 12 000 entries (Myers et al., 1995), indicates that they are rare events. Yet the magnitude of fluctuations in the deoxypurine triphosphate ratio (e.g. dGTP > dATP) would have to be very substantial, as they would need to invert the natural dATP > dGTP bias (Mathews and Ji, 1992). Furthermore, low intracellular dATP or dGTP concentrations may indirectly reflect depleted ATP or GTP pools, which could be incompatible with cell survival. Thus, the cell may not only influence the fidelity of HIV-1 replication but also restrict the mutant spectrum.

The intermediate of G → A hypermutation is a rG–dT mismatch between the RNA template and the newly synthesized (–) DNA strand. Such a stable mismatch could also be formed between (+) and (–) DNA strands during (+) strand synthesis and thus should give rise to C → T hypermutation with comparable frequency. Yet, only a single C → T hypermutant has been identified in HIV-1 to date (Vartanian et al., 1991). This inconsistency can be explained by the biased mismatch repair system operative in mammalian cells. Upon migration of an HIV provirus to the nucleus, a dG–dT mismatch between the two DNA strands would be corrected to dG–dC or dA–dT with a ratio of 19:1 (Brown and Jiricny, 1987). Thus, C → T hypermutation is strongly selected against and a single C → T transition may be the only vestige of such an event.

The genetic screening method for G → A hypermutants was used to analyse the HIV hypermutant frequency after infection of stimulated and unstimulated primary peripheral blood mononuclear cells (PBMCs). Approximately 1% of proviruses were hypermutated among activated PBMCs while around two to three times more mutants and mutations resulted from the non-productive infection of resting PBMCs (Vartanian *et al.*, 1997). Given the dNTP substrate dependence of hypermutation and knowing that the mean intracellular dTTP/dCTP ratios as determined from more than 30–50 million cells only vary about 1–10-fold in PBMC depending on the activation state, one would not expect to have found hypermutants in such cells. These observations can be reconciled if a small proportion of HIV-permissive cells, about 1–2% of activated or resting PBMCs, have highly distorted intracellular dNTP pools (Vartanian *et al.*, 1997). Whether this represents statistical fluctuations of dNTP concentrations in all cells or whether it is a small population of effete or stressed cells remains to be shown. Interestingly, a hypermutant frequency of 2% was recently found for HBV, a virus permissive for hepatocytes (Günther *et al.*, 1997). Together this may indicate that strong intracellular dTTP/dCTP biases exist for PBMCs, hepatocytes and, perhaps, many other cells in the body.

Imbalanced dNTP Pools in Genome Instability and Evolution?

Fluctuating, strongly biased intracellular dTTP/dCTP ratios may be expected to influence the genomic base composition. Due to the symmetry of chromosomal DNA replication, a bias of G → A and C → T transitions would be anticipated. Indeed, this is observed. Among germline mutations in 279 different human genes reported as the cause of inherited disease, more than half were G → A and C → T transitions (Krawczak *et al.*, 1995). The

spectrum of mutations within the p53 tumour suppressor gene was remarkably similar (Figure 5.5; Krawczak *et al.*, 1995). An equivalent bias of transitions distinguishes pseudogenes with respect to their cognate genes (Figure 5.5; Gojobori *et al.*, 1982; Li *et al.*, 1984). Furthermore, mammalian genomes' non-coding segments, which are less constrained than coding regions, are generally A+T rich. Thus, the ensemble of data highlights the important role of G:T mispairs, the most stable of all mismatches, in viral G → A hypermutation and identifies fluctuations in the dTTP/dCTP ratio as the underlying cause. Such fluctuations might also provide a link underlying oncogenesis, spontaneous and hereditary diseases as well as the long-term evolution of mammalian genomes.

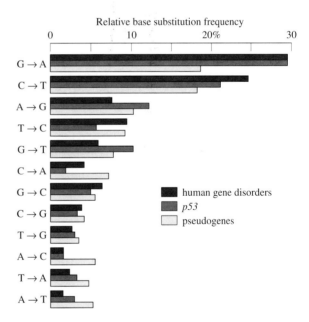

FIGURE 5.5 Relative base substitution frequencies for 229 human genes, the *TP53* tumour suppressor gene and mammalian pseudogenes. Mutation frequencies have been normalized to account for the effects of base composition. Mutation frequencies have been linearly correlated in all cases: human genes versus *TP53* ($R^2 = 0.91$), human genes versus pseudogenes ($R^2 = 0.93$); *TP53* versus pseudogenes ($R^2 = 0.89$). Data are taken from Krawczak *et al.*, 1995, and Li *et al.*, 1984.

Postreplicative A → G or U → C Hypermutations

Historically, A → G or U → C hypermutation was the first extensive monotonous substitution that was discovered in VSV defective interfering particles (O'Hara *et al.*, 1984) and subsequently in measles, respiratory syncytial virus (RSV), influenza and retroviruses (Cattaneo *et al.*, 1988; Bass *et al.*, 1989; Garcia-Barreno *et al.* 1990; Murphy *et al.*, 1991; Felder *et al.*, 1994; Rueda *et al.*, 1994; Kim *et al.*, 1996). In the case of measles infection, this form of hypermutation is invariably associated with the subacute sclerosing panencephalitis form, the effect being particularly concentrated to the M gene segment. For RSV and other viruses the effect seems not to be restricted. While the consequences for virus evolution are comparable with G → A hypermutation, the mechanism of this phenomenon seems to be mediated by a double-stranded RNA unwinding/modifying activity and not by an error-prone polymerization step (Figure 4.4B). It was suggested that adenosines within complementary RNA strands are deaminated into inosines by double-strand RNA-adenosine deaminase, which upon template copying are converted to guanosines (Cattaneo *et al.*, 1988; Bass *et al.*, 1989; Cattaneo and Billeter, 1992; Bass, 1995; Polson and Bass, 1994; Hajjar and Linial, 1995; Polson *et al.*, 1996). Depending on the strand on which the adenosine is deaminated, the mutation is identified as U → C or A → G.

Hypermutation in Virus Evolution

The mutation rate during hypermutation exceeds by orders of magnitude the inverse of the viral genome length and thus is beyond the error threshold above which the information content of a virus cannot be preserved (Eigen and Biebricher, 1988; Domingo and Holland, 1997). However, localized hypermutation might be a means by which viruses can escape being trapped in valleys of the fitness landscape. For example, the possibility exists that, by accumulation of multiple mutations in a single replica-

tion round, a virus might acquire a high level of drug resistance, which would otherwise require several rounds of adaptive mutations or even be impossible if some individual mutations were deleterious. While today this hypothesis is just a frightening possibility, there is some indication that, for instance, G → A hypermutation might have an impact on virus evolution. Mutation matrices derived from the phylogeny-based analysis of intrahost HIV-1 and SIVmac *env* gene evolution have revealed a roughly twofold excess of A → G substitutions with respect to the reverse transition (Pelletier *et al.*, 1995; Plikat *et al.*, 1997). However, lentiviruses have a characteristically elevated adenosine content of 34–38%. This apparent contradiction might be resolved if one assumes that G → A hypermutation compensates for the mutational bias. Such an assumption is well in line with the higher degree of hypermutation found in lentiviruses as compared to other retroviruses, which have an A content of only 22–30% (Karlin *et al.*, 1994).

MANIPULATION OF POLYMERASE FIDELITY FOR HYPERMUTAGENESIS

While viral hypermutation may be considered to be an accident that occurs during virus replication, targeted hypermutagenesis of nucleic acids is a way to construct very complex mutant libraries. In combination with efficient selection procedures, this bears a tremendous potential for biotechnology and basic studies in evolution. Naturally, such a combination of hypermutation and selection is exploited in the affinity maturation of antibodies after antigenic challenge *in vivo*. The variable segments of the antibody genes undergo somatic hypermutation in B lymphocytes (Berek and Milstein, 1987). Through competition for antigen deposited on the follicular dendritic cell surfaces in the germinal centres of lymph nodes, the B cells with high-affinity antibody variants receive survival signals while the other B cells die by apoptosis. Such a process increases the binding affinity of antibodies by two to three orders of magnitude (Berek and Milstein, 1987).

Studies on polymerase fidelity and the analysis of the mechanism of viral G → A hypermutation have provided the clues how to perform hypermutagenesis in practice. Most directly related to viral hypermutation are procedures that simply use highly biased nucleoside triphosphate pools during template copying (Figure 5.6) (Martinez *et al.*, 1994, 1995, 1996). Depending on the question one is interested in answering, this may be achieved *in vitro* by using an appropriate substrate mix in combination with RNA polymerases, DNA polymerases including Taq polymerase or RTases (Battula and Loeb, 1974; Leung *et al.*, 1989; Caldwell and

Joyce, 1992; Chen and Arnold, 1993; Martinez *et al.*, 1994, 1996; Rellos and Scopes, 1994; Fromant *et al.*, 1995; Vartanian *et al.*, 1996; Pezo and Wain-Hobson, 1997). From a practical point of view, hypermutagenic PCR is very promising. It has the advantage that both DNA strands may be mutated and that the mutagenized product can be obtained in sufficient quantities. By using dNTP biases with the addition of manganese cations, the PCR-mediated amplification of the dehydrofolate reductase gene (*DHFR*) of *Escherichia coli* plasmid R67 resulted in a complex mutation spectrum that included all four transitions and a sizeable proportion of

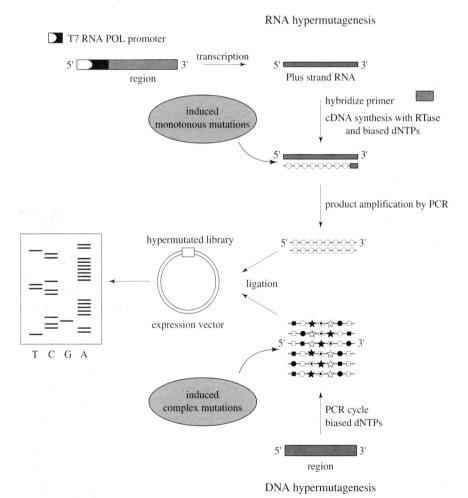

FIGURE 5.6 *In vitro* hypermutagenesis of RNA and DNA templates with imbalanced dNTPs. RNA and DNA can be used as templates for hypermutagenesis under conditions of highly biased dNTPs with RTase or Taq polymerase respectively. Amplified products may be cloned into appropriate expression vectors and analysed by DNA sequencing.

transversions (Vartanian *et al.*, 1996). Overall mutation frequencies of 10% per amplification with a range of 4–18% per clone were obtained. Instead of biased dNTP pools, nucleoside analogues like deoxyinosine triphosphate (dITP), (6-(2-deoxy-b-D-ribofuranosyl)-3,4-dihydro-8H-pyrimido-[4,5-C][1,2]oxazin-7-one triphosphate (dPTP), 8-oxo-2'-deoxyguanosine triphosphate (8-oxodGTP), and 1-(2-deoxy-b-D-ribofuranosyl)-imidazole-4-carboxamide triphosphate (dYTP) have also been used to decrease the fidelity of PCR-mediated DNA amplification (Sala *et al.*, 1996; Zaccolo *et al.*, 1996). They are used as substrates by the Taq polymerase, but have promiscuous base-pairing properties. For example, dPTP (K_m = 22μM is incorporated in place of dTTP (K_m = 9.5 μM) and with about fourfold lower efficiency in place of dCTP (Zaccolo *et al.*, 1996). After a 30-cycle PCR reaction with an equimolar mixture of the natural dNTPs and dPTP, all four transitions were obtained with high frequencies of around 4×10^{-2} (A → G and T → C) to 1×10^{-2} (G → A and C → T). In contrast, 8-oxodGTP is efficiently incorporated opposite template adenosines and can base-pair with adenosines and cytosine, giving rise to the transversions A → C and T → G with a frequency of around 1×10^{-2}. As would be expected, the combination of dPTP and 8-oxodGTP resulted in composite mutations. Because the mutant frequency was directly proportional to the number of PCR cycles, the mutational load can be easily controlled (Zaccolo *et al.*, 1996).

An alternative approach for polymerization-based hypermutagenesis relies on PCR-mediated recombination, by which template segments can be rearranged (Meyerhans *et al.*, 1990). If the template of a PCR reaction consists of a mixture of related sequences or overlapping fragments, a partially copied segment of the template derived from the premature termination of one elongation step can hybridize in a subsequent cycle to a complementary or related segment. Upon elongation by the Taq polymerase, recombinants are generated or former strand breaks repaired (Stemmer, 1994). In addition, point mutations are produced at a rate of about 0.7% (Stemmer, 1994). This approach, also known as "DNA shuffling", has been successfully used to

significantly increase the β-lactamase activity encoded by a bacterial expression plasmid against the antibiotic cefotaxime (Stemmer, 1994). The promotor and β-lactamase coding region of plasmid p182Sfi was amplified by PCR and 100–300 bp fragments generated randomly with DNAse I digestion were purified, annealed and subjected to PCR amplification. After three such rounds of DNA shuffling, mutants were generated that had a 32 000-fold increase in their β-lactamase activity. This high-resistance phenotype was due to one promotor mutation and six non-synonymous changes in the coding region (Stemmer, 1994). Thus, although this approach was impressively successful, the improved phenotype was solely due to point mutations, which are most convenient and efficiently produced by hypermutagenic PCR.

In terms of the biotechnological application of hypermutagenesis for optimizing, for instance, enzyme activities, it is important to consider the issue of optimal mutational load. How many mutations would be needed to enhance a function, and how many mutations can an enzyme accommodate without being functionally destroyed? Clearly, a comprehensive answer to this very complex issue cannot be given as it would depend on many particular features of the protein under investigation. However, going back to the natural hypermutation/selection process of antibody affinity maturation, some hints on this problem might be obtained at least for protein segments accessible on the surface. While the frequency of amino-acid changes in the variable immunoglobulin regions after somatic hypermutation can approach 10%, only few of them seem to play a role in the affinity increase (Berek and Milstein, 1987; Sharon, 1990). Other examples of the flexibility of proteins to accept amino-acid changes came, for example, from mutagenesis studies of T4 lysozyme (Rennell *et al.*, 1991), the λ repressor (Lim and Sauer, 1991) or the above-mentioned DHFR (Martinez *et al.*, 1996; Sala *et al.*, 1996; Vartanian *et al.*, 1996). For the latter, a 6% change in the total of 78 amino acids, as introduced into the respective gene by hypermutagenic PCR, was still compatible with DHFR activity. Thus the degree of changes introduced by such a mutagenesis approach may be considered an appropriate starting point for optimization studies.

Induced Hypermutation *Ex Vivo* and *In Vivo*

Hypermutation of cellular DNA in living cells or even complete organisms might also be achieved by augmenting intracellular dNTP pool imbalances through the manipulation of dNTP pool regulation. As shown in a number of eukaryotic (Phear *et al.*, 1987; Phear and Meuth, 1989) and prokaryotic systems (Barclay and Little, 1981; Kunz, 1985) and in cultured mammalian cells (Bradley and Sharkey, 1978; Meuth, 1989), mutations and chromosomal abnormalities as deletions (Seno *et al.*, 1985) and breaks (Potter, 1971) can be induced. Interestingly, DNA viruses, retroviruses and some other intracellular microbes are dependent on the cellular supply of dNTPs and, thus, any such disregulation might affect their genetic integrity as well. If the level of induced mutations can be pushed beyond the error threshold of the microorganisms, they might be driven into suicide and die. Indeed, this might well be a mechanism by which hydroxyurea, now used to treat HIV infection, exerts its antiviral effect (Meyerhans *et al.*, 1994). However, in terms of antimicrobial treatment, induced hypermutation affecting the whole cell will inevitably be cytotoxic. Consequently, promiscuous nucleoside analogues designed to be specifically utilized by microbial polymerases and thus to be selectively hypermutagenic might have therapeutic potential in the future.

CONCLUSION

Fidelity is seen as a variable with template/primer, enzyme and substrate as the principal players. For a given system, the stability of the nucleoside triphosphate pools becomes a major factor. While clearly buffered, no system is perfect, and this shows up in retroviral hypermutation. Given the frequency of such deviations and the length of cellular DNA, it is not surprising that dNTP pool biases provide a unifying hypothesis for the origin of some cancer and genetic diseases as well as genome evolution in general. Exaggeration of dNTP pool biases *in vitro* is extraordinarily mutagenic, allowing vast jumps in sequence space and freeing us from the drudgery of WORD → WORE → GORE → GONE → GENE inherent to natural selection (Maynard Smith, 1970). We are going to see some fascinating papers.

ACKNOWLEDGEMENT

We thank Monica Sala and Simon Wain-Hobson for the many interesting discussions and suggestions for the manuscript. A. M. was supported by grants from the Deutsche Forschungsgemeinschaft and J.-P. V. by grants from the Institut Pasteur and Agence Nationale de Recherche sur le SIDA.

REFERENCES

Barclay, B.J. and Little, J.G. (1981) Mutation induction in yeast by deoxythymidine monophosphate: a model. *Mol. Gen. Genet.*, **181**, 279–281.

Barnes, W.M. (1992) The fidelity of Taq polymerase catalysing PCR is improved by an N-terminal deletion. *Gene*, **112**, 29–35.

Bass, B.L., Weintraub, H., Cattenao, R. and Billeter, M.A. (1989) Biased hypermutation of viral RNA genomes could be due to unwinding/modification of double stranded RNA. *Cell*, **56**, 331.

Bass, B.L. (1995) An I for editing. *Curr. Biol.*, **5**, 598–600.

Batschelet, E., Domingo, E. and Weissman, C. (1976) The proportion of revertant and mutant phage in a growing population as a function of mutation and growth rate. *Gene*, **1**, 27–32.

Battula, N. and Loeb, L.A. (1974) The infidelity of avian myeloblastosis virus deoxyribonucleic acid polymerase in polynucleotide replication. *J. Biol. Chem.*, **249**, 4086–4093.

Bebenek, K., Abbotts, J., Roberts, J.D., Wilson, S.H. and Kunkel, T.A. (1989) Specificity and mechanism of error-prone replication by human immunodeficiency virus-1 reverse transcriptase. *J. Biol. Chem.*, **264**, 16948–16956.

Bebenek, K., Abbotts, J., Wilson, S.H. and Kunkel, T.A. (1993) Error-prone polymerisation by HIV-1 reverse transcriptase. Contribution of template-primer misalignment, miscoding, and termination probability to mutational hot spots. *J. Biol. Chem.*, **268**, 10324–10334.

Beese, L.S., Derbyshire, V. and Steitz, T.A. (1993) Structure of DNA polymerase I Klenow fragment bound to duplex DNA. *Science*, **260**, 352–355.

Berek, C. and Milstein, C. (1987) Mutation drift and repertoire shift in the maturation of the immune response. *Immunol. Rev.*, **96**, 23–41.

Boege U., Kobasa D., Onodera S., Parks G.D., Palmenberg A.C. and Scraba D.G. (1991) Mengo virus neutralization epitopes. *Virology*, **181**, 1–13.

Boosalis, M.S., Petruska, J. and Goodman, M.F. (1987) DNA polymerase insertion fidelity: gel assay for site-specific kinetics. *J. Biol. Chem.*, **262**, 14689–14696.

Borman, A.M., Quillent, C., Charneau, P., Kean, C.M. and Clavel, F. (1995) A highly defective HIV group O provirus: evidence for the role of local sequence determinants in hypermutation during negative strand DNA synthesis. *Virology*, **208**, 601–609.

Boyer, J.C., Bebenek, K. and Kunkel, T.A. (1992) Unequal human immunodeficiency virus type 1 reverse transcriptase error rates with RNA and DNA templates. *Proc. Natl Acad. Sci. USA*, **89**, 6919–6923.

Bradley, M.O. and Sharkey, N.A., (1978) Mutagenecity of thymidine to cultured chinese hamster cells. *Nature*, **274**, 607–608.

Bray, G. and Brent, T.P., (1972) Deoxyribonucleoside 5′-triphosphate pool fluctuations during the mammalian cell cycle. *Biochim. Biophys. Acta*, **269**, 184–191.

Brown, T.C. and Jiricny, J. (1987) A specific mismatch repair event protects mammalian cells

from loss of 5-methylcystosine. *Cell*, **50**, 945–950.

Burns, D.P.W. and Temin, H.M. (1994) High rates of frameshift mutations within homo-oligomeric runs during a single cycle of retroviral replication. *J. Virol.*, **68**, 4196–4203.

Caldwell, R.C. and Joyce, G.F. (1992) Randomization of genes by PCR mutagenesis. *PCR Meth. Appl.*, **2**, 28–33.

Cattaneo, R. and Billeter, M.A. (1992) Mutations and A/I hypermutations in measles virus persistent infections. *Curr. Topics Microbiol. Immunol.*, **176**, 63–74.

Cattaneo, R., Schmid, A., Eschle, D., Baczko, K., ter Meulen, V. and Billeter, M.A. (1988) Biased hypermutation and other genetic changes in defective measles virus in human brain infections. *Cell*, **55**, 255–265.

Chen, K. and Arnold, F.H. (1993) Tuning the activity of an enzyme for unusual environments: sequential random mutagenesis of subtilisin E for catalysis in dimethylformamide. *Proc. Natl Acad. Sci. USA*, **90**, 5618–5622.

Cheynier, R., Henrichwark, S., Hadida, F. *et al.* (1994) HIV and T cell expansion in splenic white pulps is accompanied by infiltration of HIV-specific cytotoxic T lymphocytes. *Cell*, **78**, 373–387.

Cheynier, R., Gratton, S., Vartanian, J.P., Meyerhans, A. and Wain-Hobson, S. (1997) G → A hypermutation does not result from polymerase chain reaction. *AIDS Res. Hum. Retroviruses*, **13**, 985–986.

Clayton, L.K., Goodman, M.F., Branscomb, E.W. and Galas, D.J. (1979) Error induction and correction by mutant and wild-type T4 DNA polymerases: kinetic error discrimination mechanisms. *J. Biol. Chem.*, **254**, 1902–1912.

Coffin, J.M., Tsichlis, P.N., Barker, C.S. and Voynow, S. (1980) Variation in avian retrovirus genomes. *Ann. NY Acad. Sci.*, **354**, 410–425.

Cohen, A., Barankiewicz, J., Lederman, H.M. and Gelfand, E.W. (1983) Purine and pyrimidine metabolism in human lymphocytes. Regulation of deoxyribonucleotide metabolism. *J. Biol. Chem.*, **258**, 12334–12340.

Condra, J.H. and Emini, E.A. (1997) Preventing HIV-1 drug resistance. *Sci. Med.*, **1**, 14–23.

Dahbo, Y. and Eriksson, S. (1985) On the mechanism of deoxyribonucleoside toxicity in human T-lymphoblastoid cells. Reversal of growth inhibition by addition of cytidine. *Eur. J. Biochem.*, **150**, 429–434.

Dalberg, M.E. and Benkovic, S.J. (1991) Kinetic mechanism of DNA polymerase I (Klenow fragment): identification of a second conformational change and evaluation of the internal equilibrium constant. *Biochemistry*, **30**, 4835–4843.

Delarue, M., Poch, O., Tordo, N., Moras, D. and Argos, P. (1990) An attempt to unify the structure of polymerases. *Prot. Eng.*, **3**, 461–467.

Delassus, S., Cheynier, R. and Wain-Hobson, S. (1992) Inhomogenous distribution of human immunodeficiency virus type 1 proviruses within spleen. *J. Virol.*, **65**, 5642–5645.

De la Torre J.C., Wimmer E. and Holland J.J. (1990) Very high frequency of reversion of guanidine resistance on clonal pools of guanidine-dependent type I poliovirus. *J. Virol.*, **64**, 664–671.

De la Torre J.C., Giachetti C., Semler B.L. and Holland J.J. (1992) High frequency of single base transitions and extreme frequency of precise, multiple-base reversion mutations in poliovirus. *Proc. Natl Acad. Sci. USA*, **89**, 2531–2535.

Domingo, E. and Holland, J.J. (1997) RNA viral mutations and fitness for survival. *Annu. Rev. Microbiol.*, **51**, 151–178.

Domingo, E., Flavell, R.A. and Weissmann, C. (1976) In vitro site-directed mutagenesis: generation and properties of an infectious extracistronic mutant of bacteriophage Qβ. *Gene*, **1**, 3–25.

Domingo, E., Sabo, D., Taniguchi, T. and Weissmann, C. (1978) Nucleotide sequence heterogeneity of an RNA phage population. *Cell*, **13**, 735–744.

Domingo, E., Davila, M. and Ortin, J. (1980) Nucleotide sequence heterogeneity of the RNA from a natural population of foot-and-mouth disease virus. *Gene*, **11**, 333–346.

Domingo, E., Escaramis, C., Martinez, M.A., Martinez-Sala, E. and Matteu, M.G. (1992) Foot-and-mouth disease virus populations are quasispecies. *Curr. Topics Microbiol. Immunol.*, **176**, 33–47.

Donlin, M.J., Patel, S.S. and Johnson, K. (1991) Kinetic partitioning between the exonuclease and polymerase sites in DNA error correction. *Biochemistry*, **30**, 538–546.

Doublié, S., Tabor, S., Long, A.M., Richardson, C.C. and Ellenberger, T. (1998) Crystal structure of a bacteriophage T7 DNA replication complex at 2.2 A resolution. *Nature*, **391**, 251–258.

Dougherty, J.P. and Temin, H.M. (1988) Determination of the rate of base pair substitution and insertion mutation in retrovirus replication. *J. Virol.*, **62**, 2817–2822.

Drake, J.W. (1991) A constant rate of spontaneous mutation in DNA-based microbes. *Proc. Natl Acad. Sci. USA*, **88**, 7160–7164.

Drake, J.W. (1993) Rates of spontaneous mutations among RNA viruses. *Proc. Natl Acad. Sci. USA*, **90**, 4171–4175.

Drake, J.W., Charlesworth, B., Charlesworth, D. and Crow, J.F. (1998) Rates of spontaneous mutation. *Genetics*, **148**, 1667–1686.

Drosopoulos, W.C. and Prasad, V.R. (1998) Increased misincorporation fidelity observed for nucleoside analogue resistance mutations M184V and E89G in human immunodeficiency virus type 1 reverse transcriptase does not correlate with the overall error rate measured in vitro. *J. Virol.*, **72**, 4224–4230.

Durbin R.K. and Stollar V. (1986) Sequence analysis of the E2 gene of a hyperglycosylated, host restricted mutant of Sindbis virus and estimation of mutation rate from frequency of revertants. *Virology*, **154**, 135–143.

Eckert, K.A. and Kunkel, T.A. (1990) High fidelity DNA synthesis by the *thermus aquaticus* DNA polymerase. *Nucleic Acids Res.*, **18**, 3739–3744.

Eigen, M. (1993) The viral quasispecies. *Sci. Am.*, **269**, 42–49.

Eigen, M. and Biebricher, C.K. (1988) Sequence space and quasispecies distribution. In: *RNA Genetics vol 3*. (eds Domingo, E., Holland, J.J. and Ahlquist, P., pp. 211–245. CRC Press, Boca Raton, FL.

Emini E.A., Jameson B.A., Lewis A.J., Larsen G.R. and Wimmer E. (1982) Poliovirus neutralization epitopes: analysis and localization with neutralizing monoclonal antibodies. *J. Virol.*, **43**, 997–1005.

Eom, S.H., Wang, J. and Steitz, T.A. (1996) Structure of Taq polymerase with DNA at the polymerase active site. *Nature*, **382**, 278–281.

Farabaugh, P.J. (1978) Sequence of the lacI gene. *Nature*, **274**, 765–769.

Felder, M.P., Laugier, D., Yatsula, B., Dezélée, P., Calothy, G. and Marx, M. (1994) Functional and biological properties of an avian variant long terminal repeat containing multiple A to G conversions in the U3 sequence. *J. Virol.*, **68**, 4759–4767.

Fersht, A.R. (1985) *Enzyme Structure and Mechanism*, p. 112. W.H. Freeman, New York.

Filler, A.G. and Lever, A.M.L. (1997) Effects of cation substitutions on reverse transcriptase and on human immunodeficiency virus production. *AIDS Res. Human Retroviruses*, **13**, 291–299.

Fitzgibbon, J.E., Mazar, S. and Dubin, D.T. (1993) A new type of G → A hypermutation affecting human immunodeficiency virus. *AIDS Res. Hum. Retroviruses*, **9**, 833–838.

Flaman, J.M., Frebourg, T., Moreau, V. *et al.* (1994) A rapid PCR fidelity assay. *Nucleic Acids Res.*, **22**, 3259–3260.

Fromant, M., Blanquet, S. and Plateau, P. (1995) Direct random mutagenesis of gene-sized DNA fragments using polymerase chain reaction. *Anal. Biochem.*, **224**, 347–353.

Galas, D.J. and Branscomb, E.W. (1978) Enzymatic determinants of DNA polymerase accuracy. Theory of coliphage T4 polymerase mechanisms. *J. Mol. Biol.*, **88**, 653.

Gao, F., Yue, L., White, A.T. *et al.* (1992) Human infection by genetically diverse SIV-sm related HIV-2 in West Africa. *Nature*, **358**, 495–499.

Garcia-Barreno, B., Portela, A., Delgado, T., Lopez, J.A. and Melero, J.A. (1990) Frame shift mutations as a novel mechanism for the generation of neutralization resistant mutants of human respiratory syncytial virus. *EMBO J.*, **9**, 4181–4187.

Gojobori, T., Li, W.-H. and Graur, D. (1982) Patterns of nucleotide substitution in pseudogenes and functional genes. *J. Mol. Evol.*, **18**, 360–369.

Goodman, M.F. (1988) DNA replication fidelity: kinetics and thermodynamics. *Mutation Res.*, **200**, 11.

Goodman, M.F. and Fygenson, K.D. (1998) DNA polymerase fidelity: from genetics toward a biochemical understanding. *Genetics*, **148**, 1475–1482.

Gorman, O.T., Bean, W.J. and Webster, R.G. (1992) Evolutionary processes in influenza viruses: divergence, rapid evolution, and stasis. *Curr. Topics Microbiol. Immunol.*, **176**, 75–97.

Günther, S., Sommer, G., Plikat, U., Wain-Hobson, S., Will, H. and Meyerhans, A. (1997) Naturally occurring hepatitis B virus subgenomes bearing the hallmarks of retroviral G → A hypermutation. *Virology*, **235**, 104–108.

Hahn Y.S., Strauss E.G., Lenches E.M. and Strauss J.H. (1989) Sindbis virus *ts* 103 has a mutation in glycoprotein E2 that leads to defective assembly of virions. *J. Virol.*, **63**, 3459–3465.

Hajjar, A.M. and Linial, M.L. (1995) Modification of retroviral RNA by double-stranded RNA adenosine deaminase. *J. Virol.*, **69**, 5878–5882.

Ho, D.D., Neumann, A.U., Perelson, A.S., Chen, W., Leonard, J.M. and Markowitz, M. (1995) Rapid turnover of plasma virions and CD4 lymphocytes in HIV-1 infection. *Nature (Lond.)*, **373**, 123–126.

Holland J.J., de la Torre J.C., Steinhauer D.A., Clarke D., Duarte E. and Domingo E. (1989) Virus mutation frequencies can be greatly underestimated by monoclonal antibody neutralization of virions. *J. Virol.*, **63**, 5030–5036.

Honess, R.W. (1984) Herpes simplex and "the herpes complex": diverse observations and a unifying hypothesis. *J. Gen. Virol.*, **65**, 2077–2107.

Honess, R.W., Bodemer, W., Cameron, K.R., Niller, H.H., Fleckenstein, B. and Randall, R.E. (1986) The A+T-rich genome of *Herpesvirus saimiri* contains a highly conserved gene for thymidylate synthase. *Proc. Nat Acad. Sci. USA*, **83**, 3604–3608.

Hsu, M., Inouye, P., Rezende, L. *et al.* (1997) Higher fidelity of RNA-dependent DNA mispair extension by M184V drug-resistant than wild type reverse transcriptase of human immunodeficiency virus 1. *Nucleic Acids Res.*, **25**, 4532–4536.

Hu, W.S. and Temin, H.M. (1990) Retroviral recombination and reverse transcription. *Science*, **250**, 1227–1233.

Hu, W.S. and Temin, H.M. (1992) Effect of gamma radiation on retroviral recombination. *J. Virol.*, **66**, 4457–4463.

Jacobo-Molina, A., Ding, J., Nanni, R.G. *et al.* (1993) Crystal structure of human immunodeficiency virus type 1 reverse transcriptase complexed with double-stranded DNA at 3.0. A resolution shows bent DNA. *Proc. Natl Acad. Sci. USA*, **90**, 6320–6324.

Johnson, P.R., Hamm, T.E., Goldstein, S., Kitov, S. and Hirsch, V.M. (1991) The genetic fate of molecularly cloned simian immunodeficiency virus in experimentally infected macaques. *Virology*, **185**, 217–228.

Jones, J.S., Allan, R.W., Seufzer, B. and Temin, H.M. (1994) Copackaging of different-sized retroviral genomic RNAs: little effect on retroviral replication or recombination. *J. Virol.*, **68**, 4097–4103.

Karlin, S., Doerfler, W. and Cardon, L.R. (1994) Why is CpG suppressed in the genomes of virtually all small eucaryotic viruses but not in those of large eucaryotic viruses? *J. Virol.*, **68**, 2889–2897.

Kerr, S.G. and Anderson, K.S. (1997) RNA dependent DNA replication fidelity of HIV-1 reverse transcriptase: evidence of discrimination between DNA and RNA substrates. *Biochemistry*, **36**, 14056–14063.

Kim, T., Mudry R.A., Jr, Rexrode C.A. II, and Pathak, V.K. (1996) Retroviral mutation rates and A-to-G hypermutations during different stages of retroviral replication. *J. Virol.*, **70**, 7594–7602.

Kinnunen, L., Pöyry, T. and Hovi, T. (1992) Genetic diversity and rapid evolution of poliovirus in human hosts. *Curr. Topics Microbiol. Immunol.*, **176**, 49–61.

Kohlstaedt, L.A., Wang, J., Friedman, J.M., Rice, P.A. and Steiz, T.A. (1992) Crystal structure at 3.5 A resolution of HIV-1 reverse transcriptase complexed with an inhibitor. *Science*, **256**, 1783–1790.

Krawczak, M., Smith-Sorensen, B., Schmidtke, J., Kakkar, V.V., Cooper, D.N. and Hovig, E. (1995) Somatic spectrum of cancer-associated single basepair substitutions in the TP53 gene

is determined mainly by endogenous mechanisms of mutation and selection. *Hum. Mutat.*, **5**, 48–57.

Kunkel, T.A. (1985) The mutational specificity of DNA, polymerase-β during in vitro DNA synthesis. Production of frameshift, base substitution, and deletion mutations. *J. Biol. Chem.* **260**, 5787–5796.

Kunkel, T.A. (1992b). DNA replication fidelity. *J. Biol. Chem.*, **267**, 18251–18254.

Kunkel, T.A. and Alexander, P.S. (1986) The base substitution fidelity of eucaryotic DNA polymerases: mispairing frequencies, site preferences, insertion preferences and base substitution by dislocation. *J. Biol. Chem.*, **261**, 160–166.

Kunkel, T.A., Silber, J.R. and Loeb, L.A. (1982) The mutagenic effect of deoxynucleotide substrate imbalances during DNA synthesis with mammalian DNA polymerases. *Mutat. Res.*, **94**, 413–419.

Kunz, B.A. (1985) Thymineless mutagenesis in bacteria. In: *Genetic Consequences of Nucleotide Pool Imbalance* (ed. de Serres, F.J.), pp. 189–209. Plenum Press, New York.

Kunz, B.A., Kohalmi, L., Kang, X.L. and Magnusson, K.A. (1990) Specificity of the mutator effect caused by disruption of the RAD1 excision repair gene of *Saccharomyces cerevisiae. J. Bactriol.*, **172**, 3009–3014.

Kunz, B.A., Kohalmi, S.E., Kunkel, T.A., Mathews, C.K., McIntosh, E.M. and Reidy, J.A. (1994) Deoxyribonucleoside triphospate levels: a critical factor in the maintenance of genetic stability. *Mutat. Res.*, **318**, 1–64.

Leider, J.M., Palese, P. and Smith, F.I. (1988) Determination of the mutation rate of the retrovirus. *J. Virol.*, **62**, 3084–3091.

Leung, D., Chen, E. and Goeddel, D. (1989) A method for random mutagenesis of a defined DNA segment using a modified polymerase chain reaction. *Technique*, **1**, 11–15.

Li, W.-H., Wu, C.-I. and Luo, C.-C. (1984) Nonrandomness of point mutátion as reflected in nucleotide sbstitutions in pseudogenes and its evolutionary implications. *J. Mol. Evol.*, **21**, 58–71.

Lim, W.A. and Sauer, R.T. (1991) The role of internal packing interactions in determining the structure and stability of a protein. *J. Mol. Biol.*, **219**, 359–376.

Loeb, L.A. and Kunkel, T.A. (1982) Fidelity of DNA synthesis. *Annu. Rev. Biochem.*, **52**, 429–457.

McIntosh, E.M. and Haynes, R.H. (1996) HIV and human endogenous retroviruses: an hypothesis with therapeutic implications. *Acta Biochim. Polon.*, **43**, 583–592.

Mansky, L.M. and Temin, H.M. (1994) Lower mutation rate of bovine leukemia virus relative to that of spleen necrosis virus. *J. Virol.*, **68**, 494–499.

Mansky, L.M. and Temin, H.M. (1995) Lower *in vivo* mutation rate of human immunodeficiency virus type 1 than that predicted from the fidelity of purified reverse transcriptase. *J. Virol.*, **69**, 5087–5094.

Mansky, L.M. (1998) Retrovirus mutation rates and their role in genetic variation. *J. Gen. Virol.*, **79**, 1337–1345.

Martinez M.A., Carrillo C., Gonzalez-Candelas F., Moya A., Domingo E. and Sobrino F. (1991) Fitness alteration of foot-and-mouth disease virus mutants: measurement of adaptibility of viral quasispecies. *J. Virol.*, **65**, 3954–3957.

Martinez, M.A., Vartanian, J.P. and Wain-Hobson, S. (1994) Hypermutagenesis of RNA using human immunodeficiency virus type 1 reverse transcriptase and biased dNTP concentrations. *Proc. Natl Acad. Sci. USA*, **91**, 11787–11791.

Martinez, M.A., Sala, M., Vartanian, J.P. and Wain-Hobson, S. (1995) Reverse transcriptase and substrate dependence of the RNA hypermutagenesis reaction. *Nucleic Acids Res.*, **23**, 2573–2578.

Martinez, M.A., Pezo, V., Marliere, P. and Wain-Hobson, S. (1996) Exploring the functional robustness of an enzyme by in vitro protein evolution. *EMBO J.*, **15**, 1203–1210.

Mathews, C.K. and Ji, J. (1992) DNA precursor asymmetries, replication fidelity, and variable genome evolution. *BioEssays*, **14**, 295–301.

Mattila, P., Korpela, J., Tenkanen, T. and Pitkänen, K. (1991) Fidelity of the DNA synthesis by the *thermococcus litoralis* DNA polymerase – an extremely heat stable enzyme with proof reading activity. *Nucleic Acids Res.*, **18**, 4967–4973.

Maynard Smith, J. (1970) Natural selection and the concept of a protein space. *Nature*, **225**, 563–564.

Mendelman, L.V., Boosalis, M.S., Petruska, J. and Goodman, M.F. (1989) Nearest neighbor influences on DNA polymerases insertion fidelity. *J. Biol. Chem.*, **264**, 14415–14423.

Mendelman, L.V., Petruska, J. and Goodman, M.F. (1990) Base mispair extension kinetics: comparison of DNA polymerase alpha and reverse transcriptase. *J. Biol. Chem.*, **265**, 2338–2346.

Meuth, M. (1989) The molecular basis of mutations induced by deoxynucleoside triphosphate pool imbalances in mammalian cells. *Exper. Cell. Res.*, **181**, 305–316.

Meyerhans, A., Cheynier, R., Albert, J. *et al.* (1989) Temporal fluctuations in HIV quasispecies *in vivo* are not reflected by sequential HIV isolations. *Cell*, **58**, 901–910.

Meyerhans, A., Vartanian, J.P. and Wain-Hobson, S. (1990) DNA recombination during PCR. *Nucleic Acids Res.*, **18**, 1687–1691.

Meyerhans, A., Vartanian, J.P., Hultgren, C. *et al.* (1994) Restriction and enhancement of human immunodeficiency virus type 1 replication by modulation of intracellular deoxynucleoside triphosphate pools. *J. Virol.*, **68**, 535–540.

Minor P.D., Schild G.C., Bootman J. *et al.* (1983) Location and primary structure of a major antigenic site for poliovirus neutralization. *Nature* **301**, 674–679.

Minor P.D., Ferguson M., Evans D.M., Almond V.W., Icenogle J.P. (1986) Antigenic structure of polioviruses of serotypes 1, 2 and 3. *J. Gen. Virol.* **67**, 1283-1291.

Monk, R.J., Malik, F.G., Stokes, D. and Evans, L.H. (1992) Direct determination of the point mutation rate of a murine retrovirus. *J. Virol.*, **66**, 3683–3689.

Murphy, D.G., Dimock, K. and Yong Kang, C. (1991) Numerous transitions in human parainfluenza virus 3 RNA recovered from persistently infected cells. *Virology*, **181**, 760–763.

Myers, G., Korber, B., Hahn, B.H. *et al.* (1995) *Human Retroviruses and AIDS*. Los Alamos National Laboratory, Los Alamos, NM.

Nassal, M. and Schaller, H. (1993) Hepatitis B

virus replication. *Trends Microbiol.*, **1**, 221–228.

O'Hara, P.J., Horodyski, F.M., Nichol, S.T. and Holland, J.J. (1984) Vesicular stomatitis virus mutants resistant to defective-interfering particles accumulate stable 5'-terminal and fewer 3'-terminal mutations in a stepwise manner. *J. Virol.*, **49**, 793–798.

Orbach, M.J., Vollrath, D., Davis, R.W. and Yanofsky, C. (1988) An electrophoretic karyotype of Neurospora crassa. *Mol. Cell. Biol.*, **8**, 1469–1473.

Oude Essink, B.B., Back, N.K.T. and Berkhout, B. (1997) Increased polymerase fidelity of 3TC-resistant variants of HIV-1 reverse transcriptase. *Nucleic Acids Res.*, **25**, 3212.

Pandey, V.N., Kaushik, N., Rege, N., Sarafianos, S.G., Yadav, P.N.S. and Modak, M.J. (1996) Role of methionine 184 in human immunodeficiency virus type 1 reverse transcriptase in the polymerase function and fidelity of DNA synthesis. *Biochemistry*, **35**, 2168–2179.

Parthasarathi, S., Varela-Echavarria, A., Ron, Y., Preston, B.D. and Dougherty, J.P. (1995) Genetic rearrangements occurring during a single cycle of murine leukemia virus vector replication: characterization and implications. *J. Virol.*, **69**, 7991–8000.

Parvin, J.D., Moscona, A., Pan, W.T., Lieder, J. and Palese, P. (1986) Measurement of the mutation rates of animal viruses: influenza-A virus and poliovirus type 1. *J. Virol.*, **59**, 377–383.

Pathak, V.K. and Temin, H.M. (1990a) Broad spectrum of *in-vitro* forward mutations, hypermutations, and mutational hotspots in a retrovital shuttle vector after a single replication cycle: substitutions, frameshifts, and hypermutations. *Proc. Natl Acad. Sci. USA*, **87**, 6019–6023.

Pathak, V.K. and Temin, H.M. (1990b). Broad spectrum of *in vitro* forward mutations, hypermutations, and mutational hotspots in a retrovital shuttle vector after a single replication cycle: deletions and deletions with insertions. *Proc. Natl Acad. Sci. USA*, **87**, 6024–6028.

Pathak, V.K. and Temin, H.M. (1992) 5-Azacytidine and RNA secondary structure

increase the retrovirus mutation rate. *J. Virol.*, **66**, 3093–3100.

Pelletier, H., Sawaya, M.R., Kumar, A., Wilson, S.H. and Kraut, J. (1994) Structures of ternary complexes of rat DNA polymerase β, a DNA template-primer, and ddCTP. *Science*, **264**, 1891–1903.

Pelletier, E., Saurin, W., Cheynier, R., Letvin, N.L. and Wain-Hobson, S. (1995) The tempo and mode of SIV quasispecies development in vivo calls for massive viral replication and clearance. *Virology*, **208**, 644–652.

Perrino, F.W., Preston, B.D., Sandell, L.L. and Loeb, L.A. (1989) Extension of mismatched 3' termini of DNA is a major determinant of the infidelity of human immunodeficiency virus type 1 reverse transcriptase. *Proc. Natl Acad. Sci. USA*, **86**, 8343–8347.

Perrino, F.W. and Loeb, L.A. (1989) Differential extension of 3' mispairs is a major contribution to the high fidelity of calf thymus DNA polymerase-alpha. *J. Biol. Chem.*, **264**, 2898–2905.

Perry, S.T., Flaherty, M.T., Kelley, M.J. *et al.* (1992) The surface envelope protein gene region of equine infectious anemia virus is not an important determinant of tropism *in vitro. J. Virol.*, **66**, 4085–4097.

Pezo, V. and Wain-Hobson, S. (1997) Biased NTP pools and manganese cations renders in vitro transcription hypermutagenic. *Gene*, **186**, 67–72.

Phear, G. and Meuth, M. (1989) A novel pathway for transversion mutation induced by dCTP misincorporation in a mutator strain of CHO cells. *Mol. Cell. Biol.*, **9**, 1810–1812.

Phear, G., Nalbantoglu, J. and Meuth, M. (1987) Next-nucleotide effects in mutations driven by DNA precursor pool imbalances at the aprt locus of Chinese hamster ovary cells. *Proc. Natl Acad. Sci. USA*, **84**, 4450–4454.

Plikat, U., Nieselt-Struwe, K. and Meyerhans, A. (1997) Genetic drift can dominate short-term human immunodeficiency virus type 1 nef quasispecies evolution in *vivo. J. Virol.*, **71**, 4233–4240.

Polson, A.G. and Bass, B.L. (1994) Preferential selection of adenosines for modification by double-stranded RNA adenosine deaminase. *EMBO J.*, **13**, 5701–5711.

Polson, A.G., Bass, B.L. and Casey, J.L. (1996) RNA editing of hepatitis delta virus antigenome by dsRNA-adenosine deaminase. *Nature*, **380**, 454–456.

Potter, C.G. (1971) Interaction of polyploidy by concentrated thymidine. *Exp. Cell. Res.* **68**, 442–448.

Preston, B.D. and Dougherty, J.P. (1996) Mechanisms of retroviral mutation. *Trends Microbiol.*, **4**, 16–21.

Preston, B.D., Poiesz, B.J. and Loeb, L.A. (1988) Fidelity of HIV-1 reverse transcriptase. *Science*, **243**, 1168–1171.

Pribnow, D., Sigurdson, D.C., Gold, L. *et al.*, (1981) rII cistrons of bacteriophage T4, DNA sequence around the intercistronic divide and positions of genetic landmarks. *J. Mol. Biol.*, **149**, 337–376.

Pusinelli, G.A. and Temin, H.M. (1991) Characterization of large deletions occurring during a single round of retrovirus replication: novel deletion mechanism involving errors in strand transfer. *J. Virol.*, **65**, 4786–4797.

Randall, S.K., Eritja, R., Kaplan, B.E., Petruska, J. and Goodman, M.F. (1987) Nucleotide insertion kinetics opposite abasic lesions in DNA. *J. Biol. Chem.*, **262**, 6864–6870.

Reichard, P. (1988) Interactions between deoxyribonucleotide and DNA synthesis. *Annu. Rev. Biochem.*, **57**, 349–374.

Reichard, P. (1993) From RNA to DNA, why so many ribonucleotide reductases? *Science*, **260**, 1773–1777.

Rellos, P. and Scopes, R.K. (1994) Polymerase chain reaction based random mutagenesis: production and characterization of thermostable mutants of *Zymomonas mobilis* alcohol dehydrogenase-2. *Protein Engin. Purific.*, **5**, 270–277.

Ren, J., Esnouf, R., Hopkins, A. *et al.* (1995) The structure of HIV-1 reverse transcriptase complexed with 9-chloro-TIBO: lessons for inhibitor design. *Structure*, **3**, 915–926.

Rennell, D., Bouvier, S.E., Hardy, L.W. and Poteete, A.R. (1991) Systematic mutation of bacteriophage T4 lysozyme. *J. Mol. Biol.*, **222**, 67–87.

Reyland, M.E. and Loeb, L.A. (1987) On the fidelity of DNA replication. *J. Biol. Chem.*, **262**, 10824–10830.

Richetti, M. and Buc, H. (1990) Reverse transcriptases and genomic variability: the accurancy of DNA replication is enzyme specific and sequence dependent. *EMBO J.*, **9**, 1583–1593.

Richetti, M. and Buc, H. (1993) E. coli DNA polymerase I as a reverse transcriptase. *EMBO J.*, **12**, 387–396.

Richetti, M. and Buc, H. (1996) A reiterative mode of DNA synthesis adopted by HIV-1 reverse transcriptase after a misincorporation. *Biochemistry*, **35**, 14970–14983.

Roberts, J.D. and Kunkel, T.A. (1988) Fidelity of a human cell DNA replication complex. *Proc. Natl Acad. Sci. USA*, **85**, 7064–7068.

Roberts, J.D., Bebenek, K. and Kunkel, T.A. (1988) The accuracy of reverse transcriptase from HIV-1. *Science*, **242**, 1171–1173.

Rodgers, D.W., Gamblin, S.J., Harris, B.A. *et al.* (1995) The structure of unligated reverse transcriptase from the human immunodeficiency virus type 1. *Proc. Natl Acad. Sci. USA*, **92**, 1222–1226.

Rubinek, T., Bakhanashvili, M. and Hizi, A. (1997) The fidelity of 3′ misinsertion and mispair extension during DNA synthesis exhibited by two drug-resistant mutants of the reverse transcriptase of human immunodeficiency virus type 1 with Leu74Val and Glu89Gly. *Eur. J. Biochem.*, **247**, 238.

Rueda, P., Garcia-Barreno, B. and Melero, J.A. (1994) Loss of conserved cysteine residues in the attachment (G) glycoprotein of two human respiratory syncytial virus escape mutants that contain multiple A–G substitutions (hypermutations). *Virology*, **198**, 653–662.

Sala, M., Wain-Hobson, S. and Schaeffer, F. (1995) Human immunodeficiency virus type 1 reverse transcriptase tG:T mispair formation on RNA and DNA templates with mismatched primers: a kinetic and thermodynamic study. *EMBO J.*, **14**, 4622–4627.

Sala, M., Pezo, V., Pochet, S. and Wain-Hobson, S. (1996) Ambiguous base pairing of the purine analogue 1-(2-deoxy-b-*scd*/*sc*-ribofuranosyl)-imidazole-4-carboxamide during PCR. *Nucleic Acids Res.* **24**, 3302–3306.

Sala, M. and Vartanian, J.P. (1998) HIV1 reverse transcriptase: an out-of-the-ordinary enzyme. *Bull. Inst. Pasteur*, **96**, 49–63.

Sauer, R.T. (1978) DNA sequence of bacterio-phage gama cI gene. *Nature*, **276**, 301–302.

Schubert, M., Harmison, G.G. and Meier, E. (1984) Primary structure of the vesicular stomatitis virus polymerase (L) gene: evidence for a high frequency of mutations. *J. Virol.*, **51**, 505–514.

Sedivy, J.M., Capone, J.P., Raj Bhandary, U.L. and Sharp, P.A. (1987) An inducible mammalian amber suppressor: propagation of a poliovirus mutant. *Cell*, **50**, 379–389.

Seno, T., Ayusawa, D., Shimizu, K., Koyama, H., Takoisha, K. and Hori, T.A. (1985) Thymineless death and genetic events in mammalian cells. *Basic Life Sci.*, **31**, 241–263.

Sharon, J. (1990) Structural correlates of high antibody affinity: three engineered aminoacid substitutions can increase the affinity of an anti p-axophenylarsonate antibody 200 fold. *Proc. Natl Acad. Sci. USA*, **87**, 4814–4817.

Sherry B., Mosser A.G., Colonno R.J. and Rueckert R.R. (1986) Use of monoclonal antibodies to identify four neutralization immunogens on a common cold picornavirus, human rhinovirus 14. *J. Virol.*, **57**, 246–257.

Shields, A., Witte, W.N., Rothenberg, E. and Baltimore, D. (1978) High frequency of aberrant expression of Moloney musine leukemia virus in clonal infections. *Cell*, **14**, 601–609.

Slabaugh, M.B., Howell, M.L., Wang, Y. and Mathews, C.K. (1991) Deoxyadenosine reverses hydroxyurea inhibition of vaccinia virus growth. *J. Virol.*, **65**, 2290–2298.

Smerdon, S.J., Jäger, J., Wang, J. *et al.* (1994) Structure of the binding site for nonnucleoside inhibitors of the reverse transcriptase of human immunodeficiency virus type 1. *Proc. Natl Acad. Sci. USA*, **91**, 3911–3915.

Smith, J. and Modrich, P. (1996) Mutation detection with MutH, MutL and MutS mismatch repair proteins. *Proc. Natl Acad. Sci. USA*, **93**, 4374–4379.

Sobrino, F., Davila, M., Ortin, J. and Domingo, E. (1983) Multiple genetic variants arise in the course of replication of foot-and-mouth disease virus in cell culture. *Virology*, **128**, 310–318.

Stec, D.S., Weddell, A., Schmaljohn, C.S., Cole, G.A. and Schmaljohn, A.L. (1986) Antibody-selected variation and reversion in Sindbis virus neutralization epitopes. *J. Virol.* **57**, 715–720.

Steinhaurer, D., de la Torre, J.C. and Holland, J.J. (1989) High nucleotide substitution error frequencies in clonal pools of vericular stomatitis virus. *J. Virol.* **63**, 2063–2071.

Steinhaurer, D. and Holland, J.J. (1986) Direct method for quantification of extreme polymerase error frequencies at selected single base sites in viral RNA. *J. Virol.* **57**, 219–228.

Stemmer, W.P.C. (1994) Rapid evolution of a protein in vitro by DNA shuffling. *Nature*, **370**, 389–391.

Suo, Z. and Johnson, K.A. (1997a). Effect of RNA secondary structure on the kinetics of DNA synthesis catalyzed by HIV-1 reverse transcriptase. *Biochemistry*, **36**, 12459–12467.

Suo, Z. and Johnson, K.A. (1997b). RNA secondary structure switching during DNA synthesis catalyzed by HIV-1 reverse transcriptase. *Biochemistry*, **36**, 14778–14785.

Tattersall, M.H.N., Ganeshaguru, K. and Hoffbrand, A.V. (1975) The effect of external deoxyribonucleosides on deoxynucleoside triphosphate concentrations in human lymphocytes. *Biochem. Pharmacol.*, **24**, 1495–1498.

Tindall, K.R. and Kunkel, T.A. (1988) Fidelity of DNA synthesis by the thermus aquaticus DNA polymerase. *Biochemistry*, **27**, 6008–6013.

Tyrsted, G. (1982) Effect of hydroxyurea and 5-fluorodeoxyuridine on deoxyribonucleoside triphosphate pools early in phytohemagglutinin-stimulated human lymphocytes. *Biochem. Pharmacol.*, **31**, 3107–3113.

Varela-Echavarria, A., Garvey, N., Preston, B.D. and Dougherty, J.P. (1992) Comparison of Moloney murine leukemia virus mutation rate with the fidelity of its reverse transcriptase *in vitro*. *J. Biol. Chem.*, **267**, 24681–24688.

Varela-Echavarria, A., Prorock, C.M., Ron, Y. and Dougherty, J.P. (1993) High rate of genetic rearrangements during replication of a Moloney murine leukemia virus-based vector. *J. Virol.*, **67**, 6357–6364.

Vartanian, J.P., Meyerhans, A., Asjo, B. and Wain-Hobson, S. (1991) Selection, recombination, and G → A hypermutation of human immunodeficiency virus type 1 genomes. *J. Virol.*, **65**, 1779–1788.

Vartanian, J.P., Meyerhans, A., Sala, M. and Wain-Hobson, S. (1994) G → A hypermutation of the human immunodeficiency virus type 1 genome: evidence for dCTP pool imbalance during reverse transcription. *Proc. Natl Acad. Sci. USA*, **91**, 3092–3096.

Vartanian, J.P., Henry, M. and Wain-Hobson, S. (1996) Hypermutagenic PCR involving all four transitions and a sizeable proportion of transversions. *Nucleic Acids Res.*, **24**, 2627–2631.

Vartanian, J.P., Plikat, U., Maheux, R., Guillemot, L., Meyerhans, A. and Wain-Hobson, S. (1997) HIV genetic variability is directed and restricted by DNA precursor availability. *J. Mol. Biol.*, **270**, 139–151.

Wain-Hobson, S. (1993) Viral burden in AIDS. *Nature*, **366**, 22.

Wain-Hobson, S., Sonigo, P., Guyader, M., Gazit, A. and Henry, M. (1995) Erratic G → A hypermutation within a complete caprine arthritis-encephalitis virus (CAEV) provirus. *Virology*, **209**, 297–303.

Wainberg, M.A., Drosopoulos, W.C., Salomon, H. *et al.* (1996) Enhanced fidelity of 3TC-selected mutant HIV-1 reverse transcriptase. *Science*, **271**, 1282–1285.

Wang, J., Sattar, A.K.M.A., Wang, C.C., Karam, J.D., Konigsberg, W.H. and Steitz, T.A. (1997) Crystal structure of a pol a replication DNA polymerase from bacteriophage RB69. *Cell*, **89**, 1087–1089.

Ward C.D. and Flanegan J.B. (1992) Determination of the poliovirus RNA poly-

merase error frequency at eight sites in the viral genome. *J. Virol.*, **66**, 3784–3793.

Weber, J. and Grosse, F. (1989) Fidelity of human immunodeficiency virus type 1 reverse transcriptase in copying natural DNA. *Nucleic Acids Res.*, **17**, 1379–1393.

Wei, X., Ghosh, S.K., Taylor, M.E. *et al.* (1995) Viral dynamics in human immunodeficiency virus type 1 infection. *Nature*, **373**, 117–122.

White B.T. and McGeoch D.J. (1987) Isolation and characterization of conditional lethal amber nonsense mutants of vesicular stomatitis virus. *J. Gen. Virol.*, **68**, 3033–3044.

Williams, K.J. and Loeb, L.A. (1992) Retroviral reverse transcriptases: error frequencies and mutagenesis. *Curr. Topics Microbiol. Immunol.*, **176**, 165–180.

Yu, H. and Goodman, M.F. (1992) Comparison on HIV-1 and avian myeloblastosis virus reverse transcriptase fidelity on RNA and DNA templates. *J. Biol. Chem.*, **267**, 10888–10896.

Zaccolo, M., Williams, D.M., Brown, D.M. and Gherardi, E. (1996) An approach to random mutagenesis of DNA using mixtures of triphosphate derivatives of nucleoside analogues. *J. Mol. Biol.*, **255**, 589–603.

Zhang, J. and Temin, H.M. (1993) Rate and mechanism of nonhomologous recombination during a single cycle of retroviral replication. *Science*, **259**, 234–238.

Zhang, J. and Temin, H.M. (1994) Retrovirus recombination depends on the length of sequence identity and is not error prone. *J. Virol.*, **68**, 2409–2414.

6

Drift and Conservatism in RNA Virus Evolution: Are They Adapting or Merely Changing?

Monica Sala and Simon Wain-Hobson

INTRODUCTION

There is no such thing as a perfect machine. Accordingly, nucleic acid polymerization is inevitably error-prone. Yet the notoriety and abundance of RNA viruses attests to their great success as intracellular parasites. Indeed some estimates suggest that 80% of viruses have RNA genomes. It follows that replication without proofreading can be a successful strategy. There is a price to pay, however. Manfred Eigen was the first to point out that without proofreading there is a limit on the size of RNA genomes. Obviously, if the mutation rate is too high, any RNA virus will collapse under mutation pressure. As it happens, RNA viral genomes are up to 32 kb long while mutation rates are 1–2 per genome per cycle or less.

Possibly, RNA viruses and retroviruses have simply not invested in proofreading, in which case mutations represent an inevitable genetic noise, to be tolerated or eliminated. Hence there would be no loss of fitness, fixed mutations being neutral. A corollary of this would be that the intrinsic life style of a virus is set in its genes. The alternative is to suppose that most fixed mutations are beneficial to the virus in allowing it to keep ahead of the host and/or host population.

By this token variation is an integral part of the viral *modus vivendi*. The twin requirements of a successful virus are replication and transmission. Under the rubric replication, a virus could vary to increase its fitness, exploit different target cells or evade adaptive immune responses. In terms of transmission, variation might allow a virus to overcome herd immunity.

These two scenarios emphasize the two sides of the molecular evolution debate; one highlights neutrality while the other puts a premium on positive selection. Purifying or negative selection is ever operative – a poor replicon invariably goes asunder. Through rounds of error and trial, positive selection is the only means of creating a novel replicon. So long as the ecological niche occupied doesn't change, the virus doesn't need to change, purifying selection being sufficient to ensure existence. This raises an important issue: we know that, over the time that we are living and loving, as well as doing experiments, writing papers and reviewing, humans are not evolving. Ernst Mayr noted that "the brain of 100 000 years ago is the same brain that is now able to design computers" (Mayr, 1997). Positive fitness selection among mammals is effectively inoperative over our lifetimes. And certainly since we have known about HIV and AIDS.

How is it that vertebrates, invertebrates, plants, fungi and bacteria, all species with a low genomic mutation rate, can control viruses which mutate so much faster – sometimes by a factor of 10^6 (Holland *et al.*, 1982; Gojobori and Yokoyama, 1985; Domingo *et al.*, 1996). Yet they do. We come to the basic question – to what extent is genetic variation exploited by an RNA virus, if at all? And if so, what is the virus adapting to? The answer invariably given to the second question is "the adaptive immune system" (Seibert *et al.*, 1995). Yet apart from the vertebrates none of the other groups mentioned above mounts antigen-specific immune responses. This chapter will argue that most fixed mutations are neutral.

MOLECULAR CLOCKS

In molecular evolution one of the remarkable observations has been the uniformity of the molecular clock. Although there has been intense debate as to what molecular clocks mean and quite how far they deviate from null hypotheses, fibronectin fixes mutations faster than alpha- or beta-globin, which do so faster than cytochrome c, etc. Rates of amino acid fixation are intrinsic to different proteins. Yet some viruses give rise to persistent infections, others to sequential acute infections. All succumb to the vagaries of transmission bottlenecks. How many rounds of infection are necessary to fix mutations? For example, the tremendous dynamics of viral replication have been described. Whether it be HIV, HBV or HCV, plasma viral turnover is of the order of 10^8–10^{12} virions per day (Ho *et al.*, 1995; Wei *et al.*, 1995; Nowak *et al.*, 1996; Zeuzem *et al.*, 1998). Between 10% and 90% of plasma virus is cleared. In the case of HIV this can involve more than 200 rounds of sequential replication per year (Wain-Hobson, 1993a,b; Ho *et al.*, 1995; Pelletier *et al.*, 1995; Wei *et al.*, 1995).

Many of these variables and unknowns can be removed by comparing the fixation of amino-acid substitutions in pairs of viral proteins from two genomes. If one assumes that the two gene fragments remain linked, through the hellfire of

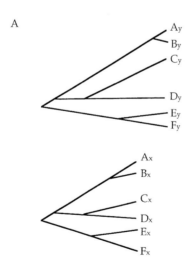
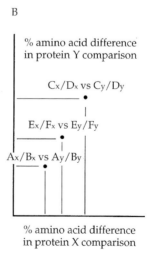

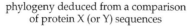

phylogeny deduced from a comparison
of protein X (or Y) sequences

FIGURE 6.1 Two different representations of distances between proteins encoded by genomes A, B, C, D, E and F. **A**. Separate phylogenetic analysis based on homologous sequences for proteins *y* and *x*. **B**. Amino acid divergence for pairs of protein *y* taken from two genomes plotted against divergence for protein *x* from the same two genomes. See footnote to Table 6.1.

immune responses and bottlenecking inherent in transmission, relative degrees of fixation should be attainable. Note that, so long as frequent recombination between highly divergent genomes is not in evidence, this assumption should be valid. This procedure is outlined in Figure 6.1.

The first example is taken from the vast primate immunodeficiency virus database (LANL, 1998). When normalized to the p66 reverse transcriptase product designated RT, amino acid sequence divergence for p17 Gag, p24 Gag, integrase, Vif, gp120, the ectodomain of gp41 and Nef all reveal highly significant linear relationships (Figure 6.2, Table 6.1). The relative rates vary by a factor of two or more. Why the hypervariable gp120 protein shows a relatively low degree of change with respect to the reverse

TABLE 6.1 Smooth drift in pairwise amino acid sequence divergence

Viruses	Paired proteins (y/x)	n	Linear relationship	r
HIV-1, -2, SIV	p17/RT	22	$y = 1.15x - 0.02$	0.98*
	p24/RT	22	$y = 0.91x + 0.03$	0.96*
	Integrase/RT	22	$y = 0.94x - 0.01$	0.99*
	gp41/RT	22	$y = 1.16x + 0.01$	0.96*
	gp120/RT	22	$y = 1.50x + 0.02$	0.96*
	vif/RT	22	$y = 1.49x + 0.01$	0.97*
	nef/RT	22	$y = 1.19x + 0.12$	0.94*
Rhinovirus	3C/3D	5	$y = 1.05x - 0.01$	0.98*
	P2A/3D	5	$y = 1.20x - 0.04$	0.91+
	P2C/3D	5	$y = 1.14x - 0.02$	0.99*
	VP1/3D	5	$y = 1.27x + 0.02$	0.99*
	VP2/3D	5	$y = 0.77x + 0.03$	0.97"
	VP3/3D	5	$y = 0.96x + 0.02$	0.95"
Potyvirus	P1/CI	21	$y = 1.42x + 0.15$	0.94*
	HC-Pro/CI	21	$y = 1.26x - 0.01$	0.99*
	P3/CI	21	$y = 1.53x + 0.06$	0.97*
	NIa/CI	21	$y = 1.13x + 0.00$	0.99*
	NIb/CI	21	$y = 0.87x + 0.01$	0.99*
	CP/CI	21	$y = 0.92x - 0.01$	0.98*
Inoviridae	Protein I/II#	15	$y = 1.06x + 0.03$	1
	Protein III/II#	15	$y = 1.02x + 0.11$	0.91*
	Protein IV/II#	15	$y = 1.02x + 0.05$	0.99*

Orthologous protein sequences were multiply aligned and divergence was calculated by the Clustalw program using protein weight matrices of the Blosum series (Henikoff and Henikoff, 1993). It is well established that protein sequence comparisons are more informative when weighted for genetic and structural biases in amino acid replacements. In the Blosum weight matrices series, the actual matrix that was used depends on how similar the sequences to be aligned are. Different matrices work differently at each evolutionary distance. For a given virus, different protein sequence sets were compared to a given reference such as RT in the case of HIV/SIV. n indicates the number of independent two-by-two comparisons. The data were checked for the possibility that a rogue genome strongly influenced the data. Only in the case of the Inoviridae were there insufficient complete sequences, six in fact, to yield satisfying analyses. Instead all pairwise comparisons were made, hence the data points reflect dependent data (#). The form of the linear regressions are given where y and x refer to the first and second protein listed in the column "Paired proteins". The correlation coefficients r were highly significant in all cases, the corresponding probabilities being: $+ < 0.02$; $" < 0.005$; $* < 0.001$.

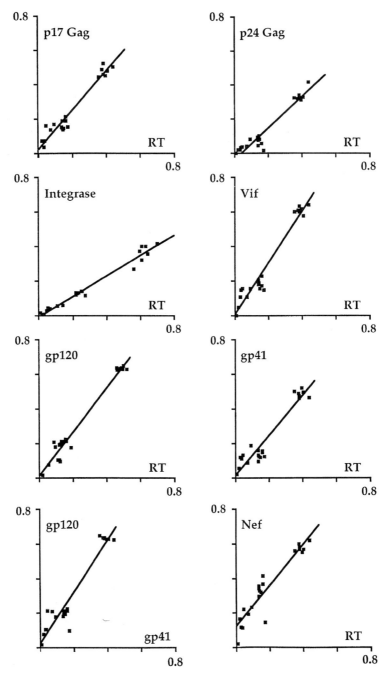

FIGURE 6.2 Graphical representation of paired divergence for orthologous proteins taken from complete HIV-1, HIV-2 and SIV genome sequences. y = different proteins, x = p66 sequence of the reverse transcriptase (RT). x and y values correspond to Blosum-corrected fractional divergence. Only non-overlapping regions were taken into account. The straight lines were obtained by linear regression analysis. Their characteristics are given in Table 6.1.

transcriptase (RT) can be explained by gap stripping, which eliminates the hypervariable regions. Consequently the gp120 data effectively reflects the conserved regions. The linearity, even out to considerable differences, indicates that multiple substitutions and back mutations, which must be occurring, do so to comparable degrees. Although these data were derived from completely sequenced primate immunodeficiency viral genomes, analyses on larger data sets, such as p17 Gag/p24 Gag or gp120/gp41, yielded relative values that differed from those given in Table 6.1 by at most 14%. The absence of points far from the linear regression substan-

tiates the assumption that recombination between highly divergent genomes is rare. This does not preclude recombination between closely related genomes.

The linear regressions passed close to the origin in nearly all cases. Only for Nef was there some deviation, suggesting that Nef was saturating to a different extent from all other proteins. However, as linear correlations involving Nef data were always statistically significant, this trend may be fortuitous. Note that the data cover the earliest phase, intrapatient variation (generally <10%), continuing smoothly to cover interclade, intertype and finally interspecies

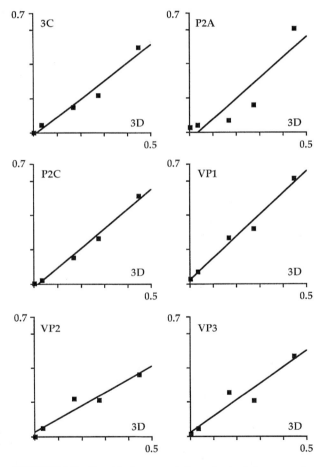

FIGURE 6.3 Graphical representation of paired divergence for orthologous proteins taken from complete rhinovirus genome sequences. y = different proteins, x = RNA-dependent RNA polymerase (3D). x and y values correspond to Blosum-corrected fractional divergence. The straight lines were obtained by linear regression analysis. Their characteristics are given in Table 6.1.

comparisons. Yet this in spite of different environments – that of an individual's immune system, different immune systems stigmatized by highly polymorphic HLA, and finally differences between humans, chimpanzees, mandrills and African green monkeys accumulated over 30 million years. The same forces were uppermost during all stages of diversification. It is remarkable that the very different proteins, such as gp120 and the gp41 ectodomain (surface glycoproteins), p17 Gag and p24 Gag (structural), RT and integrase (enzymes) and Nef and Vif (cytoplasmic), all yield linear relationships (Figure 6.2), as though fixation was an intrinsic property of the protein.

Applying the same analysis to complete rhinoviral genomes yielded comparable results, i.e. highly significant linear relationships for VP1, VP2 and VP3 (capsid proteins), P2A and 3C (proteases), P2C (cytoplasmic proteins involved in membrane reorganization) compared with the RNA-dependent RNA polymerase (3D) as reference (Figure 6.3, Table 6.1). Hence Figure 6.2 does not represent some quirk of primate lentiviruses. Of course, vertebrate viruses have a redoubtable adversary in the host adaptive immune system. The swiftness of secondary responses is reminder enough.

An analysis of proteins derived from complete potyvirus genomes, positive-stranded RNA viruses, yielded highly significant linear relationships (Table 6.1). A number of revealing points can be made. Firstly, the linear relationships hold out to very large Blosum distances (0.9). Secondly, potyviruses infect a wide variety of different plants, as their florid names betray. Finally, the linear relationships cannot result from adaptive immune pressure because plants are devoid of adaptive immune systems. They only have powerful innate immune responses.

Unfortunately there are insufficient insect RNA viral sequences to allow a comparable study. However, a glance at a few beetle nodavirus capsid sequences (Dasgupta et al., 1984; Dasgupta and Sgro, 1989) shows extensive genetic variation with a majority of synonymous base substitutions, typical of most comparisons of mammalian viral sequences (see below). For the time being there doesn't seem to be anything obviously different about insect

virus sequence variation. Although insects do not mount adaptive immune responses, the breadth and complexity of their innate immune systems is salutary (Brey and Hultmark, 1998). A final example is afforded by the inoviruses, bacteriophages of the fd group, which includes M13. Although DNA viruses fix mutations at a slower rate than RNA viruses, they too show linear relationships among comparisons of their I, II, III and IV proteins (Table 6.1). And of course bacteria are devoid of adaptive immunity as well.

Whether the comparisons were between capsid proteins versus enzymes, or secretory versus cytoplasmic molecules, significant linear relationships were obtained for pairwise comparisons in amino acid variation in all cases. Such proteins are vastly different in their three-dimensional folds and functions. Some are "seen" by humoral immunity, others are not. For the plant viruses and bacteriophages, only innate immunity is operative. It is as though the rate of amino acid sequence accumulation is an intrinsic feature of the protein, reminiscent of the differing slopes for the accumulation of substitutions by alpha-globin and cytochrome c already alluded to. Of course pairwise comparisons of these two proteins from differing organisms would yield a straight line going through the origin in a manner typical of Figures 6.2 and 6.3. Hence it is fairly safe to assume that, for viral proteins too, amino acid substitutions are accumulated smoothly over time. Indeed, this has been shown explicitly for a number of proteins from a varied group of viruses, including the influenza A, coronaviruses, HIV and herpes viruses (Hayashida et al., 1985; Gojobori et al., 1990; Querat et al., 1990; Villaverde et al., 1991; Elena et al., 1992; Sanchez et al., 1992; McGeoch et al., 1995; Yang et al., 1995; Leitner et al., 1997; Plikat et al., 1997).

The above analyses indicate that viral protein diversification is essentially a smooth process, the major parameter being the nature of the protein more than the ecological niche it finds itself in. The simplest hypothesis to explain the smoothness of protein sequence diversification is that the *majority* of fixed amino acid substitutions are neutral, being accumulated at rates intrinsic to each protein. This is not to say that positive selection is inoperative, merely that the majority of *fixed* substitutions are essentially neutral, so much so that it does not strongly distort the data from a linear relationship expected for genetic drift. In other words, neither the impact of different environments nor the ferocity of the adaptive immune response has much to do with fixation of most substitutions. This is important for the one-dimensional man in all of us sequencers who see all mutations and ask questions about genotype and phenotype – usually about genotype.

A short aside is necessary here. It is interesting that in a few areas of RNA virology much has been made of escape from the adaptive immune response, particularly cytotoxic T lymphocytes, so leading to persistence (Nowak and McMichael, 1995; McMichael and Phillips, 1997). However, it is not at all obvious that this is the case (Wain-Hobson, 1996). It must not be forgotten that it is possible to vaccinate against a number of RNA viruses such as measles, polio and yellow fever. Be that as it may, many DNA viruses, intracellular bacteria and parasites persist. In these cases *de novo* genetic variation arising from point mutations is too slow a means to thwart an adaptive immune response. For example, after 1700 generations, under experimental conditions whereby Muller's ratchet was operative, *S. typhimurium* accumulated mutations such that only 1% of the 444 lineages tested had suffered an obvious loss of fitness (Anderson and Hughes, 1996). That this number of generations could be achieved within as little as 45 days gives an idea of the time necessary to generate a mutation affecting fitness. This is more than enough time to make a vigorous immune response. Some inklings of immune system escape for the herpes virus EBV (de Campos-Lima *et al.*, 1993, 1994) came to nought (Burrows *et al.*, 1996; Khanna *et al.*, 1997). When antigenic variation is in evidence among DNA-based microbes, it invariably results from the use of cassettes and multicopy genes rather than point mutations resulting from DNA replication. And of course such complex systems could have only come about by natural selection.

Finally *de novo* genetic variation of an RNA virus has never been suggested or shown to be necessary for the course of an acute infection. For a virus to persist thanks to genetic variation the phenomenon of epitope escape must be strongly in evidence by the time of seroconversion, generally 5–6 weeks. Yet such data are not forthcoming, and not for want of trying. When viruses do play tricks with the immune system it is invariably by way of specific viral gene products that interfere with the mechanics of adaptive and innate immunity (Ploegh, 1998). In the clear cases where genetic variation is exploited by RNA viruses, it is used to overcome barriers to transmission set up by the host population, e.g. herd immunity. The obvious example is influenza A virus antigenic variation in mammals.

K_a/K_s RATIOS

Another way of assessing the contribution of positive selection to sequence variation is to compare the relative proportions of synonymous (K_s) and non-synonymous (K_a) base substitutions per site. A K_a/K_s ratio of less than 1 indicates that purifying selection is uppermost, while a ratio more than 1 is taken as evidence of an excess of positive selection. Comparisons for HIV proteins from different isolates have yielded the same result (Myers and Korber, 1994). Some mileage was made out of the fact that this ratio increased with increasing distance of SIVs with respect to HIV-1, which in turn led to a discussion of SIV pathogenesis (Shpaer and Mullins, 1993). However, this may reflect a lack of adequate correction for multiple hits. This effect is illustrated by a comparison of the set of 72 orthologous proteins encoded by herpes simplex viruses 1 and 2 (HSV-1 and HSV-2; Figure 6.4A). The more divergent the protein sequence, the greater the K_a/K_s ratio. That some proteins fix substitutions faster than others is no surprise. Yet as Figure 6.4B shows, the K_s values change little as they are near to saturation. When K_a is small, $K_s > K_a$. This suggests that reliable interpretation of K_a/K_s ratios is possible only when the degree of nucleic acid divergence is small.

Now this is the realm of viral quasispecies rendered accessible by PCR. HIV studies abound, reflecting both the phenomenal degree

A

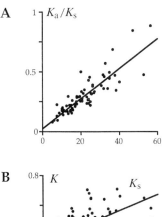

B

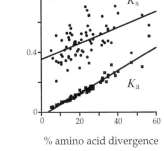

% amino acid divergence

FIGURE 6.4 Comparison of K_a and K_s values for 72 orthologous proteins encoded by herpes simplex viruses types 1 and 2 (HSV-1 and HSV-2; Dolan *et al.*, 1998). **A.** K_a/K_s ratio as a function of uncorrected percentage amino acid sequence divergence (linear regression was K_a/K_s = 1.25 divergence + 0.04, r = 0.87 (p < 0.001)). **B.** Individual K_s and K_a variation with percentage divergence (K_s = 0.53 divergence + 0.35 and K_a = 0.76 divergence – 0.03 with correlation coefficients of 0.54 and 0.97 respectively, p < 0.001 for both). Note how at small degrees of divergence, K_s>>K_a decreases as divergence increases. Basically, K_s is approaching saturation, being uncorrected for multiple and/or back mutations.

onymous substitutions. This is particularly important as the point substitution matrix is highly biased (Pelletier *et al.*, 1995; Plikat *et al.*, 1997). It turns out that when the proportions are so analysed the distributions are rarely significantly different from the neutral hypothesis (Leigh Brown, 1997). Secondly, the method for counting substitutions is highly variable, ranging from two-by-two comparisons, scoring the number of altered sites in a data set, to phylogenetic reconstruction. This latter method reflects more closely the process of genetic diversification. When so analysed, almost all of the data sets indicated proportions of synonymous to non-synonymous substitutions indistinguishable from that suggested by genetic drift and/or purifying selection (Pelletier *et al.*, 1995; Plikat *et al.*, 1997). Thirdly, prudence is called for. The fact that obviously defective sequences can be identified, occasionally accounting for large fractions of the sample (Martins *et al.*, 1991; Gao *et al.*, 1992), indicates that not all genomes have undergone the rigours of selection (Nietfield *et al.*, 1995). Indeed, in peripheral blood, HIV is invariably lurking as a silent provirus within a resting memory T-cell. Such T-cells have half lives of 3 months or more (Michie *et al.*, 1992). Hence it would be erroneous to interpret findings based on a single or clustered samples (Price *et al.*, 1997). Only when the above caveats are borne in mind is there any hope of discerning how HIV accumulates mutations. When these issues are attended to, purifying selection is dominant (Pelletier *et al.*, 1995; Leigh Brown, 1997; Plikat *et al.*, 1997).

One must not deny that positive selection is operative, merely that it is hard to pinpoint when looking at full-length sequences. Indeed it is like looking for the proverbial needle in a haystack. In the context of K_a/K_s-type analyses, the two classic cases in the literature are the HLA class I and II molecules and influenza A haemagglutinin (Hughes, 1998; Hughes and Nei, 1988; Ina and Gojobori, 1994). The peptide contact residues of both class I and II molecules have been under tremendous positive selection. Changes in the five antigenic sites on the flu A haemagglutinin help the virus overcome herd immunity set up during previous flu epidemics.

of sequence variation and its importance as a pathogen, so we'll stick to some such examples that are illustrative. Concerning K_a/K_s ratios for HIV gene segments, widely varying conclusions have been published supporting all sides (Meyerhans *et al.*, 1989, 1991; Pelletier *et al.*, 1995; Wolinsky *et al.*, 1996; Leigh Brown, 1997; Price *et al.*, 1998), so much so that three comments are in order.

Firstly, many studies have used small numbers of sequences and substitutions and even regions as small as nonameric HLA class-I-restricted epitopes. In such cases statistical analyses are essential to test the significance of the distribution of synonymous and non-syn-

Consequently, finding $K_a/K_s > 1$ in these regions was, in some ways, a pyrrhic victory because the papers needed experimental data to identify the positively selected segments in the first place.

More recently Endo *et al.* (1996) have screened the sequence data bases for proteins in which $K_a/K_s > 1$. Of 3595 homologous gene groups screened, covering about 20 000 sequences, only 17 groups came up positive, of which two were encoded by RNA viruses – the equine infectious anaemia virus envelope proteins and the reovirus σ1 (outer capsid) proteins. The former case is intriguing as there is no obvious correlation between sequence changes and neutralizing antibodies (Carpenter *et al.*, 1987). The authors noted that, when a comparable K_a/K_s analysis was restricted to small segments, the number of protein groups scoring positive rose to 5% (Endo *et al.*, 1996). Despite the explanatory power of these ratios, the number of identifiable cases of positively selected segments is small indeed. These numbers would probably shrink were phylogenic reconstruction used.

To summarize the section, synonymous changes are invariably more frequent than non-synonymous changes. Positive selection may be operative in the evolution of viral protein sequences. When it is, it apparently exploits only a small fraction of mutants.

DIVERSITY CALCULATIONS – VIROLOGICAL MAYHEM

The two rates touted by evolutionary-minded virologists are the mutation rate and the mutation fixation rate. The first describes the rate of genesis of mutations, the second attempts to describe their fixation within the population sampled over a period of time. In the case where all substitutions are neutral, the mutation rate (m) equals the fixation rate (f) per round of replication. It appears that such a situation applies to the evolution of parts of the SIV and HIV-1 genomes over 1–3 years (Pelletier *et al.*, 1995; Plikat *et al.*, 1997). If fixation rates are measured over one year, then $f = n \cdot m$, where n is the annual number of consecutive rounds of replication. It is simple to show that several hundred rounds of sequential replication are required (Wain-Hobson, 1993b; Pelletier *et al.*, 1995). Given that the proviral load of an HIV-1-positive patient ($\approx 10^7-10^9$) changes by less than a factor of 10 over 5 years or more, and given the assumption that an infected cell produces sufficient virus to generate two productively infected cells, then annual production would be something akin to 2^{200}, or 10^{60}, which is impossible. Clearly even a productive burst size of 2 is too large (Wain-Hobson, 1993a,b). This must be reduced to 1.1 to achieve a realistic proviral load ($1.1^{200} \approx 10^8$). Note that the real value for the effective burst size must be even lower, as proviral load is turning over more slowly than once a day. Yet to explain the temporal increase in proviral load, the productive burst size must be 2 or more. Thus the calculation reveals massive destruction of infected cells, precisely what was to be expected from immensely powerful innate and adaptive immune responses.

When purifying selection is in evidence, some additional factor must be introduced to couple the fixation and mutation rates. As the accumulation of most substitutions proceeds in a protein-specific linear manner for small degrees of divergence, the above equation can be modified to $f = P \cdot n \cdot m$, where $1 > P > 0$ is a constant indicating the degree of negative selection. Note immediately that, as $P < 1$, more rounds of replication are needed to produce the same percentage amino acid fixation. A corollary is an even greater degree of destruction of infected cells. Consider the example of a virus that is fixing substitutions only slowly, about 10^{-5} per site per year, something like the Ebola virus glycoprotein. The mutation rate for Ebola is not known but is probably around 10^{-4} per site per cycle (Drake, 1993). Hence $P \cdot n \approx 10^{-1}$. What is the value of n? Most mammalian viruses replicate within 24 h, while obviously outside of a body they do not replicate. Consequently a value of $n = 50-200$ is probably not unreasonable. Accordingly $P \approx 2 \times 10^{-3}$ to 5×10^{-4}. This means that most mutations generated are deleterious. Of those that are fixed, most are neutral, as has been discussed above.

VIRAL CONSERVATISM

The last two sentences describe a profoundly conservative strategy – RNA viruses are seen merely to replicate far more than giving rise to genetically distinct, even exotic, siblings. What a stultifying picture, in contrast to the shock-horror of tabloid newspaper virology and that atmospheric, yet profoundly ambiguous term, emerging viruses.

Conservative perhaps, but is there any suggestion that viruses are more or less so than other replicons? Like extrapolation, choosing examples can be problematic. However let's consider one example, the eukaryotic and retroviral aspartic proteases (Doolittle *et al.*, 1989). The former exist as a monomer with two homologous domains, while the retroviral counterpart functions as a homodimer. Despite these differences the folding patterns are almost identical, meaning that the enzymes may be considered orthologous. Between humans and chickens there is approximately 38% amino acid divergence among typical aspartic proteases (Figure 6.5). The HIV-1 and HIV-2 proteases differ by a little more, 52%. No one would doubt the considerable differences in design, metabolism and lifestyle separating us and chickens. On either side of the HIV protease coding region one finds differences: HIV-1 is *vpx⁻vpu⁺* while HIV-2 is the opposite, i.e. *vpx⁺vpu⁻*; there are differences in the size and activities of the *tat* gene product; the LTRs are subtly different. Yet both replicate in the same cells *in vivo*, produce the same disease, albeit with different kinetics: HIV-2 infection progresses more slowly. If these differences are esteemed too substantial, consider the 28% divergence between the HIV and chimpanzee SIV proteases. These two viruses are isogenic. Pig and human chromosomal aspartic proteases may differ by around 17%, the differences between these two species being, George Orwell apart, obvious to all. Even by this crude example, the AIDS viruses would seem to be more conservative than mammals in their evolution. The same argument pertains to the rhinoviral P2A and 3C serine proteases (Figure 6.5).

This conclusion is even more surprising when it is realized that HIV is fixing mutations at a rate of 10^{-2}–10^{-3} per base per year. By contrast,

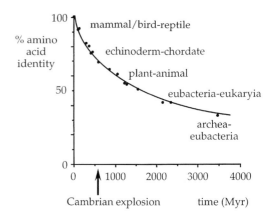

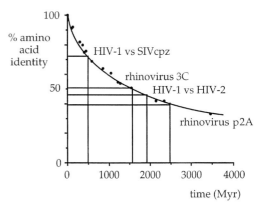

FIGURE 6.5 Mean percentage sequence divergence for an ensemble of 64 sets of proteins over geological time. The data include the trypsin family of serine proteases. Divergence for aspartic proteases, although not included, follows closely the means shown. The time of major divisions in evolution are noted, including the Cambrian explosion. The percent sequence divergence for pairs of viral proteases are projected on to the geological time scale. Adapted from Feng *et al.*, 1997.

mammals are fixing mutations approximately one million times less rapidly, i.e. approximately 10^{-8}–10^{-9} per base per year (Gojobori and Yokoyama, 1987). However, the generation times of the two are vastly different, about 1 day for HIV and about 15–25 years for humans. Normalizing for this yields a 100-fold higher fixation rate per generation for HIV than for humans. Amalgamating this with the preceding paragraph, we see that HIV is not only evolving qualitatively in a conservative manner, but it is doing so despite a 100-fold greater propensity to accommodate change. The same arguments go for almost all RNA viruses and retroviruses. Why is this? Although they mutate rapidly, their hosts are effectively invariant in an evolutionary sense. Probably sticking to the niche is all that matters, which is no mean task given the strength of innate and adaptive antiviral immune responses.

MANY WAYS TO BE A VIRUS

John Maynard Smith's argument was simply put. For organisms with a base substitution rate of less than 1 per genome per cycle, he reasoned that all intermediates linking any two sequences must be viable, otherwise the lineage would go extinct. The example used was self explanatory: WORD → WORE→ GORE → GONE → GENE (Maynard Smith, 1970). The same is true for viruses, even though their mutation rates are 6 orders higher; the rate for a given protein is still less than 1 substitution per cycle. Even for rather stable viruses like Ebola/Marburg and human T-cell leukaemia virus type 1 and 2 (HTLV-1/-2), the number of intermediates is huge. While the enormity of sequence space is basically impossible to comprehend, the amount accessible to a virus remains vast. For the lineage to exist, the probability of finding a viable mutant must be at least 1/population size within the host.

Imagine a stem-loop structure. Any replacement of a G:C base pair must proceed by a single substitution, given that the probability of a double mutation is approximately 10^{-4} that of a single mutation. Let substitution of a G:C pair pass by a G:U intermediate, finishing up as A:U. Although G:U mismatches are the most stable of all mismatches, they are less so than either a G:C or an A:U pair. There are two scenarios: either the G:U substitution is of so little consequence that it is fixed *per se*, in which case there would be no selection pressure to complete the process to A:U. Alternatively, the G:U substitution is sufficiently deleterious for selection of a secondary mutation to occur from a pool of variants, so completing the process. Yet the G:U intermediate cannot be so debilitating otherwise the process would have little chance of going to

completion. Note also that if the fitness difference is small with respect to the G:C or A:U forms, more rounds of replication are necessary to achieve fixation of G:U to A:U. A corollary is that there must be a range within which fitness variation is tolerated. This is reminiscent of nearly neutral theories of evolution and their extension to RNA viruses (Chao, 1997; Ohta, 1997). Note also that from a theoretical perspective the same secondary structure can be found in all parts of sequence space with easy connectivity (Schuster, 1995; Schuster *et al.*, 1997).

Figure 6.6 shows a number of variations on an HIV stem-loop structure, crucial for ribosomal frameshifting between the *gag* and *pol* open reading frames. There have been substitutions at positions 1, 2, 5, 8 and 11 and even an opening up of the loop. All come from viable strains, yet the environment in which these structures are operative, the human ribosome, is invariant. If the changes are all neutral the situation is formally comparable to the steady accumulation of amino acid substitutions. However, if the intermediates are less fit, it has to be understood how they can survive long enough in the face of a plethora of competitors, approximately 1/mutation rate or about 10 000 for HIV. The latter is probably the case as there are HIV-1 genomes with C:G to U:A substitutions at positions 5 and 8 (Figure 6.6). Extensions of nearly neutral theory would fit these findings well (Chao, 1997; Ohta, 1997). That there are many solutions to this stem-loop problem is clear. If HIV-2 is brought into the picture, the remarkable plurality of solutions is further emphasized (Figure 6.6).

Degeneracy in solutions found by viruses is revealed by some interesting experiments on viral revertants. The initial lesions substantially inactivated the virus. Yet with a bit of patience, sometimes more than 6 months, replication-competent variants that were not back mutations were identified (Klaver and Berkhout, 1994; Olsthoorn *et al.*, 1994; Berkhout *et al.*, 1997; Willey *et al.*, 1988; Escarmis *et al.*, 1999). As the frequencies of mutation and back mutation are not equivalent, such findings are, perhaps, not surprising. What they show is the range of possible solutions adjacent to that created by the experimentalist. Loss of fitness can be achieved

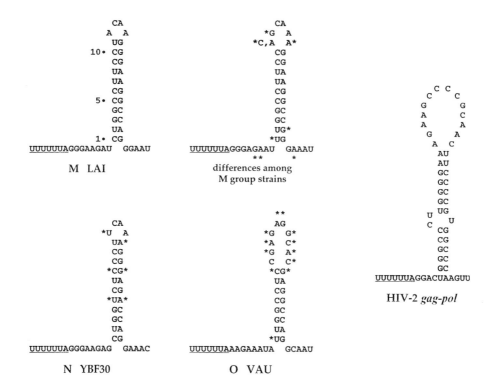

FIGURE 6.6 "Shifty" RNA stem-loop structures from HIV-1 M, N and O group strains as well as from HIV-2 Rod. This structure is part of the information that instructs the ribosome to shift from the *gag* open reading frame to that of *pol*. In addition to the hairpin is a heptameric sequence (underlined). Frameshifting occur within the *gag* UUA codon within the heptamer and continues AGG.GAA etc.* highlights differences in nucleotide sequences compared with the M reference strain LAI.

by sequential plaquing of RNA viruses, the so-called Müller's ratchet experiment, which has been analysed at the genetic level for FMDV (Escarmis *et al.*, 1996). Different lesions characterized different lineages. Recent work was aimed at characterizing the molecular basis of fitness recovery following large population passage. Not one solution was found but a variety, even in parallel experiments (Escarmis *et al.*, 1999). This reveals the impact of chance in fitness selection on a finite population of variants, which is trivially small given the immensity of sequence space.

Another example of degeneracy in viable solutions is the isolation of functional ribozymes from randomly synthesized RNA

(Bartel and Szostak, 1993; Ekland *et al.*, 1995). From a pool of approximately 10^{14} variants, through repeated rounds of positive selection, it was estimated that the frequency of the ribozyme was of the order of 10^{-8}, which is small indeed. Yet $10^{-8} \cdot 10^{14} \approx 10^6$. Even erring by four orders of magnitude, 100 distinct ribozymes could well have been present in the initial pool. Although the sequence space occupied may well represent a tiny proportion of that possible for a RNA molecule of length n, the space is so large that the number of viable solutions is large, large enough to permit a plethora of parallel solutions to the same problem. These experiments, ribozyme from dust, are cases in plurality.

Further evidence of the large proportion of viable solutions in protein sequence space comes from *in vitro* mutagenesis. For example bacteriophage T4 lysozyme can absorb large numbers of substitutions (Rennell *et al.*, 1991) with very few sites resisting replacement (Figure 6.7). Other examples include the lymphokine, interleukin 3, in which some forms with enhanced characteristics were noted (Olins *et al.*, 1995; Klein *et al.*, 1997). With modern mutagenic methods allowing mutation rates of 0.1 per base per site or less, hypermutants of the *E. coli* R67 dihydrofolate reductase (DHFR) were found by random sequencing of as little as 30 clones (Martinez *et al.*, 1996). Whatever the mutation bias, mutants with 3–5 amino acid replacements within the 78-residue protein could be attained (Figure 6.8). Other mutagenesis studies sought enzymes with enhanced catalytic constants or chemical stability. For subtilisin E variants with enhanced features for two parameters could be identified from a relatively small population of randomly mutagenized molecules (Kucher and Arnold, 1997). These data indicate that functional sequence space is probably far more dense than hitherto thought.

Most of the above examples concern maintenance or enhancement of function. An interesting example was recently afforded by engineering cyclophilin into a proline-specific endopeptidase (Quéméneur *et al.*, 1998). The proline binding pocket of cyclophilin was modified such that a single amino acid change (A91S) generated a novel serine endopeptidase with a 10^{10} mol/l proficiency with respect to cyclophilin. Addition of two further substitutions (F104H and N106D) generated a serine–aspartic-acid–histidine catalytic triad, the hallmark of serine proteases. The final enzyme proficiency was 3.5×10^{11} mol/l, typical of many natural enzymes. This shows the interconnectedness of sequence spaces for two functionally very different proteins. If sequence space were sparsely populated, the probability of observing such phenomena would be small.

Many viruses recombine, and via molecular biology more can be made, some of which are tremendously useful research tools, such as the SHIVs, chimeras between SIV and HIV (Figure 6.9). Although many groups have tried to recombine naturally HIV-1 with HIV-2 or SIV, none has succeeded. Natural and artificial recombination represent major jumps in sequence space. That one can observe such genomes means that the new site in functional sequence space must be only a few mutations

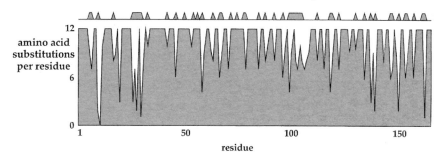

FIGURE 6.7 Systematic amino acid replacement of bacteriophage T4 lysozyme residues. Amber stop codons were engineered singly into each residue apart from the initiator methionine. The plasmids were used to transform 13 suppressor strains. Of the resulting 2015 single amino acid substitutions, 328 were found to be sufficiently deleterious to inhibit plaque formation. More than half (55%) of the positions in the protein tolerated all substitutions examined. The side chains of residues that were refractory to substitution were generally inaccessible to solvent. The catalytic residues are Glu11 and Asp20. Adapted from Rennell *et al.*, 1991.

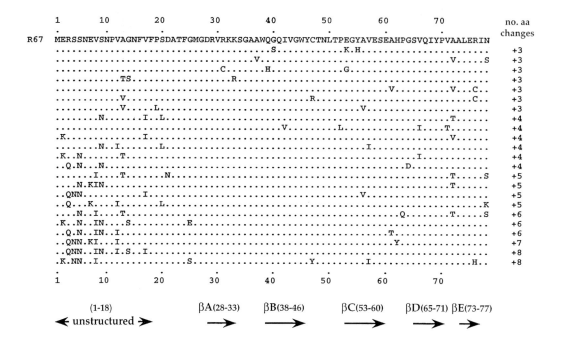

FIGURE 6.8 A modest collection of hypermutated dihydrofolate reductase protein sequences encoded by the *E. coli* R67 plasmid. All were trimethoprim resistant. Only differences with respect to the parent sequence are shown. A representation of the three-dimensional structure is shown above. Adapted from Martinez *et al.*, 1996, with permission.

from a reasonably viable solution, otherwise it would take too long to generate large numbers of cycles and, along with them, mutants. The ferocity of innate and adaptive immunity must never be forgotten.

Off on an apparent tangent, the phylogeny of Geoffrey Chaucer's *The Canterbury Tales* was recently analysed by programs tried and tested for nucleic acid sequences. The authors used 850 lines from 58 fifteenth-century manuscripts (Barbrook *et al.*, 1998). Apart from the fact that it appears that Chaucer did not leave a final version but some annotated working copy, the radiation in medieval English space is fascinating. All the versions are viable and "phenotypically" equivalent even though the "genotypes" are not so. It is ironic that William Caxton's first printed edition was far removed from the original. (N.B., printers merely make fewer errors than

scribes, tantamount to adding a 3′ exonuclease domain to an RNA polymerase).

Given the inevitability of mutation, is it possible that over the aeons natural selection has selected for proteins that are robust, those that are capable of absorbing endless substitutions? For if amino acid substitutions were very difficult to fix, huge populations would need to be explored before change could be accommodated. Recently the unstructured N-terminal segment of the *E. coli* R67 DHFR was shown to stabilize amino acid substitutions in a non-functional miniprotein devoid of this segment (Figure 6.10; Martinez *et al.*, 1996). While the mechanism by which this occurs is unknown, it suggests that there may be parts of proteins, even multiple or discontinuous segments, that may help the protein accommodate inevitable change. Formally it can be seen that such

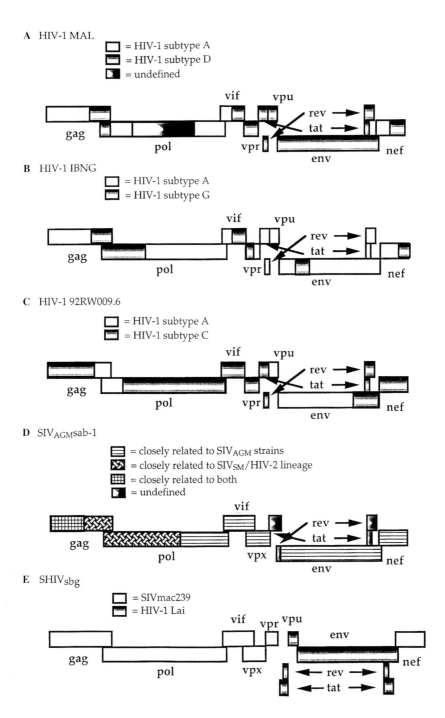

FIGURE 6.9 Genetic organization of naturally occurring HIV-1 and SIV recombinants and unnatural, genetically engineered, SIV-HIV-1 chimeras called SHIVs. Segments are hatched according to stain origin. References are HIV-1 Mal and HIV-1 IBNG (Gao *et al.*, 1996), HIV-1 92RW009.6 (Gao *et al.*, 1998), SIVagm sab-1 (Jin *et al.*, 1994) and SHIVsbg (Dunn *et al.*, 1996).

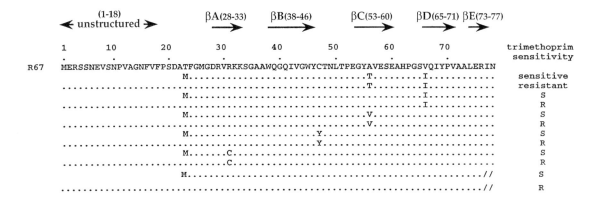

FIGURE 6.10 Recovery of R67 DHFR trimethoprim resistance by recombination of the N terminal 23 residues with otherwise drug-sensitive variants. The two slashes at the end of the last two variants indicate deletion of the COOH terminal dipeptide. Reproduced from Martinez *et al.*, 1996, with permission.

proteins would have both short- and long-term selective advantages, for they would permit the generation of larger populations of relatively viable variants as well as buffering the lineage against the effects of bottlenecking.

A HANDFUL OF RESIDUES

What fraction of amino acid residues is necessary for function? Answer – very few. A few examples taken from among the primate immunodeficiency viruses are typical. Almost all these viruses infect the same target cell using the membrane proteins CD4 and CCR5. Primary HIV-1 isolates use the chemokine receptor CCR5 and rarely the homologous molecule CXCR4, which differs by 81% in its extracellular domains. Yet two substitutions in the viral envelope protein gp120 are sufficient to allow use of the CXCR4 molecule (Hwang *et al.*, 1991). Curiously, the CCR5 chemokine receptor homologue, US28, encoded by human cytomegalovirus, can be used by HIV-1 despite the fact that US28 and CCR5 differ by 88% in the same extracellular regions (Pleskoff *et al.*, 1997). Clearly only a small set of residues are necessary for docking (Brelot *et al.*, 1997).

Another example is afforded by the Vpu protein, which is unique to HIV-1 and the chim-

panzee virus SIVcpz (Huet *et al.*, 1990). Vpu is a small protein inserted into the endoplasmic reticulum, tucked well away from humoral immunity. Despite an average amino acid sequence difference of 0.5% among orthologous human and chimpanzee proteins, HIV/SIVcpz Vpu divergence is almost beyond reliable sequence alignment (Figure 6.11): an N-terminal hydrophobic membrane anchor and a couple of perfectly conserved serine residues, which are phosphorylated, and that's about it. Among HIV-1 strains, or between SIVcpz sequences, the situation was a little better. Yet the necessity of keeping Vpu is beyond doubt. A fine final example concerns the HIV/SIV Rev proteins. These small nuclear proteins are crucial to viral replication. Despite this, only 5 residues are perfectly conserved. The situation has been taken beyond the limit, at least *ex vivo*, in that the HTLV-1 Rex protein can functionally complement for HIV-1 Rev (Rimsky *et al.*, 1989), despite the fact that they are completely different proteins.

The above is reminiscent of what is known about enzymes and surface recognition. Provided the protein fold is maintained, only a small fraction of residues actually contribute to function, a point made recently in two reviews on RNA viral proteases (Ryan and Flint, 1997; Ryan *et al.*, 1998). Insertions and deletions are generally less than 2–3 residues in length and

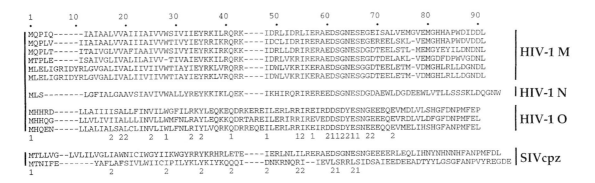

FIGURE 6.11 A selection of Vpu protein sequences from HIV-1 M, N and O group viruses as well as the two SIV chimpanzee (cpz) strains. Hyphens were used to maximize sequence alignments. Numbers 1 and 2 indicate the number of different amino acids at that position for the HIV-1 separately and HIV-1 and SIV cpz sequences together.

confined to turns, loops and coils (Pascarella and Argos, 1992). If globular proteins or at least domains are, to a first approximation, taken as spheres, then the surface area is the least for any volume. If amino acids are equally viewed as smaller, closely packed spheres, then a minimum number will be exposed on the surface, ready to partake in recognition and function.

ARE VIRUSES OPTIMIZED?

The molecular biologist frequently thinks like an engineer who can redesign from scratch. Yet replicons have been constrained by a series of historical events representing variations on a founding theme. While they are fit enough to survive, are they the best possible? This question is salutary, for we live in a society that is more and more competitive and, thanks to global communications, knows about the most successful athletes or businessmen worldwide. Yet who can remember the name of any Olympic athlete who came in fourth? Is not fourth best in any large population remarkable?

How good are viruses as machines? Once again let us look at some examples from HIV-1. Reverse transcription feeds on cytoplasmic dNTPs. Yet supplementing the culture milieu with deoxycytidine – which is scavenged and phosphorylated to the triphosphate – substan-

tially increased viral replication (Meyerhans et al., 1994). It is known that good expression of a foreign protein is frequently compromised by inappropriate codon usage. By redesigning codon usage of the jellyfish (Aequorea victoria) green fluorescent protein gene to correspond to that typical of mammalian genes, greatly improved expression was achieved in mammalian cells (Haas et al., 1996). The same group engineered codon usage of the HIV-1 gp120 glycoprotein gene segment to correspond to that of the abundantly expressed human Thy-1 surface antigen. Again expression was greatly improved (Haas et al., 1996). The coup de grâce came with the reciprocal experiment – engineering Thy-1 gene codon usage to correspond to that of gp120. Thy-1 surface expression was greatly reduced (Haas et al., 1996). Since HIV-1 was first sequenced, it has been known that its codon usage is highly biased (Wain-Hobson et al., 1985; Bronson and Anderson, 1994). Something is clearly overriding maximal envelope expression. Furthermore, gp120 codon usage is similar for all other HIV-1 genes whether they be structural or regulatory. For that matter, codon usage is comparable for most lentiviruses (Sonigo et al., 1985; Bronson and Anderson, 1994).

It was possible to show via DNA vaccination that codon-engineered gp120 elicited stronger immune responses in mice than the normal counterpart (Andre et al., 1998). Might this finding suggest that the optimum is actually away from mass production? Yet if there is a shadow of reality in this thesis, it indicates that fitness optima *in vivo* may not necessarily parallel the expectations of fitness based on *ex-vivo* models. In this context note also that HTLV-1 infects exactly the same cell as HIV, yet its codon usage is very different from that of HIV and the Thy-1 gene (Seiki *et al.*, 1983).

If fitness optimization were ever operative *in vivo*, then one would predict steady increases in virulence for those viruses that do not set up herd immunity. At some point a plateau would be reached. Yet the higgledy-piggledy way by which virulent strains come and go suggests that this is not so. Some might use the word stochastic. Whatever. If fitness selection can be overridden and we don't have a good theory for it, then we're in a sorry state.

THE (DIFFICULT) TRANSITION FROM *EX VIVO* TO *IN VIVO*

There is abundant evidence that, as a good first approximation, RNA viruses *ex vivo* perform as expected from the quasi-species model (Holland *et al.*, 1982; Eigen and Biebricher, 1988; Duarte *et al.*, 1992; Clarke *et al.*, 1993; Eigen, 1993; Novella *et al.*, 1995; Domingo *et al.*, 1996; Quer *et al.*, 1996; Domingo and Holland, 1997), which is fitness dominated. Problems arise transposing it to the *in vivo* situation, notably:

- First and foremost: how does one determine fitness *in vivo*? Should such measurements score intrahost viral titres or transmission probabilities from an index case? (If a virus doesn't spread it's dead.) For outbred populations, is it in fact virulence?
- Second: host innate immunity is hugely powerful, a fact leading Rolf Zinkernagel to remark with typical aplomb that in terms of immunity "an inferferon receptor knock-out mouse is a 1% mouse" (Huang *et al.*, 1993; van den Broek *et al.*, 1995a,b). Yet the enhanced susceptibility of SCID humans or various knock-out mice to infections indicates the part played by adaptive immunity.

For example, influenza A can persist in SCID children (Rocha *et al.*, 1991). How are innate and adaptive immune responses coupled and how are they influenced by genetic polymorphisms?
- Third: with acquired immunity rising by day 3 in an acute infection, the virus is replicating in the face of a predator whose amplitude is increasing.
- Fourth: immune responses are density-dependent. That is, the more the virus replicates the stronger the immune response. If the relationship were simply linear one could see how a virus might be able to keep just ahead, given a short lag in the immune response time. But if it were non-linear? Indeed it must be so, otherwise it would not be possible to resolve an acute infection. It is not easy to discern where optimal viral fitness would lie.
- Fifth: The wrath of combined immune responses is such that there is massive viral turnover. For the three best known cases, HIV, HBV and HCV, between 10^8 and 10^{12} virions are turning over daily, representing between 10% and 90% of the whole (Ho *et al.*, 1995; Wei *et al.*, 1995; Nowak *et al.*, 1996; Zeuzem *et al.*, 1998). Indeed, these are probably underestimates, given beautiful data from the late 1950s and 1960s showing that, for a variety of RNA viruses, plasma titres decay with a half-life of 15–20 min, whether the animal be immunologically naive or primed (Mimms, 1959; Nathanson and Harrington, 1967). From this one may conclude that any viral population is unlikely to be in equilibrium. And if a population is not in equilibrium, fitness selection is compromised.
- Sixth: A glance at any histology slide or textbook is a salient reminder of spatial discontinuities over distances of one or two cell diameters. For example the hugely delocalized immune system is characterized by a multitude of different lymphoid organs, a myriad of subtly different susceptible cell types, and a mêlée of membrane molecules. The exquisite spatial heterogeneity of HIV within the epidermis and splenic white pulps has been described (Cheynier *et al.*, 1994,

1998; Sala *et al.*, 1994). The same seems to be true for HCV-infected liver (Martell *et al.*, 1992). For HPV infiltration of skin, spatial discontinuities and gradients are also apparent. Discontinuities reduce the possibilities for competition and hence selection of the fitter forms. Indeed the Müller's ratchet experiment and clonal heterogeneity are the most vivid expressions of this.

- Seventh: much has been made of privileged sites and viral reservoirs. Basically this is reminding us of the fact that immune surveillance is modulated in some organs like the brain. There are some suggestions that cytotoxic T-cells have difficulty infiltrating the kidney. Viral reservoirs undermine fitness selection.
- Eighth: in the case of the immunodeficiency viruses, antigenic stimulation of infected yet resting memory T-cells means that variants may become amplified for reasons that have nothing to do with the fitness of the variant (Cheynier *et al.*, 1994, 1998).

Given the non-equilibrium structure of viral variants, vastly restricted population sizes in respect to sequence space, founder effects *in vivo* take on great importance. While answers for some of these issues seem far away, constraints on fitness selection cannot be so strong that a chain of infections becomes a Müller's ratchet experiment. Yet is that correct? In the experiments with phage Φ6, VSV and FMDV, most of the lineages resulted in decreased fitness. Yet for some there were no changes, while for a few there were even increases in the fitness vectors (Chao, 1990; Duarte *et al.*, 1992; Escarmis *et al.*, 1996). Could symptomatic infections reflect bottleneck transmission of those fitter clones with asymptomatic (subclinical) infection representing fitness-compromised clones?

Analysis of RNA viruses *ex vivo* is analogous to the study of bacteria in chemostats. Fitness selection dominates. Yet there is a world of difference between bacterial strains so selected and natural isolates. One of the observations frequently made upon isolation of pathogenic bacteria is the loss of bacterial virulence determinants (Miller *et al.*, 1989). Indeed, *ex-vivo* passage of RNA viruses has been used to select for attenuated strains used in vaccination.

SPACE

Mayr again: "Wherever one looks in nature, one finds uniqueness" (Mayr, 1997). As mentioned, the cardinal difference between the behaviour of RNA viruses *ex vivo* and *in vivo* is the existence of spatial discontinuities. For replicons, cloning is the ultimate separation. It allows a variant to break away from dominating competitors, disrupts or uncouples a fitter variant locked in competitive exclusion (de la Torre and Holland, 1990). The effect of bottlenecking on fitness, as well as the Müller's ratchet experiments, have been described (Chao, 1990; Duarte *et al.*, 1992; Novella *et al.*, 1995; Escarmis *et al.*, 1996). Transmission frequently involves massive bottlenecking, and is very much an exercise in cloning. All this should not surprise because allopatric speciation is omnipresent in the origin of species, Darwin's Galapagos finches being an obvious example.

AN ORIGIN OF VIRAL SPECIES?

A virus must replicate sufficiently within a host to permit infection of another susceptible host. If the new host is of the same species, differences between the two are minimal – a small degree of polymorphism being inevitable in outbred populations. Given that viruses with a small coding capacity interact particularly intimately with the host-cell machinery, it follows that infection of a host from a related species has a greater probability of succeeding if the cellular machinery is comparable. Indeed, the closer the two species, the greater the probability. In turn, if the virus gets a toehold and can generate a quasispecies, then only few mutations would probably be necessary to adapt to the new niche.

Yet species is a difficult word. What might a viral species be? Martin (1993) wrote a fascinating review on the number of extinct primate species estimated from the fossil record. Depending on the emergence time of primates of modern aspect, he was able to estimate the total number that existed as 5500–6500. The present number of 200 primates species would thus represent about 3.4–3.8%. More importantly from our viewpoint was his calculation of the average survival time of fossil primate species as a mere 1 million years (Martin, 1993). Given that RNA viruses are fixing mutations approximately 1 million times faster than mammals (Holland *et al.*, 1982; Gojobori and Yokoyama, 1985; Domingo *et al.*, 1996), a viral species would become extinct after approximately 1 year! Immediately the annual influenza A strain comes to mind. Yet rabies, polio and HTLV-1 have arguably been around for millennia. Clearly the word "species", when taken from primatology, cannot apply to the viral world. Frogs provide a more interesting example. They have been around for several hundred millions of years, and members of some lineages can interbreed despite 75 million years separation. Naturally, their protein sequences have not stood still during that time (Wilson *et al.*, 1977). Enough is conserved to allow breeding. Maybe the primate picture has undue weight in our appreciation of virology. Phenotype can be maintained despite changes in genotype – obvious to a biologist.

As usual, Holland wasn't far from the mark when he wrote:

As human populations continue to grow exponentially, the number of ecological niches for human RNA virus evolution grows apace and new human virus outbreaks will likely increase apace. Most new human viruses will be unremarkable – that is they will generally resemble old ones. Inevitably, some will be quite remarkable, and quite undesirable. When discussing RNA virus evolution, to call an outbreak (such as AIDS) remarkable is merely to state that it is of lower probability than an unremarkable outbreak. (Holland *et al.*, 1992)

New viruses can and do emerge but on a scale that is probably 15–20 logs less than the number of viral mutants generated up to that defining moment (Wain-Hobson, 1993). They will result from a small number of mutations and a dose of reproductive isolation.

CONCLUSION

The above has attempted to show that the vast majority of genetic changes fixed by RNA viruses are essentially neutral or nearly neutral in character. Positive selection exploits a small proportion of genetic variants, while functional sequence space is sufficiently dense, allowing viable solutions to be found. Although evolution has connotations of change, what has always counted is natural selection or adaptation. It is the only force for the genesis of a novel replicon. Once adapted to its niche, there is no need to change. In such circumstances an RNA

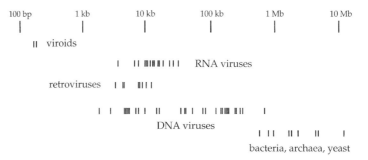

FIGURE 6.12 Latitude in microbial genome sizes. RNA viruses and retroviruses are confined to one log variation in size (3 to ≈ 32 kb). By contrast, DNA viruses span more than 2.5 logs going from the single-stranded porcine circulovirus (1.8 kb) to chlorella virus (≈330 kb, encoding at least 12 DNA endonuclease/methyltransferase genes; Zhang *et al.*, 1998) and bacteriophage G (≈670 kb). The distinction between phage DNA and a plasmid has often proven difficult (Waldor and Mekalanos, 1996). As can be seen, the genome size of the largest DNA viruses overlaps the smallest intracellular bacteria such as mycoplasmas (580 and 816 kb) and is not too far short of autonomous bacteria such as *Haemophilus influenzae* (1.83 Mb).

virus would no longer be adapting, even though it could be changing.

Why is the evolution of RNA viruses so conservative? Why do they mutate rapidly yet remain phenotypically stable? The lack of proofreading proscribes the genesis of large genomes, restricting their genome sizes to a 1 log range (Figure 6.12). Among the smallest RNA and retroviruses are MS2 and hepatitis B virus, both about 3 kb, while the largest are the coronaviruses at 32 kb or more. Most of their proteins are structural or regulatory and take up the largest part of the coding capacity of the virus. Additional proteins broadening the range of interactions with the host cell, or rendering the replicon more autonomous, are relatively few. Large, gene-sized duplications that may contribute to diversification and novel phenotypes are rare, reducing the exploration of new horizons. Thus, evolution of RNA viruses is probably conservative because they cannot shuffle domains so generating new combinations.

That the information capacity of RNA viral genomes is limited by a lack of proofreading is neither here nor there, for they are remarkably successful parasites. RNA viruses change far more than they adapt.

ACKNOWLEDGEMENTS

We would like to thank past and present members of the laboratory and numerous colleagues for endless discussions over the years. Mark Mascolini needs a special word of thanks for painstakingly going through the manuscript. This laboratory is supported by grants from the Institut Pasteur and the Agence Nationale pour la Recherche sur le SIDA.

REFERENCES

Andersson, D.I. and Hughes, D. (1996) Muller's ratchet decreases fitness of a DNA-based microbe. *Proc. Natl Acad. Sci. USA*, **93**, 906–907.

Andre, S., Seed, B., Eberle, J., Schraut, W.,

Bultmann, A. and Haas, J. (1998) Increased immune response elicited by DNA vaccination with a synthetic gp120 sequence with optimized codon usage. *J. Virol.*, **72**, 1497–1503.

Barbrook, A.C., Howe, C.J., Blake, N. and Robinson, P. (1998) The phylogeny of *The Canterbury Tales*. *Nature*, **394**, 839.

Bartel, D.P. and Szostak, J.W. (1993) Isolation of new ribozymes from a large pool of random sequences. *Science*, **261**, 1411–1418.

Berkhout, B., Klaver, B. and Das, A.T. (1997) Forced evolution of a regulatory RNA helix in the HIV-1 genome. *Nucleic Acids Res.*, **25**, 940–947.

Brelot, A., Heverker, N., Pleskoff, O., Sol, N. and Alizon, M. (1997) Role of the first and third extracellular domains of CXCR-4 in human immunodeficiency virus coreceptor activity. *J. Virol.*, **71**, 4744–4751.

Brey, P.T. and Hultmark, D. (1998) *Molecular Mechanisms of Immune Responses in Insects*. Chapman & Hall, London.

Bronson, E.C. and Anderson, J.N. (1994) Nucleotide composition as a driving force in the evolution of retroviruses. *J. Mol. Evol.*, **38**, 506–532.

Burrows, J.M., Burrows, S.R., Poulsen, L.M., Sculley, T.B., Moss, D.J. and Khanna, R. (1996) Unusually high frequency of Epstein–Barr virus genetic variants in Papua New Guinea that can escape cytotoxic T-cell recognition: implications for virus evolution. *J. Virol.*, **70**, 2490–2496.

Carpenter, S., Evans, L.H., Sevoian, M. and Chesebro, B. (1987) Role of host immune response in selection of equine infectious anemia virus variants. *J. Virol.*, **61**, 3783–3789.

Chao, L. (1990) Fitness of RNA virus decreased by Muller's ratchet. *Nature*, **348**, 454–455.

Chao, L. (1997) Evolution of sex and the molecular clock in RNA viruses. *Gene*, **205**, 301–308.

Cheynier, R., Henrichwark, S., Hadida, F. *et al.* (1994) HIV and T-cell expansion in splenic white pulps is accompanied by infiltration of HIV-specific cytotoxic T-lymphocytes. *Cell*, **78**, 373–387.

Cheynier, R., Gratton, S., Halloran, M., Stahmer, I., Letvin, N.L. and Wain-Hobson, S. (1998)

Antigenic stimulation by BCG as an *in vivo* driving force for SIV replication and dissemination. *Nat. Med.*, **4**, 421–427.

Clarke, D.K., Duarte, E.A., Moya, A., Elena, S.F., Domingo, E. and Holland, J. (1993) Genetic bottlenecks and population passages cause profound fitness differences in RNA viruses. *J. Virol.*, **67**, 222–228.

Dasgupta, R. and Sgro, J.Y. (1989) Nucleotide sequences of three Nodavirus RNA2's: the messengers for their coat protein precursors. *Nucleic Acids Res.*, **17**, 7525–7526.

Dasgupta, R., Ghosh, A., Dasmahapatra, B., Guarino, L.A. and Kaesberg, P. (1984) Primary and secondary structure of black beetle virus RNA2, the genomic messenger for BBV coat protein precursor. *Nucleic Acids Res.*, **12**, 7215–7223.

De Campos-Lima, P.O., Gavioli, R., Zhang, Q.J. *et al.* (1993) HLA-A11 epitope loss isolates of Epstein-Barr virus from a highly A11+ population. *Science*, **260**, 98–100.

De Campos-Lima, P.O., Levitsky, V., Brooks, J. *et al.* (1994) T cell responses and virus evolution: loss of HLA A11-restricted CTL epitopes in Epstein–Barr virus isolates from highly A11-positive populations by selective mutation of anchor residues. *J. Exp. Med.*, **179**, 1297–1305.

de la Torre, J.C. and Holland, J.J. (1990) RNA virus quasispecies populations can suppress vastly superior mutant progeny. *J. Virol.*, **64**, 6278–6281.

Dolan, A., Jamieson, F.E., Cunningham, C., Barnett, B.C. and McGeoch, D.J. (1998) The genome sequence of herpes simplex virus type 2. *J. Virol.*, **72**, 2010–2021.

Domingo, E. and Holland, J.J. (1997) RNA viral mutations and fitness for survival. *Annu. Rev. Microbiol.*, **51**, 151–178.

Domingo, E., Escarmis, C., Sevilla, N. *et al.* (1996) Basic concepts in RNA virus evolution. *FASEB J.*, **10**, 859–864.

Doolittle, R.F., Feng, D.F., Johnson, M.S. and McClure, M.A. (1989) Origins and evolutionary relationships of retroviruses. *Q. Rev. Biol.*, **64**, 1–30.

Drake, J.W. (1993) Rates of spontaneous mutations among RNA viruses. *Proc. Natl Acad. Sci. USA*, **90**, 4171–4175.

Duarte, E., Clarke, D., Moya, A., Domingo, E. and Holland, J. (1992) Rapid fitness losses in mammalian RNA virus clones due to Muller's ratchet. *Proc. Natl Acad. Sci. USA*, **89**, 6015–6019.

Dunn, C.S., Beyer, C., Kieny, M.P. *et al.* (1996) High viral load and CD4 lymphopenia in rhesus and cynomolgus macaques infected by a chimeric primate lentivirus constructed using the env, rev, tat, and vpu genes from HIV-1 Lai. *Virology*, **223**, 351–361.

Eigen, M. (1993) The viral quasispecies. *Sci. Am.*, **269**, 42–49.

Eigen, M. and Biebricher, C.K. (1988) Sequence space and quasispecies distribution. In: *RNA Genetics, vol. 3*, (eds Domingo, E., Holland, J.J. and Ahlquist, P.), pp. 211–245. CRC Press, Boca Raton, FL.

Ekland, E.H., Szostak, J.W. and Bartel, D.P. (1995) Structurally complex and highly active RNA ligases derived from random RNA sequences. *Science*, **269**, 364–370.

Elena, S.F., Gonzalez-Candelas, F. and Moya, A. (1992) Does the VP1 gene of foot-and-mouth disease virus behave as a molecular clock? *J. Mol. Evol.*, **35**, 223–229.

Endo, T., Ikeo, K. and Gojobori, T. (1996) Large-scale search for genes on which positive selection may operate. *Mol. Biol. Evol.*, **13**, 685–690.

Escarmis, C., Davila, M., Charpentier, N., Bracho, A., Moya, A. and Domingo, E. (1996) Genetic lesions associated with Muller's ratchet in an RNA virus. *J. Mol. Biol.*, **264**, 255–267.

Escarmis, C., Davila, M. and Domingo, E. (1999) Multiple molecular pathways for fitness recovery of an RNA virus dibilitated by operation of Müller's ratchet. *J. Mol. Biol.*, **285**, 495–505.

Feng, D.-F., Cho, G. and Doolittle, R.F. (1997) Determining divergence times with a protein clock: update and reevaluation. *Proc. Natl Acad. Sci. USA*, **94**, 13028–13033.

Gao, F., Yue, L., White, A.T. *et al.* (1992) Human infection by genetically diverse SIV-sm related HIV-2 in West Africa. *Nature*, **358**, 495–499.

Gao, F., Robertson, D.L., Morrison, S.G. *et al.* (1996) The heterosexual human immunodeficiency virus type 1 epidemic in Thailand is caused by an intersubtype (A/E) recombinant of African origin. *J. Virol.*, **70**, 7013–7029.

Gao, F., Robertson, D.L., Carruthers, C.D. *et al.* (1998) A comprehensive panel of near-full-length clones and reference sequences for non-subtype B isolates of human immunodeficiency virus type 1. *J. Virol.*, **72**, 5680–5698.

Gojobori, T. and Yokoyama, S. (1985) Rates of evolution of the retroviral oncogene of Moloney murine sarcoma virus and of its cellular homologues. *Proc. Natl Acad. Sci. USA*, **82**, 4198–4201.

Gojobori, T. and Yokoyama, S. (1987) Molecular evolutionary rates of oncogenes. *J. Mol. Evol.*, **26**, 148–156.

Gojobori, T., Moriyama, E.N. and Kimura, M. (1990) Molecular clock of viral evolution, and the neutral theory. *Proc. Natl Acad. Sci. USA*, **87**, 10015–10018.

Haas, J., Park, E.C. and Seed, B. (1996) Codon usage limitation in the expression of HIV-1 envelope glycoprotein. *Curr. Biol.*, **6**, 315–324.

Hayashida, H., Toh, H., Kikuno, R. and Miyata, T. (1985) Evolution of influenza virus genes. *Mol. Biol. Evol.*, **2**, 289–303.

Henikoff, S. and Henikoff, J.G. (1993) Performance evaluation of amino acid substitution matrices. *Proteins*, **17**, 49–61.

Ho, D.D., Neumann, A.U., Perelson, A.S., Chen, W., Leonard, J.M. and Markowitz, M. (1995) Rapid turnover of plasma virions and CD4 lymphocytes in HIV-1 infection. *Nature*, **373**, 123–126.

Holland, J., Spindler, K., Horodyski, F., Grabau, E., Nichol, S. and Vande Pol, X. (1982) Rapid evolution of RNA genomes. *Science*, **215**, 1577–1585.

Holland, J.J., de le Torre, J.C. and Steinhauer, D.A. (1992) RNA virus populations as quasispecies. *Curr. Topics Microbiol. Immunol.*, **176**, 1–20.

Huang, S., Hendriks, W., Althage, A. *et al.* (1993) Immune response in mice that lack the interferon-gamma receptor. *Science*, **259**, 1742–1745.

Huet, T., Cheynier, R., Meyerhans, A., Roelants, G. and Wain-Hobson, S. (1990) Genetic organization of a chimpanzee lentivirus related to HIV-1. *Nature*, **345**, 356–359.

Hughes, A.L. (1998) Protein phylogenies provide evidence of a radical discontinuity between arthropod and vertebrate immune systems. *Immunogenet.*, **47**, 283–296.

Hughes, A.L. and Nei, M. (1988) Pattern of nucleotide substitution at major histocompatibility complex class I loci reveals overdominant selection. *Nature*, **335**, 167–170.

Hwang, S.S., Boyle, T.J., Lyerly, H.K. and Cullen, B.R. (1991) Identification of the envelope V3 loop as the primary determinant of cell tropism in HIV-1. *Science*, **253**, 71–74.

Ina, Y. and Gojobori, T. (1994) Statistical analysis of nucleotide sequences of the hemagglutinin gene of human influenza A viruses. *Proc. Natl Acad. Sci. USA*, **91**, 8388–8392.

Jin, M.J., Hui, H., Robertson, D.L. *et al.* (1994) Mosaic genome structure of simian immunodeficiency virus from West African green monkeys. *EMBO J.*, **13**, 2935–2947.

Khanna, R., Burrows, S.R. and Burrows, J.M. (1997) The role of cytotoxic T-lymphocytes in the evolution of genetically stable viruses. *Trends Microbiol.*, **5**, 64–69.

Klaver, B. and Berkhout, B. (1994) Evolution of a disrupted TAR RNA hairpin structure in the HIV-1 virus. *EMBO J.*, **13**, 2650–2659.

Klein, B.K., Feng, Y., McWherter, C.A., Hood, W.F., Paik, K. and McKearn, I.P. (1997) The receptor binding site of human interleukin-3 defined by mutagenesis and molecular modeling. *J. Biol. Chem.*, **272**, 22630–22641.

Kucher, O. and Arnold, F.H. (1997) Directed evolution of enzyme catalysts. *Trends Biotechnol.*, **15**, 523–530.

LANL (1999) http://hiv-web.lanl.gov.

Leigh Brown, A.J. (1997) Analysis of HIV-1 *env* gene sequences reveals evidence for a low effective number in the viral population. *Proc. Natl Acad. Sci. USA*, **94**, 1862–1865.

Leitner, T., Kumar, S. and Albert, J. (1997) Tempo and mode of nucleotide substitutions in *gag* and *env* gene fragments in human immunodeficiency virus type 1 populations with a known transmission history. *J. Virol.*, **71**, 4761–4770.

McGeoch, D.J., Cook, S., Dolan, A., Jamieson, F.E. and Telford, E.A. (1995) Molecular phylogeny and evolutionary timescale for the family of mammalian herpesviruses. *J. Mol. Biol.*, **247**, 443–458.

McMichael, A.J. and Phillips, R.E. (1997) Escape of human immunodeficiency virus from immune control. *Annu. Rev. Immunol.*, **15**, 271–296.

Martell, M., Esteban, J.I., Quer, J. *et al.* (1992) Hepatitis C virus (HCV) circulates as a population of different but closely related genomes: quasispecies nature of HCV genome distribution. *J. Virol.*, **66**, 3225–3229.

Martin, R.D. (1993) Primate origins: plugging the gaps. *Nature*, **363**, 223–234.

Martinez, M.A., Pezo, V., Marliere, P. and Wain-Hobson, S. (1996) Exploring the functional robustness of an enzyme by in vitro evolution. *EMBO J.*, **15**, 1203–1210.

Martins, L.P., Chenciner, N., Asjo, B., Meyerhans, A. and Wain-Hobson, S. (1991) Independent fluctuation of human immunodeficiency virus type 1 rev and gp41 quasispecies *in vivo*. *J. Virol.*, **65**, 4502–4507.

Maynard Smith, J. (1970) Natural selection and the concept of a protein space. *Nature*, **225**, 563–564.

Mayr, E. (1997) *This is Biology*. Belknap Press, Cambridge, MA.

Meyerhans, A., Cheynier, R., Albert, J. *et al.* (1989) Temporal fluctuations in HIV quasispecies in vivo are not reflected by sequential HIV isolations. *Cell*, **58**, 901–910.

Meyerhans, A., Dadaglio, G., Vartanian, J.P. *et al.* (1991) *In vivo* persistence of a HIV-1-encoded HLA-B27-restricted cytotoxic T-lymphocyte epitope despite specific *in vitro* reactivity. *Eur. J. Immunol.*, **21**, 2637–2640.

Meyerhans, A., Vartanian, J.P., Hultgren, C. *et al.* (1994) Restriction and enhancement of human immunodeficiency virus type 1 replication by modulation of intracellular deoxynucleoside triphosphate pools. *J. Virol.*, **68**, 535–540.

Michie, C.A., McLean, A., Alcock, C. and Beverley, P.C. (1992) Lifespan of human lymphocyte subsets defined by CD45 isoforms. *Nature*, **360**, 264–265.

Miller, J.F., Mekalanos, J.J. and Falkow, S. (1989) Coordinate regulation and sensory transduction in the control of bacterial virulence. *Science*, **243**, 916–922.

Mimms, C.A. (1959) The response of mice to large intravenous injections of ectromelia virus. I. The fate of injected virus. *J. Exp. Path.*, **40**, 533–542.

Myers, G. and Korber, B. (1994) The future of human immunodeficiency virus. In: *The Evolutionary Biology of Viruses*, (ed. Morse, S.S.), pp. 211–232. Raven Press, New York.

Nathanson, N. and Harrington, B. (1967) Experimental infection of monkeys with Langat virus. *Am. J. Epidemiol.*, **85**, 494–502.

Nietfield, W., Bauer, M., Fevrier, M. *et al.* (1995) Sequence constraints and recognition by CTL of an HLA-B27-restricted HIV-1 gag epitope. *J. Immunol.*, **154**, 2189–2197.

Novella, I.S., Elena, S.F., Moya, A., Domingo, E. and Holland, J.J. (1995) Size of genetic bottlenecks leading to virus fitness loss is determined by mean initial population fitness. *J. Virol.*, **69**, 2869–2872.

Nowak, M.A. and McMichael, A.J. (1995) How HIV defeats the immune system. *Sci. Am.*, **273**, 58–65.

Nowak, M.A., Bonhoeffer, S., Hill, A.M., Boehme, R., Thomas, H.C. and McDade, H. (1996) Viral dynamics in hepatitis B virus infection. *Proc. Natl Acad. Sci. USA*, **93**, 4398–4402.

Ohta, T. (1997) The meaning of near-neutrality at coding and non-coding regions. *Gene*, **205**, 261–267.

Olins, P.O., Bauer, S.C., Braford-Goldberg, S. *et al.* (1995) Saturation mutagenesis of human interleukin-3. *J. Biol. Chem.*, **270**, 23754–23760.

Olsthoorn, R.C., Licis, N. and van Duin, J. (1994) Leeway and constraints in the forced evolution of a regulatory RNA helix. *EMBO J.*, **13**, 2660–2668.

Pascarella, S. and Argos, P. (1992) Analysis of insertions/deletions in protein structures. *J. Mol. Biol.*, **20**, 461–471.

Pelletier, E., Saurin, W., Cheynier, R., Letvin, N.L. and Wain-Hobson, S. (1995) The tempo and mode of SIV quasispecies development in vivo calls for massive viral replication and clearance. *Virology*, **208**, 644–652.

Pleskoff, O., Treboute, C., Brelot, A., Heveker, N., Seman, M. and Alizon, M. (1997) Identification of a chemokine receptor encoded by human cytomegalovirus as a cofactor for HIV-1 entry. *Science*, **276**, 1874–1878.

Plikat, U., Nieselt-Struwe, K. and Meyerhans, A.

(1997) Genetic drift can dominate short-term human immunodeficiency virus type 1 nef quasispecies evolution in vivo. *J. Virol.*, **71**, 4233–4240.

Ploegh, H.L. (1998) Viral strategies of immune evasion. *Science*, **280**, 248–253.

Price, D.A., Goulder, P.J., Klenerman, P. *et al.* (1997) Positive selection of HIV-1 cytotoxic T lymphocyte escape variants during primary infection. *Proc. Natl Acad. Sci. USA*, **94**, 1890–1895.

Price, D.A., Sewell, A.K., Dong, T. *et al.* (1998) Antigen-specific release of beta-chemokines by anti-HIV-1 cytotoxic T lymphocytes. *Curr. Biol.*, **8**, 355–358.

Quéméneur, E., Moutiez, M., Charbonnier, J.-B. and Ménez, A. (1998) Engineering cyclophilin into a proline-specific endopeptidase. *Nature*, **391**, 301–304.

Quer, J., Huerta, R., Novella, I.S., Tsimring, L., Domingo, E. and Holland, J.J. (1996) Reproducible nonlinear population dynamics and critical points during replicative competitions of RNA virus quasispecies. *J. Mol. Biol.*, **264**, 465–471.

Querat, G., Audoly, G., Sonigo, P. and Vigne, R. (1990) Nucleotide sequence analysis of SA-OMVV, a visna-related ovine lentivirus: phylogenetic history of lentiviruses. *Virology*, **175**, 434–447.

Rennell, D., Bouvier, S.E., Hardy, L.W. and Poteete, A.R. (1991) Systematic mutation of bacteriophage T4 lysozyme. *J. Mol. Biol.*, **222**, 67–88.

Rimsky, L., Dodon, M.D., Dixon, E.P. and Greene, W.C. (1989) Trans-dominant inactivation of HTLV-I and HIV-1 gene expression by mutation of the HTLV-I Rex transactivator. *Nature*, **341**, 453–456.

Rocha, E., Cox, N., Black, R.A., Harmon, M.W., Harisson, C.J. and Kendall, A. (1991) Antigenic and genetic variation in influenza A (H1N1) virus isolates recovered from a persistently infected immunodeficient child. *J. Virol.*, **65**, 2340–2350.

Ryan, M.D. and Flint, M. (1997) Virus-encoded proteinases of the picornavirus super-group. *J. Gen. Virol.*, **78**, 699–723.

Ryan, M.D., Monaghan, S. and Flint, M. (1998) Virus-encoded proteinases of the Flaviviridae. *J. Gen. Virol.*, **79**, 947–959.

Sala, M., Zambruno, G., Varianian, J.P., Marconi, A., Bertazzoni, U. and Wain-Hobson, S. (1994) Spatial discontinuities in human immunodeficiency virus type 1 quasispecies derived from epidermal Langerhans cells of a patient with AIDS and evidence for double infection. *J. Virol.*, **68**, 5280–5283.

Sanchez, C.M., Gebauer, F., Sune, C., Mendez, A., Dopazo, J. and Enjuanes, L. (1992) Genetic evolution and tropism of transmissible gastroenteritis coronaviruses. *Virology*, **190**, 92–105.

Schuster, P. (1995) How to search for RNA structures. Theoretical concepts in evolutionary biotechnology. *J. Biotechnol.*, **41**, 239–257.

Schuster, P., Stadler, P.F. and Renner, A. (1997) RNA structures and folding. From conventional to new issues in structure predictions. *Curr. Opin. Struct. Biol.*, **7**, 229–235.

Seibert, S.A., Howell, C.Y., Hughes, M.K. and Hughes, A.L. (1995) Natural selection on the gag, pol, and env genes of human immunodeficiency virus 1 (HIV-1). *Mol. Biol. Evol.*, **12**, 803–813.

Seiki, M., Hattori, S., Hirayama, Y. and Yoshida, M. (1983) Human adult T-cell leukemia virus: complete nucleotide sequence of the provirus genome intergrated in leukemia cell DNA. *Proc. Natl Acad. Sci. USA*, **80**, 3618–3622.

Shpaer, E.G. and Mullins, J.I. (1993) Rates of amino acid change in the envelope protein correlate with pathogenicity of primate lentiviruses. *J. Mol. Evol.*, **37**, 57–65.

Sonigo, P., Alizon, M., Staskus, K. *et al.* (1985) Nucleotide sequence of the visna lentivirus: relationship to the AIDS virus. *Cell*, **42**, 369–382.

Van den Broek, M.F., Muller, U., Huang, S., Aguet, M. and Zinkernagel, R.M. (1995a) Antiviral defense in mice lacking both alpha/beta and gamma interferon receptors. *J. Virol.*, **69**, 4792–4796.

Van den Broek, M.F., Muller, U., Huang, S., Zinkernagel, R.M. and Aguet, M. (1995b) Immune defence in mice lacking type I and/or type II interferon receptors. *Immunol. Rev.*, **148**, 5–18.

Villaverde, A., Martinez, M.A., Sobrino, F., Dopazo, J., Moya, A. and Domingo, E. (1991) Fixation of mutations at the *VP1* gene of foot-

and-mouth disease virus. Can quasispecies define a transient molecular clock? *Gene*, **103**, 147–153.

Wain-Hobson, S. (1993a) The fastest genome evolution ever described: HIV variation *in situ. Curr. Opin. Genet. Devel.*, **3**, 878–883.

Wain-Hobson, S. (1993b) Viral burden in AIDS. *Nature*, **366**, 22.

Wain-Hobson, S. (1996) Running the gamut of retroviral variation. *Trends Microbiol.*, **4**, 135–141.

Wain-Hobson, S., Sonigo, P., Danos, O., Cole, S. and Alizon, M. (1985) Nucleotide sequence of the AIDS virus, LAV. *Cell*, **40**, 9–17.

Waldor, M.K. and Mekalanos, J.J. (1996) Lysogenic conversion by a filamentous phage encoding cholera toxin. *Science*, **272**, 1910–1914.

Wei, X., Ghosh, S.K., Taylor, M.E. *et al.* (1995) Viral dynamics in human immunodeficiency virus type 1 infection. *Nature*, **373**, 117–122.

Willey, R.L., Smith, D.H., Lasky, L.A. *et al.* (1988) *In vitro* mutagenesis identifies a region within in the envelope gene of the human immunodeficiency virus that is critical for infectivity. *J. Virol.*, **62**, 139–147.

Wilson, A.C., Carlson, S.S. and White, T.J. (1977) Biochemical evolution. *Annu. Rev. Biochem.*, **46**, 573–639.

Wolinsky, S.M., Korber, B.T., Neumann, A.U. *et al.* (1996) Adaptive evolution of human immunodeficiency virus-type 1 during the natural course of infection. *Science*, **272**, 537–542.

Yang, Z., Lauder, I.J. and Lin, H.J. (1995) Molecular evolution of the hepatitis B virus genome. *J. Mol. Evol.*, **41**, 587–596.

Zeuzem, S., Lee, J.H., Franke, A. *et al.* (1998) Quantification of the initial decline of serum hepatitis C virus RNA and response to interferon alfa. *Hepatology*, **27**, 1149–1156.

Zhang, Y., Nelson, M., Nietfeldt, J. *et al.* (1998) Chlorella virus NY-2A encodes at least 12 DNA endonuclease/methyltransferase genes. *Virology*, **240**, 366–375.

7

Viral Quasispecies and Fitness Variations

Esteban Domingo, Cristina Escarmís, Luis Menéndez-Arias and John J. Holland

FROM EARLY REPLICONS TO PRESENT-DAY RNA VIRUSES

The quasispecies model of molecular evolution was initially proposed to describe the error-prone replication, self-organization and adaptability of primitive replicons such as those thought to have populated the earth some 4000 million years before the present (Eigen, 1971, 1992; Eigen and Schuster, 1979; Chapter 1). Error-prone replication has been maintained as a stable trait in present-day RNA viruses. This has occurred in spite of a probable origin of most extant RNA viruses in coevolution with a cellular world. Viroids and viroid-like replicons may be direct descendants of early RNA (or RNA-like) replicons that preceded an organized cellular world (Robertson, 1992; Chapter 4). Cells and viruses share a considerable number of functional domains or modules: polymerases, proteases, enzymes involved in modifications of nucleic acids, etc. Furthermore, cells and viruses display an ubiquitous ability to modify, lose or acquire new genes or gene segments through genomic rearrangements, insertions, deletions and other homologous and non-homologous recombination events. Reshuffling of functional modules among cells, viruses and other replicons (plasmids, episomes, transposons, retrotransposons) is probably a frequent occurrence through fusion, transfection, conjugation and other types of horizontal gene transfers (Botstein, 1980a,b; Hickey and Rose, 1988; Zimmern, 1988; Davis, 1997; Holland and Domingo, 1998). Sequence comparisons strongly suggest that all extant viruses have deep, ancient evolutionary roots (Gorbalenya, 1995).

ERROR-PRONE REPLICATION NECESSITATES LIMITED GENETIC COMPLEXITY

One of the critical features that distinguishes cells from viruses is the extreme difference in the complexity of their genetic material, even after accounting for repeated DNA in animal and plant cells. A typical mammalian cell includes a number of chromosomes amounting to a total of about 3×10^9 base pairs (bp) of DNA. The chromosomal DNA of *Escherichia coli* has a complexity of about 4×10^6 bp. In contrast, all known RNA viruses have genomes in the range of 3×10^3 to 3×10^4 nucleotides. Estimates of point mutation rates for eukaryotic cells are in the range of 10^{-10}–10^{-11} misincorporations per nucleotide (s/nt), while for bacterial cells values may reach up to 10^{-9} s/nt (Drake, 1993; Drake *et al.*, 1998). Mutation rates for a number of genomic sites of RNA viruses determined using both genetic and biochemical procedures are in the range of 10^{-3}–10^{-5} s/nt (Holland *et al.*, 1992; Drake, 1993; Domingo and Holland, 1994; Drake *et al.*, 1998; Chapter 5). In spite of mutation rates varying with a number of environmental parameters, the above values mean that, in the process of RNA replication or retrotranscription, each progeny genomic molecule of 10 kb will contain on average 0.1 to several mutations. These determinations of mutation rates and frequencies suggest that even the viral progeny of a single infected cell will be genetically heterogeneous. Thus many variant genomes are produced in any natural infection by an RNA virus, as extensively documented through sequence determination of molecular and biological viral clones of bacterial, animal and plant RNA viruses (Domingo *et al.*, 1978, 1980, 1988, 1995; Holland *et al.*, 1982, 1992; Temin, 1989, 1993; Holland, 1993; Domingo and Holland, 1994, 1997; Gibbs *et al.*, 1995; Roossinck, 1997; see also other chapters of this book).

The dynamic mutant distributions that compose replicating RNA viruses are termed viral quasispecies. The concept is a direct adaptation to viruses of the theoretical quasispecies model developed by Eigen, Schuster and their colleagues (Eigen, 1971; Eigen and Schuster, 1979). One of the predictions of the quasispecies structure of RNA viruses is the existence of a error-threshold relationship (reviewed in Eigen and Schuster, 1979; Eigen and Biebricher, 1988; Chapter 1). The threshold is defined by a copying fidelity value at which a sharp transition between an organized mutant spectrum and random sequences lacking information contents occurs (entry into error catastrophe). Thus the error threshold relationship establishes a limitation for the maximum complexity of genetic information that can be stably maintained by a replicon displaying a given copying accuracy. Theoretical calculations of the range of mutation rates that should be compatible with maintenance of the information carried by the simple RNA bacteriophages were compatible with the mutation rates and frequencies found experimentally (compare Domingo *et al.*, 1976, 1978, and Batschelet *et al.*, 1976, with Eigen and Schuster, 1979). This was strongly

supported by mutagenesis experiments. In DNA-based microbes, mutagenesis can increase mutation frequencies by thousands-fold (Cupples and Miller, 1989). Chemical mutagenesis could increase the mutation frequency by at most threefold at defined genomic sites of poliovirus and vesicular stomatitis virus (VSV; Holland *et al.*, 1990), and 13-fold in the case of a retroviral vector (Pathak and Temin, 1992). Also, increased mutagenesis had an adverse effect on fitness recovery of VSV clones (Lee *et al.*, 1997). These results suggest that RNA viruses replicate near the error catastrophe threshold, with a copying fidelity that allows a generous production of error copies.

Increased genetic complexity as is embodied in cells required a correspondingly higher copying accuracy of the genetic material. This appears to have been accomplished with a number of pathways for post-replicative repair mechanisms as well as with the acquisition of a 3′–5′ proofreading-repair exonuclease activity by most cellular DNA polymerases (Friedberg *et al.*, 1995; Goodman and Fygenson, 1998). No evidence of a 3′–5′ exonuclease in viral RNA polymerases and reverse transcriptases has been obtained from either biochemical or structural studies with viral enzymes (Steinhauer *et al.*, 1992; Sousa, 1996). A possible exception is an early report by Ishihama *et al.* (1986) showing that the influenza virus RNA polymerase was able to remove excess GMP residues added to a capped oligonucleotide primer.

The limited genetic complexity allows RNA viruses to exploit error copies to walk swiftly through sequence space as an adaptive strategy (Eigen and Biebricher, 1988). Mutant spectra include phenotypically relevant variant viruses (examples in: Duarte *et al.*, 1994; Nájera *et al.*, 1995; Marcus *et al.*, 1998; Morimoto *et al.*, 1998; Pawlotsky *et al.*, 1998; Quiñones-Mateu *et al.*, 1998; Wyatt *et al.*, 1998). This permits RNA viruses to refine their adaptation to a constant environment, and to have a certain probability of finding adaptive routes in response to environmental changes. Each virus population, by virtue of consisting of dynamic mutant spectra rather than a defined genomic sequence, has the potential to adapt readily to a range of environments.

FITNESS VARIATIONS IN A CONSTANT ENVIRONMENT CAN BE MODULATED BY VIRAL POPULATION SIZE

One of the consequences of the quasispecies dynamics of RNA viruses (and one that would not occur if high mutation rates and the quasispecies structure were not in operation) is fitness variations in constant environments triggered by changes in viral population size. Fitness is a complex parameter that measures the degree of adaptation of a living organism or simple replicon to a specific environment (for reviews see Williams, 1992, and Reznick and Travis, 1996). Fitness is necessarily a relative value that compares different organisms, or variants within one class of organism. For viruses, fitness values have been measured as the relative ability of two competing viruses to produce infectious progeny in either constant or changing environments (Holland *et al.*, 1991; reviewed in Domingo and Holland, 1997). Competitions are started by infecting cells or organisms with a mixture of a reference wild-type virus (given arbitrarily a fitness value of 1) and the virus to be tested, in a known proportion. The progeny are used to initiate a second round of infection, and the process is repeated a number of times (serial infections). Then the proportion of the two competing viruses at each passage is quantified relative to their proportion in the initial mixture. This defines a fitness vector, the slope of which is the fitness of the test virus relative to the reference virus (Figure 7.1; examples of fitness determinations using this procedure can be found in Duarte *et al.*, 1992, and Clarke *et al.*, 1993). The two competing viruses must be distinguishable by some phenotypic trait (e.g. a clear difference in the ability to replicate in the presence of an antibody or a drug) or genetic modification (e.g. nucleotide substitutions that allow the proportion of the two viruses to be determined by densitometry of a sequencing gel). Fitness determinations of viruses subjected to different passage regimes have established an important effect of population size on fitness evolution.

RNA virus populations subjected to severe serial bottleneck events in cell culture – such as

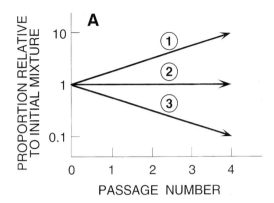

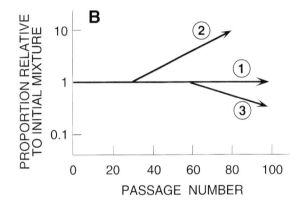

FIGURE 7.1 Fitness vectors and fitness variations. **A**. Plot of the proportion of the test virus and a reference virus, relative to their proportion in the initial mixture, as a function of passage number. The plot gives a fitness vector. The test virus can show higher relative fitness than the reference virus (line ①), equal relative fitness (neutrality, line ②) or lower fitness than the reference virus (line ③). For further details of this procedure for fitness determination and specific examples see Holland *et al.*, 1991, and Duarte *et al.*, 1992. **B**. Possible outcomes of a competition between two neutral variants. The two variants may coexist for many generations (line ①). Occasionally one variant may displace the other in a rather unpredictable manner (lines ② and ③), in agreement with the competitive exclusion principle of population genetics. For review of concepts of population biology as applied to RNA viruses, see Domingo *et al.*, 1996.

those occurring upon serial plaque-to-plaque transfers – undergo, on average, a loss in fitness (Chao, 1990; Duarte *et al.*, 1992, 1993; Escarmís *et al.*, 1996; Yuste *et al.*, 1999). This is due to the accumulation of deleterious mutations (Figure

7.2), predicted by Müller (1964) to occur for small populations of asexual organisms lacking in mechanisms, such as sex or recombination, that could eliminate or compensate for such debilitating mutations (Maynard Smith, 1976). Subjecting RNA viruses to repeated plaque-to-plaque transfers has all the ingredients (a viral population reduced to a single genome at the onset of plaque formation, high mutation rates and extreme genetic drift) to accentuate the effects of Müller's ratchet. Although RNA viruses have provided clear experimental support for the theoretical predictions of Müller (1964), evidence has been obtained recently with bacteria (Andersson and Hughes, 1996).

A study by Novella et al. (1995a) using vesicular stomatitis virus (VSV) established that the extent of fitness loss for any given bottleneck size depends on the initial fitness of the viral clone under study. The higher the initial fitness, the less severe must the bottleneck be to avoid fitness losses. Debilitated viral clones often gain fitness even when subjected to considerable bottlenecking (Novella et al., 1995a; Elena et al., 1998). Rather constant, stable fitness values could be attained by choosing the appropriate bottleneck size, although occasional fitness jumps were observed (Novella et al., 1996).

Escarmís et al. (1996) examined the genetic lesions associated with Müller's ratchet by determining genomic nucleotide sequences of foot-and-mouth disease virus (FMDV) clones prior to and after undergoing repeated plaque-to-plaque transfers. The result was that fitness loss was associated with unusual mutations that had never been seen in natural FMDV isolates or laboratory populations subjected to passages involving large viral populations. Particularly striking were an internal polyadenylate extension preceding the second functional AUG initiation codon of the FMDV genome, and amino acid substitutions at internal capsid residues. Although the presence of an internal polyadenylate tract of variable length (from 6 to about 28 residues) appeared in several FMDV clones, the majority of other mutations were unique to each bottleneck (Escarmís et al., 1996). Additions or deletions of nucleotides have been frequently observed at homopolymeric tracts, particularly on pyrimidine runs in templates

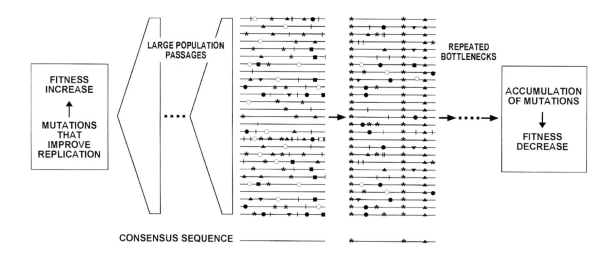

FIGURE 7.2 Schematic representation of viral quasispecies and the effect of viral population size on replicative fitness. Horizontal lines represent individual genomes, and symbols on the lines represent mutations. The random sampling of individual genomes (repeated bottlenecks; small arrows) leads to accumulation of mutations and fitness decreases. Large population passages (large arrows) lead to increases in replicative fitness. Fitness losses and gains depend on the relative fitness of the initial viral quasispecies and on the size of the bottleneck, see text for evidence with specific examples.

copied by proofreading-repair-deficient polymerases. Misalignment mutagenesis has been proposed for such frame-shift mutations at homopolymeric tracts (Kunkel, 1990; Bebenek *et al.*, 1993; see also Chapter 5). In addition to documenting the molecular basis of Müller's ratchet in FMDV, these results revealed the continuous operation of negative (or purifying) selection, a hallmark of RNA virus evolution (Domingo and Holland, 1994). In particular, the site at which adenylate residues were added qualifies as a mutational hot spot. The absence of internal oligoadenylate elongations in any FMDV population that has been examined (reviewed in Domingo *et al.*, 1990, 1992), suggests their elimination during normal RNA replication. Only when the repeated bottlenecks limit the action of negative selection can such internal polyadenylate extensions, as well as other deleterious mutations, be maintained in the FMDV genome (Escarmís *et al.*, 1996).

In contrast to bottleneck passages, large population infections generally result in fitness gains of RNA viruses (Martínez *et al.*, 1991; Clarke *et al.*, 1993; Novella *et al.*, 1995b; Escarmís *et al.*, 1998). This reflects the gradual optimization of mutant spectra when their different components are allowed unrestricted competition in a constant environment (Figure 7.2). A highly debilitated FMDV clone, which had accumulated mutations due to Müller's ratchet, including the internal polyadenylate tract extension discussed above, was studied with regard to genetic changes associated with fitness recovery when subjected to large population passages (Escarmís *et al.*, 1998). The results indicated that two true reversions and a total of 14 additional mutations in the 8.5 kb genome had been incorporated by passage 100. The fitness of this passaged clone was 30 times higher than that of the debilitated initial clone. The first genetic modification was the loss of the internal polyadenylate that was directly responsible for part of the fitness loss. Interestingly, parallel passages of subclones of the same debilitated clone revealed at least three independent molecular pathways for the loss of the internal polyadenylate tract (Escarmís *et al.*, 1999). These results with FMDV agree with previous results of analysis of revertants and pseudorevertants following infections

with altered picornaviral (reviewed by Wimmer *et al.*, 1993; Agol, 1997), retroviral (Berkhout *et al.*, 1997 and references therein) or phage (Arora *et al.*, 1996; Olsthoorn and van Duin, 1996; Licis *et al.*, 1998) genomes. A variety of genomic alterations mediated the rescuing of infectious genomes. Although in most cases fitness values were not quantified, the results support the wrightian view of evolutionary adaptation, in that multiple independent fitness peaks are available to RNA genomes (Wright, 1931, 1982). Multiple peaks, with small Hamming distances (Eigen and Biebricher, 1988), may provide similar levels of adaptation of an RNA virus to a given environment.

FITNESS PERTURBATIONS DURING VIRAL REPLICATION

When the relative fitness of the evolving quasispecies reaches a high value, even quite large population sizes can constitute an effective bottleneck and prevent continuing fitness increase. This was documented by following the fitness increase of VSV clones (Novella *et al.*, 1999). This limiting high fitness level was manifested by stochastic fluctuations in fitness values expected from random generation of mutations in a continuously evolving mutant swarm. These perturbations illustrate how difficult it is to attain a true population equilibrium even when viruses replicate in a constant environment. A rare combination of mutations – one that may occur only once over many generations of viral replication – may trigger the dominance of one viral subpopulation over another, thereby disrupting a period of population equilibrium. In competitions between two VSV clones of similar fitness coexisting at or near equilibrium, the rapid and unpredictable displacement of one VSV population by the other (Clarke *et al.*, 1994) provided support for a classical concept of population biology: the competitive exclusion principle (Gause, 1971). Furthermore, in the competition passages preceding mutual exclusion, both the winners and the losers gained fitness at comparable rates, in support of yet another concept of population genetics: the Red Queen hypothesis (van Valen, 1973; Clarke *et al.*, 1994; reviewed in Domingo *et al.*, 1996; Figure 7.1).

A multiplicity of infection (m.o.i.)-dependent fitness has been documented in competitions between two closely related subpopulations of FMDV: one of the populations showed a fitness advantage over the other only when the m.o.i. was below 1 plaque forming unit (PFU)/cell (Sevilla *et al.*, 1998). The most likely mechanism underlying this effect was a different ability of the two viruses to interact with heparan sulfate, one of the coreceptors for FMDV (Jackson *et al.*, 1996), a coreceptor that is also used by other viruses adapted to growing in cell culture (Byrnes and Griffin, 1998). The prior history of passage of the two FMDV populations in cell culture differed in viral population size relative to the number of susceptible cells (Sevilla and Domingo, 1996) and this apparently affected the degree of adaptation to heparan sulfate, as a correceptor. These results with FMDV in cell culture suggest that, even in natural infections, different viral subpopulations may be selected depending on the ratio of variant viruses to susceptible cells (Sevilla *et al.*, 1998). This possibility introduces a new element of complexity in quasispecies evolution, particularly considering that viruses often use more than one receptor and coreceptor molecule, and that the latter may be present in different amounts on the surface of potentially susceptible cells.

Another effect of the quasispecies dynamics of RNA viruses, which is dependent on a high m.o.i., is the maintenance of defective viruses by complementation (Charpentier *et al.*, 1996; Moreno *et al.*, 1997; Yamada *et al.*, 1998; Garcia-Arriaza *et al.*, unpublished results). This helper effect of a mutant spectrum is in contrast to the suppressive effects that modulate the dominance of high fitness clones depending on the surrounding mutant distribution (de la Torre and Holland, 1990; other examples reviewed in Domingo and Holland, 1997).

Results with the tripartite *Pseudomonas phaseolicala* phage Φ6 indicate that intrahost competition among coinfecting phages may suppress the advantage of sex in phage adaptation (Turner and Chao, 1998).

The main conclusion we derive from preceding paragraphs is that even in a relatively constant biological and physical environment, as is usually provided by *in-vitro* cell culture systems, the degree of adaptation of viral quasispecies may undergo remarkable quantitative variations. The present evidence strongly suggests that such variations come about from the continuous stochastic generation of mutant genomes together with different opportunities for competitive optimization of mutant spectra depending on the viral population size participating in the competition process, host cell population size, number and type of host cell receptors and coreceptors, etc.

FITNESS VARIATIONS IN CHANGING ENVIRONMENTS

The experiments on fitness variation of viruses in cell culture have been instrumental in defining some basic influences presaging fitness evolution of viral quasispecies. However, in their replication in a natural setting, viruses encounter multiple and changing environments, and they often have to cope with conflicting selective constraints. Because of polymorphisms in key proteins of cellular and humoral immune responses and in many other cell surface antigens, viruses do not face the same selective constraints in different individuals of the same host species. Biological environments are heterogeneous and vary with time within each infected individual. Furthermore, a considerable number of viruses are capable of infecting different host species, extending even further the range of environments they may face.

Arboviruses that replicate in mammalian and insect hosts constitute a classical example of obligate environmental alternacy *in vivo* (Scott *et al.*, 1994; Weaver, 1998). Early work documented that extensive replication of viruses in insect cells led to attenuation of infectivity for mammalian cells (Peleg, 1971; Mudd *et al.*, 1973). Prolonged persistence of VSV in cultured sandfly cells resulted in a 2×10^6-fold greater fitness in insect cells than in mammalian cells (Novella *et al.*, 1995c). This reflects the selective differences that VSV encounters in these two cell types. However, a single passage of the sandfly-cell-adapted VSV in mammalian cells led to an increase in fitness in mammalian cells to near original values. It would be interesting to test whether this capacity for fitness shift would be similar for non-arboviral RNA viruses able to grow in insect cells in culture. VSV adapted to sandfly cells was highly attenuated for mice. Again, a single passage in mammalian cells restored the virulence phenotype *in vivo* (Novella *et al.*, 1995c). This suggests that rapid adaptations to different cell types may have a remarkable influence on important biological traits *in vivo*.

A single specific selective constraint may often cause a quasispecies to deviate from its consensus replicating genome and lead to fitness decreases. This was clearly documented with antigenic variants of FMDV selected with neutralizing polyclonal antibodies directed to a major antigenic site of the virus (Borrego *et al.*, 1993). The selected FMDV variants showed distinct spectra of mutations at and around the antigenic site that was the target of the antibodies, and replicative fitness fell up to 10^5-fold in some cases. Fitness levels recovered upon further replication in the absence of antibodies, and sometimes even with the continuous presence of antibodies. Interestingly, such fitness gains did not involve reversion of the mutations associated with the antigenic changes. Rather, highly fit antigenically variant viruses were produced, due to additional mutations (Borrego *et al.*, 1993).

Alterations in environment produced not only by the transient presence of antibodies or antiviral agents but also by other environmental shifts can result in accelerated exploration of sequence space. This was studied by Sevilla and Domingo (1996) by analysing the evolution of genomic sequences of an FMDV clone isolated during an acute infection, adapted to cytolytic infections in cell culture and then subjected to prolonged persistence, followed by prolonged cytolytic infections, all in BHK-21 cells. The result of the study was that the alternation between persistent and cytolytic infections led to reversion of some phenotypic traits (not all), accompanied by genetic diversification (Sevilla and Domingo, 1996). Since each point of

sequence space offers a new base for further exploration of additional genotypes (and their associated phenotypes) alterations in environment, as well as transient selective pressures, may have important influences on the long-term evolution of RNA viruses.

FITNESS VARIATIONS IN NATURAL HOST ORGANISMS

Studies of fitness variations *in vivo* have been approached in three ways. A few studies have involved growth-competition experiments between two viruses replicating in host organisms. In other studies, the outcome of competitions between viruses that were isolated *in vivo* has been analyzed in primary or established cell cultures. In yet another line of research, the effects of fitness variations in cell culture on the replicative potential of viruses *in vivo* have been examined.

Carrillo *et al.* (1990) isolated two variant FMDVs in the course of replication of a clonal preparation in swine. One of the variants was a MAb-resistant (MARM) mutant, while the other was isolated from blood during the early viraemic phase of the acute infection. The ability of the two variants to compete with the parental clonal population was examined by co-infection of swine with mixtures of the parental clone and each of the two variants individually. Neither the parental FMDV clone nor either of the two variants became completely dominant in a single co-infection *in vivo*. However, fitness differences were clearly documented. The parental FMDV clone manifested a selective advantage over the MARM in that the parental clone was dominant in most lesions (vesicles) in the diseased swine. In contrast, the parental clone and the variant from the early viraemic phase were about equally represented in the lesions of the animal infected with equal amounts of the two viruses (Carrillo *et al.*, 1998).

The lentivirus equine infectious anaemia virus (EIAV) experiences continuous quasispecies fluctuations during persistent infections in horses (Clements *et al.*, 1988). EIAV quasispecies were characterized in a pony experimentally infected with a biological clone of the virus. New quasispecies were associated with recurrent episodes of disease. A large deletion in the principal neutralizing domain of the virus was identified during the third febrile episode and became dominant during the fourth febrile episode. This drastic genetic change did not appear to diminish significantly the fitness of EIAV *in vivo* and in cell culture (Leroux *et al.*, 1997).

Adaptation of VSV clones to persistent growth in sandfly cells at 28°C, which entailed strong fitness losses in mammalian cells, resulted in very low replicative fitness in mouse brain upon direct intracraneal inoculation *in vivo* (Novella *et al.*, 1995c). These insect-cell-adapted VSV populations manifested strong attenuation in a very stringent neurovirulence test. Attenuation was much stronger than for VSV clones that displayed fitness decreases in mammalian cells due to repeated bottleneck events. Evidence was also obtained that the attenuated VSV could protect against lethal doses of standard VSV in challenge experiments with mice (Novella *et al.*, 1995c). However, *in-vivo* fitness in mouse brain (and neurovirulence) was restored following a single passage in mammalian cells at 37°C.

HIV-1 DRUG RESISTANCE AND VIRUS FITNESS

An increasing number of measurements of viral fitness involve human immunodeficiency virus type 1 (HIV-1) variants isolated from quasispecies replicating *in vivo*. Particularly relevant are fitness comparisons among multiple mutants harboring amino acid substitutions related to resistance to reverse transcriptase and protease inhibitors.

HIV-1 Reverse Transcriptase Inhibitors

Since the discovery of AZT (3′-azido-3′-deoxythymidine; zidovudine) as an effective inhibitor of HIV replication (Mitsuya *et al.*,

1985), drug therapy has been widely used in the treatment of AIDS. The loss of therapeutic effect due to the acquisition of resistance was recognized for AZT in 1989, when Larder and colleagues showed that HIV isolates from patients with advanced HIV disease became less sensitive to the drug during the course of treatment (Larder *et al.*, 1989). At least six amino acid substitutions in HIV-1 reverse transcriptase have been associated with resistance to AZT: M41L, D67N, K70R, 210W, T215Y and K219Q. The first substitution arising during AZT treatment is usually K70R, followed by T215Y. These observations correlate with the results of *in-vitro* competition experiments with HIV-1 variants. Thus, in the absence of the drug, wild-type HIV-1 showed the highest fitness, followed by mutant K70R, whose estimated relative fitness was 0.97. Other tested variants, such as M41L, T215Y and the double mutant M41L/T215Y, showed a significantly reduced fitness *in vitro* (Harrigan *et al.*, 1998). The simultaneous presence of Leu-41 and Tyr-215 in the RT-coding region of the virus confers high-level resistance to AZT (for a review, see Larder, 1994). The T215Y substitution involves two nucleotide changes. The replacement of Tyr-215 by Cys, Asp or Ser has been observed *in vivo* in the absence of zidovudine treatment (Goudsmit *et al.*, 1997; Yerly *et al.*, 1998). Although 1% or 2% fitness gains can be sufficient for a significant reproductive advantage (Goudsmit *et al.*, 1996), estimations based on the frequency distribution of variants over time in an environment without zidovudine showed that the population with Tyr-215 was 10–25% less fit than the 215-Asp population (Goudsmit *et al.*, 1997). Other amino acid changes, such as T215F or L210W, may also appear during prolonged treatment with AZT (Kellam *et al.*, 1992; Harrigan *et al.*, 1996; Hooker *et al.*, 1996). The L210W substitution caused a replicative disadvantage to the virus as measured in MT-2 cells, in the context of an enzyme with Leu-41 and Tyr-215. This disadvantage was less severe when viral replication took place in the presence of AZT (Harrigan *et al.*, 1996).

Other nucleoside inhibitors of HIV-1 reverse transcriptase are listed in Table 7.1. High-level

TABLE 7.1 Amino acid substitutions in the HIV-1 reverse transcriptase and protease associated with resistance to antiretroviral drugs

Inhibitors	Amino acid substitutions associated with drug resistance*
Nucleoside analogue inhibitors of RT	
Zidovudine (AZT)	M41L, K65R, K70R, Q151M, L210W, T215Y/F, K219Q/E
Didanosine (ddI)	K65R, L74V, V75T, M184V
Lamivudine (3TC)	**K65R**, K70E, **M184V/I/T**
Stavudine (d4T)	I50T, K65R, V75T, Q151M
Zalcitabine (ddC)	K65R, T69D, V75T, Q151M, M184V, Y215C
Abacavir (1592U89)	K65R, L74V, Y115F, M184V
Adefovir	**K65R**, T69D, K70E
Non-nucleoside analogue inhibitors of RT	
Nevirapine	**K103N, V106I/A, Y181C/I, Y188C/H/L, G190A/E/Q**
Delavirdine	**L100I, K103N, Y181C, Y188H, G190E, P236L**
Efavirenz (DMP-266)	**L100I, K103N, Y188L**
Protease inhibitors	
Saquinavir	R8Q, V32I, G48V, I50V, L90M
Ritonavir	R8Q, V32I, G48V, V82F/S/A, I84V
Indinavir	R8Q, L10F, L23V, V32I, M46I, I47L, G48V, I50V, I84V, N88S
Nelfinavir	D30N, G48V, V82F, I84V, L90M
Amprenavir (VX-478)	I50V

* Only the most relevant amino acid changes are shown. Extensive compilations of resistance mutations have been published elsewhere (Schinazi *et al.*, 1997; Menéndez-Arias and Domingo, 1998). Substitutions conferring high-level resistance (here defined as greater than 50-fold increase in the 50% inhibitory concentration, IC_{50}) are shown in bold.

resistance to the nucleoside analogue 3TC (2′, 3′-dideoxy-3′-thiacytidine; lamivudine) is rapidly achieved by the substitution M184V, located at the YMDD motif, which is part of the catalytic core of the enzyme. During 3TC treatment, the substitution M184I appears first, but then disappears due to the outgrowth of the M184V-containing viruses (Keulen *et al.*, 1997). Growth competition experiments showed a selective advantage of viruses with Val-184 over those with Ile-184. This effect was accentuated in the presence of the drug. 3TC-resistant RT mutants produced significantly shorter cDNA molecules than the wild-type enzyme, because of a processivity defect that was most severe in the enzyme having Ile at position 184 (Back *et al.*, 1996). The ddI-associated substitution L74V also confers a replication disadvantage in the absence of the drug. *In-vitro* experiments have shown that this mutant produces an 11% loss of fitness compared with the wild-type virus, and quickly reverts to the wild-type sequence in the absence of the drug (Sharma and Crumpacker, 1997). It is interesting to note that a partial replication disadvantage may be associated with greater clinical benefits, since it reduces virus production and results in a viral population with less variation, which in turn may lead to a better immune response.

Drug combinations are very effective in blocking HIV replication, leading to a more than 10 000-fold reduction of viral load. However, the response of a viral quasispecies to multiple constraints (e.g. different antiviral drugs) is difficult to predict. Simultaneous treatment with AZT and ddI led to viruses with reduced sensitivity to AZT, ddC, ddI, ddG and d4T (Shirasaka *et al.*, 1995; Iversen *et al.*, 1996). The resistant viruses contained the substitutions A62V, V75I, F77L, F116Y and Q151M. The substitution Q151M, which results from two nucleotide changes, is the first to appear and confers partial resistance to AZT, ddI, ddC and d4T. *In-vitro* studies have shown that, in the presence of AZT or ddI, viruses with the substitution Q151M replicate more slowly than viruses with the additional substitutions in codons 77 and 116 (Maeda *et al.*, 1998). Single mutants with substitutions at positions 62, 75, 77 or 116 failed to propagate in the presence of AZT or ddI. A conclusion from these experiments was that HIV-1 develops drug resistance through one or more mutations, which sacrifice the replicative fitness of the virus. Optimal replication competence is then acquired by accumulating additional mutations when the multi-dideoxynucleoside-resistant mutant emerges. Multiple drug resistance to AZT and other inhibitors can be also achieved through the accumulation of mutations appearing in monotherapy (Schmit *et al.*, 1996; Shafer *et al.*, 1998). The pathway leading to multiple resistance may often depend on the pre-existence of drug resistance mutations in the HIV-1 quasispecies (Nájera *et al.*, 1995).

Resistance to one nucleoside analogue inhibitor can be suppressed by substitutions conferring resistance to another drug in molecular clones of the virus. For example, the substitutions L74V and M184V, which are selected after treatment with ddI and 3TC respectively, can suppress T215Y-mediated AZT resistance (St Clair *et al.*, 1991; Larder *et al.*, 1995). However, simultaneous resistance to 3TC and AZT has been observed (Goulden *et al.*, 1996). A recently published study has demonstrated that a substitution at codon 333 (Gly to Glu or Asp) of HIV-1 reverse transcriptase was critical in facilitating dual resistance in a complex background of AZT and 3TC resistance mutations (Kemp *et al.*, 1998). The molecular mechanism involved in the modulation of AZT resistance by Glu-333 in the presence of M184V and AZT resistance mutations is not clear. Gly-333 is located in the connection subdomain of p66 (or p51), far from the polymerase active site where residue 184 is located. Long-range effects

involving the repositioning of the template primer, and perhaps a conformational change in the RT polypeptide, could be involved in this phenomenon, but this issue awaits further investigation.

Non-nucleoside reverse transcriptase inhibitors bind to a hydrophobic cavity that is 8–10 Å away from the polymerase active site, and this cavity is lined by the side chains of Tyr-181, Tyr-188, Phe-227 and Trp-229 (Kohlstaedt *et al.*, 1992). High-level resistance appears quickly after treatment and involves amino acid changes in residues located at the inhibitor binding site (Table 7.1). Again, resistance mutations often lead to reduced *in-vitro* replicating fitness. For example, this effect has been observed in viruses selected in the presence of the quinoxaline derivative HBY 097 or the bis(heteroaryl)piperazine U-104489, which contain the substitution G190E in their reverse transcriptase (Kleim *et al.*, 1996; Olmsted *et al.*, 1996). This amino acid substitution has a detrimental effect in the polymerase and RNase H activities of the enzyme (Fan *et al.*, 1996). Interestingly, continuous exposure to the quinoxaline derivative leads to the appearance of compensatory mutations at positions 74 and 75 that mitigate some of the deleterious effects of the G190E substitution (Kleim *et al.*, 1996; Boyer *et al.*, 1998).

This trend of increasing drug resistance while simultaneously decreasing replication competence has also been observed with foscarnet. This compound is a pyrophosphate analogue, with antiviral activity against HIV-1 and herpesviruses. Foscarnet-resistance amino acid substitutions in HIV-1 reverse trancriptase include W88S, W88G, E89K, E89G, L921, S156A and Q161L/H208Y (Domingo *et al.*, 1997, and references therein). It has been recently shown that viruses harbouring the single amino acid substitutions E89K, L921 or S156A have reduced replication competence compared with the wild-type virus, in the context of different genetic backgrounds (e.g., in the sequence context of the HXB2-D or LAI strains; Tachedjian *et al.*, 1998). Foscarnet-resistant strains, as well as other drug-resistant viruses with impaired fitness, may pose a clinical benefit to the patient, by decreasing the levels of virus production and thereby delaying the emergence of highly resistant viruses.

HIV-1 Protease Inhibitors

The HIV-1 protease is a homodimeric enzyme composed of two polypeptide chains of 99 residues. The substrate binding site is located at the interface between both subunits. The side chains of Arg-8, Leu-23, Asp-25, Gly-27, Ala-28, Asp-29, Asp-30, Val-32, Ile-47, Gly-48, Gly-49, Ile-50, Phe-53, Leu-76, Thr-80, Pro-81, Val-82 and Ile-84 form the substrate binding pocket and can interact with specific inhibitors (Wlodawer and Erickson, 1993), such as those used in the clinical treatment of AIDS. It is not unexpected that many resistance mutations affect residues of the inhibitor binding pocket of the protease (Table 7.1). These substitutions often cause a deleterious effect on the replication kinetics of HIV-1 (Ho *et al.*, 1994; Markowitz *et al.*, 1995; Croteau *et al.*, 1997). Kinetic analysis of the drug-resistant proteases revealed that, in most cases, their catalytic efficiency (measured as k_{cat}/K_m) is severely reduced compared with the wild-type enzyme (Gulnik *et al.*, 1995). During virus maturation, functional proteases are required to recognize and cleave several specific processing sites in the Gag and Gag-Pol precursor polyproteins forming the immature virions. Therefore, the processing ability of the drug-resistant enzymes must be maintained above some tolerable threshold. The *in-vivo* threshold is not known but it may be well below 10% of the wild-type protease activity, particularly assuming the high concentration of the protease in the virus. It has been estimated that the number of molecules of p24 (capsid protein) is around 2500 per virion (Bourinbaiar, 1991). Since the *pol* proteins are produced at a level of about one-tenth of the *gag* proteins, we could assume that each viral particle contains around 125 protease dimers. The size of a virus is around 100–120 nm in diameter, or 7×10^{-19} liters in volume. Thus, if complete processing is achieved, the protease dimer concentration in the virion would be about 0.3 mmol/l (Tomaselli *et al.*, 1991).

The deleterious effects caused by drug resistance mutations can often be rescued by other substitutions. For example, viruses carrying the double substitution (R8Q/M46I), selected with the protease inhibitor A-77003, displayed replication kinetics similar to that of the wild-type enzyme in MT-4 cells, in contrast to the R8Q mutant alone, which appeared first and replicated less efficiently (Ho *et al.*, 1994). Interestingly, the substitutions M46I, L63P, V82T and I84V, which appear after prolonged treatment with the protease inhibitor indinavir at suboptimal dose, produce viruses that are resistant to several protease inhibitors (Condra *et al.*, 1995). Crystallographic studies of the mutant enzyme revealed that substitutions at codons 82 and 84 were critical for the acquisition of resistance, while the amino acid changes at codons 46 and 63, which are away from the inhibitor binding site, appear as compensatory mutations (Chen *et al.*, 1995; Schock *et al.*, 1996). In a similar way, saquinavir resistance implies the acquisition of substitutions G48V and L90M (Jacobsen *et al.*, 1995), where G48V exerts the major influence on resistance, and L90M, which is located away from the inhibitor binding site, contributes to increasing the stability of the HIV protease.

In-vivo evidence regarding the effect of compensatory mutations on viral fitness has been recently reported for ritonavir therapy (Eastman *et al.*, 1998). Resistance to ritonavir was achieved by the substitution V82A, whose fitness relative to the wild-type virus was estimated as 96–98%. During this study, the acquisition of additional substitutions at codons 54, 63 and 84 was associated with increases in plasma RNA levels, indicating an improvement in viral fitness.

Many resistance mutations decrease the extent of Gag and Gag-Pol cleavage, which are required for virus maturation. Mutations at residues forming the cleavage sites in the Gag and Gag-Pol polyproteins can exert a partial corrective effect on the resistance-associated loss of viral fitness (Mammano *et al.*, 1998).

Finally, it must be emphasized that the emergence of viable resistant viruses is strongly dependent on the sequence context. It has been shown that the protease substitutions G48V, A71T and V82A led to non-viable viruses in the sequence context of an NL4–3 clone, while in the context of the HIV-1 RF clone these amino acid changes resulted in viruses that replicated well in culture and displayed cross-resistance to four protease inhibitors (Rose et al., 1996). The observed growth differences between the two clones were attributed to the presence of Ile at position 10 in the protease of HIV-1 NL4-3, instead of Leu as occurs in HIV-1 RF.

DETERMINISM IN RNA VIRUS EVOLUTION: COPING WITH CONSTRAINTS

A theoretical quasispecies is a stationary, infinite distribution of error copies. This formulation is deterministic since, despite the stochastic nature of mutagenesis, all possible mutants in an infinite population arising from a constant copying fidelity per site constitute a steady distribution ranked according to the adaptive value of its components. The quasispecies as a whole, and not its individual components, is the unit upon which selection acts (Eigen, 1971; Eigen and Schuster, 1979; Eigen and Biebricher, 1988; Domingo and Holland, 1997; Chapter 1).

Despite the above, the behaviour of real viral quasispecies replicating in cells and organisms is often unpredictable. As discussed in previous sections, even under a constant biological environment, FMDV clones may move to high fitness peaks or low fitness valleys via independent molecular pathways (Escarmís et al., 1996, 1999). Intact, complex host organisms provide a complex array of selective constraints to viruses. This is due not only to immune responses, as is sometimes assumed, but also to exposure of viruses to a variety of cell types (with non-identical receptors and coreceptors and intracellular metabolites), organs with disparate cell–cell interactions, and occurrence of fluctuations in the microenvironments in which viral replication takes place (temperature, ionic composition, pH, etc.). The establishment of hepatitis C virus (HCV) persistence in human hosts follows different molecular routes, reflecting probably fitness adjustment to different environments (Manzin et al., 1998), and the highly dynamic nature of HCV quasispecies in vivo (Martell et

al., 1992; Neumann et al., 1998; Pawlotsky et al., 1998; see also Chapter 12; HIV population dynamics is reviewed in Chapters 9 and 11). Mutant distributions are constantly confronted by environmental perturbations. RNA virus evolution tends to be highly unpredictable under these circumstances. There is a range of distinct molecular pathways by which viruses escape from antibodies, CTL responses or antiviral agents during natural infections (Taboga et al., 1997; Gebauer et al., 1988; reviews in Domingo et al., 1997, and McMichael and Phillips, 1997). Similarly, a small number of different changes in haemagglutinin gene sequences were observed in both chicken and human isolates from the recent outbreak of virulent H5N1 influenza A virus in Hong Kong (Suárez et al., 1998). All contained multiple basic amino acids at the haemagglutin cleavage site, and all were highly virulent and caused similar lesions.

In some cases, however, RNA virus evolution shows a striking reproducibility. The proportion of two closely related VSV clones during extensive competition in cell culture exhibited a highly predictable, non-linear behaviour (Quer et al., 1996). More strikingly, critical points from which viral evolution could follow different trajectories were reached after nearly constant periods of time (rounds of viral replication). The proposed model for these observations was that an averaging of mutational noise signals took place due to competitive selection among variants (Quer et al., 1996). That is, by an averaging of mutational effects, in the constant environment provided by cell culture systems, some deterministic features of the theoretical quasispecies distributions were manifested in competing VSV clones.

Another example has been provided by the selection of variant FMDVs as they overcame cellular resistance to infection (Escarmís et al., 1998). The experimental design involved infection of modified BHK-21 cells that had evolved a marked resistance to FMDV during persistent infection. It has been amply documented that, during persistent viral infections in cell culture, a coevolution of virus and host cells often occurs (Ahmed et al., 1981; de la Torre et al., 1988; Pelletier et al., 1991; Dermody et al., 1993;

Martín-Hernández *et al.*, 1994; Calvez *et al.*, 1995; Chen and Baric, 1996). In the case of FMDV, prolonged persistence in BHK-21 cells selected for highly resistant cells due to an intracellular block that specifically inhibited FMDV replication (de la Torre *et al.*, 1988, 1989; Martín-Hernández *et al.*, 1994). This inhibition of FMDV replication extended to FMDV isolates quite distinct from the one that had persisted and coevolved with the cells. Infections of the modified, resistant BHK-21 cells with several biological clones of FMDV $C_3Arg/85$ resulted in late production (6–8 days postinfection rather than the usual 0.5–1 day postinfection) of a variant FMDV that had acquired the ability to infect and produce plaques on the modified BHK-21 cells with normal replication kinetics (Escarmís *et al.*, 1998). Remarkably, in five independent infections, the selected viruses showed the same two amino acid replacements at the fivefold axis of the viral capsid, and no other mutations in the entire FMDV genome. The authors interpreted this reproducibility to mean that two strongly adaptive capsid replacements were within reach of the virus in its exploration of sequence space in its early, slow replication in the modified BHK-21 cells. It is possible that one or, with lower probability, the two critical replacements pre-existed in the mutant spectrum of the initial infecting FMDV $C_3Arg/85$ clones, and were selected during the 6–8 days of infection. Adaptation would not have been feasible had replication in modified BHK-21 cells required a constellation of mutations beyond reach of the sequence space explored by the quasispecies swarm of the infecting virus (Domingo *et al.*, 1978; Domingo, 1997). What this result suggests is that, in this particular case, other molecular solutions were not immediately available to FMDV to express this particular phenotype. Structural and functional constraints contributed to a deterministic episode of RNA virus evolution.

Repeated selection of genomes carrying the same mutations following imposition of a defined selective constraint has been reported with several additional viral systems (de la Torre *et al.*, 1992; Borrego *et al.*, 1993; Chumakov *et al.*, 1994; Couderc *et al.*, 1994; Lu *et al.*, 1996; Pavio *et al.*, 1996; Sevilla and Domingo, 1996;

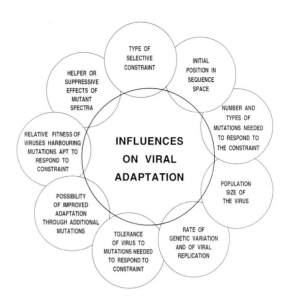

FIGURE 7.3 A number of interconnected parameters influence the adaptation of viruses to a given environment, and whether such adaptation can be achieved via one or multiple molecular pathways. Individual circles display key influences, some intrinsic to the virus and others external to the virus. The various influences and their effects on viral adaptation are discussed in sections of the chapter.

Bull *et al.*, 1997). These examples offer an interesting counterpoint to many other examples of unpredictable evolutionary change even with a simple viral clone subjected to the same environmental constraint (Escarmís *et al.*, 1999, and references therein). Figure 7.3 summarizes some key parameters that are likely to influence whether or not strong virus adaptation can be attained and whether it can be attained via one or a number of alternative sets of genetic changes. The need for a virus to maintain replicative fitness involves the same limitations on evolution that have long been recognized for all types of organisms: genetic changes are restricted by the viral genome's present position in sequence space, and by what is required to remain replication-competent (Simpson, 1949; Maynard Smith *et al.*, 1985). It must be stressed that adaptation to an environment is rarely optimal. For viruses, an extremely specialized adaptation to one single type of environment may decrease the chances for adaptation to a broad range of alternative environments, which they encounter in their natural life cycles. Nature works on the basis of imperfect approximations for adaptations among components of the biosphere.

ACKNOWLEDGEMENTS

We are indebted to many colleagues who have made important contributions to the understanding of viral quasispecies and have provided unpublished information. Work in Madrid is supported by grants from DGES PM97–0060–C02–01, FIS 98/0054–01, European Union, Comunidad Autónoma de Madrid, and Fundación Ramón Areces.

REFERENCES

Agol, V.I. (1997) Recombination and other genomic rearrangements in picornaviruses. *Sem. Virol.*, **8**, 77–84.

Ahmed, R., Canning, W.M., Kauffman, R.S., Sharpe, A.H., Hallum, J.V. and Fields, B.N. (1981) Role of the host cell in persistent viral infection: coevolution of L cells and reovirus during persistent infection. *Cell*, **25**, 325–323.

Andersson, D.I. and Hughes, D. (1996) Müller's ratchet decreases fitness of a DNA-based microbe. *Proc. Natl Acad. Sci. USA*, **93**, 906–907.

Arora, R., Priano, C., Jacobson, A.B. and Mills, D.R. (1996) Cis-acting elements within an RNA coliphage genome: fold as you please, but fold you must!! *J. Mol. Biol.*, **258**, 433–446.

Back, N.K.T., Nijhuis, M., Keulen, W. *et al.* (1996) Reduced replication of 3TC-resistant HIV-1 variants in primary cells due to a processivity defect of the reverse transcriptase enzyme. *EMBO J.*, **15**, 4040–4049.

Batschelet, E., Domingo, E. and Weissmann, C. (1976) The proportion of revertant and mutant phage in a growing population, as a function of mutation and growth rate. *Gene*, **1**, 27–32.

Bebenek, K., Abbotts, J., Wilson, S.H. and Kunkel, T.A. (1993) Error-prone polymerization by HIV-1 reverse transcriptase. Contribution of template-primer misalignment, miscoding, and termination probability to mutational hot spots. *J. Biol. Chem.*, **268**, 10324–10334.

Berkhout, B., Klaver, B. and Das, A.T. (1997) Forced evolution of a regulatory RNA helix in the HIV-1 genome. *Nucleic Acids Res.*, **25**, 940–947.

Borrego, B., Novella, I.S., Giralt, E., Andreu, D. and Domingo, E. (1993) Distinct repertoire of antigenic variants of foot-and-mouth disease virus in the presence or absence of immune selection. *J. Virol.*, **67**, 6071–6079.

Botsein, D. (1980a) A theory of modular evolution for bacteriophages. *Ann. NY Acad. Sci.*, **354**, 484–491.

Botstein, D. (1980b) A modular theory of virus evolution. In: *Animal Virus Genetics* (eds Fields, B.N., Jaenisch, R. and Fox, C.F.), pp. 11–20. Academic Press, New York.

Bourinbaiar, A.S. (1991) HIV and gag. *Nature*, **349**, 111.

Boyer, P.L., Gao, H.-Q. and Hughes, S.H. (1998) A mutation at position 190 of human immunodeficiency virus type 1 reverse transcriptase interacts with mutations at positions 74 and 75 via the template primer. *Antimicrob. Agents Chemother.*, **42**, 447–452.

Bull, J.J., Badgett, M.R., Wichman, H.A. *et al.* (1997) Exceptional convergent evolution in a virus. *Genetics*, **147**, 1497–1507.

Byrnes, A.P. and Griffin, D.E. (1998) Binding of Sindbis virus to cell surface heparan sulfate. *J. Virol.*, **72**, 7349–7356.

Calvez, V., Pelletier, I., Couderc, T., Pavio-Guédo, N., Blondel, B. and Colbére-Garapin, F. (1995) Cell clones cured of persistent poliovirus infection display selective permissivity to the wild-type poliovirus strain Mahoney and partial resistance to the attenuated Sabin 1 strain and Mahoney mutants. *Virology*, **212**, 309–322.

Carrillo, C., Plana, J., Mascarella, R., Bergadá, J. and Sobrino, F. (1990) Genetic and phenotypic variability during replication of foot-and-mouth disease virus in swine. *Virology*, **179**, 890–892.

Carrillo, C., Borca, M., Moore, D.M., Morgan, D.O. and Sobrino, F. (1998) *In vivo* analysis of the stability and fitness of variants recovered from foot-and-mouth disease virus quasi-species. *J. Gen. Virol.*, **79**, 1699 1706.

Chao, L. (1990) Fitness of RNA virus decreased by Müller's ratchet. *Nature*, **348**, 454–455.

Charpentier, N., Dávila, M., Domingo, E. and Escarmís, C. (1996) Long term, large population passage of aphthovirus can generate and amplify defective non-interfering particles deleted in the leader protease gene. *Virology*, **223**, 10–18.

Chen, W. and Baric, R.S. (1996) Molecular anatomy of mouse hepatitis virus persistence: coevolution of increased host cell resistance and virus virulence. *J. Virol.*, **70**, 3947–3960.

Chen, Z., Li, Y., Schock, H.B., Hall, D., Chen, E. and Kuo, L.C. (1995) Three-dimensional structure of a mutant HIV-1 protease displaying cross-resistance to all protease inhibitors in clinical trials. *J. Biol. Chem.*, **270**, 21433–21436.

Chumakov, K.M., Dragunsky, E.M., Norwood, L.P. *et al.*, (1994) Consistent selection of mutations in the 5′-untranslated region of oral poliovirus vaccine upon passaging *in vitro*. *J. Med. Virol.*, **43**, 79–85.

Clarke, D.K., Duarte, E.A., Moya, A., Elena, S.F., Domingo, E. and Holland, J.J. (1993) Genetic bottlenecks and population passages cause profound fitness differences in RNA viruses. *J. Virol.*, **67**, 222–228.

Clarke, D.K., Duarte, E.A., Elena, S., Moya, A., Domingo, E. and Holland, J.J. (1994) The red queen reigns in the kingdom of RNA viruses. *Proc. Natl Acad. Sci. USA*, **91**, 4821–4824.

Clements, J.E., Gdovin, S.L., Montelaro, R.C. and Narayan, O. (1988) Antigenic variation in lentiviral diseases. *Annu. Rev. Immunol.*, **6**, 139–159.

Condra, J.H., Schleif, W.A., Blahy, O.M. *et al.* (1995) *In vivo* emergence of HIV-1 variants resistant to multiple protease inhibitors. *Nature*, **374**, 569–571.

Couderc, T., Guédo, N., Calvez, V. *et al.* (1994) Substitutions in the capsid of poliovirus mutants selected in human neuroblastoma cells confer on the Mahoney type 1 strain a phenotype neurovirulent in mice. *J. Virol.*, **68**, 8386–8391.

Croteau, G., Doyon, L., Thibeault, D., McKercher, G., Pilote, L. and Lamarre, D. (1997) Impaired fitness of human immunodeficiency virus type 1 variants with high-level resistance to protease inhibitors. *J. Virol.*, **71**, 1089–1096.

Cupples, C.G. and Miller, J.H. (1989) A set of *lacZ* mutations in *Escherichia coli* that allow rapid detection of each of the six base substitutions. *Proc. Natl Acad. Sci. USA*, **86**, 5345–5349.

Davis, J.J. (1997) Origins, acquisition and dissemination of antibiotic resistance determinants. In: *Antibiotic Resistance: Origins, Evolution, Selection and Spread* (eds Chadwick, D.J. and Goode, J.), pp. 15–35. John Wiley, New York.

De la Torre, J.C. and Holland, J.J. (1990) RNA virus quasispecies can suppress vastly superior mutant progeny. *J. Virol.*, **64**, 6278–6281.

De la Torre, J.C., Martínez-Salas, E., Díez, J. *et al.* (1988) Coevolution of cells and viruses in a persistent infection of foot-and-mouth disease virus in cell culture. *J. Virol.*, **62**, 2050–2058.

De la Torre, J.C., de la Luna, S., Díez, J. and Domingo, E. (1989) Resistance to foot-and-mouth disease virus mediated by trans-acting cellular products. *J. Virol.*, **63**, 2385–2387.

De la Torre, J.C., Giachetti, C., Semler, B.L. and Holland, J.J. (1992) High frequency of single-base transitions and extreme frequency of precise multiple-base reversion mutations in poliovirus. *Proc. Natl Acad. Sci. USA*, **89**, 2531–2535.

Dermody, T.S., Nibert, M.L., Wetzel, J.D., Tong, X. and Fields, B.N. (1993) Cells and viruses with mutations affecting viral entry are selected during persistent infections of L cells with mammalian reoviruses. *J. Virol.*, **67**, 2055–2963.

Domingo, E. (1997) RNA virus evolution, population dynamics, and nutritional status. *Biol. Trace Element Res.*, **56**, 23–30.

Domingo, E. and Holland, J.J. (1994) Mutation rates and rapid evolution of RNA viruses. In: *Evolutionary Biology of Viruses* (ed. Morse, S.S.), pp. 161–184. Raven Press, New York.

Domingo, E. and Holland, J.J. (1997) RNA virus mutations and fitness for survival. *Annu. Rev. Microbiol.*, **51**, 151–178.

Domingo, E., Flavell, R.A. and Weissmann, C. (1976) *In vitro* site-directed mutagenesis: generation and properties of an infectious extracistronic mutant of bacteriophage Qβ. *Gene*, **1**, 3–25.

Domingo, E., Sabo, D.L., Taniguchi, T. and Weissmann, C. (1978) Nucleotide sequence heterogeneity of an RNA phage population. *Cell*, **13**, 735–744.

Domingo, E., Dávila, M. and Ortín J. (1980) Nucleotide sequence heterogeneity of the RNA from a natural population of foot-and-mouth disease virus. *Gene*, **11**, 333–346.

Domingo, E., Holland, J.J. and Ahlquist, P. (eds) (1988) *RNA Genetics*. CRC Press, Boca Raton, FL.

Domingo, E., Mateu, M.G., Martínez, M.A., Dopazo, J., Moya, A. and Sobrino, F. (1990) Genetic variability and antigenic diversity of foot-and-mouth disease virus. In: *Applied Virology Research, vol. II, Virus Variation and Epidemiology* (eds Kurstak, E., Marusyk, R.G., Murphy, S.A. and Van-Regenmortel, M.H.V.), pp. 233–266. Plenum Publishing Co., New York.

Domingo, E., Escarmís, C., Martínez, M.A., Martínez-Salas, E. and Mateu, M.G. (1992) Foot-and-mouth disease virus populations are quasispecies. *Curr. Topics Microbiol. Immunol.*, **176**, 33–47.

Domingo, E., Holland, J.J., Biebricher, C. and Eigen, M. (1995) Quasispecies: the concept and the word. In: *Molecular Basis of Virus Evolution* (eds Gibbs, A., Calisher, C. and García-Arenal, F.), pp. 171–180. Cambridge University Press, Cambridge.

Domingo, E., Escarmís, C., Sevilla, N. *et al.* (1996) Basic concepts in RNA virus evolution. *FASEB J.*, **10**, 859–864.

Domingo, E., Menéndez-Arias, L., Quiñones-Mateu, M.E., Holguín, A., Gutierrez-Rivas, M., Martínez, M.A., Quer, J., Novella, I.S. and Holland, J.J. (1997) Viral quasispecies and the problem of vaccine-escape and drug-resistant mutants. *Progress in Drug Research*, **48**, 99–128.

Drake, J.W. (1993) Rates of spontaneous mutations among RNA viruses. *Proc. Natl Acad. Sci. USA*, **90**, 4171–4175.

Drake, J.W., Charlesworth, B., Charlesworth, D. and Crow, J.F. (1998) Rates of spontaneous mutation. *Genetics*, **148**, 1667–1686.

Duarte, E., Clarke, D., Moya, A., Domingo, E. and Holland, J.J. (1992) Rapid fitness losses in mammalian RNA virus clones due to Müller's ratchet. *Proc. Natl Acad. Sci. USA*, **89**, 6015 6019.

Duarte, E.A., Clarke, D.K., Moya, A., Elena, S.F., Domingo, E. and Holland, J.J. (1993) Many trillionfold amplification of single RNA virus particles fails to overcome the Müller's ratchet effect. *J. Virol.*, **67**, 3620–3623.

Duarte, E.A., Novella, I.S., Weaver, S.C. *et al.* (1994) RNA virus quasispecies: significance for viral disease and epidemiology. *Infect. Agents Dis.*, **3**, 201–214.

Eastman, P.S., Mittler, J., Kelso, R. *et al.* (1998) Genotypic changes in human immunodeficiency virus type 1 associated with loss of suppression of plasma viral RNA levels in subjects treated with ritonavir (Norvir) monotherapy. *J. Virol.*, **72**, 5154–5164.

Eigen, M. (1971) Self-organization of matter and the evolution of biological macromolecules. *Naturwissenschaften*, **58**, 465–523.

Eigen, M. (1992) *Steps Towards Life*. Oxford University Press, Oxford.

Eigen, M. and Biebricher, C. (1988) Sequence space and quasispecies distribution. In: *RNA Genetics*, vol. 3 (eds Domingo, E., Holland, J.J. and Ahlquist, P.), pp. 211–245. CRC Press, Boca Raton, FL.

Eigen, M. and Schuster, P. (1979) *The Hypercycle. A Principle of Natural Self-Organization.* Springer-Verlag, Berlin.

Elena, S.F., Dávila, M., Novella, I.S., Holland, J.J., Domingo, E. and Moya, S. (1998) Evolutionary dynamics of fitness recovery from the debilitating effects of Müller's ratchet. *Evolution*, **52**, 309–314.

Escarmís, C., Carrillo, E.C., Ferrer, M. *et al.* (1998) Rapid selection in modified BHK-21 cells of a foot-and-mouth disease virus variant showing alterations in cell tropism. *J. Virol.*, **72**, 10171–10179.

Escarmís, C., Dávila, M., Charpentier, N., Bracho, A., Moya, A. and Domingo, E. (1996) Genetic lesions associated with Müller's ratchet in an RNA virus. *J. Mol. Biol.*, **264**, 255–267.

Escarmís, C., Dávila, M. and Domingo, E. (1999) Multiple molecular pathways for fitness recovery of an RNA virus debilitated by operation of Müller's ratchet. *J. Mol. Biol.*, **285**, 495–505.

Fan, N., Rank, K.B., Slade, D.E. *et al.* (1996) A drug resistance mutation in the inhibitor binding pocket of human immunodeficiency virus type 1 reverse transcriptase impairs DNA synthesis and RNA degradation. *Biochemistry*, **35**, 9737–9745.

Friedberg, E.C., Walker, G.C. and Siede, W. (1995) *DNA Repair and Mutagenesis*. American Society for Microbiology, Washington, DC.

Gause, G.F. (1971) *The Struggle for Existence*. Dover, New York.

Gebauer, F., de la Torre, J.C., Gomes, I. *et al.* (1988) Rapid selection of genetic and antigenic variants of foot-and-mouth disease virus during persistence in cattle. *J. Virol.*, **62**, 2041–2049.

Gibbs, A., Calisher, C.H. and García-Arenal, F., eds. (1995) *Molecular Basis of Virus Evolution*. Cambridge University Press, Cambridge.

Goodman, M.F. and Fygenson, D.K. (1998) DNA polymerase fidelity: from genetics toward a biochemical understanding. *Genetics*, **148**, 1475–1482.

Gorbalenya, A.E. (1995) Origin of RNA viral genomes; approaching the problem by comparative sequence analysis. In: *Molecular Basis of Virus Evolution* (eds Gibbs, A., Calisher, C. and García-Arenal, F.), pp. 49–66. Cambridge University Press, Cambndge.

Goudsmit, J., de Ronde, A., Ho, D.D. and Perelson, A.S. (1996) Human immunodeficiency virus fitness *in vivo*: calculations based on a single zidovudine resistance mutation at codon 215 of reverse transcriptase. *J. Virol.*, **70**, 5662–5664.

Goudsmit, J., de Ronde, A., de Rooij, E. and de Boer, R. (1997) Broad spectrum of *in vivo* fitness of human immunodeficiency virus type 1 subpopulations differing at reverse transcriptase codons 41 and 215. *J. Virol.*, **71**, 4479–4484.

Goulden, M.G., Cammack, N., Hopewell, P.L., Penn, C.R. and Cameron, J.M. (1996) Selection *in vitro* of an HIV-1 variant resistant to both lamivudine (3TC) and zidovudine. *AIDS*, **10**, 101–102.

Gulnik, S.V., Suvrorov, L.I., Liu, B. *et al.* (1995) Kinetic characterization and cross-resistance patterns of HIV-1 protease mutants selected under drug pressure. *Biochemistry*, **34**, 9282–9287.

Harrigan, P.R., Kinghorn, I., Bloor, S. *et al.* (1996) Significance of amino acid variation at human immunodeficiency virus type 1 reverse transcriptase residue 210 for zidovudine susceptibility. *J. Virol.*, **70**, 5930–5934.

Harrigan, P.R., Bloor, S. and Larder, B.A. (1998) Relative replicative fitness of zidovudine-resistant human immunodeficiency virus type 1 isolates *in vitro*. *J. Virol.*, **72**, 3773–3778.

Hickey, D.A. and Rose, M.R. (1988) The role of gene transfer in the evolution of eukaryotic sex. In: *The Evolution of Sex* (eds Michod, R.E. and Levin, B.R.), pp. 161–175. Sinauer Associates, Sunderland, MA.

Ho, D.D., Toyoshima, T., Mo, H. *et al.* (1994) Characterization of human immunodeficiency virus type 1 variants with increased resistance to a C_2-symmetric protease inhibitor. *J. Virol.*, **68**, 2016–2020.

Holland, J.J. (1993) Replication error, quasispecies populations, and extreme evolution rates of RNA viruses. In: *Emerging Viruses* (ed. Morse, S.S.), pp. 203–218. Oxford University Press, Oxford.

Holland, J.J. and Domingo, E. (1998) Origin and evolution of viruses. *Virus Genes*, **16**, 13–21.

Holland J.J., Spindler, K., Horodyski, F., Grabau, E., Nichol, S. and VandePol, S. (1982) Rapid evolution of RNA genomes. *Science*, **215**, 1577–1585.

Holland, J.J., Domingo, E., de la Torre, J.C. and Steinhauer, D.A. (1990) Mutation frequencies at defined single codon sites in vesicular stomatitis virus and poliovirus can be increased only slightly by chemical mutagenesis. *J. Virol.*, **64**, 3960–3962.

Holland, J.J., de la Torre, J.C., Clarke, D.K. and Duarte, E. (1991) Quantitation of relative fitness and great adaptability of clonal populations of RNA viruses. *J. Virol.*, **65**, 2960–2967.

Holland, J.J., de la Torre, J.C. and Steinhauer, D. (1992) RNA virus populations as quasispecies. *Curr. Topics Microbiol. Immunol.*, **176**, 1–20.

Hooker, D.J., Tachedjian, G., Solomon, A.E. *et al.* (1996) An *in vivo* mutation from leucine to tryptophan at position 210 in human immunodeficiency virus type 1 reverse transcriptase contributes to high-level resistance to 3′-azido-3′-deoxythymidine. *J. Virol.*, **70**, 8010–8018.

Ishihama, A., Mizumoto, K., Kawakami, K., Kato, A. and Honda, A. (1986) Proofreading function associated with the RNA-dependent RNA polymerase from influenza virus. *J. Biol. Chem.*, **261**, 10417–10420.

Iversen, A.K.N., Shafer, R.W., Wehrly, K. *et al.* (1996) Multidrug-resistant human immunodeficiency virus type 1 strains resulting from combination antiretroviral therapy. *J. Virol.*, **70**, 1086–1090.

Jackson, T., Ellard, F.M. Abu Ghazaleh, R. *et al.* (1996) Efficient infection of cells in culture by type O foot-and-mouth disease virus requires binding to cell surface heparan sulfate. *J. Virol.*, **10**, 5282–5297.

Jacobsen, H., Yasargil, K., Winslow, D.L. *et al.* (1995) Characterization of human immunodeficiency virus type 1 mutants with decreased sensitivity to protease inhibitor Ro 31–8959. *Virology*, **206**, 527–534.

Kellam, P., Boucher, C.A.B. and Larder, B.A. (1992) Fifth mutation in human immunodeficiency virus type 1 reverse transcriptase contributes to the development of high-level resistance to zidovudine. *Proc. Natl Acad. Sci. USA*, **89**, 1934–1938.

Kemp, S.D., Shi, C., Bloor, S., Harrigan, P.R., Mellors, J.W. and Larder, B.A. (1998) A novel polymorphism at codon 333 of human immunodeficiency virus type 1 reverse transcriptase can facilitate dual resistance to zidovudine and L-2′, 3′-dideoxy-3′-thiacytidine. *J. Virol.*, **72**, 5093–5098.

Keulen, W., Back, N.K.T., van Wijk, A., Boucher, C.A.B. and Berkhout, B. (1997) Initial appearance of the 184Ile variant in lamivudine-treated patients is caused by the mutational bias of human immunodeficiency virus type 1 reverse transcriptase. *J. Virol.*, **71**, 3346–3350.

Kleim, J.-P, Rösner, M., Winkler, I. *et al.* (1996) Selective pressure of a quinoxaline nonnucleoside inhibitor of human immunodeficiency virus-type 1 (HIV-1) reverse transcriptase (RT) on HIV-1 replication results in the emergence of nucleoside RT-inhibitor-specific (RT Leu-74 → Val or Ile and Val-75 → Leu or Ile) HIV-1 mutants. *Proc. Natl Acad. Sci. USA*, **93**, 34–38.

Kohlstaedt, L.A., Wang, J., Friedman, J.M., Rice, P.A. and Steitz, T.A. (1992) Crystal structure at 3.5 Å resolution of HIV-1 reverse transcriptase complexed with an inhibitor. *Science*, **256**, 1783–1790.

Kunkel, T.A. (1990) Misalignment-mediated DNA synthesis errors. *Biochemistry*, **29**, 8003–8011.

Larder, B.A. (1994) Interactions between drug resistance mutations in human immunodeficiency virus type 1 reverse transcriptase. *J. Gen. Virol.*, **75**, 951–957.

Larder, B.A., Darby, G. and Richman, D.D. (1989) HIV with reduced sensitivity to zidovudine isolated during prolonged therapy. *Science*, **243**, 1731–1734.

Larder, B.A., Kemp, S.D. and Harrigan, P.R. (1995) Potential mechanism for sustained antiretroviral efficacy of AZT–3TC combination therapy. *Science*, **269**, 696–699.

Lee, C.H., Gilbertson, D.L., Novella, I.S., Huerta, R., Domingo, E. and Holland, J.J. (1997) Negative effects of chemical mutagenesis on the adaptive behavior of vesicular stomatitis virus. *J. Virol.*, **71**, 3636–3640.

Leroux, C., Issel, C.J. and Montelaro, R.C. (1997) Novel and dynamic evolution of equine infectious anemia virus genomic quasispecies associated with sequential disease cycles in an experimentally infected pony. *J. Virol.*, **71**, 9627–9639.

Licis, N., van Duin, J., Blaklava, Z. and Berzins, V. (1998) Long-range translational coupling in single-stranded RNA bacteriophages: an evolutionary analysis. *Nucleic Acids Res.*, **26**, 3242–3246.

Lu, Z., Rezapkin, G.V., Douthitt, M.P. *et al.* (1996) Limited genetic changes in the Sabin 1 strain poliovirus occurring in the central nervous system of monkeys. *J. Gen. Virol.*, **77**, 273–280.

McMichael, A.J. and Phillips, R.E. (1997) Escape of human immunodeficiency virus from immune control. *Annu. Rev. Immunol.*, **15**, 271–296.

Maeda, Y., Venzon, D.J. and Mitsuya, H. (1998) Altered drug sensitivity, fitness, and evolution of human immunodeficiency virus type 1 with *pol* gene mutations conferring multidideoxynucleoside resistance. *J. Infect. Dis.*, **177**, 1207–1213.

Mammano, F., Petit, C. and Clavel, F. (1998)

Resistance-associated loss of viral fitness in human immunodeficiency virus type 1: Phenotypic analysis of protease and *gag* coevolution in protease inhibitor-treated patients. *J. Virol.*, **72**, 7632–7637.

Manzin, A., Solforosi, L., Petrelli, E. *et al.* (1998) Evolution of hypervariable region 1 of hepatitis C virus in primary infection. *J. Virol.*, **72**, 6271–6276.

Marcus, P.I., Rodriguez, L.L. and Sekellick, M.J. (1998) Interferon induction as a quasispecies marker of vesicular stomatitis virus populations. *J. Virol.*, **72**, 542–549.

Markowitz, M., Mo, H., Kempf, D.J. *et al.* (1995) Selection and analysis of human immunodeficiency virus type 1 variants with increased resistance to ABT-538, a novel protease inhibitor. *J. Virol.*, **69**, 701–706.

Martell, M., Esteban, J.I., Quer, J. *et al.* (1992) Hepatitis C virus (HCV) circulates as a population of different but closely related genomes: quasispecies nature of HCV genome distribution. *J. Virol.*, **66**, 3225–3229.

Martín-Hernández, A.M., Carrillo, E.C., Sevilla, N. and Domingo, E. (1994) Rapid cell variation can determine the establishment of a persistent viral infection. *Proc. Natl Acad. Sci. USA*, **91**, 3705–3709.

Martínez, M.A., Carrillo, C., González-Candelas, F., Moya, A., Domingo, E. and Sobrino, F. (1991) Fitness alteration of foot-and-mouth disease virus mutants: measurement of adaptability of viral quasispecies. *J. Virol.*, **65**, 3954–3957.

Maynard Smith, J. (1976) *The Evolution of Sex*. Cambridge University Press, Cambridge.

Maynard Smith, J., Burian, R., Kauffman, S. *et al.* (1985) Developmental constraints and evolution. *Q. Rev. Biol.*, **60**, 265–287.

Menéndez-Arias, L. and Domingo, E. (1998) Resistance tables for antirretroviral drugs. *AIDS Cyber J.*, **1**, 95–127.

Mitsuya, H., Weinhold, K.J., Furman, P.A. *et al.* (1985) 3'-azido-3'-deoxythymidine (BW A509U): an antiviral agent that inhibits the infectivity and cytopathic effect of human T-lymphotropic virus type III/lymphadenopathy-associated virus *in vitro. Proc. Natl Acad. Sci. USA*, **82**, 7096–7100.

Moreno, I.M., Malpica, J.M., Rodriguez-Cerezo,

E. and García-Arenal, F. (1997) A mutation in tomato aspermy cucumovirus that abolishes cell-to-cell movement is maintained to high levels in the viral RNA population by complementation. *J. Virol.*, **71**, 9157–9162.

Morimoto, N., Craig Hooper, D., Carbaugh, H., Fang Fu, Z., Koprowski, H. and Dietzschold, B. (1998) Rabies virus quasispecies: implications for pathogenesis. *Proc. Natl Acad. Sci. USA*, **95**, 3152–3156.

Mudd, J.A., Leavitt, R.W., Kingsbury, T. and Holland, J.J. (1973) Natural selection of mutants of vesicular stomatitis virus by cultured cells of *Drosophila melanogaster. J. Gen. Virol.*, **20**, 341–351.

Müller, H.J. (1964) The relation of recombination to mutational advance. *Mut. Res.*, **1**, 2–9.

Nájera, I., Holguín, A., Quiñones-Mateu, M.E. *et al.* (1995) The *pol* gene quasispecies of human immunodeficiency virus. Mutations associated with drug resistance in virus from patients undergoing no drug therapy. *J. Virol.*, **69**, 23–31.

Neumann, A.U., Lam, N.P., Dahari, J. *et al.* (1998) Hepatitis C viral dynamics *in vivo* and the antiviral efficacy of interferon-α therapy. *Science*, **282**, 103–107.

Novella, I.S., Elena, S.F., Moya, A., Domingo, E. and Holland, J.J. (1995a) Size of genetic bottlenecks leading to virus fitness loss is determined by mean initial population fitness. *J. Virol.*, **69**, 2869–2872.

Novella, I.S., Duarte, E.A., Elena, S.F., Moya, A., Domingo, E. and Holland, J.J. (1995b) Exponential increases of RNA virus fitness during large population transmissions. *Proc. Natl Acad. Sci. USA*, **92**, 5841–5844.

Novella, I.S., Clarke, D.K., Quer, J. *et al.* (1995c) Extreme fitness differences in mammalian and insect hosts after continuous replication of vesicular stomatitis virus in sandfly cells. *J. Virol.*, **69**, 6805–6809.

Novella, I.S., Elena, S.F., Moya, A., Domingo, E. and Holland, J.J. (1996) Repeated transfer of small RNA virus populations leading to balanced fitness with infrequent stochastic drift. *Mol. Gen. Genet.*, **252**, 733–738.

Novella, I.S., Quer, J., Domingo, E. and Holland, J.J. (1999) Exponential fitness gains of RNA virus populations are limited by bottleneck effects. *J. Virol.*, **73**, 1668–1671.

Olmsted, R.A., Slade, D.E., Kopta, L.A. *et al.* (1996) (Alkylamino)piperidine bis(hetero-aryl)piperizine analogs are potent, broad-spectrum nonnucleoside reverse transcriptase inhibitors of drug-resistant isolates of human immunodeficiency virus type 1 (HIV-1) and select for drug-resistant variants of HIV-1$_{IIIB}$ with reduced replication phenotypes. *J. Virol.*, **70**, 3698–3705.

Olsthoorn, R.C.L. and van Duin, J. (1996) Random removal of inserts from an RNA genome: selection against single-stranded RNA. *J. Virol.*, **70**, 729–736.

Pathak, V.K. and Temin, H.M. (1992) 5-Azacytidine and RNA secondary structure increase the retrovirus mutation rate. *J. Virol.*, **66**, 3093–3100.

Pavio, N., Buc-Caron, M.-H. and Corbère-Garapin, F. (1996) Persistent poliovirus infection of human fetal brain cells. *J. Virol.*, **70**, 6395–6401.

Pawlotsky, J.-M., Germanidis, G., Neumann, A.U., Pellerin, M., Frainais, P.-O. and Dhumeaux, D. (1998) Interferon resistance of hepatitis C virus genotype 1b: relationship to nonstructural 5A quasispecies mutations. *J. Virol.*, **72**, 2795–2805.

Peleg, J. (1971) Growth of viruses in arthropod cell cultures: applications. Attenuation of Semliki Forest (SF) virus in continuously cultured *Aedes aegypti* mosquito cells (Peleg) as a step in production of vaccines. *Curr. Topics Microbiol. Immunol.*, **55**, 155–161.

Pelletier, I, Couderc, T., Borzakian, S. *et al.* (1991) Characterization of persistent poliovirus mutants selected in human neuroblastoma cells. *Virology*, **180**, 729–737.

Quer, J., Huerta, R., Novella, I.S., Tsimring, L., Domingo, E. and Holland, J.J. (1996) Reproducible nonlinear population dynamics and critical points during replicative competitions of RNA virus quasispecies. *J. Mol. Biol.*, **264**, 465–471.

Quiñones-Mateu, M.E., Albright, J.L., Mas, A., Soriano, V. and Arts, E.J. (1998) Analysis of *pol* gene heterogeneity, viral quasispecies, and drug resistance in individuals infected with group O strains of human immunodeficiency virus type 1. *J. Virol.*, **72**, 9002–9015.

Reznick, D. and Travis, J. (1996) The empirical study of adaptation in natural populations. In: *Adaptation* (eds Rose, M.R. and Lauder, G.V.), pp. 243–289. Academic Press, San Diego, CA.

Robertson, H.D. (1992) Replication and evolution of viroid-like pathogens. *Curr. Topics Microbiol. Immunol.*, **176**, 213–219.

Roossinck, M.J. (1997) Mechanisms of plant virus evolution. *Annu. Rev. Phytopathol.*, **35**, 191–209.

Rose, R.E., Gong, Y.-F., Greytok, J.A. *et al.* (1996) Human immunodeficiency virus type 1 viral background plays a major role in development of resistance to protease inhibitors. *Proc. Natl Acad. Sci. USA*, **93**, 1648–1653.

Schinazi, R.F., Larder, B.A. and Mellors, J.W. (1997) Mutations in retroviral genes associated with drug resistance. *Int. Antivir. News*, **5**, 129–142.

Schmit, J.-C., Cogniaux, J., Hermans, P. *et al.* (1996) Multiple drug resistance to nucleoside analogues and nonnucleoside reverse transcriptase inhibitors in an efficiently replicating human immunodeficiency virus type 1 strain. *J. Infect. Dis.*, **174**, 962–968.

Schock, H.B., Garsky, V.M. and Kuo, L.C. (1996) Mutational anatomy of an HIV-1 protease variant conferring cross-resistance to protease inhibitors in clinical trials. Compensatory modulations of binding and activity. *J. Biol. Chem.*, **271**, 31957–31963.

Scott, T.W., Weaver, S.C. and Mallampalli, V.L. (1994) Evolution of mosquito-borne viruses. In: *The Evolutionary Biology of Viruses* (ed. Morse, S.S.), pp. 293–324. Raven Press, New York.

Sevilla, N. and Domingo, E. (1996) Evolution of a persistent aphthovirus in cytolytic infections: partial reversion of phenotypic traits accompanied by genetic diversification. *J. Virol.*, **70**, 6617–6624.

Sevilla, N., Ruiz-Jarabo, C.M., Gómez-Mariano, G., Baranowski, E. and Domingo, E. (1998) An RNA virus can adapt to the multiplicity of infection. *J. Gen. Virol.*, **79**, 2971–2980.

Shafer, R.W., Winters, M.A., Palmer, S. and Merigan, T.C. (1998) Multiple concurrent reverse transcriptase and protease mutations and multidrug resistance of HIV-1 isolates from heavily treated patients. *Ann. Intern.*

Med., **128**, 906–911.

Sharma, P.L. and Crumpacker, C.S. (1997) Attenuated replication of human immunodeficiency virus type 1 with a didanosine-selected reverse transcriptase mutation. *J. Virol.*, **71**, 8846–8851.

Shirasaka, T., Kavlick, M.F., Ueno, T. *et al.* (1995) Emergence of human immunodeficiency virus type 1 variants with resistance to multiple dideoxynucleosides in patients receiving therapy with dideoxynucleosides. *Proc. Natl Acad. Sci. USA*, **92**, 2398–2402.

Simpson, G.G. (1949) *The Meaning of Evolution*, Yale University Press, New Haven, CT.

Sousa, R. (1996) Structural and mechanistic relationships between nucleic acid polymerases. *Trends Biochem. Sci.*, **21**, 186–190.

St Clair, M.H., Martin, J.L., Tudor-Williams, G. *et al.* (1991) Resistance to ddI and sensitivity to AZT induced by a mutation in HIV-1 reverse transcriptase. *Science*, **253**, 1557–1559.

Steinhauer, D., Domingo, E. and Holland, J.J. (1992) Lack of evidence for proofreading mechanisms associated with an RNA virus polymerase. *Gene*, **122**, 281–288.

Suarez, D.L., Perdue, M.L., Cox, N. *et al.* (1998) Comparisons of highly virulent H5N1 influenza A viruses isolated from humans and chickens from Hong-Kong. *J. Virol.*, **72**, 6678–6688.

Taboga, O., Tami, C., Carrillo, E. *et al.* (1997) A large scale evaluation of peptide vaccines against foot-and-mouth disease: lack of solid protection in cattle and isolation of escape mutants. *J. Virol.*, **71**, 2606–2614.

Tachedjian, G., Mellors, J.W., Bazmi, H. and Mills, J. (1998) Impaired fitness of foscarnet-resistant strains of human immunodeficiency virus type 1. *AIDS Res. Hum. Retroviruses*, **14**, 1059–1064.

Temin, H.M. (1989) Is HIV unique or merely different? *J. AIDS*, **2**, 1–9.

Temin, H.M. (1993) The high rate of retrovirus variation results in rapid evolution. In: *Emerging Viruses* (ed. Morse, S.S.), pp. 219–225. Oxford University Press, Oxford.

Tomasselli, A.G., Howe, W.J., Sawyer, T.K., Wlodawer, A. and Heinrikson, R.L. (1991) The complexities of AIDS: an assessment of the HIV protease as a therapeutic target. *Chim. Oggi*, **9**, 6–27.

Turner, P.E. and Chao, L. (1998) Sex and the evolution of intrahost competition in RNA virus φ6. *Genetics*, **150**, 523–532.

Van Valen, L. (1973) A new evolutionary low. *Evol. Theory*, **1**, 1–30.

Weaver, S.C. (1998) Recurrent emergence of Venezuelan equine encephalomyelitis. In: *Emerging Infections 1* (eds Scheld, W.M., Armstrong, D. and Hughes, J.H.), pp. 27–42. American Society for Microbiology Press, Washington, DC.

Williams, G.C. (1992) *Natural Selection. Domains, Levels and Challenges*. Oxford University Press, Oxford.

Wimmer, E., Hellen, C.U.T. and Cao, X. (1993) Genetics of poliovirus. *Annu. Rev. Genet.*, **27**, 353–436.

Wlodawer, A. and Erickson, J.W. (1993) Structure-based inhibitors of HIV-1 protease. *Annu. Rev. Biochem.*, **62**, 543–585.

Wright, S. (1931) Evolution in Mendelian populations. *Genetics*, **16**, 97–159.

Wright, S. (1982) Character change, speciation, and the higher taxa. *Evolution*, **36**, 427–443.

Wyatt, C.A., Andrus, L., Brotnan, B., Huang, F., Lee, D.-H. and Prince, A.M. (1998) Immunity in chimpanzees chronically infected with hepatitis C virus: role of minor quasispecies in reinfection. *J. Virol.*, **72**, 1725–1730.

Yamada, K., Mori, A., Seki, M. *et al.* (1998) Critical point mutations for hepatitis C virus NS3 proteinase. *Virology*, **246**, 104–112.

Yerly, S., Rakik, A., de Loes, S.K. *et al.* (1998) Switch to unusual amino acids at codon 215 of the human immunodeficiency virus type 1 reverse transcriptase gene in seroconvertors infected with zidovudine-resistant variants. *J. Virol.*, **72**, 3520–3523.

Yuste, E., Sánchez-Palomino, S., Casado Herrero, C., Domingo, E. and López-Galíndez, C. (1999) Drastic fitness loss in HIV-1 upon serial bottleneck events. *J. Virol.*, **73**, 2745–2751.

Zimmern, D. (1988) Evolution of RNA viruses. In: *RNA Genetics*, vol. 2 (eds Domingo, E., Holland, J.J. and Ahlquist, P.), pp. 211–240. CRC Press, Boca Raton, FL.

8

The Retroid Agents:
Disease, Function and Evolution

Marcella A. McClure

INTRODUCTION

Studied for a decade now (Doolittle *et al.*, 1989; Xiong and Eickbush, 1990; McClure, 1991), the complex phylogenetic relationships of the retroid family (Fuetterer and Hohn, 1987) are still an active area of research. Retroid agents encode the ability to convert RNA into DNA by use of the reverse transcriptase (RT) function. It has been speculated that the RT function may be of ancient origin and responsible for the conversion of ancient RNA to DNA genomes (Darnell and Doolittle, 1986; McClure, 1991; Lazcano *et al.*, 1992; Doolittle and Brown, 1994). Regardless of the age of the retroid agents, they are coevolving with host genomes, some through mutualism. Recent studies on the molecular details of host gene regulation and the causes of various mutations are beginning to shed light on the intricate nature of retroid and host genome coevolution.

This chapter will provide an overview of the latest studies relevant to the complexities of retroid and host genome coevolution from the perspective of the author. The terminology used throughout this chapter is in keeping with the history of the literature. The term retroid encompasses all genetic agents that encode a RT. I prefer the term *agents* rather than *elements* to emphasize the dynamic nature of the various

roles retroid agents play in evolution. Other designations are the recent modifications (Boeke and Stoye, 1997) to the terminology suggested by Temin (Temin, 1985, 1989), with the following exceptions. The designation of long terminal repeats (LTR) retrotransposon versus non-LTR retrotransposon is misleading. Not all members of the LTR-retrotransposon lineage have LTRs (Ruiz-Perez *et al.*, 1996) and one member of non-LTR retrotransposons, DRE, has LTRs. I refer to the three distinct subfamilies of the LTR retrotransposons as *gypsy*-, DIRS1-, and *copia*-retrotransposons. Non-LTR retrotransposons are also called poly A retrotransposons; however, in keeping with the original nomenclature proposed by Temin I refer to these agents as retroposons. Contrary to some usage, sequences that have been reverse-transcribed, inserted back into genomes and lack the RT function (e.g. Alu) are retrotranscripts or retrosequences, not retroposons (Temin, 1985, 1989).

WHAT IS A RETROID AGENT?

Examples of retroid agents are found among viruses, bacteriophages, transposable elements, introns and plasmids. Relative to the host genome these agents function in two mobile classes, endogenous and exogenous. For the

Origin and Evolution of Viruses
ISBN 0–12–220360-7

most part endogenous agents are transferred vertically while exogenous agents can be both vertically and horizontally transmitted. Retroid genomes are found inserted into nuclear, organellar, bacterial, conjugative transposon and viral DNA genomes. While various retroid agents are found throughout eukaryotes, they are much less widely distributed in eubacteria (Table 8.1). Although RT activity has been detected in the Archaea, to date this activity has not been isolated and sequenced.

Retroid agents differ in several significant features: (1) the presence of LTRs; (2) the details of DNA synthesis; (3) the mechanism and location of integration; and (4) the order and conservation of genes. The first three features will be discussed briefly (Table 8.2) and the gene complement and order will be compared among retroid agents.

Retroviruses

Endogenous (ERVs) and exogenous retroviruses are found in both vertebrates and invertebrates. Among rodents additional categories of endogenous-retrovirus-like agents exist: intracisternal-A-type particles (IAPs), VL30 particles, MuRRS, GLNs, MuRVYs and ETn agents, all of which are related to various types of retrovirus. While IAP, VL30 and MuRVY genomes are most similar to retroviruses, encoding many of the same proteins, only some MuRRS and GLNs encode *gag, pol* or *env* genes. ETns are LTRs with no open-reading frames (ORFs; for a recent review see Boeke and Stoye, 1997).

All members of the retroviral lineage have LTRs at each end of their genomes that provide the regulatory sequences necessary for the viral life cycle. The RNA-dependent DNA polymerase (RDDP) uses the 3′ hydroxyl of a cellular tRNA, bound to the primer binding site (PBS) on the RNA template, to initiate DNA synthesis. While the virally encoded integrase (IN) is necessary for integration of the viral DNA, the structure of the host DNA within the nucleosome is important in determining target sites rather than any sequence or regional specificity (Brown, 1997; Table 8.2).

All retrovirus genomes encode two basic types of protein, structural and enzymatic. The structural proteins are: matrix protein (MA); capsid protein (CA), which forms an icosahedral-like shell; ribonucleoprotein (NC), which associates with the viral RNA; and envelope proteins (ENV), which insert into the host cell membrane to provide the virus coat and cell surface receptor binding. The enzymatic proteins are: aspartic acid protease (PR), which cleaves the polyprotein; the RDDP, comprising two functional domains, the RT activity, connected by a tether (T) to a segment with ribonuclease H activity (RH); and IN.

The NC and IN proteins have different histidine/cysteine (H/C) motifs, which bind zinc. In the NC this region is necessary for packaging and does not appear involved in the assembly of ribonucleoprotein (RNP) complexes. The assembly domains of basic residues in the NC are sufficient for RNP formation (Vogt, 1997). The H/C motif in the IN protein does not bind DNA directly. The motif is inferred to participate indirectly in the stabilization of the IN protein and the conserved terminal sequences of the viral DNA. At least two other regions of IN are known to bind DNA directly (Brown, 1997).

The structural and enzymatic genes are translated as a *gag-pol* polyprotein. The order of the proteins encoded by the *gag/pol* polycistron is MA/CA/NC/PR/RT-T-RH/IN. Spliced mRNAs encode the ENV proteins as well as various regulatory proteins, but the latter are not found in all retroviruses, and only evolve in some lineages. Likewise, both oncogenes and deoxyuridine triphosphatase (*dut*) genes are found among various retrovirus lineages.

Pararetroviruses

To date only three pararetroviruses have been identified; two from plants (caulimo- and badnaviruses) and one from vertebrates (hepadnaviruses). All three of these genomes are circular, partially double-stranded DNA, with a coding capacity of approximately 5000–8000 nucleotides. Replication is by way of an RNA intermediate that is reverse-transcribed by the virally encoded RDDP.

The caulimoviruses and badnaviruses differ in morphology (spherical versus bacilliform), genome organization, host range and vectors. As in retroviruses, caulimovirus DNA synthesis is initiated by the RDDP from the tRNA primer bound to a PBS of the viral polycistronic mRNA. The plant pararetrovirus genomes also carry a portion of the *pol* gene, encoding the PR, while the position of a zinc-binding domain similar to that of retrovirus NC is variable. In an upstream ORF of the petunia vein-clearing virus (PVCV) there is a sequence that encodes a putative IN and preliminary data suggest that integrated PVCV may exist in the genomes of healthy *Petunia* (Richert-Poggeler and Shepherd, 1997). There is no evidence to support any phylogenetic relationship between the capsid protein of the plant viruses and the analogous protein found in retroviruses. Plant pararetroviruses encode additional proteins unrelated to the retroid gene complement.

The hepadnaviruses have a relatively restricted host range; humans, small mammals and birds (Tiollais *et al.*, 1985). While 3′ covalently linked proteins involved in replication priming are not unknown in the viral world (adenoviruses (DNA) and picornaviruses (RNA) have them), only the hepadnaviruses use their own RDDP to provide a primer for DNA synthesis (Wang and Seeger, 1992; Table 8.2). Hepadnaviruses do not encode an IN, but they do sporadically integrate into host DNA. Integration is not necessary for viral replication. The genomes have four ORFs; only the third encodes a protein with similarity to the RT-RH of retroviruses (Miller and Robinson, 1986; Doolittle *et al.*, 1989). The second ORF encodes a capsid protein that is claimed to be related to that of the retroviruses on the one hand (Miller and Robinson, 1986) and that of the picornaviruses on the other (Argos and Fuller, 1988).

Transposons

In the 1940s Barbara McClintock discovered the mobile, changing nature of the genome and the effect of "jumping genes" on phenotypic expression (McClintock, 1987). We now know that transposable elements appear to be ubiquitous in the genomes of all life-forms and that several major classes of these elements exist (for recent discussion, see Capy *et al.*, 1998). Transposable elements that mediate their movement by an RNA intermediate and encode a RT are members of the retroid family. There are several distinct classes of RT-encoding transposable elements found throughout eukaryotes: the retrotransposons, the retroposons and retrointrons. The retrointrons are also present in eubacteria (Table 8.1).

Retrotransposons

In general, LTRs are found at the termini of retrotransposons, although there are exceptions (Ruiz-Perez *et al.*, 1996; Table 8.2). DNA replication is initiated by the RDDP, using either specific tRNA primers or various self-priming mechanisms (Lin and Levin, 1997). The site of integration varies widely among retrotransposons (Table 8.2). Some can integrate into copies of themselves, or other transposable elements, while others prefer sites upstream from genes transcribed by cellular Pol III (Chalker and Sandmeyer, 1992). In addition cellular Pol III transcription factors facilitate host DNA targeting (Kirchner *et al.*, 1995).

In general the gene order and complement of the *gypsy*-retrotransposon *pol* polycistrons are the same as extant retroviruses, PR/RT-RH/IN, while that of the *copia*-retrotransposons is PR/IN/RT-RH differing in the position of the IN. Several members of the *gypsy*-retrotransposon lineage, including the TED element inserted into a baculovirus mutant (Szatkowski Ozers and Friesen, 1996), encode putative ENV proteins. *Gypsy* can be an infectious agent and is the first retrovirus identified in invertebrates (Kim *et al.*, 1994; Song *et al.*, 1994; Table 8.1). Many members of the *gypsy*-retrotransposon lineage, however, do not encode the *env* gene. Although most *copia*-retrotransposons do not encode the *env* gene, recently a second open-reading frame in SIRE1 from soybean has been conceptually translated to reveal the presence of a putative 70 kDa protein with similarity to the ENV of retroviruses (Laten *et al.*, 1998).

TABLE 8.1 Distribution of retroid agents among eukaryotes and eubacteria (+a denotes that the *gypsy* can become an infectious virus, +b denotes that other *gypsy*- and *copia*-retrotransposons encode putative ENVs, although infectivity has not been demonstrated to date)

Retroid agents	Human	Vertebrates	Invertebrates	Plants	Fungi	Eukaryotes Protists Slime mold	Algae	Protozoa	Oomycetes	Plastids	Baculovirus	Genome	Eubacteria Conjugative transposons
Retroviruses	+	+	+a										
Pararetroviruses													
Caulimoviruses				+									
Badnaviruses				+									
Hepadnaviruses	+	+											
Transposons:													
Retrotransposons													
Gypsy-		+	+b	+	+	+		+	+	+	+		
DIRS1-			+		+	+							
Copia-		+	+b	+	+		+	+	+	+			
Retroprosons	+	+	+	+	+	+	+	+		+			
Retrointrons										+		+	+
Retroplasmids					+					+			
Retrons												+	
Retrophages												+	

TABLE 8.2 Variable features of retroid genomes. The 3′ hydroxyl required for first-strand DNA synthesis can be provided by a variety of molecules (DNA, RNA and protein) supplied by host or retroid genomes. Integration can be into copies of self, other transposable elements, specific sites, specific regions rather than sequences, and in DNA due to structural considerations (i.e. open regions of chromosomes). LTRs, long terminal repeats; PBS, primer binding site; ITRs, inverted terminal repeats; NA, not applicable; Question mark (?) indicates unknown

Retroid agent	LTRs	PBS	DNA synthesis primer			self	other	Integration specificity		
			host	self	protein			site	regional	structural
Retroviruses	+	+	tRNA	−	−	−	−	−	−	+
Pararetroviruses										
Plant	−	+	tRNA	−	−				†	
Animal	*	−	−	−	RT				‡	
Transposons:										
Retrotransposons										
gypsy-						+	+	+	+	+
gypsy	+	+	tRNA	−	−					
Tf1	+	−	−	RNA	−					
DIRS-	ITRs	−	DNA	?	?	+	?	?	?	?
copia-	+	+	tRNA	−	−	+	+	+	+	+
Retroposons	**	−	DNA	−	−	?	+	+	+	+
Retrointrons	−	−	DNA	−	−	††	?	?	?	?
Retroplasmids								NA		
Mitochondrial	−	−	−	tRNA	−					
Fungal	−	−	?	?	?					
Retrons	−	−	−	RNA	−	?	?	?	?	?
Retrophages	−	−	−	RNA	−	?	?	+	?	?

* hepadnaviruses have a retroviral LTR-like region in their genomes
† the PVCV genome encodes a putative integrase
‡ hepadnaviruses integrate sporadically and integration is not necessary for replication
** DRE elements have non-identical LTRs, other elements have 5′ and 3′ untranslated regions
‡ retrointrons can insert into intronless copies of cognate alleles

For many years the DIRS1-retrotransposon lineage had only one member, DIRS1 of *Dictyostelium discoideum* (Cappello *et al.*, 1985a,b). PAT from the nematode, *Panagrellus redivivus* (Chastonay *et al.*, 1992), and Prt1 from *Phycomyces blakesleeanus* are new members of this lineage (Ruiz-Perez *et al.*, 1996). Unlike all other retroid agents, DIRS1 and Prt1 genomes have inverted terminal repeats (ITR) of different lengths that appear unrelated to the LTRs of retroviruses or other retrotransposons. In contrast, the termini of the PAT genome are approximately one-half of a direct repeat of a 300 base internal region. This structure may have arisen by insertion into solo direct repeats that have been detected in the nematode genome (Chastonay *et al.*, 1992). DIRS1 is found randomly inserted into copies of itself and integration is thought to occur by homologous recombination (Cappello *et al.*, 1985). In contrast, only a single copy of Prt1 is found in the genomes of *P. blakesleeanus* strains (Ruiz-Perez *et al.*, 1996). While all three agents encode the two domains (RT-RH) of the RDDP, in DIRS1 the functions are adjacent, RT-RH, while in PAT and Prt 1 they are interrupted by a tether region, RT-T-RH.

Retroposons

The retroposon lineage encompasses a wide variety of agents that are variable in retroviral-like gene complement (McClure, 1991) and many genomes are truncated from the 5′ end. The reverse transcription mechanism of this diverse lineage is not well understood. Studies on the R2Bm agent of *Bombyx mori* suggest that a virally encoded endonuclease nicks the

28S rRNA gene insertion site, thereby providing a 3' hydroxyl primer for DNA synthesis (Luan *et al.*, 1993). A 250 nucleotide region at the 3' end of the R2Bm genome is also necessary for DNA synthesis and a model suggesting that R2Bm is directly reverse-transcribed into the 28S rRNA site is proposed (Luan and Eickbush, 1995).

Retroposons can be further classified into those that integrate into the host genome in a non-sequence-specific manner and those that target specific sites (McClure, 1993). Examples of the former group are the long interspersed nuclear elements (LINEs) of mammals, the I, F, G and jockey factors of *D. melanogaster*, and Tx1 of *Xenopus laevis* (Garrett *et al.*, 1989). Examples of the latter group are: Ingi3, embedded in a 512 bp ribosomal mobile element (RIME), of *Trpanosoma brucei* (Kimmel *et al.*, 1987); the spliced leader associated conserved sequences (SLACS) of RNA gene clusters in trypanosomes and trypanosomatids (Villanueva *et al.*, 1991); and the two agents R2Bm and R1Bm that insert into different positions of the 28S rRNA *B. mori* gene (Burke *et al.*, 1987; Xiong and Eickbush, 1988a).

Although retroposons encode the RT domain of the RDDP, the presence of the RH and IN genes is variable (McClure, 1991). SLACS-like retrotransposons encode a putative IN upstream from the RDDP (Villanueva *et al.*, 1991). Other retroposons encode a protein with significant similarity to the cellular apurinic/apyrimidinic endonucleases (APEs) that cleaves the DNA target sequence (Martin *et al.*, 1995; Feng *et al.*, 1996) rather than the IN found in retroviruses. Although endonucleolytic activity has been demonstrated for both R2Bm and R1Bm, only the latter encodes an APE-like protein (Feng *et al.*, 1998), while in the former the activity appears to reside in a single protein that also encodes the RT domain of the RDDP (Xiong and Eickbush, 1988b).

Many retroposons encode an additional upstream ORF that has the characteristic zinc-binding motif found in the NC protein of the retroviruses and many retrotransposons. This motif, however, is highly variable in copy number and position (McClure, 1991).

Retrointrons

Group II intron sequences of plastids can also encode the RT (Natvig *et al.*, 1984; Lang and Ahne, 1985; Kuck, 1989). These introns are self-splicing, transpose through an RNA intermediate, and can move to both allelic and non-allelic targets (Mueller *et al.*, 1993; Sellem *et al.*, 1993). The ability to move to intronless allelic sites is called retrohoming (Curcio and Belfort, 1996). In the retrointrons of the *COX1* mitochondrial gene the "integration" process occurs by reverse splicing. A dual cleavage function occurs in which the intron RNA cuts the sense strand of the host DNA and the endonuclease encoded by the intron cuts the anti-sense strand. The intron RNA (iRNA) is covalently linked to the host DNA. DNA synthesis is primed by a host 3' hydroxyl thereby directly reverse-transcribing the iRNA into a new site (Zimmerly *et al.*, 1995a,b; Yang *et al.*, 1996; Guo *et al.*, 1997).

Retrointrons are present in the chloroplast ancestor, cyanobacteria, and the probable ancestor of mitochondria, the γ subgroup of the proteobacteria (Ferat and Michel, 1993), as well as other eubacteria (Ferat *et al.*, 1994; Knoop and Brennicke, 1994). Interestingly, retrointrons have been found in several conjugative transposons (Gasson *et al.*, 1995; Mills *et al.*, 1996; Shearman *et al.*, 1996). Expression of the retrointron in pRS01 from *Lactococcus lactis* is necessary for conjugative transfer (Mills *et al.*, 1996). Retrointrons encode the PR, the RT, a region initially designated T that contains a maturase (M) activity necessary for splicing (Mohr *et al.*, 1993), and a zinc-binding region that is similar to the one found in the amino terminal region of the retrovirus/retrotransposon IN sequence (McClure, 1991).

Retroplasmids

The mitochondrial plasmids, Mauriceville-Ic, Varkud and Labelle-1b, are related (Pande *et al.*, 1989); however, their presence in different species of *Neurospora* varies (Natvig *et al.*, 1984). The mitochondrial plasmids replicate by a self-priming mechanism (Table 8.2). The priming hydroxyl is provided by a tRNA structure residing in the 3′ end of the Mauriceville RNA genome (Wang and Lambowitz, 1993). A second type of retroplasmid is found as the circular DNA plasmid of the Japanese pear fungal pathogen, *Alternaria alternata* (Kaneko *et al.*, 1997). Mitochondrial and fungal retroplasmids only encode the domains (RT-RH) of the RDDP.

Retrons and Retronphages

Among eubacteria the retroid agent called a retron appears ubiquitous in *Myxococcus*, but it is only found in a small percentage of other bacteria (Herzer *et al.*, 1990; Lampson *et al.*, 1991; Rice and Lampson, 1995). The protein gene product of a retron is the RT required for multicopy single-stranded DNA (msDNA) synthesis (Inouye *et al.*, 1989; Lim and Maas, 1989). The retron transcriptional unit comprises *msr, msd* and the RT genes. The *msr* and *msd* RNAs form a complex with the msDNA through a guanine (G) residue of the RNA molecule creating a 2′, 5′-phosphodiester linkage. Synthesis of msDNA is initiated by the RT from the 2′ hydroxyl of the internal G residue. Retrons found inserted into intergenic regions of *Myxococcus* genomes do not appear mobile (Inouye and Inouye, 1996).

The retronphages are those bacteriophages of *E. coli* that contain retrons (Sun *et al.*, 1991). In contrast to *Myxococcus*, only one retron is found inserted directly into the genome of *E. coli*; all others are inserted into cryptic phages. In at least one case, a cryptic retronphage-Ec73 produces an infectious retronphage, Φ73, by co-infection with a phage P2 helper (Inouye *et al.*, 1991). In addition, retronphages can encode other host genes (Hsu *et al.*, 1990). Although the retronphages are classified as retroid agents, perhaps they should be considered as just another viral DNA genome in which an agent, the retron, has integrated, similar to the *gypsy*-retrotransposon found in a baculovirus mutant.

The only gene encoded by retrons and retronphages related to the retroid agents is the RT. The necessary RH function, to remove the template RNA during msDNA synthesis, is provided by the host (Inouye and Inouye, 1996).

Eukaryotic Telomerase

There is only one example of a cellular gene encoding a RT function, the eukaryotic telomerase reverse transcriptase (TERT) (Lingner *et al.*, 1997; Nakamura *et al.*, 1997). All other RT activity observed to date is encoded by various retroid agents. The eukaryotic telomerase is a ribonucleoprotein enzyme with an endogenous RNA acting as the template for telomeric repeats and the protein, EST2, with RT activity. The RT ordered series of motifs common to all retroid agents is well conserved in the TERTs, although their genes are not known to be mobile

ROLES IN DISEASE AND HOST FUNCTIONS

Retroid agents are involved in a variety of roles within their hosts, including some disease processes. HIV and HTLV are well-known human retroviruses and there is a long history of the involvement of other ERVs and retroviruses in a variety of animal diseases (see Coffin *et al.*, 1997 for a recent review). Although the human ERVs (HERVs) can differentially express mRNAs in various types of tumor and teratocarcinoma cells (high), as well as in normal tissues (low) and placenta (high), their role in any aberrant cellular processes has not been established (Lower *et al.*, 1996). Among the pararetroviruses, the hepadnaviruses (type B) are the causative agents of hepatitis and hepatocellular carcinoma and are widespread in Asia and Africa (Tiollais *et al.*, 1985), while the pathogenicity of the plant pararetroviruses ranges from latency to lethal.

Some non-viral retroid agents are also implicated in disease processes. The LINEs, which comprise at least 15% of the human genome, can be found associated with several diseases: breast adenocarcinoma and teratocarcinoma, although no causal relationship has been established (Fanning and Singer, 1987). Insertional mutagenesis of a LINEs agent into specific genes can cause the human diseases hemophilia A (Kazazian *et al.*, 1988; Dombroski *et al.*, 1991) and Duchenne muscular dystrophy (Narita *et al.*, 1993).

Although the negative effects of retroid agent integration into host genes are well known, the examples of beneficial relationships are just beginning to be revealed (Britten, 1997). Retroid genomes can provide regulatory sequences for control of host genes, affect host phenotype by insertional mutagenesis (McDonald, 1995), mediate recombination, repair dsDNA breaks (Moore and Haber, 1996; Teng et al., 1996) and participate in telomere maintenance in *Drosophila* (Biessmann et al., 1990; Levis et al., 1993; Pardue et al., 1996; Biessmann and Mason, 1997). The retroposons TART and HeT-A have an exclusive target site, telomere ends of *Drosophila*. Ty-5 of *S. cerevisiae* (Zou et al., 1995) and other *copia*-retrotransposons found in *Allium cepa* (Pearce et al., 1996), and two retroposons, Zepp from *Chlorella* (Higashiyama et al., 1997) and SART1 from *B. mori*, are preferentially inserted into telomeric repeats (Takahashi et al., 1997). The generality of finding these agents in association with telomeric repeats and any role they may play in chromosome stability remains to be determined. Both retroposons and retrotransposons can mediate the repair of dsDNA breaks in eukaryotes, while retroviruses, retrotransposons and retroposons are all capable of affecting the host by insertional mutagenesis (Figure 8.1).

The first data suggesting regulatory control of a normal cellular gene by an ERV came from studies of the amylase gene cluster (Emi et al., 1988; Samuelson et al., 1988). The increase in sequencing efforts to understand the molecular basis of gene regulation and phenotypic mutation has revealed details of additional retroid agent and host gene interactions. At least one *D. melanogaster* and several different human and

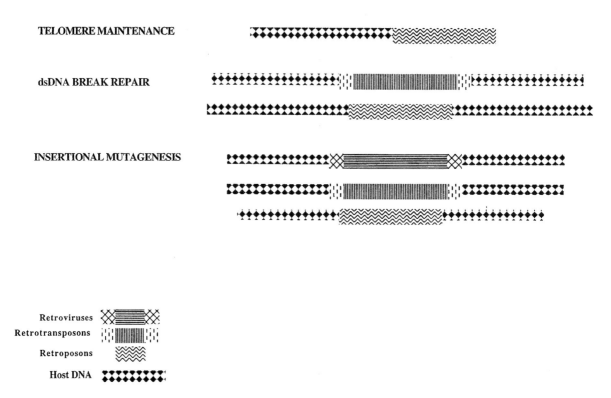

FIGURE 8.1 Retroviruses, retrotransposons and retroposons can integrate into host DNA. Retroposons maintain telomeres in *Drosophila*. Chromosomes can be repaired by the insertion of either retrotransposons or retroposons at dsDNA break points. Insertion of retroviruses, retrotransposons and retroposons can have a variety of effects on the host genome, as described in the text.

rodent genes are regulated by promoter or enhancer sequences provided by various retroid genomes (Table 8.3).

Many studies associate retroid agents with a variety of basic cellular processes. For example, ERVs appear to play a role in suppression of the mammalian female immune system during fetal implantation (Boyd *et al.*, 1993). The LTR of the HERV-H family is necessary for the transcription of a complex mRNA comprised of an Alu, a LINE-1, two ERV LTRs, and multiple copies of AUUUA proposed to play a biological role during T-cell activation (Kelleher *et al.*, 1996). HERV-mediated recombination is suggested to have played a role in the evolution of the MHC family (Svensson and Andersson, 1997).

A plethora of studies indicate that both *gypsy*- and *copia*-retrotransposons and retroposons are active in *Drosophila*. During *D. melanogaster* embryogenesis 15 different *gypsy*- and *copia*-retrotransposons exhibit conserved patterns of spatial and temporal expression, suggesting that they may be involved in a spatio-temporal regulation of surrounding host genes (Ding and Lipshitz, 1994).

The interactions between members of the *gypsy*-retrotransposon lineage and *Drosophila* are an active area of research providing a variety of examples of the coevolutionary nature of retroid agents and organismal genomes. The LTR and a repressor element within the 17.6 agent are required for lamina-specific expression in the development of *Drosophila* (Mozer and Benzer, 1994). Micropia conserves a testis-specific promoter and expresses an anti-sense RNA complementary to its own RDDP gene during primary spermatogenesis. It is postulated that this could be a mechanism for the control of germ-line expression of retrotransposition to avoid the accumulation of deleterious mutations (Lankenau *et al.*, 1994). In contrast to the potential for antisense control of germline expression of retrotransposons, it is well known that the effects of *gypsy*-induced mutations in host genes can be mitigated by binding of the host protein, *su(Hw)* (Mazo *et al.*, 1989). Furthermore, transposition of *gypsy* is controlled by the *Drosophila* gene *flamenco*. Transmission of *gypsy* to progeny is proposed to take place by infection of the germ-line in new oocytes by virus particles that have assembled in the mother's follicle cells (Pelisson *et al.*, 1997; Song *et al.*, 1997).

Drosophila genes are also involved in the

TABLE 8.3   Examples of retroid agent involvement with specific host genes. LTR-P, LTR-promoter; LRT-TSE, LTR-tissue-specific enhancer

Retroid agent	Gene	Function provided	Reference
Retroviruses			
Human			
ERV	Amylase cluster	LTR-P	Emi *et al.*, 1988; Samuelson, *et al.*, 1988
ERV-9	*HLA-DRB*	LTR-P	Mayer *et al.*, 1993; Svensson and Andersson, 1997
Rodents			
ERV-C-type	Mouse *slp*	Androgen-induced LTR-P	Stavenhagen and Robins, 1988
IAP	Rat oncomodulin	LTR-TSE	Banville and Bioe, 1989
IAP	Mouse *agouti*	LTR-P induces agouti changes	Duhl *et al.*, 1989; Michaud *et al.*, 1994
IAP	Mouse *MIPP*	LTR-TSE	VanHouten *et al.*, 1996
Retroposons			
Human			
LINEs	Thymidylate synthase	Poly-A signal	Harendza and Johnson, 1990
	Apo (a)	Enhancer sequences	Yang *et al.*, 1998
Retrotransposons			
D.*melanogaster*			
gypsy–Tom	*Om* genes	LTR-TSE induces eye morphology changes	Juni *et al.*, 1996

expression of *copia*-retrotransposons (Rabinow *et al.*, 1993). *Copia* is differentially regulated in the germ line of males and females. High expression is restricted to males in a tissue-specific manner, with highest levels detected during primary spermatogenesis (Pasyukova *et al.*, 1997). Another *copia*-retrotransposon, 1731, is spatio-temporally expressed during *Drosophila* embryogenesis, and both transcription and translation are differentially regulated in the adult ovaries and testes (Haoudi *et al.*, 1997).

Retroposon expression has been demonstrated to be tissue-specific during development in humans (Swergold, 1990) and insects (Lachume *et al.*, 1992). The Fex retroposon of *Drosophila* is inserted into the intron of the developmental control gene *point*. Fex is expressed during oogenesis and primary spermatogenesis, as well as in a variety of embryonic stages and organs (Kerber *et al.*, 1996).

Retrotransposons and retroposons are found throughout the plant world, and in some cases comprise a significant portion of plant genomic DNA (50–90%). Reversible methylation inactivation of retrotransposon sequences appears to control transposition frequency (Wessler *et al.*, 1995). In general, retroid agents are only expressed under stressful conditions in plants (Wessler, 1996).

Data are accumulating revealing the intricate relationship between retroid and host genomes. One of the most revealing findings regarding retroid agents and their hosts is their mobilization in the undermethylated genome of progeny resulting from interspecific hybridization between *Macropus eugenii* and *Wallabia bicolor* (Australian wallabies). This mobilization occurred in one generation, illustrating that retroid agents play a role in the restructuring of genomes and are no doubt involved in speciation (Kidwell and Lisch, 1998; Waugh O'Neill *et al.*, 1998).

RETROID AGENT EVOLUTION

The evolution of the retroid agents can be studied by analysis of both individual genes and entire genomes. The following discussion will cover several aspects of tracing the history of individual protein sequence relationships and the acquisition or loss of genes in the construction of retroid genomes.

Protein Evolution

The basic assumption in tracing the evolutionary history of a protein is that analogous functions actually share common ancestry and are, therefore, homologs. Statistical confidence supporting homology is based on the Dayhoff criteria (Schwartz and Dayhoff, 1978). Protein sequences that share more than 25% identity are usually homologous. While there is no doubt that common ancestry exists within the major lineages of retroid agents, there is no direct statistical evidence for protein homology between these lineages. Monte Carlo simulations of sequence similarities among and between retroid and cellular sequences for the PR, RT and RH fail the Dayhoff criteria for homology (McClure, 1992, unpublished). There is no support for significant phylogenetic signals between many of the RT sequences and the trees derived from such data are not statistically supported (Zanotto *et al.*, 1996). Given these facts I refer to the sequences that fail homology criteria as functionally equivalent proteins (FEPs). The use of both homologous and FEP sequences in "phylogenetic reconstruction" naturally leads to the idea of expressing this relationship as a mixture of supported phylogenetic lineages (within major lineages) and a functionally equivalent network, FEN (between major branches and lineages). I will use the term FEN to denote the deep relationships among the genes common to the retroid agents.

Homologous and Functionally Equivalent Proteins

The only function common to all retroid agents is the RT. Only the retrons and retronphages, and cellular TERTs, encode this single function. In most other retroid agents the RT function is present in the amino portion of the RDDP while the RH function resides in the carboxyl portion (Johnson *et al.*, 1986; Figure 8.2). Initial studies

RNA-dependent DNA Polymerase

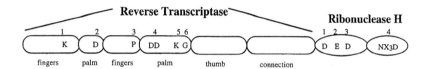

Aspartic Acid Protease

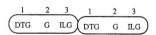

Integrase

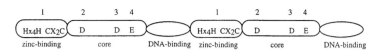

FIGURE 8.2 Enzymes of the retroid genome. The RNA-dependent DNA polymerase (RDDP) is a monomer comprised of two functional domains, reverse transcriptase (RT) and ribonuclease H (RH). The most highly conserved residues of the ordered series of motifs of the RT functional domain (McClure, 1993) are placed within the structural domains (fingers, palm, fingers, palm, thumb and connection) defined by the HIV-1 RT crystal (Kohlstaedt *et al.*, 1992). The most highly conserved residues of the ordered series of motifs of the RH domain (McClure, 1991) are placed within the two structural RH domains based upon comparison of the HIV-I and *E. coli* RH crystal structures (Davies *et al.*, 1991). The retroid protease is an aspartic acid protease that functions as a dimer. The most highly conserved residues of the ordered series of motifs are positioned within each half of the dimer. The integrase (IN) is divided into three domains: a zinc-binding region; the catalytic core that conserves the DDE residues of the polynucleotidyl transferases in three motifs; and a DNA-binding region (adapted from Brown, 1997). IN functions as a dimer. The most highly conserved residues of the ordered series of motifs of the IN protein (McClure, 1991) are placed in each domain within each half of the dimer. Amino acids are in the single letter code. X indicates invariant numbers of residues within indicated motifs. Each motif of the ordered-series of-motifs in each protein is labeled numerically.

on the evolutionary history of retroid agents were based on the analysis of the RT relationships (Doolittle *et al.*, 1989; Xiong and Eickbush, 1990; McClure, 1991). Independent FENs representing the relationships for the other proteins common among retroid agents generally support RT FEN topology (Doolittle *et al.*, 1989; McClure, 1991, 1993).

The rooting of the RT FEN has been attempted by using various RNA-dependent RNA polymerases (RDRPs) as outgroups (Xiong and Eickbush, 1990; Nakamura *et al.*, 1997; Nakamura and Cech, 1998). The reasoning behind this approach is that the canonical motif, a hydrophobically embedded Asp-Asp (DD), of the RDRPs of positive-strand RNA viruses is

also present in the RT domain of the RDDPs. This observation led to the notion that the RT domain and all RDRPs (size range approximately 200–2000 amino acids) share a common ancestry (Argos, 1988; Poch *et al.*, 1989; Delarue *et al.*, 1990). There is no statistically significant sequence relationship between the RT domain and the RDRPs to support the idea that RNA-dependent polymerases share a common ancestry (Zanotto *et al.*, 1996).

It is suggested that all DNA polymerases share ancestry. Relationships based on structural similarities, however, may be due to mechanistic analogy or convergence, rather than ancestry (Steitz, 1998). Recent analysis of DNA polymerase structures indicates that there are four unrelated DNA polymerase families, one of which is the RT domain family (Joyce and Steitz, 1994; Brautigam and Steitz, 1998).

Within all known RT domains there is an ordered series of motifs that contains only seven conserved residues (Figure 8.2). The degree of sequence relationship between the RT domain and other RDRPs is far less than seven residues and there is no significant sequence similarity between RT domains and other DNA polymerases.

The RH domain of retroviral RDDPs was initially identified by comparison to the *E. coli* RH sequence (Johnson *et al.*, 1986). Most, but not all retroid agents encode the RH domain. In addition to the retrons and retronphages, and TERTs mentioned above, neither the R1Bm and R2Bm retroposons nor the retrointrons encode the RH domain. Cellular RH genes appear ubiquitous in eubacteria and eukaryotes.

Among the retroid agents only the Moloney murine leukemia virus (MoMLV) RH domain can function as an independent module, separate from the RT domain. A comparison of the MoMLV and *E. coli* RH sequences reveals a shared stretch of basic amino acids not present in other RH sequences (Davies *et al.*, 1991). Comparison among and between retroid agent and *E. coli* RH sequences indicates that there are only five conserved residues within the ordered series of motifs (Figure 8.2). This level of sequence relationship cannot be demonstrated to support common ancestry (McClure, 1992).

Early sequence comparisons suggested that the retroviral PR was a member of the aspartic acid protease family (e.g. pepsin; Toh *et al.*, 1985). The pepsin family sequence underwent an ancient tandem gene duplication, thereby physically linking the two sets of ordered series of motifs required for catalysis. Retroid agent PR gene segments have not undergone this duplication event and sequence comparisons predicted that the retrovirus PR would be a dimer (Pearl and Taylor, 1987; Doolittle *et al.*, 1989; Figure 8.2). These predictions were confirmed by X-ray crystal determination for the Rous sarcoma virus PR (Miller *et al.*, 1989).

Comparisons between retroid PR sequences to either the amino or carboxyl pepsin analogs from several eukaryotes did not identify any potential

cellular ancestral sequence. The retroid PR is equidistant from representative pepsin sequences. A straightforward interpretation of the PR FEN is that the retroid PRs diverged from one another prior to the duplication event that gave rise to the pepsin family. The time of the duplication event, estimated from the rates of change of cellular proteases, is approximately 2 billion years ago. These data suggest that the viral protease could be of ancient origin (Doolittle *et al.*, 1989). Given the accelerated rate of evolution exhibited by some retroid agents one could expect all similarity to any host ancestral PR to be lost by now. Viruses, however, exist as populations under strong selection pressures that act to control extreme rates of change (Domingo and Holland, 1997). Such pressures could very well act to maintain critical residues necessary for optimal enzymatic function in the retroid PR over a long evolutionary time span. Alternatively all sequence similarity to an ancestral pepsin could have eroded over time, leaving only the ordered series of motifs that confers function. As in the RDDP sequences, little phylogenetic signal remains among and between the retroid PRs and cellular aspartic acid proteases. Only seven amino acids of the retroid and cellular PR sequences are conserved in the ordered series of motifs that folds into the catalytic site (Figure 8.2).

Analogous Functions

There are two cases of analogous function, endonuclease (IN and APE) and ribonucleoprotein (NC), found within various retroid genomes. In the endonuclease example the IN protein has three domains as determined by sequence and functional analyses. The first domain of the IN is a zinc-binding region (Burke *et al.*, 1992). The core domain is the catalytic region, which contains three motifs, each of which conserves one residue of the canonical aspartic acid, aspartic acid, glutamic acid (DDE) pattern of the polynucleotidyl transferases (Fayet *et al.*, 1990; Dyda *et al.*, 1994). The third domain binds DNA. As revealed by the crystal structure, IN functions as a dimer (Dyda *et al.*, 1994; Figure 8.2).

The encoding of a retroviral-like IN is variable among retroid agents. In the four definitive cases where IN is found – *gypsy*- and *copia*-retrotransposons, one member of the Caulimoviruses, PVCV, and the SLACS-like retroposons – the genes' position in these genomes varies (Villanueva *et al.*, 1991).

Among the retroposons the source of the endonuclease activity is variable. R1Bm and many LINEs-like agents encode a cellular-like APE protein (Figure 8.3). In contrast, other retroposons (CZAR, SLAC and CRE1) do not encode the APE protein. They encode an IN just upstream from the RDDP.

Retroposons R2Bm and Ingi3 are two unusual examples regarding the source of the endonuclease function. R2Bm is known to encode an endonuclease function within the protein with RT activity and it does not encode an APE function (Feng *et al.*, 1998). The residues responsible for the endonuclease function in R2Bm, however, have not been experimentally identified. Early sequence analysis identified IN-like sequences in the retroposons Ingi3 and R2Bm, but not in their relatives, I factor and R1Bm respectively (McClure, 1991). In the Ingi3 genome the IN-like sequence is immediately adjacent to the RH domain. Read through of a stop codon would allow this sequence to be translated. In contrast, in R2Bm, the IN-like sequence is immediately adjacent to the RT in the same ORF (Figure 8.3). While the zinc-binding domain is conserved in Ingi3 and most mitochondrial, bacterial and conjugative transposon retrointrons, it is not well conserved in R2Bm. Within the putative zinc-binding motif ($HX_5HX_{31}CX_2H$, where X_n represents the number of variable amino acid residues) of the *COXI* gene al1 and al2 retrointrons, however, similarity to one of the motifs ($HX_{14-24}NX_{10-24}H$, N is asparagine) found in bacterial endonucleases can be recognized. Although the terminal

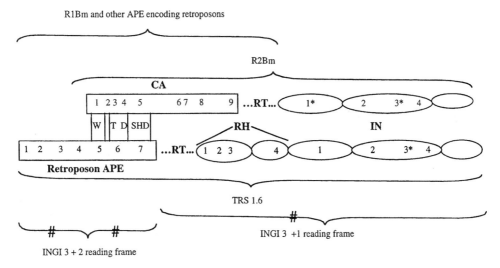

FIGURE 8.3 Schematic representation of different coding strategies for the two different endonuclease activities found in retroposons. The APE region is usually found in the same reading frame upstream of the RT domain in R1Bm-like agents. In TRS 1.6, however, the APE is found upstream and adjacent to the RT domain. In Ingi3 it is in the +2 reading frame with two stop codons. Analysis of the R2Bm sequence, which does not encode an APE, suggests that a region upstream of the RT domain is similar to the CA sequence of retroviruses (McClure, 1991). Motifs five, six and seven of the retroposon APEs share similarity with motifs one, three, four and five of the CA respectively (see text for details). Both TRS 1.6 and Ingi3 encode IN-like sequences immediately downstream from the RH domain. In R2Bm the IN-like sequence is found adjacent and downstream from the RT domain. Each motif of the ordered-series-of-motifs in each protein is labeled numerically (see Figure 8.2 for the highly conserved amino acids of each motif). Asterisks (*) indicate less conserved residues within specific motifs as described in the text. The hash sign (#) indicates the position of stop codons.

histidines are common to the two motifs there is no statistical support for common ancestry (Gorbalenya, 1994; Shub *et al.*, 1994). Mutagenesis studies on the zinc-binding domain of the retrointrons confirm that this region contains the endonuclease activity (Zimmerly *et al.*, 1995b). The retrointrons do not encode the other two domains of IN. The IN-like sequences in both Ingi3 and R2Bm, however, conserve four motifs in common with retroviral IN sequences in addition to the zinc-binding motif (McClure, 1991). Three of these motifs comprise the IN core domain and coincide with the highly conserved residues DDE (Figure 8.2). In the Ingi3 and R2Bm sequences, residues surrounding the second D residue of motif 3 are conserved, although the D residue itself is not (Corro and McClure, unpublished).

A second Ingi3-like agent, TRS 1.6, encodes an RDDP (Murphy *et al.*, 1987). The genome of this agent also encodes a putative APE (Olivares *et al.*, 1997). In contrast to Ingi3, the TRS 1.6 genome has a longer ORF that encodes the APE, the RDDP and IN-like sequences (Figure 8.3). Given this finding, all six reading frames of the Ingi3 agent were examined. The relationship among cellular and viral APEs consists of an ordered series of seven motifs. Six of these motifs conserve one to three identical amino acid residues (Feng *et al.*, 1996; Blesa and Martinez-Sebastian, 1997). In the +2 reading frame of the Ingi3 genome all seven motifs of the APE can be found, interrupted by two stop codons (Corro and McClure, unpublished; Figure 8.3).

While the endonuclease activity of R2Bm resides in the RT protein (Xiong and Eickbush, 1988b), and the region proposed to have IN-like sequences is downstream and adjacent to the RT domain, it was of interest to re-examine the sequences upstream of the RT domain. This region of the R2Bm genome, as well as several other retroposons, is proposed to encode a retroviral capsid-like sequence with nine motifs (McClure, 1991). In APE+ retroposons the amino portion of the proposed CA is now known to be the carboxyl end of the APE (Figure 8.3). The R2Bm sequence in this region indicates that the APE and CA have conserved motifs in common. Motif two in the CA is found in an unaligned

region of APE sequences between motifs five and six. The threonine (T) and aspartic acid (D) residues of the APE motif six are conserved in CA motifs three and four, respectively. APE motif seven corresponds to the CA motif five. In the motifs common to APE and CA sequences, R2Bm only conserves the D residue of motifs seven and five, respectively. In APE motif five and CA motif one trytophan (W) is conserved among many of the retroposons that is not present in cellular APEs. In searching upstream of this region in the R2Bm genome the remaining APE motifs could not be identified. Interestingly, some of the retrovirus CA sequences conserve more residues with the APE carboxyl terminus than does the R2Bm sequence. In addition, between the end of the APE and the beginning of the RT sequences APE+ retroposons share additional sequence similarity to the retroviral CA sequence (McClure, 1991).

The lack of conservation of the APE sequence and detection of endonucleolytic activity in the R2Bm RT protein on the one hand, and the presence of IN-like sequences on the other, suggests that the endonuclease activity is located in the IN-like sequences. The mechanism of action may differ from that of retroviruses and retrotransposons. The genomes of Ingi3 and TRS 1.6 suggest that some retroposons may encode both an APE and an IN, although these activities have yet to be demonstrated for these agents. The overlap in APE and CA motifs could be a coincidence, vestigial, or perhaps alternative proteolytic cleavage of the polyprotein can give rise to a CA protein in these cases.

The second case of analogous function is the NC, the protein component of the RNP complex. Many retroposons encode an additional upstream ORF that has a putative zinc-binding motif characteristic of the NC protein of the retroviruses and many retrotransposons (McClure, 1991). This motif, however, is highly variable in copy number and position. The protein encoding this motif varies in size. In caulimoviruses an upstream ORF has a small region that can be aligned to the H/C motif of the NC (Covey, 1986), while in badnaviruses and PVCV this region is found in frame, upstream from the PR (Richert-Poggeler and

Shepherd, 1997). In retroviruses and retrotransposons the NC varies from approximately 10 to 15 kDa, although the analogous protein found in the RNP of retroposon Tx1L in *Xenopus* oocytes is an approximately 105 kDa protein that is not protolytically cleaved (Pont-Kingdon *et al.*, 1997). It is the basic residues of the NC, called assembly domains, that are sufficient for RNP complex formation (Vogt, 1997). There is no information regarding the presence of such domains in the 105 kDa protein. The similarity of the 10–15 kDa and 105 kDa NC proteins is limited to the zinc-binding motif and, given the size difference between these two types of protein, there is little support for common ancestry. In contrast to the zinc-binding motif NCs, the human LINE-1 encodes a 40 kDa protein with a leucine zipper in an upstream ORF that forms a RNP complex with the LINE-1 RNA (Hohjoh and Singer, 1996).

Retroid agents provide a new source of genomes to study the acquisition of genes with analogous functions (endonucleases and ribonucleoproteins). Two types of putative endonuclease are encoded in the ingi-like TRS 1.6 genome in the same ORF as the RDDP. In the Ingi3 genome the IN sequence is in the same reading frame as the RDDP and the APE is in an upstream reading frame. Although the coding sequences for each of these endonucleases are interrupted by stop codons (Figure 8.3), it is well known that retroviral mRNA stop codons can be read through and frame shifts can occur during translation to produce polyproteins. Whether or not this can occur with the Ingi3 genome to produce IN and/or APE proteins remains to be determined, but there is nothing to impede the production of both these proteins in the TRS 1.6 agent. All other retroposon genomes analysed to date, however, only encode the IN, the zinc-binding motif of IN or the APE. Are the ingi-like agents the descendants of a genome that once encoded both putative functions or were they independently acquired?

In the case of the NC, there are three distinct types; large and small zinc-binding proteins and a medium-sized leucine zipper protein, involved in RNP formation of different retroid agents. There are no data to support common ancestry of these proteins, suggesting that they have been independently acquired.

Genome Evolution

Reconstructing the relationships of retroid agents cannot be based on the RT domain alone. The remaining gene complement and order of the genomes must be considered. The RT domain, however, has the slowest relative rate of change among all retroid proteins (McClure *et al.*, 1988). The topology of the RT FEN, therefore, is an appropriate choice upon which to map the relationships from other regions of the genome (McClure, 1991, 1993). The actual path by which various functions were acquired or lost by retroid genomes cannot be known. Parsimony is used to provide a network of evolutionary events that accounts for the gene complement of extant retroid agents (Figure 8.4).

There are two major branches of the retroid agents based on reconstruction of the relationship between the homologous and FEP RT sequences. Compared to retroviruses the remaining gene complement of retroid genomes is variable among lineages (Figure 8.4). Eukaryotic branch I could have arisen from a hypothetical CA/NC/PR/RT-RH/IN/ENV ancestral genome. Allowing transposition of the IN portion of the *pol* gene along some lineages prior to subsequent divergence, the hypothetical ancestor would account for the genome arrangement found in retroviruses, *gypsy*- and *copia*-retrotransposons. The *env* gene may have been lost by many of the *gypsy*- and *copia*-retrotransposons over evolutionary time.

Among the remaining branch I agents, only the plant pararetroviruses encode additional retroid proteins. All caulimo- and badnaviruses encode the NC and PR. To date only one member of this group, PVCV, encodes a potential IN protein. The FEN topology for the pararetrovirus RH domains and PR proteins are congruent with the RT FEN (Doolittle *et al.*, 1989) suggesting that these regions came from a common ancestor.

The plant pararetrovirus RDDP sequences are most similar to those of the *gypsy*-retrotransposons/retroplasmids, and DIRS1 retrotransposons, suggesting that portions of the polycistron were transferred among these agents (Figure 8.4). Given that the *gypsy*-like retroplasmid only encodes the RDDP, perhaps it acts as a vector for dissemination of *gypsy*-like RDDP sequences. The hepadnavirus (Hep B) RDDP sequences are most similar to those of the *copia*-retrotransposons, suggesting that one branch acquired this part of the genome from the other (Figure 8.4). Alternatively, the smaller retroid genomes along the *gypsy*- and *copia*-retrotransposon lineages could have lost ancestral genes.

Branch II bifurcates into the eukaryotic retroposons, and a lineage that includes the retrointrons (eubacteria and plastid), retrons and retronphages (eubacteria), retroplasmids of mitochondria and TERTs (eukaryotes). The two major lineages of the retroposons are SLACS-like genomes (IN/RT-RH) and the LINEs-like

lineage that can be accounted for by the hypothetical ancestral genome APE-CA/RT-T-RH/IN (Figure 8.4). This ancestor encoded two endonuclease functions (APE and IN), as suggested by the TRS 1.6 and Ingi3 genomes (Figure 8.3). In general, subsequent loss of either APE or IN and T/RH along various sublineages accounts for the LINEs-like retroposon genomes. The source of the transposed IN sequence in the SLACS-like genomes has not been determined.

Among the eubacterial retroid agents, the retrointrons encode a PR-like sequence that may be vestigial, the RT domain, the M region and the zinc-binding domain of the IN. The RT domains of the mitochondrial retroplasmid RDDP and the retrons/retronphages RT proteins share more sequence similarity to one another than to either the retrointrons or TERTs.

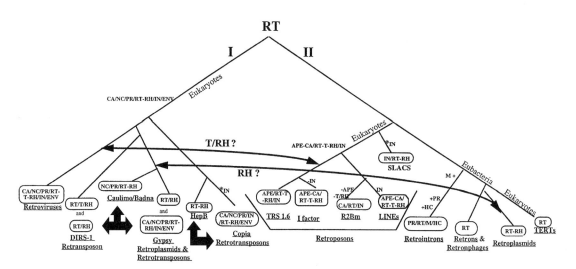

FIGURE 8.4 The functionally equivalent network (FEN) of the retroid agents. The topology of the FEN is based on the pairwise distance values of RT sequences representative of the retroid agents and is unrooted. Since the actual path by which various genomes acquired or lost functions cannot be known, a parsimony approach is used to account for the gene compliment of extant genomes (adapted from McClure, 1991, 1993). Two major branches are indicated (I and II). CA/NC/PR/RT-RH/IN/ENV is the hypothetical ancestor of branch I. The RDDP sequence similarity among *gypsy*-retrotransposons and retroplasmids, DIRS1 retrotransposons and the plant pararetroviruses (caulimo/badna), suggests that either these genomes acquired various subsets of genes from one another or lost genes from a hypothetical ancestor (three-way arrow). The PR and RDDP sequence similarity between hepadnaviruses (Hep B) and *copia*-retrotransposons suggests that one lineage acquired this part of the genome from the other or that Hep B lost some of the gene complement of the hypothetical ancestor (two-way arrow). Branch II bifurcates into two lineages, the eukaryotic specific retroposons, and the eubacterial lineage, which also includes the eukaryotic TERTs. Independent FEN topologies for RT and RH sequences indicate that the RH sequence of the eubacterial retroplasmids is more similar to that of both *gypsy*-retrotransposons and plant pararetroviruses. The RH sequences of the Ingi3 and I factor branch are more similar to that of the retroviruses (McClure, 1991). Lines with double arrowheads indicate these relationships. Abbreviations for various proteins and domains, and the names of retroid agents, are defined in the text. Asterisks(*) indicate transpositions of IN along various lineages.

The Argument for Common Ancestry

As discussed in the section on protein evolution, the sequence similarity remaining between any one domain or protein (RT-RH, PR or IN) that performs analogous function in the retroid life cycle is not significant enough to support common ancestry for all retroid agents. How then can common ancestry be asserted? The argument for common ancestry is based on positing the question of divergence versus convergence at different levels of gene and genome relationships. First, the PR, RT-RH and IN sequences have each conserved a specific ordered series of motifs that confers catalysis. Second, in general, the gene complement and order of retroid genomes is conserved (Figure 8.4). The gene order for the enzymatic core of the genome is PR/RT-RH/IN, although there are exceptions for the IN position as mentioned. The conservation of both the ordered series of motifs in the enzymatic core genes and the gene complement in retroid genomes strongly suggests ancestry rather than convergence.

Common ancestry is also supported by the congruent topology of the different phylogenetic branches and FENs representing the similarity or distance between PR, RT-RH and IN sequences. The only exceptions to these congruent topologies are two cases of suspected xenologous recombination. Xenologous recombination is the replacement of a gene in a genome with a foreign copy of the gene. The RH domain sequences of Ingi3 from *T. bruci* and I factor from *D. melanogaster* are significantly more similar to the RH of retroviruses, while the RH sequences of mitochondrial retroplasmids, and caulimoviruses and *gypsy*-retrotransposons are most similar. The FEN topology for RH sequences suggests that in these two cases xenologous recombination played a role in the evolution of some retroid genomes (McClure, 1991; Figure 8.4).

Rates of Change and Horizontal Transfer

In RNA virus replication systems the error-prone polymerase, the lack of an RNA correc-tion mechanism and the high rate of replication can give rise to RNA genomes with mutation rates of 10^{-3}–10^{-6} mutations/nucleotide/replication cycle. In contrast, the mutation rate of DNA genomes is 10^{-8}–10^{-11} mutations/nucleotide/replication cycle (Preston, 1996, and references therein). The high rate of accumulated mutation in RNA genomes leads to the existence of a single virus population as a dynamic distribution of non-identical genomes, or quasispecies (for a recent review of RNA virus population dynamics, see Domingo and Holland, 1997). It is well known that the HIV exists as a quasispecies (Delassus *et al.*, 1991; Sala *et al.*, 1994; Nájera *et al.*, 1995), but can other retroid agents rapidly mutate and create quasispecies? One of the rare, active *copia*-retrotransposons of plants, Tnt1 of tobacco, is demonstrated to exist as a quasi-species-like population (Casacuberta *et al.*, 1995, 1997). Although the heterogeneous nature of some retroid agents is well documented there is little direct measurement of mutation rates. The existing data suggest that different retroid agents can evolve at different rates. Among the retroposons, two different LINE-1 lineages in *Peromyscus* evolve at different rates within the same genome (Casavant *et al.*, 1996), while the R1 and R2 agents of *Drosophila* appear to be changing at the same rate as nuclear genes (Eickbush *et al.*, 1995). In sharp contrast, the rate of mutation has been directly measured for the *copia*-retrotransposon Ty1, demonstrating that this agent can mutate as rapidly as a retrovirus (Gabriel *et al.*, 1996). This variability in the rate of evolution is not unexpected. The frequency of transposition and, therefore, the opportunity for mutation is correlated with retroid genome copy number: some host genomes have only one copy, while others have hundreds of thousands (Kumar *et al.*, 1997).

Does this variation in rate require a re-examination of data claiming horizontal transfer? It has been demonstrated that some retroid agents are only transmitted vertically and suggested that claims of horizontal transfers based on unexpectedly similar RT sequences found in distantly related organisms may be due to purifying selection rather than horizontal transfer (McAllister and Werren, 1997). If either phylogenetic inference or claims of horizontal transfer are only based on analysis of the highly conserved RT ordered series of motifs and surrounding regions, there is a risk of misinterpreting similarities that are more probably due to parallel mutations and selection rather than ancestry. Although the RT is the most conserved of all retroid proteins, it should be recalled that the protein sequence relationships between various retroid lineages are, in general, less than 25% identical and in many cases far less. Most of this identity is in the ordered series of motifs that are functionally constrained. This low-level similarity no doubt reflects the purifying selection aspect of RT evolution over long time spans. In cases where there is highly significant identity (e.g. greater than 50%) between the retroid genes found in organisms that have been separated for hundreds of millions of years it is difficult to invoke such selection. Examples of probable horizontal transfer of retroid agents are accumulating that are based on more than just sequence similarity.

One of the best arguments for horizontal transfer of retroid agents is the discontinuous distribution of a specific agent among species or organisms. Although loss of a specific agent in many organisms could account for some of the observed distributions, it appears that loss is a rare event (Eickbush, 1997). Sequence similarity, not only of the RT but of entire genomes including the LTRs, is also used to support horizontal transfer.

Examples of probable horizontal transfer of retrotransposons, retroposons and retronphages are beginning to accumulate (Table 8.4). Among *gypsy*-retrotransposons, the sequence analysis of the complete genome of the *C. elegans* Cer1 agent suggests that it was recently acquired (Britten, 1995a). There are only two *gyspy*-retrotransposons of vertebrates. Cpr1 of *Clupea pallasi* (herring; Britten, 1995b); and Hsr1 from the terrestrial salamander *Hydromantes* (Marracci *et al.*, 1996). Although Hsr1 is found throughout the genus *Hydromantes* in Europe it is not present in the American relative *H. shastae* or in two other urodeles examined. These data suggest that the Hrs1 agent was acquired prior to the European *Hydromantes* speciation (Marracci *et al.*, 1996).

To data DIRS1-retrotransposons have only been found in slime molds (DIRS1), nematodes (PAT) and fungi (Prt1). Although PAT is present in multiple copies in the *P. redivivus* genome it is not present in the closely related *C. elegans*

TABLE 8.4   Examples of horizontal transfer of retroid agents

Retroid agent	Organism(s) involved	Reference
Retrotransposons		
Gypsy-		
Cer 1	*C. elegans*	Britten, 1995a
Hsr 1	European *Hydromantes*	Marracci *et al.*, 1996
DIRS-		
PAT	*P. redivivus*	Chastonay *et al.*, 1992
Prt1	*P. blakesleeanus*	Ruiz-Perez *et al.*, 1996
Copia-		
Tch1	*C. harengus*	Flavell and Smith, 1992
Tpm1	*Colubridae*	Flavell *et al.*, 1995
Tst1	*S. tubersosum, P. infestans*	Tooley and Garfinkel, 1996
Copia	*Melanogaster subgroup*	Jordan and McDonald, 1998
Retroposons		
LINE	*Viperidae, Ruminantia*	Kordis and Gubensek, 1997, 1998
Retrophages	*E. coli*	Inouye and Inouye, 1993

genome (Chastonay *et al.*, 1992). In contrast, only a single copy of Prt1 is found in the genomes of geographically isolated *P. blakesleeanus* strains and it is absent in other *Phycomyces* species (Ruiz-Perez *et al.*, 1996). Both these cases support the horizontal acqusition of DIRS1-retrotransposons from an unknown source.

There are several probable cases of horizontal transmission of *copia*-retrotransposons. Tch1 was the first agent of this class found in vertebrates. It is present in *C. harengus* (herring), but not in zebrafish, salmon or humans (Flavell and Smith, 1992). A second study extended the finding of *copia*-retrotransposons in vertebrates and several agents have been found and characterized among amphibians and reptiles. A survey of Old and New World snakes for the Tpm1 agent indicates that its presence is discontinuous among the descendants of the ancestral *Colubridae* snake. Massive losses of this agent from the *Crotalinea* and *Elpidae* lineages would have to be invoked to account for its absence. It is more likely that Tpm1 has been transferred horizontally among some Old World snakes (Flavell *et al.*, 1995). Sequence similarity between the Tst1 agent of the host, *Solanum tuberosum* (potato), and some strains of the *Phytophthora infestans* pathogen carrying Tst1 also supports horizontal transfer (Tooley and Garfinkel, 1996). Most recently, a study on the distribution of *copia* 5′ LTR-ULR size variants in the *Melanogaster* subgroup indicates that both vertical and horizontal transfer play a role in the evolution of these agents (Jordan and McDonald, 1998).

One of the most compelling examples of retroid agent horizontal transfer is the study of the distribution of the retroposon bovine LINE agent (Bov-B). Among mammals Bov-B is restricted to *Ruminantia*, and its distribution in *Squamata* (snakes) is discontinuous. It is not present in a sampling of other vertebrates, invertebrates and plants. The data are consistent with Bov-B transferring from the snakes to the ancestral *Ruminantia* (Kordis and Gubensek, 1997, 1998).

All of the examples mentioned above are restricted to the eukaryotic retroid agents. Is there any evidence for the horizontal transfer of the retrointrons, retrons or retronphages? The retrointron in the pRS01 conjugative transposon from *L. lactis* is more similar to the mitochondrial retrointron sequences of yeast than to other bacterial retrointrons (Mills *et al.*, 1996). The current analysis of these data cannot distinguish if this similarity is due to purifying selection or horizontal transfer.

As mentioned above, retrons are ubiquitous in *Myxococcus* and only present in a small percentage of other bacteria. This discontinuous distribution suggests that retrons are being acquired by various bacteria. Retrons are inserted into different locations in various *E. coli* strains, and two different retrons are found in the same position in a prophage. These findings are consistent with retron horizontal transfer among bacteria and retronphages may be responsible for their dissemination (Inouye and Inouye, 1993).

Are Retroid Agents Descendents of the Ancient RNA World?

It has been suggested that the entities encoding the RT function were necessary for the transfer of genetic information from ancient RNA to DNA (Darnell and Doolittle, 1986; McClure, 1991; Lazcano *et al.*, 1992). This assertion implies that some form of translation mechanism evolved prior to the divergence of the eucarya, archaea and eubacteria (Doolittle and Brown, 1994). Evidence suggesting an RNA genome as the last common ancestor to the eucarya, archaea and eubacteria has come from comparative analysis of the genomes of *S. cerevisiae*, *Haemophilus influenzae* and *Mycoplasma genitalium*. This study indicates that the minimal, common gene set is missing specific functional categories of conserved genes, i.e. the major DNA replication genes, while retaining the genes necessary for the translation machinery (Koonin and Mushegian, 1996). Other sequence analyses indicate that, while transcription and translation components are universally conserved, DNA replication is not (Olsen and Woose, 1997).

Whether the genome of the last universal common ancestor (LUCA) was DNA or RNA is unknown. A reasoned hypothetical model of

LUCA, a ribo-organism that existed before the encoding of protein synthesis, is proposed (Jeffares *et al.*, 1997; Poole *et al.*, 1998). This hypothesis posits that the LUCA was an eukaryl-like set of linear RNA genomes, and that the tendency was for genetic information to increase early in the evolution of life. Furthermore the prokaryotes were derived from LUCA by a RT process of plasmid transfer. As RNA genomes were reverse-transcribed into intronless dsDNA they also circularized and became associated with one another (Poole *et al.*, 1998). Although this model is quite radical in proposing that prokaryotes evolved from an eukaryl-like ancestor, it does explain the apparently ubiquitous distribution of retroid agents and introns in eukaryotes and their paucity in eubacteria.

Are retroid agents the descendents of the LUCA responsible for the transformation from RNA to DNA? If they are, which are the most ancient? It has been suggested that mitochondr-

ial retroplasmids (Wang and Lambowitz, 1993) and retrons (Inouye and Inouye, 1996) are each the most ancient. The distribution of mitochondrial retroplasmids is not ubiquitous, although these agents do provide a link to the ribo-world because they encode a tRNA at the 3′ end of their genome that provides the primer for dsDNA synthesis. The retrons are not uniformly distributed in eubacteria, they are not present in eukaryotes and they are not known to be mobile. Their distribution supports invasion rather than a feature of antiquity.

Retrointrons have only been identified in eubacteria and plastids to date. The discovery of their reverse splicing ability brings together three important features: splicing, cleavage and reverse transcription in the intron RNP (iRNP) complex (Zimmerly *et al.*, 1995a,b; Yang *et al.*, 1996; Guo *et al.*, 1997; Figure 8.5). The three functions encoded by retrointrons suggest coevolution of a ribozyme and the PR/RT/M/HC protein (Zimmerly *et al.*, 1995a,b). Although only

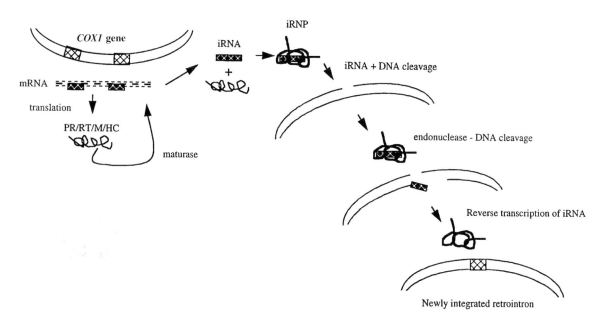

FIGURE 8.5 Schematic representation of the pathway of retrointron integration. The *COX1* has two retrointrons (white stippled boxes). The maturase function (M) of the PR/RT/M/HC (corkscrew) protein mediates the ribozyme splicing of the Cox1 mRNA to release the iRNA (black stippled box). The PR/RT/M/HC and iRNA form the RNP complex (black stippled box with corkscrew). Within this complex the iRNA cleaves the plus-strand of the host DNA and the HC domain cleaves the minus DNA strand. The iRNA becomes covalently linked to the host DNA and the RT function uses the host 3′ OH to prime DNA synthesis. By reverse splicing the retrointron is integrated into a new position in the host. All other abbreviations are as in the text. Adapted from Curcio and Belfort, 1996.

the zinc-binding motif of the IN is found in the retrointrons, it is involved in endonuclease activity (Zimmerly et al., 1995b). The ability to move is clearly dependent on the ribozyme component of the iRNA and the RT/M/HC portion of the RNP complex (Figure 8.5). It has been suggested that splicing and mobility may have evolved from a primitive RT-associated RNA that later gave rise to the world of retroid agents (Curcio and Belfort, 1996). However, one critical function, that which removes the iRNA template, has not been identified in retrointrons to date. Without the RH or an analogous function how could the retrointrons have transformed the ancient RNA world into DNA?

The recent characterization of the eukaryotic TERTs provides yet another possible source for the conversion of ancient RNA to DNA. The TERT gene is not known to be mobile, however, and appears to be vertically transmitted in eukaryotes. Interestingly, not all TERT genes are interrupted by introns: they are present in *Tetrahymena thermophlia* and *Saccharomyces pombe* but absent from *Oxytricha trifallax*, *Euplotes aediculatus* and *S. cerevisiae* (Bryan et al., 1998). Did a retroid agent reverse-transcribe TERT mRNAs and reinsertion intronless TERT genes back into some genomes or are the observed introns due to a recent invasion of this gene? The widespread eukaryotic distribution of TERTs is consistent with the hypothesis that eukaryl-like sequences gave rise to all other genomes, but again the critical RH function has not been identified in TERTs.

The functions necessary for the transformation of RNA into DNA, and subsequent DNA genome evolution, are RT-RH and some type of endonuclease. To date all three functions have not been found in any one of the most distant retroid agents (retrointrons, retrons, retroplasmids and TERTs). The current status of relationships of the RT sequences among this subset of agents cannot resolve this matter either. While statistical confidence in the FEN topology of the relationships of these agents represented in Figure 8.4 can be generated using some methods it is not as well supported by others (McClure, unpublished). Perhaps the ancestor no longer exists, or

maybe a single ancestor never did exist. Could the branch II retroid agents represent the descendants of an ancient RNA gene pool that complemented one another to allow the transformation of RNA to DNA?

CONCLUSIONS AND PERSPECTIVE

There was little direct evidence to support evolutionary roles for retroid agents prior to the 1990s. They were considered "junk DNA" or selfish parasites without benefits to their hosts. Early studies suggested retroid agent involvement in a variety of events ranging from phenotypic expression to increased host fitness and speciation (for a review of early literature, see McDonald, 1993). The recent finding of the activation of retroid agents in the interspecific hybridization between two Australian wallaby species supports a role in speciation (Kidwell and Lisch, 1998; Waugh O'Neill et al., 1998). As we approach the 21st century there is no doubt that retroid agents are ubiquitous in eukaryotes and many have coevolved with their respective hosts in a mutual manner. There has been a plethora of recent reviews on this topic with one common theme: retroid agents are involved in generating genetic diversity in eukaryotes (Daboussi, 1997; Flavell et al., 1997; Kidwell and Lisch, 1997; Labrador and Corces, 1997). Recently, the work of Moran et al. demonstrates that LINE agents can transduce host DNA to new locations, thereby creating a source of genomic diversity (Moran et al., 1999). It is now accepted that various retroid agents have been vertically transmitted since the radiation of animals and plants, and some are mobile within and between hosts. The variety of host functions involving retroid agents is an active area of research. The extent to which these agents have participated in genome evolution is just beginning to be revealed.

Among the eubacteria only three types of retroid agent have been detected: retrointrons, retrons and retronphages. While the distribution of retrons is ubiquitous in *Myxococcus*, it is very narrow within other eubacteria, suggesting acquisition rather than ancestry. Only the genomes of eukaryotic plastids have both eukaryotic (*gypsy-* and *copia-*retrotransposons, retroposons and retroplasmids) and eubacterial (retrointrons) retroid agents (Table 8.1). Until recently, retroplasmids were thought to be unique to plastids and related to other eubacterial retroid agents. The fungal retroplasmid of *A. alternata*, however, clearly belongs to the *gypsy-*retrotransposon lineage (Kaneko *et al.*, 1997; Figure 8.4).

The variety of gene complement and order found among various retroid agents is not restricted to a single branch or lineage of the RT functionally equivalent network (Figure 8.4). Contrary to some analyses, retroviruses and *gypsy-* and *copia-*retrotransposons are not monophyletic when RT sequences from the DIRS1-retrotranspsons, fungal retroplasmid and pararetroviruses are considered. In addition analyses of other retroid protein sequences from these agents do not support a monophyletic origin of retroviruses and *gypsy-*retrotransposons.

Current statistical methods do not generate support for common ancestry between all retroid lineages based on the sequence analysis of any one of the enzymatic functions or the trees generated from some published RT alignments (Zanotto *et al.*, 1996). Given the conservation of both the ordered series of motifs in the catalytic protein sequences (PR, RT-RH and IN) and the genomic gene complement and order it is unlikely that convergent evolution, rather than common ancestry, explains the similarities observed among retroid agents. In addition, retroid genomes can encode analogous functions: different types of endonucleases and ribonucleoproteins.

The exogenous and endogenous nature of the retroviruses is well documented. Although the *gypsy-* and *copia-*retrotransposons were thought to only be endogenous, the identification of the *env* gene in some of these genomes indicates that they are retroviruses of invertebrates. Infectivity has not been demonstrated for any *env+ copia-*retrotransposon to date. While some *env+ gypsy* strains are infectious, others have a degenerate *env* gene and are non-infectious (Alberola *et al.*, 1997). Although there is no direct evidence for the widespread assumption that retroviruses came directly from the *gypsy-*

retrotransposons, they may have. Many *gypsy-* and *copia-*retrotransposons, however, are endogenous, lack the *env* gene and are non-infectious. Recall that one of the first genes to erode in ERVs is the *env* gene. ERVs can be activated only through co-infection with a virus that can provide a missing function. Can *gypsy-* and *copia-* retrotransposon ENV proteins act in *trans* to allow exogenous mobilization of otherwise genome-restricted agents?

The presence of both retroviruses and *gypsy-*retrotransposons in fish lead to speculation that perhaps the piscine retroviruses would be more similar to the *gypsy-*retrotransposons, thereby proving the link between the two lineages (Britten, 1995b). A recent study on the diversity and distribution of retroviruses in vertebrates indicates that retroviruses are monophyletic (Herniou *et al.*, 1998). Different reptilian, amphibian and piscine retroviruses cluster with three groups of mammalian and avian sequences. In a small sampling of chordates from lampreys to mollusks, retroviruses were not detected. In the nonvertebrate/chordate, *Mya arenaria* (soft-shelled claim), however, it has been demonstrated that an exogenous type B retrovirus causes seasonal neoplasia (Oprandy and Chang, 1983). There is no doubt, however, that there are three distinct lineages of retroid agents in amphibians, reptiles and fish: retroviruses and *gypsy-* and *copia-* retrotransposons. The distribution of retroviruses among primitive chordates is an area of future study well worth pursuing and may provide clues as to when the retroviruses first appeared in the chordates.

The distribution of *gypsy-* and *copia-*retrotransposons appears to be restricted in vertebrates. They are only sporadically found in amphibians, reptiles and fish. Does this distribution imply that these genomes are the remnants of an ancient retroviral infection with few remaining active agents? Interestingly, none of the *gypsy-*retrotransposons isolated from marine animals encodes the *env* gene (Britten, 1995b). Does the vertebrate distribution of retroviruses represent the still active component of an ancestral infection? Whether or not an initial transmission of an ancestral infectious retroid agent (and its direction of transfer) between invertebrates and vertebrates actually took place, giving rise to the branch I lineages, may be unknowable. There is no doubt, however, that members of branch I are more similar to one another than branch II retroid agents (Figure 8.4).

Branch II agents include both eukaryotic and eubacterial retroid agents. Since the discovery of TERTs, the interpretation of studies on the evolutionary history of the RT function has become somewhat of a controversy (Nakamura and Cech, 1998). Most RT evolutionary trees that are representative of the diversity of retroid agents are only based on the ordered series of motifs and surrounding regions (Xiong and Eickbush, 1990; Nakamura et al., 1997), because of the difficulty in aligning the distantly related motif-intervening regions. Phylogenetic reconstruction based only on functionally constrained regions may reflect evolutionary parallelism rather than true ancestry and descent. Some studies use the RDDPs as an outgroup to root the RT tree (Xiong and Eickbush, 1990; Nakamura et al., 1997), even though there is no statistically significant evolutionary signal between RDDPs and RDRPs (Zanotto, et al., 1996). Only with a rooted tree can the evolutionary pathway leading to extant retroid genomes be unambiguously discerned. One study makes an arbitrary decision to divide the retroid agents strictly between eukaryotic and eubacterial hosts (Eickbush et al., 1997). This division is not supported by any of the distance calculations or trees for any of the proteins common to the retroid family (Doolittle et al., 1989; McClure, 1991, 1993). Although the statistical confidence in the topology of the network of relationships of these agents varies, based on the distribution and number of sequences analyzed as well as the methods employed, recent RT sequence studies (Nakamura et al., 1997; Nakamura and Cech, 1998) support the topology in Figure 8.4. Using a small sample representing over 600 RT sequences the topology presented in Figure 8.4 is better supported than any other (McClure, unpublished). Given the mobile nature of retroid agents it would not be surprising that some of these genomes have found their way across evolutionary domains at various evolutionary stages.

Issues regarding the rate of accumulated mutation and horizontal transfer (Capy et al., 1994), and sequence similarity due to ancestry, recombination and convergence of retroid genomes provides an exciting area of evolutionary research. A study of the complexity of the evolutionary history of the catalytic core functions, PR, RT-RH and IN, of retroid agents is under way (McClure and Fitch, unpublished). The actual rate of accumulated mutation over long evolutionary time spans cannot be known. The retroid sequence sample size is now sufficiently large, however, to apply new analytical methods to the data and determine the relative rate of protein evolution among and between different retroid lineages. There are well over 600 unique RT sequences available, with fewer sequences for the PR, RH and IN functions, given their presence or absence in various retroid genomes (Figure 8.4).

The basic problem, however, in reconstructing the evolutionary history of highly divergent sequences is the generation of an adequate representation of the sequence data. Regardless of the alignment method employed, although the ordered series of motifs of highly divergent protein sequences may be correctly aligned, the motif-intervening regions are often poorly aligned or not aligned at all. The ordered series of motifs are functionally constrained, while the motif-intervening regions provide information on neighboring relationships within subclasses of a protein family. The alignment of both types of regions is important to maximize the amount of information obtained from sequence analyses for use in phylogenetic inference. How can an informative multiple alignment of data as divergent as the RT sequences be obtained? In recent years I have investigated the use of stochastic production models, specifically the hidden Markov model (HMM) approach, to generate multiple alignments of highly divergent sequence data (McClure and Raman, 1995; McClure et al., 1996). Basically, a HMM is generated by training a model with the sequence data. Using the initial model and all training sequences, all possible paths for each sequence through the model are evaluated to obtain new estimates of the parameters that will increase the likelihood of the model. This process is repeated until the model converges. Sequences are never aligned to one another directly. A multiple alignment is generated by computing the negative logarithm of the probability of the single most likely path through the model for a particular sequence, given all the possible paths generated by the training sequences (Krogh *et al.*, 1994; Hughey and Krogh, 1996).

The HMM approach has a variety of advantages over more classical multiple alignment methods. It is grounded in probability theory and the computational cost of aligning a set of sequences to a HMM is directly proportional to the number of sequences to be aligned. Unlike classical alignment methods, knowledge of phylogenetic history or pairwise ordering is not required, and insertions and deletions are treated probabilistically in a variable, position-dependent manner. Experimentally derived information can be incorporated into the model *a priori*. To date there are no other methods with the flexibility of design allowed by the HMM approach.

Based on attempts to construct a HMM that can adequately-represent the RT sequences a new strategy of HMM generation has been developed. This approach independently models both the ordered series of motifs and motif-intervening regions in a sequence subclassification manner (McClure and Kowalski, 1999; McClure *et al.*, 1998). Refinements to the HMM representing over 600 RT sequences are nearly complete and the construction of HMMs representing the other retroid enzymatic proteins is in process. The new alignments generated from these models will allow for more robust measurements for determining phylogenetic relationships of the core enzymes. Although the generation of HMMs representing retroid protein sequences will provide data relevant to issues of academic interest, these models will have a variety of practical uses that are just beginning to be realized. For example, data on the frequency and type of amino acid substitutions on a per-site basis is also generated by the HMM representing natural retroid agent populations. This information can be used to study and classify the various types of mutations arising from chemotherapeutic regimes used in the AIDS epidemic. The model can be easily refined as more sequences become available and a separate model representing chemotherapeutic mutants will be generated when enough data are available.

We may never be able to unambiguously resolve issues pertaining to the origin(s) and ancestry of the retroid agents. Given existing data it is plausible that they are the descendants of the agent(s) responsible for the conversion of RNA to DNA. While modern invasions of retroid agents may be "parasitic", this term is inappropriate for the established lineages that have coevolved with their hosts and are now integral to host function. Evolution is ongoing, however, and retroid agents will continue to play a dynamic role in this process. To paraphrase a paraphrase of Dobzhansky's famous phrase, "Nothing about retroid agents makes sense except in the light of the origins of life and genome evolution" (Kidwell and Lisch, 1997).

ACKNOWLEDGEMENTS

This work is supported by NIH grant AI 28309 and a Research Career Development Award. I thank Seanna Corro and Micah Potter for scientific technical assistance and Tiffany Cabral and Megan Smith for library reference assistance. I also thank Angela Baldo, King Jordan, Seanna Corro and Julianna Hudak for discussion and critical reading of the manuscript.

REFERENCES

Alberola, T.M., Bori, L. and de Frutos, R. (1997) Structural analysis of *Drosophilia subobscura gypsy* elements (*gypsyDs*). *Genetica*, **100**, 39–48.

Argos, P. (1988) A sequence motif in many polymerases. *Nucl. Acids Res.*, **16**, 9909–9916.

Argos, P. and Fuller, S.D. (1988) A model for the hepatitis B core protein: prediction of antigenic sites and relationship to RNA virus capsid proteins. *EMBO J.*, **7**, 819–824.

Banville, D. and Bioe, Y. (1989) Retroviral long terminal repeat is the promoter of the gene encoding the tumor-associated calcium-binding protein oncomodulin in the rat. *J. Mol. Biol.*, **207**, 481–497.

Biessmann, H. and Mason, J.M. (1997) Telomere maintenance without telomerase. *Chromosoma*, **106**, 63–69.

Biessmann, H., Mason, J.M., Ferry, K. *et al.* (1990) Addition of telomere-associated HeT DNA sequences "heals" broken chromosome ends in *Drosophila*. *Cell*, **61**, 663–673.

Blesa, D. and Martinez-Sebastian, M.J. (1997) *bilbo*, a non-LTR retrotransposon of *Drosophila subobscura*: a clue to the evolution of LINE-like elements in *Drosophilia*. *Mol. Biol. Evol.*, **14**, 1145–1153.

Boeke, J.D. and Stoye, J.P. (1997) Retroposons, endogenous retroviruses, and the evolution of retroelements. In: *Retroviruses* (ed. Coffin, J.M. Hughes, S.H. and Varmus, H.E.), pp. 343–435. Cold Spring Harbor Laboratory Press, Cold Spring Harbor, NY.

Boyd, M.T., Bax, C.M.R., Bax, B.E., Bloxam, D.L. and Weiss, R.A. (1993) The human endogenous retrovirus ERV-3 is upregulated in differentiating placental trophoblast cells. *Virology*, **196**, 905–909.

Brautigam, C.A. and Steitz, T.A. (1998) Structural and functional insights provided by crystal structures of DNA polymerases and their substrate complexes. *Curr. Opin. Struct. Biol.*, **8**, 54–63.

Britten, R.J. (1995a) Active *gypsy/Ty3* retrotransposons or retroviruses in *Caenorhabditis elegans*. *Proc. Natl Acad. Sci. USA*, **92**, 599–601.

Britten, R.J. (1995b) Gypsy/Ty3-class retrotransposons integrated in the DNA of herring, tunicates and echinoderms. *J. Mol. Evol.*, **40**, 13–24.

Britten, R.J. (1997) Mobile elements inserted in the distant past have taken on important functions. *Gene*, **205**, 177–182.

Brown, P.O. (1997) Integration. In: *Retroviruses* (ed Coffin, J.M., Hughes, S.H. and Varmus, H.E.), pp. 161–203. Cold Spring Harbor Laboratory Press, Cold Spring Harbor, NY.

Bryan, T.M., Sperger, J.M., Chapman, K.B. and Cech, T.R. (1998) Telomerase reverse transcriptase genes identified in *Tetrahymena thermophila* and *Oxytricha trifallax*. *Proc. Natl. Acad. Sci. USA*, **95**, 8479–8484.

Burke, W.D., Calalang, C.C. and Eickbush, T.H. (1987) The site-specific ribosomal insertion element type II of *Bombyx mori* (R2Bm) contains the coding sequence for a reverse transcriptase-like enzyme. *Mol. Cell. Biol.*, **7**, 2221–2230.

Burke, C.J., Sanyal, G., Bruner, M.W. *et al.* (1992) Structural implications of spectroscopic characterization of a putative zinc-finger peptide from HIV-1 integrase. *J. Biol. Chem.*, **267**, 9639–9644.

Cappello, J., Handelsman, K., Cohem, S.M. and Lodish, H.F. (1985a) Structure and regulated transcription of DIRS-1: an apparent retrotransposon of *Dictyostelium discoideum*. Cold Spring Symposium on Quantitative Biology, Cold Spring Harbor Laboratory Press, Cold Spring Harbor, NY.

Cappello, J., Handelsman, K. and Lodish, H.F. (1985b) Sequence of dictyostelium DIRS-1: an apparent retrotransposon with inverted terminal repeats and an internal circle junction sequence. *Cell*, **43**, 105–115.

Capy, P., Anxolabehere, D. and Langin, T. (1994) The strange phylogenies of transposable elements: are horizontal transfers the only explanation? *Trends Genet.*, **10**, 339–344.

Capy, P., Bazin, C., Higuet, D. and Langin, T. (1998) *Dynamics and Evolution of Transposable Elements*. Landes Bioscience, Austin, Tx.

Casacuberta, J.M., Vernhettes, S. and Grandbastien, M.-A. (1995) Sequence variability within the tobacco retrotransposon Tnt1 population. *EMBO J.*, **14**, 2670–2678.

Casacuberta, J.M., Vernhettes, S., Audeon, C. and Grandbastien, M.-A. (1997) Quasispecies in retrotransposons: a role for sequence variability in Tnt1 evolution. *Genetica*, **100**, 109–117.

Casavant, N.C., Sherman, A.N. and Wichman, H.A. (1996) Two persistent LINE-1 lineages in peromyscus have unequal rates of evolution. *Genetics*, **142**, 1289–1298.

Chalker, D.L. and Sandmeyer, S.B. (1992) Ty3 integrates within the region of RNA polymerase III transcription initiation. *Genes Dev.*, **4**, 117–128.

Chastonay, Y. de, Felder, H., Link, C., Aeby, P., Tobler, H. and Muller, F. (1992) Unusual features of the retroid element PAT from the nematode *Panagrellus redivivus*. *Nucl. Acids Res.*, **20**, 1623–1628.

Coffin, J.M., Hughes, S.H. and Varmus, H.E. (eds) (1997) *Retroviruses*. Cold Spring Harbor Laboratory Press, Cold Spring Harbor, NY.

Covey, S.N. (1986) Amino acid sequence homology in gag region of reverse transcribing elements and the coat protein gene of cauliflower mosaic virus. *Nucl. Acids Res.*, **14**, 623–633.

Curcio, M.J. and Belfort, M. (1996) Retrohoming: cDNA-mediated mobility of group II introns requires a catalytic RNA. *Cell*, **84**, 9–12.

Daboussi, M.J. (1997) Fungal transposable elements and genome evolution. *Genetica*, **100**, 253–260.

Darnell, J.E. and Doolittle, W.F. (1986) Speculations on the early course of evolution. *Proc. Natl Acad. Sci. USA*, **83**, 1271–1275.

Davies, J.F., Hostomska, Z. *et al.* (1991) Crystal structure of the ribonuclease H domain of HIV-1 reverse transcriptase. *Science*, **252**, 88–95.

Delarue, M., Poch, O., Tordo, N., Moras, D. and Argos, P. (1990) An attempt to unify the structure of polymerases. *Prot. Eng.*, **3**, 461–467.

Delassus, S., Cheynier, R. and Wain-Hobson, S. (1991) Evolution of human immunodeficiency virus type 1 nef and long terminal repeat sequences over 4 years *in vivo* and *in vitro*. *J. Virol.*, **65**, 225–231.

Ding, D. and Lipshitz, H.D. (1994) Spatially regulated expression of retrovirus-like transposons during *Drosophilia melanogaster* embryogenesis. *Genet. Res. Camb.*, **64**, 167–181.

Dombroski, B.A., Mathias, S.L., Nanthakumar, E., Scott, A.F. and Kazazian, H.H. Jr (1991) Isolation of an active human transposable element. *Science*, **254**, 1805–1807.

Domingo, E. and Holland, J.J. (1997) RNA virus mutations and fitness for survival. *Annu. Rev. Microbiol.*, **51**, 151–178.

Doolittle, W.F. and Brown, J.R. (1994) Tempo, mode, the progenote, and universal root. *Proc. Natl Acad. Sci. USA*, **91**, 6721–6728.

Doolittle, R.F., Feng, D.-F., Johnson, M.S. and McClure, M.A. (1989) Origins and evolutionary relationships of retroviruses. *Rev. Bio.*, **64**, 1–30.

Duhl, D.M.J., Vrieling, K.A., Miller, G.L., Wolff, G.L. and Barsch, G.S. (1994) Neomorphic *agouti* mutations in obese yellow mice. *Nat. Genet.*, **8**, 59–65

Dyda, F., Hickman, A.B., Jenkins, T.M., Engelman, A., Craigie, R. and Davies, D.R. (1994) Crystal structure of the catalytic domain of HIV-1 integrase: similarity to other polynucleotidyl transferases. *Science*, **266**, 1981–1985.

Eickbush, T.H. (1997) Telomerase and retrotransposons: which came first? *Science*, **277**, 911–912.

Eickbush, D.G., Lathe, W.C.I., Francino, M.P. and Eickbush, T.H. (1995) R1 and R2 retrotransposable elements of *Drosophila* evolve at rates similar to those of nuclear genes. *Genetics*, **139**, 685–695.

Eickbush, T.H., Burke, W.D., Eickbush, D.G. and Lathe III, W.C. (1997) Evolution of R1 and R2 in the rDNA units of the genus *Drosophila*. *Gentica*, **100**, 49–61.

Emi, M., Horii, A., Tomita, N. *et al.* (1988) Overlapping two genes in human DNA: a salivary amylase gene with a gamma-actin pseudogene that carries an integrated human endogenous retroviral DNA. *Gene*, **62**, 229–235.

Fanning, T.G. and Singer, M.F. (1987) Line-1: a mammalian transposable element. *Biochim. Biophys. Acta*, **910**, 203–212.

Fayet, O., Ramond, P., Polard, P., Prere, M.F. and Chandler, M. (1990) Functional similarities between retroviruses and the IS3 family of bacterial insertion sequences. *Mol. Microbiol.*, **4**, 1771–1777.

Feng, Q., Moran, J.V., Kazazian, H.H.J. and Boeke, J.D. (1996) Human L1 retrotransposons encodes a conserved endonuclease required for retransposition. *Cell*, **87**(5), 905–916.

Feng, Q., Schumann, G. and Boeke, J.D. (1998) Retrotransposon R1Bm endonuclease cleaves the target sequence. *Proc. Natl. Acad. Sci. USA*, **95**, 2083–2088.

Ferat, J.-L. and Michel, F. (1993) Group II self-splicing introns in bacteria. *Nature*, **364**, 358–361.

Ferat, J.-L., Le Gouar, M. and Michel, F. (1994) Multiple group II self-splicing introns in mobile DNA from *Escherichia coli*. *C.R. Acad. Sci. Paris*, **317**, 141–148.

Flavell, A.J. and Smith, D.B. (1992) A ty1-copia group retrotransposon sequence in a vertebrate. *Mol. Gen. Genet.*, **233**, 322–326.

Flavell, A.J., Jackson, V., Iqbal, M.P., Riach, I. and Waddell, S. (1995) Ty1-*copia* group retrotransposon sequences in amphibia and reptilia. *Mol. Gen. Genet.*, **246**, 65–71.

Flavell, A.J., Pearce, S.R., Heslop-Harrison, J.S.P. and Kumar, A. (1997) The evolution of Ty1-*copia* group retrotransposons in eukaryote genomes. *Genetica*, **100**, 185–195.

Fuetterer, J. and Hohn, T. (1987) Involvement of nucleocapsids in reverse transcription: a general phenomenon? *Trends Biochem. Sci.*, **12**, 92–95.

Gabriel, A., Willems, M., Mules, E.H. and Boeke, J.D. (1996) Replication infidelity during a single cycle of Ty1 retrotransposition *Proc. Natl Acad. Sci. USA*, **93**, 7767–7771.

Garrett, J.E., Knutzon, D.S. and Carroll, D. (1989) Composite transposable elements in the *Xenopus laevis* genome. *Mol. Cell. Biol.*, **9**, 3018–3027.

Gasson, M.J., Gordon, J.-J., Pillidge, C.J., Eaton, T.J., Jury, K. and Shearman, C.A. (1995) Characterization and exploitation of conjugation in *Lactococcus lactis*. *Int. Dairy J.*, **5**, 757–762.

Gorbalenya, A.E. (1994) Self-splicing group I and II introns encode homologous (putative) DNA endonucleases of a new family. *Prot. Sci.*, **3**, 1117–1120.

Guo, H., Zimmerly, S., Perlman, P.S. and Lambowitz, A.M. (1997) Group II intron endonucleases use both RNA and protein subunits for recognition of specific sequences in double-stranded DNA. *EMBO J.*, **16**, 6835–6848.

Harendza, C.J. and Johnson, L.F. (1990) Polyadenylation signal of the mouse thymidylate synthase gene was created by insertion of an L1 repetitive element downstream of the open reading frame. *Proc. Natl Acad. Sci. USA*, **87**, 2531–2535.

Haoudi, A., Rachidi, M., Kim, M.-H., Champion, S., Best-Belpomme, M. and Maisonhaute, C. (1997) Developmental expression analysis of the 1731 retrotransposon reveals an enhancement of *gag-pol* frameshifting in males of *Drosophila melanogaster*. *Gene*, **196**, 83–93.

Herniou, E., Martin, J., Miller, K., Cook, J., Wilkinson, M. and Tristem, M. (1998) Retroviral diversity and distribution in vertebrates. *J. Virol.*, **72**, 5955–5966.

Herzer, P.J., Inouye, S., Inouye, M. and Whittam, T.S. (1990) Phylogenetic distribution on branched RNA-linked multicopy single-stranded DNA among natural isolates of *Escherichia coli*. *J. Bacteriol.*, **172**, 6175–6181.

Higashiyama, T., Noutoshi, Y., Fujie, M. and Yamada, T. (1997) Zepp, a line-like retrotransposon accumulated in the *chlorella* telomeric region. *EMBO J.*, **16**, 3715–3723.

Hohjoh, H. and Singer, M.F. (1996) Cytoplasmic ribonucleoprotein complexes containing human LINE-1 protein and RNA. *EMBO J.*, **15**, 630–639.

Hsu, M.-Y., Inouye, M. and Inouye, S. (1990) Retrons for the 67-base multicopy single-stranded DNA from *Escherichia coli*: a potential transposable element encoding both reverse transcriptase and Dam methylase functions. *Proc. Natl Acad. Sci. USA*, **87**, 9454–9458.

Hughey, R. and Krogh, A. (1996) Hidden Markov models for sequence analysis: extension and analysis of the basic method. *CABIOS*, **12**, 95–107.

Inouye, S. and Inouye, M. (1993) The retron: a bacterial retroelement required for the synthesis of msDNA. *Curr. Opin. Genet. Devel.*, **3**, 713–718.

Inouye, S. and Inouye, M. (1996) Structure, function, and evolution of bacterial reverse transcriptase. *Virus Genes*, **11**, 81–94.

Inouye, S., Hsu, M.-Y., Eagle, S. and Inouye, M. (1989) Reverse transcriptase associated with the biosynthesis of the branched RNA-linked msDNA in myxococcus xanthus. *Cell*, **56**, 709–717.

Inouye, S., Sunshine, M.G., Six, E.W. and Inouye, M. (1991) Retronphage fR73: an *E. coli* phage that contains a retroelement and integrates into a tRNA gene. *Science*, **252**, 967–971.

Jeffares, D.C., Poole, A.M. and Penny, D. (1997) Relics from the RNA world. *J. Mol. Evol.*, **45**, 18–36.

Johnson, M.S., McClure, M.A., Feng, D.-F., Gray, J. and Doolittle, R. F. (1986) Computer analysis of retroviral pol genes: assignment of enzymatic functions. *Proc. Natl Acad. Sci. USA*, **83**, 7648–7652.

Jordan, I.K. and McDonald, J.F. (1998) Evolution of the *copia* retrotransposon in the *Drosophila melanogaster* species subgroup. *Mol. Biol. Evol.*, **15**, 1160–1171.

Joyce, C.M. and Steitz, T.A. (1994) Function and structure relationships in DNA polymerases. *Annu. Rev. Biochem.*, **63**, 777–822.

Juni, N., Awaskai, T., Yoshida, K. and Hori, S.H. (1996) The om (1E) mutation in *Drosophila annassae* causes compound eye overgrowth due to tom retrotransposon-driven overexpression of a novel gene. *Genetics*, **143**, 1257–1270.

Kaneko, I., Katsuya, S. and Tsuge, T. (1997) Structural analysis of the plasmid pAAT56 of the filamentous fungus *Alternaria alternata*. *Gene*, **203**, 51–57.

Kazazian, H. Jr, Wong, C., Youssoufian, H., Scott, F., Phillips, D. and Antonarakis, S. (1988) Haemophilia A resulting from *de novo* insertion of L1 sequences represents a novel mechanism for mutation in man. *Nature*, **332**, 164–166.

Kelleher, C.A., Wilkinson, D.A., Freeman, J.D., Mager, D.L. and Gelfand, E.W. (1996) Expression of novel transposon-containing mRNAs in human T cells. *J. Gen. Virol.*, **77**, 1101–1110.

Kerber, B., Fellert, S., Taubert, H. and Hock, M. (1996) Germ line and embryonic expression of fex, a member of the *Drosophila* f-element retrotransposon family, is mediated by an internal cis-regulatory control region. *Mol. Cell. Biol.*, **16**, 2998–3007.

Kidwell, M.G. and Lisch, D. (1997) Transposable elements as sources of variation in animals and plants. *Proc. Natl Acad. Sci. USA*, **94**, 7704–7711.

Kidwell, M.G. and Lisch, D.R. (1998) Transposons unbound. *Nature*, **393**, 22–23.

Kim, A.I., Terzian, C., Snatamaria, P., Pelisson, A., Prud'homme, N. and Bucheton, A. (1994) Retroviruses in invertebrates: the gypsy retrotransposon is apparently an infectious retrovirus of *Drosophila melanogaster*. *Proc. Natl Acad. Sci. USA*, **91**, 1285–1289.

Kimmel, B.E., Ole-Moiyoi, O.K. and Young, J.R. (1987) Ingi, a 5.2-kb dispersed sequence element from *Trypanosoma brucei* that carries half of a smaller mobile element at either end and has homology with mammalian lines. *Mol. Cell. Biol.*, **7**, 1465–1475.

Kirchner, J., Connolly, C.M. and Sandmeyer, S.B. (1995) Requirement for RNA polymerase III transcription factors for *in vitro* position-specific integration of a retrovirus-like element. *Science* **267**, 1488–1491.

Knoop, V. and Brennicke, A. (1994) Evidence for a group II intron in *Escherichia coli* inserted into a highly conserved reading frame associated with mobile DNA sequences. *Nucl. Acid Res.*, **22**, 1167–1171.

Kohlstaedt, L.A., Wang, J. *et al.* (1992) Crystal structure at 3.5 A resolution of HIV-1 reverse transcriptase complexed with an inhibitor. *Science*, **256**, 1783–1790.

Koonin, E.V. and Mushegian, A.R. (1996) Complete genome sequences of cellular life forms: glimpses of theoretical evolutionary genomics. *Curr. Opin. Genet. Devel.*, **6**, 757–762.

Kordis, D. and Gubensek, F. (1997) Bov-B long interspersed repeated DNA (LINE) sequences are present in *Vipera ammodytes* phospholipase A2 genes and in genomes of Viperidae snakes. *Eur. J. Biochem.*, **246**, 772–779.

Kordis, D. and Gubensek, F. (1998) Unusual horizontal transfer of a LINE element between distant vertebrate classes. *Proc. Natl Acad. Sci. USA*, **95**, 10704–10709.

Krogh, A., Brown, M., Mian, I.S., Sjolander, K. and Haussler, D. (1994) Hidden Markov models in computational biology: applications to protein modeling. *J. Mol. Biol.*, **235**, 1501–1531.

Kuck, U. (1989) The intron of a plastid from a green alga contains an open reading frame for a reverse transcriptase-like enzyme. *Mol. Gen. Genet.*, **218**, 257–265.

Kumar, A., Pearce, S.R., McLean, K. *et al.* (1997) The Ty1-*copia* group of retrotransposons in plants: genomic organization, evolution, and use as molecular markers. *Genetica*, **100**, 205–217.

Labrador, M. and Corces, V.G. (1997)

Transposable element-host interactions: regulation of insertion and excision. *Annu. Rev. Genet.*, **31**, 381–404.

Lachume, P., Bouhidel, K., Mesure, M. and Pinon, H. (1992) Spatial and temporal expression of the I factor during oogenesis in *Drosophilia melanogaster. Development*, **14**, 729–735.

Lampson, B., Inouye, M. and Inouye, S. (1991) Survey of multicopy single-stranded DNAs and reverse transcriptase genes among natural isolates of *Myxococcus zanthus. J. Bact.*, **173**, 5363–5370.

Lang, B.F. and Ahne, F. (1985) The mitochondrial genome of the fission yeast *Schizosaccharomyces pombe*. The cytochrome b gene has an intron closely related to the first two introns in the *Saccharomyces cerevisiae cox1* gene. *J. Mol. Biol.*, **184**, 353–366.

Lankenau, S., Corces, V.G. and Lankenau, D.-H. (1994) The *Drosophila* micropia retrotransposon encodes a testis-specific antisense RNA complementary to reverse transcriptase. *Mol. Cell. Biol.*, **14**, 1764–1775.

Laten, H.M., Majumdar, A. and Gaucher, E.A. (1998) SIRE-1, a *copia/Ty1*-like retroelement from soybean, encodes a retroviral envelope-like protein. *Proc. Natl Acad. Sci. USA*, **95**, 6897–6902.

Lazcano, A., Valverde, V., Hernandez, G., Gariglio, P., Fox, G.E. and Oro, J. (1992) On the early emergence of reverse transcriptase: theoretical basis and experimental evidence. *J. Mol. Evol.*, **35**, 524–536.

Levis, R.W., Ganesan, R., Houtchens, K., Tolar, L.A. and Sheen, F.-M. (1993) Transposons in place of telomeric repeats at a *Drosophila* telomere. *Cell*, **75**, 1083–1093.

Lim, D. and Maas, W.K. (1989) Reverse transcriptase-dependent synthesis of a covalently linked, branched DNA-RNA compound in *E. coli B. Cell*, **56**, 891–904.

Lin, J.-H. and Levin, H.L. (1997) A complex structure in the mRNA of Tf1 is recognized and cleaved to generate the primer of reverse transcription. *Genes Devel.*, **11**, 270–285.

Lingner, J., Hughes, T.R., Shevchenko, A., Mann, M., Lundblad, V. and Cech, T.R. (1997) Reverse transcriptase motifs in the catalytic subunit of telomerase. *Science*, **276**, 561–567.

Lower, R., Lower, J. and Kurth, R. (1996) The viruses in all of us: characteristics and biological significance of human endogenous retrovirus sequences. *Proc. Natl Acad. Sci. USA*, **93**, 5177–5184.

Luan, D.D. and Eickbush, T.H. (1995) RNA Template requirements for target DNA-primed reverse transcriptase by the R2 retrotransposable element. *Mol. Cell. Biol.*, **15**, 3882–3891.

Luan, D.D., Korman, M.H., Jakubczak, J.L. and Eickbush, T.H. (1993) Reverse transcription of R2B mRNA is primed by a nick at the chromosomal target site: a mechanism for non-LTR retrotransposition. *Cell*, **72**, 595–605.

McAllister, B.F. and Werren, J.H. (1997) Phylogenetic analysis of a retrotransposon with implications for strong evolutionary constraints on reverse transcription. *Mol. Biol. Evol.*, **14**, 69–80.

McClintock, B. (1987) *The Discovery and Characterization of Transposable Elements: the Collected Papers of Barbara McClintock*. Garland, New York.

McClure, M.A. (1991) Evolution of retroposons by acquisition or deletion of retrovirus-like genes. *Mol. Biol. Evol.*, **8**, 835–856.

McClure, M.A. (1992) Sequence analysis of eukaryotic retroid proteins. *Math. Comp. Model. Int. J.*, **16**, 121–136.

McClure, M.A. (1993) Evolutionary history of reverse transcriptase. In: *Reverse Transcriptase* (eds Skalka, A.M. and Goff, S.P.), pp. 425–444. Cold Spring Harbor Laboratory Press, Cold Spring Harbor, NY.

McClure, M.A. and Kowalski, J. (1999) The effects of ordered-series-of-motifs anchoring and subclass modeling on the generation of HMMs representing highly divergent protein sequences. In: *Proceedings of the Pacific Symposium on Biocomputing, '99* (eds Altman, R.B., Dunker, A.K., Hunter, L., Klein, T.E. and Lauderdale, K.), pp. 162–170. World Scientific, New Jersey.

McClure, M.A. and Raman, R. (1995) Parameterization studies of Hidden Markov Models representing highly divergent protein sequences. In: *28th Annual Hawaii International Conference on System Sciences* (eds Hunter, L. and Shriver, B), *IEEE* Computer Society Press, Hawaii.

McClure, M.A., Johnson, M.S., Feng, D.-F. and Doolittle, R.F. (1988) Sequence comparisons of retroviral proteins: relative rates of change and general phylogeny. *Proc. Natl Acad. Sci. USA*, **85**, 2469–2473.

McClure, M.A., Smith, C. and Elton, P. (1996) Parameterization studies for the SAM and HMMER methods of Hidden Markov Model generation (D. States). In: *Proceedings of the Fourth International Conference on Intelligent Systems for Molecular Biology*. AAAI Press, Menlo Park, CA.

McClure, M.A., Hudak, J. and Kowalski, J. (1998) Low identity, low similarity protein sequences: independent modeling of the ordered-sense-of-motifs and motif-intervening regions. In: *Genome Informatics 1998* (eds Miyano, S. and Takagi, T.), pp. 183–192. Universal Academy Press, Tokyo.

McDonald, J.F. (1993) Evolution and consequences of transposable elements. *Curr. Opin. Genet. Devel.*, **3**, 855–864.

McDonald, J.F. (1995) Transposable elements: possible catalysts of organismic evolution. *Trends Ecol. Evol.*, **10**, 123–126.

Marracci, S., Batistoni, R., Pesole, G., Citti, L. and Nardi, I. (1996) Gypsy/ty3-like elements in the genome of the terrestrial salamander hydromantes (amphilbia, urodela). *J. Mol. Evol.*, **43**, 584–593.

Martin, F., Maranon, C., Olivares, M., Alonso, C. and Lopez, M.C. (1995) Characterization of a non-Long terminal repeat retrotransposon cDNA (L1Tc) from *Trypanosoma cruzi*: homology of the first ORF with the ape family of DNA repair enzymes. *J. Mol. Biol.*, **247**, 49–59.

Mayer, W.E., O'hUigin, C. and Klein, J. (1993) Resolution of the HLA DRB6 puzzle: a case of grafting a *de novo* generated exon on an existing gene. *Proc. Natl Acad. Sci. USA*, **90**, 10720–10724.

Mazo, A.M., Mizrokhi, L.J., Karavanov, A.A., Sedov, Y.A. and Krichievska, A.A. (1989) Suppression in *Drosophilia*: su(Hw) and su(f) gene products interact with a region of *gypsy* (*mdg4*) regulating its transcriptional activity. *EMBO J.*, **8**, 903–911.

Michaud, E.J., van Vugt, M.J., Bultman, S.J., Sweet, H.O., Davisson, M.T. and Woychik, R.P. (1994) Differential expression of a new

dominant agouti allele (*Aiapy*) is correlated with methylation state and is influenced by parental lineage. *Genes Devel.*, **8**, 1463–1472.

Miller, R.H. and Robinson, W.S. (1986) Common evolutionary origin of hepatitis B virus and retroviruses. *Proc. Natl Acad. Sci. USA*, **83**, 2531–2535.

Miller, M., Jaskolski, M., Rao, J.K., Leis, J. and Wlodawer, A. (1989) Crystal structure of a retroviral protease proves relationship to aspartic protease family. *Nature*, **337**, 576–579.

Mills, D.A., McKay, L.L. and Dunny, G.M. (1996) Splicing of a group II intron involved in the conjugative transfer of pRSO1 in *Lactococci*. *J. Bacteriol.*, **178**, 3531–3538.

Mohr, D.A., McKay, L.L. and Lambowitz, A.M. (1993) Evolutionary relationships among group II intron-encoded proteins and identification of a conserved domain that may be related to maturase function. *Nucl. Acids Res.*, **21**, 4991–4997.

Moore, J.K. and Haber, J.E. (1996) Capture of retrotransposon DNA at the sites of chromosomal double-strand breaks. *Nature*, **383**, 645–646.

Moran, J.V., De Barardinis, R.J. and Kazazian Jr., H.H. (1999) Exon Shuffling by L.1. retrotransposition *Science*, **283**, 1530–1534.

Mozer, B.A. and Benzer, S. (1994) Ingrowth by photoreceptor axons induces transcription of a retrotransposon in a developing *Drosophila* brain. *Development*, **120**, 1049–1058.

Mueller, M.W., Allmaier, M., Eskes, R. and Schweyen, R.J. (1993) Transposition of group II intron all in yeast and invasion of mitochondrial genes at new locations. *Nature*, **366**, 174–178.

Murphy, N.B., Pays, A., Tebabi, P. *et al.* (1987) *Trypanosome brucei* repeated element with unusual structural and transcriptional properties. *J. Mol. Biol.*, **195**, 855–871.

Najera, I., Holguin, A., Quinones-Mateu, M. *et al.* (1995) Pol gene quasispecies of human immunodeficiency virus: mutations associated with drug resistance in virus from patients undergoing no drug therapy. *J. Virol.*, **69**, 23–31.

Nakamura, T.M. and Cech, T.R. (1998) Reversing time: origin of telomerase. *Cell*, **92**, 587–590.

Nakamura, T.M., Morin, G.B., Chapman, K.B. *et al.* (1997) Telomerase catalytic subunit homologs from fission yeast and human. *Science*, **277**, 955–959.

Narita, N., Nishio, H., Kitoh, Y. *et al.* (1993) Insertion of a 5′ truncated L1 element into the 3′ end of exon 44 of the dystrophin gene resulted in skipping of the exon during splicing in a case of Duchenne muscular dystrophy. *J. Clin. Invest.*, **91**, 1862–1867.

Natvig, D.O., May, G. and Taylor, J.W. (1984) Distribution and evolutionary significance of mitochondrial plasmids in *Neurospora* spp. *J. Bact.*, **159**, 288–293.

Olivares, M., Alonso, C. and Lopez, M.C. (1997) The open reading frame 1 of the L1Tc retrotransposon of *Trypanosoma cruzi* codes for a protein with apurinic-apyrimidinic nuclease activity. *J. Biol. Chem.*, **272**, 25224–25228.

Olsen, G.J. and Woose, C.R. (1997) Archeal genomics: an overview. *Cell*, **89**, 991–994.

Oprandy, J.J. and Chang, P.W. (1983) 5-bromodeoxyurdine induction of hematopoietic neoplasia and retrovirus activation in the soft-shelled claim, *Mya arenaria*. *J. Invertebr. Pathol.*, **42**, 196–206.

Pande, S., Lemire, E.G. and Nargang, F.E. (1989) The mitochondrial plasmid from *Neurospora intermedia* strain labelle-1b contains a long open reading frame with blocks of amino acids characteristic of reverse transcriptases and related proteins. *Nucl. Acids Res.*, **17**, 2023–2042.

Pardue, M.L., Danilevskaya, O.N., Lowenhaupt, K., Slot, F. and Traverse, K.L. (1996) *Drosophila* telomeres: new views on chromosome evolution. *Trends Genet.*, **12**, 48–52.

Pasyukova, E., Nuzhdin, S., Li, W. and Flavell, A. (1997) Germ line transposition of the copia retrotransposon in *Drosophila melanogaster* is restricted to males by tissue-specific control of copia RNA levels. *Mol. Gen. Genet.*, **255**, 115–124.

Pearce, S.R., Pich, U., Harrison, G. *et al.* (1996) The Ty1-copia group retrotransposons of *Alluim cepa* are distributed throughout the chromosomes but are enriched in the terminal heterochromatin. *Chromo. Res.*, **4**, 357–364.

Pearl, L.H. and Taylor, W.R. (1987) A structural model for the retroviral proteases. *Nature*, **329**, 351–354.

Pelisson, A., Teysset, L., Chalvet, F. *et al.* (1997) About the origin of retroviruses and the co-evolution of the *gypsy* retrovirus with the *Drosophila flamenco* host gene. *Genetica*, **100**, 29–37.

Poch, O., Sauvaget, I., Delarue, M. and Tordo, N. (1989) Identification of four conserved motifs among the RNA-dependent polymerase encoding elements. *EMBO J.*, **8**, 3867–3874.

Pont-Kingdon, G., Chi, E., Christensen, S. and Carroll, D. (1997) Ribonucleoprotein formation by the ORF1 protein of the non-LTR retrotransposon Tx1L in xenopus oocytes. *Nucl. Acids Res.*, **25**, 3088–3094.

Poole, A.M., Jeffares, D.C. and Penny, D. (1998) The path from the RNA world. *J. Mol. Evol.*, **46**, 1–17.

Preston, B.D. (1996) Error-prone retrotransposition: rime of the ancient mutators. *Proc. Natl Acad. Sci. USA*, **93**, 7427–7431.

Rabinow, L., Chiang, S.L. and Birchler, J.A. (1993) Mutations at the Darkerer of apricot locus modulate transcript levels of *copia* and *copia*-induced mutations in *Drosophila melanogaster*. *Genetics*, **134**, 1175–1185.

Rice, S.A. and Lampson, B.C. (1995) Phylogenetic comparison of retron elements among the *myxobacteria*: evidence for vertical inheritance. *J. Bacteriol.*, **177**, 37–45.

Richert-Poggeler, K.R. and Shepherd, R.J. (1997) Petunia vein-cleaning virus: a plant pararetrovirus with core sequences for an integrase function. *Virology*, **236**, 137–146.

Ruiz-Perez, V.L., Murillo, F.J. and Torres-Martinez, S. (1996) Prt1, an unusual retrotransposon-like sequence in the fungus *Phycomyces blakesleeanus*. *Mol. Gen. Genet.*, **253**, 324–333.

Sala, M., Zambruno, G., Leceller G. *et al.* (1994) Spatial discontinuities in human immunodeficiency virus type 1 quasispecies derived from epidermal Langerhans cells of a patient with AIDS and evidence for double infection. *J. Virol.*, **68**, 5280–5283.

Samuelson, L.C., Wiebauer, K., Gumcio, D.L. and Meisler, M.H. (1988) Expression of the human amylase genes: recent origin of a salivary amylase promoter from an actin pseudogene. *Nucl. Acids Res.*, **16**, 8261–8276.

Schwartz, R.M. and Dayhoff, M.O. (1978) Matrices for detecting distant relationships. *Atlas Prot. Seq. Struct.*, 353–358.

Sellem, C.H., Lecelher, G. and Belcour, L. (1993) Transposition of a group II intron. *Nature*, **366**, 176–178.

Shearman, C., Gordon, J. and Gasson, M. (1996) Splicing of a group II intron in a functional transfer gene of *Lactocuccus lactis*. *Mol. Microbiol.*, **21**, 45–53.

Shub, D.A., Goodrich-Blair, H. and Eddy, S.R. (1994) Amino acid sequence motif of group I endonucleases is conserved in open reading frames of group II introns. *Trends Biochem. Sci.*, **19**, 402–404.

Song, S.U., Gerasimova, T., Kurkulos, M., Boeke, J.D. and Corces, V. G. (1994) An env-like protein encoded by a *Drosophila* retroelement: evidence that *gypsy* is an infectious retrovirus. *Genes Devel.*, **8**, 2046–2057.

Song, S.U., Kurkulos, M., Boeke, J.D. and Corces, V.G. (1997) Infection of the germ line by retroviral particles produced in the follicle cells: a possible mechanism for the mobilization of the *gypsy* retroelement of *Drosophila*. *Development*, **124**, 2789–2798.

Stavenhagen, J.B. and Robins, D.M. (1988) An ancient provirus has imposed androgen regulation on the adjacent mouse sex-limited protein gene. *Cell*, **55**, 247–254.

Steitz, T.A. (1998) A mechanism for all polymerases. *Nature* **391**, 231–232.

Sun, J., Inouye, M. and Inouye, S. (1991) Association of a retroelement with a P4-like cryptic prophage (retronphage phiR73) integrated into the selenocystyl tRNA gene of *Escherichia coli*. *J. Bacteriol.*, **173**, 4171–4181.

Svensson, A.-C. and Andersson, G. (1997) Presence of retroelements reveal the evolutionary history of the human DR haplotypes. *Hereditas*, **127**, 113–124.

Swergold, G.D. (1990) Identification, characterization, and cell specificity of a human LINE-1 promoter. *Mol. Cell. Biol.*, **10**, 6718–6729.

Szatkowski Ozers, M. and Friesen, P.D. (1996) The Env-like open reading frame of the *Baculovirus*-integrated retrotransposon TED encodes a retrovirus-like envelope protein. *Virology*, **226**, 252–259.

Takahashi, H., Okazaki, S. and Fujiwara, H. (1997) A new family of site-specific retrotransposons, START1, is inserted into telomeric repeats of the silkworm, *Bombyx mori*. *Nucl. Acids Res.*, **25**, 1578–1584.

Temin, H.M. (1985) Reverse transcription in eukaryotic genome: retroviruses, pararetroviruses, retrotransposons, and retrotranscripts. *Mol. Biol. Evol.*, **2**, 455–468.

Temin, H.M. (1989) Retrons in bacteria. *Nature*, **339**, 252–255.

Teng, S.-C., Kim, B. and Gabriel, A. (1996) Retrotransposon reverse-transcriptase-mediated repair of chromosomal breaks. *Nature*, **383**, 641–644.

Tiollais, P., Pourcel, C. and Dejean, A. (1985) The hepatitis B virus. *Nature*, **317**, 489–495.

Toh, H., Ono, M., Saigo, K. and Miyata, T. (1985) Retroviral protease-like sequence in the yeast transposon Ty1. *Nature*, **315**, 691.

Tooley, P.W. and Garfinkel, D.J. (1996) Presence of ty1-copia group retrotransposon sequences in the potato late blight pathogen *Phytophthora infestans*. *MPMI*, **9**, 305–309.

VanHouten, J.N., Natoli, F. and Asch, B.B. (1996) Frequent ectopic expression of a placenta-specific gene at high levels in BALB/c mouse mammary carcinomas. *Oncogene*, **12**, 2241–2245.

Villanueva, M.S., Williams, S.P., Beard, C.B., Richards, F.F. and Aksoy, S. (1991) A new member of a family of site-specific retrotransposons is present in the spliced leader RNA genes of *Trypanosoma cruzi*. *Molec. Cell. Biol.*, **11**, 6139–6148.

Vogt, V.M. (1997) Retroviral virions and genomes. In: *Retroviruses* (eds Coffin, J.M., Hughes, S.H. and Varmus, H.E.), pp. 27–69. Cold Spring Harbor Laboratory Press, Cold Spring Harbor, NY.

Wang, H. and Lambowitz, A.M. (1993) The Mauriceville plasmid reverse transcriptase can initiate cDNA synthesis *de novo* and may be related to the progenitor of the reverse transcriptases and DNA polymerases. *Cell*, **75**, 663–670

Wang, G.-H. and Seeger, C. (1992) The reverse transcriptase of hepadna B virus acts as a protein primer for viral DNA synthesis. *Cell*, **71**, 663–670

Waugh O'Neill, R.J., O'Neill, M.J. and Marshall Graves, J.A. (1998) Undermethylation associated with retroelement activation and chromosome remodelling in an interspecific mammalian hybrid. *Nature*, **393**, 68–72.

Wessler, S.R. (1996) Plant retrotransposons: turned on by stress. *Curr. Biol.*, **6**, 959–961.

Wessler, S.R., Bureau, T.E. and White, S. E. (1995) LTR-retrotransposons and MITEs: important players in the evolution of plant genomes. *Curr. Opin. Genet.*, **5**, 814–821.

Xiong, Y. and Eickbush, T.H. (1988a) The site-specific ribosomal DNA insertion element R1Bm belongs to a class of non-long-terminal-repeat retrotransposons. *Molec. Cell. Biol*, **8**, 114–123.

Xiong, Y. and Eickbush, T.H. (1988b) Functional expression of a sequence-specific endonuclease encoded by the retrotransposon R2Bm. *Cell*, **55**, 235–246.

Xiong, Y. and Eickbush, T.H. (1990) Origin and evolution of retroelements based upon their reverse transcriptase sequences. *EMBO J.*, **9**, 3353–3362.

Yang, J., Zimmerly, S., Perlman, P.S. and Lambowitz, A.M. (1996) Efficient integration of an intron RNA into double-stranded DNA by reverse splicing. *Nature* **381**, 332–335.

Yang, Z., Boffelli, D., Boonmark, N., Schwartz, K. and Lawn, R. (1998) Apolipoprotein (a) gene enhancer resides within a LINE element. *J. Biol. Chem.*, **273**, 891–897.

Zanotto, P., Gibbs, M.J., Gould, E.A. and Holmes, E.C. (1996) A reevaluation of the higher taxonomy of viruses based on RNA polymerases. *J. Virol.*, **70**, 6083–6093.

Zimmerly, S., Huatao, G., Perlman, P.S. and Lambowitz, A.M. (1995a) Group II intron mobility occurs by target DNA-primed reverse transcription. *Cell*, **82**, 545–554.

Zimmerly, S., Guo, H., Eskes, R., Yang, J., Perlman, P.S. and Lambowitz, A.M. (1995b) A group II intron RNA is a catalytic component of a DNA endonuclease involved in intron mobility. *Cell*, **83**, 529–538.

Zou, S., Wright, D.A. and Voytas, D.F. (1995) The *Saccharomyces* Ty5 retrotransposon family is associated with origins of DNA replication at the telomeres and the silent mating locus HMR. *Proc. Natl Acad. Sci. USA*, **92**, 920–924.

Dynamics of HIV Pathogenesis and Treatment

Dominik Wodarz and Martin A. Nowak

INTRODUCTION

Recent quantitative virological data from HIV-1-infected patients receiving antiviral therapy, in conjunction with mathematical models, have provided for the first time precise information on the kinetics of virus turnover *in vivo* (Ho *et al.*, 1995; Loveday *et al.*, 1995; Nowak *et al.*, 1995a; Schuurman *et al.*, 1995; Wei *et al.*, 1995; Perelson *et al.*, 1996, 1997; Bonhoeffer *et al.*, 1997a). It was shown that the half-life of productively infected cells is around 2 days (Ho *et al.*, 1995; Wei *et al.*, 1995). Free virus particles in the plasma have an even shorter half-life. The current estimate is about 6 h, but it is likely that the actual figure is lower than this, which suggests that close to 100% of the free virus population in the plasma of a patient turns over every day (Perelson *et al.*, 1996). These studies provide conclusive evidence that the asymptomatic phase of HIV-1 infection is not a period of latency but a dynamic process involving continuous rounds of *de novo* replication and infection, and that HIV is not a "slow virus" in terms of its daily replication potential. These findings are crucial for understanding the driving forces underlying disease progression in HIV-1-infected patients: the rapid viral turnover during the asymptomatic phase indicates a great potential for the virus to evolve in response to selection pressures exerted either by the immune system or by drug treatment (Nowak *et al.*, 1991; Wain-Hobson 1993; Ho *et al.*, 1995; Wei *et al.*, 1995). The potential for rapid viral evolution is demonstrated by findings that wild-type virus can be replaced by drug-resistant strains within 14–28 days of drug therapy (Wei *et al.*, 1995; Nowak *et al.*, 1997).

Such infection dynamics involve non-linear interactions of many different components and this makes the use of mathematical models necessary to provide a correct interpretation of empirical results, to generate new insights and hypotheses as well as to guide further experimental work. In this review we will demonstrate how mathematical models have been used to identify evolutionary mechanisms that may drive disease progression in HIV-infected patients as well as to achieve a better understanding of antiviral therapy.

NATURAL HISTORY OF HIV INFECTION

In the primary phase of the disease, after an individual has been infected with HIV, the virus replicates to relatively high levels for a short but variable period of time. The subsequent rise of the immune response suppresses

Origin and Evolution of Viruses
ISBN 0–12–220360-7

this primary viremia and the patient enters the asymptomatic phase. During this phase, the virus is kept at low levels and one observes a slow decline in the CD4$^+$ cell count of the patient. The duration of the asymptomatic phase varies a lot between individuals. Fast progressors develop AIDS within a few years, whereas slow or non-progressors remain without symptoms for 10–15 years. The final development of AIDS is often characterized by an upsurge in virus load, accompanied by a fall in the number of CD4$^+$ cells found in the blood. It is now common to define AIDS disease if the number of CD4$^+$ cells has fallen below 200 per ml (a healthy, uninfected person has about 1000 CD4$^+$ cells per ml).

At the beginning of the infection, there is little sequence variation in the HIV population isolated from a patient (Kuiken et al., 1992; Zhang et al., 1993; Zhu et al., 1993; Safrit et al., 1994). In the course of infection, HIV mutates in the CTL and antibody epitopes, resulting in an increase in antigenic diversity until the patient develops AIDS (Saag et al., 1988; Eigen, 1989; Balfe et al., 1990; Phillips et al., 1991; McMichael et al., 1995, 1996; Borrow et al., 1997; Goulder et al., 1997; Price et al., 1997).

Virus strains isolated in the asymptomatic phase tend to use the CCR5 coreceptor (R5 viruses; Berger et al., 1998; Doms and Moore, 1998). Therefore, they not only infect primary T cells but also have the ability to infect macrophages. These isolates tend to show a slow rate of replication (Connor and Ho, 1994) and a low degree of cytopathogenicity, and are usually associated with the non-syncytium-inducing (NSI) phenotype (Fenyo et al., 1988; Roos et al., 1992; Schuitemaker et al., 1992; Zhu et al., 1993; Van't Wout et al., 1994; Schuitemaker, 1994; Rudensey et al., 1995; Fouchier et al., 1996). During the asymptomatic phase of the infection, HIV evolves to show stronger tropism for T cells (Schuitemaker, 1994). These virus strains also tend to replicate at a faster rate (Connor and Ho, 1994), and are characterized by higher degrees of cytopathogenicity (Rudensey et al., 1995; Fouchier et al., 1996). Such isolates may either use both the CCR5 and the CXCR4 coreceptors (termed R5X4 strains; Berger et al., 1998; Doms and Moore, 1998) or, in about 50% of the

patients, specialise in using the CXCR4 receptor only (termed X4 viruses; Berger et al., 1998; Doms and Moore, 1998). X4 strains have lost the ability to infect macrophages, are associated with the syncytium-inducing phenotype and are markers of disease progression (Tersmette et al., 1988, 1989; Schuitemaker et al., 1992; Schuitemaker, 1994).

In the following sections we will show how the evolutionary dynamics of HIV may help us understand the transition from the asymptomatic period of the infection to full-blown AIDS. In addition, we discuss the implications of these dynamical interactions for antiviral therapy.

THE BASIC MODEL OF VIRUS DYNAMICS

The basic model of virus dynamics describes the interactions between the populations of uninfected target cells (x), infected target cells (y) and free virus particles (v). Here we add a CTL response (z). The model is given by the following system of ordinary differential equations:

$$
\begin{aligned}
\dot{x} &= \lambda - dx - \beta xv \\
\dot{y} &= \beta xv - ay - pyz \\
\dot{v} &= ky - uv \\
\dot{z} &= cyz - bz.
\end{aligned} \tag{1}
$$

The model is explained schematically in Figure 9.1. Uninfected target cells are produced at a rate λ. They die at a rate dx and become infected by the virus at a rate βxv. Infected cells die at a rate ay and produce free virus particles at a rate ky. Free virus particles decay at a rate uv. CTLs proliferate in response to antigen at a rate cyz, die at a rate bz and lyse infected cells at a rate pyz.

An important concept is the basic reproductive ratio of the virus, denoted by R_0 (Figure 9.2). It describes the average number of secondary infected cells produced by each infected cell at the beginning of the infection (Anderson and May, 1979a,b, 1991) and is given by $R_0 = \beta \lambda k / adu$. If the basic reproductive ratio of the virus is less than unity, each infected cell produces on average less than one newly infected cell, making it impossible for the virus popula-

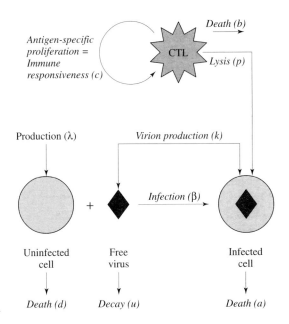

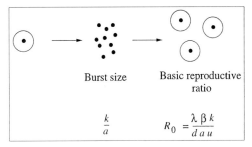

FIGURE 9.2 The concept of the basic reproductive ratio of the virus. An infected cell produces on average k/a virions (burst size). The basic reproductive ratio of the virus (R_0) is defined as the average number of secondary infections produced by a single infected cell at the beginning of the infection, i.e. if $x = \lambda/d$. For the basic model of virus dynamics it is given by $R_0 = \lambda\beta k/adu$.

FIGURE 9.1 Basic model of virus dynamics including a CTL response. Uninfected cells are produced at a rate λ, die at a rate d and become infected at a rate β. Infected cells produce free virus particles at a rate k and die at a rate a. Free virus decays at a rate u. Upon recognition of antigen, CTLs proliferate at a rate c, and kill infected cells at a rate p. CTLs die at a rate b.

tion to maintain itself. This outcome is described by (E1):

$$x^{(1)} = \lambda/d, \ y^{(1)} = 0, \ v^{(1)} = 0, \ z^{(1)} = 0.$$

On the other hand, if $R_0 > 1$, the virus can successfully establish an infection. The infection may be limited by target cell availability alone without the presence of a CTL response. This is described by (E2):

$$x^{(2)} = au/\beta k, \ y^{(2)} = \lambda/a - du/\beta k,$$
$$v^{(2)} = \lambda k/au - d/\beta, \ z^{(2)} = 0.$$

On the other hand, if the immune responsiveness of the host is high enough (i.e. $cy^{(2)} > b$), the infection is controlled by a combination of target cell availability and the CTL response, described by (E3):

$$x^{(3)} = \frac{\lambda cu}{cdu + \beta bk}, \ y^{(3)} = \frac{b}{c}, \ v^{(3)} = \frac{bk}{cu}, \ z^{(3)} = \frac{1}{p}\left(\frac{\lambda\beta ck}{cdu + \beta bk}\right).$$

This model provides the basis for studying viral dynamics and evolution in HIV-infected patients. Important insights gained by such models are set out as follows.

ANTIGENIC VARIATION AND DIVERSITY

Diversity Threshold

Since there is no proofreading mechanism during reverse transcription, HIV is characterized by a relatively high mutation rate of about one base per genome per replication cycle (Preston *et al.*, 1988; Roberts et al. 1988). This high mutation rate, together with the selection pressure exerted by the antiviral immune response, provide ideal conditions for the evolution of antigenic escape mutants (Phillips *et al.*, 1991; Borrow *et al.*, 1997; Goulder *et al.*, 1997). In fact, the error rate of the reverse transcriptase of HIV seems to maximize the chance of producing antigenic escape mutants (Nowak, 1990). These escape mutants may in turn be controlled by rising immune responses specific for these newly evolved virus strains. However, there is an inherent asymmetry in the interaction between the evolving virus population and the antiviral immune response. While a given immune response may have to be specific for a given virus strain in order to kill it, each virus strain can infect and kill any T helper cell regardless of its specificity (Figure 9.3). These basic assumptions can be captured in simple mathematical

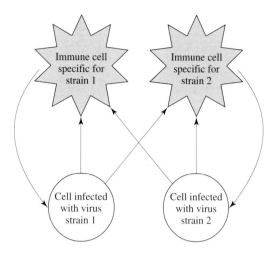

FIGURE 9.3 Asymmetry in the dynamics between HIV and the immune response. A given immune cell may only be able to fight the virus strain it is specific for. However, the HIV population may impair the immune response regardless of its specificity.

models (Nowak *et al.*, 1991; Nowak and May, 1991, 1993; Nowak, 1992a, 1995). Such models take into account the evolving virus population, strain-specific immune responses and cross-reactive immune responses. The evolutionary dynamics also involve a stochastic process, where the probability that a new mutant emerges is either constant or, more realistically, proportional to the total virus load, since the number of mutations is proportional to the number of replication events.

Such models are characterized by three parameter ranges, depending on the relationship between viral replication and/or immune impairment and the strength of the strain-specific and cross-reactive immune responses.

1. If viral replication together with immune impairment outruns the combined effect of strain-specific and cross-reactive immune responses, the immune system immediately collapses, resulting in uncontrolled viral replication and the development of AIDS without the presence of a prolonged asymptomatic period. No antigenic variation will be observed and the fastest replicating strain will emerge to dominate the virus population. Such dynamics are observed in rapid progressors (Wolinsky *et al.*, 1996).

2. If the cross-reactive immune response alone is sufficiently large compared to virus replication and the amount of immune impairment, the immune system may effectively control the virus population without the development of disease. In this case, the level of virus load controlled by the immune response depends on the rate of viral replication, the amount of immune impairment by the virus, the efficacy of the immune response and the number of antigenic variants present. However, uncontrolled viral replication is not possible in this parameter region. Although some HIV-infected patients are long-term non-progressors and may fall in this parameter region, it is not clear whether they will develop symptoms at a later stage. Chronic infection without development of disease is, however, the rule for simian immunodeficiency virus (SIV) infection in its natural host (Hartung *et al.*, 1992). The lack of disease progression in this case may be due to strong cross-reactive immune responses as well as slower replication rates of the virus.

3. The most interesting case predicting the course of HIV infection in humans arises when the combination of cross-reactive and strain-specific immune responses may control the virus, but the strain-specific response alone is unable to do so. In this case, the dynamics depend on the amount of antigenic diversity. If antigenic diversity is low, the virus is controlled, corresponding to the asymptomatic phase of the infection. Increasing antigenic diversity results in an increase in virus load. The immune system keeps the virus population in check as long as the antigenic diversity lies below a threshold. Crossing the diversity threshold leads to uncontrolled virus growth accompanied by a collapse of the immune system.

A typical course of disease progression predicted by the model is illustrated in Figure 9.4. The length of the asymptomatic period is determined by the magnitude of the diversity threshold which in turn depends on the efficacy of the immune response as well as on the amount of viral replication and immune impairment.

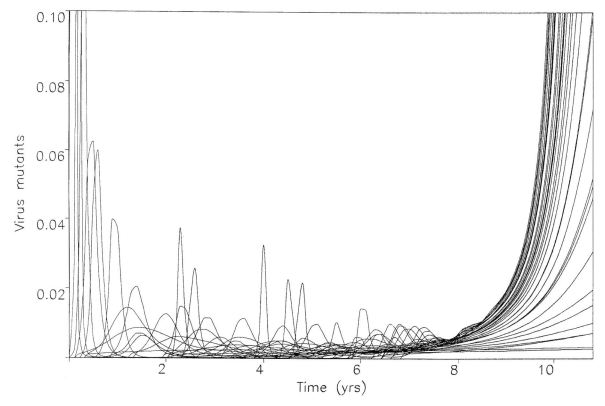

FIGURE 9.4 Evolution of many different virus strains during the time of infection. Initially, the strains grow to high levels, which may cause the clinical symptoms observed during primary HIV infection. The subsequently emerging escape mutants are suppressed at a faster rate, because of the action of cross-reactive immune responses. Different virus strains grow to different levels according to their growth rates. The accumulation of viral diversity breaches the diversity threshold after approximately 7 years in this simulation. In the final phase, the fastest growing strains dominate the virus population. The *y*-axis indicates the relative concentration of different virus mutants.

Variation in these parameters among patients may therefore account for the variability observed in the disease process. In addition, the stochastic nature of the emergence of escape mutants may also contribute to the variability of the disease process. Whether the first few escape mutants arise sooner or later than average may result in marked differences in the pattern of disease progression (Nowak and May, 1993).

In summary, viral evolution towards increased antigenic diversity may be the cause rather than the consequence of disease progression and may contribute to the transition from the asymptomatic period of the disease to AIDS.

Wolinsky *et al.* (1996) interpreted their finding that rapid progressors can have low *genetic* vari-

ation as being at variance with the diversity threshold theory. This conclusion is incorrect. As explained in the first papers on the diversity threshold idea, rapid progression is expected to occur with low antigenic diversity if the patient has weak anti-HIV immune responses.

CTL Responses Against Multiple Epitopes

In most viral infections, including HIV, there are immune responses against several epitopes; this complicates viral dynamics and the effect of antigenic variation. Therefore, a theoretical framework was developed to explore the dynamics of an antigenically variable virus population and an array of CTL responses directed against multiple epitopes (Nowak et al., 1995b, c; Nowak, 1996). This multiple epitope theory has also been applied to other infectious diseases (Gupta et al., 1996, 1998)

Central to the understanding of the effect of CTL responses against multiple epitopes is that these CTL clones are in competition with one another. The different CTL clones can be viewed as species of predators that proliferate in response to a common food source (the virus population) with different efficiencies.

Consider first the simple assumption of an antigenically homogeneous virus population. In such a scenario, the model predicts that the CTL response with the strongest immunogenicity will outcompete all other CTL responses. This is analogous to the competitive exclusion principle in ecology: the strongest competitor will reduce virus load to levels that are not sufficiently high to stimulate the competitively inferior CTL responses. This may explain the concept of immunodominance, i.e. that *in vivo* the CTL response is predominantly directed against one or only a few epitopes (Buus et al., 1987; Adorini et al., 1988; Schaeffer et al., 1989; Liu et al., 1993; Sercarz et al., 1993).

For a heterogeneous virus population, the situation is more complicated than this. One has to distinguish between two cases. Either all antigenic variants have the same replication rate, or their replication kinetics differ. Consider the simpler case of equal replication rates first. The model still predicts the occurrence of immunodominance, but now the competitive ability of a CTL response against a given epitope depends not only on its immunogenicity but also on viral diversity in this epitope. Viral diversity essentially reduces immunogenicity. Assume that CTL proliferate equally well in response to all variants of a given epitope. Denoting the immune responsiveness against epitope A as c, the immune responsiveness against epitope B as k and the number of antigenic strains in the respective epitopes as n_A and n_B, the immune response against epitope A will dominate if $c/n_A > k/n_B$.

On the other hand, if the antigenic variants replicate at different rates, the model predicts

the coexistence of CTL responses directed against the different epitopes. In other words, a heterogeneous virus population characterized by concomitant differences in the viral replication kinetics induces the lack of immunodominance.

Antigenic variation in multiple epitopes can lead to complicated "antigenic oscillations" (Figure 9.5; Nowak et al., 1995b), i.e. distinct peaks in viral abundance, often dominated by a single genotype, that arise when the CTL response against a given variant declines to low levels due to temporary lack of stimulation. This leads to concomitant fluctuations in immunodominance. In contrast to the occurrence of peaks in viral abundance upon the emergence of new variants (antigenic drift), antigenic oscillations are the result of the non-linear dynamics between the existing heterogeneous virus population and CTL responses against multiple epitopes and do not require the emergence of new mutants.

Next, consider the emergence of a new mutant in a given epitope. Suppose the existence of a homogeneous virus population and an immunodominant CTL response against one of two epitopes (A, Figure 9.6). The emergence of an escape mutant in epitope A may lead to one of four possible outcomes depending on the replication rates and the immunogenicities of the mutant relative to the wild-type (Figure 9.6). Denote the replication rate of wild-type and mutant as r_w and r_m respectively and denote the immunogenicity of wild-type and mutant epitope A as c_w and c_m respectively. The immunogenicity of epitope B is described by k. The four outcomes are as follows.

1. A new specific response against epitope A is induced by the mutant without affecting epitope B. This represents diversification in epitope A and will be observed if $1/c_w + 1/c_m > 1/k$.
2. No new response against epitope A is induced by the mutant, but the response against epitope B is enhanced. This is a partial shift in immunodominance and will occur if $1/c_w + 1/c_m < 1/k$ and $r_w > r_m$.
3. The mutant may induce a new response against epitope A, which outcompetes the

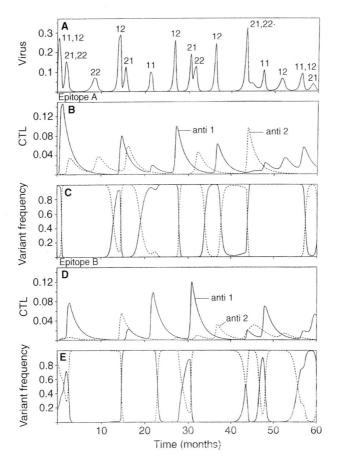

FIGURE 9.5 Antigenic oscillations and fluctuating immunodominance in a model with two epitopes. Peaks of viral abundance that consist of antigenically different variants are accompanied by oscillations in the size and specificity of the CTL responses. All virus variants are present at the beginning of the simulation and there are no additional mutational events.

response against the wild-type, consequently causing a partial shift in immunodominance. The condition for this outcome is given by $1/c_w + 1/c_m > 1/k$ and $r_w < r_m$.

4. Finally, if $1/c_m > 1/k$ and $r_w < r_m$, the mutant virus outcompetes the wild-type, resulting in a complete shift in immunodominance. Thereby, the selective advantage of the escape mutant becomes negligible and it may not reach fixation even if there is no CTL response against the variant peptide.

In terms of HIV infection, this analysis extends and reinforces the idea that viral evolution towards increased antigenic diversity may drive progression of the disease. A stable response against an invariant epitope will lead to efficient control of the virus and slow progression. Viral diversity may lead to unstable dynamics, and the evolution of new antigenic variants may result in a shift in immunodominance to weaker epitopes, leading to less efficient control of HIV and an increase in virus load. These notions have been confirmed

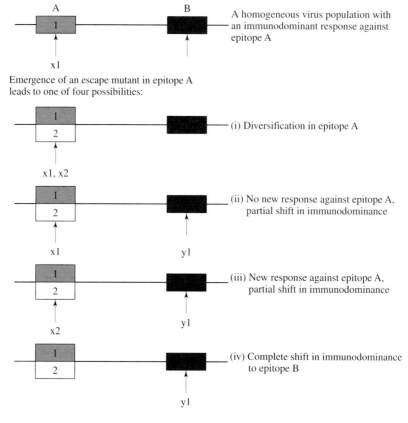

FIGURE 9.6   Antigenic variation may shift immunodominance. The emergence of an escape variant in an immunodominant epitope (A) leads to one of four possible outcomes. For details see text.

experimentally by Borrow *et al.*, (1997), Goulder *et al.* (1997) and Price *et al.* (1997).

REPLICATION KINETICS

The notion that changes in the replication rate of HIV may be important for disease progression has often been pointed out (e.g. Asjo *et al.*, 1986; Fenyo *et al.*, 1988; Tersmette *et al.*, 1989; Gruters *et al.* 1991; Ferbas *et al.* 1996).

Connor and Ho (1994) analysed sequential HIV-1 isolates from a patient who progressed to AIDS within 5 years. Virus strains isolated at the beginning of the infection were characterized by relatively slow replication kinetics. During the asymptomatic phase they observed a strong and steady increase in the replication rate of the virus until the onset of AIDS. The increasing abundance of faster replicating variants during disease progression has been described mathematically by Nowak and May (1991, 1992), deBoer and Boerlijst (1994) and Schenzle (1994). More specifically, Wodarz *et al.* (1998) showed how viral evolution towards increased replication kinetics in the asymptomatic phase may contribute to the transition to full-blown AIDS. They considered a mathematical model (Figure 9.7) where the target cells are antigen presenting cells or T helper cells (which are both infected by HIV). They distinguished between precursor and effector CTL and assumed that the proliferation of precursor CTL requires the help of uninfected CD4[+] T cells, e.g. via cytokine production.

Virus Dynamics:

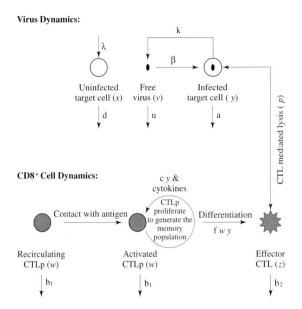

CD8⁺ Cell Dynamics:

FIGURE 9.7 Schematic representation of a model describing virus dynamics with immune impairment. The dynamics between the virus and its target cells are identical to the basic virus infection model. In this model, the target cells are assumed to be T-helper cells or antigen presenting cells. With the CTL response, we distinguish between CTL precursor and effector cells. The precursor cells circulate through the body, and upon contact with antigen, proliferate at a rate c. This proliferation is enhanced by cytokines secreted by CD4⁺ cells. The precursors die at a rate b_1. On further contact with antigen, precursor cells from the expanded CTLp pool differentiate into effector cells at a rate f. Effector cells die at a rate b_2 and lyse infected cells at a rate p.

These assumptions led to the result that the virus may induce the exhaustion of the virus-specific CTL response. In terms of HIV, two parameters are important in determining the fate of the CTL response. There is a replication rate threshold beyond which the HIV-specific CTL response will be exhausted. The higher the immune responsiveness of the host, the higher the replication rate of the virus required to induce CTL exhaustion. CTL exhaustion has been proposed as a possible mechanism for the eventual breakdown of the immune system upon progression to AIDS (e.g. Doherty, 1993; Moss *et al.*, 1995; Rinaldo *et al.*, 1995a). Figure 9.8 demonstrates how evolution towards increased replication rates

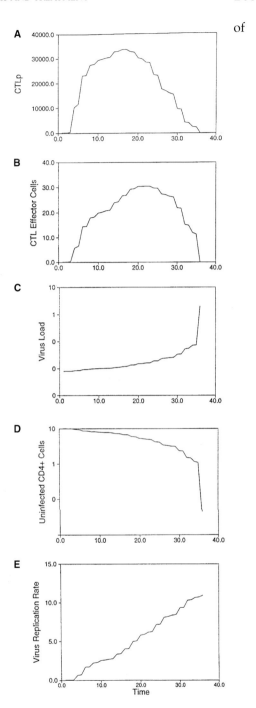

FIGURE 9.8 Implications of CTL exhaustion for HIV disease progression. The graphs show the effect of virus evolution towards higher replication kinetics on the number of: (**A**) CTL precursor cells; (**B**) CTL effector cells; (**C**) virus load; and (**D**) the number of uninfected target cells. At each time interval the viral replication rate (**E**) is increased by a random amount and the equilibrium values of the respective variables are calculated. See text for details.

HIV may lead to a pattern of disease progression similar to that observed in HIV-infected patients. As long as the replication rate of the virus is sufficiently low for the CTL response to persist, evolution towards faster replication results in a relatively slow decline of the uninfected T helper cells and a slow rise in virus load. When the virus evolves beyond the replication rate threshold, it induces the exhaustion of the specific CTL response, allowing virus load to shoot up to relatively high levels and causing a sharp drop in the number of T helper cells found in the patient. As was the case with the diversity threshold model, variation in the immune responsiveness of the host may account for variability in the duration of the asymptomatic phase of the infection, with a stronger responder remaining asymptomatic for a longer period of time. As can be seen in Figure 9.8, the model also predicts that the CTL precursor population starts to decline earlier and declines at a faster rate than the CTL effector population. Thus, when the CTL precursors have largely vanished, one might still expect to see considerable CTL effector activity. Such observations have been made by Rinaldo et al. (1995a,b). They found that there was a correlation between CTL effector activity and time, and that, towards the end stage of the disease, a significant CTL effector response was still present, possibly arising from residual memory.

Evolution towards higher replication rates is related to evolution of antigenic diversity in that mutation in the CTL epitopes may allow effectively faster viral replication. This may occur either through simple escape mutations, or through the evolution of altered peptide ligands, which may act as TCR antagonists (Jameson et al., 1993; Bertoletti et al., 1994; Klenerman et al., 1994; Meier et al., 1995) or promote responses with inappropriate specificities (Jameson and Bevan, 1995; Kalams and Walker, 1995; Klenerman et al., 1995, 1996; McAdam et al., 1995; Borrow et al., 1997). Moreover, evolution of resistance to immune responses inhibiting the overall replication kinetics of the virus, such as interferon or antibodies, may also contribute to the exhaustive process.

CELL TROPISM

The differential tropism of HIV for macrophages and T cells may be a key element of HIV pathogenesis (Crowe and Kornbluth, 1994; Karlsson et al., 1994; Mosier and Sieburg, 1994; Schuitemaker, 1994; Crowe, 1995; McKnight and Clapham, 1995; Rudensey et al., 1995; Fauci, 1996; Koot et al., 1996; Connor et al., 1997; Dittmar et al., 1997). Strains using the CCR5 coreceptor (R5 viruses) infect primary T cells and macrophages. They are slowly replicating and relatively non-virulent (NSI phenotype). As disease progresses, HIV becomes increasingly T-cell tropic, using the CXCR4 coreceptor. Virus strains may use both the CCR5 and the CXCR4 coreceptors (termed R5X4 virus; Doms and Moore 1998), or they evolve to specialize on the CXCR4 coreceptor (X4 virus, Doms and Moore 1998), which happens in about 50% of patients. Such isolates have lost the ability to infect macrophages, replicate at a faster rate and tend to show the syncytium-inducing (SI) phenotype. The evolution of virulent X4 strains is associated with progression to AIDS and mathematical models can help us identify factors that lead to the emergence of virulent CXCR4 tropic mutants.

Mathematical models have so far focused on two basic concepts concerning cell tropism in HIV infection. (1) First, there is a difference in the biology of macrophages and T cells: while HIV can only replicate in activated T cells, activation and cell division does not seem to be required for macrophage infection (Stevenson and Gendelman, 1994; Stevenson, 1996). The reason for this difference rests in the mechanism underlying the transport of the HIV genome into the nucleus of the two cell types (Stevenson and Gendelman, 1994; Emerman et al., 1994; Heinzinger et al., 1994; Stevenson et al., 1994, 1995; Bukrinskaya et al., 1996; Stevenson, 1996). While the HIV preintegration complex may reach the macrophage nucleus by active transport, this mechanism does not work in T cells, where breakdown of the nuclear envelope is required for infection. (2) Second, there is a difference in the viral phenotypes between strains using the CCR5 and the CXCR4 receptor. While

R5 viruses are supposed to be non-cytopathic and slowly replicating, X4 viruses are thought to be strongly cytopathic and faster replicating.

Macrophage Versus T-cell Infection

Macrophage infection by HIV can be modeled by the basic equations describing virus dynamics set out at the beginning of this review. However, in order to study T-cell infection in a more realistic way, Wodarz *et al.* (1999) modified the basic virus infection model to take into account resting target cells that cannot be infected and susceptible target cells that are generated from the resting cell population through activation in response to antigen (Figure 9.9). These assumptions complicate the dynamics of HIV infection. If the rate of background activation of T cells due to other pathogens being present at the time of HIV infection is relatively low, the outcome may depend on a complex interplay between host and viral parameters as well as the initial conditions. An important host factor is the immune responsiveness. If the immune responsiveness lies above a certain threshold, the infection is cleared. The reason for this is that a relatively strong CTL response will depress levels of virus load to very low levels, not enough to maintain a significant number of activated T helper cells. Since these are the target cells of the virus, the infection will vanish. Among the viral parameters, the replication rate of the virus is required to lie above a certain threshold for infection to be possible. In the parameter region where the virus may establish a persistent infection, the outcome of the dynamics depends on the initial abundance of virus and virus-specific CTLs (Figure 9.10). The

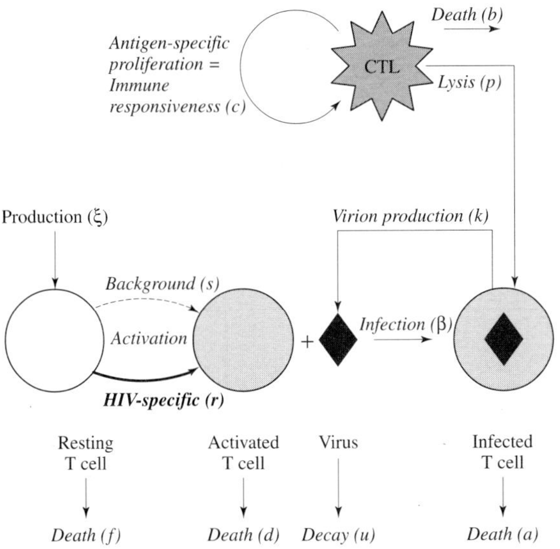

FIGURE 9.9 Model of T-cell infection by HIV. We distinguish between resting and activated T cells. Only activated cells may be infected. Resting T cells are produced at a rate ξ and die at a rate f. We assume that activation occurs mainly in response to HIV-replication (r), but there may also be a certain amount of background activation (s), e.g. due to other pathogens being present in the host at the time of HIV infection. Activated T cells die at a rate d and become infected at a rate β. Infected cells produce free virus particles at a rate k and die at a rate a. Free virus decays at a rate u. Upon contact with antigen, CTLs proliferate at a rate c and lyse infected cells at a rate p. CTLs die at a rate b.

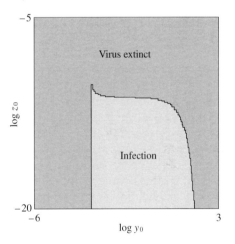

FIGURE 9.10 Dependence of the outcome of T-helper-cell infection on the initial conditions. Initial virus load (y_0) needs to lie above a threshold for persistent infection to be possible. However, in the presence of a CTL response, the virus population also goes extinct if virus load lies above a certain threshold. This is because a high initial virus load increases the initial growth rate of the CTL response, which in turn quickly suppresses viral abundance. This reduces the number of activated target cells below the levels required for persistent replication. Therefore, initial virus load needs to be intermediate for establishment of persistent infection to be successful. The initial number of CTLs (z_0) also plays an important role. The higher the initial levels of CTLs, the lower the upper virus load threshold, leading to virus extinction. Above a certain initial abundance of CTLs, the establishment of persistent infection becomes impossible.

initial virus load must lie above a certain thresh-old level in order to activate a sufficient number of target cells for invasion to be possible. Given that the initial virus load is sufficiently high to allow invasion, the CTL response may either clear the infection or control a persistently repli-cating virus. If the initial virus load lies above a threshold, the CTL response can grow fast enough to quickly reduce the virus population by a significant amount. This in turn leads to a decline in the number of activated target cells, making it impossible to support persistent repli-cation. Consequently, the virus population can-not recover and goes extinct. On the other hand, intermediate initial values of virus load allow virus replication to outrun the CTL response, which can only grow at a slower rate. Thus, per-sistent replication can be established. The high-er the initial abundance of CTLs, the lower the threshold level of virus load required to achieve clearance of the infection. Above a certain initial CTL abundance, the establishment of persistent infection becomes impossible.

This model has been extended to take into account both types of virus strain: R5 viruses capable of infecting macrophages and X4 virus-es capable of infecting T cells only. We assumed that T helper cells are activated by both types of virus. The presence of strains capable of infect-ing macrophages significantly enhances the invasion of CXCR4 tropic strains, eliminating the complex dependence on initial conditions. The model identifies three parameter regions (Wodarz *et al.*, 1999). If the immune responsive-ness to macrophage-tropic HIV is above a cer-tain threshold, CXCR4 tropic strains cannot invade the host. On the other hand, if the immune responsiveness to macrophage tropic strains is sufficiently low so that invasion of CXCR4 tropic mutants is possible, one has to distinguish between two situations (Figure 9.11). If the immune responsiveness to both virus strains is not low enough, CXCR4 tropic strains may initially grow to a peak but this rise is only temporary, since the CTL response is still strong enough to suppress these mutants to very low levels, indicating extinction. If CXCR4 tropic strains are continuously produced, this may result in the appearance of "blips" of CXCR4 tropic HIV, which immediately goes

extinct (Figure 9.11). Finally, if the immune responsiveness to both strains is reduced below a threshold, the immune system is not strong enough to suppress these HIV variants, leading to the permanent rise of CXCR4 tropic HIV (Figure 9.11).

To summarize, these studies indicate that infection of macrophages may be essential for maintaining a persistent infection and for facili-tating the eventual rise of CXCR4 tropic mutants once the immune system has been suf-ficiently weakened by the virus, e.g. due to anti-genic variation. These results are in line with the observation that individuals with a deletion in the CCR5 coreceptor, making macrophage infec-tion impossible, are unlikely to become infected

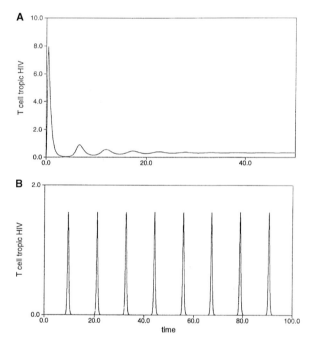

FIGURE 9.11 Two types of dynamical behavior can occur if CXCR4-tropic HIV has a positive initial growth rate. Either (**A**) the equilibrium is stable, reached by damped oscillations or (**B**) the occurrence of stable limit cycles is observed. The limit cycles are characterized by very low troughs. This indicates that, in the parameter space where limit cycles occur, the emergence of a CXCR4-tropic mutant leads to strong initial growth up to a peak. Subsequently, the immune response suppresses these mutants towards extinc-tion. If CXCR4-tropic HIV is continuously generated, this would lead to the occurrence of "blips" of such mutants, which rapidly go extinct.

by HIV (Dean *et al.*, 1996; Dragic *et al.*, 1996; Huang *et al.*, 1996; Liu *et al.*, 1996; Samson *et al.*, 1996; Balotta *et al.*, 1997; Biti *et al.*, 1997; Rana *et al.*, 1997; Theodorou *et al.*, 1997).

R5 Versus X4 Viruses

While CCR5 tropic HIV persists throughout the infectious process, the virus specializes in the use of the CXCR4 receptor in about 50% of patients, which is associated with the syncytium-inducing phenotype and marks progression to the end stage of the disease. Mathematical models suggest a possible reason for the rise of such specialist HIV in only about half of the patients. Wodarz and Nowak (1999a) devised an evolutionary model describing the dynamics of cell tropism (Figure 9.12). They considered *n* virus strains, which could infect two alternative target cell types with different efficiencies. The rate of target cell entry of the successive mutants was assigned in a way that it monotonically increased in one cell type while it monotonically decreased in the alternative cell type (Figure 9.13). As can be seen in Figure 9.13, the change in the rate of target cell entry in the successive mutants can either be greater than

linear or less than linear. In biological terms, a greater than linear change in the viral replication kinetics may occur if, for instance, the accumulation of mutations leads to an increasing effect on viral replication. Similarly, a less than linear change may occur if the initial mutations have a relatively large effect on viral replication but subsequent mutations exert only a weaker effect. Wodarz and Nowak (1999a) found that the virus population will only specialize in both target cell types if this change is greater than linear, otherwise generalism will evolve (Figure 9.13).

The correlation between coreceptor usage and tropism for macrophages and T cells is complex. While macrophage infection is dependent on the CCR5 receptor, both the CCR5 and the CXCR4 receptors may promote T-cell infection. However, it has been reported that, in T cells, expression of the CCR5 and the CXCR4 receptors is subset-dependent and tends to be mutually exclusive (Bleul *et al.*, 1997; Rowland-Jones and Tan, 1997; Unutmaz and Littman, 1997; Loetscher *et al.*, 1998). Therefore, we can distinguish between two "target cell types": those susceptible mainly to R5 strains and those susceptible mainly to X4 virus. Consequently, our model for the evolution of cell tropism may be applied and predicts that it depends on the exact fitness landscape of the virus whether HIV evolves to specialize on the CXCR4 coreceptor in a given patient. Full X4 strains will evolve only if the fitness landscape of the virus is greater than linear, otherwise R5X4 strains will evolve, using both coreceptors with similar efficiencies. The fitness landscape may differ between patients. This may simply be due to chance in the mutations occurring and thus in the course of evolution taken by the virus Moreover, a greater than linear change in the replication kinetics of the successive mutants may also be promoted by escape from mechanisms limiting viral replication.

Assuming that the fitness landscape of the viral mutants allows the evolution of specialism, Wodarz and Nowak (1998a) adopted this model specifically for HIV and investigated the evolution of X4 strains in the absence and presence of different immune responses. The fitness landscape of HIV mutants assumed in these

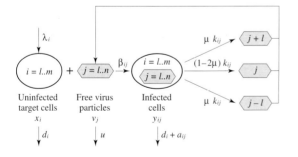

FIGURE 9.12 Model describing the evolutionary dynamics of cell tropism. We distinguish between target cells of type i ($i = 1 .. m$) and virus particles of strain j ($j = 1 .. n$). Uninfected target cells (x_i) are produced at a rate λ_i and die at a rate d_i. They become infected by free virus of strain j (v_j) at a rate β_{ij}. Cells of type i infected with strain j (y_{ij}) die at a rate $d_i + a_{ij}$ and produce free virus at a rate k_{ij}. Free virus decays at a rate u. During replication, virus strain j mutates at a rate μ resulting in strains $j + 1$ as well as $j - 1$.

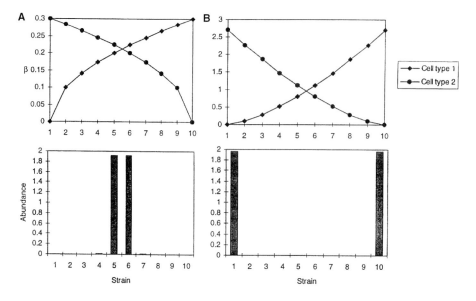

FIGURE 9.13 Evolution of generalism versus specialism assuming the existence of two target cell types. We also assume that the viral replication kinetics monotonically increase from strain 1 to *n* in one cell type, while they monotonically decrease in the alternative cell type. The resulting fitness landscape of the virus mutants can be divided into two categories. The change in the replication rate of the successive mutants may be (**A**) less than linear or (**B**) greater than linear. If the change is less than linear, the virus population evolves to utilize both cell types with similar efficiency (generalism), while a greater than linear change leads to the evolution of specialism.

studies is shown in Figure 9.14. In accord with empirical findings (e.g. O'Brien, 1994; Rudensey *et al.*, 1995; Fouchier *et al.*, 1996), we assume that CXCR4 tropic HIV may evolve to higher replication kinetics and at a faster rate than CCR5 tropic strains, and that the half-life of cells infected with X4 virus is shorter than that of cells infected with R5 strains. This fitness landscape leads to the dominance of CCR5 tropic HIV with CXCR4 tropic strains being suppressed to relatively low levels (Figure 9.15A). We explored the effect of four types of immune response on the course of evolution. These included an antibody response, CTL-mediated lysis, CTL-mediated inhibition of viral entry into target cells, and CTL-mediated inhibition of virion production. The evolutionary outcome of these models is shown in Figure 9.15B–E. Clearly, different types of immune response have different effects on the rise of CXCR4 tropic mutants. Immune responses acting on the

virus before it has integrated into the host genome (antibodies, CTL-mediated inhibition of infection) select against the rise of CXCR4 tropic HIV, suppressing these strains to relatively low levels, while immune responses acting on infected cells (CTL-mediated lysis and inhibition of virion production) favor the rise of CXCR4 tropic strains, leading to the coexistence of R5 and X4 virus with similar abundances.

These results are the consequence of the finding that, without an immune response or under the pressure of immune responses acting on the virus before integration into the host genome, non-cytopathic viruses attain significantly higher levels of virus load than cytopathic ones, while this is not the case under immune responses acting on infected cells (Wodarz and Nowak, 1999b). Moreover, these studies have also shown that the viral replication kinetics do not significantly influence the equilibrium number of infected cells. Thus, while in certain cir-

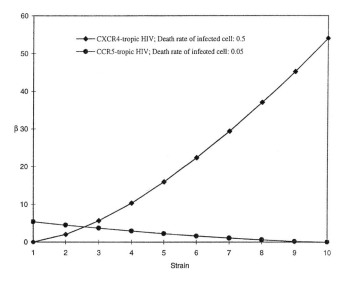

FIGURE 9.14 Fitness landscape assumed for CCR5 and CXCR4 tropic HIV. The change in the replication kinetics of the successive viral mutants is greater than linear. Therefore, the virus population may evolve towards specialism. According to empirical findings, we assume that X4 strains may evolve to faster replication kinetics and at a faster rate than R5 variants, and that cells infected with CXCR4 tropic mutants have a shorter half-life than those infected with CCR5 tropic virus.

cumstances, non-cytopathic R5 variants may have a selective advantage over cytopathic X4 strains; this advantage is not cancelled out by the relatively slow replication kinetics of CCR5 tropic HIV.

Directly after infection, HIV can replicate freely before the rise of any immune responses. Since the models have demonstrated that such conditions select against the rise of virulent CXCR4 tropic mutants, these dynamics may also help to understand why at the beginning of HIV infection only non-virulent, slowly replicating R5 viruses are found and why individuals carrying a deletion in the CCR5 receptor are unlikely to become successfully infected by HIV (Dean *et al.*, 1996; Dragic *et al.*, 1996; Huang *et al.*, 1996; Liu *et al.*, 1996; Samson *et al.*, 1996; Balotta *et al.*, 1997; Biti *et al.*, 1997; Rana *et al.*, 1997; Theodorou *et al.*, 1997). Experimental studies investigating the rise of virulent X4 virus also support the result that different immune responses may exert different selection pressure on virulent CXCR4 tropic strains. Thus, it has been found that an efficient neutralizing antibody response may lead to the suppression of X4 mutants, while evolution of viral resistance to neutralizing antibodies correlates with the rise of virulent X4 strains (Tsang *et al.*, 1994; Cornelissen *et al.*, 1995; Lathey *et al.*, 1997; Nkengasong *et al.*, 1997).

It is interesting to consider CTL-mediated inhibition of viral entry in HIV infection. This is mainly due to the beta-chemokines macrophage inflammatory proteins MIP 1α and 1β or RANTES (Cocchi *et al.*, 1995; Zanussi *et al.*, 1996; Gallo and Lusso, 1997). They inhibit only the less cytopathic R5 strains while the more virulent X4 mutants are not affected by these chemokines (Jansson *et al.*, 1996). According to our models, more cytopathic virus strains may have a significant selective disadvantage when susceptible to CTL-mediated inhibition of virus entry, while this is not the case for less cytopathic strains. Therefore, the evolution of more

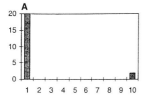

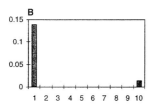

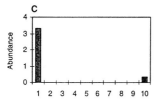

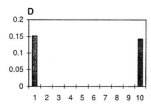

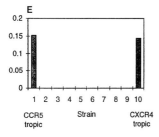

FIGURE 9.15 Effect of different immune responses on the rise of virulent CXCR4 tropic HIV. **A**. No immune response. **B**. Antibody response. **C**. CTL-mediated inhibition of virus entry. **D**. CTL-mediated inhibition of virion production. **E**. CTL-mediated lysis. The graphs show that no immune response as well as immune responses acting on the virus before integration into the host genome (**B** and **C**) suppress virulent CXCR4 tropic strains to low levels. On the other hand, immune responses acting on infected cells (**D** and **E**) allow such strains to reach similar abundances compared to CCR5 tropic ones.

cytopathic properties of the virus may require the evolution of resistance to the CTL-secreted chemokines, and therefore the use of the CXCR4 coreceptor. This may help to interpret the observation that the SI phenotype is usually associated with the use of the CXCR4 coreceptor even if the syncytium-inducing phenotype may also potentially arise in strains using the CCR5 receptor (Doms and Moore 1998).

IMPLICATIONS FOR ANTIVIRAL THERAPY

In the previous section we have considered the effect of various aspects of viral evolution on HIV disease progression. A common theme in all the models examined was that a diminished efficacy of the immune system as well as high replication kinetics of the virus enhance the rate of progression to AIDS. Consequently, a boost of the antiviral immune response by immunotherapy as well as suppression of viral replication by drug treatment may be important means of delaying or preventing the development of the disease.

Virus Dynamics During Drug Treatment

In recent years, a number of potent drugs have been developed inhibiting HIV-1 replication *in vivo* and inducing a decline in virus load by about two orders of magnitude as well as a significant rise in the CD4+ T-cell count. They come in two classes. Reverse transcriptase inhibitors prevent infection of new cells while protease inhibitors interfere with the production of new infectious virions by infected cells.

Mathematical modeling has shed light on the basic dynamics of drug treatment (McLean and Nowak, 1992; Ho *et al.*, 1995; Wei *et al.*, 1995; Perelson *et al.*, 1996, 1997; Bonhoeffer and Nowak, 1997; Herz *et al.*, 1996; Bonhoeffer *et al.*, 1997a, b; Wodarz and Nowak 1998b). In terms of the basic model of virus dynamics, the action of reverse transcriptase inhibitors is expressed as a reduction in the parameter β to low values

(Bonhoeffer *et al.*, 1997a). The outcome of drug treatment obviously depends on the efficacy of the drug. If the rate of infection (β) is sufficiently reduced so as to push the basic reproductive ratio (R_0) of the virus below unity, then the virus population sensitive to the drug is eradicated. In the extreme case of 100% efficacy ($\beta = 0$), the decline of infected cells and free virus over time is described by $y(t) = y^{(2)} e^{-at}$ and $v(t)=v^{(2)} (ue^{-at} - ae^{-ut})/(u - a)$ respectively. Thus, the number of infected cells diminishes exponentially with time. The form of decline of free virus particles is different (Figure 9.16). Given the reasonable assumption that the half-life of free virus is significantly shorter than that of infected cells ($u \gg a$), initially, a shoulder phase with a relatively small decay in plasma virus is observed, the length of which is determined by the half-life of free virus particles. After a time delay of approximately $1/u$, the decline in plasma virus follows an exponential decay.

Herz *et al.*, (1996) analysed the effect of the intracellular phase of the virus life cycle and showed that the general shape of decline of infected cells and free virus particles is very similar in the case of protease inhibitors. That is, they again observed the initial shoulder phase the duration of which was determined by the half-life of free virus particles ($\approx 2/u$), followed by exponential decay.

If the drug does not completely prevent infection ($\beta \neq 0$), its effectiveness may be represented by $s\beta$ ($s < 1$). Bonhoeffer *et al.* (1997a) demonstrated that the general patterns of drug-induced virus decay remain unchanged with the additional feature that the half-life of infected cells now becomes a function of the efficacy of the drug. Thus, a more effective treatment should lead to a faster decay and a shorter estimate for the half-life of virus-producing cells.

If, on the other hand, the drug is not sufficiently strong to reduce the basic reproductive ratio of the virus below 1, the model suggests that the new equilibrium virus load (during treatment) is essentially the same as before treatment. In other words, as long as $R_0 > 1$, the equilibrium virus load is relatively independent of the rate of infection β. This is not in agreement with observation: a combination of zidovudine and lamivudine leads to a sustained 10–100

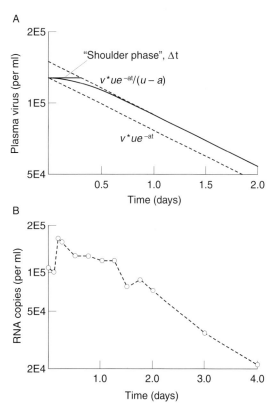

FIGURE 9.16 Short-term dynamics of virus decline during drug therapy. **A.** Under the reasonable assumption that the half-life of infected cells is longer than that of free virus particles ($u > a$), the model predicts that the decline of plasma virus load upon drug treatment is characterized by two phases. Initially, a shoulder phase with a relatively small decay in plasma virus is observed, the length of which is determined by the half-life of free virus particles. After a time delay of approximately $1/u$, the decline in plasma virus follows an exponential decay. **B** Plasma virus decline in a patient treated with the protease inhibitor ritonavir. (Reproduced from Bonhoeffer *et al.*, 1997a, with permission). Virus load starts to fall exponentially around 30–40 hours after initiating treatment. This time span is a combination of the shoulder phase described above, a pharmacological delay of the drug and the intracellular phase of the virus life cycle (Perelson *et al.*, 1996; Herz *et al.*, 1996).

times reduction in equilibrium virus load. How can this observation be explained? Including a relatively strong CTL response into the dynamics between the virus and their target cells showed that the effectiveness of the immune response correlated with the amount of decay of virus upon drug treatment (Bonhoeffer *et al.*, 1997b). Thus, if the immune responsiveness is relatively high, one might observe a linear relationship between levels of virus load and the rate of infection of target cells. The importance of the presence of an immune response for drug treatment to result in the observed reduction in virus load was also shown in a model taking into account immune impairment due to CD4+ T-cell infection (Wodarz *et al.*, 1999a). In such a scenario, the replication rate of the virus becomes a major determinant of virus load at equilibrium. Alternative mechanisms that might account for the observed reduction in virus load upon administration of the reverse transcriptase inhibitors may include bystander killing of uninfected cells, e.g. via the formation of syncytia or shedding of gp120 molecules, or differential effectivity of the drug on infected cells in different tissues (Bonhoeffer *et al.*, 1997b).

Drug Resistance

A major problem of antiviral therapy is the rapid emergence of drug-resistant virus (Larder *et al.*, 1989; Richman, 1994; Condra *et al.*, 1995). HIV-1 can develop resistance against all individual drugs in use. In monotherapy, resistant virus can emerge within weeks, but the key to success is to use several drugs simultaneously (Eron *et al.*, 1995). Combining reverse transcriptase inhibitors and protease inhibitors leads in many patients to a rapid reduction in plasma virus abundance. Often virus load stays below detection limits for the duration of treatment (which is now 2–3 years in some patients). In other patients, however, multiple drug therapy is not as successful. In some cases virus load never declines below detection limit, in other cases virus load resurges during therapy. Treatment failure is both a consequence of viral resistance and lack of compliance. Mathematical models help to define the concept of drug resistance and give us an understanding of its evolution.

The basic model of virus dynamics provides a clear functional definition of resistance. Therapy reduces the basic reproductive ratio of viral mutants. A drug-resistant mutant is defined by having a basic reproductive ratio of greater than

unity during therapy. Therefore, *in-vitro* measurements of viral inhibition by drugs do not define resistance. Resistance is not a property of the virus mutant alone but a combined property of virus and host. A mutant may be resistant in one patient but not in another.

A further important issue when evaluating the success of drug therapy is to be clear about the concept of "benefit of treatment". Bonhoeffer and Nowak (1997) defined this as "the total gain of uninfected CD4 cells or the total reduction of virus load over the duration of treatment compared to baseline before treatment". Interestingly, the basic model suggests that this benefit of treatment is independent of the degree of inhibition of sensitive strains. Thus, in the extreme case where the drug has no effect on resistant virus, the benefit of treatment is independent of the efficacy or dosage of the drug. A dependence on the efficacy and dosage of the drug will only be observed if the drug also has a negative effect on resistant strains.

Mathematical models (McLean and Nowak, 1992; Frost and McLean, 1994) have suggested that the eventual upsurge of virus load during drug treatment is the result of the dynamical interactions between the virus and its target cells with the rise of viremia being due to the presence of an increased number of CD4+ cells during treatment. On the other hand, competition between virus strains is the important factor governing the emergence of drug-resistant mutants. A very important question is whether drug-resistant mutants pre-exist within the virus population before the start of therapy (Bonhoeffer *et al.*, 1997a; Bonhoeffer and Nowak 1997; Ribeiro *et al.*, 1998; Coffin, 1995). Riberio *et al.* (1998) used standard quasispecies equations (Eigen and Schuster, 1977; Holland *et al.*, 1982; Domingo *et al.*, 1985, 1996; Domingo, 1991; Nowak, 1992b) to show that the pretreatment frequency of drug-resistant HIV strains depends on the mutation rate, the selection coefficient for the individual strains and the number of point mutations required to confer resistance. Importantly, they showed that the probability of a drug-resistant strain to be present before treatment drastically falls with an increasing number of point mutations required to evolve resistance. Roughly speaking, most strains requiring a single point mutation, many of the strains requiring two point mutations and only very few or none of the strains requiring more than two point mutations will be present before the start of therapy.

Assume that drug-resistant mutants exist before the start of treatment and that the basic reproductive ratio of the sensitive strains (R_s) is significantly higher than that of the resistant strains (R_r) such that the resistant strains will be maintained by a selection mutation balance $(R_s \gg R_r > 1)$. Denoting the basic reproductive ratios after treatment by R'_s and R'_r respectively, mathematical models (Bonhoeffer *et al.*, 1997a) define four possible scenarios depending on the efficacy of the drug and the competitive interactions between sensitive and resistant strains. In the most favorable case, a very effective drug will reduce the basic reproductive ratio of both sensitive and resistant strains below unity $(R'_s < 1 > R'_r)$, thereby completely eradicating the infection. At the other extreme, a very weak drug may only reduce the basic reproductive ratios by a relatively small amount without altering the competitive hierarchy, i.e. $R'_s > R'_r > 1$. In this case, the situation stays the same as before treatment, i.e. the rise of resistant mutants will not be selected for. On the other hand, a drug strong enough to reverse the competitive hierarchies, such that $R'_r > R'_s > 1$, will select for the rise of resistant virus after a certain period of time. Finally, an even stronger drug, reducing the basic reproductive ratio of the sensitive strains below 1, such that $R'_s < 1 < R'_r$, will induce an exponential decay of the population of sensitive virus and a rise of the resistant strains. For the last two cases, the models provide the important insight that the stronger the action of the drug the faster the rise of resistant strains.

In contrast, when drug-resistant mutants do not exist before the start of therapy, mathematical models may result in the opposite prediction (Bonhoeffer *et al.*, 1997a; Bonhoeffer and Nowak, 1997). One has to distinguish between two cases. If the drug reduces the basic reproductive ratio of the sensitive strains below unity, it is unlikely that any resistant mutants will be generated because the number of infected cells produced during treatment is less than the number of infected cells produced before therapy (Bonhoeffer and Nowak, 1997). If, on the other hand, the drug is less effective and does not reduce the basic reproductive ratio of the sensitive strains below unity, such that $R'_r > R'_s > 1$, drug-resistant strains will be generated. However, interestingly, and contrary to the prediction for the pre-existence of resistant mutants, a strong and effective drug will delay the emergence of resistant virus.

Thus, in summary, mathematical models predict that the rise and selection in favor of drug-resistant mutants is a consequence of (1) the existence of resistant strains before the onset of therapy, in which case a strong drug will speed up the emergence of resistance, and (2) the drug not being able to eliminate the sensitive strain, leading to a delay in the emergence of resistance by a strong drug.

Can HIV be Eradicated?

One of the main messages emerging from the above discussion is that treatment should be applied as early and as hard as possible in the infectious process. The earlier treatment is started the more likely it is that resistant mutants do not pre-exist, since virus load will still be relatively low. Using a combination of a variety of drugs also decreases the chances of resistant strains being present in the population before therapy. In general, any treatment regimes facilitating the existence of drug-resistant mutants, such as adding drugs sequentially during therapy or cycling between drugs, as well as irregular drug consumption behavior by the patient, should be avoided.

Given that an optimal treatment regime is adopted, the question arises whether it is possible to eradicate HIV from patients and how long the drug would have to be applied. Obviously, the models suggest that the drug cocktail used must reduce the basic reproductive ratio of *all* virus strains below unity such that extinction is possible. However, the *in-vivo* scenario of HIV infection is more complex than the situation captured in the simple models discussed above. *In vivo*, three classes of infected cells can be distinguished according to their half-life and the cell type infected (Schockmel and Perrin, 1997). Acutely infected CD4+ T cells have a half-life of about 2 days. They comprise only a small fraction of the total pool of infected cells, but contribute most of the plasma virus particles found in a patient (Klenerman *et al.*, 1996; Bonhoeffer *et al.*, 1997a). Chronically infected cells have a longer half-life, produce virus particles over a longer period of time and are thought to be mainly macrophages. Finally, the virus may enter a dormant stage after infection, leading to latently infected cells, and may become reactivated at a later stage.

Perelson *et al.* (1997) and Nowak *et al.*, (1997) estimated the half-life of latently infected cells to be around 10–20 days. On the basis of these figures a minimum time period of 2–3 years was estimated to be necessary to eradicate the virus. However, it is unlikely that the 10–20-day half-life, which was estimated using data from the first few weeks of therapy, can be extrapolated to such a long time (Finzi *et al.*, 1997; Schockmel and Perrin, 1997; Wong *et al.*, 1997). In other words, it is more likely that the latently infected cell population is heterogeneous and that some of these cells live much longer (McLean and Mitchie, 1995). Indeed, data by Chun *et al.* (1997), which are based on a measurement 1 year after initiating combination therapy, suggest a half-life of latently infected cells of about 40 days. Wei *et al.* (1995) estimated the half-life of HIV provirus to be about 100 days. Most of this provirus is defective, but ultimately this half-life defines the kinetics of the longest-lived replication competent virus. Thus, eradication by inhibiting virus replication alone will not be practical.

CONCLUSION

This review has shown how mathematical models may help us in understanding virus dynamics in individual infections. The principle of mathematical modeling is to follow a small set of essential assumptions rigorously to their logical conclusions. Infection dynamics often involve non-linear interactions of many different components. In such a scenario, mathematical models are necessary to provide a correct interpretation of empirical results, to provide a framework for parameter estimation and to further our understanding of process and concept. Mathematical models play an important role in estimating kinetic parameters of infection dynamics *in vivo* and in determining what needs to be measured to improve understanding (Wei *et al.*, 1995; Ho *et al.*, 1995; Nowak and Bangham, 1996; Perelson *et al.*, 1997; Bonhoeffer *et al.*, 1997a). Furthermore, mathematical models also generate new insights and create new hypotheses (Murray, 1989; Anderson and May, 1991; Nowak *et al.*, 1991; Levin *et al.*, 1997).

ACKNOWLEDGEMENTS

We acknowledge support by The Florence Gould Foundation; J. Seward Johnson, Sr. Charitable Trusts; The Ambrose Morell Foundation; The Alfred P. Sloan Foundation.

REFERENCES

Adorini, L., Appella, E., Doria, G. and Nagy, Z.A. (1988) Mechanisms influencing the immunodominance of T-cell determinants. *J. Exp. Med.*, **168**, 2091–2104.

Anderson R.M. and May R.M. (1979a) Population biology of infectious diseases. I. *Nature*, **280**, 361–367.

Anderson R.M. and May R.M. (1979b) Population biology of infectious diseases. II. *Nature*, **280**, 455.

Anderson, R.M. and May, R.M. (1991) *Infectious Diseases of Humans.* Oxford University Press, Oxford.

Asjo, B., Morfeldt-Manson, L., Albert, J. *et al.* (1986) Replicative capacity of human immunodeficiency virus from patients with varying severity of HIV infection. *Lancet*, **20**, 660–662.

Balfe, P., Simmonds, P., Ludlam, C.A., Bishop, J.O. and Brown, A.J.L. (1990) Concurrent evolution of human-immunodeficiency-virus type-1 in patients infected from the same source – rate of sequence change and low-frequency of inactivating mutations. *J. Virol.*, **64**, 6221–6233.

Balotta, C., Bagnarelli, P., Violin, M. *et al.* (1997) Homozygous Delta 32 deletion of the CCR-5 chemokine receptor gene in an HIV-1 infected patient. *AIDS*, **11**, F67–F71.

Berger, E.A., Doms, R.W., Fenyo, E.M. *et al.* (1998) A new classification for HIV-1. *Nature*, **391**, 240.

Bertoletti, A., Sette, A., Chisari, F.V. *et al.* (1994) Natural variants of cytotoxic epitopes are T cell receptor antagonists for antiviral cytotoxic T cells. *Nature*, **369**, 407–110.

Biti, R., French, R.F., Young, J., Bennetts, B., Stewart, G. and Liang, T. (1997) HIV-1 infection in an individual homozygous for the CCR-5 deletion allele. *Nature Med.*, **3**, 252–253.

Bleul, C.C., Wu, L., Hoxie, J.A., Springer, T.A. and Mackay, C.R. (1997) The HIV coreceptors CXCR4 and CCR5 are differently expressed and regulated on human T lymphocytes. *Proc. Natl. Acad. Sci.*, USA **94**, 1925–1930.

Bonhoeffer S. and Nowak M.A. (1997) Pre-existence and emergence of drug resistance in HIV-1 infection. *Proc. Roy. Soc. Lond. B*, 1997, **264**, 631–637.

Bonhoeffer, S., May, R.M., Shaw, G.M. and Nowak, M.A. (1997a) Virus dynamics and drug therapy. *Proc. Natl Acad. Sci. USA*, **94**, 6971–6976.

Bonhoeffer S., Coffin J.M. and Nowak M.A. (1997b) Human immunodeficiency virus drug therapy and virus load. *J Virol*, **71**, 3275–3278.

Borrow, P., Lewicki, H., Wei, X.P. *et al.* (1997) Antiviral pressure exerted by HIV-1-specific cytotoxic T lymphocytes (CTLs) during primary infection demonstrated by rapid selection of CTL escape virus. *Nature Med.*, **3**, 205–211.

Bukrinskaya, A.G., Ghorpade, A., Heinzinger, N.K., Smithgall, T.E., Lewis, R.E. and Stevenson, M. (1996) Phosphorylation-dependent human immunodeficiency virus type-1 infection and nuclear targeting of viral DNA. *Proc. Natl. Acad. Sci. USA*, **93**, 367–371.

Buus, S., Sette, A., Colon, S.M., Miles, C. and Grey, H.M. (1987) The relation between major

histocompatibility complex (MHC) restriction and the capacity of IA to bind immunogenic peptides. *Science*, **235**, 1353–1358.

Chun, T.W., Stuyver, L., Mizell, S.B. *et al.* (1997) Presence of an inducible HIV-1 latent reservoir during highly active antiretroviral therapy. *Proc. Natl Acad. Sci. USA*, **94**, 13193–13197.

Cocchi, F., Devico, A.L, Garzinodemo, A., Arya, S.K., Gallo, R.C. and Lusso, P. (1995) Identification of RANTES, Mip-1-alpha, and MIP-1-beta as the major HIV suppressive factors produced by CD8[+] T-cells. *Science*, **270**, 1811–1815.

Coffin, J.M. (1995) HIV population dynamics *in vivo*: implications for genetic variation, pathogenesis and therapy. *Science*, **267**, 483–489.

Condra, J.H., Schleif, W.A., Blahy, O.M. *et al.* (1995) In-vivo emergence of HIV-1 variants resistant to multiple protease inhibitors. *Nature*, **374**, 569–571.

Connor, R.I. and Ho, D.D. (1994) Human immunodeficiency virus type 1 variants with increased replicative capacity develop during the asymptomatic stage before disease progression. *J. Virol.*, **68**, 4400–4408.

Connor, R.I., Sheridan, K.E., Ceradini, D., Choe, S. and Landau, N.R. (1997) Change in coreceptor use correlates with disease progression in HIV-1 infected individuals. *J. Exp. Med.*, **185**, 621–628.

Cornelissen, M., Nulderkampinga, G., Veenstra, J. *et al.* (1995) Syncyticum inducing (SI) phenotype suppression at seroconversion after intramuscular inoculation of a non-syncyticum-inducing/SI phenotypically mixed human immunodeficiency virus population. *J. Virol.*, **69**, 1810–1818.

Crowe, S.M. (1995) Role of macrophages in the pathogenesis of human immunodeficiency virus (HIV) infection. *Aust. NZ. J. Med.*, **25**, 777–783.

Crowe, S.M. and Kornbluth, R.S. (1994) Overview of HIV interactions with macrophages and dendritic cells – the other infection in AIDS. *J. Leukocyte Biol.*, **56**, 215–217.

Dean, M., Carrington, M., Winkler, C. *et al.* (1996) Genetic restriction of HIV-1 infection and progression to AIDS by a deletion allele of the CKR5 structural gene. *Science*, **273**, 1856–1862.

DeBoer, R.J. and Boerlijst, M.C.B. (1994) Diversity and virulence thresholds in AIDS. *Proc. Natl. Acad. Sci. USA*, **91**, 544–548.

Dittmar, M.T., McKnight, A., Simmons, G., Clapham, P.R., Weiss, R.A. and Simmonds, P. (1997) HIV tropism and coreceptor use. *Nature*, **385**, 495–496.

Doherty, P.C. (1993) Immune exhaustion: driving virus-specific CD8+ T cells to death. *Trends Microbiol.*, **1**, 207–209.

Domingo, E. (1991) Quasispecies structure and dynamics of RNA virus populations. *Biol. Chem. Hoppe-Seyler*, **372**, 646.

Domingo, E., Martínez Salas, E., Sobrino, F. *et al.* (1985) The quasispecies (extremely heterogeneous) nature of viral RNA genome populations – biological relevance – a review. *Gene*, **1**, 1–8.

Domingo, E., Escarmís, C., Sevilla, N. *et al.* (1996) Basic concepts in RNA virus evolution. *FASEB J.*, **10**, 859–864.

Doms, R.W. and Moore, J.P. (1998) HIV-1 coreceptor use: a molecular window into viral tropism.   *WWW*:http:/Perec/Desktop%20Folder/HIV-1% 20Coreceptor%20Use

Dragic, T., Litwin, V., Allaway, G.P. *et al.* (1996) HIV-1 entry into CD4(+) cells is mediated by the chemokine receptor CC-CKR-5. *Nature*, **381**, 667–673.

Eigen, M. (1989) The AIDS debate. *Naturwissenschaften*, **76**, 341–350.

Eigen, M. and Schuster P. (1977) The hypercycle. A principle of natural self-organisation. Part A: Emergence of the hypercycle. *Naturwissenschaften*, **64**, 541–565.

Emerman, M., Bukrinsky, M. and Stevenson, M. (1994) HIV infection of non-dividing cells – reply. *Nature*, **369**, 108.

Eron, J.J., Benoit, S.L., Jemsek, J. *et al.* (1995) Treatment with lamivudine, zidovudine, or both in HIV-positive patients with 200–500 CD4+ cells per cubic millimeter. *N. Engl. J. Med.*, **333**, 1662–1669.

Fauci, A.S. (1996) Host factors and the pathogenesis of HIV-induced disease. *Nature*, **384**, 529–534.

Fenyo, E.M., Morfeldt-Manson, L., Chiodi, F. *et al.* (1988) Distinct replicative and cytopathic characteristics of human immunodeficiency virus isolates. *J. Virol.*, **62**, 4414–4419.

Ferbas, J., Daar, E.S., Grovit-Ferbas, K. *et al.* (1996) Rapid evolution of human immunodeficiency virus strains with increased replicative capacity during the seronegative window of primary infection. *J. Virol.*, **70**, 7285–7289.

Finzi, D., Hermankova, M., Pierson, T. *et al.* (1997) Identification of a reservoir of HIV-1 in patients on highly active antiretroviral therapy. *Science*, **278**, 1295–1300.

Fouchier, R.A.M., Meyaard, L., Brouwer, M., Hovenkamp, E. and Schuitemaker, H. (1996) Broader tropism and higher cytopathicity for CD4(+) T-cells of a syncytium-inducing compared to a non-syncytium-inducing HIV-1 isolate as a mechanism for accelerated CD4+ T cell decline *in vivo*. *Virology*, **219**, 87–95.

Frost, S.D.W. and McLean, A.R. (1994) Quasispecies dynamics and the emergence of drug resistance during zidovudine therapy of HIV infection. *AIDS*, **8**, 323–332.

Gallo, R.C. and Lusso, P. (1997) Chemokines and HIV infection. *Curr. Opin. Infect. Dis.*, **10**, 12–17.

Goulder, P.J.R., Phillips, R.E., Colbert, R.A., *et al.* (1997) Late escape from an immunodominant cytotoxic T-lymphocyte response associated with progression to AIDS. *Nature Med.*, **3**, 212–217.

Gruters, R.A., Terpstra, F.G., de Goede, R. *et al.* (1991) Immunological and virological markers in individuals progressing from seroconversion to AIDS. *AIDS*, **5**, 837–844.

Gupta, S., Maiden, M.C.J., Feavers, I.M., Nee, S., May, R.M. and Anderson, R.M. (1996) The maintenance of strain structure in populations of recombining infectious agents. *Nature Med.*, **2**, 437–442.

Gupta, S., Ferguson, N. and Anderson, R. (1998) Chaos, persistence, and evolution of strain structure in antigenically diverse infectious agents. *Science*, **280**, 912–915.

Hartung, S., Boller, K., Cichutek, K., Norley, S.G. and Kurth, R. (1992) Quantitation of a lentivirus in its natural host – simian immunodeficiency virus in African-green monkeys. *J. Virol.*, **66**, 2143–2149.

Heinzinger, N.K., Burkrinsky, M.I., Haggerty, S.A. *et al.* (1994) The VPR protein of human immunodeficiency virus type-1 influences nuclear localisation of viral nucleic acids in non-dividing host cells. *Proc. Natl Acad. Sci. USA*, **91**, 7311–7315.

Herz, A.V.M., Bonhoeffer, S., Anderson, R.M., May, R.M. and Nowak, M.A. (1996) Viral dynamics *in vivo*: limitations on estimates on intracellular delay and virus decay. *Proc. Natl Acad. Sci. USA*, **93**, 7247–7251.

Ho, D.D., Neumann, A.U., Perelson, A.S., Chen, W., Leonard, J.M. and Markowitz, M. (1995) Rapid turnover of plasma virions and CD4 lymphocytes in HIV-1 infection. *Nature*, **373**, 123–126.

Holland, J., Spindler, K., Horodyski, F., Grabeu, E., Nichol, S. and Vandepol, S. (1982) Rapid evolution of RNA genomes. *Science*, **215**, 1577–1585.

Huang, Y.X., Paxton, W.A., Wolinsky, S.M. *et al.* (1996) The role of a mutant CCR5 allele in HIV-1 transmission and disease progression. *Nature Med.*, **2**, 1240–1243.

Jameson, S.C. and Bevan, M.J. (1995) T cell receptor antagonists and partial agonists. *Immunity*, **2**, 1–11.

Jameson, S.C., Carbone, F.R. and Bevan, M.J. (1993) Clone specific T-cell receptor antagonists of major histocompatibility complex class-1 restricted cytotoxic T cells. *J. Exp. Med.*, **177**, 1541–1550.

Jansson, M., Mikulas, P., Karlsson, A. *et al.* (1996) Sensitivity to inhibition by beta chemokines correlates with biological phenotype of primary HIV-1 isolates. *Proc. Natl Acad. Sci. USA*, **93**, 15382–15387.

Kalams, S.A. and Walker, B.D. (1995) Cytotoxic T lymphocytes and HIV-1 related neurologic disorders. *Curr. Topics Micobiol. Immunol.*, **202**, 79–88.

Karlsson, A., Parsmyr, K., Sandstrom, E., Fenyo, E.M. and Albert, J. (1994) MT-2 celtropism as prognostic marker for disease progression in human immunodeficiency type-1 infection. *J. Clin. Microbiol.*, **32**, 364–370.

Klenerman, P., Rowland-Jones, S., McAdam, S. *et al.* (1994) Cytotoxic T cell activity antagonized by naturally occurring HIV-1 *gag* variants. *Nature*, **369**, 403–407.

Klenerman, P., Meier, U.C., Phillips, R.E. and McMichael, A.J. (1995) The effects of natural altered peptide ligands on the whole blood cytotoxic T lymphocyte response to human immunodificiency virus. *Eur. J. Immunol.*, **25**, 1927–1931.

Klenerman, P., Phillips, R.E., Rinaldo, C.R. *et al.* (1996) Cytotoxic T lymphocytes and viral turnover in HIV type 1 infection. *Proc. Natl Acad. Sci. USA*, **93**, 15323–15328.

Koot, M., van't Wout, A.B., Koostra, N.A., deGoede, R.E.Y., Tersmette, M. and Schuitemaker, H. (1996) Relation between changes in cellular load, evolution of viral phenotype, and the clonal composition of virus populations in the course of human immunodeficiency virus type-1 infection. *J. Infect. Dis.*, **173**, 349–354.

Kuiken, C.L., Dejong, J.J., Baan, E., Keulen, W., Tersmette, M. and Goudsmit, J. (1992) Evolution of the V3 envelope domain in proviral sequences and isolates of human immunodeficiency virus type-1 during transition of the viral biological phenotype. *J. Virol.*, **66**, 4622–4627.

Larder, B.A., Darby, G. and Richman, D.D. (1989) HIV with reduced sensitivity to zidovudine (AZT) isolated during prolonged therapy. *Science*, **243**, 1731–1734.

Lathey, J.L., Pratt, D. and Spector, S.A. (1997) Appearance of autologous neutralising antibody correlates with reduction in virus load and phenotype switching during primary infection with human immunodeficiency virus type 1. *J. Infect. Dis.*, **175**, 231–232.

Levin, S.A., Grenfell, B., Hastings, A. and Perelson, A.S. (1997) Mathematical and computational challenges in population biology and ecosystem science. *Science*, **275**, 334–343.

Liu, Z.R., Williams, K.P., Chang, Y.H. and Smith, J.A. (1993) Immunodominance – a single amino-acid substitution within an antigenic site alters intramolecular selection of T-cell determinants. *J. Immunol.*, **151**, 1852–1858.

Liu, R., Paxton, W.A., Choe, S. *et al.* (1996) Homozygous defect in HIV-1 coreceptor accounts for resistance of some multiply exposed individuals to HIV-1 infection. *Cell*, **86**, 367–377.

Loetscher, P., Uguccioni, M., Bordoli, L. *et al.*

(1998) CCR5 is characteristic of Th1 lymphocytes. *Nature*, **391**, 344–345.

Loveday C., Kaye S., Tenant-Flowers M. *et al.* (1995) HIV-1 RNA serum load and resistant viral genotypes during early zidovudine therapy. *Lancet*, **345**, 820–824.

McAdam, S.N., Klenerman, P., Tussey, L.G. *et al.* (1995) Immunogenic HIV variant peptides that bind to HLA-B8 but fail to stimulate cytotoxic T lymphocyte responses. *J. Immunol.*, **155**, 2729–2736.

McKnight, A. and Clapham, P.R. (1995) Immune escape and tropism for HIV. *Trends Microbiol.*, **3**, 356–361.

McLean A.R. and Mitchie C.A. (1995) *In vivo* estimates of division and death rates of human T lymphocytes. *Proc. Natl Acad. Sci. USA*, **92**, 3707–3711.

McLean A.R. and Nowak M.A. (1992) Competition between zidovudine-sensitive and zidovudine-resistant strains of HIV. *AIDS*, **6**, 71–79.

McMichael, A., Rowland-Jones, S., Klenerman, P. *et al.* (1995) Epitope variation and T-cell recognition. *J. Cell. Biochem.*, **60–60**.

McMichael, A., Goulder, P., Rowland-Jones, S., Nowak, M. and Phillips, R. (1996) HIV escapes from cytotoxic lymphocytes. *Immunology*, **89**, SE111.

Meier, U.C., Klenerman, P., Griffin, P. *et al.* (1995) Cytotoxic T lymphocyte lysis inhibited by viable HIV mutants. *Science*, **270**, 1360–1362.

Mosier, D. and Sieburg, H. (1994) Macrophage-tropic HIV – critical for AIDS pathogenesis. *Immunol. Today*, **15**, 332–339.

Moss, P.A.H., Rowland-Jones, S.L., Frodsham, P.M. *et al.* (1995) Persistent high frequency of human immunodeficiency virus-specific cytotoxic T-cells in peripheral-blood of infected donors. *Proc. Natl Acad. Sci. USA*, **92**, 5773–5777.

Murphy, B.R. and Chanock, R.M. (1990) Immunication against viruses. In *Virology* (eds Fields, R.M., Knipe, D.M., Chanock, R.M. *et al.*), pp. 463–502. Raven Press, New York.

Murray, J.D. (1989) *Mathematical Biology*, 2nd edn. Biomathematics Texts. Springer-Verlag, New York.

Nkengasong, J.N., Fransen, K., Willems, B. *et al.* (1997) Virologic, immunologic, and clinical follow-up of a couple infected by the human immunodeficiency virus type one, group O. *J. Med. Virol.*, **51**, 202–209.

Nowak, M. (1990) HIV mutation-rate. *Nature*, **347**, 522.

Nowak, M.A. (1992a) Variability of HIV infections. *J. Theor. Biol.*, **155**, 1–20.

Nowak, M.A. (1992b) What is a quasi-species? *Trends Ecol. Evol.*, **7**, 118–121.

Nowak, M.A. (1995) AIDS pathogenesis: from models to viral dynamics in patients. *J. AIDS Hum. Retrovirol.*, **10**, S1–S5.

Nowak, M.A. (1996) Immune-responses against multiple epitopes – a theory for immunodominance and antigenic variation. *Sem. Virol.*, **7**, 83–92.

Nowak, M.A. and Bangham, C.R.M. (1996) Population dynamics of immune responses to persistent viruses. *Science*, **272**, 74–79.

Nowak, M.A. and May, R.M. (1991) Mathematical biology of HIV infections – antigenic variation and diversity threshold. *Math. Biosci.*, **106**, 1–21.

Nowak, M.A. and May, R.H. (1992) Coexistence and competition in HIV infections. *J. Theor. Biol.*, **159**, 329–342.

Nowak, M. and May, R.M. (1993) AIDS pathogenesis – mathematical models of HIV and SIV infections. *AIDS*, **7**, S3–S18.

Nowak, M.A., Anderson, R.M., McLean, A.R., Wolfs, T.F.W., Goudsmit, J. and May, R.M. (1991) Antigenic diversity thresholds and the development of AIDS. *Science* **254**, 963–969.

Nowak, M.A., Bonhoeffer, S., Loveday, C. *et al.* (1995a) HIV results in the frame – results confirmed. *Nature*, **375**, 193–193.

Nowak, M.A., May, R.M., Phillips, R.E. *et al.* (1995b) Antigenic oscillations and shifting immunodominance in HIV-1 infections. *Nature*, **375**, 606–611.

Nowak, M.A., May, R.M. and Sigmund, K. (1995c) Immune-responses against multiple epitopes. *J. Theor. Biol.* **175**, 325–353.

Nowak, M.A., Bonhoeffer, S., Shaw, G.M. and May, R.M. (1997) Anti-viral drug treatment: dynamics of resistance in free virus and infected cell populations. *J. Theor. Biol.*, **184**, 205–219.

O'Brien, W.A. (1994) HIV-1 entry and reverse transcription in macrophages. *J. Leukocyte Biol.*, **56**, 273–277.

Perelson, A.S., Neumann, A.U., Markowitz, M., Leonard, J.M. and Ho, D.D. (1996) HIV-1 dynamics *in-vivo* – virion clearance rate, infected cell life-span, and viral generation time. *Science*, **271**, 1582–1586.

Pereison, A.S., Essunger, P. and Ho, D.D. (1997) Dynamics of HIV-1 and CD4⁺ lymphocytes *in vivo*. *AIDS*, **11**, S17–S24.

Phillips, R.E., Rowland-Jones, S., Nixon, D.F. *et al.* (1991) Human immunodeficiency virus genetic variation that can escape cytotoxic T-cell recognition. *Nature*, **354**, 453–459.

Preston, B.D., Poiesz, B.J. and Loeb, L.A. (1988) Fidelity of HIV-1 reverse-transcriptase. *Science*, **242**, 1168–1171.

Price, D.A., Goulder, P.J.R., Klenerman, P. *et al.* (1997) Positive selection of HIV-1 cytotoxic T lymphocyte escape variants during primary infection. *Proc. Natl Acad. Sci. USA*, **94**, 1890–1895.

Rana, S., Besson, G., Cook, D.G. *et al.* (1997) Role of CCR5 in infection of primary macrophages and lymphocytes by macrophage-tropic strains of human immunodeficiency virus: resistance to patient-derived and prototype isolates resulting from the Delta CCR5 mutation. *J. Virol.*, **71**, 3219–3227.

Ribeiro, R.M., Bonhoeffer, S. and Nowak, M.A. (1998) The frequency of resistant mutant virus before anti-viral therapy. *AIDS*, **12**, 461–465.

Richman, D.D. (1994) Resistance, drug failure, and disease progression. *AIDS Res. Hum. Retroviruses*, **10**, 901–905.

Rinaldo, C.R., Beltz, L.A., Huang, X.L., Gupta, P., Fan, Z. and Torpey, D.J. (1995a) Anti-HIV type I cytotoxic T lymphocyte effector activity and disease progression in the first 8 years of HIV type 1 infection of homosexual men. *AIDS Res. Hum. Retroviruses*, **11**, 481–489.

Rinaldo, C.R., Huang, X.L., Fan, Z. *et al.* (1995b) High levels of anti-human immunodeficiency virus type 1 (HIV-1) memory cytotoxic T-lymphocyte activity and low viral load are associated with lack of disease in HIV-1 infected long-term nonprogressors. *J. Virol.*, **69**, 5838–5842.

Roberts, J.D., Bebenek, K. and Kunkel, T.A. (1988) The accuracy of reverse transcriptase from HIV-1. *Science*, **242**, 1171–1173.

Roos, M.T.L., Lange, J.M.A., deGoede, R.E.Y. *et al.* (1992) Viral phenotype and immune response in primary human immunodeficiency virus type-1 infection. *J. Infect. Dis.*, **165**, 427–432.

Rowland-Jones, S. and Tan, R. (1997) Control of HIV coreceptor expression: implications for pathogenesis and treatment. *Trends Microbiol.*, **5**, 300–303.

Rudensey, L.M., Kimata, J.T., Benveniste, R.E. and Overbaugh, J. (1995) Progression to AIDS in Macaques is associated with changes in the replication, tropism, and cytopathic properties of the simian immunodeficiency virus variant population. *Virology*, **207**, 528–542.

Saag, M.S., Hahn, B.H., Gibbons, J. *et al.* (1988) Extensive variation of human immunodeficiency virus type-1 *in vivo*. *Nature*, **334**, 440–444.

Safrit, J.T., Lee, A. and Koup R.A. (1994) Characterisation of HLA-B7-restricted cytotoxic T lymphocyte clones specific for the 3rd variable region of the HIV GP120, isolated from 2 patients during acute seroconversion. *AIDS Res. Hum. Retroviruses*, **10**, S2–S33.

Samson, M., Libert, F., Doranz, B.J. *et al.* (1996) Resistance to HIV-1 infection in caucasian individuals bearing mutant alleles of the CCR-5 chemokine receptor gene. *Nature*, **382**, 722–725.

Schaeffer, E.B., Sette, A., Johnson, D.L. *et al.* (1989) Relative contribution of determinant selection and holes in the T-cell repertoire to T-cell responses. *Proc. Natl Acad. Sci. USA*, **86**, 4649–4653.

Schenzle, D. (1994) A model for AIDS pathogenesis. *Stat. Med.*, **13**, 2067–2079.

Schockmel, G.A. and Perrin, L. (1997) Could treatment of primary HIV infection eradicate HIV? *J. HIV Comb. Ther.*, **2**, 57–61.

Schuitemaker, H. (1994) Macrophage-tropic HIV variants – initiators of infection and AIDS pathogenesis. *J. Leukocyte Biol.*, **56**, 218–224.

Schuitemaker, H., Koot, M., Kootstra, M.A. *et al.* (1992) Biological phenotype of human immunodeficiency virus type-1 clones at different stages of infection–progression of disease is associated with a shift from monocytotropic to T-cell tropic virus populations. *J. Virol.*, **66**, 1354–1360.

Schuurman, R., Nijhuis, M., Vanleeuwen, R. *et al.* (1995) Rapid changes in human-immunodeficiency-virus type-1 RNA load and appearance of drug-resistant virus populations in persons treated with lamivudine (3TC). *J. Infect. Dis.*, **171**, 1411–1419.

Sercarz, E.E., Lehmann, P.V., Ametani, A., Benichou, G., Miller, A. and Moudgil, K. (1993) Dominance and crypticity of T-cell antigenic determinants. *Annu. Rev. Immunol.*, **11**, 729–766.

Stevenson, M. (1996) Portals of entry–uncovering HIV nuclear transport pathways. *Trends Cell Biol.*, **6**, 9–15.

Stevenson, M. and Gendelman, H.E. (1994) Cellular and viral determinants that regulate HIV-1 infection in macrophages. *J. Leukocyte Biol.*, **56**, 278–288.

Stevenson, M., Bukrinski, M.I., Gulizia, J. (1994) Identification of factors which govern HIV-1 replication in non-dividing host cells. *AIDS Res. Hum. Retroviruses*, **10**, S82.

Stevenson, M., Bukrinskaya, A., Ghorpade, A. *et al.* (1995) Molecular basis for cell cycle dependent HIV replication. *AIDS Res. Hum. Retroviruses*, **11**, S87.

Tersmette, M., deGoede, R.E.Y., Al, B.J.M. *et al.* (1988) Differential syncyticum-inducing capacity of human immunodeficiency virus isolates – frequent detection of syncyticum-inducing isolates in patients with acquired immunodeficiency syndrome (AIDS) and AIDS related complex. *J. Virol.*, **62**, 2026–2032.

Tersmette, M., Gruters, R.A., de Wolf, F. *et al.* (1989) Evidence of a role of virulent human immunodeficiency virus (HIV) variants in the pathogenesis of acquired immunodeficiency syndrome: studies on sequential HIV isolates. *J. Virol.*, **63**, 2118–2125.

Theodorou, I., Meyer, L., Magierowska, M., Katlama, C. and Rouzioux, C. (1997) HIV-1 infection in an individual homozygous for CCR5 Delta 32. *Lancet*, **349**, 1219–1220.

Tsang, M.L., Evans, L.A., McQueen, P. *et al.* (1994) Neutralising antibodies against sequential autologous human immunodefi-

ciency virus type-1 isolates after seroconversion. *J. Infect. Dis.*, **170**, 1141–1147.

Unutmaz, D. and Littman, D.R. (1997) Expression pattern of HIV coreceptors on T cells: implications for viral transmission and lymphocyte homing. *Proc. Natl Acad. Sci. USA*, **94**, 1615–1618.

Van't Wout, A.B., Kootstra, N.A., Mulder-Kampinga, G.A. *et al.* (1994) Macrophage tropic variants initiate human immunodeficiency virus type-1 infection after sexual, parenteral, and vertical transmission. *J. Clin. Invest.*, **94**, 2060–2067.

Wain-Hobson, S. (1993) The fastest genome evolution ever described: HIV variation *in situ*. *Curr. Opin. Genet. Deval.*, **3**, 878–883.

Wei, X.P., Ghosh, S.K., Taylor, M.E. *et al.* (1995) Viral dynamics in human-immunodeficiency-virus type-1 infection. *Nature*, **373**, 117–122.

Wodarz, D. and Nowak, M.A. (1998a) The effect of different immune responses on the evolution of virulent CXCR4 tropic HIV. *Proc. Roy. Soc. Lond. B. Biol. Sci.*, **265**, 2149–2158.

Wodarz, D. and Nowak, M.A. (1998b) Mathematical models of virus dynamics and resistance. *J. HIV Ther.*, **3**, 36–41.

Wodarz, D. and Nowak, M.A. (1999a) Virus dynamics and cell tropism: competition and the evolution of specialism. *Math. Biosci.*, submitted.

Wodarz, D. and Nowak, M.A. (1999b) Virus dynamics under different immune responses. *J. Theor. Med.*, in press.

Wodarz, D. Klenerman, P. and Nowak, M.A. (1998) Dynamics of cytotoxic T-lymphocyte exhaustion. *Proc. Roy. Soc. Lond. B.*, **265**, 191–203.

Wodarz, D., Lloyd, A.L., Jansen, V.A.A. and Nowak, M.A. (1999) Dynamics of macrophage and T cell infection by HIV. *J. Theor. Biol.*, **196**, 101–113.

Wolinsky, S.M., Korber, B.T.M., Neumann, A.U. *et al.* (1996) Adaptive evolution of human immunodeficiency virus type-1 during the natural course of infection. *Science*, **272**, 537–542.

Wong, J.K., Hezareh, M., Gunthard, H.F. *et al.* (1997) Recovery of replication-competent HIV despite prolonged suppression of plasma viraemia. *Science*, **278**, 1291–1295.

Zanussi, S., Dandrea, M., Simonelli, C., Tirelli, U. and Depaoli, P. (1996) Serum levels of RANTES and MIP-1 alpha in HIV positive long-term survivors and progressor patients. *AIDS*, **10**, 1431–1432.

Zhang, L.Q., Mackenzie, P., Cleland, A., Holmes, E.C., Brown, A.J.L. and Simmonds, P. (1993) Selection for specific sequences in the external envelope protein of human immunodeficiency virus type-1 upon primary infection. *J. Virol.*, **67**, 1772–1777.

Zhu, T.F., Mo, H.M., Wang, N. *et al.* (1993) Genotypic and phenotypic characterisation of HIV-1 in patients with primary infection. *Science*, **261**, 1179–1181.

10

Interplay Between Experiment and Theory in Development of a Working Model for HIV-1 Population Dynamics

I. M. Rouzine and J. M. Coffin

INTRODUCTION

Eradication of HIV from infected individuals and development of more effective therapeutic methods for its control remain a high priority of HIV research. Elucidating the dominant biological mechanisms responsible for persistence and stability of HIV infection could greatly assist with these goals. Given the complex behavior of this system, with its many different cell types, the task requires combination of modern experimental techniques with the advanced mathematical tools of kinetic modeling and evolution theory.

In this chapter, we review our current understanding of the dynamics of HIV and its host cells in infected individuals. We start with an overview of the recent, widely accepted findings concerning the distribution and dynamics of virus and, particularly, of infected cells. We then venture off this familiar ground into a large uncharted realm of host-cell kinetics. We are attempting this quest not with faith that our models and predictions will ultimately prove correct; rather, we hope to illustrate how such modeling can work for experiment and to challenge you, our reader, to test our predictions.

The predicament of a mathematician who is trying to produce practically useful work on real experimental systems is that it is not known in advance which biological factors (among hundreds reported in the literature or not even studied yet) are of greatest importance to the main features of the system. Taken at face value, approximations are often counterintuitive: apparently safe assumptions distort predictions and seemingly crude simplifications work surprisingly well. The quality of an approximation depends on the parameters one is trying to predict and cannot be evaluated until the analysis is over. We thus face a "time paradox": on the one hand a well-defined set of approximations must be introduced in the beginning to make the analysis possible and intuitively clear. On other hand, finding the right set of approximations (a biological model) is the final aim of such research. The only escape from the "time paradox" we know of is a dynamic interplay between experiment and theory. One starts from a preliminary analysis of the most important experimental results and formulates a starting model. After calculating its predictions, one searches aggressively for a consistent contradiction with experiment and, when such is found, checks initial assumptions by changing them, one by one, and estimating what has changed in the predictions. The time it takes for the process to converge to a satisfactory solution is never known in advance; it depends on the complexity of the system, on the amount of experimental information and, most important, on the luck of the worker.

To approach the problem of cell dynamics in HIV infection, we analyse a score of experimental features in a typical patient, both in the presence and in the absence of drug therapy, and use these facts as tests for a panel of mathematical models. In this work, our search is restricted to models that do not postulate that HIV impairs the homeostatic replenishment mechanism of T-cells and that do not include interaction of HIV-specific effector cells with uninfected permissive cells or interaction of infected cells with each other. Within this group, the simplest model that agrees with data, and thus can be considered as the working model, requires that:

1. Most T-cell replenishment is due to non-antigen-specific activation of these cells by an unknown mechanism specific for HIV.
2. The virus infects and kills T-cells when they are cycling.
3. Killing of T-cells is independent of their infection.
4. Additional, virus-independent sources of CD4[+] T-cells (probably, of homeostatic origin) account for the slow decline of CD4[+] cell count.
5. The virus causes fratricide among HIV-specific effector cells.

From the many models we considered, we present only a few, which represent most important steps in our search for the working model. We restrict mathematical derivations to an appendix, using qualitative explanations and diagrams in the main text.

BASICS OF HIV POPULATION DYNAMICS

Course of Infection and Drug Therapy

The course of an average HIV-1 infection (Coffin, 1996) is schematically shown in Figure 10.1A. It can be divided into three phases. In the acute phase, viremia increases rapidly, and then declines but does not disappear altogether. Rather, it drops to a level around two orders of magnitude smaller than its peak value. The acute phase is usually accompanied by mild influenza-like symptoms. The maximum viremia correlates with a deep, but transient depletion in CD4[+] T-cell count, presumably reflecting massive infection and killing of cells of this type. After the decline in viremia, the CD4[+] T-cell count rebounds, but returns only partially to its preinfection level. The long asymptomatic phase, which then ensues, is characterized by constant or slowly climbing viremia, and the CD4[+] T-cell count slowly decreases until almost vanishing in an average of 7 years (ranging from 2 to 15 for individual patients; the rate of progression to AIDS correlates positively with the virus load in the asymptomatic phase). The final phase of acute immunodeficiency is characterized by an explosion in virus load, rapid disappearance of the few remaining CD4[+] T-cells, spread of opportunistic infections, malignancies, neurological symptoms and death. A small fraction of HIV-infected individuals, termed long-term non-progressors, establish a very low virus load in the asymptomatic phase, do not exhibit decline in CD4[+] T-cells and do not develop AIDS.

The grim outcome of HIV infection is postponed in patients receiving HAART, a mixture of several antiviral drugs, which efficiently suppresses virus replication and, in many cases, does not allow emergence of drug-resistant virus variants for a number of years. The dynamics of virus load and of CD4[+] T-cell count in the asymptomatic phase, before and after onset of HAART, is shown in Figure 10.1B (Kelleher et al., 1996; Autran et al., 1997; Kelleher, 1997; Zhang et al., 1998b; Lederman et al., 1998; Connick et al., 1999). The onset of therapy causes a rapid decrease in the virus load and a partial return of CD4[+] T-cells.

An expanded view of Figure 10.1B corresponding to the first few weeks of HAART is shown in Figure 10.2A. After a small initial delay of about 1 day (during which it slightly increases for reasons unknown), the viremia experiences a rapid decline, consistent with a very short average lifetime for most (95–99%) infected cells and

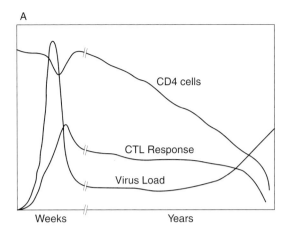

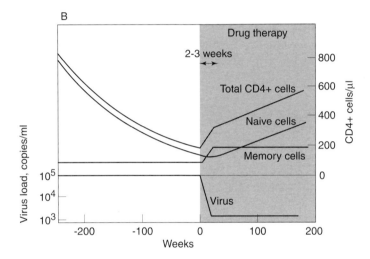

FIGURE 10.1 **A**. Typical course of HIV infection in the absence of drug intervention. Lines show the virus load, the CD4⁺ cell count in the peripheral blood and the frequency of HIV-specific CD8 cytotoxic cells at different times after transmission. **B**. Long-term effect of highly active antiretroviral therapy (HAART) on the virus load and the two subcompartments of CD4 cells. Time 0 is onset of HAART.

an even shorter average virus-producing phase (Ho *et al.*, 1995; Wei *et al.*, 1995). Using an exponential decay to fit the experimental decline, the average lifetime of virus-producing cells was estimated as 1–2 days (Perelson *et al.*, 1996; Haase, 1999).* The clearance time of free infectious par-

*In fact, the observed decay of plasma RNA during the first week, in a typical patient, is not a simple exponential. In recent HAART trials, a faster decay at days 1–3 is followed by a more flat exponential slope at day 4–6 (Polis *et al.*, 1999). Neither phase is a certain exponential.

ticles from lymphoid tissue is even shorter, a few hours (Perelson *et al.*, 1996). After 10–14 days, the decline slows down significantly, implying the existence of a relatively small, but still very large long-lived compartment of infected cells (Cavert *et al.*, 1997; Perelson *et al.*, 1997). These cells have not been directly identified. They may be latently infected cells randomly activated to produce virus with a half-life of 2–3 weeks or some other cell type, such as macrophages, that might be chronic producers of virus. Alternatively, they

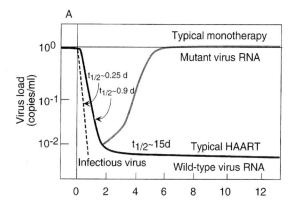

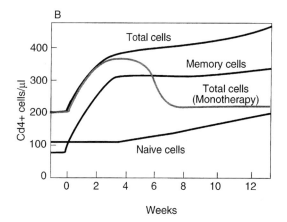

FIGURE 10.2 Typical response to different types of antiviral therapy on a short time scale. **A.** Time dependence of virus load during HAART (black line) and during monotherapy (gray line). Both parameters are normalized to their base (steady-state) levels. **B.** Changes in the memory, naive and total CD4 cells during HAART and monotherapy.

might be cells infected with virus of very low replication capability. (Ongoing selection for diversity due to immune pressure could maintain a high load of such mutants.)

If these two classes of infected cell were the only possible cell types that could produce infectious virus, HAART therapy for a few years would suffice to eradicate the infection in an average individual (Perelson *et al.*, 1996). However, recent evidence indicates the presence of a very small fraction (10^{-5}–10^{-6}) of resting T-cells that do not produce infectious virus when isolated but can be induced to do so by culture conditions that lead to cell activation (Chun *et al.*, 1997a; Finzi *et al.*, 1997; Wong *et al.*, 1997). The half-life of such cells *in vivo* is not known

with any certainty, and it is not clear whether they give any noticeable contribution to the long-lived virus load under HAART; they are also unlikely to be of any importance to the natural course of disease. However, they may well foil attempts to eradicate the virus from infected individuals in a reasonable period of time, dooming patients to decades of difficult and expensive therapy. The possible existence of other reservoirs of long-lived infected cells has also not been effectively addressed.

It should be noted that the frequency of production of such cells must be very low. If they are derived from the same pool as productively infected cells, represent 1% of such cells and their half-life is at least 100-fold greater, then less than one infection event in 10^4 will give rise to them. It can be hypothesized that their appearance might be due to infection of activated cells just as they are entering a resting phase, shutting down virus expression until a later activation event. Unfortunately, the low frequency of appearance of these cells will make experimental tests very difficult.

Before HAART, monotherapy using single reverse transcriptase or protease inhibitors was in use. Monotherapy almost inevitably fails because of the emergence of virus variants resistant to the inhibitor used. The virus and CD4+ T-cell count during a typical monotherapy trial are shown schematically in Figure 10.2 (Coffin, 1995; Danner *et al.*, 1995; Wei *et al.*, 1995). The resistant virus, in this example, replaces the wild type in 2 weeks (the timing depends on the drug and the dosage) and returns rapidly to the predrug level. Correspondingly, the initial increase in the CD4+ T-cell count is followed by a slow decline. Note that both CD4+ cells and viremia return approximately to their predrug levels (at different times), without overshooting these levels. We will return to this apparently trivial feature in the next section.

Topography of Virus and Infected Cells

Recent improvements in labeling and imaging techniques advanced our knowledge on how free virus and infected cells are distributed in the body (Haase, 1999). Of 2×10^{11} CD4+ T-cells in a normal young adult, 98% are found in the lymphoid tissue; 2% are in the peripheral blood. Not surprisingly, most HIV-infected cells are found in the lymphoid tissue as well. Late in infection, a fraction of CD4+ T-cells as large as 1% may harbor virus DNA (Chun *et al.*, 1997b). A much smaller subset of CD4+ T-cells, 0.05% or 4×10^7 cells in the whole body, express viral RNA (> 20 copies/cell at one time) in an average patient with a CD4+ T-cell count of 400/μl blood (Haase *et al.*, 1996). Calculations of total body burden of infected cells assume that the tissues sampled – tonsils and peripheral lymph nodes – are representative of all lymphoid tissue, most of which is in the gut (gut-associated lymphoid tissue, or GALT). The error is unlikely to be as much as an order of magnitude, however.

Most HIV RNA in the body, which amounts to some 10^{11} copies, is found neither in infected cells nor in blood. Rather, it is located in virions attached to the surface of the antigen-presenting follicular dendritic cells (FDC) in lymphoid tissue. Most of the remaining 2–3% is dispersed inside the tissue or associated with productively infected cells, and only a small fraction (10^8–10^9 copies) is present in the blood at any one time. Productively infected cells contain an average of 70 RNA copies per cell at one time (Haase *et al.*, 1996; Haase, 1999).

Dynamics of HIV Infection

The static figures presented above are quite distinct from the dynamic picture of HIV infection. FDCs "store" most of the virus load at any one time but are probably of little consequence for the overall dynamics of the virus. Although virus remains attached to an FDC surface for an average of 2–3 days (Cavert *et al.*, 1997; Haase, 1999), decline of infectious virus in blood is much more rapid, with the average time 6 hours (Perelson *et al.*, 1996), implying that release of the virus load from FDC is not the dominant contributor of virion drain into blood, and that most of virions on FDC undergo endocytosis ($\sim 5 \times 10^{10}$/day). In principle, the observed decay of infectious virus could reflect the loss of infectivity of virus particles before they are released from FDC, rather than their release rate

(Dimitrov, 1999). But in this case, were FDC a major contributor to the virion flux into blood, one would expect a divergence between the decay of plasma RNA and decay of infectious titer of PBMC in blood. Significant difference between the two decays is not detected within the first week of drug therapy, although later PMBC titer falls below plasma RNA (Perelson *et al.*, 1997). Therefore, most virions leaving tissue for blood must flow around the FDC.

The dynamic flow of virus particles filtered daily from blood (and, therefore, produced daily in lymphoid tissue) is very intense. Uninfected monkeys inoculated with virus directly into blood in clear it in an average time of 5 min (Zhang *et al.*, 1999a). In an average HIV-infected person with 10^5 HIV RNA copies per milliliter of blood, the total HIV RNA flux from tissue to blood can be estimated as $10^5 \times 5 \cdot 10^3$ ml $\times 288/\text{day} = 1 \cdot 4 \times 10^{10}$ copies/day. This corresponds to 4000 RNA copies per a lifetime of productively infected cell. Although it is possible that the number of infected cells is underestimated in these studies, the estimate of 2000 virions per cell agrees with the principle that a cell has to produce many infectious particles to ensure a steady state, and with the fact that the ratio of infectious units per virus particle (as measured *in vitro*) is very low. The static value of 70 RNA copies per cell (Haase *et al.*, 1996) implies that these RNA molecules turn over about 50 times during the productive phase of cell infection (≈ 1 day).

CD4+ T-cells in blood increase rapidly after the onset of therapy, at about 5% per day (Figure 10.2B). This increase led some workers (Ho *et al.*, 1995; Wei *et al.*, 1995) to postulate a turnover time of T-cells in HIV+ individuals of a few weeks. This assertion was challenged by others, who noted that altered trafficking of lymphocytes due indirectly to decreased infection could cause their apparent increase in the peripheral blood (Mosier, 1995; Spreut and Tough, 1995; Dimitrov and Martin, 1995). Later, the rapid turnover was confirmed by direct measurements of turnover time of T-cells in blood and lymphoid tissue by BrdU labeling in SIV-infected animals (Mohri *et al.*, 1998; Rosenzweig *et al.*, 1998). Although the controversy continues and other arguments ensue (see Summary of T-cell kinetics, below), the latter experiments are convincing and our modeling will assume rapid turnover of CD4+ T-cells.

Initial observations on the rebound of CD4+ T-cells under HAART were quite disappointing in that the number seemed to saturate after about 2 weeks at values substantially less than normal (Kelleher *et al.*, 1996; Autran *et al.*, 1997; Kelleher, 1997; Figure 10.2B). However, more detailed analysis revealed that the two important subsets of CD4+ T-cells react to drug in different ways. The memory (CD45RA⁻CD45RO⁺) compartment expands rapidly in the beginning but soon saturates while the total count is still well below normal, and the return of naive (CD45RA⁺CD45RO⁻) cells starts in a few weeks after the onset of HAART and proceeds at a linear rate, 20–80 times lower than the initial return of memory cells (Kelleher *et al.*, 1996; Autran *et al.*, 1997; Lederman *et al.*, 1998; Connick *et al.*, 1999). The net increase in CD4+ T-cell count after 6 months of HAART is mostly due to naive cells (Zhang *et al.*, 1998) but is still a long way from normal values.

The rapid flow of virus particles into blood suggests that most virions travel far, as opposed to being trapped near producing cells. From the decay of infectious virus load under a protease inhibitor (Perelson *et al.*, 1996), we can conclude that a typical virus particle is drained into blood in a few hours after it was produced, or even sooner. This conclusion is consistent with the visualization of separate HIV genetic variants in spleen by selective labeling (Reinhart *et al.*, 1998). Infected cells are spread in the tissue in a macroscopically uniform fashion. At a more detailed level, infected cells concentrate in separate islands. Since most islands are shared by different genetic variants (Reinhart *et al.*, 1998), the speckled pattern is likely to result from a non-uniform supply of uninfected host cells, rather than from the spread of infection by local cell-to-cell transfer, as suggested by some authors (Haase, 1999). It especially well agrees with the idea (which we propose below on different grounds) that HIV activates uninfected resting cells to divide and thus generates its own host cells. To conclude, on a macroscopic level, the infected tissue can be viewed as a "well-stirred pot" that contains uninfected and infected cells and free ambient virus, which spreads infection between cells. For more experimental details, the reader is referred to a very informative recent review (Haase, 1999), but is warned that our conclusions do not necessarily coincide with the conclusions of its author.

HIV INFECTION AS A QUASI STEADY STATE

In the previous section, we observed that, although infected and uninfected CD4+ T-cells turn over many times during the asymptomatic phase of HIV infection, their numbers stay approximately constant. Therefore, replenishment and depletion of each cell type nearly compensate each other (Coffin, 1995). Clearly, mere coincidence between some biological parameters cannot explain the almost exact balance, not for so many individual cases, and over such a wide range of virus loads. The collective of cells somehow arrives at a steady state and then maintains it automatically, through interaction between its parts. The method of kinetic modeling that we use in this work allows us to test whether a given model predicts a steady state. To define a model, we have to specify which cell types it includes and how they interact with

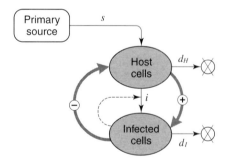

FIGURE 10.3 A simple model illustrating the steady state concept. Replenishment of CD4 cells occurs from a constant homeostatic source. Uninfected host cells either die naturally or become infected, infected cells are cleared within a constant average time. Broad gray arrows denote the negative and the positive feedback between the host cell and the infected cell compartments. For reasons discussed in the text, this cannot be the correct model of HIV infection.

each other. Making different assumptions, one can construct a large number of models even for, say, three cell phenotypes. However, most models do not predict a stable steady state, or contradict other experimental facts, and can be excluded, thus saving experimental work to test them.

We illustrate the concept of steady state using the simplest model of HIV kinetics with a constant replenishment source of CD4+ T-cells (Figure 10.3). The model, although simple and popular in mathematical works on HIV, cannot be the actual model of HIV kinetics, since it contradicts some important facts (see next section). We consider this model only in this section, to prepare for the subsequent discussion, and to introduce the notation used.

The different cell compartments in Figure 10.3 are shown as shaded ovals, one representing uninfected but infectible CD4+ T-cells, which we will refer to as "host cells", and another representing infected cells. (The host cells are usually referred to as "target cells". We use the "host cell" designation to avoid confusion with cells that are targets for CTL-mediated killing.) In general, the number of cells in each compartment may change with time. According to the model, uninfected cells are replenished from a homeostatic source, die from old age or become infected, and infected cells die. The cell fluxes in and out of the two compartments are shown by solid arrows. We show by the dashed (controlling) arrow that the number of cells infected per day depends on the number of existing infected cells. Unless specified otherwise, we will imply that each cell flux is linearly proportional to the size of the compartment it is coming from and the size of the controlling compartment, if any. For example, the number of cells becoming infected per day is proportional both to the number of infected cells (i.e. the amount of virus) and to the number of host cells existing at that time. Letters near solid arrows are biological parameters: s is the source intensity; i is the infection efficiency parameter proportional to the number of infectious particles per infected cell per day and to the virus entry efficiency; d_H (d_I) is the inverse average time of the host (infected) cell death. The parameters are unknown in advance and all have to be found

from fitting to experimental data. Such notation contains all the information about properties of the model system that there is to know. It can be translated into equations, which we list in Appendix, together with the main mathematical results. There is one equation for each compartment in the diagram.

Note that the model in Figure 10.3 does not treat free virus as a separate compartment. An average infectious virus particle is cleared from lymphoid tissue in a few hours (Perelson *et al.*, 1996). Therefore, at time scales of 1 day or longer, the number of virus particles can, for all practical purposes, be treated as directly proportional to the number of productively infected cells. Below, we use the terms "virus load" and "number of infected cells" as synonyms.

When the infection efficiency is sufficiently high (or the natural life span of host cells is long), the system in Figure 10.3 can be shown to have a steady state. The main idea is that, at very large virus loads the number of host cells infected per day exceeds the replenishment rate, and at a very small virus load the imbalance is in the opposite direction. Therefore, at a certain intermediate value of virus load, host cells reach an exact balance between their production and infection rates. Similarly, the infected cell number reaches an exact balance at a certain number of host cells. Formulas for steady-state numbers of both compartments are given in the appendix.

It is not enough for a model system to have a steady state. To match the experimental observations of persistence of HIV *in vivo*, the steady state must be stable: it has to resume after any possible transient perturbation, whether accidental or intentional. Such perturbations include increasing virus production by treatment with interferon-γ, depleting virus by antiviral drug therapy, depleting or activating T-cells by anti-CD3 antibodies, and increasing the host cell population by immune response to a vaccine or IL-2. A necessary condition for the stability is the presence of a "negative feedback loop" in the system. That is, a perturbation must create a compensating response in the opposite direction. Below we show that the steady state in the model shown in Figure 10.3 is stable.

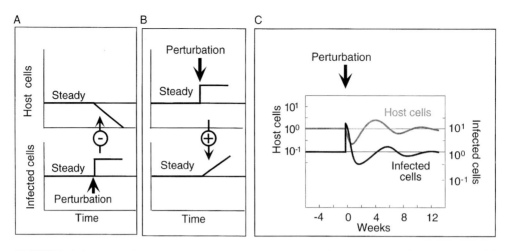

FIGURE 10.4 A negative feedback mechanism ensures stability of steady state predicted by the model in Figure 10.3. **A**. An increase in virus load causes depletion of host cells (negative feedback). **B**. An increase in host cell number causes expansion of infected cells (positive feedback). **C**. Combined, these two effects result in stability of the steady state: both cell compartments gradually resume their steady-state levels after an initial perturbation. Note oscillations around the steady-state levels due to delay in the negative feedback response.

Suppose that there is a small accidental increase in the virus load. The increase will cause depletion of host cells since more of them become infected (Figure 10.4A). The depletion in host cells will cause a decrease in the virus load since there are now fewer cells to infect, and the death of infected cells now prevails over infection *de novo* (Figure 10.4B). The negative feedback from the virus to host cells and the positive feedback from the host cells back to the virus (two shaded wide arrows in Figure 10.3) together make a negative feedback loop: any increase in the virus load eventually causes it to decrease. The resulting time courses of the two compartments exhibit oscillations converging gradually to the base levels (Figure 10.4C). The oscillations are caused by a delay in the host cell number. In the general case, when the delay is too large, oscillations can become divergent and the steady state will be lost. In this particular model, this does not happen. Note that the fact of stability or instability is a property of the model, not of the type of perturbation: a stable system, by definition, resists small perturbations of any kind (strong perturbations are discussed below, under A second mechanism of replenishment). A rigorous stability analysis, based on finding eigenvalues of the dynamic matrix, is discussed in the appendix.

The model in Figure 10.3 predicts an exact steady state, while the actual CD4+ T-cell count in HIV infection slowly drifts towards zero. Many hypotheses have been proposed to explain the downward decline ending in AIDS (Feinberg and McLean, 1997). Can we explain the decline without using any additional ideas but the simple kinetic arguments already introduced? An answer is shown in Figure 10.5A: two independent compartments of host cells, which are activated to divide, at two very different rates, by a homeostatic replenishment mechanism. The cells from the slowly replenished group are also assumed to be infected at a much lower rate (e.g. because of a different receptor or coreceptor expression). Although the two host-cell compartments can be of a similar size, cells turn over more rapidly in the compartment fed from the stronger source. Such a model system does not have an exact steady-state. To be in balance, each of the two host cell compartments requires a certain amount of the virus. The two required amounts, in general, differ and cannot be met together. However, if the activation rate of compartment 2 is small enough, the virus load will be forced close to the steady-state value of the "fast" compartment. Intuitively, the negative feedback from each host-cell compartment is trying to bring the

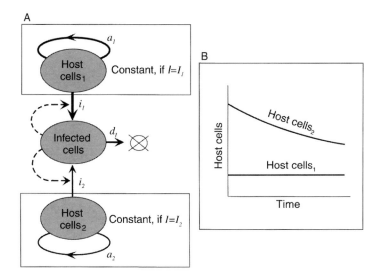

FIGURE 10.5   A. The simplest model with two compartments of host cells illustrating the concept of quasisteady state. Both compartments are activated to divide by a homeostatic mechanism, but the first compartment is (according to the model) activated much more frequently and is infected much more efficiently than the second compartment. **B**. Predicted schematic time dependencies for the two host-cell compartments. Replenishment and death of the "slow" compartment are far out of balance. For reasons discussed in the text, this cannot be the correct model of HIV infection.

virus to its own steady-state level, and the feedback from the fast compartment acts faster and wins out. As a result, the virus load and the "fast" compartment will be almost constant in a long run, while the "slow" compartment will be far out of balance and either decline or increase slowly. The decline occurs over a broad range of model parameters (Table 10.1, in the appendix, and Figure 10.5B).

An important feature of the model shown in Figure 10.5A is that it predicts a slow decline in CD4 cell count without postulating that any model parameters depend on time. For example, the clearance rate of infected cells, d_1, may be constant, decrease, or even increase slowly with time after infection, but the CD4 cell count, according to this model, will still decline. (On average, d_1 is approximately constant; Ho et al., 1995.) This model also implies that the short-term and the long-term kinetic properties of HIV infection are, in general, separate. In particular, it is hardly possible to infer the clearance mechanism of infected cells ("CTL versus virus-induced death") from the correlation between

the clearance rate and the CD4 count in individual patients, as is sometimes suggested (Feinberg and McLean, 1997).

The models shown in Figure 10.3 and 10.5A disagree with several important features of HIV kinetics. For example, the oscillations that, according to both models, are predicted to follow a transient perturbation are not observed in most patients when the rebound of drug-resistant viremia during monotherapy trials is monitored (Coffin, 1995; Danner et al., 1995; Wei et al., 1995; Figure 10.2B). The oscillations are rather stubborn: one cannot eliminate them and explain the experimental results by merely adjusting the model parameters (see simulation in Figure 10.6 and Table 10.2, in Appendix). The oscillations survive even if the drug resistance is only partial; i.e. if the new virus variant has a smaller fitness than the wild-type virus prior to drug therapy. In this case, the CD4 cell count is predicted to converge to a level higher than the predrug level, but the dramatic oscillations of virus load would remain (simulation results not shown).

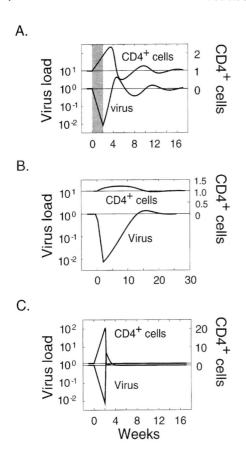

FIGURE 10.6 Kinetics of CD4 cells and virus load under monotherapy as predicted by the model in Figure 10.3. Both quantities are normalized to their steady-state levels. It is assumed that virus resistant to drug appears at week 2. Parameter values: **A**. Matching increase in CD4+ count. This choice of parameters matches the first two weeks of experimental kinetics (Ho *et al.*, 1995; Wei *et al.*, 1995; Kelleher *et al.*, 1996): $is = 0.02/day^2$, $d_1 = 0.33/day$, $d_H << is/d_1$. **B**. Low source intensity. Near the threshold of stability of persistent infection: $is = 0.02/day^2$, $d_1 = 0.33/day$, $d_H = 0.8\ is/d_1 = 0.048/day$. **C**. Too-high source intensity eliminates oscillations: $d_1 = 0.33/day$, $is = 4d_1^2 = 0.44/day^2$, $d_H << d_1$.

Oscillations can be eliminated from the model by assuming that there are two types of CD4 cell: short-lived proliferating cells permissive for virus replication and resting cells which are not permissive, but are activated at some rate to become proliferating (de Jong *et al.*, 1996). However, this assumption alone is not sufficient, since it creates a new sort of trouble: the model (de Jong *et al.*, 1996) predicts a slowly decaying overshoot of base level by the drug-resistant virus, since the CD4 level at the time of

rebound is still elevated and there are more cells for virus to infect. The overshoot is typically not observed in monotherapy trials (Coffin, 1995; Danner *et al.*, 1995; Wei *et al.*, 1995). This is confirmed by data obtained from SIV-infected animals who received an antiviral drug during 2 weeks, and then the treatment was stopped: the overshoot of viremia did not exceed about 50% (Nowak *et al.*, 1997). De Jong *et al.* (1997) published data on four AIDS patients who showed a considerable overshoot of viremia after therapy was discontinued, for different reasons. However, of the four patients, three were in final stages of AIDS (base-level CD4 counts 20/µl blood or below), which hardly qualifies as a steady-state HIV infection, which the model describes. The model also does not explain saturation of CD4 cell number and some other facts discussed in the next section.

Therefore, we have to change the model drastically: either replenishment of T-cells is not from a constant source, or their killing is not due to direct infection, or there are more players in the field, i.e. more cell types to consider. As we shall see, we will have to postulate all the above: HIV infection appears to be a slightly more complex process than depicted by any model composed from an intuitive standpoint. More detailed analysis of experimental information will be given in the next section. However, ideas of steady state, quasisteady state and stability, which these models helped us to illustrate, are still very useful. Later we will return to the idea of two separate host cell types, and will identify the long-lived compartment with "naive" cells and the short-lived with "memory" cells.

SUMMARY OF T-CELL KINETICS: MAIN POSTULATES

Let us obtain a more complete and, at the same time, a more condensed picture of T-cell kinetics in a typical patient than we did in the previous sections. We do not propose to explain here the entire body of data available on HIV kinetics. Rather, we will carefully select the facts that turn out to be most useful for segregating the possible kinetic models. For example, we will not consider individual variation between patients and mostly not consider different correlations, such as the correlation between the slope of virus decay under therapy and the virus load. We do not include this material because it is usually hard to determine which parameters of a model vary more strongly between individuals and because we wish to explain the typical experimental pattern first. The acute phase of HIV infection is beyond the scope of the present work, since it is likely to be influenced by even more factors than affect the asymptomatic phase and since experimental data on acute phase kinetics are sparse. We also will not discuss issues related to the existence of the relatively small long-lived virus compartment, which is important for evolution of drug-resistant variants and for seeding of HIV re-infection after prolonged drug therapy. We also do not consider the relatively large compartment of cells containing HIV DNA in some sort of defective state that cannot be induced to produce infectious virus (Chun *et al.*, 1997b). Instead, we will focus on the short-lived 95–99% of the virus population. Therefore, the model we obtain in the end has to be treated as a first approximation, a working model subject to further tests and refinements.

When listing below some important experimental features, we will group them by relevance, concluding each group with an observation. We start from the facts pertinent to the behavior of HIV population on short time scales, from days to weeks.

1. *Short lifetime of infected cells.* After onset of a potent antiviral drug therapy which suppresses infection of new cells by virus, viremia in HIV-infected patients exhibits a (roughly) exponential decay with a half-time of 1.2 days (Ho *et al.*, 1995; Wei *et al.*, 1995; Haase, 1999). After 10–14 days of decay, the decrease slows down significantly.

2. *Rapid turnover rates and rapid drug kinetics of "memory" T-cells.* The onset of drug therapy in HIV-infected patients is followed by an increase in "memory" (CD45RO+CD45RA−) T-cell count, with an average rate 5% of the predrug CD4+ T-cell count per day (for patients with counts around 200/μl blood; Ho *et al.*, 1995; Wei *et al.*,

1995; Kelleher *et al.*, 1996; Lederman *et al.*, 1998; Connice *et al.*, 1999). As determined by BrdU labeling experiments in SIV-infected animals (Mohri *et al.*, 1998; Rosenzweig *et al.*, 1998), most memory T-cells in both lymphoid tissue and peripheral blood have an average lifetime of 1–2 months, a few times shorter than in uninfected controls.

3. *Almost constant cell numbers.* In an average HIV-infected person both the virus load and the T-cell count are approximately constant during the course of a year, except for some random fluctuations (Coffin, 1995, 1996, and references therein). The virus load at steady state is often referred to as the set point.

4. *Absence of oscillations/overshoot.* In a typical monotherapy trial the emerging drug-resistant virus variant saturates at the predrug level without significant (> 50%) overshoot or oscillation around it (Coffin, 1995; Danner *et al.*, 1995; Wei *et al.*, 1995). Similar behavior is observed in SIV-infected animals whose antiviral treatment is experimentally discontinued (Nowak *et al.*, 1997).

Uninfected and infected T-cell populations turn over dozens and thousands of times, respectively, during an asymptomatic HIV infection. For both populations, cell death is nearly compensated, on average, by cell replenishment, exhibiting existence of a (quasi-) steady state (Coffin, 1995). After a transient perturbation, the steady state restores itself by monotonous changes in cell numbers, without overshooting base levels. In other words, the steady state is not only stable it is also non-oscillatory.

5. *A small number of productively infected cells.* An average HIV-infected person with a CD4+ T-cell count of 400/μl blood harbors ≈ 4×10^7 HIV-RNA-expressing cells in his/her lymphoid tissue, about 0.05% of total CD4+ T-cells (Haase *et al.*, 1996).

Combining 1, 2 and 5, it follows that less than 1% of CD4+ T-cells that die do so after being infected. The remaining 99% of cells must die as a result of some indirect mechanism (Haase *et al.*, 1996).

6. *Saturation of memory cells.* In drug-treated HIV+ patients, return of "memory" T-cells satu-

rates after about 2 weeks of therapy, while the total count of CD4[+] T-cells is still below the normal level (Kelleher *et al.*, 1996; Kelleher, 1997; Autran *et al.*, 1997; Lederman *et al.*, 1998; Connick *et al.*, 1999; Haase, 1999), creating an impression of "impaired homeostasis".

7. *Dominance of virus strains using CCR4 co-receptor.* In the asymptomatic phase of HIV infection, most virus variants use the CCR4 co-receptor expressed by "memory" T-cells and macrophages for entry into a host cell, as opposed to variants using the CXCR5 coreceptor expressed by "naive" T-cells (Bleul *et al.*, 1997). CXCR5-using strains appear in only a fraction of patients and only at late stages of HIV infection. The short-term kinetics and the dominant host cell type are associated with the "memory" phenotype as opposed to the "naive" phenotype of T-cells.

8. *Similarity between CD4[+] and CD8[+] T-cell kinetics.* Features 2,6,9 of T-cell kinetics on time scales of a few weeks are strikingly similar between the two compartments (Kelleher, 1997; Mohri *et al.*, 1998; Rosenzweig *et al.*, 1998; Lederman *et al.*, 1998; Connick *et al.*, 1999), even though CD8[+] T-cells are not infectible by HIV and are elevated above normal levels until late stages of infection.

Combining 6, 8 and 14 (below), the mechanism of memory T-cell replenishment in HIV infection is independent of CD4/CD8 molecule expression and is, most likely, not of homeostatic origin (unless the homeostatic mechanism of T-cell replenishment is somehow damaged).

9. *Rapid dilution of BrdU label in SIV-infected animals.* After the BrdU labeling of animals has stopped, the decline rate in labeled memory T-cells is similar to the increase rate during the labeling phase (Mohri *et al.*, 1998).

The rate at which a cell population becomes labeled in the presence of BrdU is determined by the cell turnover time. In the absence of BrdU, unlabeled cells divide into unlabeled cells and labeled cells generate labeled progeny. Each of these subcompartments is in a steady state of its own. Therefore, rather than decreasing rapidly as observed (Mohri *et al.*, 1998), the fraction of BrdU-positive cells is expected to stay almost constant for a period much longer than

the cell turnover time, until the label is gradually diluted below the detection threshold. This discrepancy could be explained assuming that a resting "memory" T-cell *in vivo*, upon receiving an activation signal, goes rapidly through many (5–6 or more) cycles of division in less than a week, before returning to the resting state. This rapid series of sequential divisions rapidly dilutes the label below the detectable level.

10. *Virions travel far and clear rapidly.* Estimates based on studies of the clearance rate of free virus from the peripheral blood (Zhang *et al.*, 1999), on the HIV RNA quantitation in the lymphoid tissue (Haase *et al.*, 1996), and on the decay rate of infectious virus titer under HAART (Perelson *et al.*, 1996), suggest that most virus particles produced in the tissue drain within a few hours or less into the blood, from where they are removed by liver cells within a few minutes.

A typical infectious virus particle infects a cell situated far from the cell that produced this particle. Therefore, an infected lymphoid organ can be approximately viewed, at least on macroscopic level (Reinhart *et al.*, 1998), as a "well-stirred pot" of infected and uninfected cells and free virus particles.

We will now turn to the course of infection on a longer time scale: from months to years.

11. *Slow decline of CD4[+] cell count.* In an average patient, the CD4[+] T-cell count declines from 1000 cells/μl blood to almost zero in 8 years: $\approx 0.2\%$ per day at a count of 200 cells/μl. The progression time is distributed almost uniformly between 2 and 15 years among patients (Smith *et al.*, 1997).

12. *Slow turnover rate and return of naive CD4[+] T-cells.* The return of naive CD4[+] T-cells is 20–80 times slower than the return of memory T-cells during the first week of drug therapy (Kelleher *et al.*, 1996; Autran *et al.*, 1997; Connick, 1999; Zhang *et al.*, 1998). In one study, the rate of increase in tissue was 7×10^7 cells/day, or $\approx 0.1\%$ of the total CD4[+] T-cell count for patients with about 200 cells/μl blood (Zhang *et al.*, 1998). The average lifetime of naive T-cells in SIV-infected animals is not detectably less than in uninfected controls (Mohri *et al.*, 1998).

13. *Timing of the return of naive CD4⁺ T-cells.* In one study the return of naive CD4⁺ T-cells is delayed with respect to the onset of antiviral therapy by a few weeks (Kelleher *et al.*, 1996; Kelleher, 1997).

The naive T-cell phenotype is closely associated with the long-term kinetics. A steady state does not exist for naive cells in the absence of drug: since the predrug decline and the postdrug increase in the naive CD4⁺ T-cell numbers occur at comparable rates, one can conclude that even an approximate balance between the production and replenishment of naive cells is absent from infected individuals.

14. *Elevated level of CD8⁺ T-cells.* In an average patient, the CD8⁺ T-cell count stays above the normal level almost until the onset of AIDS (Margolick *et al.*, 1995). There is no similarity between the CD4⁺ and CD8⁺ T-cell kinetics on a long time scale. (Connick *et al.*, 1999).

15. *An approximately constant half-life of infected cells.* In spite of declining CD4 cell count, the half-life of infected cells, on average, does not increase or decrease with progression to AIDS (Ho *et al.*, 1995; Wei *et al.*, 1995).

To summarize, within both CD4⁺ and CD8⁺ T-cell compartments there are two groups of cells with strikingly different behavior: "primed" or "memory" cells, which participate in fast processes; and "naive" cells, which are associated with long-term features. Memory cells are in a steady state and exhibit similar kinetics for CD4⁺ and CD8⁺ subsets. Naive cells are not even in an approximate steady state, and their CD4⁺ and CD8⁺ subsets behave differently. Memory cells have a replenishment mechanism that appears to be independent of homeostatic activation. Most are killed without being infected. The slow replenishment of naive cells agrees with the rate of homeostatic reconstitution of lymphocytes, 6×10^7 per day as measured in young adult cancer patients after radiation therapy (Haase, 1999). The fact that only the CD4 subset of naive cells declines during the asymptomatic phase of HIV infection implies direct infection (or indirect CD4-dependent killing) as the killing mechanism. To account for these features, we will conduct our further search for a working model of HIV population within three assumptions, which can be loosely stated as follows.

- The homeostatic mechanism of T-cell replenishment is not seriously impaired in HIV infection.
- There are two distinct groups of host cells that differ in how they are replenished and killed and in their average turnover times.
- Proliferation of host cells in the short-lived group is induced by the virus itself via a non-antigen-specific signaling pathway, possibly unique for HIV. (This postulate is necessary to account for the similarity between CD4 and CD8 memory cells.)

This set of postulates leaves out alternative explanations, whose plausibility is hard to estimate at this point. For example, some researchers suggest that the homeostatic mechanism is impaired by HIV or/and is "blind" with respect to CD4/CD8 phenotype (Margolick *et al.*, 1995). Since we did not find sufficient experimental information on the mechanism of homeostasis, it would be difficult for us to speculate on how it could be impaired, leave alone constructing any closed models based on such an assumption. Therefore, we had no choice but to leave out this possibility for the future, when more information on this topic will be available.

Another idea is that the rapid initial expansion of memory CD4 T-cells under HAART is due to redistribution of these cells between different organs (Mosier, 1995; Spreut and Tough, 1995; Dimitrov and Martin, 1995). However, similar drug kinetics were observed in several peripheral lymphoid organs and in the blood (Zhang *et al.*, 1998), and the direct BrdU labeling experiments in the blood and lymphoid tissue of SIV-infected animals confirm the accelerated turnover of T-cells (Mohri *et al.*, 1998; Rosenzweig *et al.*, 1998). Proponents of the trafficking hypothesis believe the redistribution between the (not yet studied) GALT, the lungs and the studied organs to be responsible for the observed kinetics (Haase, 1999). Experiments in which turnover of T-cells is estimated by antibody staining for cell-surface protein Ki67 show that the proportion of Ki67⁺ CD4⁺ T-cells in tissue early in infection is similar to uninfected controls (Fleury *et al.*, 1998; Haase, 1999) and is elevated threefold later in infection (Zhang *et al.*, 1998). Unfortunately, these experiments are harder to interpret than BrdU-labeling experiments since expression of Ki67 by proliferating cells may depend on specific conditions *in vivo*, and since the relationship between the frequency of proliferating cells and the turnover rate depends on a kinetic model. Interpretation of telomere length measurements, which do not suggest an increased turnover of CD4$^+$ T-cells in HIV infection (Wolthers *et al.*, 1996), also remains controversial (Wolthers and Miedema, 1998).

Note that turnover times for CD4 cells observed in SIV-infected animals (Mohri *et al.*, 1998; Rosenzweig *et al.*, 1998) are two to three times longer (Haase, 1999) than the estimate which can be obtained from the initial rate of increase in CD4$^+$ count (Ho *et al.*, 1995; Wei *et al.*, 1995), under the assumption that the killing of T-cells stops under HAART, but that their proliferation rate is not affected by the therapy. The working model we arrive at in the next section explains this discrepancy, as well as other quoted facts of HIV kinetics. According to this model, proliferation of T-cells is indirectly affected by therapy, since it is the virus that activates T-cells to divide. The initial expansion of memory T-cells observed under HAART is due to a continuing proliferation of cells activated by the virus before onset of the therapy.

SEARCHING FOR A MODEL OF SHORT-TERM KINETICS

In this section, we consider the behavior of different cell types in HIV infection at relatively short time scales, from days to weeks. On these scales, the quasisteady state can be viewed, approximately, as a true steady state. The three major cell groups that interact with each other are infected cells, host cells and HIV-specific effector cells. The virus load, as we have already discussed, is not an independent compartment, since it follows the changes in the number of infected cells. In principle, all three groups of cells must be considered together, within the same model (Figure 10.7A). However, if HIV-

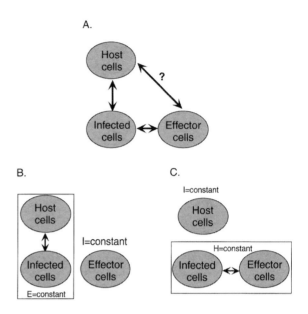

FIGURE 10.7 Three kinetic models including host cells, HIV-specific effector cells, and infected cells. **A.** General model. **B.** Host-cell-controlled model. **C.** Immune-cell-controlled model. **B** and **C** Show particular cases when the steady-state number of infected cells is determined by the balance condition for the host effector cell compartment alone. In the first approximation, kinetics of effector host cells may be studied separately from two other cell types.

specific effector cells and host cells do not interact directly (arrow with question mark in Figure 10.7A), but only via the infected cell compartment, and infected cells do not interact with each other, it may be possible to consider host cells and anti-HIV immune response separately and thus significantly simplify our task.

All models that predict a steady state can be classified into one of three families (appendix), which can be called "host-cell-controlled", "immune-cell-controlled" (De Boer and Perelson, 1998), and "host-immune-cell-controlled" models, respectively. In the first family (Figure 10.7B), the virus load in the steady state is determined by the balance between death and replenishment of the host-cell compartment alone. The HIV-specific immune cells in such a model can reach an exact balance at any (or within an interval of) virus load. In the immune-cell-controlled models (Figure 10.7C), the situation is the exact opposite: host cells can reach an exact balance at any value of the virus load but HIV-specific immune cells require a specific

A. Model 1: Linear activation and killing ...

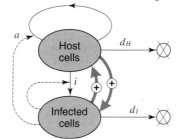

B. ... gives an unstable steady rate

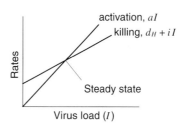

C. Model 2: Non-linear activation or killing ...

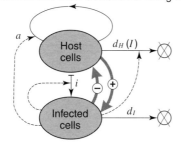

D. ...gives a steady state with undecaying oscillations

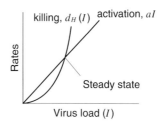

FIGURE 10.8 Two models with HIV-induced replenishment of host cells. **A.** Activation and killing rate of host cells are linear in the virus load. **C.** The killing rate is non-linear in virus load. **B.** and **D.** Dependence of activation and killing rates on virus load for the models shown in panels **A** and **C** respectively. The point at which the two curves intersect corresponds to an unstable (**B**) and an oscillatory (**D**) steady state.

virus load for their balance. In models of the third family, one has to consider the host-cell and the immune-cell kinetics together to find the steady-state value of virus load.

The idea of this approach is that a host-cell-controlled model may be (approximately) separated into two independent parts. One part deals with the kinetics of host cells and the virus load, assuming that the clearance rate of infected cells is constant (e.g. the model in Figure 10.3), and the second part considers HIV-specific immune cells, assuming a fixed virus load (Figure 10.7B). One can obtain steady-state expressions for both parts separately and match them to each other later. The method is an approximation, since it neglects coupling between the kinetics of HIV-specific effector cells and virus. Therefore, in the end, the two components of the full model have to be reunited and verified against experimental evidence. Similarly, in an immune-cell-controlled model (Figure 10.7C), one studies HIV-specific effector cells and virus load at a fixed number of host cells, and then host cells alone at a fixed virus load. As we show in the immune response against

HIV (below), the immune-response-controlled models predict unrealistically short lifetimes of HIV-specific effector cells. Therefore, in this section, we consider only the host-cell-limited models, assuming that the clearance rate of infected cells and of free virus is constant. We return to HIV-specific cells in the later section. The method of separation of a model into two parts will not work for the third family, which, together with models including collective effects among infected cells, will be studied elsewhere.

According to our third postulate (see above), T-cells are activated to proliferate by HIV itself. The mechanism cannot be normal, antigen-specific activation, since the frequency of antigen-specific naive cells is extremely small, $\approx 10^{-5}$. The HIV-specific CD4+ T-cell effector compartment is very small and often cannot be detected at all (Rosenberg *et al.*, 1997). The unknown activation mechanism of T-cells by HIV has to be non-antigen-specific and independent of CD4/CD8 expression. The simplest model of such type is shown in Figure 10.8A. It includes only two compartments: host cells and infected cells.

Host cells are infected by virus and, in addition, die at a constant rate (e.g. as a result of the illegitimate activation). The death and activation rates are linear to the virus load. Although the model, formally speaking, has a steady state, i.e. the point where the death rate of host cells is equal to their activation rate (Figure 10.8B), this state is not stable. Indeed, since the activation rate increases faster with the virus load than the killing rate, an accidental decrease in the virus will cause an excess of killing over the activation, which will cause a depletion of host cells, which will cause a further decrease in the virus and so on, until the virus is fully cleared. In other words, the model system contains a positive feedback loop and is, therefore, unstable.

To obtain a negative feedback, the killing rate should increase with the virus faster than the activation rate. This means that at least one of two rates must be non-linear in the virus load: either the killing rate must increase faster than the first power with the virus load increasing, or the activation rate must increase slower than the first power with the virus load. Both assump-

tions are equally non-trivial in the biological sense, because infected cells are very dilute (below). We will assume henceforth that it is the killing rate that has this strange feature (Figure 10.8C,D), although the reader has to keep in mind that it may be otherwise as well.

As it turns out, the model shown in Figure 10.8C is still not satisfactory, since it predicts that, once perturbed, the virus load and the host cell number will oscillate forever without decay. As we discussed, such oscillations would be observable in the drug-resistant viremia rebound (Coffin, 1995; Danner *et al.*, 1995; Wei *et al.*, 1995). The mechanism behind the predicted oscillations is that the response of host cells to the initial perturbation in virus is delayed because of a long lifetime of host cells. As a result, the virus load is expected to overshoot its steady-state level and then do it again and again (Figure 10.9A,B).

The easiest way to eliminate oscillations is to assume that the T-cell compartment consists of two connected compartments: resting cells, which constitute the majority of T-cells, and

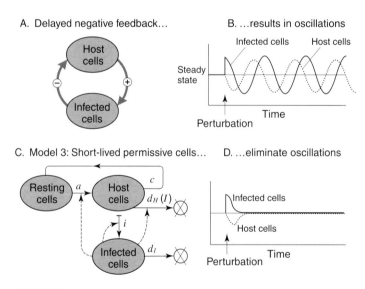

FIGURE 10.9 **A**. Time delay in the negative feedback response. **B**. Oscillations in the two cell compartment numbers following a perturbation of steady state in the model (Figure 10.8C). **C**. A model with two subcompartments of CD4 cells. Proliferating cells are the (short-lived) host cells. Host cells are killed indirectly, with the killing rate $d_H(I)$ increasing with the virus load I faster than a square. **D**. Non-oscillatory convergence of all compartments to steady state after a transient perturbation.

short-lived proliferating cells (de Jong *et al.*, 1996). Resting cells are activated by the virus to proceed through a cell cycle, during which they can be killed indirectly or infected by virus, or return naturally into the resting state (Figure 10.9C). Such a model has a non-oscillatory steady state, if the cycling time is even (a little) shorter than the lifetime of infected cells (which is, on average, 1.2 days; Haase, 1999). In this case, the host-cell compartment reacts quickly to changes in the virus load, so that the initial perturbation decays monotonously (Figure 10.9D). It is critical that the virus kills and infects short-lived proliferating cells selectively, rather than both compartments of CD4+ T-cells; otherwise, the steady state acquires oscillations or becomes unstable. Figuratively speaking, the model in Figure 10.9C portrays HIV infection as an extremely wasteful method of agriculture: the farmer (virus) breeds rabbits (host cells) for food, but has to destroy most of them just to keep their number under control.

What could the mathematical prediction that the killing (or activation) rate of host cells must be non-linear in the virus load mean biologically? Infected cells in the lymphoid tissue are very dilute (Haase *et al.*, 1996) and all the virus-derived factor concentrations are linearly proportional to the number of infected cells. Naturally, one would expect only linear dependencies on the number of infected cells. One possibility is that the death of host cells is caused by a combined action of several soluble factors secreted by infected cells. The number of these factors will be the power of the virus load dependence: two factors yield the square, three factors the cubic dependence of virus load, etc. Another possibility, suggested by neuron networks, is that the death signal is provided by a single factor but cell response to the signal is non-linear in the frequency of separate pulses.

On a qualitative level, the model shown in Figure 10.9C also explains the increase and the saturation of the T-cell number under HAART (Kelleher *et al.*, 1996; Kelleher, 1997; Autran *et al.*, 1997; Lederman *et al.*, 1998; Connick *et al.*, 1999; Haase, 1999). Suppose that infection of host cells is completely suppressed by a potent drug. Since the clearance rate does not change, the virus load starts an exponential descent. As a result, the killing and activation of resting T-cells, both induced by virus, cease rapidly. The already activated host cells complete division, causing an expansion of the total number of T-cells (Figure 10.10). The only problem with this explanation of the saturation effect is that, since the cell cycle time has to be short (about 1 day), the predicted magnitude of the total increase is very small, a few percent of the base count of CD4+ T-cells. Therefore, we have to change the model again. However, this time we do not need to make any additional assumptions. The rapid dilution of BrdU label in SIV-infected animals (Mohri *et al.*, 1998) implies that the T-cell replenishment occurs as a rapid series of, at least, five to six divisions following an initial activation event (see above). The modified model including multiple proliferating compartments is shown in Figure 10.11A.

As a result of this change in the model, the predicted delay of proliferation increases, and the size of the final clone produced by an initial activation event increases even more. Suppose that we have five divisions of 1 day each. At steady state, the negative feedback from the host cell compartment (see above) adjusts the virus load, so that, on average, half the cells in each proliferating subcompartment are killed by the virus and another half complete division and pass to the next subcompartment. As a result, each resting progenitor cell, after all cycles of division, generates, on average, a single cell, which resumes the resting phenotype (Figure 10.11A). The feedback ensures exact

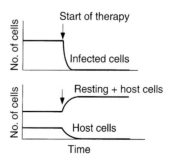

FIGURE 10.10 Schematic time dependence of three cell compartments during HAART in the model shown in Figure 10.9C.

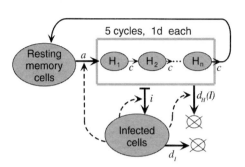

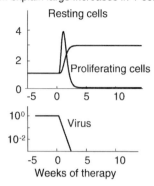

FIGURE 10.11   **A**. A model with multiple divisions of proliferating CD4 cells (host cells). Compartments H_1 through H_n correspond to consecutive cell cycles following activation of a resting cell. **B**. Kinetics predicted by the model for different cell compartments after onset of HAART. Parameter values: $d_I = 0.5$/day (Perelson *et al.*, 1996); the values $n = 5$, $c = 1$/day are chosen to resemble the average experiment on virus and CD4 cell kinetics (Kelleher *et al.*, 1996). The activation rate, $aI_0 = 0.025$/day, is found from the average turnover rate of CD4 cells in SIV-infected animals (Mohri *et al.*, 1998). The infection efficiency parameter, $i = 1.6 \times 10^{-10}$/day/cell, is estimated from the steady-state value $R_0 = 5 \times 10^{10}$ (Table 10.1, assuming total CD4 cell count ≈ 200/ml blood, half-and-half split between memory and naive cells (Lederman *et al.*, 1998; Connick *et al.*, 1999), and a 99:1% tissue-to-blood ratio (Haase, 1999)). From Table 10.1, one obtains the proliferating cell number in steady state, $5H_{10} = 3 \times 10^9$ cells. The killing rate dependence is put to be cubic: $d(I) = \text{const} \cdot I^3$.

balance of the resting cell compartment. After the onset of HAART, cell killing ceases rapidly and so does cell activation. However, an already activated cell expands into a large clone. As a result, we obtain a fair semblance of a typical experimental time course for patients with pre-drug CD4[+] T-cell counts of around 200 cells/µl blood (Kelleher *et al.*, 1996; Kelleher, 1997; Lederman *et al.*, 1998; Connick *et al.*, 1999) (cf. Figure 10.11B and Figure 10.2B). As with the single-division model, the absence of oscillations/overshoot requires a non-linear dependence of the killing rate on the virus load (Table 10.2, in Appendix).

A SECOND MECHANISM OF REPLENISHMENT: STABILITY AND PROGRESSION TO AIDS

In this section, we discuss the slow decline of the CD4[+] cell level in the absence of antiviral treatment and the rebound of the drug-resistant virus under treatment. We link both of these apparently unrelated phenomena to a second, virus-independent mechanism of T-cell replenishment.

We postulated above that the replenishment of naive T-cells is provided by a homeostatic mechanism, much slower than the virus-induced replenishment of memory cells. In principle, replenishment of naive cells can occur both from primary lymphoid tissue such as thymus and by reversion of memory cells into the naive phenotype (Haase, 1999). For the purposes of this discussion, we assume that the replenishment of naive cells occurs, mostly, from a primary source. The intensity of cell flux from the primary source is expected to be upregulated by a homeostatic mechanism when CD4[+] cells become depleted. For the sake of simplicity, we assume that the cell flux intensity is fixed and equal to its maximum values, $(6–7) \times 10^8$ cells/day (Haase, 1999). Given the normal number of naive cells in young adults, approximately 2×10^{11}, and the fact that changes in memory/naive ratio in healthy adults occur over a decade time scale, this implies a very long natural lifespan of naive cells, more than 8 years.

As we proposed above, there is no balance between production and death of naive cells, since the virus load adjusts itself to provide the balance for memory cells, the main host-cell compartment. Naive cells will be gradually depleted by one of two mechanisms. They may be slowly eliminated by virus infection or they may have a finite lifespan and their precursors are infected and killed by the virus. To explain why CD4+ but not CD8+ T-cells are depleted, it is tempting to invoke the direct infection of naive cells as a mechanism of killing. Indeed, the number of productively infected cells dying per day (Smith *et al.*, 1997) and the number of naive cells dying per day (Haase *et al.*, 1996) are in the same range, 10^7–10^8 cells/day. However, the dominant virus variants use for entry CC-type chemokine coreceptors weakly expressed by naive cells (Bleul *et al.*, 1997). Given that the sizes of the naive and memory compartments of uninfected CD4 cells are comparable, most of the directly infected cells must be memory cells. In other words, the rate of naive cell depletion seems too high to be explained by direct infection of these cells. There are two possible explanations. First, perhaps precursors of naive cells, unlike mature naive cells, can be infected efficiently. Naive cells gradually become extinct, since their homeostatic pipeline is cut. Second, naive resting cells may be killed indirectly by a mechanism that discriminates between CD4 and CD8 phenotypes, such as apoptosis induced by soluble gp120 protein (Westendorp *et al.*, 1995). Remember that we are discussing a very small effect, with a rate about 100 times less than the indirect killing of memory cells proposed in the last section. We are not aware of any direct data that allow us to decide between these two mechanisms of naive cell depletion. However, the fact that, in half of all patients, the CD4+ T-cell depletion occurs in less than 7 years and, in some patients, as early as year 2 (Smith *et al.*, 1997) suggests the indirect killing of resting naive cells rather than their natural extinction due to altered homeostatic replenishment. On the other hand, to explain the rebound of drug-resistant strains (Coffin, 1995; Danner *et al.*, 1995; Wei *et al.*, 1995) we will need to assume direct infection of naive precursors by HIV. Therefore, killing of both precursors and of mature naive cells is essential.

A model with two mechanisms of permissive cell supply is shown in Figure 10.12A. The upper part of the diagram is taken from the memory host model (Figure 10.11A). By analogy with the memory host cell model, it is assumed that naive precursors are both killed indirectly and infected by the virus. In the absence of HAART, the homeostatic pipeline is cut and naive cells decline as a result of indirect killing by the virus. After the initiation of HAART and suppression of the virus load, precursor cells start to expand and replenish the main naive compartment, in agreement with the experimental picture (Figure 10.1B). Expansion of the precursor compartment cannot happen instantaneously. One can roughly estimate the lifetime of precursor cells from the delay in the naive cell return (Kelleher *et al.*, 1996; Kelleher, 1997) and obtain a value of about 4 weeks. Below, we shall confirm this estimate from a different kind of argument.

Postulating the existence of a second source of T-cells is also necessary to account for the stability of the HIV infection. HIV infection is notoriously robust. After being significantly suppressed by monotherapy for a few weeks, a drug-resistant virus appears and the virus load rebounds to the predrug level (Danner *et al.*, 1995). Clearly, however, any model system based on HIV-induced replenishment alone cannot be stable against a sufficiently strong perturbation. The steady states in the models discussed in the last section are stable with respect to small perturbations only ("metastable"). In fact, these models have two metastable states: with and without virus (points #1 and #2 in Figure 10.12B). The zero-virus state is also metastable since, in the absence of virus, there are no permissive cells and a small initial amount of virus added to a system will be cleared. If the system is initially in the steady state with virus, a potent drug will switch it to the state without virus (arrow in Figure 10.12B). The virus will not return even after the treatment is lifted. Adding a second, virus-independent source of host cells to the model allows us

A.

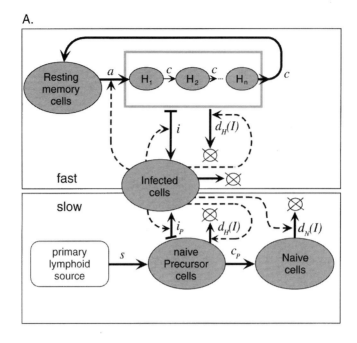

B.

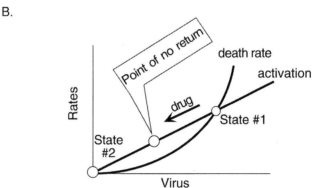

FIGURE 10.12   **A**. A model with a second, virus-independent source of host cells (precursors of naive CD4 cells). Both precursor and mature naive cells are killed by virus indirectly, although mature cells are killed much more slowly. The model is introduced to explain the long-term features of HIV infection, with and without HAART. Parameter values of the second source, $d_N = 10^{-3}$/day, $c_P = ¼$ weeks, are chosen to fit the typical HAART kinetics of naive cells shown in Figures 10.1B and 10.2B is $s = 7 \times 10^7$/day (Haase, 1999). **B**. In the absence of a second source (model in Figure 10.11A), the system has two steady states, state 1 with virus and state 2 without, both states stable against a small perturbation. A potent drug causes the system to switch from state 1 to state 2.

to destabilize the extra state #2 and, thereby, to explain the drug resistant virus rebound observed in experiment (Coffin, 1995; Danner *et al.*, 1995; Wei *et al.*, 1995).

Figure 10.13 shows the simulated kinetics of monotherapy with a single replication inhibitor for the model shown in Figure 10.12A. The efficiency of infection of naive precursors is chosen to resemble the results obtained by Wei *et al.*

(1995). Following this experiment, we assume that a drug-resistant variant of virus starts to dominate during the second week of therapy and has a fitness similar to the fitness of wild-type virus in the absence of drug. (Here, we consider only the kinetic aspect of drug resistance, as opposed to the origin or evolution of drug-resistant strains.) The lifespan of naive precursor cells has to be, for the best fit, about 2 weeks

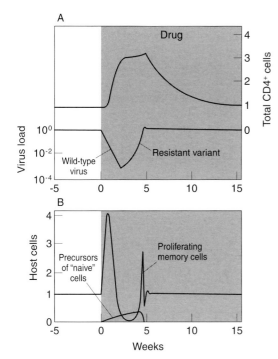

FIGURE 10.13 Simulated monotherapy kinetics for the model shown in Figure 10.12A. Drug-resistant virus is assumed to appear at week 2. The infection efficiency of naive precursors, $i_p = 0.9 \times 10^{-9}$, is chosen to simulate the exponential growth rate of the resistant virus in a typical experiment, approximately 0.5 of the virus decline rate (Figure 10.2A). Values of other parameters are given in legends for Figures 10.11 and 10.12. Steady-state values obtained from equations in Table 10.1: $P_0 = 6 \times 10^7$ cells, $N_0 = 2 \times 10^9$ cells. Note that the contribution of naive precursors to steady-state host cells is small, $i_p P_0 / d_1 = 0 \times 1$, and that the steady-state value of naive cells N_0 is very low, corresponding to 4 cells/µl blood.

or more, which agrees with the estimate above. A qualitative interpretation of the complex kinetics shown for different cell types in Figure 10.13 is as follows.

Initially, the system is in steady state. Almost all precursor cells coming from the primary source are indirectly killed by the virus. Since the source is weak, only a small proportion of infected cells are due to infection of naive precursor cells. Then a drug combination suppresses virus replication. The virus starts to decay and naive precursors, correspondingly, start to expand slowly (Figure 10.13A,B). Proliferating cells expand transiently too, but then convert into the non-permissive resting state. At week 2, when the drug-resistant vari-

ant starts to dominate, naive precursors dominate the host-cell compartment. The most critical point here is that, in order for virus to grow rapidly, the naive precursors have to exceed, by a factor of 1.5–2 (Wei et al., 1995), the level at which the virus replication rate is equal to its clearance rate, at about 2 weeks. (The replication rate is, of course, proportional to the host cell number when infection spreads by ambient virions, as discussed earlier.) For this to happen, we had to assume that naive precursors, on a per-cell basis, are infected by virus a few times more often than proliferating memory cells (see caption to Figure 10.13).

THE IMMUNE RESPONSE AGAINST HIV: FRATRICIDE

"Normal" CTL Responses: The LCMV Model

Most viral infections are cleared from the body in a matter of days. Despite considerable experimental and mathematical effort, the full scenario of the cytotoxic response is not known. One of the best studied animal models is lymphocytic choriomeningitis virus (LCMV). LCMV strain Armstrong causes an acute infection in mice that either kills the animal or is cleared within a week (Christensen et al., 1994; Matloubian et al., 1994; Moskophidis et al., 1995; Tishon et al., 1995). MHC-II, IFN-γ genes and CD4+ T-cells are not required for the rapid virus clearance, which implies a modest role of antibodies and helper cells in this infection (Christensen et al., 1994; Matloubian et al., 1994; Tishon et al., 1995). The virus is not lytic. Consequently, LCMV-Arm is considered a good animal model for studies of a CTL-mediated, helper-free immune response. Some other strains of LCMV (Clone 13, Docile) cause persistent infection in mice and (Docile) exhaustion of CTL effector cells (Moskophidis et al., 1993).

The current paradigm of functional immune response to a "normal" viral infection, such as LCMV-Arm, includes three phases: expansion of effector cells, their death and the memory phase (Asano and Ahmed, 1996). As a correlation between the time course of CTL response (Gallimore *et al.*, 1998; Murali-Krishna *et al.*, 1998) and the antigen level in LCMV-infected mice (Murali-Krishna *et al.*, 1998) shows, CTL effector cells divide each time they are re-activated with antigen. The critical point is that effector cells cannot die until the antigen falls below an extremely low level, which means the existence of an additional, "stand-by" phase of the immune response. The general argument supporting this point is that, otherwise, any virus infection would develop into a chronic infection, if the antigen dynamics were transiently slowed down by some factor, such as depletion of host cells or a non-specific immune response. (Experimentally, a standby phase is also evident in HIV-specific CTL effector cells under HAART (Kalams, 1999). After the virus is suppressed by orders of magnitude, effector cells stay at the predrug level for months.) A schematic representation of a "normal" primary CTL response including the standby phase is shown in Figure 10.14A.

The HIV-specific Immune Response: Some Facts

1. HIV-specific immune effector cells are present throughout the asymptomatic phase of HIV infection (Yang *et al.*, 1996, 1997a,b; Lieberman *et al.*, 1997; Rosenberg *et al.*, 1997). As shown by the use of tetramer assay data, their frequency is much lower in a typical patient (0.1–1%; Altman *et al.*, 1996; Ogg *et al.*, 1998) than during an acute virus infection (50–70%; Murali-Krishna *et al.*, 1998). In general, all effector cell types can be present, although CD4$^+$ helper T-cells are not detectable in individuals with a high virus load (Rosenberg *et al.*, 1997).
2. When isolated from patients, cells appear to be functional *in vitro*. They show normal cytolytic activity and secrete the expected cytokines (Klenerman *et al.*, 1996; Yang *et al.*, 1996, 1997a,b; Rosenberg *et al.*, 1997).
3. In patients undergoing HAART, the number of CTL effector cells (both direct effector cells and their precursors) stays almost constant

for a period of time ranging, among patients, from 1 to 12 months. Then their numbers drop rapidly (Kalams, 1999; Figure 10.14C).

4. The rate of genetic evolution in some recognized epitopes and Ab-binding regions is significantly higher than in antigenically inactive regions of HIV genome (Wolfs *et al.*, 1991; Holmes *et al.*, 1992; Burns and Desrosiers, 1994; Wolinsky *et al.*, 1996; Goulder *et al.*, 1997; Brander *et al.*, 1998), implying selective pressure from the immune response on the virus population. However, the shortest genetic turnover half-life observed for antigenically important bases is more than 1 month (Borrow *et al.*, 1997), indicating that such pressure is not very strong (Wolinsky *et al.*, 1996; Brander *et al.*, 1998).

Hypothesis: HIV Causes Death of Effector Cells

Although the immune response clears most viral infections in animals, some viruses manage to survive the ordeal and establish a persistent infection (Janeway and Travers, 1996). A virus can escape immune eradication using a variety of tricks.

- It may become invisible to the immune system by establishing a latent state, as with herpesviruses.
- It may replicate in sites not accessible to effector cells or antibodies.
- As with many retroviruses, it may infect individuals at a sufficiently young age to induce tolerance.
- It may escape recognition by mutation.
- It may express a product or products that impair or counteract effector cell function.
- It may prevent or impair expansion of immune cells, as with some strains of LCMV.
- It may kill effector cells.

Which of these methods is the central tool of the HIV persistence? Although the answer is unknown, we can, at least, try to guess by eliminating explanations that do not work. HIV infection, as a whole, is not latent: many infected cells in asymptomatic HIV infection actively express virus RNA and virus proteins.

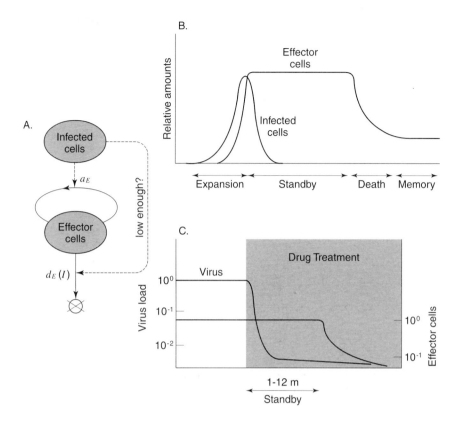

FIGURE 10.14 **A**. A model for cytotoxic immune response to an acute viral infection in which helper cells are not very important (acute LCMV infection; Christensen *et al.*, 1994; Matloubian *et al.*, 1994; Tishon *et al.*, 1995). Effector cells are allowed to die only if the antigen level is extremely low. **B**. Schematic time course of unimpaired cytotoxic response to an acute primary virus infection. **C**. Schematic time dependence of HIV-specific CTL effector cell frequency after onset of HAART (Kalams, 1999).

HIV does not hide: most infected cells are located in the lymphoid tissue and should be exposed to HIV-specific effector cells proliferating in the same tissue (Haase, 1999). The antigenic escape model does not work out: the rate of antigenic escape (Borrow *et al.*, 1997) is too slow. The half-time of genetic turnover at the critical base would have to be as short as the infected cell lifetime: 1–2 days (we will return to this point later). HIV does employ a variety of mechanisms to impair function of different effector cell types, such as the internalization of MHC-I molecules from the surface of infected cells (Le Gall *et al.*, 1997). Extensive glycosylation of the envelope protein may effectively impair binding of virions by antibodies (Reitter *et al.*, 1998), which is equivalent to rendering B-cells ineffective. However, if the impaired function of effector cells alone were the cause of persistence, the number of effector cells during the asymptomatic phase would be unusually high rather than low, as observed (Altman *et al.*, 1996; Ogg *et al.*, 1998). Indeed, effector cells would expand until the antigen is removed, as they do in the case of LCMV-Arm infection (Murali-Krishna *et al.*, 1998). Recent experiments on CD8[+] cell depletion in SIV infected animals show that the viremia level is efficiently controlled by these cells (Schmitz *et al.*, 1999).

A possibility consistent with the low level of effector cells is that HIV limits the expansion of immune cells by interfering with their proliferation. (To our knowledge, a mechanism that could prevent proliferation of CD8[+] T-cells has not been reported for HIV.) However, the impaired proliferation alone cannot explain the stable level of effector cells in the asymptomatic infection: since the virus level fluctuates randomly throughout asymptomatic infection, the effector cell level would slightly move upward each time the virus load accidentally drops.

The remaining possibility (also not reported) is that HIV selectively kills HIV-specific effector cells. According to this idea, the killing compensates the antigenically driven proliferation and results in a steady state for effector cells. Below we find out the simplest model of the killing that agrees with the observed drug kinetics (Kalams, 1999). We consider only a single type of effector cell, CTL, in the absence of helper cells and at a fixed rate of infectious virion clearance. To complement the two previous sections devoted to the host cell kinetics, here we assume that either the infected cell number or the host cell number is fixed, corresponding to the host-cell-controlled and the immune-cell-controlled models (Figure 10.7) respectively. The model of immune response against HIV, obtained in this section, can be generalized for the case of four effector cell types (CTL, Th1, Th2, B cells), which we will describe elsewhere.

Models with a Fixed Virus Load: Fratricide

The simplest model including virus-induced proliferation and virus-induced death of effector cells is shown in Figure 10.15A. Both proliferation and death rates are determined, from all possible cell compartments, by the number of infected cells only. Since the virus load is assumed to be fixed (to keep the balance between depletion and proliferation of host cells), the activation and death rates are fixed as well. Therefore, such a model cannot have a steady state: the activation of effector cells is not expected to compensate for their death. To obtain a steady state, we must change the model, assuming that either the death rate or the activation rate depends on the number of pre-existing effector cells. In this case, by adjust-

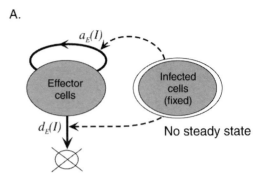

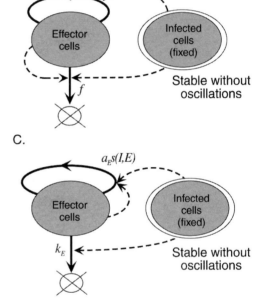

FIGURE 10.15 Three models for CTL effector cell kinetics in HIV infection. The number of infected cells is fixed. **A**. Killing of individual effector cells by virus. **B**. Virus-induced fratricide among effector cells. **C**. Virus-induced collective sabotage of antigenic activation by effector cells combined with killing of individual effector cells by virus.

ing the effector cell number, we can achieve an exact balance between proliferation and death. For example, the activation rate is proportional only to the virus load (remember that both infected cells and HIV-specific effector cells are very dilute; Altman *et al.*, 1996; Haase *et al.*, 1996), while the death rate is proportional to both the effector cell number and the virus load (Figure 10.15B). Biologically, this model implies that the death is a collective effect, i.e. that HIV causes fratricide of effector cells. In another version, the activation rate negatively correlates with the number of effector cells, and the death rate depends on the virus load only (Figure 10.15C, Table 10.1). This means that effector cells are killed individually but interfere with each other's proliferation. This "collective sabotage" model is more complex than the "fratricide" model since it assumes the existence of two separate mechanisms specific for HIV: first, killing HIV-specific effector cells and second, making them sabotage each other's proliferation. Therefore, we focus on the simpler model, although the second option has to be kept in mind.

The fratricide model (Figure 10.15B) meets all the important selection criteria used in previous sections for the host-cell kinetics models. First, it predicts the existence of a stable steady state, since it has a negative feedback curtailing fluctuations of the effector cell number. Second, the steady state is also non-oscillatory, as observed after transient depletion of CD8$^+$ cells in SIV infected animals (Schmitz et al., 1999). A kinetic system must have, at least, two time-dependent variables to be able to oscillate, and the fratricide model has only one – the effector cell number. Third, the fratricide model predicts the trivial, but correct, drug kinetics: effector cells are predicted to remain at the predrug level after

the drug is administered (Figure 10.14; Kalams, 1999). Drug treatment can be introduced into the model by assuming that the virus load is no longer fixed but depends on time: initially it is constant but then, at some time point, it starts to decline rapidly (Figure 10.16). The number of effector cells stays at the predrug level since, according to the model, both the proliferation and the killing of effector cells almost cease when the virus load declines to very low levels. The virus decrease slows down in 1–2 weeks to that of the long-lived virus compartment (Cavert et al., 1997; Perelson et al., 1997). Only after a few months does the virus reach a very low level at which effector cells are allowed to die (Figure 10.14). This delay in virus clearance allows us to visualize the standby phase of the CTL response.

Models with a Fixed Number of Host Cells

Consider now the number of infected cells (proportional to the virus load) as a time-dependent variable and assume a constant number of host cells. The dominant mechanism of infected cell killing *in vivo* is still unknown. It may be either direct killing by the virus itself (Terai et al., 1991; Corbeil et al., 1996) or recognition and killing by CTLs (Yang et al., 1996, 1997a). If the virus is directly responsible for cell killing, as some workers believe (Feinberg and McLean, 1997), its kinetics are independent of CTL and we again arrive at the host-cell-controlled models we considered above. Therefore, in this section, we assume a CTL-mediated mechanism of infected-cell clearance. Analysis of this case is more complex than analysis of the fixed-antigen models, and we list only the final results here. As it turns out, the simplest model satisfying the usual requirements is, again, the fratricide model, except with virus kinetics included (Figure 10.17). However, the existence of an additional variable (the virus load) imposes severe restrictions on the model parameters to ensure stability and absence of oscillations in the steady state. First, either the

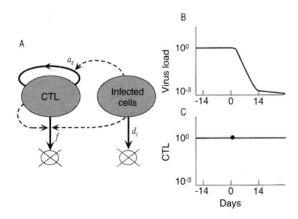

FIGURE 10.16 **A**. The fratricide model of effector cell kinetics in the presence of HAART. **B**. Decay of the virus load after onset of HAART assumed in the model in panel **A**. **C**. Effector-cell kinetics predicted by the model.

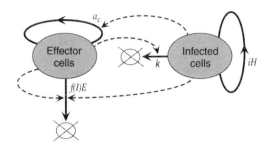

FIGURE 10.17 A model of CTL response when the host-cell number (rather than the virus load) is fixed. The model predicts a stable, non-oscillatory steady state for effector cells, but only if the effector cell population turns over extremely rapidly, as given by $a_E I > d_1$

death rate of effector cells must increase with the virus load more slowly, or the activation rate must be faster than the first power. Second, the lifespan of all effector cells in the steady state must be shorter than the lifespan of infected cells (i.e. about 1 day). The latter condition is improbable for the asymptomatic phase of the HIV infection. Such a high death rate of effector cells has been observed only in acute infection in one individual (Borrow *et al.*, 1997), when both the effector cell number and the virus load are 1–2 orders of magnitude higher. Since, in the model discussed, the death rate of effector cells is proportional to both these variables, the prediction contradicts the data.

Correspondence between the immune-cell-controlled and host-cell-controlled models is not restricted to the fratricide model. In the next level of complexity, we found a version of "collective sabotage model" (Figure 10.15C) with the virus kinetic part attached, as in Figure 10.17 (diagram not shown). It has similar (unrealistic) requirements for the lifetime of effector cells as the fratricide model. A particular case of this model, when the death rate of effector cells is constant, was suggested earlier by De Boer and Perelson (1998) who noted that effector cell competition for targets (infected cells) can interfere with each other's activation. However, as we discussed, the low frequency of HIV-specific CTL effector cells in lymphoid tissue of HIV-infected individuals, 0.1–1%, implies that the competition effect is rather weak. One would need some special virus-induced effect ("collective sabotage") to have the

decrease of the activation rate with the effector cell number. In addition, the model (De Boer and Perelson, 1998) predicts oscillations in monotherapy trials and a fast decay of effector cells under HAART. Neither effect is observed in a typical experiment.

Therefore, we do not consider the immune-cell-controlled models beyond this point.

EFFECTOR AND HOST CELL KINETICS TOGETHER

Following our plan, we now consider the host cell model (Figure 10.11A) and the HIV-specific effector model (Figure 10.15B) together. The resulting full model is presented in Figure 10.18. In principle, we must reapply our verification tests (stability near the steady state, drug kinetics with and without drug-resistance) to the full model, and test for possible changes in predictions as compared to its separate parts. Such changes could be expected due to interaction between fluctuations in virus load and in effector cell number. Accidentally, this is not necessary for the listed tests, since, on the one hand, a change in the virus load in this particular model does not perturb the number of effector cells provided the latter is initially at the steady-state level (see last section). On the other hand, the constant level of effector cells ensures a constant clearance rate of infected cells. In other words, the two approximations we made (assuming that the virus load is constant when considering the HIV-specific block and that the virus clearance rate is constant when considering the host cell block) happen to be exact statements. This peculiar feature of the fratricide model of the HIV-specific immune response simplifies our analysis.

Although the above speculations and the model shown in Figure 10.18 imply that the clearance of infected cells is mediated by cytotoxic immune cells (CTL), the same conclusions follow if a part (or almost all) of infected cell clearance does not depend on the effector cell number and is constant.

The virus and the effector cell kinetics, however, are not necessarily independent when the HIV population is not initially in steady state.

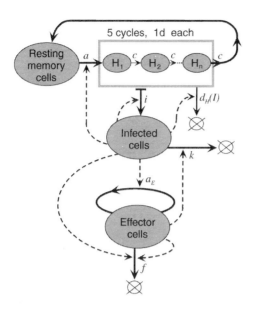

FIGURE 10.18 A full model of HIV infection (including host cells, infected cells and HIV-specific effector cells) obtained by connecting diagrams in Figures 10.15B and 10.11A. The second, "slow" compartment of host cells (bottom part of Figure 10.12A) is implied but not shown in the figure.

For example, in the acute phase of an HIV infection, the frequency of effector cells starts from a very small value and then increases by many orders of magnitude, and the rate of this change, according to the fratricide model (Figures 10.4–10.6), should depend on the virus load. Therefore, in this case, one would have to calculate the time dependence of the virus and the effector cell number, solving the entire set of equations together.

POSSIBLE TESTS AND CONCLUSION

General Conclusions

Let us review briefly the logic of the previous sections. To explain some features of short-term T-cell kinetics, such as the similarity between CD4+ and CD8+ cell behavior and the association with the "memory" phenotype, we postulated that HIV induces replenishment of its host cells and kills indirectly, independent of CD4 versus CD8 subset. Pressured by the absence of oscilla-

tions in the resistant virus rebound, we had to assume that the virus infects and kills T-cells in the short-lived proliferation phase. The stability of the steady state against small perturbations suggested that either the killing of T-cells or their activation is not proportional to the number of infected cells, in spite of the fact that the latter are very dilute in the infected tissue. The very rapid BrdU delabeling in SIV-infected animals implied that an HIV-activated T-cell expands into a clone of 30–60 cells without additional activation, which fact also explains the effect of premature saturation of T-cells under the antiviral therapy. The model we obtained portrays HIV as a wasteful farmer who breeds rabbits (host cells) in enormous quantities, eats a small part and destroys the rest.

The model was incomplete since, if the only source of host cells were virus-induced activation, an infection would never return after a transient antiviral treatment, as it actually does. We found that a relatively weak, but HIV-independent, source of T-cells, such as a homeostatic replenishment source, would suffice to ensure the virus rebound. The existence of an additional source also explained why CD4+ T-cell count declines slowly, why the slow component of T-cell kinetics is associated with the "naive" phenotype of T-cells, and the possible advantage of using two separate entry coreceptors. Assuming that we are biologically close to the original primate host, the coreceptor restriction prevents a premature death of the host caused by an early depletion of the TCR repertoire. Eventually, the depletion does occur, causing AIDS.

Our starting point in the immunological part of modeling was the understanding that effector cells in other, "normal" acute infections do not die until after the antigen is cleared. To explain why HIV-specific effector cells persist in the asymptomatic phase at a small but stable level, we needed to postulate, in the simplest version, fratricide among HIV-specific effector cells induced by the virus. The fratricide model agreed with the observed HIV-specific effector cell kinetics in drug therapy, which shows a virtual absence of changes in their frequency for a few months after the virus has been significantly suppressed.

An important conclusion of this work is that it

is possible to explain the slow decline in CD4 cell count ending in AIDS within a closed model without postulating that any model parameters depend on time. This implies that the short-term and the long-term kinetic properties of HIV infection are, in general, not related. In particular, the (absence of) correlation between the clearance rate of infected cells and the CD4 count in individual patients (Ho *et al.*, 1995; Wei *et al.*, 1995) does not contain any information about the clearance mechanism of infected cells ("CTL versus virus-induced death"), as sometimes suggested (Feinberg and McLean, 1997). For example, in the model shown in Figure 10.18, the clearance rate, d_I is constant because there is a constant number of CTL effector cells; nevertheless, CD4 cell count declines slowly as a result of depletion in naive cells.

Scenarios of Fratricide

One could imagine three different versions of fratricide (Figure 10.19). A chemical secreted by infected cells can (1) force an effector cell to secrete a factor that causes death of other effector cells; (2) become a death signal for an effector cell when combined with another factor normally secreted by effector cells; (3) sensitize an effector cell so that a factor normally secreted by other effector cells becomes lethal afterwards. Cases 2 and 3 differ in their timing. In addition, all interactions can be either by contact or via soluble factors.

The third, sensitization, version of fratricide may be an explanation of the relatively slow rate of antigenic escape observed in HIV infection. The average time of genetic turnover is at least 20 times longer than the lifespan of an average infected cell (Borrow *et al.*, 1997). If the immune response worked normally and CTLs were the factor that cleared infected cells, both times would have to be the same. A possible explanation is that CTLs kill only 5% of infected cells; the rest die from virus-derived factors (McLean, 1999). Although we cannot exclude this possibility, here is another explanation that applies even if 100% of the infected cell clearance is due to CTLs.

Let effector cells be engaged in fratricide that is serotype-specific with respect to the victim-

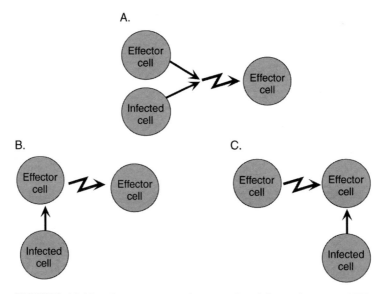

FIGURE 10.19 Three scenarios of virus-induced fratricide among HIV-specific effector cells. **A**. "Adding". An effector cell dies upon receiving two independent signals, from an effector cell and an infected cell. **B**. "Arming". An effector cell is induced by a virus-derived factor to secrete a death signal for other effector cells. **C**. "Sensitizing". Upon receiving a sensitization signal from an infected cell, an effector cell becomes sensitive to an otherwise harmless factor secreted by other effector cells.

ized CTLs: each virus serotype can induce death only of its cognate CTL clone (whether the killing CTL is from the cognate clone or any other clone is irrelevant). Suppose, a CTL cell attached itself to a cell infected with the cognate virus serotype. Before being lysed, the infected cell secretes a signal, that makes the cytotoxic cell a target for the death signals from other CTLs (Figure 10.20). Consider now two CTL clones and their cognate virus serotypes: serotype 1, which we call "wild type", and serotype 2, which we call "antigenic escape mutant" (Figure 10.20). The two subsystems, each consisting of a virus serotype and its cognate CTL clone, will be independent, since all interactions between cells are antigen-specific. Both subsystems will have independent steady states, as discussed above for the system consisting of a single serotype and cognate CTL clone. The rate of antigenic escape will be exactly zero. One obtains a finite rate of antigenic escape in this model by introducing any weak interaction between the two subsystems (such as a weak cross-serotype fratricide).

In SIV-infected animals, effector cell death induced by the virus could be detected by double-labeling cells with SIV-specific soluble tetramer and BrdU. This would allow a test of whether the effector cell turnover slows down under HAART, as the fratricide model predicts.

The collective nature of fratricide could be tested *in vitro* only if a steady-state system of effector cells could be set up, adding infected cells afterwards. This might be rather challenging, since it would require fine tuning of the culture conditions. Yielding here to experts on immunology, we only note that, since the mechanism of hypothetical effect is likely to be specific for HIV and novel, testing the existence of this effect first would be more efficient than testing for different biochemical mechanisms. If and after the existence of fratricide is confirmed, the general scenarios of fratricide as classified above could be segregated, allowing for a more detailed research.

So far, the only mechanism of CTL fratricide in the literature is the sensitization of CTLs by extracellular HIV peptides by binding to MHC-I on the CTL cell surface (Siliciano *et al.*, 1988; Walden and Eisen, 1990; Su *et al.*, 1993). There are a few reasons why we think that this is not the mechanism of fratricide for HIV. First, there are too many ($\approx 10^2 / 10$ kb genome) different types of peptide binding a particular MHC-I subtype set, and only several CTL clones are present at any time *in vivo*. Only a tiny fraction of MHC-I–peptide complexes on the cell surface would be recognized by these clones. (The legitimate, intracellular loading of MHC-I molecules has a mechanism that ensures that a few

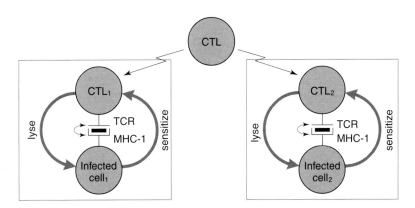

FIGURE 10.20 A hypothetical mechanism of fratricide among HIV-specific effector cells. This model is the antigen-specific version of sensitization scenario (Figure 10.19C) and may be an explanation of the very slow antigenic escape in HIV infection (Wolinsky *et al.*, 1996; Borrow *et al.*, 1997; Brander *et al.*, 1998). Two or more subsystems, each consisting of a virus serotype and a CTL clone that recognizes it, coexist and maintain independent steady states for an indefinite period of time.

peptides are much more efficiently presented than the rest.) Second, even when a single peptide is used for pre-pulsing of a CTL, the concentration necessary for recognition by other CTLs is unreasonable large. Finally, even assuming for a moment that the prepulsing is much higher than it is, such a fratricide would not be specific for HIV infection.*

A Sea of Models

From a vast, uncharted ocean of kinetic models of HIV infection, we have fetched but a glassful. We restricted our search to models that did not postulate an impaired homeostatic mechanism, nor did they include interaction between host cells and HIV-specific effector cells or interaction among different infected cells. Even within this, following our analogy, relatively small bay, we considered only the two particular cases when the infected cell number in steady state can be found either from the host-cell or from the immune-cell balance condition alone. From a few dozen models we scanned within this group, we discussed 13, including the simplest model that agrees with experiment. We hope to expand our search and explore other families of models (especially those with interaction between infected cells) elsewhere.

The motive behind our search for a simplest model of HIV infection within a certain class is not that we hope that the first model thus found is likely to be the correct model of HIV infection. This simplistic interpretation of the "Occam's razor" principle certainly does not work for biological systems. Rather, we believe that, given that each finite group of facts can be matched by an infinite number of models, the search on the stormy sea of models is dangerous and has to be carefully organized. On the one hand, eliminating simplistic explanations which can be shown

to disagree with already well-established facts by a rigorous analytic procedure saves a lot of experimental work on testing them *in vitro* or *in vivo*. On the other hand, a relatively simple model that agrees with existing data and has striking testable predictions is easier to refute in follow-up experiments than a complex model with 100 parameters, which can be adjusted to "explain" any data whatsoever. After a simpler model is proven to be wrong (but not sooner), one can move on to the next level of complexity, and so on. We hope to find support for this, practically very important philosophy among mathematicians interested in HIV and among experimentalists, who would be daring and practical enough to risk their time and precious grant money to test someone else's predictions.

ACKNOWLEDGMENTS

We would like to thank our colleagues for providing information prior to publication: D.D. Ho (SIV clearance rate), A.T. Haase (HIV distribution in tissues) and S.A. Kalams (HIV immunology). We also thank N.L. Letvin, H.N. Eisen, J. Lieberman, S.A. Kalams, D.D. Ho, D.S. Dimitrov, R.P. Johnson, O.O. Yang and Å Öhagen, for interesting discussions, and E. Goldberg for useful suggestions on organizing this chapter. This work was supported by research grant R35 CA 44385 of the National Cancer Institute. J.M.C. was a Research Professor of the American Cancer Society.

APPENDIX

Notation

We show the different cell types (compartments) as shaded ovals in the figures. The current state of each compartment is characterized by its size only. Cell flux to/from a compartment is shown by solid arrows. A dashed arrow from an oval to a solid arrow indicates that the compartment size controls the intensity of the cell flux. By default, a cell flux is linearly proportional to the

---

*Recent study of CTL kinetics using MHC-I–peptide–tetramer complexes (Ogg, 1999) showed existence of a non-exponential decay of HIV-specific CD8⁺ cells after onset of drug therapy, on both a short (days) and a long time scale (months). As will be discussed elsewhere, the decay in CTL can be accounted for in the model (Figure 10.20) by assuming that sensitization of a CTL is not transient but persists beyond its disengagement from the infected cell.

size of the compartment it is coming from and to the size of the controlling compartment, if any. Letters near solid arrows are the proportionality coefficients, which represent parameters of the model. A general, non-linear form of the dependence of a cell flux on the controlling compartment size is implied when a parameter is shown as a function with an argument in parentheses. For example, $d_H(I)$ means that the death rate of host cells need not be linearly proportional to the number of infected cells. A solid arrow separated by a perpendicular bar from the compartment it comes from means that the corresponding outcoming cell flux can be neglected or incorporated into other outcoming flux (e.g. the direct killing flux in Figure 10.8C can be included in the indirect killing flux).

A diagram defined according to the above rules can be unambiguously translated into a system of mast er equations (Table 10.1). The compartment sizes, denoted by the initial letters of the corresponding compartment names in the diagram (I for Infected cells, R for Resting cells, H for Host cells, etc.), represent time-dependent variables. A dot above a variable denotes the time derivative. The remaining mathematical notations are standard or self-explanatory. We applied a set of mathematical procedures/tests, described below, to the 13 models discussed in the main text. Tables 10.1 and 10.2 show their master equations and summarize results we obtained on stability and drug kinetics. We refer to the separate models using figure labels of their diagrams.

Steady State

The steady-state values of variables are found from the condition that the right-hand side of each equation is zero (the balance condition). If the resulting system of algebraic equations has a solution (or solutions), the system has an exact steady state (or states). We denote steady-state values by the subscript 0 on the corresponding variables (Table 10.1). If the balance condition is met for some variables only, and the remaining variables, as found from the corresponding master equations, change relatively slowly (as

compared to rates at which the first set of variables would change if far out of balance), we refer to a "quasisteady state".

Stability Against Small Perturbations

Given the facts of HIV infection discussed in the main text, the steady state we are interested in must be stable and non-oscillatory, i.e. it must gradually resume after any transient perturbation, without having the variables oscillate around their steady-state levels. To test for stability against perturbations of small magnitude, we obtain the eigenvalue equation for the linearized master equation set. We illustrate this procedure on the example of model 9C (Table 10.1). We express each variable near its steady state value, as given by $R = R_0 (1 + x)$, $H = H_0 (1 + y)$, $I = I_0 (1 + z)$, where $x, y, z \ll 1$. Replacing the right-hand side of each master equation in Table 10.1 by its linear expansion in x, y, z, we rewrite the master equation set in the matrix form, $\vec{r} = \hat{D}\vec{r}$, where vector $\vec{r} \equiv (x, y, z)$, and $\hat{D}$ is the dynamic matrix (not shown). The condition $\det (\hat{D} - \lambda) = 0$ represents the equation for eigenvalues $\lambda_1, \lambda_2, \lambda_3$ of the dynamic matrix $\hat{D}$ (fourth column of Table 10.1). The steady state is stable and non-oscillatory if all (in this example, three) eigenvalues are real and negative: $\mathrm{Im}\,(\lambda_{1,2,3}) = 0$, $\lambda_{1,2,3} < 0$ (first column of Table 10.2). For those models in Table 10.1 for which obtaining the analytic form of the eigenvalue equation was difficult, we assessed the stability/absence of oscillations numerically, simulating a small perturbation around the steady state.

Drug Kinetics

To simulate potent antiviral therapy (HAART), we assumed that $i = 0$ or $i = i_p = 0$ starting at some time point following infection. We used t_{infec} and t in Tables 10.1 and 10.2 to denote the time period elapsed since the transmission event and since the onset of therapy, respectively. To simulate the emergence of a drug-resistant virus at week 2, we restored the predrug value(s) of the infectivity parameter(s). Most

TABLE 10.1 Summary of models: master equations and eigenvalues

Figure	Master equations	Steady state	Eigenvalue equation
3	$\dot{H} = s - H(iI + d_H)$ $\dot{I} = I(iH - d_I)$	$H_0 = d_I/i$ $I_0 = \dfrac{d_H}{i}\left(\dfrac{si}{d_I d_H} - 1\right)$	$\lambda^2 + d_H(\rho - 2)\lambda + d_I d_H(\rho - 1) = 0$ where $\rho \equiv \dfrac{is}{d_I d_H}$
5	$\dot{H}_1 = H_1(a_1 - iI)$ $\dot{H}_2 = H_2(a_2 - iI)$ $\dot{I} = I[(i_1 H_1 + i_2 H_2 - d_I)]$ $i_1 \gg i_2, a_1 \gg a_2, a_1/i_1 > a_2/i_2$	Quasi-steady state: $I_0 \simeq a_1/i_1, H_1 \simeq d_I/i_1$ $H_2(t) = const \cdot exp[-(a_1 i_2/i_1 - a_2)t]$	Ignoring H_2: $\lambda^2 + a_1 d_I = 0$
8A	$\dot{H} = H[(a - i)I - d_H]$ $\dot{I} = I(iH - d_I)$	First: $H_0 = d_I/i, I_0 = d_H/(a - i)$ Second: $H_0 = I_0 = 0$	$\lambda^2 - d_I d_H = 0$
8C	$\dot{H} = H[aI - d_H(I)]$ $\dot{I} = I(iH - d_I)$	First: $H_0 = d_I/i, I_0 : I_0 = d_H(I_0)/a$ Second: $H_0 = I_0 = 0$	$\lambda^2 + d_I d_H(I_0)(\eta - 1) = 0$ $\eta \equiv \dfrac{d \ln d_H(I_0)}{d \ln I_0}$
9C	$\dot{R} = 2cH - aRI$ $\dot{H} = aRI - H[c + d_H(I)]$ $\dot{I} = I(iH - d_I)$	First: $H_0 = d_I/i, I : d_H(I) = c, R_0 = 2cd_I/ai I_0$ Second: $H_0 = 0, I_0 = 0, R_0 = const$	$\lambda^3 + (aI_0 + 2c)\lambda^2 + (\eta - 2)cd_I\lambda$ $+ \eta a I_0 c d_I = 0$
11A	$\dot{R} = 2cH_n - aRI$ $\dot{H}_1 = aRI - H_1[c + d_H(I)]$ $\dot{H}_2 = 2cH_1 - H_2[c + d_H(I)]$ $\vdots$ $\dot{H}_n = 2cH_{n-1} - H_n[c + d_H(I)]$ $\dot{I} = I[(H_1 + H_2 + \ldots + H_n) - d_I]$	First: $H_{10} = H_{20} = \ldots = H_{m0} = \dfrac{d_I}{ni}$ $I_0 : d_H(I_0) = c, R_0 = \dfrac{2cd_I}{nai I_0}$ Second: $H_{10} = \ldots = H_{m0} = 0$ $I_0 = 0, R_0 = const$	
12	$\dot{R}, \dot{H}_1, \ldots, H_n$ - see model 11A $\dot{I} = I[i(H_1 + H_2 + \ldots + H_n) + i_P P$ $- d_I]$ $\dot{P} = s - P[d_H(I) + c_P]$ $\dot{N} = c_P P - N d_N(I)$	R_0, I_0 - see model 11A; $P_0 = \dfrac{s}{c_P + c}$ $H_{10} = \ldots = H_{m0} = (1/ni)(d_I - i_P P_0)$ Exact s.s.: $N_0 = \dfrac{s}{d_N(1+c/c_P)}$ Quasi s.s. at $N \gg N_0$: $N(t_{infec}) = N_{normal} exp[-t_{infec} d_N(I_0)]$	

14	$\dot{E} = E[a_E I - d_E(I)]$, where $d_E(I) \neq 0$ only if $I < I_{\text{thresh}}$ I – arbitrary kinetics	No steady state	
15A	$\dot{E} = E[a_E(I) - d_E(I)]$, $I = \text{const}$	No steady state	
15B	$\dot{E} = EI(a_E - fE)$, $I = \text{const} = I_0$	$E_0 = a_E/f$	$\lambda + a_E I_0 = 0$
15C	$\dot{E} = EI(a_E s(I,E) - k_E)$ where $s(I,E) \leq 1$, $s(0,E) = s(I,0) = 1$ $I = \text{const} = I_0$	$E_0 : s(I_0, E_0) = k_E/a_E$	$\lambda - aI_0^2 E_0^2 \dfrac{\partial s(I_0, E_0)}{\partial E_0} = 0$
17	$\dot{E} = E[a_E I - f(I)E]$ $\dot{I} = I(iH - kE)$ $H = \text{const} = H_0$	$E_0 = iH_0/k$ $I_0 : I_0 = \dfrac{iH_0}{ka_E} f(I_0)$	$\lambda^2 + a_E I_0 \lambda + iH_0 a_{E_0} I_0(1 - \xi) = 0$ $\xi \equiv \dfrac{d \ln f(I_0)}{d \ln I_0}$
18	$\dot{R}, \dot{H}_1, \ldots, \dot{H}_n$ – see model 11A $\dot{I} = I[i(H_1 + \ldots + H_n) - kE]$ $\dot{E} = EI(a_E - fE)$	$R_0, I_0, H_{10}, \ldots, H_{n0}$ – see model 11A with $d_i \equiv ka_E/f$ $E_0 = a_E/f$	

TABLE 10.2 Summary of models: stability and kinetics*

Figure	Stability/no oscillation condition	HAART kinetics	Monotherapy kinetics
3	$\dfrac{d_H}{d_I} > \dfrac{4(\rho-1)}{(\rho-2)^2}$ and $\rho > 2$	$I(t) = I_0 \exp[-d_H t]$ $H(t) = H_0 [1 + (\rho-1)(1 - \exp[-d_H t])]$, at $t >> 1/d_I$	Numerically in Figure 10.6
5	Non-decaying oscillations		
8A	Unstable		
8C	$\eta < 1$: unstable $\eta > 1$: non-decaying oscillations		
9C	$\eta > 2$ and $c/d_I > \eta - 2$	Schematically in Figure 10.10	
11A	No oscillations at $aI_0 << d_I \lesssim c$ Broad overshoot of s.s. level by virus by < 50% at $\eta > 2$	Numerically in Figure 10.11B	
12	See model 10.11A	$P(t) \simeq P_0 + \dfrac{s}{c_p}\,(1 - \exp[-c_p t])$ $N(t) \simeq N(t_{\text{infec}}) + \dfrac{s}{c_p}\,(c_p t - 1 + \exp[-c_p t])$ $s/c_p >> P_0,\ t >> 1/d_I$	Numerically in Figure 10.13
15B	Always stable	$E(t) \equiv E_0$	$E(t) \equiv E_0$
15C	$\dfrac{\partial s(I_0, E_0)}{\partial E_0} < 0$	$E(t) \simeq E_0$ if $k_E I_0 << d_I$	$E(t) \simeq E_0$ if $k_E I_0 << d_I$
17	$a_E I_0 > 4iH_0\,(1 - \xi)$ and $\xi < 1$		
18	See model 11A	$E(t) \equiv E_0$ The rest numerically in Figure 10.11B	

*Notation is defined in the corresponding rows of Table 10.1.

drug kinetics calculations were performed numerically.

Families of Models

Suppose that infected cells do not interact with each other, and host cells do not interact with HIV-specific immune cells directly. For any model from this group, the master equations and the exact-balance (steady-state) conditions have a general form:

Master equation   Balance condition

$$\dot{\vec{H}} = \vec{\phi}(\vec{H}, I) \qquad \vec{\phi}(\vec{H}, I) = 0 \qquad (1)$$

$$\dot{\vec{E}} = \vec{\chi}(\vec{E}, I) \qquad \vec{\chi}(\vec{E}, I) = 0 \qquad (2)$$

$$\dot{I} = I\psi(\vec{H}, \vec{E}) \qquad \psi(\vec{H}, \vec{E}) = 0, \qquad (3)$$

where $\vec{H} = (H_1, \ldots, H_n)$ is the vector whose components are several (n) host-cell (or related) compartments, and $\vec{E} = (E_1, \ldots, E_m)$ is the vector of m effector-cell compartments. All such models can be classified in three families, depending on the order in which steady-state values of different variables are specified by Equations 1–3.

Equation	Family I	Family II	Family III
Eq. 1 $\Rightarrow$	$I = I_0, H_1$ undefined	$\vec{H} = [\vec{H}_0$ or $\vec{H}_0(I)]$	$\vec{H} = [\vec{H}_0$ or $H_0(I)]$
Eq. 2 $\Rightarrow$	$\vec{E} = [\vec{E}_0$ or $\vec{E}_0(I)]$	$I = I_0, E_1$ undefined	$\vec{E} = [\vec{E}_0$ or $\vec{E}_0(I)]$
Eq. 3 $\Rightarrow$	$H_1 = H_{10}$	$H_1 = E_{10}$	$I = I_0$

In other words, the infected cell number can be specified by host-cell kinetics alone, by immune-cell kinetics alone or by both together. We call the first family "host-cell-controlled models" and the second family "immune-cell-controlled models". For these two families, immune-cell kinetics or host-cell kinetics, respectively, may be studied, in the first approximation, separately from other variables. These partial models can be compared with the experimental facts cited in the main text. In the end this approach has to be verified by joining the partial models which survived all the tests together and repeating these tests again. Models of family III require evaluation of the entire set of Equations 1–3.

REFERENCES

Altman, J.D., Moss, P.A.H., Goulder, P.J.R. *et al.* (1996) Phenotypic analysis of antigen-specific T lymphocytes. *Science*, **274**, 94–96.

Asano, M.S. and Ahmed, R. (1996) CD8 T cell memory in B cell-deficient mice. *J. Exp. Med.*, **183**, 2165–2174.

Autran, B., Carcelain, G., Li, T.S. *et al.* (1997) Positive effects of combined antiretroviral therapy on CD4+ T cell homeostasis and function in advanced HIV disease. *Science*, **277**, 112–116.

Bleul, C.C., Wu, L., Hoxie, J.A., Springer, T.A. and Mackay, C.R. (1997) The HIV coreceptors CXCR4 and CCR5 are differentially expressed and regulated on human T lymphocytes. *Proc. Natl Acad. Sci. USA*, **94**, 1925–1930.

Borrow, P., Lewicky, H., Wei, X. *et al.* (1997) Antiviral pressure exerted by HIV-1-specific T lymphocytes (CTLs) during primary infection demonstrated by rapid selection of CTL escape virus. *Nature Med.*, **3**, 205–211.

Brander, C., Hartman, K.E., Trocha, A.K. *et al.* (1998) Lack of strong immune selection pressure by the immunodominant, HLA-A*0201-restricted cytotoxic T lymphocyte response in chronic human immunodeficiency virus-1 infection. *J. Clin. Invest.*, **101**, 2559–2566.

Burns, D.P. and Desrosiers, R.C. (1994) Envelope sequence variation, neutralizing antibodies, and primate lentivirus persistence (review). *Curr. Topics Microbiol. Immunol.*, **188**, 185–219.

Cavert, W., Notermans, D.W., Staskus, K. *et al.* (1997) Kinetics of response in lymphoid tissues to antiretroviral therapy of HIV-1 infection. *Science*, **276**, 960–964.

Christensen, J.P., Marker, O. and Thomsen, A.R. (1994) The role of CD4+ T-cells in cell-mediated immunity to LCMV: studies in MHC class I and class II deficient mice. *Scand. J. Immunol.*, **40**, 373–382.

Chun, T.-W., Stuyver, L., Mizell, S.B. *et al.* (1997a) Presence of an inducible HIV-1 latent reservoir during highly active antiretroviral therapy. *Proc. Natl Acad. Sci. USA*, **94**, 13193–13197.

Chun, T.W., Carruth, L., Finzi, D. *et al.* (1997b) Quantification of latent tissue reservoirs and total body viral load in HIV-1 infection. *Nature*, **387**, 183–188.

Coffin, J.M. (1995) HIV population dynamics *in vivo*: implications for genetic variation, pathogenesis, and therapy. *Science*, **267**, 483–488.

Coffin, J.M. (1996) HIV viral dynamics. *AIDS*, **10**, S75–S84.

Connick, E., Lederman, M.M., Kotzin, B.L. *et al.* (1999) Immunologic effects of 48 weeks of zidovudine, lamivudine, and ritonavir therapy of HIV infection. In press.

Corbeil, J., Tremblay, M. and Richman, D.D. (1996) HIV-induced apoptosis requires the CD4 receptor cytoplasmic tail and is accelerated by interaction of CD4 with p561ck. *J. Exp. Med.*, **183**, 39–48.

Danner, S.A., Carr, A., Leonard, J.M. *et al.* (1995) A short-term study of the safety, pharmacokinetics, and efficacy of ritonavir, an inhibitor of HIV-1 protease. *N. Engl J. Med.*, **333**, 1528–1533.

De Boer, R.J. and Perelson, A.S. (1998) Target cell limited and immune control models of HIV infection: a comparison. *J. Theor. Biol.*, **190**, 201–214.

De Jong, M.D., Veenstra, J., Stilianakis, N.I. *et al.* (1996) Host–parasite dynamics and outgrowth of virus containing a single K70R amino acid change in reverse transcriptase are responsible for the loss of human immunodeficiency virus type 1 RNA load suppression by zidovudine. *Proc. Natl Acad. Sci. USA*, **93**, 5501–5506.

De Jong, M.D., de Boer, R.J., de Wolf, F. *et al.* (1997) Overshoot of HIV-1 viraemia after early discontinuation of antiretroviral treatment. *AIDS*, **11**, F79–F84.

Dimitrov, D.S. and Martin, M.A. (1995) CD4+ T cell turnover. *Nature*, **375**, 194-195.

Dimitrov, D.S., Willey, R., Martin, M. and Blumenthal, R. (1992) Kinetics of HIV-1 interactions with sCD4 and CD4+ cells. *Virology*, **187**, 398–406.

Feinberg, M.B. and McLean, A.R. (1997) AIDS: decline and fall of immune surveillance? *Curr. Biol.*, **7**, R136–R140.

Finzi, D., Hermankova, M., Pierson, T. *et al.* (1997) Identification of a reservoir for HIV-1 in patients on highly active antiretroviral therapy. *Science*, **278**, 1295–1300.

Fleury, S., de Boer, R.J., Rizzardi, G.P. *et al.* (1998) Limited CD4+ T-cell renewal in early HIV-1 infection: effect of highly active antiretroviral therapy. *Nature Med.*, **4**, 794–801.

Gallimore, A., Glithero, A., Godkin, A. *et al.* (1998) Induction and exhaustion of lymphocytic choriomeningitis virus-specific cytotoxic T lymphocytes visualized using soluble tetrameric major histocompatibility complex class I-peptide complexes. *J. Exp. Med.*, **187**, 1383–1393.

Goulder, P., Price, D., Nowak, M., Rowland-Jones, S., Phillips, R. and McMichael, A. (1997) Co-evolution of human immunodeficiency virus and cytotoxic T-lymphocyte responses. *Immunol. Rev.*, **159**, 17–29.

Haase, A.T. (1999) Population biology of HIV-1 infection: viral and CD4+ T cell demographics in lymphatic tissues. *Ann. Rev. Immunol.*, in press.

Haase, A.T., Henry, K., Zupancic, M. (1996) Quantitative image analysis of HIV-1 infection in lymphoid tissue. *Science*, **274**, 985–989.

Ho, D.D., Neumann, A.U., Perelson, A.S., Chen, W., Leonard, J.M. and Markowitz, M. (1995) Rapid turnover of plasma virions and CD4 lymphocytes in HIV infection. *Nature*, **373**, 123–126.

Holmes, E.C., Zhang, L.Q., Simmonds, P., Ludlam, C.A. and Brown, A.J.L. (1992) Convergent and divergent sequence evolution in the surface envelope glycoprotein of human immunodeficiency virus type 1 within a single infected patient. *Proc. Natl. Acad. Sci. USA*, **89**, 4835–4839.

Janeway, C.A. and Travers, P. (1996) *Immunobiology. The Immune System in Health and Disease.* Current Biology Garland Publishing, London/New York.

Kalams, S.A., Goulder, P.J., Shea, A.K., *et al.* (1999) Levels of HIV-1-specific CTL effector and memory responses are in equilibrium with viral burden and decline with HAAR. In press.

Kelleher, A.D. (1997) Alterations in the immune response of HIV-infected subjects treated with an HIV-specific protease inhibitor, ritonavir. PhD Thesis, Ch. 5.

Kelleher, A.D., Carr, A., Zaunders, J. and Cooper, D.A. (1996) Alterations in the immune response of HIV-infected subjects treated with an HIV-specific protease inhibitor, ritonavir. *J. Infect. Dis.*, **163**, 321–329.

Klenerman, P., Phillips, R.E., Rinaldo, C.R. (1996) Cytotoxic T lymphocytes and viral turnover in HIV type 1 infection. *Proc. Natl Acad. Sci. USA*, **93**, 15323–15328.

Lederman, M.M., Connick, E., Lunday, A. *et al.* (1998) Immunologic responses associated with 12 weeks of combination antiretroviral therapy consisting of zidovudine, lamivudine and ritonavir. Results of AIDS clinical trials group protocol 315. *J. Infec. Dis.* **178**, 70–7.

Le Gall, S., Heard, J.M. and Schwartz, O. (1997) Analysis of Nef-induced MHC-1 endocytosis. *Res. Virol.*, **148**, 43–47.

Lieberman, J., Fabry, J.A., Fong, D.M. and Parkerson, G.R. III. (1997) Recognition of a small number of diverse epitopes dominates the cytotoxic T lymphocytes response to HIV type 1 in an infected individual. *AIDS Res. Hum. Retroviruses*, **13**, 383–392.

McLean, A. (1998) To be published.

Margolick, J.B., Munoz, A., Donnenberg, A.D. *et al.* (1995) Failure of T-cell homeostasis preceding AIDS in HIV-1 infection. *Nature Med.*, **1**, 674–680.

Matloubian, M., Concepcion, R.J. and Ahmed, R. (1994) CD4⁺ T-cells are required to sustain CD8⁺ cytotoxic T-cell responses during chronic viral infection. *J. Virol.*, **68**, 8056–8063.

Mohri, H., Bonhoeffer, S., Monard, S., Perelson, A.S. and Ho, D.D. (1998) Rapid turnover of T lymphocytes in SIV-infected Rhesus macaques. *Science*, **279**, 1223–1227.

Mosier, D.E. (1995) CD4⁺ cell turnover. *Nature*, **375**, 193–194.

Moskophidis, D., Lechner, F., Pircher, H. and Zinkernagel, R.M. (1993) Virus persistence in acutely infected immunocompetent mice by exhaustion of antiviral cytotoxic effector T-cells. *Nature*, **362**, 758–761.

Moskophidis, D., Battegay, M., van den Broek, M., Laine, E., Hoffmann-Rohrer, U. and Zinkernagel, R.M. (1995) Role of virus and host variables in virus persistence or immunopathological disease caused by a non-cytolytic virus. *J. Gen. Virol.*, **76**, 381–391.

Murali-Krishna, K., Altman, J.D., Suresh, M. *et al.* (1998) Counting antigen-specific CD8 T-cells: a reevaluation of bystander activation during viral infection. *Immunity*, **8**, 177–187.

Nowak, M.A., Lloyd, A.L., Vasquez, G.M. *et al.* (1997) Viral dynamics of primary viremia and antiretroviral therapy in simian immunodeficiency virus infection. *J. Virol.*, **71**, 7518–7525.

Ogg, G.S., Jin, X., Bonhoeffer, S. *et al.* (1998) Quantitation of HIV-1-specific cytotoxic T lymphocytes and plasma load of viral RNA. *Science*, **2792**, 103–2106.

Ogg, G.S., Jin, X., Bonhoeffer, S *et al.* (1999) Decay kinetics of human immunodeficiency virus-specific effector cytotoxic T lymphocytes after combinations antiretroviral therapy. *J. Virol.* **73**, 797–800.

Perelson, A.S., Neumann, A.U., Markowitz, M., Leonard, J.M. and Ho, D.D. (1996) HIV-1 dynamics *in vivo*: virion clearance rate, infected cell life-span, and viral generation time. *Science*, **271**, 1582–1586.

Perelson, A.S., Essunger, P., Cao, Y. *et al.* (1997) Decay characteristics of HIV-1-infected compartments during combination therapy. *Nature*, **387**, 188–191.

Polis, M., Yoder, C., Boon, J *et al.* (1999) Increased initial rate of HIV-1 elimination with a four-drug anti-retroviral regimen in treatment-naive patients. Presented at 38th Interscience conference on Antimicrobial agents and Chemotherapy. San Diego, 1998. Abstract No. LB-3.

Reinhart, T.A., Rogan, M.J., Amedee, A.M. *et al.* (1998) Tracking members of the simian immunodeficiency virus deltaB670 quasi-species population in vivo at single-cell resolution. *J. Virol.*, **72**, 113–120.

Reitter, J.N., Means, R.E. and Desrosiers, R.C. (1998) A role for carbohydrates in immune evasion in AIDS. *Nature Med.*, **4**, 679–684.

Rosenberg, E.S., Billingsley, J.M., Caliendo, A.M. *et al.* (1997) Vigorous HIV-1-specific CD4⁺ T cell responses associated with control of viremia. *Science*, **278**, 1447–1450.

Rosenzweig, M., DeMaria, M.A., Harper, D.M., Friedrich, S., Jain, R.K. and Johnson, R.P. (1998) Increased rates of CD4(+) and CD8(+) T lymphocyte turnover in simian immunodeficiency virus-infected macaques. *Proc. Natl Acad. Sci. USA*, **95**, 6388–6393.

Schmitz, J.E., Kuroda, M.J., Samtra, S *et al.* (1999) Control of viremia in simian immunodeficiency virus infections by CD8+ lymphocytes. *Science* **283**, 857–860.

Siliciano, R.F., Lawton, T., Knall, C. *et al.* (1988) Analysis of host-virus interactions in AIDS with anti-gp120 T cell clones: effect of HIV sequence variation and a mechanism for CD4+ cell depletion. *Cell*, **54**, 561–575.

Smith, M.W., Dean, M., Carrington, M. *et al.* (1997) Contrasting genetic influence of CCR2 and CCR5 variants on HIV-1 infection and disease progression. *Science*, **277**, 959–965.

Sprent, J. and Tough, D (1995) CD4+ T cell turnover. *Nature*, **375**, 194.

Su, M.W., Walden, P.R., Eisen, H.N. and Golan, D.E. (1993) Cognate peptide-induced destruction of CD8+ cytotoxic T lymphocytes is due to fratricide. *J. Immunol.*, **151**, 658–667.

Terai, C., Kornbluth, R.S., Pauza, C.D., Richman, D.D. and Carson, D.A. (1991) Apoptosis as a mechanism of cell death in cultured T lymphoblasts acutely infected with HIV. *J. Clin. Invest.*, **87**, 1710–1715.

Tishon, A., Lewicki, H., Rall, G., Von Herrath, M. and Oldstone, M.B. (1995) An essential role for type 1 interferon-gamma in terminating persistent viral infection. *Virology*, **212**, 244–250.

Walden, P.R. and Eisen, H.N. (1990) Cognate peptides induce self-destruction of CD8+ cytolytic T lymphocytes. *Proc. Natl Acad. Sci. USA*, **87**, 9015–9019.

Wei, X., Ghosh, S., Taylor, M.E. *et al.* (1995) Viral dynamics in human immunodeficiency virus type 1 infection. *Nature*, **373**, 117–122.

Westendorp, M.O., Frank, R., Ochsenbauer, C. *et al.* (1995) Sensitization of T-cells to CD95-mediated apoptosis by HIV-1 tat and gp120. *Nature*, **375**, 497–500.

Wolfs, T.F.W., Zwart, G., Bakker, M., Valk, M., Kuiken, C.L. and Gousmit, J. (1991) Naturally occurring mutations within HIV-1 V3 genomic RNA leads to antigenic variation dependent on a single amino acid substitution. *Virology*, **185**, 195–205.

Wolinsky, S.M., Korber, B.T.M., Neumann, A.U. *et al.* (1996) Adaptive evolution of human immunodeficiency virus-type 1 during the natural course of infection. *Science*, **272**, 537–542.

Wolthers, K.C. and Miedema, F. (1998) Telomeres and HIV-1 infection: in search of exhaustion. *Trends Microbiol.*, **6**, 144–147.

Wolthers, K.C., Bea, G., Wisman, A. *et al.* (1996) T cell telomere length in HIV-1 infection: no evidence for increased CD4+ T cell turnover. *Science*, **274**, 1543–1547.

Wong, J.K., Hezareh, M., Günthard, H.F. *et al.* (1997) Recovery of replication-competent HIV despite prolonged suppression of plasma viremia. *Science*, **278**, 1291–1295.

Yang, O.O., Kalams, S.A., Rosenzweig, M. *et al.* (1996) Efficient lysis of human immunodeficiency virus type 1-infected cells by cytotoxic T lymphocytes. *J. Virol.*, **70**, 5799–5800.

Yang, O.O., Kalams, S.A., Trocha, A. *et al.* D. (1997a). Suppression of human immunodeficiency virus type 1 replication by CD8+ cells: evidence for HLA class I-restricted triggering of cytolytic and noncytolytic mechanisms. *J. Virol.*, **71**, 3120–3128.

Yang, O.O., Trans, A.C., Kalams, S.A., Johnson, R.P., Roberts, M.R. and Walker, B.D. (1997b). Lysis of HIV-1-infected cells and inhibition of viral replication by universal receptor T-cells. *Proc. Natl Acad. Sci. USA*, **94**, 11478–11483.

Zhang, Z.Q., Notermans, D.W., Sedgewick, G. *et al.* (1998) Kinetics of CD4+ T cell repopulation of lymphoid tissues after treatment of HIV-1 infection. *Proc. Natl Acad. Sci. USA*, **95**, 1154–1159.

Zhang, L., Dailey, P.J., Gettie, A., *et al.* (1999) Rapid clearance of simian immunodeficiency virus particles from plasma of Rhesus macaques. *J. Virol.*, **73**, 855–860.

11

Plant Virus Evolution: Past, Present and Future

A. J. Gibbs, P. L. Keese, M. J. Gibbs and F. García-Arenal

INTRODUCTION

Evolution is the process by which the genetic structure of the population of an organism changes with time. Experimental evidence showing that plant viruses could adapt to new hosts was first obtained in the 1920s. Over the following 40 years much was learned of the morphology and antigenic relationships of the virions of viruses. Using features of the virions, vectors and symptoms as characters, isolates were clustered into species and "groups" (now mostly genera), and these groupings were mostly confirmed by the earliest molecular data on the composition and amino-acid sequences of the viral proteins (Gibbs, 1968). The relationships, if any, between the genera were unknown, and their origins and modes of evolution were merely a subject of speculation. However, in the 1960s, methods for determining the nucleotide sequences of genes were invented and the study of virus origins and evolution became an attainable goal.

Most of the early studies of viral gene sequences focused on the genomes of single species. They revealed "the remarkable diversity of plant RNA virus genomes" (Zaccomer *et al.*, 1995) and, to a lesser extent, of plant DNA virus genomes. Analysis of these sequences showed that many of the viral genera fell into larger sets that were totally unexpected. So, in

this chapter, we start by briefly reviewing the current understanding of plant virus phylogenetics and give, as contrasting examples, the results of analyses of the tobamoviruses and the luteoviruses. Next we outline our rapidly expanding state of knowledge of plant virus population genetics and mechanisms of evolutionary change. Finally we speculate both about some of the special features of plant viruses that might soon be explained by evolutionary studies, and also what the future may hold for plant viruses.

PHYLOGENETICS OF PLANT VIRUSES

The discovery of the –GDD– sequence motif in many viral polymerases (Kamer and Argos, 1984; Argos, 1988) showed that viruses, previously thought be to quite unrelated, had seemingly related genes. Haseloff *et al.* (1984) showed that tobacco mosaic tobamovirus (TMV), alfalfa mosaic alfamovirus, cucumber mosaic cucumovirus and Sindbis alphavirus had related replication proteins, even though they were previously not known to be related in any way other than that their genomes were single-stranded RNA. Most surprising was that the first three of these only infected plants and the fourth, Sindbis alphavirus, replicated only in vertebrates and invertebrates! Haseloff and his colleagues concluded that "reassortment of functional modules of coding and regulatory sequence from pre-existing viral or cellular sources, perhaps via RNA recombination, may be an important mechanism in RNA virus evolution". This phenomenon was originally described and called "modular evolution" by Botstein (1980), who detected it in the bacteriophages of coliform bacteria, and involves genetic recombination (Lai, 1992), which has been shown to be one of the dominant features of viral evolution.

After these pioneering efforts many virus genomes were sequenced, and gene and viral phylogenies inferred (Habili and Symons, 1989; Morozov *et al.*, 1989; Bruenn, 1991; Gorbalenya *et al.*, 1990; Koonin and Dolja, 1993; Mushegian and Koonin, 1993). These studies, especially those of the RNA polymerases, produced a clear consensus on most of the closer groupings, but not the distant ones. Koonin and Dolja (1993) suggested, for example, "that all positive-strand RNA viruses and some related double-stranded RNA viruses could have evolved from a common ancestor virus that contained genes for RNA-dependent RNA polymerase, a chymotrypsin-related protease that also functioned as the capsid protein, and possibly an RNA helicase". However Zanotto *et al.* (1996), noting the lack of consensus on the higher groupings, re-examined the data used in the major polymerase studies by three independent methods: Monte Carlo randomization, bootstrap sampling and phylogenetic signal analysis. They concluded that "[Although] clear relationships between some viral taxa were identified, overall the sequence similarities and phylogenetic signals were insufficient to support many of the proposed evolutionary groupings of RNA viruses. Likewise, no support for the common ancestry of RNA-dependent RNA polymerases and reverse transcriptases was found." In fact Zanotto *et al.* (1996) found support for a dozen or more separate polymerase groups and found no statistical evidence of relatedness between those groups. They concluded that "it is more appropriate to present the evolutionary relationships between RNA viruses as a set of dis-

tinct subtrees, the links between which are unclear, rather than as a single and resolved phylogenetic tree". Further, "[The] lack of conservation in primary sequence and size among polymerases suggests that the requirement for polymerase function can be fulfilled by diverse means, supporting the notion of multiple origins."

The results of the analyses of polymerase relationships by Zanotto and his colleagues implies that some of the groupings claimed for other genes based on similar or lesser confidence levels must also be treated with caution. For example, although the 30K movement proteins of different tobamoviruses are clearly related to one another (Melcher, 1990), the suggestion that the movement proteins of different viral genera form larger groupings (Mushegian and Koonin, 1993) is not convincing. The movement proteins of plant viruses are probably polyphyletic in origin. One possibility is that these genes are overprinted genes (Keese and Gibbs, 1992, 1993; Gibbs and Keese, 1995) that have arisen *de novo* during the establishment of each genus; the slight similarity of motifs found in some of them may reflect selection for similar function or composition rather than shared ancestry.

The Phylogeny of the Tobamoviruses

Many different "strains", "types" or "forms" of TMV were recognized in early studies. Some had different host preferences but all could be grouped by the shape and size of their virions and by the antigenic relationships (van Regenmortel, 1986) or the amino acid sequence and composition of their virion proteins (Gibbs, 1986). Many have subsequently been designated as distinct tobamovirus species as they have separate evolutionary histories and infect different hosts.

The nucleotide sequences of 20 complete tobamovirus genomes are now (late 1998) known. Classifications of these sequences by distance matrix methods or by parsimony methods (Lartey *et al.*, 1996) show that the 20 isolates are from perhaps 12 tobamovirus species. The relationships between all, except odontoglossum ringspot virus (ORSV) isolates, are closely similar (Figure 11.1) whether the replicase, movement or virion protein genes or encoded proteins are used for the comparisons. The replicase genes of the ORSVs are closest to those of the brassica tobamoviruses but their movement and coat protein genes place them among those isolated from Solanaceae; thus the ancestral ORSV was, most probably, a recombinant between species of two major tobamovirus lineages (Lartey *et al.*, 1996; Gibbs *et al.*, 1997). There is also evidence that TMV-Ob, which was isolated from pepper, is also a recombinant but between closely related species of the same lineage (M.J. Gibbs, J. Armstrong and A.J. Gibbs, unpublished data). The relationships between the other tobamoviruses can be shown to be a mostly coherent set (Gibbs, 1999) with remarkably uniform rates of evolution in their replicase and movement protein genes, although there was some evidence of rate differences among the coat protein genes (Lartey *et al.*, 1996).

The gene sequence taxonomy of the tobamoviruses (Figure 11.1) correlates well with groupings based on other criteria. For example, Fukuda *et al.* (1980) proposed that there are two major groups of tobamoviruses: those infecting cucurbits and legumes have their origin of virion assembly region within the coat protein gene, whereas the virion assembly region of all others is within the movement protein gene. These groups are the two primary lineages defined by the nucleotide sequence classification. This grouping also agrees with the conclusion of Lartey *et al.* (1996) that the brassica-infecting tobamoviruses are a sublineage of the tobamoviruses in that they have replaced a small deleted 3'-terminal portion of their movement protein gene by "overprinting" the 5'-terminal part of their coat protein gene. Lartey *et al.* (1996) deduced from the overprinting that the brassica-infecting tobamoviruses were the "derived" rather than the "ancestral" group. The gene sequence taxonomy also correlates with the observed differences in the structure of the virion proteins of five of the viruses (Wang and Stubbs, 1994; Wang *et al.*, 1997, 1998). The TMV, TMGMV and ORSV proteins are most similar, but distinct from that of the ribgrass mosaic virus (RMV; synonym crucifer-TMV),

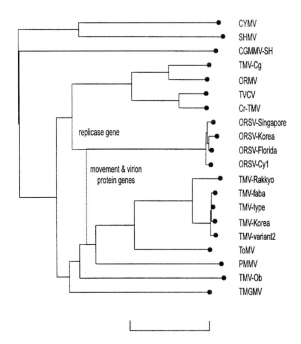

FIGURE 11.1 Neighbor-joining tree calculated from the pairwise percentage nucleotide differences (excluding gaps) between the aligned replicase protein genes of 20 tobamoviruses. The dotted lines indicate the approximate position of the ORSV lineage in similar trees calculated from the pairwise percentage nucleotide differences (excluding gaps) for the movement and virion protein genes. Acronyms as in Table 11.1; the bottom bar corresponds to a 10% difference (uncorrected for multiple mutations). The genomic sequences of CYMV (clitoria yellow mottle virus; K. Wei, A.M. Mackenzie and A.J. Gibbs) and TMV-faba (X. Zhou, personal communication) are as yet unpublished. The others came from the public databases and had the following accession codes: CGMMV-SH (cucumber green mottle mosaic virus), D12505; cr-TMV (crucifer TMV), Z29370; ORMV (oilseed rape mosaic virus), U30944; ORSV-Cy1 (odontoglossum ringspot virus-Cy1), S83257; ORSV-Florida, U89894; ORSV-Korea, X82130; ORSV-Singapore, U34586; PMMV (pepper mild mottle virus), M81413; SHMV (sunnhemp mosaic virus), J02413; TMGMV (tobacco mild green mosaic virus), M34077; TMV-Cg, D38444; TMV-Korea, X68110; TMV-Ob (TMV-Ob), D13438; TMV-Rakkyo, D63809; TMV-type, J02415; TMV-variant 2, U01409; ToMV (tomato mosaic virus), X02144; TVCV (turnip vein clearing virus), U03387.

and the most distinct of all is the coat protein of cucumber green mottle mosaic virus (CGMMV).

It would, of course, be of great interest to estimate the age of the tobamoviruses, i.e. to place a time scale on the ordinate of Figure 11.1. This could be done by estimating the current rate of evolution of tobamoviruses and extrapolating to obtain the time scale for the tobamovirus dendrogram. However there is clear evidence, discussed below, that most tobamovirus populations are very stable and do not evolve at a measurable rate. Gene sequences of TMGMV isolates obtained from herbarium specimens of *Nicotiana glauca* collected in eastern Australia over the past century (Fraile *et al.*, 1997a) were found not to have changed in a time-related way. Thus the data provide no estimate of the rate of TMGMV evolution, except to indicate that it will minimally require millenia, not centuries, to measure it.

An alternative is to extrapolate from evolutionary rates established for other organisms. Twenty years ago the virion protein sequences of seven tobamoviruses had been reported, and differed by about 120% in sequence, after correction for multiple mutations. Gibbs (1980a) argued that, if they had evolved at a rate of 1% every 2–10 million years, like many other proteins (Wilson *et al.*, 1977), then the prototobamovirus probably arose about 120–600 million years ago.

A safer way to infer the age of a fossil-less group of organisms, like tobamoviruses, is to link its phylogeny with that of a group whose fossil history is known, in this instance their hosts. There are two lines of evidence of this sort, and both indicate that the tobamoviruses may be more than 100 million years old. First there is Holmes's (1950) observation that the species of *Nicotiana* that respond to infection by TMV in a hypersensitive manner are all natives of the Americas: *N. glutinosa*, which is a native of Peru, *N. repanda* of Mexico, *N. rustica* of Ecuador and Peru and *N. langsdorfii* of Brazil. Several species of other genera of the Solanaceae native to South America behave in the same way, including *Solanum capsicastrum* of Brazil and *S. tuberosum* of Bolivia and Peru. By contrast, *Nicotiana* species that respond to TMV infection with bright chlorosis and mottling, and accumulate the greatest concentrations of virions, are mostly found in North America, southern South America and Australia. Holmes (1950), arguing that the hypersensitive response may reflect exposure to and selection by TMV, stated that this "would seem to imply that the original habitat of tobacco-mosaic virus was

within an area of the New World, centering about some part of Peru, Bolivia, or Brazil." He also noted that in this region there are now three species of *Nicotiana* (*N. glauca*, *N. raimondii* and *N. wigandioides*) that tolerate TMV infection, show few or no symptoms and may be the long-term niche of TMV; *N. tabacum* itself is an amphidiploid species found only in crops or as a "crop fugitive", and is unlikely to have been the original preferred host of TMV.

The Solanaceae is a mostly tropical family and its earliest fossils are from the Cretaceous, 65 million years ago (D'Arcy, 1991). However, the present distribution of the Solanaceae has some Gondwanan features (Symon, 1991): the major centre of diversity is Central and South America and there are minor centres in Eurasia, especially around the Himalayas, and also Australia. Hence the Solanaceae may have originated earlier than 65 million years ago, as the Indian subcontinent separated from Gondwana around 80 million years ago (Raven, 1983). Furthermore, sequence analysis of the chloroplast genome of 21 *Nicotiana* species (Olmstead and Palmer, 1991), including some tested by Holmes, shows that the species hypersensitive to TMV are in two basal lineages. Thus hypersensitivity to TMV is either the ancestral condition, has arisen more than once, or spreads by hybridization between species. Whichever is correct, it suggests that the sort of processes that might have been responsible for the development of the American "centre of resistivity" observed by Holmes have involved tens of millions, not thousands, of years.

The other reason to believe that the tobamoviruses coevolved with their hosts is that their taxonomy (Figure 11.1) mostly correlates with the taxonomy of the hosts from which they have been isolated (Gibbs, 1986; 1999; Lartey *et al.*, 1996). All the tobamoviruses isolated from solanaceous plants fall into one large cluster; other pairs of species (on long branches) infect brassicas and legumes; and similarly distinct are those isolated from cucurbits and orchids. A major division of the eudicotyledonous angiosperms (Figure 11.2), first identified by Young and Watson (1970) in an early computed taxonomic study of the phenotypic characters of plants and recently confirmed by molecular

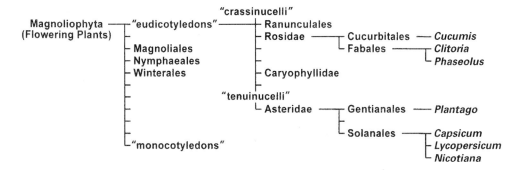

FIGURE 11.2   The topological relationships of the hosts of some tobamoviruses based on data from Young and Watson (1970) and the NCBI taxonomy database (http:www.ncbi.nlm.nih.gov/ Taxonomy/taxonomyhome.html)

sequence comparisons (http:www.ncbi.nlm. nih.gov/Taxonomy/taxonomyhome.html), separates more ancient crassinucellate plants (Rosidae, Caryophyllidae and others) from the more modern tenuinucellate plants (Asteridae). Cucurbitaceae and Fabaceae are crassinucellate, and Solanaceae are tenuinucellate. Thus the pattern of relationships of the hosts correlate well with that of the major tobamovirus groups. The simplest explanation for this correlation is that most of the tobamoviruses have coevolved with their hosts. The possibility that the viruses have moved between host lineages and have then been selected by those hosts to appear similar is less parsimonious. Furthermore, tobamovirus taxonomies calculated from synonymous nucleotide differences are the same as those calculated from all differences, indicating that changes to encoded proteins are not the source of the phylogenetic signal revealed in the trees.

At variance with the correlation between tobamovirus taxonomy and that of their hosts is that those isolated from brassicas, which are crassinucellate, are most closely related to the tenuinucellate (solanaceous) tobamoviruses. However, one of the brassica-infecting tobamoviruses is best known as ribgrass mosaic virus (Holmes, 1941), as it is common worldwide in weed species of *Plantago*, a tenuinucellate genus. Further work on ribgrass mosaic virus isolates from around the world may show that plantains are the natural hosts of brassica-infecting tobamoviruses.

The correlations outlined above indicate that

the tobamoviruses have coevolved with their hosts. The angiosperms first appeared in the fossils from 120–140 million years ago, although most modern families did not appear before 60–80 million years ago (Raven, 1983). Thus the proto-tobamoviruses infecting the earliest angiosperms are probably represented by the basal "edges" of the CGMMV and sunn-hemp mosaic (SHMV) lineages in the taxonomy in Figure 11.1. The major tobamovirus radiations resulting in the clusters of species now found in the Solanaceae, the legumes and cucurbits may have occurred when these modern plant families radiated 60–80 million years ago. This period includes both the final stages of dismemberment of Gondwana, and also the Cretaceous–Tertiary extinction (Schultz and Dhondt, 1996), and either of these events may also account for the deep branches within some of these tobamovirus lineages.

Tobamovirus genes have clear sequence homologies (Koonin and Dolja, 1993; Zanotto *et al.*, 1996) with those of a large set of viruses known as the "alpha-like" virus group (ALVG; Goldbach and de Haan, 1994). This group now has many extant genera. In the "tobamo lineage", the tobamoviruses are closest to hordeiviruses and tobraviruses and also soilborne wheat mosaic virus, and this cluster is a sister to other plant-infecting genera that include the alfamoviruses, bromoviruses, cucumoviruses, ilarviruses and idaeoviruses. The "alpha lineage" is also supported by most analyses and includes the rubiviruses, hepatitis

E virus and beet necrotic yellow vein "furovirus" and, maybe, the alphaviruses; and the "potex lineage" includes the potexviruses, tymoviruses, capilloviruses and closteroviruses. However, the relationships between these lineages is uncertain. The same relationships are also shown by the 5' terminal methyl transferases and helicases of the tobamoviruses (Koonin and Dolja, 1993), indicating that these enzyme genes have formed a module with a long phylogenetic history. The tobamovirus coat proteins are also clearly related in sequence and structure (Dolja *et al.*, 1991) to those of other viruses with rod-shaped and filamentous virions, especially the tobraviruses (Goulden *et al.*, 1992) and also the hordeiviruses and furoviruses. Thus all the species of the ALVG share an enzyme module and some of them, notably the tobamoviruses and others with rod-shaped virions, also share the coat protein gene. In the multicomponent ALVG viruses the enzyme module is now divided among separate genome segments, and the "alpha lineage" viruses have acquired a papain-like serine protease gene, which is inserted between the methyl transferase and helicase genes. If it is correct that the tobamoviruses coevolved with the angiosperms, then their links with other viruses

of the ALVG occurred before 120–140 million years ago.

The Phylogeny of the Luteoviruses

The luteoviruses and several other genera of viruses with isometric virions form the "carmo-like" and "sobemo-like" supergroups. The phylogeny of the luteoviruses, in contrast with that of the tobamoviruses, does not correlate with that of their hosts (Figure 11.3). Host switching rather than virus–host coevolution has dominated their evolution. For example, whereas most luteoviruses infect dicotyledonous plants, two species (barley yellow dwarf virus – MAV/PAV – and barley yellow dwarf virus – RPV) infect cereals, but these are not a subgroup within the luteovirus genus. Instead, each is more closely related to a luteovirus that infects dicotyledonous plants. Similarly the three luteovirus species that infect legumes (groundnut rosette, pea enation mosaic and soybean dwarf viruses) are not clustered within the luteovirus taxonomy. Thus luteoviruses seem to have switched between very different kinds of hosts much more frequently than tobamoviruses.

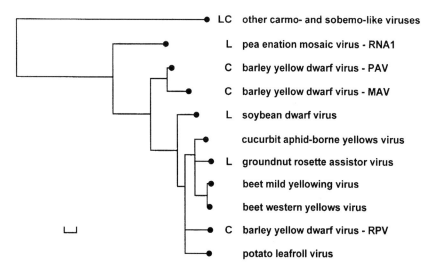

FIGURE 11.3 Maximum likelihood tree (heuristic) showing the relationships of the virion protein (22K) genes of luteoviruses and some related viruses. C, those infecting cereals; L, those infecting legumes. The bar represents 0.1 changes per nucleotide position.

These differences indicate that there may be conflicting evolutionary pressures which favor either host switching or virus–host coevolution. All viruses are probably selected to some degree to interact efficiently with a particular host species or family of hosts, and hence coevolve with them, but it is likely that variants of virus able to infect new host species will have the advantage of avoiding the competition of their parental population.

Luteoviruses also differ from tobamoviruses in that many more luteovirus lineages have evolved through interspecies recombination. One major event replaced the RNA-dependent RNA polymerase (RdRp) of an ancestral luteovirus with an unrelated RdRp, so that extant luteoviruses can now be divided into two sets that have different RdRps (Miller *et al.*, 1988). These RdRps, one set related to those of the carmoviruses and the other to those of the sobemoviruses, clearly have different origins and have little or no sequence homology (Zanotto *et al.*, 1996). By contrast, all the luteovirus virion proteins are encoded by unequivocally related members of a single gene family. The virion proteins determine the mode of luteovirus transmission by aphids (i.e. "circulative non-propagative") and it is this that distinguishes luteoviruses from other carmo- and sobemo-like viruses. At least two other instances of recombination between luteoviruses with the same kinds of RdRp gene can be detected in their phylogeny (Gibbs and Cooper, 1995; Guilley *et al.*, 1995). The fact that interspecies recombination has been so common among luteoviruses is probably related to their ecology, as several of them are found in lasting associations, probably symbiotic, with other luteoviruses or with other carmo-like viruses.

It is possible that the clear pattern of host switching, rather than coevolution, among the luteoviruses is linked with their history of interspecies recombination (Gibbs, 1995). Interspecies recombination probably changes genomes much more radically than point mutation. It is likely to produce viruses with novel combination of properties, which may include host preferences. In this context it is noteworthy that ORSV has hosts that are atypical for a tobamovirus (i.e. orchids, which are mono-cotyledons), and is a recombinant of parents that probably had unrelated dicotyledonous hosts, whereas TMV-Ob is a recombinant with parents from the same tobamovirus lineage, and has retained the host family preferences of those parents.

THE EVOLUTION OF PLANT VIRUS POPULATIONS

Measures of Genetic Diversity

Transmission experiments showed at an early date that viral characteristics are mutable. Passage experiments showed that host preferences could be changed, a phenomenon described as host adaptation (Yarwood, 1979). Variants were also obtained by other experimental methods, such as temperature shift passaging (Van Vloten-Doting and Bol, 1988). Gene sequence analysis has shown that passage adaptation results from selection among variants present, or newly generated, in the original virus stock (Donis-Keller *et al.*, 1981). Laboratory stocks of plant viruses are also heterogeneous mixtures of variants. Estimates of population heterogeneity for these stocks are scanty. For TMV a 0.2–0.3% of symptom mutants has been reported (Gierer and Mundry, 1958). For TMGMV, nucleotide heterogeneity was estimated as 0.0025 (Rodríguez-Cerezo and García-Arenal, 1989), which is less than that estimated for bacterial and animal RNA viruses. Isolates obtained from single lesion passage were less heterogeneous but not homogeneous. However, stocks obtained from biologically active cDNA clones of TMV, the satellite RNA of cucumber mosaic virus (CMV-satRNA) and the satellite of TMV (STMV) are initially homogeneous, but mutations can accumulate quickly and, as a result, there may be genetic changes in the virus population because of selection and/or random founder effects (Aldaoud *et al.*, 1989; Kurath and Palukaitis, 1989, 1990; Kurath and Dodds, 1995). The fast appearance and accumulation of mutants has also been described for bacterial or animal RNA viruses (Domingo and Holland, 1997) and has been associated with large error rates of RNA-dependent RNA polymerases. No accurate estimates of the error rates of RNA polymerases of plant viruses have been reported (those described by Kearney *et al.*, 1993, were inconclusive), but data indicate that the rate is similar to that reported for other RNA viruses (Domingo and Holland, 1997). Heterogeneous population and host passage effects have been described also for viroids (e.g. Fagoaga *et al.*, 1995; Visvader and Symons, 1985) and for DNA viruses (Kunkel, 1947; Daubert and Routh, 1990; Sanger *et al.*, 1991; Al-Kaff *et al.*, 1994). Thus, no significant differences appear to exist in the capacity of viroids, or of plant viruses with either RNA or DNA genomes, to vary.

A potential to vary need not result in variability, as the latter depends on the balance between mutation and selection, and indeed there are plenty of examples of genetic stability in plant viruses. For instance, stocks of TMV (strains *vulgare* and U1) with the same origin, passaged for over 50 years in different laboratories, are very similar in nucleotide sequence (Goelet *et al.*, 1982; Dawson *et al.*, 1986). The nucleotide difference of isolates of the same virus usually is less than 5%, regardless of the virus genetic material (e.g. Kruse *et al.*, 1994; Murao *et al.*, 1994; Ward *et al.*, 1995; Sanz *et al.*, 1999). Nuclotide sequence differences between isolates of wound tumor virus obtained 50 years apart in the eastern USA, or between isolates of TMGMV sampled in eastern Australia over a 100-year period, were less than 3% (Hillman *et al.*, 1991; Fraile *et al.*, 1997b). Furthermore, isolates of turnip yellow mosaic tymovirus (TYMV) from Europe and Australia that differed by only 3–5% in nucleotide sequence represent populations that may have been isolated from one another for at least 13 000 years (Blok *et al.*, 1987; Skotnicki *et al.*, 1993).

Many factors may ensure that a virus that can vary does not. To identify these factors and to understand their role in virus evolution, it is necessary to analyse the genetic structure of natural populations of the virus. This work has been done with only a few viruses. The genomic nucleic acids of individual isolates are typed to assess the genetic diversity of the population. Different methods vary in the precision with which they estimate genetic diversity. Genetic diversities are most accurately estimated from nucleotide sequence data, but can also be obtained, with larger errors, from the analysis of restriction fragment length, or RNase T1 fragment, polymorphisms (Nei, 1987, pp. 64–107). Nucleotide diversities cannot be estimated from ribonuclease protection assay (RPA) data unless the method has been calibrated directly using known sequences (Aranda *et al.*, 1995) and cannot be estimated from some methods, such as single-strand conformation polymorphism analysis. It is important to measure variability as genetic distances as, although a population may contain many variants, it may not be very variable. For instance, RPA analysis of an Australian population of TYMV revealed a large number of haplotypes/isolate for most probes, resulting in an heterozygosity index of about 0.9, but the population was not very diverse, as the isolates only differed about 2% in nucleotide sequence (Skotnicki *et al.*, 1993).

Studies have established that plant virus populations differ in structure, presumably reflecting the wide range of influencing factors and the likelihood that they have been observed at different stages in their evolution and speciation. A virus population may be a single, mostly undifferentiated one, as in TYMV infecting wild *Cardamine lilacina* in the Australian Alps (Skotnicki *et al.*, 1993; Nolan *et al.*, 1996), wheat streak mosaic whestrevirus in wheat in western USA (McNeil *et al.*, 1996), pepper mild mottle virus (PMMV) in pepper in Spain (Rodríguez-Cerezo *et al.*, 1989), TMGMV in wild *Nicotiana glauca* in Spain (Rodríguez-Cerezo *et al.*, 1991) or cotton leaf curl begomovirus in cotton in Pakistan (Sanz *et al.*, 1999). Populations may differ geographically, as described for TMGMV in different regions of the world (Fraile *et al.*, 1996) or beet curly top curtoviruses (BCTV) in the western USA (Stenger and McMahon, 1997), or there may be host and geographical differences, as in kennedya yellow mosaic tymoviruses in east-coast Australia (Skotnicki *et al.*, 1996). Some populations show no consistent pattern but randomly fluctuate in time and space. This metapopulation structure has been reported for CMV in Spain (Fraile *et al.*, 1997a). The virus population structure may be similar to that of its host plant or its insect vectors, as reported for TYMV and its host plant *C. lilacina* (Skotnicki *et al.*, 1993; Nolan *et al.*, 1996) and for CMV and its aphid vectors *Myzus persicae* and *Aphis gossypii* (Fraile *et al.*, 1997a; Martinez-Torres *et al.*, 1998). But a biological link need not result in similar population structures: the population structure and dynamics of CMV and its satellite RNA are different, indicating that the satellite RNA spreads as an independent, but linked, epidemic within the population of its helper virus (Alonso-Prados *et al.*, 1998). The frequency distribution of genetic types in the population may also differ. All genetic types in the population may be equally common (as for BCTV; Stenger and McMahon, 1997) or, more often, their frequency may show a gamma distribution, in which there are a few common types plus several infrequent ones (Skotnicki *et al.*, 1993, Rodríguez-Cerezo *et al.*, 1989, 1991; McNeil *et al.*, 1996; Sanz *et al.*, 1999). Gamma distributions, also called quasispecies distributions (Domingo *et al.*, 1995), are commonly found in laboratory stocks (Rodríguez-Cerezo and García-Arenal, 1989; Kurath *et al.*, 1992). However, the genetic structure of a virus population seems not to correlate with whether the viral genome is RNA or DNA, nor indeed with any other known trait of its life history.

As previously mentioned, the diversity of a population does not depend only on the number of variants it contains but also on the frequency with which the variants occur and on the genetic distances between them. Table 11.1 shows the population diversity, estimated as nucleotide diversity per site, of those viruses for which information is available. It shows that none of the viruses analysed so far, irrespective of whether their genome is RNA or DNA, are genetically diverse. Thus all the available diversity data suggest that considerable genetic stability (rather than diversity) is the rule for plant viruses. It could be that the spectacularly variable animal viruses, influenza A orthomyxovirus, hepatitis C flavivirus (Ina and Gojobori, 1994; Ina *et al.*, 1994) and the immunodeficiency lentiviruses (Hahn *et al.*, 1986) are the exception, even for animal viruses (Moya and García-Arenal, 1995), or that very variable plant viruses have not yet been identified and studied. Populations of geminiviruses, especially begomoviruses, may be more variable, and much of their biological variability may result from their populations being complex mixtures of recombinants (Harrison *et al.*, 1997; Liu *et al.*, 1998; Zhou *et al.*, 1998). If this is so, then measures of genetic diversity that include indices of recombination may have to be devised. Obviously, the

TABLE 11.1 Intrapopulation nucleotide diversities of some plant viruses; diversities expressed as nucleotide diversity per site (Nei, 1987, p. 276)

Virus	Population	Diversity value	Analysis of
PMMV	Spain	0.018	RNase T1 fragments
TMGMV	Spain	0.022	RNase T1 fragments
	Spain	0.020	Sequences
	Australia	0.022	Sequences
	World	0.057	Sequences
CMV-sat RNA	Spain	0.079	RPA
WSMV	USA	0.031	RFLPs
BCTV	USA, CHF	0.026	RFLPs
	USA, Worland	0.021	RFLPs
CLCuV	Pakistan	0.019	RFLPs

Data from Rodríguez-Cerezo *et al.*, 1989, for PMMV; Rodríguez-Cerezo *et al.*, 1991, and Fraile *et al.*, 1996, for TMGMV; Alonso-Prados *et al.*, 1998, for CMV-sat RNA; Sanz *et al.*, 1999, for CLCuV; McNeil *et al.*, 1996, for WSMV; and from Stenger and McMahon, 1997, for BCTV used to calculate estimates.

natural populations of more viruses will have to be analysed in detail before general conclusions can be drawn.

Population Evolutionary Processes

The two main processes that may modify the genetic structure of the population of an organism are selection and random genetic drift. Selection is the process by which fitter variants increase their frequency in populations (i.e. positive selection) and less fit ones are eliminated (i.e. negative selection) from it. The effect of selection is directional and results in decreased population diversity. It may also increase the differences between populations. Selection is the process most frequently invoked in the virus literature to explain population structures and evolutionary processes, although in many instances the effects ascribed to selection could also result from other processes (see below). Selection pressures can be associated with the intrinsic properties of the virus, such as the maintenance of structures important for the viability of the virus. For example, the structure of replication signals in RNA and DNA viruses is strongly conserved (Kurath et al., 1993; Argüello-Astorga et al., 1994; Bacher et al., 1994), as too are non-coding RNAs (Fraile and García-Arenal, 1991; Kofalvi et al., 1997; Ambros et al., 1998). Selection pressures can also be associated with the obligate interactions of a plant virus with its host plants and vectors; however, not all reports of host adaptation are clear proof of host selection, but some are (Donis-Keller et al., 1981; Kurath and Palukaitis, 1990). Differentiation of field isolates according to host plant species (e.g. Kofalvi et al., 1997), and the avoidance of host resistance genes (Harrison, 1981) can also be explained by host-associated selection. Host-associated selection may have an important role in virus evolution when a new host is colonized and a virus adapts to it, or when a virus coevolves with its hosts, as has been proposed for the tobamoviruses (Lartey et al., 1996; Gibbs, 1998). The coevolutionary link between viral and host macromolecules (Parker and Coleman, 1997) could synchronize their evolutionary rates, even though those of the virus were RNA-encoded and those of the host DNA-encoded (see below). Phenomena resulting from vector-associated selection have been less often reported. Examples are loss of vector transmissibility upon repeated passage and the preferential transmission of variants from mixtures (Perry and Francki, 1992; Suga et al., 1995; Uyeda et al., 1995; Fraile et al., 1997a). The behavior of phytophagous animals, including vector species, may differ on virus-infected and virus-free plants. For example, Baker (1960) found that aphid species commonly found on sugar beet in the UK preferred beet plants infected with beet yellows closterovirus rather than comparable virus-free plants. The aphids preferred leaves showing the severest symptoms and, on these, the species Myzus persicae lived half as long again and produced three times as many offspring as on virus-free leaves. Thus it is likely, but not proven, that the severest variants of beet yellows will induce the optimal transmission by vectors and will be selected preferentially, at least in crops when virus-infected and virus-free plants are not competing. Virus infection may also influence the behavior of herbivores (Gibbs, 1980b, 1983) and this too could also result in selection pressures that, to our knowledge, have not been analysed.

Sequence analyses have shown that negative selection limits the accumulation of mutations. This is also indicated by the infrequency with which point insertions and deletions interrupt ORFs. One way to assess the extent of negative selection in protein-encoding genes is to measure the proportion of nucleic acid changes that produce changes in the encoded protein. This can be estimated from the ratio between nucleotide diversities at non-synonymous (dNS) and synonymous (dS) positions. Table 11.2 shows some of these values for different plant viral genes. Both the degree of variation and the evolutionary constraint to variation (indicated by the dS/dNS ratio) vary largely according to the gene and the virus and, as might be expected, different genes are differently constrained. Interestingly the degree of constraint falls within the range reported for DNA-encoded genes of cellular organisms (Nei, 1987, Table 5.4). Thus, viral proteins, irrespective of whether they are encoded by RNA or DNA

TABLE 11.2 Nucleotide diversities of some virus genes; all estimates made by the method described by Pamilo and Bianchi (1993) and Li (1993)

Virus	Gene	dNS	dS	dNS/dS
TMGMV	183k	0.012	0.052	0.230
CMV	1a	0.006	0.301	0.020
	2a	0.007	0.222	0.032
	Coat protein	0.004	0.086	0.050
TLCV	C1	0.157	0.477	0.329
CLCuV	AC1	0.076	0.505	0.150
	Coat protein	0.021	0.556	0.048

Data are from: Fraile *et al.*, 1996, for TMGMV; Fraile *et al.*, 1997b, for CMV; Ooi *et al.*, 1997, for TLCV; Sanz *et al.*, 1998, for CLCuV; for CaMV estimates made from seven sequences and for PVY from 39 sequences, all from the EMBL databank.

genomes, are under similar levels of constraint to those of their hosts. This must be relevant to virus evolution and leads us to suggest that host, or vector, proteins may constrain change in the virus-encoded proteins with which they interact (Parker and Coleman, 1997) and, as a result, will evolve at similar rates.

Negative selection may also restrict genetic exchange by reassortment of genomic segments or by recombination. Evidence for this comes from data showing that reassortment of genetic segments is not random, either in experimental conditions (Uyeda *et al.*, 1995; Fraile *et al.*, 1997a; Qiu *et al.*, 1998) or in nature (Fraile *et al.*, 1997a); there is selection against particular segment combinations. Data from natural populations indicate that the frequency of occurrence of recombinants in the population varies greatly with different viruses (Fraile *et al.*, 1997a; Zhou *et al.*, 1997, 1998). Also, that some reassortants or recombinants are more fit than their parents and may be important in virus evolution (Chenault and Melcher, 1994; White *et al.*, 1995; Lartey *et al.*, 1996; Revers *et al.*, 1996). The strength of selection may be less in non-coding regions of viral genomes and in non-coding pathogenic RNAs, as shown by the large number of point mutations, deletions and duplications, and of recombinants in satellite RNAs and viroids (Visvader and Symon, 1985; Dáros and Flores, 1995; Aranda *et al.*, 1997).

Positive selection is more difficult to demonstrate than negative selection. It has been suggested that positive selection might have occurred in the Spanish population of TMGMV (Moya *et al.*, 1993; Fraile *et al.*, 1996). It should be pointed out that, while data showing selection are relatively abundant, in few instances can the selection factor be identified, and more work is needed in this important aspect of virus evolution.

Complementation of less fit variants in a population by fully functional ones masks deleterious alleles and modifies the effect of selection; however this has been mostly ignored in analyses of virus evolution. Its importance in shaping the genetic structure in virus populations is shown by a study in which complementation of a movement-defective mutant of TAV RNA3 was analysed (Moreno *et al.*, 1997). The degree of complementation (i.e. the probability that the mutant functioned as well as the functional variant) was estimated to be 0.13. As a result, the usually defective mutant accumulated and became 76% of the population.

Genetic drift results from founder effects associated with population bottlenecks, and this may be important during virus transmission from plant to plant (Pirone and Blanc, 1996) or to new hosts or areas. Population bottlenecks result in random changes in the genetic composition of populations, as opposed to the directional changes resulting from selection. They have been found to decrease within-population diversity and increase between-population diversity in many organisms, including viruses (Duarte *et al.*, 1992; Novella *et al.*, 1995). Random changes in genetic composition observed in passage experiments (Aldaoud *et al.*, 1989; Kurath and Palukaitis, 1989; Kurath and Dodds, 1995), as well as some reported cases of small within-population diversity and subdivision in natural populations (Fraile *et al.*, 1996, 1997a), can be explained by founder effects. It is important to stress that founder effects may be misinterpreted as due to selection.

All the factors discussed above may act in populations of individual viruses. However, in nature, mixed infections are common and interactions between the component viruses may affect the genetic structure of their populations. One documented case is that of the interaction of TMV and TMGMV in *N. glauca* plants in Australia (Fraile *et al.*, 1997b), where both viruses were present in specimens collected between the 1890s and about 1950, and only TMGMV was found after this date. The probable reason for the disappearance of TMV was revealed in experiments, which showed that mixed infection of TMGMV and TMV does not affect the accumulation level of TMGMV but depressed that of TMV to a tenth of what it was in the absence of TMGMV. The concentration of TMV was then probably less than that required to ensure the transmission of the most fit genotype, so that the population became progressively dominated by less fit genomes, and succumbed. This process is known as Müller's ratchet (Lynch *et al.*, 1993). Another well-studied type of interaction is that between helper viruses and their satellites. The strain of helper virus has been shown in passage experiments to affect the genetic structure of CMV-satRNA (Palukaitis and Roossinck, 1995; Roossinck and Palukaitis, 1995). Such an effect, however, is not detected in field populations (Grieco *et al.*, 1997; Alonso-Prados *et al.*, 1998). Constraints on evolution in nature and in experiments may be different, and one should be cautious when extrapolating from experimental conditions to explain observations in natural populations.

"NOTHING MAKES SENSE IN BIOLOGY EXCEPT IN THE LIGHT OF EVOLUTION" (Dobzhansky, 1964)

There have been great advances in our understanding of viruses, especially those of plants, since 1898, when Beijerinck's insight first revealed their world. Nonetheless, many questions remain and some of these will only be answered by studies of their evolution. Here we discuss some such questions that interest us.

Why Do Such a Large Proportion of the Viruses of Flowering Plants have RNA Genomes?

Over 1000 plant viruses have been described (Brunt *et al.*, 1996). Most of them have single-stranded RNA genomes that are replicated in the cytoplasm via RNA intermediates. There are no plant retroviruses or plant viruses with large double-stranded DNA genomes, as commonly found in bacteria and animals. Indeed there are only two groups of plant viruses with double-stranded DNA genomes (badnaviruses and caulimoviruses) and both replicate via RNA intermediates rather than exploiting the standard cellular DNA replication machinery, as comparable animal viruses with DNA genomes do. This domination of plants by viruses with RNA genomes is in marked contrast to the pattern found in other organisms. For example, bacteria are most commonly invaded by viruses with double-stranded DNA genomes of greater than 20 kb; fungi are host to a multitude of viruses with double-stranded RNA genomes; whereas viruses of vertebrates have genomes of every type imaginable. One further conundrum with which to grapple is that most of the described viruses of plants are restricted to flowering plants. There are few reports of viruses in gymnosperms, ferns or bryophytes, although several have been found in algae, but most of these have large, double-stranded DNA genomes. So what are the possible explanations for these highly biased distributions? Are there any features peculiar to plants that favor infection by RNA viruses? For example, plants (unlike animals and single-celled organisms) have a much more restricted zone of replicating cells, the meristem. This may inhibit exploitation by DNA viruses that parasitize the host DNA replication machinery. In addition, plants have a distinctive cell–cell communication system through the plasmodesmata, and this appears to be the primary pathway for spreading viral infection throughout the plant. These channels may be too small to allow the transport of large DNA viruses. Another possibility is that the answer is historical. Perhaps the first viruses to invade flowering plants were RNA viruses; they coevolved rapidly and in some way dominated before other viruses had an opportunity to gain an effective toe-hold.

However the idea that appeals to us is that the bias may reflect the host's defence mechanisms. Thus most viruses of bacteria have large double-stranded DNA genomes and bacteria combat them using complex DNA restriction/modification systems. By contrast, vertebrate animals are host to a very wide range of viruses with genomes of all types and sizes, and they counter these with a highly developed immune system that targets the proteins produced by all virus types. Recently it has emerged that flowering plants control viruses with a cytoplasmic RNA degradation system (Lindbo *et al.*, 1993; Smith *et al.*, 1994; Baulcombe, 1996; Covey *et al.*, 1997), which was revealed by the appearance of post-transcriptional gene silencing in studies of transgenic plants. This cytoplasmic system recognizes foreign RNA sequences and targets them for degradation in a sequence-specific manner. In contrast to vertebrates' immune systems, which are prearmed with a vast array of antibodies, the plants' defence system responds *de novo* to RNA pathogens in a sequence-specific way. This correlation between the dominant virus type and the host's defence mechanism may seem paradoxical. However we suggest that the only host:pathogen pairs that survive are those able to establish a *modus vivendi* in which neither wins, because when that happens the relationship succumbs.

Have flowering plants benefited from their probable "incessant evolutionary dance" (Haldane, 1949) with RNA viruses? Strong selection for a subtle and innovative RNA defence mechanism may have inadvertently produced mechanisms that allow plants to generate developmental novelty quickly too. This may have been one of the prime factors leading to the ecological domination of much of the land surface of the planet by the flowering plants.

Finally, why have so many viruses been isolated from flowering plants, whereas so few have from non-flowering plants? Could this be because the explosion of diversity of flowering plants, especially over the past 80 million years, coincided with a parallel diversification of leaf eaters, mostly insects, many of which are also virus vectors. During most of their prehistory only the root parasites of non-flowering plants behaved in a way that made them possible virus vectors. Thus the rarity of viruses of non-flowering plants may be more apparent than real. The viruses may be lurking in roots, transmitted through soil or by root parasites and pathogens, and hence gaining no selective advantage from

spreading to leaves and causing symptoms to attract the attention of modern leaf-eating vectors (and virologists).

Why Are There so Many Virus/Virus Associations Among Plant Viruses?

Why are plants so often co-infected by more than one virus? Some of these co-infections are clearly not coincidental, and they occur far more frequently than would be expected by chance. Some of the associations are obligatory, others are not. In some, one or both of the partners clearly benefits from the association, they are symbioses, whereas in some of the satellite/helper associations one virus is growing at the expense of the other. Nine different kinds of symbiotic association have been recognized, each involving viruses from a different combination of genera (Chin *et al.*, 1993; Murant, 1993). By contrast, there are only two similar associations between vertebrate-infecting viruses (Berns, 1990; Taylor, 1991).

Does this difference between animal and plant associations result from differences in the defence mechanisms of these hosts, or in the dynamics of transmission, or mode of spread through tissues, or some combination of such factors? Do these differences reflect the fact that most virus infections of plants are systemic and persistent, whereas those of animals are not, so there are more opportunities for developing plant virus associations? Do these differences perhaps reflect contrasting consequences of multiple infection during spread? Present evidence does not provide sensible clues about this possibility, for, whereas stable populations of some plant viruses seem to require spread by inocula that contain a large and representative sample of the viral population (see TMGMV studies described above), vector-borne viruses seem to survive happily despite frequent "bottlenecking". Furthermore, large inocula of animal viruses often produce "defective interfering variants" of animal viruses, but these are uncommon among plant viruses (Kong *et al.*, 1997; Li and Simon, 1991). Finally, are any of these factors linked with the fact that most plant viruses with divided genomes have those components separately encapsidated, whereas animal viruses with divided genomes encapsidate the genomic components together?

Are There Different Modes of Evolution?

Neo-Darwinian Evolution?

In this chapter we have viewed evolution merely as "the process by which the genetic structure of the population of an organism changes with time", and above we have used comparisons calculated from such changes to map short- and long-term changes in viral populations. Fortunately, it is likely that most of the nucleotide changes upon which these calculations were based were "neutral", "evolutionary noise", seen by Campbell (1993) as "nothing but a soap opera. Its actors . . . forever changing and adapting to crisis after crisis but never getting anywhere", but used by tree builders to build dendrograms! But Campbell (1993) also exhorts us to search for signs of "progressive" evolution, namely "those changes in gene structure that have advanced a gene, gene product or organism", and also for signs of meta-evolution (Balch, 1989), namely "the advancement of evolution . . . New evolutionary mechanisms and strategies (that) come into being as life advanced"!

Progressive Evolution?

Probably the simplest form of progressive or functional evolution shown by viruses is their ability to switch hosts. The phylogenies of viruses and of their hosts are often, in parts, similar, but never the same. The simplest explanation of this pattern is that it reflects a mixture of coevolution and host switching, the relative proportions of which differ in different virus groups.

Some analyses have shown that a single nucleotide change may permit host switching (e.g. Boulton *et al.*, 1991; Fujita *et al.*, 1996), but most experiments suggest that the host range of a virus is determined by a complex interaction of parts of one or more of its genes, and that many pleiotropic effects may be involved

(Schoelz *et al.*, 1986; De Jong *et al.*, 1994; Fenczik *et al.*, 1995). Host switching is rarely observed in the field: new viruses emerge, but proof that they have switched host rather than extended their geographic range is difficult to obtain. One clear example is the appearance in Australia in 1991 of a variant of watermelon mosaic potyvirus 1 (WMV-1) able to infect and cause severe disease in papaya crops. WMV-1, otherwise known as PRSV-W (papaya ringspot potyvirus – watermelon), was first recorded in Australia in 1978 and is unable to infect papaya. Phylogenetic analysis of partial gene sequences of the papaya-infecting variant (PRSV-P) showed that it had most probably evolved from Australian PRSV-W (Bateson *et al.*, 1994). Similar analyses have shown subsequently that a similar papaya-infecting variant appeared in Thai papaya crops some time ago, and that it arose quite independently from Thai WMV-1 (Worawan Chaleeprom, personal communication). The genetic changes that permitted this host switch have not yet been determined, but it is clear that they occur infrequently.

The "progressive" evolutionary events that produce major successful new lineages of plant viruses can be deduced from comparisons of viral genera and families. They can be seen to involve both genetic recombination to produce new "decks" of the basic enzyme and virion protein genes of existing lineages, but also the appearance of entirely novel genes. The former seem to provide the core viral functions and the latter the ecological adaptations of the new lineages. Some, if not most, of the novel genes are probably derived *de novo* by overprinting existing genes (Keese and Gibbs, 1992; 1993; Gibbs and Keese, 1995); however the conditions that would permit novel genes to arise are unknown. It is possible that chronic long-term infections of clonally propagated plants would provide the conditions for mutational and recombinational experimentation to generate new genes.

Meta-Evolution?

Campbell (1993) noted that "Now, *Homo sapiens* has developed the facility to design his and other species' evolution towards preconceived goals deliberately . . . Ask not what evolution can do for a species but what a species can do for evolution!"

The significant changes produced by human activity have obviously already affected plant virus populations. The past 5000 years, and especially the past 100 years, have seen the spread of agriculture and horticulture. Crops usually consist of large uniform populations of genetically identical plants, often surrounded by similarly selected populations of weeds, many species of which are found worldwide. These conditions select the viruses and vectors able to exploit them. Thus, present broad-acre cropping practices encourage vector insects, especially whitefly and aphids, and hence the present is probably an all-time heyday of the geminiviruses, luteoviruses and potyviruses. Intensive cropping, where plants are grown in dense stands, handled or wounded by tools, fosters other viruses, such as the tobamoviruses, and also viroids. Thus evolution is no longer being driven solely by the interaction of the environment with the phenotypes encoded by genetically stored information, but nowadays also by information stored and manipulated in the human brain. Brain-driven intervention in the "natural order" is likely to continue and increase.

It is widely believed that increased levels of carbon dioxide and methane in the atmosphere caused by human activity are likely to increase global temperatures and change weather patterns. Organisms with short life cycles, like viruses and their vectors, will probably respond more quickly to change than other organisms, although in what way is difficult to predict. For example, increased temperatures may stimulate the growth rate of phytophagous vectors, but increased carbon dioxide concentrations in the atmosphere may increase the carbohydrate content of plants and, by altering the relative nutrient concentration, decrease their growth rate (Tripp *et al.*, 1992).

Perhaps the most direct way in which the evolution of plant viruses may soon become increasingly "brain-driven" is by the growing of crops of plants containing transgenes, especially those derived from viruses (MacLean *et al.*, 1997; Tepfer and Balázs, 1997a). Virus genes are being used in this way in attempts worldwide to pro-

duce plants with "cross-protection" to virus infection (Hamilton, 1980; Gibbs, 1982). Plant virus genes are being introduced for this purpose into cultivars of every major crop plant. The viral mRNAs or proteins expressed in the plants interfere with homologous invading viruses, sometimes making these cultivars partially or completely resistant to the virus species or strain from which the gene was obtained; a combination of "gene silencing" and "coat protein resistance" (Baulcombe, 1996), respectively, seems to be involved in viral pathogen-derived resistance. Some of the new cultivars are already being grown in commercial crops in the USA (Fuchs and Gonsalves, 1997).

However, it has been shown that plant viruses infecting plants containing viral transgenes can recombine with viral mRNAs expressed in the plants, and in this way acquire the transgene or a part of it (Greene and Allison, 1996; Frischmuth and Stanley, 1998). In some experiments, recombination between viral transgene mRNA and superinfecting viruses has been detected only when there is strong selection pressure against the parental virus, thus favoring the recombinant; the recombinants arose only in plants infected with a mutant virus and the recombinational event restored the function of the mutant. It might be thought that these conditions would not occur in crops; however, mutants of poor viability and fitness occur in populations of viruses in the field and as most plant viruses are transmitted by animal vectors, and transmission is a stochastic process, it is likely that some transgenic plants will be infected by these mutants alone. Thus, recombination between naturally occurring mutant viruses and the transgenic mRNA may well occur in the absence of competition from the virus from which the transgene was derived, and so they may be more likely to survive.

It is also possible that transgenic plants will indirectly influence the rate at which recombinants arise or survive. The first firm evidence of this has come from experiments by Jakab *et al.* (1997), who used transgenic plants carrying a gene from potato Y potyvirus. A line of these plants was partially resistant to one strain but not another, but when they infected the plants with both strains they detected several recombinants. Surprisingly, they did not detect recombinants in non-transgenic plants that were similarly co-infected. They concluded that the recombinants were outcompeted by one strain in the non-transgenic plants but that in the transgenic plants the replication of that strain was suppressed and the recombinants fostered. The recombinants identified by Jakab *et al.* (1997) were generated by recombination between the two virus strains, not between virus and transgene; however, these experiments have important implications as they demonstrate another unexpected way in which transgenes may influence the interactions between different virus species.

Thus the widespread cultivation of transgenic plants carrying viral genes may increase the pace of virus evolution and increase the frequency with which damaging new viral diseases arise in crops. The speed with which this evolutionary adventure will impact on agriculture is uncertain because, although most large plant virus laboratories are involved in developing crop plants with viral transgenes, very few laboratories are funded to study the possible effects of such plants on the environment and on natural viral populations (Tepfer and Balázs, 1997b). Thus the evolutionary effects of viral transgenes may well come as a surprise to mankind!

REFERENCES

Aldaoud, R., Dawson, W.O. and Jones, G.E. (1989) Rapid, random evolution of the genetic structure of replicating tobacco mosaic virus populations. *Intervirology*, **30**, 227–233.

Al-Kaff, N. and Covey, S.N. (1994) Variation in biological properties of cauliflower mosaic virus clones. *J. Gen. Virol.*, **75**, 3137–3145.

Alonso-Prados, J.L., Aranda, M.A., Malpica, J.M., García-Arenal, F. and Fraile, A. (1998) Satellite RNA of cucumber mosaic cucumovirus spreads epidemically in natural populations of its helper virus. *Phytopathology*, **88**, 520–524.

Ambros, S., Hernandez, C., Desvignes, J.C. and Flores, R. (1998) Genomic structure of three phenotypically different isolates of peach latent mosaic viroid: implications of the existence of constraints limiting the heterogeneity of viroid quasispecies. *J. Virol.*, **72**, 7397–7406.

Aranda, M.A., Fraile, A., García-Arenal, F. and Malpica, J.M. (1995) Experimental evaluation of the ribonuclease protection assay method for the assessment of genetic heterogeneity in populations of RNA viruses. *Arch. Virol.*, **140**, 1373–1383.

Aranda, M.A., Fraile, A., Dopazo, J., Malpica, J.M. and García-Arenal, F. (1997) Contribution of mutation and RNA recombination to the evolution of a plant pathogenic RNA. *J. Mol. Evol.*, **44**, 81–88.

Argos, P. (1988) A sequence motif in many polymerases. *Nucl. Acids Res.*, **16**, 9909–9916.

Argüello-Astorga, G., Herrera-Estrella, L. and Rivera-Bustamante, R. (1994) Experimental and theoretical definition of geminivirus origin of replication. *Plant Mol. Biol.*, **26**, 553–556.

Bacher, J.W., Warkentin, D., Ramsdell, D. and Hancock, F. (1994) Selection versus recombination: what is maintaining identity in the 3′ termini of blueberry leaf mottle nepovirus RNA1 and RNA2? *J. Gen. Virol.*, **75**, 2133–2137.

Balch, S.H. (1989) Metaevolution and biochemical history. *J. Soc. Biol. Struct.*, **12**, 303–318.

Baker, P.F. (1960) Aphid behaviour on healthy and yellows-virus-infected sugar-beet. *Ann. Appl. Biol.*, **48**, 384–391.

Bateson, M.F., Henderson, J., Chaleeprom, W., Gibbs, A.J. and Dale, J.L. (1994) Papaya ringspot potyvirus: isolate variability and the origin of PRSV type P (Australia) *J. Gen. Virol.*, **75**, 3547–3553.

Baulcombe, D. C. (1996) Mechanisms of pathogen-derived resistance to viruses in transgenic plants. *Plant Cell*, **8**, 1833–1844.

Berns, K.I. (1990) Parvovirus replication. *Microbiol. Rev.* **54**, 316–329.

Blok, J., Mackenzie, A., Guy, P. and Gibbs, A.J. (1987) Nucleotide sequence comparisons of turnip yellow mosaic virus isolates from Australia and Europe. *Arch. Virol.*, **97**, 283–295.

Botstein, D. (1980) A theory of modular evolution for bacteriophages. *Ann. NY Acad. Sci.*, **354**, 484–491.

Boulton, M.I., King, D.I., Donson, J. and Davies, J.W. (1991) Point substitution in a promoter-like region and the V1 gene affect the host range and symptoms of maize streak virus. *Virology*, **183**, 114–121.

Bruenn, J.A. (1991) Relationships among the positive strand and double-strand RNA viruses as viewed through their RNA-dependent RNA polymerases. *Nucl. Acids Res.*, **19**, 217–226.

Brunt, A.A., Crabtree, K., Dallwitz, M.J., Gibbs, A.J. and Watson, L. (1996) *Viruses of Plants.*, C.A.B. International, London.

Campbell, J.H. (1993) A tilt at cladism or Let's contemplate evolution instead of our belly buttons. *Mem. Assoc. Australas. Palaeontol.*, **15**, 43–50.

Chenault, K.D. and Melcher, U. (1994) Phylogenetic relationships reveal recombination among isolates of cauliflower mosaic virus. *J. Mol. Evol.*, **39**, 496–505.

Chin, L.S., Foster, J.L. and Falk, B.W. (1993) The beet western yellows virus ST9-associated RNA shares structural and nucleotide sequence homology with carmo-like viruses. *Virology*, **192**, 473–482.

Covey, S.N., Al-Kaff, N.S., Lángera, A. and Turner, D.S. (1997) Plants combat infection by gene silencing. *Nature (Lond.)*, **385**, 781–782.

D'Arcy, W.G. (1991) The Solanaceae since 1976, with a review of its biogeography. In: *Solanaceae III: Taxonomy, Chemistry, Evolution* (eds Hawkes, J.G., Lester, R.N., Nee, M. and Estrada, N.), pp. 75–137. Royal Botanic Gardens and Linnaean Society of London, London.

Dáros, J.A. and Flores, R. (1995) Characterization of multiple circular RNAs derived from a plant viroid-like RNA by sequence deletions and duplications. *RNA*, **1**, 734–744.

Daubert, S. and Routh, G. (1990) Point mutations in cauliflower mosaic virus gene VI confer host-specific symptoms changes. *Mol. Plant–Microbe Interact.*, **3**, 341–345.

Dawson, W.O., Beck, D. L., Knorr, D.A. and Grantham, G.L. (1986) cDNA cloning of the complete genome of tobacco mosaic virus

and production of infectious trancripts. *Proc. Natl Acad. Sci. USA*, **83**, 1832–1836.

De Jong, W., Mise, K. and Ahlquist, P. (1994) The multigenic nature of RNA virus adaptation to plants. *Trends Microbiol.*, **2**, 29–31.

Dobzhansky, T.G. (1964) *Heredity and the Nature of Man.* Harcourt, Brace and World, New York.

Dolja, V.V., Boyko, V.P., Agranovsky, A.A. and Koonin, E.V. (1991) Phylogeny of capsid proteins of rod-shaped and filamentous RNA plant viruses: two families with distinct patterns of sequence and probably structure conservation. *Virology*, **184**, 79–86.

Domingo, E. and Holland, J.J. (1997) RNA virus mutations and fitness for survival. *Annu. Rev. Microbiol.*, **51**, 151–178.

Domingo, E., Holland, J., Beibricher, C. and Eigen, M. (1995) Quasi-species: the concept and the word. In: *Molecular Basis of Virus Evolution* (eds Gibbs, A.J., Calisher, C.H. and García-Arenal, F.), pp. 181–191. Cambridge University Press, Cambridge.

Donis-Keller, H., Browning, K.S. and Clarck, J.M. (1981) Sequence heterogeneity in satellite tobacco necrosis virus RNA. *Virology*, **110**, 43–54.

Duarte, E.A., Clarke, D.K., Moya, A., Domingo, E. and Holland, J. J. (1992) Rapid fitness losses in mammalian RNA virus clones due to Müller's ratchet. *Proc. Natl Acad. Sci. USA*, **89**, 6015–6019.

Fagoaga, C., Semancik, J.S. and Durán-Vila, N. (1995) A citrus exocortis viroid variant from broad bean (*Vicia faba* L): infectivity and pathogenesis. *J. Gen. Virol.*, **76**, 2271–2277.

Fenczik, C.A., Padgett, H.S., Holt C.A., Casper, S.J. and Beachy, R.N. (1995) Mutational analysis of the movement protein of odontoglossum ringspot virus to identify a host-range determinant. *Mol. Plant–Microbe Interact.*, **8**, 666–673.

Fraile, A. and García-Arenal, F. (1991) Secondary structure as a constraint on the evolution of a plant viral satellite RNA. *J. Mol. Biol.*, **221**, 1065–1069.

Fraile, A., Malpica, J. M., Aranda, M.A., Rodríguez-Cerezo, E. and García-Arenal, F. (1996) Genetic diversity in tobacco mild green mosaic tobamovirus infecting the wild plant *Nicotiana glauca*. *Virology*, **223**, 148–155.

Fraile, A., Alonso-Prados, J.L., Aranda, M.A., Bernal, J.J., Malpica, J.M. and García-Arenal, F. (1997a) Genetic exchange by recombination or reassortment is infrequent in natural populations of a tripartite RNA plant virus. *J. Virol.*, **71**, 934–940.

Fraile, A., Escriu, F., Aranda, M.A., Malpica, J.M., Gibbs, A.J. and García-Arenal, F. (1997b) A century of tobamovirus evolution in an Australian population of *Nicotiana glauca*. *J. Virol.*, **71**, 8316–8320.

Frischmuth, T. and Stanley, J. (1998) Recombination between viral DNA and the transgenic coat protein gene of African cassava mosaic geminivirus. *J. Gen. Virol.*, **79**, 1265–1271.

Fuchs, M. and Gonsalves, D. (1997) Risk assessment of gene flow associated with the release of virus-resistant transgenic crop plants. In: *Virus-resistant Plants: Potential Ecological Impact* (eds Teper, M. and Balázs, E.), pp. 114–120. Springer-Verlag, Berlin.

Fujita, Y., Mise, K., Okuno, T., Ahlquist, P. and Furusawa, I. (1996) A single codon change in a conserved motif of a bromovirus movement protein gene confers compatibility with a new host. *Virology*, **223**, 283–291.

Fukuda, M., Okada, Y., Otsuki, Y. and Takebe, I. (1980) The site of initiation of rod assembly on the RNA of a tomato and a cowpea strain of tobacco mosaic virus. *Virology*, **101**, 493–502.

Gibbs, A. (1968) Plant virus classification. *Adv. Virus Res.*, **14**, 263–328.

Gibbs, A.J. (1980a) How ancient are the tobamoviruses? *Intervirology*, **14**, 101–108.

Gibbs, A.J. (1980b) A plant virus that partially protects its wild legume host against herbivores. *Intervirology*, **13**, 42–47.

Gibbs, A.J. (1982) Competition between viruses; a transposable phenomenon? In: *Genetic Engineering for Agriculture, the Substance behind the Promise* (eds Sutton, B.G. and Williams, P.), Australian Institute of Agricultural Science Occasional Publication 2, pp. 65–68. Australian Institute of Agriculture, Sydney, NSW.

Gibbs, A.J. (1983) Virus ecology – "struggle" of the genes. In: *Encyclopaedia of Plant Physiology*, vol. 12C (eds Lange, O.L., Nobel, P.S.,

Osmond, C.B. and Ziegler, H.), pp. 537–558. Springer-Verlag, Berlin.

Gibbs, A. (1986) Tobamovirus classification. In: *The Plant Viruses. 2. The Rod-shaped Plant Viruses* (eds van Regenmortel, M.H.V. and Fraenkel-Conrat, H.), pp. 167–180. Plenum Press, New York.

Gibbs, A.J. (1999) Evolution and origins of tobamoviruses. *Proc. Roy. Soc. B.*, in press.

Gibbs, M.J. (1995) The luteovirus supergroup: rampant recombination and persistent partnerships. In: *Molecular Basis of Virus Evolution* (eds Gibbs, A.J., Calisher, C.H. and García-Arenal, F.), pp. 351–368. Cambridge University Press, Cambridge.

Gibbs, M.J. and Cooper J.I. (1995) A recombinational event in the history of luteoviruses probably induced by base-pairing between the genomes of two distinct viruses. *Virology*, **206**, 1129–1132.

Gibbs, A. and Keese, P. (1995) In search of the origins of viral genes. In: *Molecular Basis of Virus Evolution* (eds Gibbs, A.J., Calisher, C.H. and García-Arenal, F.), pp. 76–90. Cambridge University Press, Cambridge.

Gibbs, M.J., Armstrong, J., Weiller, G.F. and Gibbs, A.J. (1997) Virus evolution; the past a window on the future? In: *Virus-resistant Plants: Potential Ecological Impact* (eds Tepfer, M. and Balázs, E.), pp. 1–19. Springer-Verlag, Berlin.

Gierer, A. and Mundry, K.W. (1958) Production of mutants of tobacco mosaic virus by chemical alteration of its ribonucleic acid *in vitro*. *Nature*, **182**, 1457–1458.

Goelet, P., Lomonossoff, G.P., Butler, P.J.G., Akam, M.E., Gait, M.J. and Karn, J.N. (1982) Nucleotide sequence of tobacco mosaic virus RNA. *Proc. Natl Acad. Sci. USA*, **79**, 5818–5822.

Goldbach, R. and de Haan, P. (1994) RNA viral supergroups and the evolution of RNA viruses. In: *The Evolutionary Biology of Viruses* (ed. Morse, S.), pp. 161–184. Raven Press, New York.

Gorbalenya, A.E., Koonin, E.V. and Wolf, Y.I. (1990) A new superfamily of putative NTP-binding domains encoded by genomes of small DNA and RNA viruses. *FEBS Lett.*, **262**, 145–148.

Goulden, M.G., Davies, J.W., Wood, K.R. and Lomonossoff, G.P. (1992) Structure of tobraviral particles: a model suggested from sequence conservation in tobraviral and tobamoviral coat proteins. *J. Mol. Biol.*, **227**, 1–8.

Greene, A.E. and Allison, R.F. (1994) Recombination between viral RNA and transgenic plant transcripts. *Science*, **263**, 1423–1425.

Grieco, F., Lanave, C. and Gallitelli, D. (1997) Evolutionary dynamics of cucumber mosaic virus satellite RNA during natural epidemics in Italy. *Virology*, **229**, 166–174.

Guilley, H., Richards, K. and Jonard, G. (1995) Nucleotide sequence of beet mild yellowing virus. *Arch. Virol.*, **140**, 1109–1118.

Habili, N. and Symons, R.H. (1989) Evolutionary relationship between luteoviruses and other RNA plant viruses based on sequence motifs in their putative RNA polymerases and nucleic acid helicases. *Nucl. Acids Res.*, **17**, 9543–9555.

Hahn, B.H., Shaw, G.M., Taylor, M.E. *et al.* (1986) Genetic variation in HTLV-III/LAV over time in patients with AIDS or at risk from AIDS. *Science*, **232**, 1548–1553.

Haldane, J.B.S. (1949) Disease and evolution. *Ric. Sci. Suppl.*, **19**, 68–76.

Hamilton, R.I. (1980) Defenses triggered by previous invaders: viruses. In: *Plant Disease: An Advanced Treatise, vol. 5. How Plants Defend Themselves* (eds Horsfall, J.G. and Cowling, E.B.), pp. 279–303. Academic Press, New York.

Harrison, B.D. (1981) Plant virus ecology: ingredients, interactions, and environment influences. *Ann. Appl. Biol.*, **99**, 195–209.

Harrison, B.D., Zhou, X., Otim-Nape, G.W., Liu, Y. and Robinson, D.J. (1997) Role of a novel type of double infection in the geminivirus-induced epidemic of severe cassava mosaic in Uganda. *Ann. Appl. Biol.*, **131**, 437–448.

Haseloff, J., Goelet, P., Zimmern, D., Ahlquist, P., Dasgupta, R. and Kaesberg, P. (1984) Striking similarities in amino acid sequence among nonstructural proteins encoded by RNA viruses that have dissimilar genomic organization. *Proc. Natl Acad. Sci. USA*, **81**, 4358–4362.

Hillman, B.I., Anzola, J.V., Halpern, B.T., Cavileer, T.D. and Nuss, D.L. (1991) First field isolation of wound tumor virus from a plant host: minimal sequence divergence from the type strain isolated from an insect vector. *Virology*, **185**, 896–900.

Holmes, F.O. (1941) A distinctive strain of tobacco-mosaic virus from *Plantago*. *Phytopathology*, **31**, 1089–1098.

Holmes, F.O. (1950) Indications of a New-World origin of tobacco-mosaic virus. *Phytopathology*, **41**, 341–349.

Ina, X. and Gojobori, T. (1994) Statistical analysis of nucleotide sequences of the hemagglutinin gene of human influenza A viruses. *Proc. Natl Acad. Sci. USA*, **91**, 8388–8392.

Ina, Y., Mizokami, M., Ohoba, K. and Gojobori, T. (1994) Reduction of synonymous substitutions in the core protein gene of hepatitis C virus. *J. Mol. Evol.*, **38**, 50–56.

Jakab, G., Vaistij, F.E., Droz, E. and Malnoë, P. (1997) Transgenic plants expressing viral sequences create a favourable environment for recombination between viral sequences. In: *Virus-resistant Plants: Potential Ecological Impact* (eds Tepfer, M. and Balázs, E.), pp. 45–51. Springer-verlag, Berlin.

Kamer, G. and Argos, P. (1984) Primary structural comparisons of RNA-dependent polymerases from plant, animal and bacterial viruses. *Nucl. Acids Res.*, **12**, 7269–7282.

Kearney, C.M., Donson, J., Jones, G.E. and Dawson, W.O. (1993) Low level genetic drift in foreign sequences replicating in an RNA virus in plants. *Virology*, **192**, 11–17.

Keese, P. and Gibbs, A. (1992) Origins of genes: big bang or continuous creation? *Proc. Natl Acad. Sci. USA*, **89**, 9489–9493.

Keese, P. and Gibbs, A. (1993) Plant viruses: master explorers of evolutionary space. *Curr. Opin. Genet. Devel.*, **3**, 873–877.

Kofalvi, S.A., Marcos, J.F., Cañizares, M.C., Pallás, V. and Candresse, T. (1997) Hop stunt viroid (HSVd) sequence variants from *Prunus* species: evidence for recombination between HSVd isolates. *J. Gen. Virol.*, **78**, 3177–3186.

Kong, Q., Oh, J.W., Carpenter, C.D. and Simon, A.E. (1997) The coat protein of turnip crinkle virus is involved in subviral RNA-mediated symptom modulation and accumulation. *Virology*, **238**, 478–85.

Koonin, E.V. and Dolja, V.V. (1993) Evolution and taxonomy of positive-strand RNA viruses: implications of comparative analysis of amino acid sequences. *Crit. Rev Biochem. Mol. Biol.*, **28**, 375–430.

Kruse, M., Koenig, P., Hoffmann, A. *et al.* (1994) Restriction fragment length polymorphism analysis of reverse transcription-PCR products reveals the existence of two major strain groups of beet necrotic yellow vein virus. *J. Gen. Virol.*, **75**, 1835–1842.

Kunkel, L.O. (1947) Variation in phytopathogenic viruses. *Annu. Rev. Microbiol.*, **1**, 85–100.

Kurath, G. and Dodds, J.A. (1995) Mutation analyses of molecularly cloned satellite tobacco mosaic virus during serial passage in plants: evidence for hotspots of genetic change. *RNA*, **1**, 491–500.

Kurath, G. and Palukaitis, P. (1989) RNA sequence heterogeneity in natural populations of three satellite RNAs of cucumber mosaic virus. *Virology*, **173**, 231–240.

Kurath, G. and Palukaitis, P. (1990) Serial passage of infectious transcripts of a cucumber mosaic virus satellite RNA clone results in sequence heterogeneity. *Virology*, **176**, 8–15.

Kurath, G., Rey, M.E.C. and Dodds, A. (1992) Analysis of genetic heterogeneity within the type strain of satellite tobacco mosaic virus reveals several variants and a strong bias for G to A substitution mutations. *Virology*, **189**, 233–244.

Kurath, G., Heick, J.A., Dodds, J.A. (1993) RNase protection analyses show high genetic diversity among field isolates of satellite tobacco mosaic virus. *Virology*, **194**, 414–418.

Lai, M.M. (1992) RNA recombination in animal and plant viruses. *Microbiol. Rev.*, **56**, 61–79.

Lartey, R.T., Voss, T.C. and Melcher, U. (1996) Tobamovirus evolution: gene overlaps, recombination, and taxonomic implications. *Mol. Biol. Evol.*, **13**, 1327–1338.

Li, W.-H. (1993) Unbiased estimates of the rates of synonymous and nonsynonymous substitutions. *J. Mol. Biol.*, **36**, 96–99.

Li, X.H. and Simon, A.E. (1991) *In vivo* accumulation of a turnip crinkle virus defective inter-

fering RNA is affected by alterations in size and sequence. *J. Virol.*, **65**, 4582–90.

Lindbo, J.A., Silva-Rosales, L., Proebsting, W.M. and Dougherty, W.G. (1993) Induction of a highly specific antiviral state in transgenic plants: implications for regulation of gene expression and virus resistance. *Plant Cell*, **5**, 1749–1759.

Liu, Y., Robinson, D.J. and Harrison, B.D. (1998) Defective forms of cotton leaf curl virus DNA-A that have different combinations of sequence deletion, duplication, inversion and rearrangement. *J. Gen. Virol.*, **79**, 1501–1508.

Lynch, M.R. Burger, R., Butcher, D. and Gabriel, W. (1993) The mutational meltdown in asexual populations. *J. Hered.*, **84**, 339–344.

MacLean, G.D., Waterhouse, P.M., Evans, G. and Gibbs, M.J. (eds) (1997) *Commercialization of Transgenic Crops: Risk, Benefit and Trade Considerations*. Australian Government Publishing Service, Canberra.

McNeil, J.E., French, R., Hein, G.L., Baezinger, P.S. and Eskridge, K.M. (1996). Characterization of genetic variability among natural populations of wheat streak mosaic virus. *Phytopathology*, **86**, 1222–1227.

Martinez-Torres, D., Carrio, R., Latorre, A., Simon, J.C., Hermoso, A. and Moya, A. (1998) Assessing the nucleotide diversity of three aphid species by RAPD. *J. Mol. Evol.*, **10**, 459–477.

Melcher, U. (1990) Similarities between putative transport of plant viruses. *J. Gen. Virol.*, **71**, 1009–1018.

Miller, W.A., Waterhouse, P.M. and Gerlach, W.L. (1988) Sequence and organisation of barley yellow dwarf virus genomic RNA. *Nucl. Acids Res.*, **16**, 6097–6111.

Moreno, I., Malpica, J.M., Rodríguez-Cerezo, E. and García-Arenal, F. (1997) A mutation in tomato aspermy cucumovirus that abolishes cell-to-cell movement is maintained to high levels in the viral RNA population by complementation. *J. Virol.*, **71**, 9157–9162.

Morozov, S.Y., Dolja, V.V. and Atabekov, J.G. (1989) Probable reassortment of genomic elements among elongated RNA-containing plant viruses. *J. Mol. Evol.*, **29**, 52–62.

Moya, A. and García-Arenal, F. (1995) Population genetics of viruses: an introduction. In: *Molecular Basis of Virus Evolution* (eds Gibbs, A.J., Calisher, C.H. and García-Arenal, F.), pp. 213–223. Cambridge University Press, Cambridge.

Moya, A., Rodríguez-Cerezo, E. and García-Arenal, F. (1993) Genetic structure of natural populations of the plant RNA virus tobacco mild green mosaic virus. *Mol. Biol. Evol.*, **10**, 449–456.

Murant, A.F. (1993) Complexes of transmission-dependent and helper viruses. In: *Diagnosis of Plant Virus Diseases* (ed. Matthews, R.E.F.), pp. 333–357. CRC Press, Boca Raton, FL.

Murao, K., Suda, N., Uyeda, I. *et al.* (1994) Genomic heterogeneity of rice dwarf phytoreovirus field isolates and nucleotide sequences of variants of genome segment 12. *J. Gen. Virol.*, **75**, 1843–1848.

Mushegian, A.R. and Koonin, E.V. (1993) Cell-to-cell movement of plant viruses. *Arch. Virol.*, **133**, 239–257.

Nei, M. (1987) *Molecular Evolutionary Genetics*. Columbia University Press, New York.

Nolan, M.F., Skotnicki, M.L. and Gibbs, A.J. (1996) RAPD variation in populations of *Cardamine lilacina* (Brassicaceae). *Austral. Systemat. Bot.*, **9**, 291–299.

Novella, I.S., Elena, S.F., Moya, A., Domingo, E. and Holland, J.J. (1995) Size of genetic bottlenecks leading to virus fitness loss is determined by mean of initial population fitness. *J. Virol.*, **69**, 2869–2872.

Olmstead, R. and Palmer, J.D. (1991) Chloroplast DNA and systematics of the Solanaceae. In: *Solanaceae III: Taxonomy, Chemistry, Evolution* (eds Hawkes, J.G., Lester, R.N., Nee, M. and Estrada, N.), pp. 161–168. Royal Botanic Gardens and Linnaean Society of London, London.

Ooi, K., Ohshita, S., Ishii, I. and Yahara, T. (1997) Molecular phylogeny of geminivirus infecting wild plants in Japan. *J. Plant Res.*, **110**, 247–257.

Palukaitis, P. and Roossinck, M.J. (1995) Variation of the hypervariable region of cucumber mosaic virus satellite RNAs is affected by the helper virus and the inial helper context. *Virology*, **206**, 765–768.

Pamilo, P. and Bianchi, N.O. (1993) Evolution of the Zfx and Zfy genes: rates and interdependence between the genes. *Mol. Biol. Evol.*, **10**, 271–281.

Parker, J.E. and Coleman, M.J. (1997) Molecular intimacy between proteins specifying plant–pathogen recognition. *TIBS*, **22**, 291–296.

Perry, K.L. and Francki, R.I.B. (1992) Insect-mediated transmission of mixed and reassorted cucumovirus genomic RNAs. *J. Gen. Virol.*, **73**, 2105–2114.

Pirone, T.P. and Blanc, S. (1996) Helper-dependent vector transmission of plant viruses. *Annu. Rev. Phytopathol.*, **34**, 227–247.

Qiu, W.P., Geske, S.M., Hickey, C.M. and Moyer, J.W. (1998) Tomato spotted wilt tospovirus genome reassortment and genome segment-specific adaptation. *Virology*, **244**, 186–194.

Raven, P.H. (1983) The migration and evolution of floras in the southern hemisphere. *Bothalia*, **14**, 325–328.

Revers, F., Legall, O., Candresse, T., Leromancer, M. and Dunez, J. (1996) Frequent occurrence of recombinant potyvirus isolates. *J. Gen. Virol.*, **77**, 1953–1965.

Rodríguez-Cerezo, E. and García-Arenal, F. (1989) Genetic heterogeneity of the RNA genome population of the plant virus U5-TMV. *Virology*, **170**, 418–423.

Rodríguez-Cerezo, E., Moya, A. and García-Arenal, F. (1989) Variability and evolution of the plant RNA virus pepper mild mottle virus. *J. Virol.*, **63**, 2198–2203.

Rodríguez-Cerezo, E., Elena, S.F., Moya, A. and García-Arenal, F. (1991) High genetic stability in natural populations of the plant RNA virus tobacco mild green mosaic virus. *J. Mol. Evol.*, **32**, 328–332.

Roossinck, M.J. and Palukaitis, P. (1995) Genetic analysis of helper virus-specific selective amplification of cucumber mosaic virus satellite RNA. *J. Mol. Evol.*, **40**, 25–29.

Sanger, M., Daubert, S. and Goodman, R.M. (1991) The regions of sequence variation in caulimovirus gene VI. *Virology*, **182**, 830–834.

Sanz, A.I., Fraile, A., Gallego, J.M., Malpica, J.M. and García-Arenal, F. (1999) Genetic structure and evolution of natural populations of cotton leaf curl geminivirus, a single-stranded DNA virus. Submitted.

Schoelz, J., Shepherd, R.J. and Daubert, S. (1986)

Region VI of cauliflower mosaic virus encodes a host range determinant. *Mol. Cell Biol.*, **6**, 2632–2637.

Schultz, P.H. and Dhondt, S. (1996) Cretaceous–Tertiary (Chicxulub) impact angle and its consequences. *Geology*, **24**, 963–967.

Skotnicki, M.L., Mackenzie, A.M., Ding, S.W., Mo, J.Q. and Gibbs, A.J. (1993) RNA hybrid mismatch polymorphisms in Australian populations of turnip yellow mosaic tymovirus. *Arch. Virol.*, **132**, 83–99.

Skotnicki, M.L., Mackenzie, A.M. and Gibbs, A.J. (1996) Genetic variation in populations of kennedya yellow mosaic tymovirus. *Arch. Virol.*, **141**, 99–110.

Smith, H.A., Swaney, S.L., Parks, D. Wernsman, E.A. and Dougherty, W.G. (1994) Transgenic plant virus resistance mediated by untranslatable sense RNAs: expression, regulation, and fate of nonessential RNAs. *Plant Cell*, **6**, 1441–1453.

Stenger, D.C. and McMahon, C.L. (1997) Genotypic diversity of beet curly top virus populations in the Western United States. *Phytopathology*, **87**, 737–744.

Suga, H., Uyeda, I., Yan, J. *et al.* (1995) Heterogeneity of rice ragged stunt oryzavirus genome segment 9 and its segregation by insect vector transmission. *Arch. Virol.*, **140**, 1503–1509.

Symon, D.E. (1991) Gondwanan elements of the Solanaceae. In: *Solanaceae III: Taxonomy, Chemistry, Evolution* (eds Hawkes, J.G., Lester, R.N., Nee, M. and Estrada N.), pp. 139–150. Royal Botanic Gardens and Linnaean Society of London, London.

Taylor, J.M. (1991) Human hepatitis delta virus. *Curr. Topics Microbiol. Immunol.*, **168**, 141–166.

Tepfer, M. and Balázs, E. (eds) (1997a) *Virus-resistant Plants: Potential Ecological Impact*. Springer-Verlag, Berlin.

Tepfer, M. and Balázs, E. (1997b) Concluding remarks and recommendations. In: *Virus-resistant Plants: Potential Ecological Impact* (eds Tepfer, M. and Balázs, E.), pp. 121–123. Springer-Verlag, Berlin.

Tripp, K.E., Kroen, W.K., Peet, M.M. and Willits, D.H. (1992) Fewer whiteflies found on CO_2-enriched greenhouse tomatoes with high C:N ratios. *Hort. Sci.*, **27**, 1079–1080.

Uyeda, I., Ando, Y., Murao, K. and Kimura, I. (1995) High resolution genome typing and genomic reassortment events of rice dwarf phytoreovirus. *Virology*, **212**, 724–727.

Van Regenmortel, M.H.V. (1986) Tobacco mosaic virus: antigenic structure. In: *The Plant Viruses. 2. The Rod-shaped Plant Viruses* (eds van Regenmortel, M.H.V. and Fraenkel-Conrat, H.), pp. 79–104. Plenum Press, New York.

Van-Vloten-Doting, L. and Bol, J.F. (1988) Variability, mutant selection, and mutant stability in plant RNA viruses. In: *RNA Genetics*, vol. 3 (eds Domingo, E., Holland, J. and Ahlquist, P.), pp. 37–52, CRC Press, Boca Raton, FL.

Visvader, J.E. and Symons, R.H. (1985) Eleven new sequence variants of citrus exocortis viroid and the correlation of sequence with pathogenicity. *Nucl. Acids Res.*, **13**, 2907–2920.

Wang, H. and Stubbs, G. (1994) Structure determination of cucumber green mottle mosaic virus by X-ray fiber diffraction. *J. Mol. Biol.*, **239**, 371–384.

Wang, H. Culver, J.N. and Stubbs, G. (1997) Structure of ribgrass mosaic virus at 2.9A resolution: evolution and taxonomy of tobamoviruses. *J. Mol. Biol.*, **269**, 769–779.

Wang, H., Planchart, A. and Stubbs, G. (1998) Caspar carboxylates: the structural basis of tobamovirus disassembly. *Biophys. J.*, **74**, 633–638.

Ward, C.W., Weiller, G.F., Shukla, D.D. and Gibbs, A.J. (1995) Molecular systematics of the Potyviridae, the largest plant virus family. In: *Molecular Basis of Virus Evolution* (eds Gibbs, A.J., Calisher, C.H. and García-Arenal, F.), pp 477–500. Cambridge: Cambridge University Press.

Weiller, G.F. and Gibbs A.J. (1995) DIPLOMO: the tool for a new type of evolutionary analysis. *CABIOS*, **11**, 535–540.

White, P.S., Morales, F.J. and Roossinck, M.J. (1995) Interspecific reassortment in the evolution of a cucumovirus. *Virology*, **207**, 334–337.

Wilson, A.C., Carlson, S.S. and White, T.J. (1977) Biochemical evolution. *Annu. Rev. Biochem.*, **46**, 573–639.

Yarwood, C.E. (1979) Host passage effects with plant viruses. *Adv. Virus Res.*, **25**, 169–190.

Young, D.J. and Watson, L. (1970) The classification of dicotyledons: a study of the upper levels of the hierarchy. *Austral. J. Bot.*, **18**, 387–433.

Zaccomer, B., Haenni, A.-L. and Macaya, G. (1995) The remarkable variety of plant RNA virus genomes. *J. Gen. Virol.*, **76**, 231–247.

Zanotto, P.M. de A., Gibbs, M.J., Gould, E.A. and Holmes, E.C. (1996) A reevaluation of the higher taxonomy of viruses based on RNA polymerases. *J. Virol.*, **70**, 6083–6096.

Zhou, X, Liu, Y., Calvert, L. *et al.* (1997) Evidence that DNA-A of a geminivirus associated with severe cassava mosaic disease in Uganda has arisen by interspecific recombination. *J. Gen. Virol.*, **78**, 2101–2111.

Zhou, X., Liu, Y., Robinson, D.J. and Harrison, B.D. (1998) Four DNA-A variants among Pakistani isolates of cotton leaf curl virus and their affinities to DNA-A of geminivirus isolates from okra. *J. Gen. Virol.*, **79**, 915–923.

12

Genetics, Pathogenesis and Evolution of Picornaviruses

Matthias Gromeier, Eckard Wimmer and Alexander E. Gorbalenya

INTRODUCTION

The discovery of viruses heralded an exciting new era for research in the medical and biological sciences. Many contemporary virologists do not know, however, that the first animal virus described was a picornavirus, the etiological agent of the dreaded foot-and-mouth disease in cloven-footed animals. The discovery of foot-and-mouth disease virus (FMDV) by F. Loeffler and P. Frosch in 1898 (Loeffler and Frosch, 1898) occurred at the same time as M.W. Beijerinck described the amazing "contagium vivum fluidum" in 1898. This liquid was a filtered leaf extract derived from tobacco plants suffering from tobacco mosaic disease. Free of bacteria, it was yet able to transmit the disease to uninfected plants. Already in 1892, I. Ivanovski had made a similar observation with tobacco mosaic virus but apparently he was unable to fully convince his peers of the significance of his discovery (Waterson and Wilkinsen, 1978).

Research on viruses, now formally in its hundred-and-first year, has yielded an immense harvest of biochemical and biological information. The studies were driven not only by an urgent need to understand, and possibly prevent, viral disease; they were also fueled by a strong curiosity about the minute biologicals called viruses, which we can view as chemicals, on the one hand and as "living" entities on the other. Poliovirus is an exquisite example of a chemical with a known empirical formula (Molla *et al.*, 1991) that can be crystallized (Schaffer and Schwerdt, 1955) yet causes a devastating disease in humans. Poliovirus was discovered 90 years ago by Landsteiner and Popper (1909) to be the causative agent of poliomyelitis. The current knowledge of its chemical and three-dimensional structure and of its life cycle and pathogenesis is second to none. Indeed, the intense research efforts on poliovirus over a period of nine decades will lead to its demise in the near future: global eradication of poliovirus is considered possible by the year 2000.

Following the identification of FMDV and poliovirus, a deluge of other viruses with similar properties were uncovered. These viruses have now been classified as Picornaviridae, a large family of small (Lat. *pico*) RNA (rna) viruses. Currently, Picornaviridae consists of six genera: *Enterovirus, Rhinovirus, Hepatovirus, Parechovirus, Cardiovirus* and *Aphthovirus* (Table 12.1). The first four genera include predominantly human pathogens, which cause a bewildering array of disease syndromes. Although a disease syndrome may be considered characteristic for a specific picornavirus group, the same syndrome can possibly be also produced by

Origin and Evolution of Viruses
ISBN 0–12–220360–7

TABLE 12.1 Picornaviridae (nd, no data; italics denote a non-human pathogen)

Genus	Clusters	Major associated syndromes	Receptor	References
Enterovirus	A			
	Coxsackieviruses A 2,3,5,7,8,10,12,14,16	Herpangina, hand-foot-and-mouth disease, respiratory disease, meningitis, poliomyelitis (CAV7)	nd	
	Enterovirus 71	Hand-foot-and-mouth disease, meningitis, paralysis	nd	
	B			
	Coxsackieviruses A9	(See CAV above), poliomyelitis, myocarditis, pleurodynia, meningitis, hand-foot-and-mouth disease, respiratory disease, neonatal, infections	$\alpha_v\beta_3$*	(1)
	Coxsackieviruses B1–6		HCAR**, DAF	(2)
	Echoviruses 1–9, 11–21, 24–27, 29–33	Meningitis, encephalitis, pleurodynia, exanthema	VLA-2(=$\alpha_2\beta_1$) DAF (=CD55)	(3)
	Enterovirus 69		nd	(4)
	C			
	Poliovirus types 1–3	Poliomyelitis, meningitis	CD155	(5)
	Coxsackieviruses 1, 11, 13, 15, 17, 18–22, 24	Common cold, infantile diarrhea	ICAM-1†	(6)
	Coxsackievirus 24v‡	Acute hemorrhagic conjunctivitis		
	D			
	Enteroviruses 70 (68)	Acute hemorrhagic conjunctivitis	nd	
	E			
	Bovine enterovirus types 1 and 2	Diarrhea (cattle)	nd	
*Parechovirus***	Human echovirus types 22 and 23	Respiratory disease, encephalitis	nd	
Rhinovirus	Major receptor group rhinov. (>90 serotypes)	Common cold	ICAM-1	(7)
	Minor receptor group rhinov. (>10 serotypes)	Common cold	LDL receptor	(8)
Hepatovirus	Hepatitis A virus	Type A hepatitis	HAVcr-1	(9)
Aphthovirus	Foot-and-mouth-disease virus FMDV A12 O1	Foot-and-mouth disease (cloven-footed livestock)	$\alpha_v\beta_3$ Heparan sulfate	(10) (11)
Cardiovirus	Encephalomyocarditis virus	Encephalitis, myocarditis (hoofed livestock)	VCAM-1	(12)
	Mengovirus	Encephalitis (murine)	nd	
	Theiler's murine encephalomyocarditis virus	Encephalitis	nd	
	Vilyuisk virus			

The following viruses have been recognized as picornaviruses on the basis of their genome sequences and physico-chemical properties as well as the result of comparative sequence analyses (see the section on Evolution): equine rhinovirus types 1 and 2, Aichi virus, porcine enterovirus, avian encephalomyelitis virus, infectious flacherie virus of silkworm

Clusters of enteroviruses refer to groups of enteroviruses arranged predominantly according to genotypic kinship (Hyypia et al., 1997). More clusters, including mainly animal enteroviruses, have been proposed.

List of human syndromes adapted from Melnick, 1996. Common syndromes in humans caused predominantly by one and/or other member(s) of the cluster but member viruses of other clusters may cause the same syndrome.

Receptors may be specific for specific serotypes. For details, see text.

References describing the identification of receptors: (1) Roivainen et al., 1994; (2) Tomko and Philipson, 1997; Shafren et al., 1997; (3) Bergelson et al., 1992; (4) Bergelson et al., 1994; Ward et al., 1994; (5) Mendelsohn et al., 1989; (6) Shafren et al., 1997; (7) Staunton et al., 1989; Greve et al., 1989; (8) Hofer et al., 1994; (9) Feigelstock et al., 1998; (10) Neff et al., 1998; Berinstein et al., 1995; (11) Jackson et al., 1996; (12) Huber, 1994.

* shared with adenovirus type 2.
† DAF (decay accelerating factor) may function as non-essential (infection-augmenting) coreceptor.
‡ Coxsackie virus A24v is a genetic variant of coxsackie virus A24.
** Pringle, 1996.

other picornaviruses. It has been realized that the cellular receptor guiding a virus to a target cell cannot be the sole determinant of a virus's pathogenic potential. Indeed, it is a major challenge of the day to decipher the molecular mechanism(s) that determine viral tissue tropism and disease.

What is the identity of picornaviruses? It relates to ancestral viruses whose identity we will never know. Comparative analyses of the structures of genomes and their products, however, have placed the picornaviruses into a large "picorna-like" virus family, in which they occupy a prominent place (discussed in the section on Evolution). These same analyses have led to an evolutionary tree of picornaviruses that reveals the extent of kinship (Figure 12.1A).

One result of these phylogenetic investigations was a radical reorganization of the taxonomy of *Enterovirus*, a genus of Picornaviridae comprising numerous members infecting the gastrointestinal tract. The enteroviruses have now been divided into clusters (Table 12.1; Figure 12.1B) grouping the viruses mainly corresponding to genotypes (Hyypia *et al.*, 1997). Earlier classifications were based (1) on specific properties of the virions, (2) on disease patterns, (3) on the apparent absence of pathogenesis (ECHO is an acronym for "enteric cytopathic human orphan" because no disease was originally correlated with these viruses), or (4) in reference to the site of discovery (e.g. the town of Coxsackie in New York State) and pathogenesis in suckling mice. As the number of known enteroviruses increased and the properties of these new isolates were elucidated, the need for a modified classification became apparent. However, even the latest dendrograms are likely to be modified again.

Principally, viruses that have been classified as belonging to a specific genus may be further divided into serotypes. A serotype is defined by the virus's ability to elicit a set of neutralizing antibodies ("antiserum") in a host animal; this set of neutralizing antibodies will generally *not* neutralize any other virus, regardless of the origin of the antiserum. Neutralizing antibodies, in turn, are elicited by structures specific for a virus's capsid, and they have been referred to as neutralization antigenic determinants (or sites). The poliovirion

carries at its surface four distinct neutralization antigenic determinants (Nomoto and Wimmer, 1987; Minor, 1990). However, poliovirus expresses only three unique sets of these four determinants; hence poliovirus occurs in three serotypes. Hepatitis A virus, on the other hand, expresses only one set of neutralization antigenic determinants; hence, it occurs in only one serotype. In contrast, human rhinoviruses (HRV) can express more than 100 unique sets of four antigenic determinants. Rhinoviruses, therefore, occur in more than 100 serotypes. It should be noted that a poliovirus has been constructed that expresses neutralization antigenic determinants of all three serotypes. This virus, which is severely handicapped in proliferation, is trivalent as it can be neutralized by all three serotype-specific antibodies (Murdin *et al.*, 1992).

A genus consisting of viruses that cause the same disease syndrome can be subdivided further on the basis of receptor use. For example, all member viruses of the genus *Rhinovirus* cause the common cold, yet they use two different receptors (ICAM-1 and LDL receptor; Table 12.1). On the basis of genotypes, however, this division no longer holds up (Figure 12.1).

As mentioned, the enteroviruses have now been subdivided into clusters based on genotypes (Table 12.1, Figure 12.1B). For example, the C-cluster consists of the three serotypes of poliovirus and of serotypes 1, 11, 13, 15, 17, 18–22, 24 of coxsackie A virus (C-CAV). Originally the C-CAVs were not considered related to polioviruses because of the profound difference in pathogenesis (common cold and poliomyelitis, respectively) and the different use of receptor (ICAM-1 and CD155, respectively). However, their very close kinship was revealed by genome sequence. This proximity has led to the interesting question of whether the C-CAVs are genetic variants of poliovirus (Harber *et al.*, 1995) or *vice versa* (discussed in detail in Evolution

An interesting recent variant of CAV24 is CAV24v, an agent that emerged in the early 1970s and that causes acute hemorrhagic conjunctivitis. This syndrome is also associated with a new variant of enterovirus 70, a D-cluster enterovirus (Table 12.1; Yin-Murphy, 1973). The phenomenon of the sudden appearance of enterovirus strains causing human diseases not previously associated with picornaviruses is of

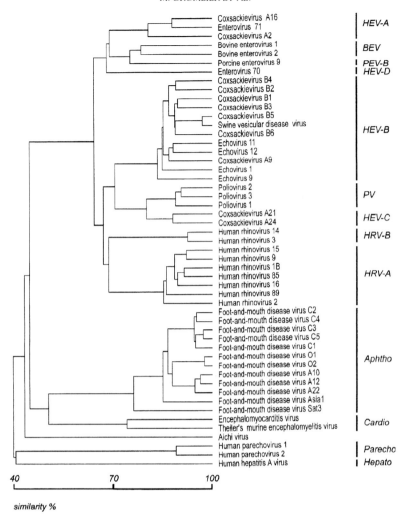

FIGURE 12.1 Evolutionary relationships of Picornaviridae. This part is a dendrogram of Picornaviridae showing amino-acid similarity in the capsid region. Sequences were aligned and the dendrogram was generated using the PILEUP program of the Genetics Computer Group software package. (*continued on facing page*)

greatest interest with respect to the dynamics of picornavirus diversification, particularly in view of the eradication of poliovirus.

What are the mechanisms by which the picornaviruses and other RNA virus families have diversified? Clearly, the genetic program inscribed into the viral genome is being changed as the viruses acquire new genetic traits. The predominant driving force of the changes in the genotype is largely an adaptation to new opportunities to proliferate. In the following, we will discuss some mechanisms and rules of genetic diversification and evolution of picornaviruses.

GENETICS OF PICORNAVIRUSES

The Picornavirus Genome

The picornavirus genome is single-stranded RNA of plus strand polarity, roughly 7500 nt long (for an overview, see Rueckert, 1996). The genome can be divided into three parts: the 5' non-translated region (5'NTR), the open reading frame of the polyprotein and the 3' non-translated region (3'NTR) (Figure 12.2).

FIGURE 12.1 (*continued*) An unrooted tree based on the region for the RNA polymerase 3D^{pol}. The tree was constructed using the maximum likelihood method (DNAML from the PHYLIP package). The branch lengths are proportional to genetic distance. In this tree the clusters are designated as Human Enterovirus-A, -B, -C, etc. (HEV-A; HEV-B; HEV-C, etc.). The groups boxed (PV and HEV-C, HEV-A, and HEV-D) are considered statistically significant by bootstrapping analysis. Figures kindly provided by Leena Kinnunen.

The 5'NTR

A signature of all picornaviruses is their unusually long 5'NTR (733–1199 nt; >10% of genome length), which seems to defy the notion that RNA viruses strive to keep their genome at minimum length (Wimmer *et al.*, 1993; see below). These 5'NTRs carry important structures: the terminal protein VPg, adjacent *cis*-acting sequences involved in RNA replication, and the internal ribosomal entry site (IRES), controlling translation.

The virus-encoded 5'-terminal protein, VPg (viral protein genome-linked) is covalently linked to the 5'-terminal uridylic acid via a O^4-(5'-uridylyl)tyrosine bond (Lee *et al.*, 1976, 1977; Flanegan *et al.*, 1977; Nomoto *et al.*, 1977b; Rothberg *et al.*, 1978). Picornavirus VPgs are

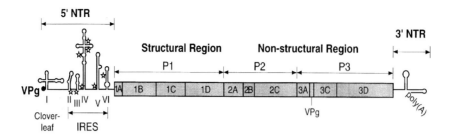

FIGURE 12.2 Structure of the poliovirus genome. The single-stranded RNA is shown with the genome-linked protein VPg (3B) at the 5′ end of the non-translated region (5′ NTR, single line) and the 3′ NTR (single line) connected to the poly(A) tail. The boxed region shows the polyprotein and vertical lines within the box indicate proteinase cleavage sites. The locations of the structural (P1) and non-structural (P2, P3) region are shown on top. RNA structural domains within the 5′ non-translated region are shown by Roman numerals, cloverleaf (I), IRES (II–VI). Stars indicate the positions of non-initiating AUG triplets.

22–24 amino acids long; their third amino acid (from the N-terminus) is always a tyrosine, the residue linking VPg to the genome. Genome-linked proteins are quite common amongst viruses belonging to the picorna-like super family (see Figure 12.10)

Picornavirus VPgs are attached to 5′-terminal nucleotide sequences that form complex structures typical for entero- and rhinoviruses on the one hand, or cardio-, aphtho- and hepatoviruses on the other. These sequences are important signals in genome replication. Entero- and rhinoviruses share a cloverleaf structure (Rivera *et al.*, 1988; Andino *et al.*, 1990) that has been subject to intense studies (see below). Relatively little is known about the role of corresponding structural elements (which do not form cloverleaves) of cardio, aphtho- and hepatovirus genomes.

The cloverleaf is followed by the internal ribosomal entry site (IRES), arguably the most complex *cis*-acting element in any RNA virus genome known (Figures 12.2, 12.3; Wimmer *et al.*, 1993). Picornavirus IRES elements, which are approximately 400 nt long, regulate the initia-

tion of polyprotein synthesis. In deviation to cap-dependent "scanning", IRESes promote internal ribosomal entry, i.e. they allow initiation of translation independently of a capping group and even a free 5′ end (Jang *et al.*, 1988, 1989; Pelletier and Sonenberg, 1988; Molla *et al.*, 1992; Chen and Sarnow, 1995). Remarkably, IRES elements are defined by their function, not by their sequences or apparent higher-order structure(s). This is illustrated in Figure 12.3, which depicts the sequence and folding pattern of the IRES elements of poliovirus and encephalomyocarditis virus (EMCV; Pilipenko *et al.*, 1989a,b). In spite of these differences, the poliovirus IRES has been exchanged with that of EMCV, leading to a novel chimeric virus with excellent growth properties (Alexander *et al.*, 1994). Similarly, the IRES of hepatitis C virus (HCV), a flavivirus, was found to functionally substitute for the poliovirus IRES, yielding a polio/HCV chimeric virus (Lu and Wimmer, 1996; Zhao *et al.*, 1999). Finally, a construct in which the IRES of human rhinovirus type 2 (HRV2) replaced that of poliovirus yielded a PV/HRV chimeric virus (PV1(RIPO)) that is

FIGURE 12.3 Sequences and secondary structures of IRES elements of poliovirus and encephalomyocarditis virus. **A.** Poliovirus IRES; individual domains have been labeled with Roman numerals. **B.** Encephalomyocarditis virus (EMCV) IRES; domains have been labeled with capital letters. Both IRESes contain a conserved YnXmAUG motif, of which the oligopyrimidine stretch (Yn) and the AUG triplet are indicated by solid bars. Note that in the EMCV IRES, the AUG triplet of the YnXmAUG motif is the initiating codon of the polyprotein. In the poliovirus IRES, this AUG triplet is silent and is separated from an AUG codon initiating the synthesis of the polyprotein by a "spacer sequence" of 154 nt (Jang *et al.*, 1990). Single attenuating mutations in the poliovirus vaccine strains map to domain V (Wimmer *et al.*, 1993).

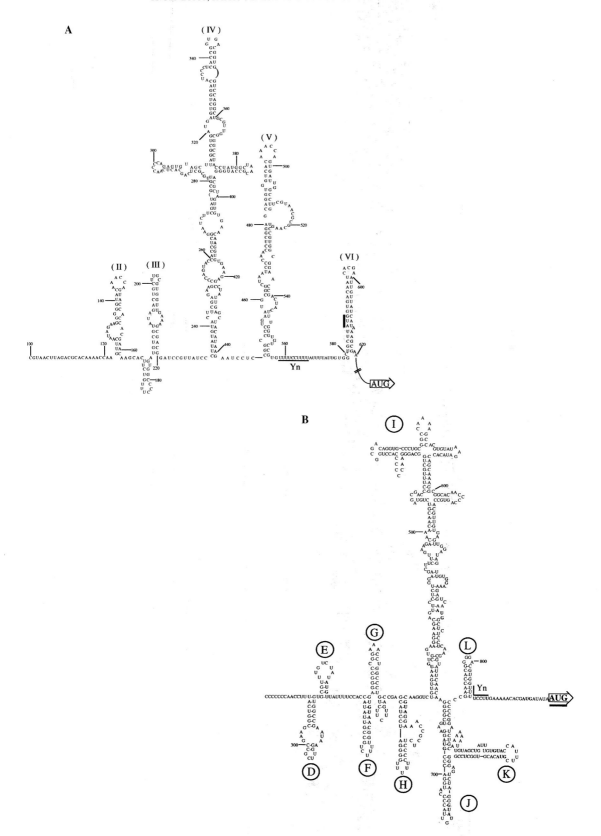

indistinguishable from wt poliovirus with respect to replication in HeLa cells yet is highly attenuated in poliovirus-receptor-transgenic mice and in monkeys (Gromeier *et al.*, 1996, 1999a; discussed in the section on Pathogenesis). The properties of this interesting novel virus will be discussed in a later section.

The mechanism by which IRES elements function is still obscure.

The Polyprotein

Translation of picornavirus mRNA is initiated downstream of the IRES to yield an unstable "polyprotein" that is rapidly cleaved by virus-encoded proteinases to proteins involved in viral proliferation (Figure 12.4; see also Evolution and Figure 12.10). It is important to note that the mRNA found in viral polyribo-somes that encodes the polyprotein differs from virion RNA in one important aspect: it is termi-nated with pUp ... (Hewlett *et al.*, 1976; Nomoto *et al.*, 1976). Apparently, the terminal protein VPg has been cleaved from incoming or from newly synthesized RNA. It has been sug-gested that the enzyme cleaving the VPg–pUp phosphodiester bond is of cellular origin but the reason for the removal of the protein and the nature of the enzyme catalyzing it remain unknown. Moreover, it is not clear whether the incoming VPg-linked virion RNA will be processed immediately after entry or whether the removal of VPg will occur only after the first round(s) of viral protein synthesis.

Entero- and rhinoviruses encode the two pro-teinases 2A^{pro} and 3C/3CD^{pro}, aphthoviruses the two proteinases L^{pro} and 3C^{pro}, and cardioviruses only the proteinase 3C^{pro}. Interestingly, both car-dioviruses and aphthoviruses have evolved a peculiar cleavage mechanism between 2A and 2B that occurs only in *cis* and is an enzyme-inde-pendent reaction (reviewed by Ryan and Flint, 1997). A similar as yet unknown mechanism of proteolytic cleavage is that between VP4 and VP2 (Figure 12.4D), which occurs only during maturation of the virion (maturation cleavage) and appears also to be proteinase-independent (Harber *et al.*, 1991; see below). The origin of these fascinating enzymes and of specific cleav-age events are discussed in the section on

Evolution. Since most details of proteolytic pro-cessing have been accumulated for poliovirus, much of the following discussion will center on this viral system.

The two poliovirus proteinases 2A^{pro} and 3C/3CD^{pro} cleave at different sites, as deter-mined by the sequences of the scissile bond (Figure 12.4B, C). Theoretically, the poliovirus polyprotein could give rise to 77 different cleav-age products if proteolytic processing by these enzymes and the maturation cleavage were entirely random (Wimmer *et al.*, 1993). In fact, only roughly 29–30 cleavage products have been identified in poliovirus-infected cells (Nicklin *et al.*, 1986). It has thus been concluded that processing of the picornavirus polyproteins is not random but follows a pathway that is determined by protein folding (masking of cleavage sites) and by the amino-acid sequences surrounding the scissile bond (Figure 12.4B, C; Harris *et al.*, 1990). For example, the precursor 3CD^{pro} can be cleaved into 3C^{pro} and 3CD^{pol} by a (*cis*?) cleavage in which the 3C/3CD^{pro} pro-teinase is involved. Both 3C^{pro} and 3D^{pol} are quite stable end-products of processing. However, in the case of poliovirus type 1 (Mahoney) (PV1 (M)), 3CD^{pro} can also be efficiently processed in *trans* by 2A^{pro} to 3C' and 3D' (Figure 12.4C), two polypeptides with no apparent function in viral proliferation (Lee and Wimmer, 1988). Just like 3C^{pro} and 3D^{pol}, 3C' and 3D' are quite stable end-products of processing, even though 3D^{pol} har-bors a perfect cleavage site for 2A^{pro} and 3C' har-bors a cleavage site for 3C/3CD^{pro} (Figure 12.4). Indeed, in PV1(M)-infected cells, nearly equal amounts of the four cleavage products of pre-cursor 3CD^{pro} are observed. It is assumed that structural constraints mask one or the other cleavage site from recognition and processing once the cleavage product has been formed (Lee and Wimmer, 1988).

The preferred cleavage sequence for 3C/3CD^{pro} in the poliovirus polyprotein is AxxQ*G; hence, cleavage sites with this sequence are usually rapidly processed. Numerous mutational studies have supported the identity of this 3C/3CD^{pro} cleavage motif (reviewed in Dougherty and Semler, 1993, and Wimmer *et al.*, 1993). An intriguing genetic analysis has made use of a viral construct that

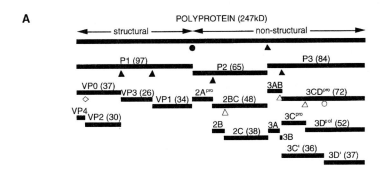

Site	P6	P5	P4	P3	P2	P1	P1'	P2'	P3'	P4'	P5'
B VP0/VP3	T	L	P	R	L	Q	G	L	P	V	M
VP3/VP1	Q	K	A	L	A	Q	G	L	G	Q	M
2A/2B	E	E	A	M	E	Q	G	I	T	N	Y
2B/2C	P	Y	V	I	K	Q	G	D	S	W	L
2C/3A	M	E	A	L	F	Q	G	P	L	Q	Y
3A/3B	L	F	A	G	H	Q	G	A	Y	T	G
3B/3C	R	T	A	K	V	Q	G	P	G	F	D
3C/3D	Y	F	T	Q	S	Q	G	E	I	Q	W
C VP1/2A	K	D	L	T	T	Y	G	F	G	H	Q
3C'/3D'	K	L	L	D	T	Y	G	I	N	L	P
D VP4/VP2	T	A	P	M	L	N	S	P	N	I	E

FIGURE 12.4 Processing scheme and cleavage sites of the poliovirus polyprotein. **A.** Proteolytic cleavages of the polyprotein. Triangles indicate cleavage by 3C^{pro} and/or 3CD^{pro}. Note that both enzymatic entities can efficiently cleave the non-structural proteins. In contrast, the P1 capsid precursor can be processed by 3CD^{pro} only. Solid triangles represent efficient cleavage sites, whereas open triangles represent slowly cleaved sites resulting in stable precursor proteins. The 2A^{pro}-mediated cleavages are depicted with circles. Only the cleavage between P1 and P2–P3 (solid circle) is essential, whereas the cleavage of 3CD^{pro} to 3C' and 3D' is dispensable (open circle). The maturation cleavage is indicated by the open diamond. The mechanism by which this cleavage occurs is unknown. Numbers in brackets indicate the molecular weight in kDa. **B–D.** Amino-acid residues at sites cleaved by **(B)** 3C^{pro} and/or 3CD^{pro}, **(C)** by 2A^{pro} and **(D)** during the maturation cleavage are shown in a single-letter code. The positions of the amino-acid residues are designated P1, P2, P3, . . . at the newly generated C-termini, or P1', P2', P3', . . . at the newly generated N-termini. The fastest cleavages catalyzed by 3C^{pro}/3CD^{pro} occur at sites in which the P4 position is a small aliphatic amino acid (e.g. AxxQ*G). Cleavage at TQSQ*G between 3C and 3D is slow, giving rise to the 3CD^{pro} cleavage intermediate with a long half-life (Cao and Wimmer, 1996).

mutated this AxxQ*G cleavage motif at a specific site in order to avoid proteolytic processing. The amino acids placed by the mutants into the motif confirmed the proposed interaction between substrate and enzyme during cleavage (Cao and Wimmer, 1996, and references therein). As will be discussed later, poliovirus is a purist with respect to cleavage signals, since the scissile bond in all cleavages,

catalysed by 3C/3CD^{pro}, is Q*G (Kitamura *et al.*, 1981; Semler *et al.*, 1981a, 1981b). In other picornaviruses, or viruses of the large picorna-like superfamily, the cleavage site may differ from the canonical Q*G signal.

A most important observation in studies of picornavirus proliferation is that cleavage intermediates may have important functions that in some cases may even be distinct from that of their end-products (e.g. 3CD^{pro} yielding 3C^{pro} and 3D^{pol}).

The structure of picornavirus 3C^{pro} enzymes has been accurately predicted by Gorbalenya *et al.* (1986), leading to the genetic analyses alluded to above. The structures were proved to be correct by X-ray crystallographic studies of 3C^{pro} of human hepatitis A virus (Allaire *et al.*, 1994) and human rhinovirus 14 (Matthews *et al.*, 1994).

The 3′NTR

Following the ORF, there is a heteropolymeric region that may be different with respect to length (72–126 nt) and structure in different picornavirus genomes (Xiang *et al.*, 1997). However, all picornavirus genomes terminate with poly(A), as was shown first for poliovirus (Yogo and Wimmer, 1972). The role these sequences play in replication will be discussed below.

The Polarity of Picornavirus RNA and Reverse Genetics

The genomic RNA of picornaviruses can serve as mRNA and, consequently, it is of the same polarity as cellular mRNA. By convention, this polarity has been designated plus-strand polarity (Baltimore, 1971). Fortunately, the genomic RNA of picornaviruses is infectious; that is, upon transfection into suitable host cells, virion RNA will initiate a complete infectious cycle (Wimmer *et al.*, 1993). Interestingly, poliovirus and its purified genome will replicate even in enucleated cells (Morgan-Detjen *et al.*, 1978), an observation suggesting that the nucleus does not contribute factors essential for viral proliferation.

Using reverse transcriptase, Racaniello and Baltimore (1981) generated full-length "complementary" DNA (cDNA) that contained the entire genetic information of the viral genome (currently, cDNA refers to double-stranded DNA generated from the original complementary DNA strands). Transfections into HeLa cells of the cDNA that contained heterologous DNA sequences at either end of the virus-specific sequences generated, surprisingly, poliovirus.

With this experiment, "reverse genetics" of RNA viruses was born as the RNA genome was now amenable to manipulations developed for DNA.

The efficiency with which the original cDNA clones induced an infectious cycle in HeLa cells was very low (about 10 PFU/µg DNA; Racaniello and Baltimore, 1981). Construction of plasmids that could replicate in transfected cells dramatically increased the specific infectivity to 10^3 PFU/µg DNA; Semler *et al.*, 1984). However, reverse genetics was made more practical when the cDNA was cloned downstream of the phage T7 RNA transcriptase promoter and, using purified T7 transcriptase, virtually unlimited amounts of highly infectious transcript RNA could be produced in a simple test-tube experiment (>10^5 PFU/µg of transcript RNA; van der Werf *et al.*, 1986). This was important because mutant genomes with highly debilitating replication phenotypes could not be recovered by the inefficient cDNA transfection method. It was known before that VPg is not required to be at the 5′ end for poliovirus RNA to be infectious (Nomoto *et al.*, 1977a). The 5′ end of the T7 transcripts is *ppp*GGUUAAAA . . . whereas that of virion RNA is VPg-pUUAAAA . . . The extra G residues do not prevent transfection but they reduce the specific infectivity of the transcript. In any event, picornavirus RNA is quite tolerant of modifications of the 5′ end of its genome and, in all cases, the virion RNAs isolated after transfections have the authentic terminus restored (Wimmer *et al.*, 1993).

Infectious cDNAs have now been generated from members of all picornaviruses. The method of choice to generate virus remains transfection of T7 transcripts. Recently developed methods of RT/PCR allow researchers to generate infectious cDNA clones in less than 1 month (Tellier *et al.*, 1996).

The Mechanism of Genome Replication

In general terms, genome replication proceeds in two steps: synthesis of a complementary RNA strand (–strand) that then serves as template for plus RNA strands (+strands; Figure

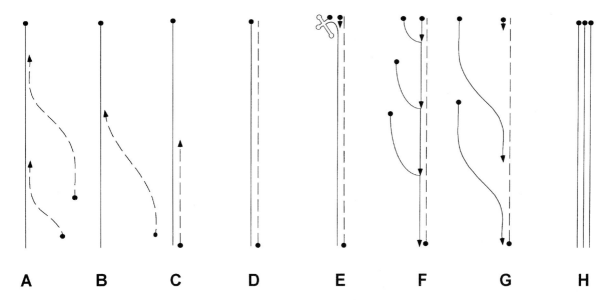

A   B   C   D   E   F   G   H

FIGURE 12.5 Steps in the replication of the poliovirus genome. Parental, positive-stranded virion RNA (solid line) is transcribed, yielding – RNA (broken line) after protein (VPg)-priming by the viral RNA-dependent RNA polymerase 3D^{pro} (enzyme or any other proteins involved are not shown). A replicative intermediate (RI) form consisting of a single +strand template and multiple nascent – RNA strands (**A**) has not been detected, so that, more probably, intermediates in – RNA synthesis are either mainly single-stranded (**B**) or double-stranded (**C**). Elongation of the nascent – RNA (**C**) yields the replicative form (RF) double-stranded RNA (**D**). Available evidence suggests that the RF is an intermediate in genome replication (discussed in Xiang *et al.*, 1997). Accordingly, a cloverleaf/RNP is formed at the end of the RF that promotes VPg-primed synthesis of +RNA (**E**). The structures formed after multiple initiation could either be "closed" (entirely base-paired; **F**) or "open" (**G**). Available evidence suggests that structure **F** is the correct intermediate (note that 3D^{pol} is an unwindase). For details, see Wimmer *et al.*, 1993, and Xiang *et al.*, 1997. Modified from Wimmer *et al.*, 1993.

12.5). The validity of this scheme has been known for almost three decades yet only very few details of the individual steps have been elucidated (Agol *et al.*, 1999). Because the vast majority of studies have been carried out with poliovirus, this review will concentrate predominantly on this viral system.

Viral and Cellular Polypeptides Involved in Genome Replication

With the exception of the capsid proteins, all viral non-structural proteins and even processing intermediates have been implicated in genome replication (Xiang *et al.*, 1997). The evidence for the involvement of these proteins (2A^{pro}, 2B, 2BC, 2C, 3A, 3AB, VPg, 3C/3CD^{pro}, 3D^{pol}) is based largely on genetic data or on biochemical experiments assumed to be indicative of genome replication (Wimmer *et al.*, 1993; Xiang *et al.*, 1997). For example, genetic and bio-

chemical analyses of 3AB strongly suggest that this protein, a non-specific RNA binding protein, and the proteinase 3CD^{pro} participate in the formation of an initiation complex for +strands (Xiang *et al.*, 1997). Another example is the involvement of 2C in RNA replication. Briefly, poliovirus RNA synthesis is highly sensitive to the presence of 2 mM guanidine hydrochloride (Gua HCl); poliovirus mutants resistant to 2 mM Gua HCl harbor a single amino-acid exchange (N179A/G) in polypeptide 2C. It has recently been established that 2C is an ATPase (and not a GTPase) and we now refer to it as 2C^{ATPase} (Pfister and Wimmer, 1999). The ATPase activity of purified 2C^{ATPase} is inhibited by 2 mM Gua HCl, whereas that of purified 2C^{ATPase} with a N179/G mutation is resistant to this concentration of the drug (Pfister and Wimmer, 1999). On the basis of these considerations, it can be assumed that the ATPase activity of 2C^{ATPase} is essential for genome replication. Just as with

3AB or 3CD^{Pro}, however, the step(s) by which 2C^{ATPase} is exerting its essential function are still unknown.

The only proteins whose role in genome replication has been firmly established are VPg and 3D^{pol}. The crystal structure of 3D^{pol} has recently been solved (Hansen and Schultz, 1997), a result that will greatly advance our (limited) understanding of this important enzyme. Importantly, 3D^{pol} was established already in 1977 as being a primer-dependent and RNA-dependent RNA polymerase (Flanegan and Baltimore, 1977). Although a deluge of circumstantial evidence suggested that a uridylylated form of VPg might serve as primer for 3D^{pol} (Nomoto et al., 1977b; Wimmer, 1982; Takeda et al., 1986; Toyoda et al., 1987), direct evidence for this mechanism has been obtained only very recently (Paul et al., 1998b). Briefly, VPg is being uridylylated to VPg-pU(pU) by the viral RNA polymerase 3D^{pol} in the presence of template (poly(A)). VPg-pU(pU) then primes the transcription of poly(A), leading to the synthesis of poly(U), which is the 5' terminus of –strands (Paul et al., 1998b).

In spite of these seemingly simple experiments (Paul et al., 1998b), the mechanism of initiation of RNA synthesis was a matter of controversy for almost two decades. Baltimore's and Flanegan's groups presented evidence favoring "hairpin priming", whereas Wimmer's group accumulated data suggesting "protein priming" (reviewed by Richards and Ehrenfeld, 1990). The controversy has finally been settled in favor of protein priming.

At low concentration of enzyme, poliovirus polypeptide 3AB stimulates the transcriptional activity of 3D^{pol} up to 100-fold (Lama et al., 1994; Plotch et al., 1989; Paul et al., 1994). Indeed, biochemical and genetic evidence suggests that 3D^{pol} and 3AB form a complex in solution (Molla et al., 1994). The significance of these observations is not yet known.

An important additional property of 3D^{pol} is its ability to unwind double-stranded RNA. That is, the enzyme, while transcribing a template, can replace a dormant RNA strand that is hybridized to the template with the new strand that is just being synthesized (Cho et al., 1993). It should be noted, however, that 3D^{pol} is not a helicase as it will not separate two strands without transcribing one of them (Cho et al., 1993).

The participation in picornavirus replication of cellular proteins, referred to by investigators as "host factors", has also had a history of controversies. Several polypeptides were proposed to be involved in replication (e.g. a kinase or a uridylic acid transferase) but these proteins have disappeared after further analysis (Richards and Ehrenfeld, 1990). Ehrenfeld's and Semler's groups have recently identified a cellular 38 kDa RNA binding protein, poly(rC) binding protein 2 (PCBP2), that is not only required for the function of the poliovirus IRES but it has also the propensity to bind, together with 3CD^{Pro}, to the poliovirus 5'-terminal cloverleaf (Blyn et al., 1996). PCBP2 (or PCBP1, a protein related to PCBP2; Gamarnik and Andino, 1997) is undoubtedly the "host factor p36" that was originally proposed by Baltimore's group to effect the binding of 3CD^{Pro} to the poliovirus cloverleaf (Andino et al., 1990, 1993).

Andino et al. (1993) provided first evidence suggesting that the formation of a specific protein/cloverleaf RNP complex consisting of viral protein 3CD, a cellular protein ("p36") and the viral RNA is required for the initiation of +strand synthesis (Andino et al., 1993). This hypothesis has been further supported by the discovery of PCBP2 (Gamarnik and Andino, 1997; Parsley et al., 1997). PCBP2 is therefore a sensible candidate for a "host factor" involved in poliovirus RNA replication. However, poliovirus protein 3AB can replace PCBP2 in all biochemical reactions characteristic of the formation of a 5' terminal RNP. Moreover, 3AB and 3CD^{Pro}, both cleavage products of the P3 precursor (Figure 12.4A), are associated in solution (Molla et al., 1994). Finally, the phenotypes of mutants of 3AB in vivo and in vitro support the conjecture that 3AB is involved in the formation of a cloverleaf/3CD^{Pro}/3AB complex (Harris et al., 1994; Xiang et al., 1995a,b). Currently, there is no compelling evidence in favor of the cloverleaf/3CD^{Pro}/PCBP2 complex over that of cloverleaf/3CD^{Pro}/3AB with respect to poliovirus genome replication (see a discussion in Xiang et al., 1997).

Cis-*acting RNA Signals*

Recognition of RNA signals located somewhere in the RNA genome is a prerequisite for specificity in genome replication. This review will concentrate only on *cis*-acting elements of entero- and rhinoviruses because, as mentioned earlier, the overwhelming number of experiments deal with these viruses.

Currently, only the 5'-terminal cloverleaf has been firmly established as a *cis*-acting signal in enterovirus genome replication (see previous section), although the mechanism by which it functions is still obscure. Clearly, the formation of a specific RNP plays a role the significance of which will be discussed below.

More complicated is the recognition of the +strand template for the initiation of –strands. Since replication of picornavirus RNAs commences at the 3'-terminal poly(A), a homopolymeric sequence found also in most cellular mRNAs, poly(A) alone cannot be a determinant for virus-specific –strand synthesis. VPg-pU(pU) can be synthesized in the presence of poly(A), and VPg-poly(U), the 5' end of –strands, will follow the synthesis of the primer (Paul *et al.*, 1998). This reaction, however, does not reveal the mechanism of specificity.

Mutational analysis of the heteropolymeric sequence of the 3'NTR of enteroviruses indicated that this region was critically important for replication (Pilipenko *et al.*, 1996; Melchers *et al.*, 1997). However, the poliovirus 3'-terminal heteropolymeric sequence can be replaced with that of HRV14, a hairpin with no apparent homology with the poliovirus structure, and the resultant poliovirus/HRV14 hybrid genome replicated with wt kinetics (Rohll *et al.*, 1995). Even more startling was a report from Semler's group that presented evidence that the heteropolymeric region could be deleted altogether without loss of viability (Todd *et al.*, 1997). Currently, the paradox intrinsic to these findings remains unsolved (Agol *et al.*, 1999).

It is possible that the 3' heteropolymeric region plays an important role in the efficient formation of an initiation complex for replication but to a much lesser extent in +strand template recognition. The authentic recognition signal may reside in RNA-internal sequences, as

proposed by McKnight and Lemon (1996). These authors reported that, surprisingly, a stem loop structure mapping to the coding region of the HRV14 capsid proteins was absolutely necessary for genome replication. Fittingly, a stem-loop RNA structure that has been uncovered in poliovirus RNA also appears to play a role in genome replication; it maps to the coding region of 2C^{ATPase} (Goodfellow *et al.*, 1998). The mechanism by which these new elements influence replication has yet to be resolved.

Finally, evidence has been presented suggesting that sequences within the IRES play a role in genome replication (Borman *et al.*, 1994; Shiroki *et al.*, 1995). This is difficult to comprehend if one considers chimeric IRES viruses. As mentioned above, the cognate IRES of poliovirus can be replaced with IRES elements from different viruses whose IRES are merely related (HRV2, HRV14, CBV4, CAV9, CAV24, EV71; Gromeier *et al.*, 1996, 1999a and unpublished results) or entirely different (EMCV; Alexander *et al.*, 1994; HCV, Lu and Wimmer, 1996; Zhao *et al.*, 1999) without loss of genome replication.

Cis-*dominance of Translation in Genome Replication*

Defective interfering particles (DI particles; see below) of poliovirus are naturally occurring variants with deletions (of varying sizes) in the P1 region, encoding the capsid proteins. DI particles can replicate their RNA without helper function but they need wt virus for encapsidation. Sequence analyses of genomic RNAs of DI particles led Nomoto and his colleagues to the surprising observation that in all cases the deletions were in-frame of the polyprotein coding sequence. On the other hand, artificial genomes engineered with out-of-frame deletions were unable to replicate their RNA, even in the presence of wt helper virus (Kuge *et al.*, 1986; Hagino-Yamagushi and Nomoto, 1989). It was concluded that translation was necessary for the cognate genome to replicate. That is, translation had a *cis* effect on replication that could not be complemented in *trans* by a helper genome. These observations were later confirmed and extended (Wimmer *et al.*, 1993; Novak and Kirkegaard, 1994; Agol *et al.*, 1999).

There are several hypotheses that are used to explain the phenomenon. The least likely is that certain replication proteins can only function in *cis*. If so, only viral mRNA could serve as template in RNA synthesis. Since viral mRNAs lack VPg (Hewlett *et al.*, 1976; Nomoto *et al.*, 1976; see above), every +strand RNA that functions as template in RNA synthesis should also lack VPg. Available evidence suggests that all RNA templates involved in replication are terminated with VPg (Nomoto *et al.*, 1977b; Petterson *et al.*, 1977; Wu *et al.*, 1978; Larsen *et al.*, 1980). Furthermore, RNA replication occurs in a tight membranous environment (Bienz *et al.*, 1992). Thus, it is unlikely that these genome replicating membranous complexes also harbor viral polysomes (Wimmer *et al.*, 1993). Indeed, crude, membranous replication complexes can be isolated from infected cells that can replicate poliovirus RNA yet they are free of ribosomes (Takegami *et al.*, 1983; Takeda *et al.*, 1986; Toyoda *et al.*, 1987).

An alternative explanation is that the observed *cis* effect is operating only during the very first round of translation at the onset of infection. Clearly, translation of an infecting genome will have to be somehow arrested to allow the template to switch from translation to transcription. It is possible that, once the switch has been made, replication can proceed independently of translation. This does not exclude the possibility that viral proteins, perhaps intermediates with a short half-life or short-lived protein complexes, must be continuously supplied to the RNA synthesizing machinery.

The question of the switch from translation to RNA synthesis of infecting + stranded genomic RNA has been subject of much speculation. The classical study of Kolakofsky and Weissmann (1971) on phage Qβ replication solved the dilemma by showing that the phage replicase (a complex of four proteins) can repress translation of viral mRNA. A similar model has been proposed for poliovirus by Gamarnik and Andino (1998): the formation of an RNP consisting of cloverleaf/3CD^{pro}/PCBP2 at the 5' end of the viral mRNA inhibits further translation, thereby switching the template to replication. One problem with this model is that at the peak of poliovirus replication, translation and RNA synthesis occur concomitantly in the presence of an excess of 3CD^{pro} molecules (note that for each virus particle, 60 molecules of 3CD^{pro} are synthesized; the ratio of viral +strand RNA to unprocessed 3CD^{pro} may be 1:100 through most of the replicative cycle). Moreover, if the genome has to be translated for replication to occur, how can inhibition of translation promote RNA synthesis?

Genome Replication: An Overview

A very schematic representation of steps in genome replication is shown in Figure 12.5 (Wimmer *et al.*, 1993). The possible RNA structures involved in replication have been divided into three categories: (1) single-stranded +RNA (continuous line; H), −strand RNA (dotted lines); (2) closed replicative form (RF: D) and open replicative form (B); and (3) closed (C, F) and open (G) replicative intermediate RNA (RI). The structures differ in an important aspect: in one case the nascent RNA strands are always fully base-paired to the template ("closed form"; C, F), in the other case, the nascent strands are only base-paired at the replication fork ("open form"; A, B, G). All strands are VPg-linked (indicated by closed circles; Nomoto *et al.*, 1977b).

We have argued before that the cumulative evidence favors the "closed forms" for RF and RI but this view may not be shared by others (Wimmer *et al.*, 1993; Xiang *et al.*, 1997). Since 3D^{pol} is an "unwindase" (Cho *et al.*, 1993; see above), the scheme does not necessarily require a helicase. Indeed, so far no picornaviral helicase has been identified, and purified 2C^{ATPase} has stubbornly refused to exhibit such activity (Pfister and Wimmer, 1999).

Briefly, VPg will be uridylylated at the 3'-terminal poly(A). VPg-pU(pU), in turn, will then prime synthesis of −strands (Figure 12.5C). It is unlikely that multiple initiation of −strands on the same template (prior to completion of the first −strand) occurs, since an RI with multiple −strands (such as in Figure 12.5A) has not been found in infected cells (Bishop and Koch, 1969). It is even possible that initiation at the poly(A) tail of poliovirus RNA occurs only once. Completion of the −strand will thus yield RF

(Figure 12.5D), which we consider an intermediate in replication and not a byproduct (Wimmer *et al.*, 1993). One compelling argument in favor of this assumption is that in the RF the 5′ end of +strands is in the close vicinity of the 3′ end of −strands, a prerequisite first proposed by Baltimore and his colleagues (Andino *et al.*, 1993; Harris *et al.*, 1994). Destabilization of this end of the RNA will lead to the formation of an RNP consisting either of cloverleaf/3CD^{Pro}/3AB or cloverleaf/3CD^{Pro}/PCBP2, which, in turn, will free the 3′ end of the −strand for VPg-primed +strand synthesis to occur (Figure 12.5E). Multiple initiation at this end will lead to the multistranded RI (Figure 12.5F), the nascent or full-length +strands being replaced during transcription by the 3D^{Pol} unwindase.

Initiation of +strands may be more efficient than initiation of −strands; hence the large excess of +strands over −strands in infected cells. Note that a reconstituted replication system of purified viral and cellular components capable of synthesizing +strands from input +strands has not been achieved; thus many of the hypotheses put forward in this scheme have not yet been tested.

Replication of Poliovirus in Oocytes

Gamarnik and Andino (1996) have described a novel system to study poliovirus replication in *Xenopus* oocytes by injecting poliovirus RNA into these cells. However, virus will replicate only if a HeLa cell S10 extract was co-injected with the RNA. Interestingly, the authors have been able to separate the HeLa supporting activities (S10) into two factors, one necessary for poliovirus IRES-driven translation, the other for poliovirus RNA synthesis. This system offers an excellent opportunity to separate and characterize viral and cellular factors involved in virus replication.

Cell-free, De Novo Synthesis of Poliovirus

Viruses, lacking the genetic information as well as the tools to provide most of the essential components to replicate, are obligatory intracellular parasites. The complexity of viral proliferation – macromolecular synthesis of polypeptides and genomic nucleic acid, and encapsidation – has led to the text book wisdom that viruses are obligatory intracellular parasites unable to proliferate outside living cells.

However, poliovirus RNA (obtained either from virions or by transcription with phage T7 RNA polymerase from plasmid DNA), when incubated in an extract of uninfected HeLa cells void of nuclei, mitochondria and cellular mRNA, will direct translation, genome replication and genome encapsidation such that infectious particles are formed. These newly synthesized virions are indistinguishable from poliovirus isolated from tissue cultures. Thus, a picornavirus (poliovirus) is the first virus that has been synthesized *de novo* in a cell-free extract of mammalian cells (Molla *et al.*, 1991).

This experiment has nullified the notion that viruses can proliferate exclusively in living cells. Moreover, the novel approach can be used to study individual steps of viral replication in the absence of cell-membrane barriers. Several interesting observations regarding protein–protein interactions, the role of membranes, of cellular membranous components or soluble cellular factors, or of inhibitors of viral RNA synthesis, have been published (Barton and Flanegan, 1993; Molla *et al.*, 1993c, 1994; Barton *et al.*, 1995; Parsley *et al.*, 1997; Cuconati *et al.*, 1998; Towner *et al.*, 1998). The use of the cell-free cellular extract for studies of poliovirus RNA replication, however, is still in the early stages of exploitation. Nevertheless, it has been possible to even achieve genetic recombination of poliovirus in cell-free HeLa extracts (Duggal *et al.*, 1997; Tang *et al.*, 1997; Duggal and Wimmer, 1999; see below).

Error Rate in Genome Replication

General Observations

In the course of transcription, all template-dependent nucleic acid polymerases make errors in incorporating nucleotides with roughly the same frequency (10^{-3}–10^{-4}). As is discussed in Chapter 7, this phenomenon has profound biological consequences for RNA viruses. Because RNA viruses have chosen not to develop mechanisms by which misincorporations of nucleotides can be recognized and corrected, the average number of "spontaneous" mutations per replication of the genome, referred to as *error rate*, is around 10^{-4}.

The high error rate in the absence of mechanisms of proofreading and editing has several consequences.

First, the average genome length of animal RNA viruses is small (12 000 nt). Notwithstanding the genome of the exceptional coronavirus (30 000 nt), RNA viruses with genomes exceeding 150 kb (e.g. the DNA viruses, herpes viruses, poxviruses, iridoviruses) are inconceivable because of the high probability that each genome would carry multiple mutations after each round of synthesis. It should be noted that these considerations by no means imply that DNA viruses with very small genomes do not exist. In fact, the animal virus with the smallest known genome is hepatitis B virus (3.2 kb). As to picornaviruses, their average genome length is 8000 nt (see also Wimmer *et al.*, 1993).

Second, RNA viruses replicate near the threshold of error catastrophe (Holland *et al.*, 1990). That is, the artificial increase of misincorporation of nucleotides (e.g. by chemical mutagens) may lead to a rapid decline of the viability of the entire virus population.

Third, plaque-purified clones of RNA viruses are not homogeneous but populations of many different, albeit very closely related genotypes; hence the term "quasispecies" (Eigen, 1993).

Fourth, the genetic heterogeneity allows an RNA to rapidly adapt to a changing environment.

Mutations and "Quasi-infectious" Genomes

A simple example should demonstrate the ease with which a drug-resistant mutant of poliovirus can be isolated. As mentioned, poliovirus RNA replication is highly sensitive to the presence of 2 mM guanidine hydrochloride (Gua HCl). After plating a stock of plaque-purified poliovirus on a monolayer of HeLa cells in the presence of 2 mM Gua HCl, a few plaques will arise corresponding to resistant variants (*gr*) with mutations mapping to 2C^{ATPase} (Pincus *et al.*, 1986; Tolskaya *et al.*, 1994).

In the case of the selection of *gr* poliovirus mutants, it should be noted that the resistant variants already existed in the population of the inoculating virus. If the virus inoculum had been entirely free of *gr* variants, no selection of *gr* mutants could have occurred since the drug inhibits RNA synthesis; hence, there would have been no misincorporation of nucleotides to generate the *gr* mutations in 2C^{ATPase}. Although it may be a trivial thing to repeat, it is important to remember that genetic variation by misincorporation of nucleotides (just as recombination) requires replication. No replication, no mutants.

The genetic plasticity of genotypes and the dynamics of genetic variation can be studied conveniently when transcript RNA, produced by transcription of cDNA with T7 RNA polymerase, is transfected on to HeLa cell monolayers and the corresponding plaque phenotype of progeny virus is analysed. In the case of wt virus, the plaques are, by convention, "large". If mutant RNAs are analysed in plaque assays, one may observe only "small plaques" with a rare "large" plaque emerging on the plate. This rare large plaque may signal a reversion (either directly or through suppresser mutations) to a fast-replicating genotype. Passage of the population of small and large plaque phenotype viruses (at multiplicities of infection of more than 5) will rapidly yield populations of only the faster-growing virus because the impaired genomes are eliminated by competition. An example of this phenomenon has been described by Lu *et al.*, (1995b), who analysed a hybrid poliovirus in which the cognate 2A^{pro} coding sequence was exchanged to that of coxsackie B4 virus.

A special case of a genetic phenomenon is that of a "quasi-infectious" genome. This term was originally introduced by Agol and his colleagues (Gmyl *et al.*, 1993) to describe the following phenomenon. Genetically engineered poliovirus variant RNA was transfected into HeLa cells. Progeny virus was harvested, sometimes only after prolonged incubations of the transfected tissue cell cultures. Analysis of the genotypes of progeny virus genomes (by RT/PCR) revealed only revertant or pseudorevertant RNAs. None of the original mutant genotypes were detectable. This phenomenon can be explained if the original mutant genotype was able to replicate its RNA, albeit only at levels too low for virus production or even for the development of CPE. Nevertheless, the slowly replicating mutant genome allowed for mutation (either misincorporation or deletions, insertions), eventually leading to fast-growing genotypes. By definition, the progeny of quasi-infectious genomes will not yield virus with the parental genotype.

If a mutation (point mutation, linker insertion, etc.) engineered into the genome RNA is lethal, the lesion may effect complete abrogation of genome replication. Hence, reversion to viability cannot be expected. An interesting example of quasi-infectious versus lethal mutations in the poliovirus genome was described when mutations in VPg were studied (Kuhn et al., 1988; Reuer et al., 1990; Cao and Wimmer, 1995). As mentioned, VPg is linked to the genome via a O^4-(5'-uridylyl) tyrosine (the tyrosine in position three of VPg). A mutation of tyrosine to phenylalanine (Y3F) was originally described as being lethal (Reuer et al., 1990). This conclusion made sense, since phenylalanine lacks a O^4 hydroxyl group for phosphodiester formation. However, Cao and Wimmer (1995) later observed that cells transfected with VPg(Y3F) variant RNA produced viable virus, albeit only at very low frequency and only after prolonged incubation of the cultures. All of the progeny genomes carried a F3Y reversion. The possibility of contamination of the cultures with wt virus was excluded. The only explanation for this surprising result was that the VPg(Y3F) variant was quasi-infectious, presumably in that the threonine residue in position four of VPg may have served as a (poor) surrogate acceptor for uridylylation and protein priming of RNA synthesis (it should be noted that genome-linked terminal proteins are often attached to serine residues; Salas, 1991). Further analyses supported this hypothesis. A VPg(T4A) variant was found to be viable, expressing good growth kinetics. In contrast, VPg(F3Y, T4A) variant RNA never yielded progeny virus and was, therefore, considered unable to replicate its RNA. This mutation then can be considered to be lethal.

Genetic analysis of mutant genomes and their revertants has been an invaluable tool to study the structure and function of picornavirus genetic elements and picornavirus proteins (Wimmer et al., 1993).

Genetic Recombination

Genetic recombination of picornaviruses is the exchange of genetic elements between two viruses that may occur during replication in the same cell. Discovered by Hirst (1962), and used first by Cooper and his colleagues (Cooper, 1977) to map poliovirus genetic units, lingering skepticism about the phenomenon was dispelled through biochemical analyses of poliovirus recombinant proteins (Romanova et al., 1980; Tolskaya et al., 1983) or FMDV recombinant genomes (King et al., 1982). (For a detailed review of recombination, the reader is referred to Wimmer et al., 1993.)

Detection of Recombinants

Picornavirus recombinants have been detected because (1) they acquired genetic traits from the parental strains allowing them to proliferate under conditions restricting the growth of either parent and (2) they arose in excess over (replication competent) revertants of the parental strains (Cooper, 1977). Restricting conditions for the selection of recombinants can include specific drugs, such as 2 mM Gua HCl, monoclonal neutralizing antibodies (Emini et al., 1984) or host-cell specificity (Duggal et al., 1997). An elegant method to study poliovirus recombination under normal growth conditions (without selection) has been developed by Jarvis and Kirkegaard (1992).

Mechanism of Recombination

A wealth of experimental data has shed light on the most important steps in recombination. Details of individual steps in recombination, however, remain to be elucidated. The current knowledge can be summarized as follows.

1. Recombination is homologous and it occurs by copy choice; i.e. an incomplete (nascent) RNA strand may switch template strands

during genome replication. The probability of crossover depends strongly upon the degree of homology between the two recombining viral genomes.

2. Genetic analyses have indicated that template switching occurs (predominantly?) during –strand synthesis (Kirkegaard and Baltimore, 1986).

3. Recombination is precise: no deletions or insertions at sites of homologous recombination have been observed. This is true even if crossover occurred in the 154 nt long noncoding region (spacer region) between IRES and initiating AUG of poliovirus (Jarvis and Kirkegaard, 1992) even though the sequence of this "spacer" is not conserved amongst the three poliovirus serotypes (Toyoda et al., 1984). Indeed, deletions in the "spacer" could conceivably be tolerated in view of the observations by Kuge and Nomoto (1987) and others that the spacer can be partially or completely deleted without loss of viability.

4. Template switching requires that the replication complex pauses, allowing a heterologous (invading) +strand to offer its service as template. An unsolved question is whether pausing and crossover is random (Jarvis and Kirkegaard, 1991) or non-random (Romanova et al., 1986). The latter is in all probability true.

Agol and his colleagues have proposed that higher order structures formed on template RNAs may favor pausing of RNA synthesis and crossover (Romanova et al., 1986) In addition, King (1988) suggested that there are preferred sites of recombination in poliovirus RNA, and that crossover may be favored immediately after synthesis of two uridylate residues (UU) in the nascent strand.

Duggal and Wimmer (1999) observed that crossover patterns changed significantly when recombination occurred at different temperatures. Specifically, crossover between two genetically marked RNA strands at 34°C occurred over a wide range of the genome with preference for sequences coding for structural proteins in the 5'-terminal half of the genome. In contrast, recombination in vivo at 37°C and 40°C yielded crossover patterns that had shifted dramatically to a region encoding nonstructural proteins (Duggal and Wimmer, 1999). Preferential selection of recombinants at 37°C and 40°C was ruled out by analyses of the growth kinetics of the recombinants. The reason for the temperature effect is unknown. Temperature-dependent stability of higher order RNA structures seems possible.

Recombination Frequencies

Recombination frequencies are calculated by dividing the yield of the recombinant virus by the sum of the yield of the parental virus. For picornaviruses with linear genomes, the distance between genetic markers used to determine recombination is proportional to the recombination frequency. As mentioned above, the degree of homology between the parental genomes strongly influences the probability for crossover.

The frequency of recombination between homologous genomes is remarkably high (2×10^{-3} between markers only 600 nt apart; Jarvis and Kirkegaard, 1992). It has been estimated that 10–20% of the homologous viral genomes may undergo genetic recombination within a single growth cycle (King, 1988). This would mean an unprecedented genetic shuffling between genotypes of which the fittest retain the "wt" phenotype.

Experimental results that support a high frequency of recombination between sibling strands were obtained with engineered, quasi-infectious poliovirus genomes carrying two adjacent VPg sequences (Cao and Wimmer, 1996). After transfection, all progeny viruses had lost the downstream VPg, most probably by homologous recombination during –strand synthesis. On studying recombination using genetically marked genomes, 90% of the recombination events occurred between sibling strands.

Finally, Jarvis and Kirkegaard (1992) have demonstrated that the frequency of recombination increases with the progression of the infectious cycle; i.e. the larger the concentration of intracellular viral RNA the higher the probability of recombination.

Cell-free Recombination of Poliovirus

Recombination in a cell-free extract of uninfected HeLa cells (Molla *et al.*, 1991) has recently been reported by two groups. Recombination of parental viruses in the cell-free medium was detected either by RT/PCR in the absence of selection (Tang *et al.*, 1997), or by plating the progeny virus under conditions that were restricted for either parent (Duggal *et al.*, 1997). The recombination frequencies were found to be roughly the same as that *in vivo* (in tissue culture cells).

The crossover pattern of recombination *in vitro* and *in vivo* at 34°C was the same, lending credibility to the cell-free system as reflecting an *in-vivo* environment (Duggal and Wimmer, 1999). The *in-vitro* approach has the potential to decipher some basic steps in recombination, as, for example, the invasion of heterologous template strands into the replication complex. As mentioned before, the pattern of recombination changed significantly when recombination was carried out *in vivo* at 37°C or 40°C. Unfortunately, this effect at higher temperature cannot be analysed *in vitro* because cell-free synthesis of poliovirus is highly inefficient or completely absent at temperatures above 36°C (Molla *et al.*, 1993c).

Poliovirus Vectors, Defective Interfering Particles and Illegitimate Recombination

Picornavirus genomes are extremely "plastic" in that any change in their genotype can lead to unexpected nucleotide rearrangements. This has been observed, for example, in analyses of IRES elements, where deletions or insertions lead to unexpected new genotypes with excellent growth properties (see, for example, Dildine and Semler, 1989; Gmyl *et al.*, 1993; Pilipenko *et al.*, 1992; Alexander *et al.*, 1994; Charini *et al.*, 1994; Cao and Wimmer, 1995). Genetic plasticity is particularly apparent when poliovirus genomes are constructed harboring foreign sequences. The analysis of genetic rearrangements and deletions is especially important when picornavirus genomes are to be used as vectors for the delivery of foreign genes.

Picornavirus vectors and illegitimate recombination Generally, polioviruses respond to the insertion of foreign sequences by rapidly deleting these sequences either partially or completely. The driving force behind the selection of deletion variants may be: (1) excessive length of the genome, restricting encapsidation; (2) interference with efficient processing of the polyprotein; (3) interference with initiation of translation; (4) alteration of RNA structures necessary for replication; or others. There are examples, however, where poliovirus, or deletion mutants thereof, appear to tolerate a foreign sequence inserted into the genome.

An interesting approach to studying IRES elements was the insertion of the EMCV IRES into the ORF of poliovirus, thereby making unnecessary the primary cleavage between P1*P2 catalyzed normally by 2A^pro (Figure 12.6B; Molla *et al.*, 1992). The resulting RNA transcripts proved highly infectious (Molla *et al.*, 1992). Although this dicistronic virus expressed a small plaque phenotype, neither the plaque size nor the genotype surrounding the insertion changed over six passages. These observations suggested that the insertion was stable, at least under the conditions studied. The EMCV IRES has no apparent sequence homology with any sequence of the poliovirus genome, the poliovirus IRES included. This eliminated the possibility that the EMCV IRES was rapidly removed by homologous recombination. Perhaps, illegitimate recombination (see below) to delete the IRES without debilitating the virus was a very rare event and not apparent in progeny virus. It should be noted in parenthesis that the viability of the dicistronic virus shown in Figure 12.6B proved for the first time the function of the EMCV IRES as a true internal ribosomal entry site (Molla *et al.*, 1992).

On the basis of these experiments, a specific version of novel expression vectors was constructed (Figure 12.6C,D) that included, in addition to the foreign IRES, a foreign ORF (Alexander *et al.*, 1994; Lu *et al.*, 1995a). None of these vectors was genetically stable over extended numbers of passages, and in some cases the deletion occurred during first passage (Lu *et al.*, 1995a). Nevertheless, the insertion of a foreign ORF between two IRESes in the 5'NTR,

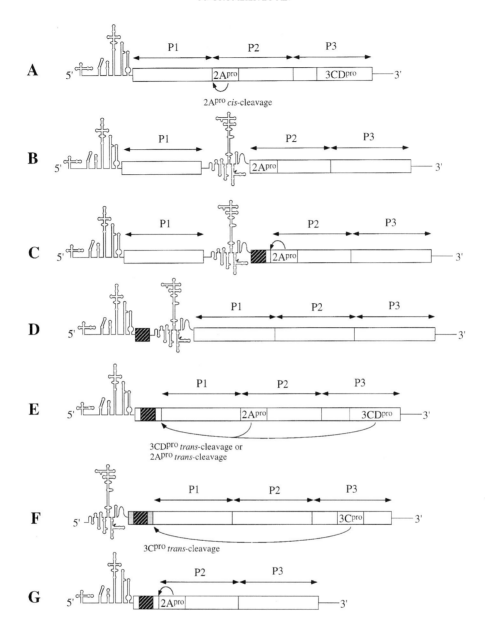

FIGURE 12.6 Picornavirus genomes and expression vectors. **A.** The poliovirus genome. **B.** Dicistronic poliovirus RNA in which the EMCV IRES (Figure 12.3) has been inserted into the genome at the site specifying primary 2A^{pro} cleavage. In this case, the function of the 2A^{pro} has been rendered superfluous. **C.** Expression vector based on the dicistronic genome depicted in **B**. The foreign gene (dark stippled box) was inserted upstream of 2A^{pro}, which now delivers the foreign protein by a *cis* cleavage. **D.** Dicistronic poliovirus generated by inserting a foreign gene and the EMCV IRES into the 5'NTR. In this case the foreign gene is synthesized independently from the polyprotein. **E.** Generation of an expression vector by fusing the coding sequence of a foreign gene to the N-terminus of the poliovirus polyprotein. The foreign gene product is liberated through *trans* cleavage, by either 3D^{pro}/3CD^{pro} or 2A^{pro}. **F.** Expression vector based on mengovirus, a cardiovirus that carries a small leader sequence preceding the P1 region of the polyprotein (see Figure 12.9). In this case, the foreign gene is inserted into the L coding sequence. Note that the organization of the genomes in **E** and **F** is identical. **G.** Encapsidation-incompetent poliovirus expression vector in which a portion of the P1 coding sequence has been replaced by a foreign gene. This genome can be encapsidated in *trans* but, by itself, it can only go through one cellular cycle of replication.

as shown in Figure 12.6D, yielded a replicating poliovirus vector that efficiently expressed the CAT gene over several passages (Alexander *et al.*, 1994). No deletion was apparent after the first passage. Remarkably, the genome of this construct is 17% larger than that of the wt genome, an observation indicating that the capsid of naturally occurring polioviruses is not "full". However, attempts failed to encapsidate and express the larger luciferase gene instead of the CAT gene in the context of the dicistronic virus. The luciferase activity was clearly detectable in cells transfected with the appropriate dicistronic transcript, but the genome harboring luciferase was not encapsidated (Alexander *et al.*, 1994). Apparently, an increase of genome length to 31% (luciferase gene plus EMCV IRES) was not tolerable for encapsidation (Alexander *et al.*, 1994).

A different strategy to convert poliovirus to an expression vector was the fusion of a foreign ORF directly to the poliovirus polyprotein (Figure 12.6E; Andino *et al.*, 1994). In these experiments, the strategy of Altmeyer *et al.* (1994) was mirrored, which made use of the genetic make-up of cardioviruses (mengovirus or EMCV). Specifically, the cardiovirus polyprotein is preceded by a small leader protein (67 aa) that is cleaved from the capsid region P1 by the viral 3C^{pro} proteinase, thereby allowing maturation and encapsidation of the virion. Altmeyer *et al.* (1994) inserted into the leader sequence of the mengovirus genome a foreign gene (Figure 12.6F) and expressed the product of this fusion protein over several passages in tissue culture (Altmeyer *et al.*, 1994, 1995). Andino *et al.*, (1994) generated a similar "leader" protein in front of the poliovirus polyprotein. In this case, however, it was necessary to engineer a novel 3C/3CD^{pro} cleavage site between the foreign ORF (the new "leader") and the viral polyprotein such that the foreign polypeptide can be cleaved from the poliovirus capsid precursor (Figure 12.6E). Although these poliovirus constructs were originally claimed to express excellent growth properties and, more importantly, were reported to be genetically highly stable (Andino *et al.*, 1994), the poliovirus-based vectors proved, in fact, impaired in replication and prone to rapid deletions, at least if the insert was

more than 500 nt in size (Mueller and Wimmer, 1998, and references therein; see below). It should be noted that the cardiovirus-based vectors also suffered from loss of the inserts of a foreign gene upon repeated passage, an observation suggesting that even cardioviruses do not tolerate an extended leader protein for the purpose of gene therapy (Altmeyer *et al.*, 1994, 1995).

In a third strategy of the construction of picornavirus vectors, the P1 capsid region of the picornavirus genome is partially replaced with a foreign ORF (Figure 12.6G), yielding proliferation-incompetent replicons that appear to be genetically quite stable (see, for example, Porter *et al.*, 1993). For a possible application as vectors in gene therapy, the replicons are *trans*-encapsidated via a vaccinia virus-based P1 expression vector, with relatively low yields of proliferation-incompetent virions. The apparent genetic stability of these replicons may be due to the fact that the RNAs are similar in size when compared to the wt genome, and that the naturally occurring *cis* cleavage (between P1*P2) catalyzed by 2A^{pro} is highly efficient, placing no restriction on this step of polyprotein processing. However, the rapid selection of faster growing variants that lost the foreign gene is unlikely, since the *trans*-encapsidated replicons can only proceed to a one-step infectious cycle. This is very different from the selection pressure in proliferation-competent vectors, which engage in second-round infections.

What is the mechanism by which poliovirus may eliminate foreign sequences? Homologous recombination cannot function because there is not enough sequence homology to engage in crossover. Pilipenko *et al.* (1995) have nevertheless proposed that short sequences may serve as parting and anchoring sites for template switching in illegitimate (non-homologous) recombination. An alternative mechanism is "loop-out" deletion, in which the nascent strand skips endogenous sequences, jumping to an upstream sequence that serves as anchoring sequence. Given the high frequency by which recombination occurs among sibling strands, a crossover mechanism may be favored, but decisive experiments to decide between these two mechanisms are lacking. In any case, a detailed study of genetic variations of polyprotein fusion vectors (Figure 12.6D) strongly supports the model of parting and anchoring sites for template switching or loop-out deletion (Mueller and Wimmer, 1997; see below).

Briefly, when expression vectors (Figure 12.6E) consisting of a *gag* gene (encoding p17-p24; 1161 nt) of human immunodeficiency virus that was fused to the N-terminus of the poliovirus polyprotein (Andino *et al.*, 1994; Mueller and Wimmer, 1998) were analysed after transfection into HeLa cells, the genomes were not only found to be severely impaired in viral replication but they were also genetically unstable (Mueller and Wimmer, 1997). Upon replication, the inserted sequences were rapidly deleted as early as the first growth cycle in HeLa cells. Interestingly, the vector viruses did not readily revert to wt sequences but rather retained some of the insert plus the artificial 3C/3CD^(pro) cleavage site (to allow processing at the N-terminus of the polyprotein). Thus, variants of different genotypes that replicated nearly as well as wt poliovirus had followed an evolutionary pathway towards the genetic organization of cardioviruses (Mueller and Wimmer, 1998). That is, the poliovirus polyprotein of these variants was preceded by *gag*-derived "leader" proteins of different but distinct sizes (predominantly between 20 and 50 aa long), the most prominent leader size reflecting the length of that in cardioviruses (67 aa). In the immediate vicinity of the deletion borders of several isolates, short direct sequence repeats were observed that are likely to allow alignment of RNA strands for non-homologous (illegitimate) recombination during –strand synthesis (Figure 12.7; Mueller and Wimmer, 1998). Interestingly, the selection of the leader size occurred during the very first rounds of replication of the transfected RNA; in most cases, as sequential shortening of the leader sequence was not observed.

Defective interfering particles An interesting phenomenon of naturally occurring deletion mutants of picornaviruses are defective interfering particles (DI particles) that can be (rarely) discovered in laboratory stocks of virus or generated (with difficulty) by passage of virus at high multiplicities (reviewed in Wimmer *et al.*, 1993). All naturally occurring DI particles carry deletions in the P1 capsid precursor region (Cole et al., 1971; Cole and Baltimore, 1973; Lundquist *et al.*, 1979; Nomoto *et al.*, 1979; Kajigaya *et al.*, 1985; Kuge *et al.*, 1986). As mentioned before, Nomoto and his colleagues have found that the deletions in all DI particles are in-frame (Kuge *et al.*, 1986). Since genetically engineered DI genomes (replicons) with an out-of-frame deletion in the P1 region are unable to replicate, Nomoto and his colleagues (Kuge *et al.*, 1986; Hagino-Yamagushi and Nomoto, 1989) correctly concluded that poliovirus RNA replication requires, at some stage of the replicative cycle, translation of the replicating RNA (*cis* requirement of translation), and that the DI particles with out-of-frame deletion cannot be complemented in *trans*. Novak and Kirkegaard (1994) confirmed this hypothesis in that replicons with translation termination codons downstream of the P1 coding region could not be rescued in *trans*. For hypotheses to explain this phenomenon, see above.

Genetic Recombination by Non-replicative Mechanisms

Chetverin *et al.* (1997) have recently made the startling observation that certain RNA fragments can join to one another via a molecular pathway determined by the intrinsic chemical properties of the RNA molecules (Chetverin *et al.*, 1997). The fragments that formed chemically stable duplexes were selected by the replicase of phage Qβ: only those dimers that had acquired signals from two different RNA molecules were able to replicate.

Gmyl *et al.* (1999) have now reported that viable recombinants could also be generated from non-replicating and non-translatable segments of the poliovirus genome. These fragments by themselves were unable to generate the viral RNA-dependent RNA polymerase necessary for a replication-dependent recombination event. The "crossovers" were targeted to the highly variable segment of 154 nucleotides located within the 5′NTR upstream of the initiating AUG codon for the polyprotein. A great number of recombinants have been obtained by transfection of mixtures of RNA fragments. Analyses of viruses that evolved after the mix-

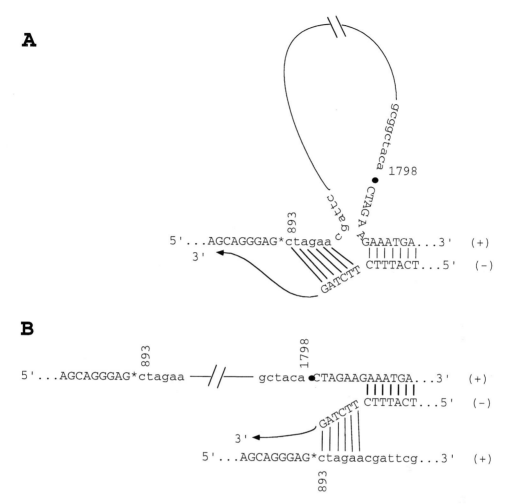

FIGURE 12.7 Two models of illegitimate recombination during – RNA synthesis as was observed with an expression vector shown in Figure 12.6E (Mueller and Wimmer, 1998). Both models require a partial dissociation of the nascent – RNA from the template +RNA, caused presumably by pausing of the RNA polymerase. The free 3′ end of the nascent – RNA can re-anneal to a short complementary sequence further upstream on the same template strand, thereby looping out the intervening sequence (**A**), or it can re-anneal to the same complementary sequence but on a sibling +strand, and complete synthesis on this second template (**B**; strand switching). In both cases, the resulting strands would have excised the same sequence and could now, in turn, give rise to truncated +RNA genomes. Note that this deletion event can occur even during the first round of replication of the expression vector leading to partial or complete deletion of the foreign coding sequences. Reproduced, with permission, from Mueller and Wimmer, 1998.

ture of the RNA species strongly suggested that the connection between two fragments was the result of chemical reactions between the fragments rather than of template-switching (Gmyl *et al.*, 1999).

The mechanism by which the chemical linkage between two fragments is formed is obscure but it could involve structures reminiscent of ribozyme-like activities in the viral RNAs. Nevertheless, this observation could have pro-

found implications for the generation of novel genotypes in nature (Gmyl *et al.*, 1999).

Genetic Complementation

Is the Picornavirus Genome Monocistronic?

Genetic complementation is the compensatory action of gene products of two homologous genetic systems to alleviate defects of mutant genes. Genetic complementation has been firmly established in picornavirus replication. However, because of the complexity of diverse function(s) of precursor proteins and their cleavage products, it has not been possible to define complementation groups (Wimmer *et al.*, 1993). Complementation groups are indicative of genetic elements that can function independently, and they have been the basis of the definition of cistrons (Benzer, 1957). A cistron, therefore, may be equated with a gene, i.e. a functional unit of genetic material specifying a single protein. On the basis of these definitions, the picornavirus genome, encoding only the polyprotein whose products function in many cases in overlapping or even opposing fashions, cannot be called multicistronic.

In general genetics, mutations affecting the same polypeptide can occasionally complement each other, a phenomenon referred to as intracistronic complementation (Schlesinger and Levinthal, 1963). Based on these considerations, we have suggested that the picornavirus genome be considered "monocistronic" (Wimmer *et al.*, 1993). It follows that the genome encodes only one gene product, the polyprotein. The polyprotein, in turn, contains multiple genetic units whose products may or may not be capable of intracistronic complementation. If this definition is accepted, one should avoid referring to individual coding regions of the picornavirus genome as "genes". Thus, there would be no "3D^{pol} gene". This convention makes good sense if one considers that a "gene for 3D^{pol}" is for the most part also the gene for 3CD^{pro}, a proteinase with properties unrelated to the polymerase 3D^{pol}. It should be noted that Theiler's virus is an exception to the monocistronic nature of picornaviruses in that it encodes a small protein in a separate reading frame, mapping towards the N-terminus of the polyprotein (Chen *et al.*, 1995).

It is interesting to consider that there is no absolute requirement that picornaviruses must exist as monocistronic (single-polyprotein-producing) entities. For example, the insertion of a second IRES into the genome would represent a viable dicistronic virus (Figure 12.6B), an entity artificially generated by Molla *et al.* (1992,

1993b). Apparently, during the evolution of picornaviruses, the elimination of genetic elements regulating the expression of different picornavirus proteins was favored over retaining them. Similarly, there was no pressure to generate such regulatory sequences and insert them into the genome. In other words, proteolytic processing of a single polyprotein evolved not only to be highly efficient but also as a means to regulate the temporal appearance of viral proteins (e.g. precursor proteins versus end-products of proteolytic cleavage). In contrast, in prokaryotic RNA phages or in –strand RNA viruses, the expression of proteins is regulated by sequence elements located between different cistrons.

Interestingly, the dicistronic poliovirus depicted in Figure 12.6B resembles the genetic composition of cow pea mosaic virus (CPMV), a plant virus (Hellen *et al.*, 1989; Molla *et al.*, 1992). Indeed, CPMV and the dicistronic poliovirus shown in Figure 12.6B have similar gene order and amino acid sequences, the main difference being that the genome of CPMV is bipartite. That is, rather than inserting a sequence such as an IRES into the genome, CPMV preferred to divide the genome into two portions, one coding for the capsid proteins the other for the replication proteins. Such a genetic arrangement, which requires two particles to initiate a complete infectious cycle, may be suitable for a plant virus (where the yield of virus per host can be extremely high and, equally importantly, the local concentration of host organisms can be high) but it would be highly disadvantageous for an animal virus.

The fascinating topic of the structure and evolution of polyproteins within the RNA-like virus superfamily is discussed later.

General Observations of Picornavirus Complementation

Evidence for genetic complementation *in vivo* has existed for decades, the best-known involving guanidine-generated mutants. Complementation of guanidine mutants seemed unidirecional (Wimmer *et al.*, 1993). Bernstein *et al.* (1986) provided the first conclusive evidence for symmetric complementation, using mutants that were generated in 2A^{pro} and 3A. These authors, however, also made the unexpected observation that mutants mapping to 2B or 3D^{pol} (which they had also generated by genetic engineering) could not be complemented (Bernstein *et al.*, 1986). On the other hand, Charini *et al.* (1991) clearly showed that mutations in 3D^{pol} mapping to a different region of the coding sequence could be rescued in *trans*. This example and many others (Wimmer *et al.*, 1993) support the notion that the polyprotein is a single genetic unit that does not consist of non-overlapping genes whose functions can be separated by complementation grouping.

A special case of complementation was tested using dicistronic polioviruses. Briefly, Cao and Wimmer (1995) constructed a virus with a genotype shown in Figure 12.6B, the extra cistron being the coding region for poliovirus 3AB. The dicistronic construct yielded a virus expressing a small plaque phenotype but it was genetically unstable, losing its inserts after several passages. Nevertheless, if the lethal mutation VPg(Y3F, T4A) was engineered into the second cistron of the dicistronic virus (see above), the first cistron (3AB) could rescue the genome, albeit inefficiently.

A much more efficient rescue of lesions in the 3AB coding sequence was reported by Towner *et al.* (1998), who used the cell-free system of poliovirus replication developed by Molla *et al.* (1991). Apparently, the supply in *trans* of P3 polypeptides *in vitro* is more efficient that *in vivo*, a phenomenon that remains as yet unexplained. Interestingly, it appeared as if a mutation in 3AB could be rescued only if the complementing polypeptide was a precursor of 3AB, preferably P3 (see Figure 12.4). Perhaps, efficient complex formation between 3AB/3CD^{pro} (see above) in this system depends on the cleavage of the P3 precursor *in situ*.

PATHOGENESIS OF PICORNAVIRUSES

Picornaviridae combines species that infect animals with exceedingly varied pathogenic features affecting almost every organ system (Table 12.1). In the following, we will concentrate only on human picornavirus infections, which range in severity from protean symptoms associated with the common cold (e.g. rhinoviruses), mild gastroenteritis (e.g. echoviruses), hepatitis (e.g. hepatitis A virus), to fatal CNS manifestations (e.g. PV, enterovirus 71) or lethal myocarditis (coxsackievirus B group). Despite the enormous variety in organ tropism observed with different species of the picornavirus group, pathogenic features of every single species are recognized in the form of highly distinct disease syndromes (exceptionally, coxsackieviruses can cause fatal disseminated infections in neonates with widespread viral propagation in multiple organs).

.We will define pathogenic properties of picornaviruses as a combination of different viral traits: (1) those that affect tropism (determining the target cell type of, and influencing spread in, the host); (2) those that affect virulence (determining kinetics of particle propagation); and (3) those determining the progression of a disease syndrome ("pathogenicity proper", the propensity to cause clinical symptoms). There is a fourth parameter, which is strictly related to a condition of the host. An example is injury-provoked ("provocation") poliomyelitis, which will also be discussed below.

Surprisingly, the disparity in pathogenic properties may be contrasted with a high degree of sequence conservation among certain groups of picornaviruses. This is most evident with the cluster C enteroviruses (Figure 12.1B). For example, on the basis of sequences of the 3D^{pol}, poliovirus serotype 2 (Lansing) (PV2(L)) shares more than 90% sequence homology with its close relative coxsackievirus A24 (CAV24). Indeed, their sequence similarity exceeds that between PV2(L) and PV1 (Mahoney) or PV3 (Leon) (Figure 12.1B). Yet, whereas all PV serotypes are associated with poliomyelitis, a severe and frequently fatal infection of the CNS, CAV24 causes mild upper respiratory tract infections only. Since minor sequence variations of picornaviruses can account for drastically different disease syndromes it may be assumed that the pathogenic phenotype of picornaviruses is encrypted within a few crucial genetic determinants.

These basic determinants of pathogenic

features appear to be dynamic, leading occasionally to the emergence of novel virus variants causing clinical syndromes not previously observed with their ancestors. This was the case when widespread epidemics of acute hemorrhagic conjunctivitis ravaged Africa and the pacific rim (Yin-Murphy, 1973) and quickly expanded worldwide. Two picornaviruses were associated with this previously unknown clinical syndrome, coxsackievirus A24 variant (CAV24v) and enterovirus 70 (EV70; Table 12.1). The former evolved from its ancestral CAV24, causing mild upper respiratory tract infections, whereas EV70 was primarily recognized for its association with a poliomyelitis-like neurological disorder (Melnick et al., 1974). The deviation of tropism toward ocular tissues resulting in acute hemorrhagic conjunctivitis suggests a switch or an expansion in receptor specificity. This hypothesis, however, awaits confirmation, since the cellular receptor(s) of EV70 is unknown (Table 12.1). Available evidence suggests that CAV24v can still use ICAM-1 as receptor but may have extended its cellular binding activity to a second unknown protein. CAV24v and EV70 do not share receptors (N. Takeda, personal communication; see later).

The circumstances and conditions that may favor a switch in host cell tropism with resulting changes in the pathogenic phenotype are unknown. A detailed discussion of our current view of the evolution of enteroviruses, however, is presented in the following section on Evolution. This is particularly relevant in the context of the imminent global eradication of poliovirus. It is known that coxsackieviruses A7 and A9 (CAV7, CAV9) as well as enteroviruses 70 and 71 (EV71) occasionally cause a clinical syndrome with striking resemblance to poliomyelitis (Melnick et al., 1974). Occurrence of poliomyelitis caused by these virus species has only rarely been reported in epidemic proportions (Voroshilova and Chumakov, 1959; Melnick et al., 1980); generally they occur as isolated incidents. Fortunately, preliminary evidence (da Silva et al., 1996) suggests that to date no surge in non-PV-caused poliomyelitis has occurred in response to the eradication of poliovirus in Latin America. However, the time elapsed since the eradication of wt polioviruses

in the western hemisphere is too short in terms of evolution to conclude that the incidence of non-PV-caused poliomyelitis and poliovirus eradication are unrelated (see section on Evolution).

The observation of diverse specific clinical syndromes caused by closely related picornaviruses (particularly enteroviruses) has sparked interest amongst virologists in identifying those factors that may determine the clinical outcome of picornaviral infections. Generally, signals for pathogenic phenotypes can be found in all parts of the viral genome. However, factors that determine cell and tissue tropism are not necessarily the same as factors determining virulence or attenuation. For example, the capsid mutations of the live attenuated strains of PV (the Sabin strains) have an attenuating effect without altering the tropism of the Sabin strains for the prime target of PV: spinal anterior horn motor neurons.

A different effect of capsid proteins on an extended host tropism has been reported for several poliovirus type 2 strains, e.g. PV2(L). A small segment in capsid protein VP1 of PV2(L) (the B-C loop) has been identified as carrying determinants of host range extension from primates to rodents (Murray et al., 1988). However, whereas PV2(L) infection causes poliomyelitis in primates and in CD155 tg mice, normal mice developed histopathology indicative of panencephalomyelitis, which was radically distinct from poliomyelitis observed in primates and CD155 tg mice (Gromeier et al., 1995). This observation suggests PV2(L) tropism toward a cell type in mice that is not targeted in primates. It is likely that the mouse-adapted PV2(L) acquired additional receptor specificity but the nature of the receptor for PV2(L) in normal mice is unknown. It should be noted that although PV2(L) causes disease in mice after intracerebral injection, cultivated mouse L cells cannot be infected with this strain (Gromeier et al., 1995).

The fact that single determinants of pathogenicity (e.g. the PV capsid) can carry signals that influence either tropism (e.g. PV2(L)) and/or virulence (Sabin strains of poliovirus) indicates different dimensions of picornaviral pathogenesis. Taking multiple determinants of tissue tropism and virulence (shared between the capsid, non-structural viral proteins and non-coding sequences) into account, the enormous complexity of the molecular basis of picornaviral pathogenesis comes into perspective.

Capsid protein structure determines the interaction of a virus with its cellular receptor. As pointed out for the poliovirus Sabin strains and PV2(L), small differences in the structure of viral capsids are critical for cell and organ tropism as they ultimately determine the pathognomic features of the resulting infection. Similarly, the capsid was determined to harbor sequences critical for cardiotropism (Tracy *et al.*, 1995; Cameron-Wilson *et al.*, 1998) as well as diabetogenicity (Kang *et al.*, 1994) of coxsackie B viruses (CBV). Moreover, diabetogenicity of EMCV in mice mapped to the capsid (Jun *et al.*, 1998).

As pointed out, the mechanism by which the changes in the capsid may affect pathogenesis may be related to differences in the interaction between virus and receptor. Apart from direct receptor switching (or extending receptor specificity to more than one cell surface molecule), virus capsid alterations may affect the kinetics of virus/receptor binding or particle stability. This would influence the virulence of that virus without a concurrent change in host cell tropism. Reduced particle integrity has been proposed to participate in the attenuation phenotype of the Sabin strains of poliovirus (Filman *et al.*, 1989). However, theories linking capsid mutations within the Sabin strains with structural elements important for protomer cohesion and capsid integrity remain inconclusive.

In contrast to a likely relationship between capsid structure and pathogenic phenotype, the role of non-structural viral gene products in the determination of disease has been less obvious. Non-structural viral proteins are cell-internal and, hence, do not influence tropism in a strict sense. Cell-type-specific restrictions in viral replication regulated by non-coding regions of the viral proteins or by non-structural proteins will not change the spectrum of target cells infected but may critically influence virulence. Viruses normally evolve to adapt to host cells offering adequate portal of entries (receptors), thereby exposing the viral particles to an intracellular milieu supportive of particle propagation. Thus, favorable cell-internal conditions for

viral replication would ideally be matched by a suitable viral receptor to avoid virus entry into cells that do not permit replication. For most viruses invading a host organism, the match is not perfect and, hence, the viruses are restricted to replication in fewer cells or organs than the distribution of receptor molecules would suggest.

Cell-internal determinants mapping to the viral genome of various picornaviruses have been suggested to influence virulence. These can be divided into loci mapping to viral proteins or to non-coding regions (e.g. the 5'NTR). For example, mutations within the coding region for the RNA-dependent RNA polymerase 3D^{pol} of the PV1 Sabin vaccine have been implicated in the attenuation phenotype (Toyoda *et al.*, 1987; Tardy-Panit *et al.*, 1993). It was found that these mutations contributed to the *ts* phenotype of this Sabin vaccine strain (Toyoda *et al.*, 1987). Mutations in the P2 region of hepatitis A virus have been correlated with an attenuated phenotype of HAV (Raychaudhuri *et al.*, 1998). In many instances the genetic loci of pathogenesis mapping to viral non-structural proteins have been identified through sequence comparison. This approach, of course, did not reveal mechanisms to account for reduced virulence.

The non-coding genetic elements of picornaviruses have also been shown to carry signals determining virulence (Evans *et al.*, 1985; Duke *et al.*, 1990; Gromeier *et al.*, 1996; Tracy *et al.*, 1996). As has been discussed in the section on Genetics, the 5'NTR harbors the internal ribosomal entry site (IRES) that, on the basis of the early observation by Evans *et al.* (1985) with the Sabin type 3 vaccine strain, has been identified as a major determinant of virulence for a number of picornaviruses. Sequence comparison of the Sabin strains of poliovirus with their *wt* progenitors revealed point mutations within a confined region of the 5'NTR in all three serotypes, known as domain V (Figure 12.3; reviewed in Wimmer *et al.*, 1993). In analyses of viral strains recovered from patients who acquired paralytic polio after vaccination, a point mutation at position 472 (direct reversion in domain V) of the IRES of PV3(Sabin) was proposed to contribute to the neurovirulence of the isolate (Evans *et al.*, 1985). Recent analyses led to a different hypothesis stressing a co-operative attenuating effect of capsid mutations with mutations in the IRES and other locations in the Sabin vaccine genomes (McGoldrick *et al.*, 1995). The mechanism responsible for IRES-mediated attenuation of neurovirulence remains obscure. Analyses of cell-type-specific growth restrictions in cell lines of neuronal origin and biochemical studies of cell-type-specific IRES function suggested impairment of initiation of translation in a cell-type-specific manner (Haller *et al.*, 1996; Agol *et al.*, 1989; La Monica and Racaniello, 1989). Following the example of poliovirus, IRES elements or surrounding sequences of a large number of picornaviruses were found to contain genetic markers with a role in the determination of a pathogenic phenotype. This was most impressively demonstrated by the drastic attenuating effect of a deletion within the poly(C) tract of the 5'NTR of EMCV (Duke *et al.*, 1990).

How do the multiple mechanisms alluded to interlace to produce a specific picornavirus disease syndrome? The intricacies of dual cell external and internal determinants of viral pathogenic features are best illustrated using the example of poliovirus. This most thoroughly studied prototype picornavirus is characterized by pathogenic properties of specificity untypical of viral pathogens of the CNS. Poliovirus host range is limited to primates only. Within the primate organism poliovirus replicates within an unknown site in the gastrointestinal tract and within associated lymphatic structures, leading to viremia (Bodian, 1972). At this stage, the virus causes hardly any disease symptoms. However, viremia may, in only about 1% of infections, lead to CNS invasion, presumably through passive passage of the blood–brain barrier (Yang *et al.*, 1997). On muscle injury, the virus may also reach the CNS via retrograde axonal transport (Gromeier and Wimmer, 1998, 1999). Within the CNS, poliovirus uniquely targets spinal cord anterior horn motor neurons. Lytic destruction of anterior horn motor neurons results in flaccid paralysis, the hallmark clinical sign of paralytic poliomyelitis (Bodian, 1972).

Whereas efficient poliovirus proliferation occurring in the human gastrointestinal tract produces few or no symptoms, this virus's pathological potential is expressed in a small and relatively inaccessible subpopulation of neurons in the CNS. This peculiar restriction has been the subject of research interest for many years. Available evidence clearly suggests that the restriction of the host range of poliovirus to primates is determined predominantly by the receptor. In humans, this receptor is the immunoglobulin superfamily molecule CD155 (Mendelsohn *et al.*, 1989) and two proteins closely related to CD155 function as poliovirus receptor in simians (Koike *et al.*, 1990). Clearly, the observed organ and cell tropism are co-determined by the virus's dependence on the cellular receptor. We believe that, at least in part, the expression of CD155 must also determine the highly restrictive cell tropism of poliovirus within the CNS, because mice transgenic for the *CD155* gene develop a neurological condition with pathologic and clinical features identical to those observed in primates (Ren *et al.*, 1990; Koike *et al.*, 1991; Gromeier *et al.*, 1996). Furthermore, support for this hypothesis comes from studies of transcriptional control (Solecki *et al.*, 1999) and developmental expression of the *CD155* gene (Gromeier *et al.*, 1999b). It has been reported that *CD155* expression is restricted to structures in close anatomical and functional relationship with spinal cord anterior horn neurons during embryonic development of the CNS (Gromeier *et al.*, 1999b). It is thus likely that the restrictive expression pattern of *CD155* may indeed direct poliovirus tropism toward a specific cellular compartment of the CNS.

Analyses of pathogenesis related to genetic determinants mapping to the capsid, proteins, IRES, or 3D^{pol} have recently been extended using poliovirus hybrid viruses. Specifically, polioviruses have been constructed in which the cognate IRES was replaced by that of other picornaviruses (Gromeier *et al.*, 1996). By exchanging the cognate IRES element of PV by that of other picornavirus species, it could be shown that neuropathogenicity of PV can be eliminated without affecting growth properties in non-neuronal cell types normally susceptible to poliovirus (Gromeier *et al.*, 1996). Thus, it was determined that the IRES of rhinovirus type 2, a virus species never associated with neurological disease, confers the attenuation phenotype to poliovirus. Significantly, this chimeric virus, called PV1(RIPO), did not carry any attenuating mutations in the poliovirus-specific sequence of its genome.

The neuropathogenic potential of a picornavirus IRES cannot be predicted but it is innate, of course, in its sequence. Certainly, IRES elements of enteroviruses known to cause poliomyelitis (CAV7, CAV9, EV70, and EV71) are candidates for "neurovirulent IRESes" and, indeed, a corresponding PV/CAV chimera has proved this hypothesis (Gromeier *et al.*, unpublished results). C-cluster enteroviruses, on the other hand, never cause poliomyelitis. However, because of their close genetic kinship, the IRESes of C-cluster coxsackieviruses confer a highly neurovirulent phenotype to the PV/CAV chimeras (Gromeier *et al.*, unpublished). Finally, IRES elements of the genus *Rhinovirus* ablate neuropathogenesis in poliovirus chimeric viruses (Gromeier *et al.*, 1996).

These observations indicate that, indeed, cell-external restriction in cell tropism as well as cell-internal factors exert powerful limitations toward enterovirus pathogenesis. Many relatives of poliovirus of the *Enterovirus* genus (particularly the C cluster; Table 12.1) presumably would equal the neuropathogenic properties of PV if their capsid structure allowed interaction with neuronal cells. Thus, non-neurovirulent C-cluster enteroviruses with high sequence homology to PV and "neurovirulent" IRES elements may gain tropism for neurons in the future (see under Evolution).

In addition to virus-encoded factors of PV neuropathogenicity, the circumstances within the host organism at the time of infection or shortly thereafter may influence the outcome of poliovirus infection. Trivial muscle injury has been shown to increase the probability of neurological complications of concurrent poliovirus infection (McCloskey, 1950). A proposed pathogenic mechanism for provocation polio identified a deviation of the route of CNS invasion toward retrograde axonal transport to account for the increased risk of polio among individuals who received intramuscular injections (Gromeier and Wimmer, 1998, 1999).

Picornaviruses have adapted to a wide variety of cellular components of their hosts. The diverse spectrum of disease syndromes associated with human picornaviruses provides an excellent field of study to examine the factors that determine the clinical outcome of a viral infection. The enormous amount of sequence information, combined with a broad knowledge of the molecular biology of many of these agents, have sparked hopes of a rapid elucidation of the molecular basis for their pathogenic properties. Initial optimism that sequence comparison of virulent strains with their attenuated variants alone would rapidly identify those elements responsible for a pathogenic phenotype and unravel mechanisms of pathogenesis is, however, unjustified. This is particularly true for poliovirus, the most thoroughly studied picornavirus. Progress in the analysis of poliovirus neuropathogenicity has revealed that the interactions of poliovirus with the host are characterized by a degree of complexity not previously appreciated. Mechanistic concepts of viral pathogenesis, combined with one-dimensional views of virus replication and its relation to the host organism, have helped little in increasing our understanding of the selective susceptibility to poliovirus of motor neurons. Viral infections result in complex clinical syndromes that are difficult to explain in terms of single viral genetic elements. Using poliovirus as an example, picornavirus-induced disease is the result of an intricate interplay of numerous factors, of both viral and host origin, that coordinately affect the ability of the virus to propagate in any particular cell type or organ.

EVOLUTION

General Remarks

Numerous investigations, particularly those of J.J. Holland, E. Domingo and their colleagues (see Chapter 7), have led to the realization that the rapid evolution of RNA viruses results from high mutation rates combined with exceedingly large heterogenic populations. This is true for Picornaviridae that encode variants of conserved protein folds as well as catalytic systems not found in the cellular world and, hence, have explored an enormous evolutionary space. It is less appreciated, although equally true, that it is the host environment that has provided new opportunities for viruses to proliferate and select new variants out of a huge number of mutants. Without this "co-operation" between host and parasites, picornavirus evolution would not have resulted in the tremendous diversity of genotypes whose number, now counting in the hundreds, is currently biased towards those infecting mammals.

The key role of virus–host interaction is evident in the evolution of RNA viruses. Indeed, the relatively slow pace by which the cellular environment is changing imposes severe restrictions on the mode of RNA virus evolution. A model has been proposed suggesting that, in each moment of protein evolution, mutations could only be accepted in a very limited number of positions of a polypeptide chain; otherwise, protein structure and function would have been severely compromised. These limited places of acceptable variation have been called covarions (Fitch, 1971). It has been argued that the fast evolution of RNA viruses in a constrained environment has to proceed through the exploitation of constantly emerging but vastly overlapping covarions (Koonin and Gorbalenya, 1989). This model of virus evolution provides a sensible hypothesis as to how picornaviral proteins have managed to accept a heavy load of mutations that are quite frequently unique and rarely seen at sites in cellular proteins, while still not entirely losing some discernible similarity with other viral and cellular homologs. This has important practical consequences, since structural similarities can be used to reconstruct the evolution of RNA viruses, an undertaking impossible just 20 years ago.

Apart from the biological properties generally associated with all RNA viruses, the current end-product of evolution makes each virus species unique. This is engraved in the virus' genetic plan: the organization of the genome and the mode of gene expression. During the past years we have learned that +strand RNA viruses employ variations of surprisingly few basic genetic plans. The genetic organization and genetic expression of picornaviruses has

been outlined in detail above. In the following, the genetic organization of different picornaviruses will be put into an evolutionary perspective. In addition, we will discuss in a rather speculative manner some hypothetical implications of evolutionary consequences of the eradication of poliovirus.

Classification and Phylogenetic Relationships of Picornaviridae

Numerous RNA viruses have been combined into a picorna-like supergroup of which the Picornaviridae comprise a rather compact domain (Figure 12.8). The viruses of the Picorna-like supergroup, a taxon not yet recognized as higher-than-family rank, share a conserved array of replicative proteins (see below). They infect plants, insects, birds and mammals. Most of the established members of Picornaviridae are mammalian viruses, and they fall principally into six genera (Table 12.1). The newly characterized avian encephalomyelitis virus (AEV), which is a member of Picornaviridae, is most closely related to hepatitis A virus (Marvil *et al.*, 1999).

The latest revision of the classification of picornaviruses, although closely related to the original version, has a clear evolutionary flavor since it tends to combine viruses in accord with phylogenetic kinship rather than relying on phenotypic properties (see the Introduction). The largest number of picornaviruses fall into the most closely related genera, *Enterovirus* and *Rhinovirus* (Table 12.1). *Cardiovirus* and *Aphthovirus* comprise two other genera that have most probably emerged from a common ancestor. It appears that the two pairs of picornavirus genera diverged after hepato- and parechoviruses split from the main trunk of the picornavirus tree (Figures 12.1, 12.8). The exact phylogenetic interrelationship between hepato- and parechoviruses remains somewhat uncertain and may be different from that shown in Figure 12.1 (see also below). In addition, three other picornaviruses, equine rhinovirus types 1 and 2 (ERV1 and 2 respectively) and Aichi virus (AiV), remain to be classified. ERV1 was shown

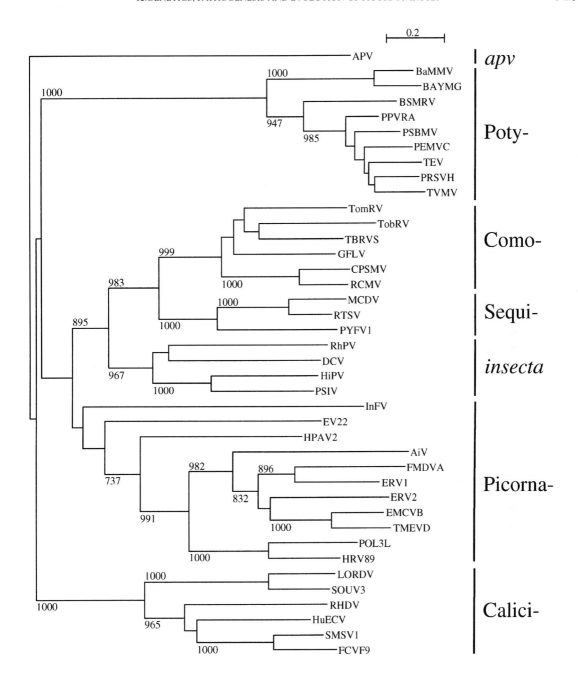

FIGURE 12.8 Tree of 3D^{pol} and its homologs encoded by viruses of the picorna-like supergroup. The tree was generated for 3D (-like) sequences of a representative set of picorna-like viruses, including the five established families Potyviridae, Comoviridae, Sequiviridae, Caliciviridae and Picornaviridae, and two provisional groups of insect viruses, one dubbed *insecta*, the other *apv* (A.E. Gorbalenya, unpublished results). The neighbor-joining method with the Kimura correction for multiple substitutions, as implemented in the ClustalX program, was used to calculate the tree. Those radiations that have persisted in more than 700 out of 1000 bootstrap replications are marked at the bifurcations.

to be a distant relative of FMDV (Wutz *et al.*, 1996), while ERV2 (Li *et al.*, 1996; Wutz *et al.*, 1996) and AiV (Yamashita *et al.*, 1998) appear to be related to the cardio- and cardio/aphthovirus branches respectively. They are not recognized, however, as cardio- or aphthoviruses.

In addition to mammalian viruses, a number of insect viruses have been previously included into the Picornaviridae on the basis of phenotypic criteria. During the last 2 years, however, the complete genome structure of six picorna-like insect viruses has been reported (van der Wilk *et al.*, 1997; Isawa *et al.*, 1998; Johnson and Christian, 1998; Moon *et al.*, 1998; Nakashima *et al.*, 1998; Sasaki *et al.*, 1998). Some of these viruses feature different genome organizations, and only infectious flacherie virus of silkworm (InFV; Isawa *et al.*, 1998) was shown to possess a genotype and gene organization that may justify placing it in the Picornaviridae (A.E. Gorbalenya, unpublished results). It is likely that an ancestor of InFV separated from the main branch of Picornaviridae before the radiation of mammalian viruses (Figure 12.8).

The rapid accumulation of new virus genotypes has not been matched by an understanding of its evolutionary meaning. Therefore, the basis of picornavirus classification may need to be revisited. Moreover, the relationship between Picornaviridae and other genetic systems may have to be defined within a new classification. Although confusing for virologists on first sight, any reclassification into hierarchically organized taxa will ultimately aid our understanding of evolution, host range of viruses and pathogenesis. It should be kept in mind, however, that any classification is, at best, an approximation of true phylogenetic relationships, and the current classification of Picornaviridae should be treated as such.

The Conserved Backbone and Hot Spots of Genome Organization

Picornaviridae have evolved by speciation from a common ancestor. This plausible statement has been supported by computer analyses of nucleotide and protein sequences as well as by studies of the tertiary structure of capsid proteins and 3C^{pro} proteinases. There is every reason to believe that the putative ancestral viral entity had a genetic organization that has been conserved largely in contemporary picornaviruses. Its signature is a long 5'NTR-long open reading frame-3'NTR-poly(A) (Figure 12.2). With the sole exception of a strain of Theiler's virus (TMEV; see below), all picornavirus proteins are generated by autocatalytic processing of the gigantic polyprotein (Figure 12.3).

Proteins

The backbone of the polyprotein is formed by a set of polypeptides conserved in all known picornaviruses. In addition, the backbone may be decorated with a few optional proteins unique to a particular virus or virus group. Among the proteins, the conservation increases in the order (Figures 12.4, 12.9):

$$L < 2A^{(pro)} < VP4 < 3A, 2B < 3B^{VPg} < VP1,VP2 < VP3 < 2C^{ATPase} < 3C^{pro} < 3D^{pol}.$$

This order can be deduced from analyses of virus groups belonging to different phylogenetic ranks – clusters of closely related viruses, of distinct genera, of an entire family, or of the entire picorna-like supergroup (Figure 12.10; A.E. Gorbalenya, unpublished results). Slight deviations can only be seen upon analysis of some small taxonomic groups. As already noted, TMEV encodes a small, unique L* polypeptide outside the main reading frame at a locus overlapping the L reading frame. Additionally, the insect picornavirus InFV was predicted to encode unique domains as part of the polyprotein (Isawa *et al.*, 1998; A.E. Gorbalenya, unpublished results). A closer look at these proteins reveals the following.

The L polypeptide preceding the capsid coding region (Figure 12.9) is encoded by all picornaviruses except the entero- and rhinoviruses. The L protein is the most variable of all picornavirus proteins, and it exists in five different versions. L proteins with proteinase activity are encoded only by FMDV, ERV1 and ERV2 (Skern, 1998). Cardioviruses and AiV encode three different versions of L polypeptides containing a putative Zn finger (Chen *et al.*, 1995; A.E.

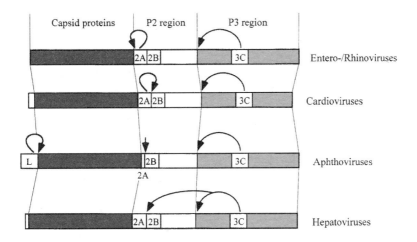

FIGURE 12.9 Primary cleavage in picornavirus polyproteins. Open boxes at the left end depict L proteins, of which only that of aphthoviruses is a proteinase. Of the 2A coding sequences, only 2A^{pro} of entero- and rhinoviruses is a proteinase. In cardio-and aphthoviruses, processing at the C-terminus of 2A is strictly a *cis* cleavage event. In hepatoviruses, even this cleavage is catalyzed by 3C^{pro}. Modified from Ryan and Flint, 1997.

Gorbalenya, unpublished observations), while hepato-, parechoviruses and InFV appear to encode unique L proteins (Najarian *et al.*, 1985; Hyypia *et al.*, 1992; Isawa *et al.*, 1998).

Some evolutionary characteristics of the 2A polypeptides parallel those of the L proteins. Entero- and rhinoviruses encode 2A of the same protein family, known as 2A cysteine chymotrypsin-like proteinases (Bazan and Fletterick, 1988), whereas each of the three groups of hepato- and parechoviruses and InFV encodes a unique 2A with unknown function(s). The other picornaviruses, cardio- and aphthoviruses and AiV, encode a 2A protein having a characteristic C-terminal motif (or a derivative thereof) that has been implicated in the spontaneous separation of 2A and 2B proteins during polyprotein synthesis (reviewed in Ryan and Flint, 1997).

Besides L and 2A, the only other protein of the family not conserved at the primary structure is VP4 (Palmenberg, 1989; A.E. Gorbalenya, unpublished observation). It may therefore not come as a surprise that VP4 has been poorly resolved in X-ray analyses of picornavirions whose structure has been solved (Lentz *et al.*, 1997). It is interesting to note that this small pro-tein, which occupies a position upstream of VP2, has moved its position to between VP2 and VP3 in insect InFV (Sasaki *et al.*, 1998).

Other picornavirus proteins are conserved, albeit to varying degrees. These polypeptides therefore may play similar role(s) in the life cycle of different picornaviruses. For example, 2B and 3A are of variable sizes but they contain hydrophobic regions thought to be involved in the anchoring of these proteins to membranes in RNA replication complexes. Only the hydrophobic patches of 2B and 3A polypeptides, however, have been conserved (A.E. Gorbalenya, unpublished results). In capsid proteins, the most pronounced conservation is evident in residues critically important for fold maintenance. Finally, in the key replicative enzymes 2C^{ATPase}, 3C^{pro} and 3D^{pol}, as well as in 3B^{VPg}, the active site residues are amongst the most highly conserved (Gorbalenya and Koonin, 1993a).

The majority of proteins of picornaviruses, regardless of how well they have been conserved within their own family, have homologs among cellular and other viral proteins. First, the three capsid proteins, VP1, VP2 and VP3, have adopted different versions of an eight-

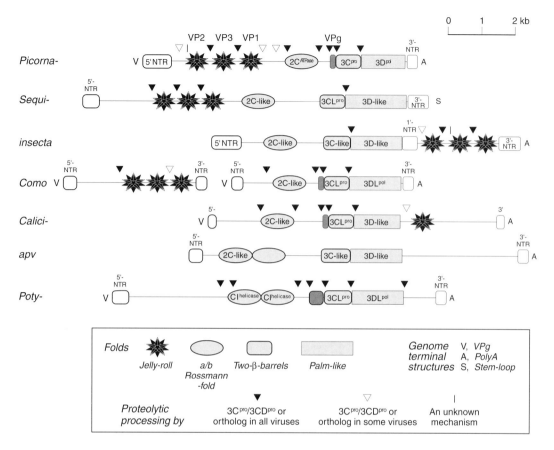

FIGURE 12.10 Comparison of the genome organizations of the main groups of the picorna-like supergroup. For each picorna-like family group, excluding APV, the conserved organization of an "averaged" genome typical for this group is shown and compared with that of picornaviruses. The genomes are aligned with respect to the position of the 3D (-like) locus (the RNA polymerase). "Averaging" was carried out with respect to genome size so that most conserved genome features could be shown. Note that bymoviruses, comprising a genus of Potyviridae, have a bipartite genome. It is believed that all picorna-like viruses contain VPg at the 5'-end, although a VPg has not yet been demonstrated for every group of viruses. In picorna-like viruses, proteins were designated so as to reflect their similarity to the prototype picornavirus enzymes, although other nomenclatures may be in use by other investigators. Apart from studies with picornaviruses, enzymatic activities have been ascribed to some proteins of como-, poty-, calici- and sequiviruses, but the complete processing map of polyproteins has been established only for como- and potyviruses. The conserved α/β Rossmann-fold and palm-like fold comprise only one of the domains of 2C and 3D, respectively, or their homologs. For further details, see legends to Figures 12.2 and 12.4, and the text.

stranded antiparallel beta-barrel fold, dubbed "jelly-roll" (Rossmann and Johnson, 1989). Amongst RNA and DNA viruses of different families, this fold is the most common to build icosahedral capsids (Rux and Burnett, 1998). It is also conserved in a number of cellular proteins (Rossmann, 1987; Orengo *et al.*, 1997). Second, the core domain of 3D^{pol}, containing several highly conserved sequence motifs, is related to a number of polynucleotide polymerases includ-

ing RNA-dependent RNA polymerases of RNA viruses, reverse transcriptases of viral and cellular origins, and DNA-dependent DNA polymerases (Hansen *et al.*, 1997). An analysis of the crystal structure of PV 3D^{pol} has also identified a (palm) subdomain adopting a RRM-like fold conserved among a number of functionally different proteins, including ribosomal proteins L7/L12 and S6 as well as the U1A splicing factor (Hansen *et al.*, 1997). Third, two picornavirus proteinases, the ubiquitous 3C^{pro} and entero/rhinovirus-specific 2A^{pro}, have adopted 12-stranded antiparallel two beta-stranded barrel folds, conserved in cellular serine proteases with chymotrypsin as the prototype (reviewed in Skern, 1998). These picornavirus proteinases have also relatives that are encoded by (+)RNA viruses belonging to dozens of different species (Gorbalenya and Snijder, 1996; Ryan and Flint, 1997). Unlike cellular proteases, the picornaviruses 3C^{pro} and 2A^{pro} employ cysteine as the principal catalytic nucleophile and, in some lineages, have another unique replacement – instead of the catalytic Asp they use a Glu (Gorbalenya *et al.*, 1989). The other small family of picornavirus proteinases, L^{pro} of aphthoviruses, ERV1 and ERV2, is related to cellular-papain-like proteases (Gorbalenya *et al.*, 1991; Skern, 1998) whose homologs have been identified in many animal and plant RNA viruses as well (Gorbalenya and Snijder, 1996). Finally, 2C^{ATPase}, whose structure is yet to be solved, belongs to the so-called helicase superfamily III. This protein group includes polynucleotide-stimulated ATPases, some with helicase activity, which are encoded by (+)RNA and small DNA viruses as well as proteins of cellular origin (Gorbalenya and Koonin, 1990, 1993b). The 2C^{ATPase} has been predicted to be a three-domain protein. Two α/α domains flank an ATP-binding domain adopting a variation of the α/β "Rossmann" fold, which is widespread in the protein world (Teterina *et al.*, 1997).

With respect to details, our current understanding of the function of picornavirus proteins is rather fragmentary. Nevertheless, a preliminary functional profile of picornavirus proteins fits patterns of conservation evident at the structural level. The most conserved non-structural proteins provide the basic enzymatic activities needed for the synthesis and expression of viral RNAs inside the cell. The three conserved capsid proteins form the scaffold of virions shielding virus RNA from the detrimental environment outside the cell. All these activities appear to be virus-specific, although they may be modulated by cell-encoded components. In contrast, non-conserved viral proteins seem to sense and modify the host environment in addition to serving basic biosynthetic processes pro-grammed by the viral genomes (for instance, see Piccone *et al.*, 1995; Zoll *et al.*, 1996; Svitkin *et al.*, 1998; Ventoso *et al.*, 1998). Virions also have host-dependent functions, such as the recognition of the cellular receptor, entry and, possibly, virion maturation. Different lines of evidence have shown that the least conserved regions of three capsid proteins, as well as VP4, may mediate these early activities of host cell entry (for recent work, see Hadfield *et al.*, 1997, and Lentz *et al.*, 1997).

NTRs

Much of what has been said about proteins applies also to the terminal NTRs of the picornavirus genomes. These regions are conserved within related genera but they may diverge when groups are compared (e.g. *Enterovirus/Rhinovirus* versus *Cardiovirus/Aphthovirus*) even although they play identical roles in viral proliferation (discussed in the section on Genetics). Variants of two very different conserved secondary structure organizations of IRES elements are shown in Figure 12.3, prototyped by those of PV and EMCV. It is unclear what type of 5'NTR was encoded by an ancestor of picornaviruses – that resembling one of the contemporary prototypes or rather a "consensus" one (Le and Maizel, 1998). In contrast, the 3'NTR region has diverged profoundly amongst picornaviruses and it is not conserved even within the otherwise closely related entero- and rhinoviruses (Poyry *et al.*, 1996).

Conservation of the Organization of Proteins Within the Polyprotein

The polyprotein of numerous +strand RNA viruses has evolved such that its organization reveals an additional level of conservation – the order of mature proteins in this large precursor (Figure 12.10; see also the order of protein domains in the prototype PV in Figure 12.4). This order is inflexible and none of the picornaviruses violates it, although entero- and rhinoviruses do not encode L proteins (see above). Despite a near absolute conservation of the order of protein domains, there is some plasticity (Figure 12.10). Upon computer sequence analyses of the picorna-like viruses, it has become evident that the polyprotein can be divided into two parts, one comprising capsid proteins and the other the non-structural proteins. These parts are expressed rather independently. In a group of picorna-like insect viruses, *Rhopalosiphum padi* virus (RhPV), *Drosophila* C virus (DCV), *Plautia stali* intestine virus (PSIV) and cricket paralysis virus (CrPV), non-structural and capsid proteins are encoded by two ORFs, separated by a NTR (Koonin and Gorbalenya, 1992; Johnson and Christian, 1998; Moon *et al.*, 1998; Nakashima *et al.*, 1998; Sasaki *et al.*, 1998). In Comoviridae, a family of plant viruses, the capsid and non-structural proteins are encoded by two distinct RNAs, RNA2 and RNA1 (Goldbach, 1986). Remarkably, in the dendrogram shown in Figure 12.8, Comoviridae, Sequiviridae, a plant virus family having the same polyprotein organization as Picornaviridae (Turnbull-Ross *et al.*, 1993) and picorna-like insect viruses form a division immediately adjacent to the Picornaviridae.

It is important to stress that comparison of the sequences of many viruses of the picorna-like supergroup has revealed a profile of sequence conservation that parallels that observed for Picornaviridae. Thus, two groups of highly conserved clusters can be distinguished in polyproteins. The first group comprises the capsid proteins VP2–VP3–VP1, the second the non-structural proteins $2C^{ATPase}$–(VPg)–$3C^{pro}$–$3D^{pol}$ (or equivalents). The functions assigned to the individual members of the group of non-structural proteins remain provisional for the majority of known viruses. They have been inferred largely on the basis of sequence similarities with proteins of well-characterized viruses like poliovirus. Among the positionally highly conserved non-structural proteins, the genome-linked protein VPg has a special standing (highlighted by bracketing) since it is conserved functionally rather than structurally in the picorna-like viruses (Figure 12.10; Gorbalenya and Koonin, 1993a).

The combination of conserved non-structural proteins of picorna-like viruses has been termed "replicative module" (Goldbach, 1986). Such module of related proteins has been recognized also as "capsid modules" built of three "jelly-

roll" proteins. Animal Caliciviridae, plant Potyviridae and insect *Acyrthosiphon pisum* virus, all of which are distantly related to Picornaviridae, encode a distinct variety of the replicative module that is associated with one of three unique sets of capsid protein(s) encoded in the 3'-region of viral genomes (Domier *et al.*, 1986; Meyers *et al.*, 1991; van der Wilk *et al.*, 1997; Figure 12.10).

Conservation of Proteins is Paralleled by the Mechanism of Processing

The conservation of protein order in the picornavirus polyprotein and the patterns of expression (proteolytic processing) have been conserved. Pairs of neighboring proteins are separated at scissile bonds cleaved by a virus proteinase or, in case of the VP4/VP2 junction, by an unknown mechanism. It could be hypothesized that the position of protein domains could be changed as long as the corresponding proteins were released more or less independently from the precursor. This, however, is not the case, as the pathway of proteolytic processing in picornavirus polyproteins is not random. Furthermore, at least some intermediate precursors, e.g. 2BC, 3AB and $3CD^{pro}$, have essential functions that differ from those of the end-product of processing (see the section on Genetics). These considerations provide a biological reasoning for the observed conservation of the protein order in polyproteins.

We have already pointed out that the order of two conserved units within the polyprotein, the capsid precursor and replicative modules, is flexible. The two least conserved proteins, L and 2A, flank the capsid precursor at the N- and C-termini, respectively, and bring additional plasticity to the organization of the polyprotein. This is reflected also in terms of expression, i.e. the mechanism of proteolytic processing. Processing of the capsid precursors as well as of the replicative module at junctions separating conserved proteins involves exclusively the conserved $3C/3CD^{pro}$ proteinases, a mechanism functioning not only in picornaviruses but also in other picorna-like viruses (Figure 12.10; Ryan and Flint, 1997). In contrast, the three cleavages separating the poorly conserved L and 2A proteins from the neighboring polypeptide chains (L/VP4, VP1/2A and 2A/2B) are processed by a range of mechanisms in a genus-specific manner. Furthermore, whereas picornaviruses use two general pathways of cleavages – 3C^{pro}/3CD^{pro} versus distinct mechanisms involving L and 2A – this genetic repertoire may be further diversified in some picorna-like viruses. For instance, in Comoviridae and insect viruses, capsid precursor and non-structural proteins are encoded by distinct ORFs (Figure 12.10), which eliminates the need for cleavages separating these polypeptide chains.

3C^{pro}/3CD^{pro} have emerged as the major enzymatic factors in the regulation of protein expression in all picorna- and related viruses. Interestingly, the primary structure of sites recognized by these proteases is virus-specific rather than position-specific. Among picornaviruses, entero- and rhinoviruses employ sets of structurally uniform sites while viruses of the other genera use more diversified sets. Poliovirus and HAV exemplify the most extreme diversity. In poliovirus, all eight cleavage sites have the same ("canonical") Q/G structure (Figure 12.4), whereas in HAV, six variations of this structure were described in different sites (Palmenberg, 1990).

Poliovirus proteins produced from its replicative module seem to have been exceptionally strongly constrained not only with respect to the type of the terminal amino acids but also with respect to size. Mature poliovirus proteins (except 3C^{pro}), as well as processing intermediates, have sizes that can be divided by 11 without remainder or with only a small remainder (Gorbalenya et al., 1986). This feature separates poliovirus proteins from the overwhelming majority of cellular and viral proteins. The latter are heterogeneous both in size and sequence, particularly at their termini, because of a relative abundance of mutations, including insertions and deletions. Structural regularities documented for poliovirus can be visualized in a form of weak primary structure periodicities with the common denominator of 11 comprising the major portion of the replicative module. On the basis of these observations, it has been proposed that the replicative module of picornaviruses has originated from a primitive self-

replicating RNA molecule through consecutive multistep duplications (Gorbalenya, 1995; Gorbalenya et al., 1986).

Does the History of a Polyprotein Determine How it is Likely to Evolve in the Future?

We have briefly described different levels of evolutionary conservation in picornaviruses by using results of comparative sequence analyses. The conservation of different properties is the result of a long evolutionary process, accompanied by numerous radiations. Does the history of the polyprotein determine how picornaviruses may evolve in the future? We are unaware that this question has ever been directly addressed in experimental studies, although many results obtained by using genetic engineering seem to be quite relevant. These data can technically be separated into two sets – those obtained in studies using site-directed mutagenesis and those aimed at constructing chimeras.

In numerous studies of the first type, it has been observed that different regions of the picornavirus genome express a differential tolerance to replacements (Wimmer et al., 1993). It can be predicted that a profile of the "accepted" mutability, drawn over the entire genome, would fit the conservation profiles described above. Such a result would support the hypothesis that the past of picornaviruses influences their future in terms of evolution. However, mutagenesis saturating the genome has never been systematically carried out. Therefore, the available "mutagenesis profile" can only be used as a rough approximation of the yet-to-be-defined "accepted" mutability profile in relation to the conservation of modules. The "resolution" of the mutagenesis studies that remained unresolved is potentially relevant to an understanding of the evolution of contemporary picornaviruses, regardless of whether this relates to recent evolutionary events or to the complete historical past.

The second group of data involving genome engineering complements the mutagenesis studies and helps to address the question posed above. In wt genomes of entero- and rhinoviruses, the ORF of the capsid precursor is preceded by the cognate 5′NTR (Figure 12.2). As was observed in studies of poliovirus expression vectors (Figure 12.5), genetically stable variants of poliovirus have been selected (Mueller and Wimmer, 1997) in which an additional leader peptide is encoded that is fused to the N-terminus of the polyprotein, just downstream from the 5′NTR. This organization may look unique on first sight, but in fact it resembles that of all other picornaviruses distantly related to entero- and rhinoviruses. These terminal appendices in the poliovirus variants resemble the L proteins and, hence, these poliovirus chimeras have a "cardio-like" organization (Figure 12.6E). We can speculate that PV has "accepted" an artificial L peptide because a similar event has already happened in the past history of its ancestors.

In a different set of experiments, several poliovirus chimeras have been generated in which the heterologous EMCV IRES was placed into the sequences specifying scissile bonds of the polyprotein, thereby dividing the polyprotein into two parts (Figure 12.6B). This insertion radically modified the conserved protein expression mechanism of picornaviruses, since it functionally replaced a proteolytic cleavage event by an event of internal initiation of translation directed by the alien IRES. In all, poliovirus genomes were constructed in which the EMCV IRES was placed between the Y*G cleavage site of 2A^{pro} (Figure 12.6B) or all possible Q*G cleavage sites involving the 3C^{pro}/3CD^{pro} proteinase (Molla *et al.*, 1992, 1993b; Paul *et al.*, 1998a). Only two poliovirus-EMCV dicistronic chimeras, specifically those carrying EMCV IRES between VP1 and 2A and between 2A and 2B, have given rise to viable and stable virus progeny (Molla *et al.*, 1992, 1993b; Paul *et al.*, 1998a). Although the genome organizations of these chimeras do not match anything found in nature, immediate parallels come to mind with genomes of picorna-like viruses in which capsid and replicative modules are encoded by different ORFs (for example Comoviridae, see above and the section on Genetics).

These considerations imply that the conserved and non-conserved features in organization and structure of genomes of picornaviruses and even picorna-like viruses are indicative of an evolutionary plasticity and of possible future changes of a picornavirus. Perhaps an "evolutionary space" of a picornavirus can be approximated from the past. Mechanistically, this can be seen as if the past evolution of the entire family has been "imprinted" in the organization of the genome of each of the contemporary picornaviruses.

Mechanism of Evolution

Phylogenetic trees that have been built for different picornaviral proteins (most often VP3, 2C^{ATPase} and 3D^{pol}) by employing parsimonious and maximum-likelihood methods proved roughly topologically equivalent even though different regions of the polyproteins have definitely evolved at different rates (Stanway, 1990; Rodrigo and Dopazo, 1995; Hyypia *et al.*, 1997). These observations strongly favor a concerted evolution of (the majority of) the picornavirus proteins. This conclusion is not compromised by some incongruity in the tree topology of closely related viruses, e.g. the C cluster of the enteroviruses (Poyry *et al.*, 1996), or very distantly related groups, e.g. hepato- and parechoviruses. It is likely that some trees generated for different regions look different, as a result of technical limitations related to phylogenetic and biopolymer sequence analyses as well as a biased representation of some groups. Also, possible recombination events between closely related viruses may have complicated phylogenetic analyses. We shall analyse sequence alignments of picornavirus proteins and polynucleotides aimed at deducing the mechanisms functioning in picornavirus evolution.

Uniform sizes of each of the VP3, 2C^{ATPase}, 3C^{pro}, or 3D^{pol} polypeptides have been maintained in all picornaviruses. The diversity of the proteins is therefore most probably the result of numerous in-frame mutations. For the other proteins, some additional mechanism of diversification may have been functioning in the course of evolution. Among the viruses encoding 2A proteins sharing the NPGP motif, the two viruses FMDV and ERV1 encode a 2A consisting of only 16 amino acids, whereas the cardioviruses ERV2 and AiV encode a 2A ranging between 67 and 150 residues. It can be speculated that deletion events in the 2A coding region of FMDV and ERV1 are the result of "jumping" of 3D^{Pol}, perhaps by loop-out deletion or by illegitimate recombination (Figure 12.7). On the other hand, the three adjacent coding regions for VPg uniquely found in all strains of FMDV suggest duplication events. In other viruses, e.g. ERV2 or TMEV, genetic events such as local duplication and deletions may have occurred, leading to considerable size heterogeneity of the corresponding VPgs and adjacent sequences (Wutz *et al.*, 1996; A.E. Gorbalenya, unpublished results). Duplications have also been discovered in the 5'NTR of enteroviruses (Pilipenko *et al.*, 1989a). Picornavirus genomic redundancy, known as duplications, may have been generated by intragenomic recombination. After duplications, however, the sequences must have undergone some variation so as to avoid elimination by homologous recombination. Indeed, the nucleotide sequences (and to a small extent also the amino acid sequences) of the three VPgs of FMDV differ such that homologous recombination at this locus is unlikely (Cao and Wimmer, 1996).

In spite of lack of evidence, duplications by intragenomic recombination might have been involved in the production of large differences in size found in capsid proteins VP1 and VP2, or in non-structural 3A and 2B proteins of some picornaviruses. The capsid proteins contain long extra loops while the 2B protein of ERV1 has an enormous size relative to the 2B proteins of all other picornaviruses (283 versus 100–150 amino acids; Wutz *et al.*, 1996). On the other hand, Charini *et al.* (1994) have reported that, surprisingly, a viable poliovirus isolate they selected from a swarm of revertants had captured a short segment of cellular ribosomal RNA. Thus, capture of entirely foreign RNA sequences, although very rare, cannot be excluded from the mechanisms of diversification.

At least two different mechanisms could have given rise to the contemporary diversity of 2A and L protein families. The diversity includes, amongst others, chymotrypsin-like proteinase and NPGP motif-containing polypeptides for 2A and papain-like proteinase and Zn-finger proteins for L. Phylogenetic analyses suggest that "new" unrelated 2A and L proteins have emerged in the course of evolution of picornaviruses on several occasions, following the split of the major groups of the picornavirus tree. It is logical to assume that, following each split, one of the two descendants has arisen from an ancestral viral source, the other from an "independent" source. As to the latter, the coding sequence of either 2A or L could have recombined with a gene of either another virus or of the cell, leading to the replacement of the ancestral coding sequence. For example, this replacement mechanism could have resulted in the capture of cellular chymotrypsin-like (2A^{pro}) or papain-like (L^{pro}) activities. This hypothesis is, of course, purely speculative since no potential partners in recombination have been identified as yet.

Alternatively, the diversity of the 2A and L families may be the result of frame-shifting events. For example, enteroviruses have a "spacer sequence" between the IRES and the ORF of the polyprotein. This spacer commences with an unused ("silent") AUG at the 3' border of the IRES. In poliovirus, it is 154 nt long and represents a small out-of-frame ORF terminating inside the polyprotein ORF. If the silent AUG at the 5' end of the spacer were to trigger initiation of translation and, in addition, a frame-shift mutation connected the small ORF with the main ORF, a small "leader" peptide would be created fused to the polyprotein. All that is then necessary is a 3C^{pro} cleavage site to sever the "leader" from VP4 – and a genetic arrangement would have been created resembling that of cardio- and aphthoviruses (Jang *et al.*, 1990). Indeed, the silent AUG of poliovirus can be turned on by changing its Kozak context (Pestova *et al.*, 1994), and stable poliovirus variants can be isolated that carry short foreign leaders (see above; Mueller and Wimmer, 1998). Thus, the conversion of an enterovirus to a cardiovirus genotype with respect to an L protein can be envisioned by relatively simple genetic changes. Similarly, it should be possible to convert a cardiovirus genotype in this region into an enterovirus genotype by silencing its L ORF.

It is relevant that, as already mentioned, a

strain of Theiler's virus has been identified that, just like the normal cardioviruses, synthesizes a polyprotein-fused L protein and, in addition, a polypeptide L* in a separate ORF. L* synthesis is initiated at its own AUG initiation codon (Takata *et al.*, 1998). Apparently, the synthesis of L* may present the virus with an advantage in the natural host, a fact that may have contributed to its selection.

By comparison with the L protein region in TMEV, two 2A proteins may have existed in the ancestral picornavirus genome, one active in the polyprotein, the other "silent". In the course of subsequent speciation, each of these 2A variants may have been used in separate picornavirus lineages. The activation of the "silent" 2A may have led to a concomitant inactivation of the other 2A.

It should be mentioned that the presence of multiple alternative ORFs in ancestral picornavirus genomes may have been the rule rather than the exception, particularly if the polyprotein evolved by amplification of 11-mers (Gorbalenya, 1995; see also above). Ohno (1984) has demonstrated that periodicity-organized polynucleotides with a period that cannot be divided by 3 (11-long periodicity included) have an identical coding capacity in each frame. In other words, if one ORF is open the two other frames are open also. In the course of evolution, two out of three reading frames may have deteriorated or may have given rise to genetic variation as speculated for the generation of the diversity in 2A and L proteins.

Driving Force of Evolution

Numerous studies attest to a remarkable stability of the picornavirus genotype if grown under identical conditions (Wimmer *et al.*, 1993). On the other hand, if exposed to altered conditions in the environment, a shift to new variants can be readily observed. Just like other biological systems, it can be assumed that picornavirus speciation has been driven by a changing environment.

Circumstances upon which a picornavirus may encounter a "new" environment include:

(1) horizontal or vertical transfer to a new (different) host; (2) entering a natural host through a non-natural gate; (3) infecting immunized (natural) hosts previously exposed to the same virus. Although there is no proof, it is intuitively highly likely that all three scenarios have played a role in picornavirus speciation. In the following, a speculative reconstruction of forces will be presented that may have contributed to the evolution of picornaviruses.

Picornaviruses belonging to a genus or a cluster may have almost identical phenotypes with respect to growth properties and even in regard to pathogenic potential. A most important characteristic, however, does further divide a group of very closely related picornaviruses (e.g. polioviruses): the susceptibility to activation by different neutralizing antibodies and, hence, the separation into serotypes (see the Introduction). It is logical to assume that the (negative) pressure of the immune system may be largely accountable for serotype diversification of picornaviruses. That is, the immune response can lead to the selection of viral variants resistant to the neutralizing immune response produced by the surviving host. Such variants would form a pool from which a new serotype could be further selected. In fact, such mechanism of virus evolution seems to dominate in the case of influenza A virus or immunodeficiency virus (HIV). However, the sheer unlimited degree of serotype diversification observed in influenza viruses or HIV is an exception rather than the rule amongst viruses. Indeed, not all picornaviruses seem to be able to easily produce new serotypes. For example, the genus *Hepatovirus* encompasses only one serotype while others are restricted to a few serotypes (e.g. poliovirus).

New viral variants that have escaped the immune surveillance must, of course, interact with multiple host components at virtually every stage of their reproduction in order to survive. This includes virus entry into the host cell, translation and processing, genome replication, encapsidation and maturation, spread in the host. Each of these steps are checkpoints and every new viral variant must be fit to pass these barriers.

The earliest events in the infectious cycle –

receptor interaction, uptake, uncoating – and the mechanisms of neutralization are amongst the least understood in the molecular biology of picornaviruses. The crystal structures of some member viruses of four picornavirus genera have been solved; examples are: *Enterovirus*, poliovirus 1 and 3 (Hogle *et al.*, 1985); *Rhinovirus*, human rhinoviruses 2, 3, 14, 16 (Rossman *et al.*, 1985); *Cardiovirus*, mengovirus (Luo *et al.*, 1987), Theiler's virus (Luo *et al.*, 1996); *Aphthovirus*, FMDV (Acharya *et al.*, 1989). (For a complete list, see Lentz *et al.*, 1999). However, the precise localization and structures of different neutralization antigenic sites (the structures interacting with neutralizing antibodies) is known only for polioviruses, rhinoviruses and aphthoviruses. For aphthoviruses and for polioviruses, the available evidence suggests that the same structures that determine in part the serotype identity are also involved in receptor recognition (Domingo *et al.*, 1993; Mason *et al.*, 1994; Harber *et al.*, 1995). Thus, immune-escaping viral mutants are likely to be enriched in those variants that have maintained the ability to efficiently interact with the cognate receptor and follow the pathway of uptake and uncoating. This is, of course, only speculative but, if correct, it would explain in part serotype restriction (Harber *et al.*, 1995).

In this respect it may be informative to compare receptor specificities with serotype diversities of human enteroviruses, on the one hand and rhinoviruses on the other. These two genera encompass viruses that have diverged from an immediate common ancestor and radiated during the same time period (Figure 12.1). In the course of evolution, different serotypes in roughly the same numbers have been generated in these two picornavirus branches: there are about 66 enterovirus and over 100 rhinovirus serotypes. This implies that viruses of the two genera are similarly prone to accumulation of changes in those capsid structures giving rise to new serotypes. But what about receptor specificity of these viruses?

At the time of writing, two receptors have been assigned for human rhinoviruses (which is probably all that will be found) and six receptors for human enteroviruses (at least four more are awaiting identification; Table 12.1). Thus, in contrast to the quite similar extent of serotype diversification in both genera, adaptation to new receptors is significantly more restricted in rhinoviruses than in the closely related enteroviruses. Importantly, there is an overlap between the two receptor patterns and, taken together, the ICAM-1 receptor specificity appears to be dominant among entero- and rhinoviruses. This can be interpreted to mean that the immediate common ancestor of both entero- and rhinoviruses may have used a receptor related to ICAM-1. Regardless of whether this is true or not, the subsequent evolution of ICAM-1-recognizing picornaviruses has proceeded differently, as seen in the disparity of the current use of this cellular receptor (>90 for rhinoviruses versus 11 for C-cluster coxsackieviruses). Given that the serotype diversification has proceeded at a similar pace in entero- and rhinoviruses, enteroviruses may have had greater opportunities – or a greater need – to adapt to new receptors in order to initiate an infection. This may be related to the function(s) of receptors in viral docking and uncoating: whereas rhinoviruses may need the receptor only for docking and uptake (because of their inherent sensitivity to the acidic pH inside late endosomes), the exceedingly stable enteroviruses do need the receptor (and possibly a co-receptor) for docking, uptake *and* uncoating. With poliovirus, a particle stable to detergents, proteases and low pH (pH 2), this is exemplified in the formation of A-particles, a labile product of receptor/virion interaction and an intermediate in uncoating (Wimmer *et al.*, 1994). A-particle formation appears to involve also sequences of neutralization antigenic sites (Harber *et al.*, 1995). Thus, the intercourse between receptor and enterovirion may be much more complex than that between receptor and rhinovirion. Consequently, a change in the serotype may have forced enteroviruses to search for new receptors to retain the uncoating capacity of the cellular receptor.

The unusually large serotype diversity of the major receptor group human rhinoviruses may then be explained as follows. It seems possible that the initiation of an infectious cycle of HRV *does not critically require* an interaction between structures of the neutralization antigenic sites of the virion and ICAM-1. That is, the N-terminal domain of ICAM-1, by inserting itself into the virion's canyon, can effect docking, uptake and uncoating of the particle. Progression through any of these events is *not critically dependent* on sequences of the neutralization antigenic sites. If correct, it follows that variation of the antigenic sites does not restrict viral proliferation and serotype evolution. Consistently, in other picornaviruses the neutralization antigenic sites and the determinants recognizing the receptor would be much more overlapping and mutually dependent.

It is likely that an initial immune-driven selection might also finally result in a virus variant with changed or extended tissue tropism. This might have happened with CAV24v, a C-cluster human enterovirus. Immune pressure might have initiated the selection of the CAV24v mutant derived from a CAV24 swarm. As mentioned before, CAV24v is a very recent variant of CAV24 and, unlike its parent and the other members of the C-cluster, it can cause acute hemorrhagic conjunctivitis. Apart from the possibility that CAV24v emerged through immune selection, it could also have been selected from a swarm when the parental CAV24 was accidentally inoculated into the eye.

Another type of selection might have been responsible for the emergence of swine vesicular disease virus (SVDV). Phylogenetic analysis of genomes of human enteroviruses identified SVDV as being interleaved with human viruses comprising the CBV-like cluster (Poyry *et al.*, 1996). This observation is strongly indicative of selection of SVDV from a mutant of a human coxsackie-B virus entering the new host through frequent contacts of these domestic animals with (infected) humans.

Picornaviridae are a Prosperous Virus Family

We have discussed different aspects relevant to picornavirus evolution, but we did not address one crucial question: Are picornaviruses a successful family? We believe that the answer is: yes. In discussing this issue, we will also formu-

late considerations regarding the worldwide eradication of poliovirus.

One of the strongest criteria of biological prosperity is the diversity of a taxonomic group. Despite some bias inherent in current analyses, phylogenetic studies of picornaviral genomes suggest that Picornaviridae have radiated densely over the course of evolution, at both early and late stages (Figure 12.1). Furthermore, picornaviruses are members of a superfamily with numerous distant relatives (Figure 12.8) that infect a wide range of organisms, including both plants and animals. Some of these viruses, like Sequiviridae, employ a genetic plan that is basically a variation of the genetic plan used by picornaviruses (Figure 12.10).

Prosperity of the host is another prerequisite for a virus to be successful. By this criterion also, picornaviruses are successful, since the majority of them, representing different branches of the picornavirus tree, infect humans. Humans are arguably one of the most successful species in the biological world. In truth, picornaviruses are relatively harmless even though few humans, if any, can escape picornavirus infections. This too can be viewed as evidence that these viruses have adapted well to their host, as they have not significantly undermined human affairs. This is true even for poliovirus, an agent that is commonly regarded as a deadly virus following epidemics of poliomyelitis. However, prior to this century, poliovirus did not cause epidemics, even though it infected humans at rates approaching 100%. Epidemics emerged because human behavior changed through the invention of modern hygiene. Hygiene broke the chain of natural immunization through infant infection combined with infant protection by maternal antibodies. Even in this century's devastating epidemics, however only 1–2% of infected individuals developed poliomyelitis.

A Special Case of Picornavirus Evolution – the Human Battle with Poliovirus

The poliovirus–human relationship alluded to above deserves to be discussed in more detail. Humans, who occupy a unique niche in the biological world (because they care about *each* human life), did not accept their potential defeat as poliomyelitis became an epidemic. Unprecedented efforts combining medical research with modern technologies led to the development of two highly effective poliovirus vaccines, the inactivated poliovirus vaccine by Jonas Salk and the live attenuated vaccine by Albert Sabin (Wimmer *et al.*, 1993). Through education of the populace and advanced health-care measures, mass vaccinations have gradually eliminated wild-type poliovirus, first in the developed countries and later in most of the world. Incredibly, the few cases of poliomyelitis in the western hemisphere now result from vaccination with the live Sabin strains. Overall, polio vaccination is a success story of greatest consequence. Indeed, through worldwide efforts led by the World Health Organization, it is likely that wild-type polioviruses will be eradicated globally by the turn of the century (WHO, 1985).

Do these considerations allow us to safely conclude that, after its global eradication, poliovirus will have no chance to re-emerge through enterovirus evolution? For discussion of this issue, we will first summarize hypotheses about the possible origin of polioviruses and their closest relatives, the C-cluster coxsackieviruses.

The three serotypes of poliovirus belong to the C-cluster of enteroviruses (Table 12.1; Figure 12.1B). The most comprehensive analysis of the C-cluster has been performed with sequences of the VP4-VP2 capsids and with sequences of the 3D^{pol} RNA polymerase (Pulli *et al.*, 1995). Results of these analyses are consistent with data obtained in a study of the other regions of the viral genome using a less representative set of sequences (Poyry *et al.*, 1996). Therefore, these relationships shown in Figure 12.1B can be assumed to be quite reliable. A phylogenetic analysis of the capsid VP4-VP2 region of C-cluster viruses indicated that the tree has split at least twice, perhaps before the emergence of an immediate ancestor of polioviruses. The first split led to the separation of a branch encompassing CAV1, CAV21 and CAV24 from the main C-cluster trunk, and the second, more recent one resulted in the separation of the

ancestor for PV and the ancestor for CAV11, CAV13, CAV17, CAV18, CAV20 and CAV20b. The results obtained with sequences of 3D^{pol} favor an even more complex evolutionary history of poliovirus, including more than five intermediate steps (Pulli *et al.*, 1995). Consistent with the results of the analysis of the capsid region, CAV21 and CAV24 were among those viruses that diverged from the main trunk relatively early in evolution while the three poliovirus serotypes clustered together with CAV11, CAV13, CAV17 and CAV18. Remarkably, in the tree based on 3D^{pol} sequences the latter four coxsackieviruses (as well as several other coxsackieviruses) are interleaved with, rather than separated from, the three poliovirus serotypes (Figure 12.1). This stands in contrast to the tree of the capsid region. Assuming the most parsimonious scenario of evolution, the combination of these results strongly implies that coxsackieviruses that recognize the ICAM-1 receptor formed a pool from which polioviruses, interacting with the CD155 receptor, have evolved. This conclusion is compatible with a hypothesis of the immune-driven evolution of entero- and rhinoviruses presented above. Furthermore, the analyses do not indicate that three polioviruses comprise a monophyletic subgroup within the C-cluster enteroviruses and, hence, have emerged from an ancestral virus by speciation, as one could expect from a distinct phenotypic profile of these viruses.

We have previously hypothesized that the coxsackiviruses may have derived from polioviruses by switching receptors from CD155 to ICAM-1 (Harber *et al.*, 1995). This possibility may be supported from the fact that the IRESes of C-cluster coxsackieviruses are highly 'neuropathogenic". On the other hand, the assessment presented above favors an evolutionary relationship in the opposite direction. Regardless of the direction in which these viruses emerged, the receptor switch has profound consequences for their pathogenic properties: whereas the C-cluster coxsackieviruses cause respiratory disease, poliovirus can cause deadly neurological disease. These considerations may also have important practical implications. For the sake of the argument, we will assume that the poliovirus eradication campaign has been successfully completed and no more poliovirus particles, including those of the vaccine strains, are circulating worldwide. Furthermore, we will assume that all vaccination against poliovirus (including vaccination by inactivated vaccines) has been terminated, a scenario that has been envisioned to be a reality by the end of the next decade. These measures would mark the beginning of a new era in the history of mankind: there will be no human exposure to polioviruses and their antigens.

Generations of humans will be born that have not been infected with wild-type or vaccine polioviruses and, gradually, they will replace the older generations who carry anti-poliovirus antibodies. At that point, the world will not only be free of poliovirus, but its human population will also no longer carry anti-poliovirus antibodies. Thus, a new environment will emerge for human viruses, in particular for C-cluster coxsackieviruses, which are the closest genetic relatives of poliovirus. These C-CAVs" are expected to circulate widely in the human population, exploring a new evolutionary space.

Within the human space populated by the C-CAVs, there will then exist also a free space that was previously occupied by the three (extinct) poliovirus serotypes. It is possible that mutations in antigenic sites of the C-CAVs may (re)generate affinity to CD155. Prior to eradication, C-CAVs carrying such mutations could conceivably be eliminated by anti-poliovirus antibodies (Harber *et al.*, 1995) but in the poliovirus-free world they may remain unchecked. This means that, once emerged, these new viruses carrying poliovirus-like neutralization antigenic sites with CD155 receptor affinity are less likely to be eliminated from the human population after eradication than before. Since all enteroviruses, the variants included, lead to enteric infections, these variants may find a passage to the CNS and, mediated by their affinity to CD155, may cause neurological disease. It is relevant to point out (Gromeier and Wimmer, unpublished results) that poliovirus chimeric viruses in which the poliovirus IRES has been replaced with that of C-cluster IRES elements have been found to be highly neurovirulent in *CD155* tg mice (see the section on Pathogenesis). Thus, there is reason to fear that

in a poliovirus-free world new coxsackievirus-related, poliovirus-like pathogens that can cause poliomyelitis may emerge in the course of natural viral evolution. The time frame, however, cannot be predicted. It could be one generation or 1000 years.

The human condition favors an increasing rate of diversity of human viruses simply because of the increasing size of the human population (estimated to stabilize at 8–12 billion during the next century). This population explosion will lead to a dramatic increase of human contacts, either in cities, particularly megacities (harboring more than 50% of the world's population), through travel or otherwise. Clearly, this presents a fertile ground for proliferation and diversification of the highly infectious human picornaviruses. Thus, the possibilities of genetic variation of picornaviruses leading to new or renewed human pathogens, such as CAV24v, must always be kept in mind.

At this point, however, our considerations of the possible re-emergence of poliovirus-like pathogens in the post-eradication era pale in the face of mankind's heroic attempt to eradicate an RNA virus for the first time. After all, poliovirus has caused, and is still causing, terrible human suffering.

CONCLUDING REMARKS

Picornaviruses have been discovered because they cause diseases in animals and humans. Fortunately, most human picornavirus infections are self-limiting. Yet the enormously high rate of picornavirus infections in the human population can lead to a significant incidence of disease complications that may be permanently debilitating or even fatal. The case of poliovirus has taught us that a change of human behavior, which, paradoxically, was the invention of modern hygiene, has greatly aggravated the impact of infection by this specific agent. Clearly, this scenario could repeat itself with other human picornavirus species. The terror of this century's poliomyelitis epidemics has driven picornavirus research forward more than any other factor. This work has led to a wealth of discoveries in biology in general, and to an abundance of data describing the unique biology of picornaviruses and their evolution in particular.

Picornaviruses employ one of the simplest imaginable genetic systems: they consist of single-stranded RNA that encodes only a single multidomain polypeptide, the polyprotein. The RNA is packaged into a small, rigid, naked, icosahedral virion whose proteins are unmodified except for a myristate at the N-termini of VP4. The RNA itself does not contain modified bases. Thus, picornaviruses travel with light baggage. On first sight, the replication of picornaviruses is exceedingly simple. After having chosen a receptor from a large menu of cell-surface proteins, the virion enters the cytoplasm and immediately translates its genome, controlled by its IRES element. Thereafter, the polyprotein is processed by its own proteinases. RNA replication occurs by a unique, protein-primed mechanism catalyzed by the RNA-dependent RNA polymerase. Assembly appears to be linked to RNA synthesis, and release of the progeny virions follows a passive mechanism. There is no need for a cellular nucleus. Indeed, the entire replication cycle can occur in a cell-free system free of nuclei, mitochondria and perhaps of all other cellular organelles.

Yet as of now we understand only a small fraction of these viruses' life cycle, and we are awed by the sophistication with which the viruses express their genetic information. The IRES, arguably one of the most complex cis-acting signals known in RNA systems, has freed picornaviruses from the cellular constraint of cap-dependent translation. This, in turn, allows the primer-dependent RNA polymerase, an enzyme with properties generally ascribed only to DNA polymerases or reverse transcriptase, to prime with VPg and leave the RNA uncapped. Polyprotein processing proceeds in a controlled manner yielding cleavage intermediates and end-products that can be used for different functions. Thus, the menu of gene products is expanded through the temporal regulation of proteolytic processing. Details of all of these steps in replication are still obscure (Agol et al., 1999).

The key to ultimately understanding picornaviruses may be to rationalize the huge amount of information about these viruses from the perspective of evolution. It is possible that the replicative apparatus of picornaviruses originated in the precellular world and was subsequently refined in the course of thousands of generations in a slowly evolving environment. Picornaviruses cultivated the art of adaptation, which has allowed them to "jump" into new niches offered in the biological world. Also, by having chosen humans as an additional host, they were offered an abundance of opportunities to proliferate in different tissues, which has contributed to their diversification. These opportunities have further increased through the human population explosion and through changes in human behavior.

We suggest that, in addition to drastic and expansive measures such as global eradication, strategies should be developed that aim at predicting the possible evolution of new picornavirus pathogens and preparing for their control. The results reviewed in this article may contribute to achieving this tantalizing and desirable goal.

ACKNOWLEDGMENTS

We are indebted to Leena Kinnunen for providing Figure 12.1, and to Steffen Mueller for Figure 12.6. We thank Astrid Wimmer for editing parts of the manuscript. Work by M.G. and E.W. described here has been supported in part by grants from the National Institutes of Health, the National Cancer Institute, and the Centers for Disease Control.

REFERENCES

Acharya, R., Fry, E., Stuart, D., Fox, G., Rowlands, D. and Brown, F. (1989) The three-dimensional structure of foot-and-mouth disease virus at 2.9 Å resolution. *Nature*, **337**, 709–716.

Agol, V.I., Drozdov, S.G., Ivannikova, T.A., Kolesnikova, M.S., Korolev, M.B. and Tolskaya, E.A. (1989) Restricted growth of attenuated poliovirus strains in cultured cells of a human neuroblastoma. *J. Virol.*, **63**, 4034–4038.

Agol, V.I., Paul, A.V. and Wimmer, E. (1999) Paradoxes of the replication of picornaviral genomes. *Virus Research.*, in press.

Alexander, L., Lu, H.H. and Wimmer, E. (1994) Polioviruses containing picornavirus type 1 and/or type 2 internal ribosomal entry site elements: genetic hybrids and the expression of a foreign gene. *Proc. Natl Acad. Sci. USA*, **91**, 1406–1410.

Allaire, M., Chernaia, M.M., Malcolm, B.A. and James, M.N. (1994) Picornaviral 3C cysteine proteinases have a fold similar to chymotrypsin-like serine proteinases. *Nature*, **369**, 72–76.

Altmeyer, R., Escriou, N., Girard, M., Palmenberg, A. and van der Werf, S. (1994) Attenuated Mengo virus as a vector for immunogenic human immunodeficiency virus type 1 glycoprotein 120. *Proc. Natl Acad. Sci. USA*, **91**, 9775–9779.

Altmeyer, R., Girard, M., van der Werf, S., Mimic, V., Seigneur, L. and Saron, M.F. (1995) Attenuated Mengo virus: a new vector for live recombinant vaccines. *J. Virol.*, **69**, 3193–3196.

Andino, R., Rieckhof, G.E. and Baltimore, D. (1990) A functional ribonucleoprotein complex forms around the 5′ end of poliovirus RNA. *Cell*, **63**, 369–380.

Andino, R., Rieckhof, G.E., Achacoso, P.L. and Baltimore, D. (1993) Poliovirus RNA synthesis utilizes an RNP complex formed around the 5′-end of viral RNA. *EMBO J.*, **12**, 3587–3598.

Andino, R., Silvera, D., Suggett, S.D. *et al.* (1994) Engineering poliovirus as a vaccine vector for the expression of diverse antigens. *Science*, **265**, 1448–1451.

Baltimore, D. (1971) Expression of animal viral genomes. *Bacteriol. Rev.*, **35**, 235–241.

Barton, D. and Flanegan., J.B. (1993) Coupled translation and replication of poliovirus RNA *in vitro*: synthesis of functional 3D polymerase and infectious virus. *J. Virol.*, **67**, 822–831.

Barton, D.J., Black, P.A. and Flanegan, J.B. (1995) Complete replication of poliovirus *in vitro*: preinitiation RNA replication complexes

require soluble cellular factors for the synthesis of VPg-linked RNA. *J. Virol.*, **69**, 5516–5527.

Bazan, J.F. and Fletterick, R.J. (1988) Viral cysteine proteases are homologous to the trypsin-like family of serine proteases: structural and functional implications. *Proc. Natl Acad. Sci. USA*, **85**, 7872–7876.

Benzer, S. (1961) Genetic fine structure. In: *Harvey Lectures*, series 56. Academic Press, New York.

Bergelson, J.M., Chan, M., Solomon, K.R., St. John, N.F., Lin, H. and Finberg, R.W. (1994) Decay-accelerating factor (CD55), a glycosylphosphatidyylinsitol-anchored complement regulatory protein, is a receptor for several echoviruses. *Proc. Natl Acad. Sci. USA*, **91**, 6245–6249.

Bergelson, J.M., Shepley, M.P., Chan, B.M., Hemler, M.E. and Finberg, R.W. (1992) Identification of the integrin VLA-2 as a receptor for echovirus 1 [see comments]. *Science*, **255**, 1718–1720.

Bernstein, A., Roivainen M., Hovi, T., Mason, P.W. and Baxt, B (1995) Antibodies to the vitronectin receptor (integrin alpha V beta 3) inhibit binding and infection of foot-and-mouth disease virus to cultured cells. *J Virol.*, **69**, 2664–2666.

Bernstein, H.D., Sarnow, P. and Baltimore, D. (1986) Genetic complementation among poliovirus mutants derived from an infectious cDNA clone. *J. Virol.*, **60**, 1040–1049.

Bienz, K., Egger, D., Pfister, T. and Troxler, M. (1992) Structural and functional characterization of the poliovirus replication complex. *J. Virol.*, **66**, 2740–2747.

Bishop, J.M. and Koch, G. (1969) Infectious replicative intermediate of poliovirus: purification and characterization. *Virology*, **37**, 521–534.

Blyn, L.B., Swiderek, K.M., Richards, O., Stahl, D.C., Semier, B.L. and Ehrenfeld, E. (1996) Poly(rC) binding protein 2 binds to stem-loop IV of the poliovirus RNA 5′ noncoding region: identification by automated liquid chromatography-tandem mass spectrometry. *Proc. Natl Acad. Sci. USA*, **93**, 11115–11120.

Bodian, D. (1972) Poliomyelitis. In: *Pathology of*

the Nervous System (ed. Minckler, J.), pp. 2323–2344. McGraw-Hill, New York.

Borman, A.M., Deliat, F.G. and Kean, K.M. (1994) Sequences within the poliovirus internal ribosome entry segment control viral RNA synthesis. *EMBO J.*, **13**, 3149–3157.

Cameron-Wilson, C.L., Pandolfino, Y.A., Zhang, H.Y., Pozzeto, B. and Archard, L.C. (1998) Nucleotide sequence of an attenuated mutant of coxsackievirus B3 compared with the cardiovirulent wildtype: assessment of candidate mutations by analysis of a revertant to cardiovirulence. *Clin. Diagn. Virol.*, **9**, 99–105.

Cao, X. and Wimmer, E. (1995) Intragenomic complementation of a 3AB mutant in dicistronic polioviruses. *Virology*, **209**, 315–326.

Cao, X. and Wimmer, E. (1996) Genetic variation of the poliovirus genome with two VPg coding units. *EMBO J.*, **15**, 23–33.

Charini, W.A., Burns, C.C., Ehrenfeld, E. and Semler, B.L. (1991) *Trans* rescue of a mutant poliovirus RNA polymerase function. *J. Virol.*, **65**, 2655–2665.

Charini, W.A., Todd, S., Gutman, G.A. and Semler, B.L. (1994) Transduction of a human RNA sequence by poliovirus. *J. Virol.*, **68**, 6547–6552.

Chen, C.Y. and Sarnow, P. (1995) Initiation of protein synthesis by the eukaryotic translational apparatus on circular RNAs. *Science*, **268**, 415–417.

Chen, H.H., Kong, W.P., Zhang, L., Ward, P.L. and Roos, R.P. (1995) A picornaviral protein synthesized out of frame with the polyprotein plays a key role in a virus-induced immune-mediated demyelinating disease. *Nature Med.*, **1**, 927–931.

Chetverin, A.B., Chetverina, H.V., Demidenko, A.A. and Ugarov, V.I. (1997) Nonhomologous RNA recombination in a cell-free system: evidence for a transesterification mechanism guided by secondary structure. *Cell*, **88**, 503–513.

Cho, M.W., Richards, O.C., Dmitrieva, T.M., Agol, A. and Ehrenfeld., E. (1993) RNA duplex unwinding activity of poliovirus RNA-dependent RNA polymerase 3D^pol. *J. Virol.*, **67**, 3010–3018.

Cole, C. and Baltimore, D. (1973) Defective interfering particles of poliovirus. *J. Mol. Biol.*, **76**, 325–343.

Cole, C.N., Smoler, D., Wimmer, E. and Baltimore, D. (1971) Defective interfering particles of poliovirus. 1. Isolation and physical properties. *J. Virol.*, **7**, 478–485.

Cooper, P.D. (1977) Genetics of picornaviruses. In: *Comprehensive Virology*, (ed. Fraenkel-Conrat, W.E.), pp. 133–207. Plenum Press, New York.

Cuconati, A., Molla, A. and Wimmer, E. (1998) Brefeldin A inhibits cell-free, *de novo* synthesis of poliovirus. *J. Virol.*, **72**, 6456–6464.

Da Silva, E.E., Winkler, M.T. and Pallansch, M.A. (1996) Role of enterovirus 71 in acute flaccid paralysis after the eradication of poliovirus in Brazil. *Emerg. Infect. Dis.*, **2**, 231–233.

Dildine, S.L. and Semler, B.L. (1989) The deletion of 41 proximal nucleotides reverts a poliovirus mutant containing a temperature-sensitive lesion in the 5′ noncoding region of genomic RNA. *J. Virol.*, **63**, 847–862.

Domier, L.L., Franklin, K.M., Shahabuddin, M. *et al.* (1986) The nucleotide sequence of tobacco vein mottling virus RNA. *Nucl. Acids Res.*, **14**, 5417–5430.

Domingo, E., Diez, J., Martinez, M.A. *et al.* (1993) New observations on antigenic diversification of RNA viruses: antigenic variation is not dependent on immune selection. *J. Gen. Virol.*, **74**, 2039–2045.

Dougherty, W.G. and Semler, B.L. (1993) Expression of virus-encoded proteinases: functional and structural similarities with cellular enzymes. *Microbiol. Rev.*, **57**, 781–882.

Duggal, R. and Wimmer, E. (1999) Temperature-dependent alteration of cross-over sites in poliovirus recombination. *Virology*, submitted.

Duggal, R., Cuconati, A., Gromeier, M. and Wimmer, E. (1997) Genetic recombination of poliovirus in a cell-free system. *Proc. Natl Acad. Sci. USA*, **94**, 13786–13791.

Duke, G.M., Osorio, J.E. and Palmenberg, A.C. (1990) Attenuation of mengo virus through genetic engineering of the 5′ noncoding poly(C) tract. *Nature*, **343**, 474–476.

Eigen, M. (1993) The origin of genetic information: viruses as models. *Gene*, **135**, 37–47.

Emini, E.A., Leibovitz, J., Diamond, D.C., Bonin, J. and Wimmer, E. (1984) Recombinants of Mahoney and Sabin strain poliovirus type 1: analysis of *in vitro* phenotypic markers and evidence that resistance to guanidine maps in the nonstructural proteins. *J. Virol.*, **137**, 74–85.

Evans, D.M., Dunn, G., Minor, P.D. *et al.* (1985) Increased neurovirulence associated with a single nucleotide change in a noncoding region of the Sabin type 3 poliovaccine genome. *Nature*, **314**, 548–550.

Feigelstock, D., Thompson, P., Mattoo, P., Zhang, Y. and Kaplan, G.G. (1998) The human homolog of HA Vcr-1 codes for a hepatitis A virus cellular receptor. *J Virol.*, **72**, 6621–6628.

Filman, D.J., Syed, R., Chow, M., Macadam, A.J., Minor, P.D. and Hogle, J.M. (1989) Structural factors that control conformational transitions and serotype specificity in type 3 poliovirus. *EMBO J.*, **8**, 1567–1579.

Fitch, W.M. (1971) Rate of change of concomitantly variable codons. *J. Mol. Evol.*, **1**, 84–96.

Flanegan, J.B. and Baltimore, D. (1977) Poliovirus-specific primer-dependent RNA polymerase able to copy poly(A). *Proc. Natl Acad. Sci. USA*, **74**, 3677–3680.

Flanegan, J., Pettersson, R., Ambros, V., Hewlett, M. and Baltimore, D. (1977) Covalent linkage of a protein to a defined nucleotide sequence at the 5′-terminus of virion and replicative intermediate RNAs of poliovirus. *Proc. Natl Acad. Sci. USA*, **74**, 961–965.

Gamarnik, A.V. and Andino, R. (1996) Replication of poliovirus in Xenopus oocytes requires two human factors. *EMBO J*, **15**, 5988–5998.

Gamarnik, A.V. and Andino, R. (1997) Two functional complexes formed by KH domain containing proteins with the 5′ noncoding region of poliovirus RNA. *RNA*, **3**, 882–892.

Gamarnik, A.V. and Andino, R. (1998) Switch from translation to replication in a positive-stranded RNA virus. *Genes and Dev.*, **12**, 2293–2304.

Gmyl, A.P., Pilipenko, E.V., Maslova, S.V., Belov, G.A. and Agol, V.I. (1993) Functional and genetic plasticities of the poliovirus genome: quasi-infectious RNAs

modified in the 5′-untranslated region yield a variety of pseudorevertants. *J. Virol.*, **67**, 6309–6316.

Gmyl, A.P., Belousov, E.V., Maslova, S.V., Khitrina, E.V., Chetverin, A.B. and Agol, V.I. (1999), Submitted.

Goldbach, R.W. (1986) Molecular evolution of plant RNA viruses. *Annu. Rev. Phytopathol.*, **24**, 289–310.

Goodfellow, I., Meredith, J., Richardson, A., Barclay, W., Alm, O.J. and Evans, D.J. (1998) Abstract. Europic'98 (Jena, Germany).

Gorbalenya, A.E. (1995) Origin of RNA viral genomes: approaching the problem by comparative sequence analysis. In: *Molecular Basis of Virus Evolution* (eds Gibbs, A.J., Calisher, C.H. and Garcia-Arenal, F.) pp. 49–66. Cambridge University Press, Cambridge, UK.

Gorbalenya, A.E. and Koonin, E.V. (1990) Superfamily of UvrA-related NTP-binding proteins. Implications for rational classification of recombination/repair systems. *J. Mol. Biol.*, **213**, 583–591.

Gorbalenya, A.E. and Koonin, E.V. (1993a) Comparative analysis of the amino acid sequences of the key enzymes of the replication and expression of positive-strand RNA viruses. Validity of the approach and functional and evolutionary implications. *Sov. Sci. Rev. D. Physiochem. Biol.*, **11**, 1–84.

Gorbalenya, A.E. and Koonin, E.V. (1993b) Helicases: amino acid sequence comparisons and structure-function relationships. *Curr. Opin. Struct. Biol.*, **3**, 419–429.

Gorbalenya, A.E. and Snijder, E.J. (1996) Viral cysteine proteinases. *Persp. Drug Discov. Design*, **6**, 64–86.

Gorbalenya, A.E., Blinov, V.M. and Donchenko, A.P. (1986) Poliovirus induced proteinase 3C: a possible evolutionary link between cellular serine and cysteine proteinase families. *FEBS Lett.*, **194**, 253–257.

Gorbalenya, A.E., Donchenko, A.P., Blinov, V.M. and Koonin, E.V. (1989) Cysteine proteases of positive strand RNA viruses and chymotrypsin-like serine proteases: a distinct protein super-family with a common structural fold. *FEBS Lett.*, **243**, 103–114.

Gorbalenya, A.E., Koonin, E.V. and Lai, M.M.

(1991) Putative papain-related thiol proteases of positive-strand RNA viruses. Identification of rubi- and aphthovirus proteases and delineation of a novel conserved domain associated with proteases of rubi-, alpha- and coronaviruses. *FEBS Lett.*, **288**, 201–205.

Greve, J.M. and Rossman, M. (1994) Interaction of rhinovirus with its receptor, ICAM-1. In: *Cellular Receptors for Animal Viruses*, (ed. Wimmer, E.), pp. 195–213. Cold Spring Harbor Laboratory Press, Cold Spring Harbor.

Gromeier, M. and Wimmer, E. (1998) Mechanism of injury-provoked poliomyelitis. *J. Virol.*, **73**, 5056–5060.

Gromeier, M. and Wimmer, E. (1999) Prophylactic injections and the onset of paralytic poliomyelitis. *Rev. Med. Vir.*, in press.

Gromeier, M., Lu, H.-H. and Wimmer, E. (1995) Mouse neuropathogenic poliovirus strains cause damage in the central nervous system different from poliomyelitis. *Microb. Pathog.*, **18**, 253–267.

Gromeier, M., Alexander, L. and Wimmer, E. (1996) Internal ribosomal entry site substitution eliminates neurovirulence in intergeneric poliovirus recombinants. *Proc. Natl Acad. Sci. USA.*, **93**, 2370–2375.

Gromeier, M., Bossert, B., Arita, M., Nomoto, A. and Wimmer, E. (1999a) Dual stem loops within the poliovirus internal ribosomal entry site control neurovirulence. *J. Virol.*, **73**, 958–964.

Gromeier, M., Solecki, D. and Wimmer, E. (1999b) The human poliovirus receptor/CD155 promoter directs reporter-gene expression in floor plate and optic nerve of transgenic mice. *Genes Devel.*, submitted.

Hadfield, A.T., Lee, W., Zhao, R. *et al.* (1997) The refined structure of human rhinovirus 16 at 2.15 A resolution: implications for the viral life cycle. *Structure*, **5**, 427–441.

Hagino-Yamagushi, K. and Nomoto, A. (1989) In vitro construction of poliovirus defective interfering particles. *J. Virol.*, **66**, 5386–5392.

Haller, A.A., Stewart, S.R. and Semler, B.L. (1996) Attenuation stem-loop lesions in the 5′ noncoding region of poliovirus RNA: neuronal cell-specific translation defects. *J. Virol.*, **70**, 1467–1474.

Hansen, J., Long, A.M. and Schultz, S. (1997) Structure of the RNA-dependent RNA polymerase of poliovirus. *Structure*, **15**, 1109–1122.

Harber, J.J., Bradley, J., Anderson, C.W. and Wimmer, E. (1991) The catalysis of the poliovirus VPO maturation cleavage is not mediated by serine 10 of VP2. *J. Virol.*, **65**, 326–334.

Harber, J., Bernhardt, G., Lu, H.-H., Sgro, J.-Y. and Wimmer, E. (1995) Serotype polymorphism of poliovirus-cellular receptor interaction: separation of events of viral attachment and uptake. *Virology*, **214**, 559–570.

Harris, K.S., Hellen, C.U.T. and Wimmer, E. (1990) Proteolytic processing in the replication of picornaviruses. *Sem. Virol.*, **1**, 323–333.

Harris, K.S., Xiang, W., Alexander, L., Paul, A.V., Lane, W.S. and Wimmer, E. (1994) Interaction of the poliviral polypeptide 3CD^{pro} with the 5′ and 3′ termini of the poliovirus genome: identification of viral and cellular cofactors necessary for efficient binding. *J. Biol. Chem.*, **269**, 27004–20014.

Hellen, C.U.T., Kräusslich, H.-G. and Wimmer, E. (1989) Proteolytic processing of viral polyproteins in the replication of RNA viruses. *Biochemistry*, **28**, 9881–9890.

Hewlett, M., Rose, J. and Baltimore, D. (1976) 5′-terminal structure of poliovirus polyribosomal RNA is pUp. *Proc. Natl Acad. Sci. USA*, **73**, 327–330.

Hirst, G. (1962) Genetic recombination with Newcastle disease virus, polioviruses and influenza. Cold Spring Harbor Symp. *Quant. Biol.*, **27**, 303–308.

Hofer, F., Gruenberger, M., Kowalski, H., Machat, H., Huettinger, M., Kuechler, E. and Blass, D. (1994) Members of the low density lipoprotein receptor family mediate cell entry of a minor-group common cold virus. *Proc. Natl Acad. Sci. USA*, **91**, 1839–1842.

Hogle, J.M., Chow, M. and Filman, D.J. (1985) The three dimensional structure of poliovirus at 2.9 A resolution. *Science*, **229**, 1358–1365.

Holland, J.J., Domingo, E., de la Torre, J.C. and Steinhauer., D.A. (1990) Mutation frequencies at defined single codon sites in vesicular stomatitis virus and poliovirus can be increased only slightly by chemical mutagenesis. *J. Virol.*, **64**, 3960–3962.

Huber, S.A. (1994) VCAM-1 is a receptor for encephalomyocarditis vitrus on murine vascular endothelial cells. *J Virol.*, **68**, 3453–3458.

Hyypia, T., Horsnell, C., Maaronen, M. *et al.* (1992) A distinct picornavirus group identified by sequence analysis. *Proc. Natl Acad. Sci. USA*, **89**, 8847–8851.

Hyypia, T., Hovi, T., Knowles, N. and Stanway, G. (1997) Classification of enteroviruses based on molecular and biological properties. *J. Gen. Virol.*, **78**, 1–11.

Isawa, H., Asano, S., Sahara, K., Iizuka, T. and Bando, H. (1998) Analysis of genetic information of an insect picorna-like virus, infectious flacherie virus of silkworm: evidence for evolutionary relationships among insect, mammalian and plant picorna(-like) viruses. *Arch. Virol.*, **143**, 127–143.

Jackson, T., Ellard, F.M., Ghazaleh, R.A., Brookes, S.M., Blakemore, W.E., Corteyn, A.H., Stuart, D.I., Newman, J.W. and King, A.M. (1996) Efficient infection of cells in culture by type O foot-and-mouth disease virus requires binding to cell surface heparan sulfate. *J. Virol.*, **70**, 5282–5287.

Jang, S.K., Kräusslich, H.-G., Nicklin, M.J.H., Duke, G.M., Palmenberg, A.C. and Wimmer, E. (1988) A segment of the 5′ nontranslated region of encephalomyocarditis virus RNA directs internal entry of ribosomes during in vitro translation. *J. Virol.*, **62**, 2636–2643.

Jang, S.K., Davies, M.V., Kaufman, R.J. and Wimmer, E. (1989) Initiation of protein synthesis by internal entry of ribosomes into the 5′ nontranslated region of encephalomyocarditis virus RNA in vitro. *J. Virol.*, **63**, 1651–1660.

Jang, S.K., Pestova, T., Hellen, C.U.T., Witherell, G.W. and Wimmer, E. (1990) Cap-independent translation of picornavirus RNAs: structure and function of the internal ribosomal entry site. *Enzyme*, **44**, 292–309.

Jarvis, T.C. and Kirkegaard, K. (1991) The polymerase in its labyrinth: mechanisms and implications of RNA recombination. *Trends Genet.*, **7**, 186–191.

Jarvis, T.C. and Kirkegaard, K. (1992) Poliovirus RNA recombination: mechanistic studies in the absence of selection. *EMBO J.*, **11**, 3135–3145.

Johnson, K.N. and Christian, P.D. (1998) The novel genome organization of the insect picorna-like virus *Drosophila* C virus suggests this virus belongs to a previously undescribed virus family. *J. Gen. Virol.*, **79**, 191–203.

Jun, H.S., Kang, Y., Yoon, H.S., Kim, K.H., Notkins, A.L. and Yoon, J.W. (1998) Determination of encephalomyocarditis viral diabetogenicity by a putative binding site of the viral capsid protein. *Diabetes*, **47**, 576–582.

Kajigaya, S., Arakawa, H., Kuge, S., Koi, T., Imura, N. and Nomoto., A. (1985) Isolation and characterization of defective-interfering particles of poliovirus Sabin 1 strain. *Virology*, **142**, 307–316.

Kang, Y., Chatterjee, N.K., Nodwell, M.J. and Yoon, J.W. (1994) Complete nucleotide sequence of a strain of coxsackie B4 virus of human origin that induces diabetes in mice and its comparison with nondiabetogenic coxsackie B4 JBV strain. *J. Med. Virol.*, **44**, 353–361.

King, A.M.Q. (1988) Preferred sites of recombination in poliovirus RNA: an analysis of 40 intertypic cross-over sequences. *Nucl. Acids Res.*, **16**, 11705–11723.

King, A., McKahon, D., Slade, W. and Newman, J. (1982) Recombination in RNA. *Cell*, **29**, 921–928.

Kirkegaard, K. and Baltimore, D. (1986) The mechanism of RNA recombination in poliovirus. *Cell*, **47**, 433–443.

Kitamura, N., Semler, B.L., Rothberg, P.G. *et al.* (1981) Primary structure, gene organization and polypeptide expression of poliovirus RNA. *Nature*, **291**, 547–553.

Koike, S., Horie, H., Ise, I. *et al.* (1990) The poliovirus receptor protein is produced both as membrane-bound and secreted forms. *EMBO J.*, **9**, 3217–3224.

Koike, S., Taya, C., Kurata, T. *et al.* (1991) Transgenic mice susceptible to poliovirus. *Proc. Natl. Acad. Sci. USA*, **88**, 951–955.

Kolakovsky, D. and Weissmann, C. (1971) Qβ replicase as repressor of Qβ RNA-directed protein synthesis. *Biochim. Biophys. Acta*, **246**, 596–599.

Koonin, E.V. and Gorbalenya, A.E. (1989) Evolution of RNA genomes: does the high mutation rate necessitate high rate of evolution of viral proteins? *J. Mol. Evol.*, **28**, 524–527.

Koonin, E.V. and Gorbalenya, A.E. (1992) An insect picornavirus may have genome organization similar to that of caliciviruses. *FEBS Lett.*, **3**, 81–86.

Kuge, S. and Nomoto, A. (1987) Construction of viable deletion and insertion mutants of the Sabin strain type 1 poliovirus: function of the 5′ noncoding sequence in viral replication. *J. Virol.*, **61**, 1478–1487.

Kuge, S., Saito, L. and Nomoto, A. (1986) Primary structure of poliovirus defective interfering particle genomes and possible generation mechanism of the particles. *J. Mol. Biol.*, **192**, 473–487.

Kuhn, R.J., Tada, H., Ypma-Wong, M.F., Semler, B.L. and Wimmer, E. (1988) Mutational analysis of the genome-linked protein VPg of poliovirus. *J. Virol.*, **62**, 4207–4215.

Lama, J., Paul, A.V., Harris, K.S. and Wimmer, E. (1994) Properties of purified recombinant poliovirus protein 3AB as substrate for viral proteinases and as co-factor for viral polymerase 3D^{pol}. *J. Biol. Chem.*, **269**, 66–70.

La Monica, N. and Racaniello, V.R. (1989) Differences in replication of attenuated and neurovirulent poliovirus in human neuroblastoma cell line SH-SY5Y. *J. Virol.*, **63**, 2357–2360.

Landsteiner, K. and Popper, E. (1909) Ubertragung der Poliomyelitis acuta auf Affen. *Z. Immunitaetsforsch. Orig.*, **2**, 377–390.

Larsen, G., Dorner, J., Harris, T. and Wimmer, E. (1980) The structure of poliovirus replicative form. *Nucl. Acids Res.*, **8**, 1217–1229.

Le, S.Y. and Maizel, J.V., Jr. (1998) Evolution of a common structural core in the internal ribosome entry sites of picornavirus. *Virus Genes*, **16**, 25–38.

Lee, C.-K. and Wimmer, E. (1988) Proteolytic processing of poliovirus polyproteins: elimination of 2A pro-mediated, alternative cleavage of polypeptide 3CD by in vitro mutagenesis. *Virology*, **166**, 405–414.

Lee, Y., Nomoto, A. and Wimmer, E. (1976) The genome of poliovirus is an exceptional eukaryotic mRNA. *Prog. Nucl. Acid. Res. Mol. Biol.*, **19**, 89–95.

Lee, Y., Nomoto, A., Detjen, B. and Wimmer, E. (1977) The genome-linked protein of picornaviruses. I. A protein covalently linked to poliovirus genome RNA. *Proc. Natl Acad. Sci. USA*, **74**, 59–63.

Lentz, K.N., Smith, A.D., Geisler, S.C. *et al.* (1997) Structure of poliovirus type 2 Lansing complexed with antiviral agent SCH48973: comparison of the structural and biological properties of three poliovirus serotypes. *Structure*, **5**, 961–978.

Li, F., Browning, G.F., Studdert, M.J. and Crabb, B.S. (1996) Equine rhinovirus 1 is more closely related to foot-and-mouth disease virus than to other picornaviruses. *Proc. Natl Acad. Sci. USA*, **93**, 990–995.

Loeffler, F. and Frosch, P. (1898) Berichte der Kommission zur Erforschung der Maul- und Klauenseuche bei dem Institut fuer Infektionskrankheiten in Berlin. *Zbl. Bakter. Abt. 1. Orig.*, **23**, 371–391.

Lu, H.H. and Wimmer, E. (1996) Poliovirus chimeras replicating under the translational control of genetic elements of hepatitis C virus reveal unusual properties of the internal ribosomal entry site of hepatitis C virus. *Proc. Natl Acad. Sci. USA*, **93**, 1412–1417.

Lu, H.H., Alexander, L. and Wimmer, E. (1995a) Construction and genetic analysis of dicistronic polioviruses containing open reading frames for epitopes of human immunodeficiency virus type 1 gp 120. *J. Virol.*, **69**, 4797–4806.

Lu, H.H., Li, X., Cuconati, A. and Wimmer, E. (1995b) Analysis of picornavirus 2A(pro) proteins: separation of proteinase from translation and replication functions. *J. Virol.*, **69**, 7445–7452.

Lundquist, R., Sullivan, M. and Maizel, J.V. Jr (1979) Characterization of a new isolate of poliovirus defective interfering particles. *Cell*, **18**, 759–769.

Luo, M., Vriend, G., Kamer, G. (1987) The atomic structure of mengo virus at 3.0 Å resolution. *Science*, **235**, 182–191.

Luo, M., Toth, K.S., Zhou, L., Pritchard, A. and Lipton, H.L. (1996) The structure of a highly virulent Theiler's murine encephalomyelitis virus (GDVII) and implications for determinants of viral persistence. *Virology*, **220**, 246–250.

McCloskey (1950) The relation of prophylactic inoculations to the onset of poliomyelitis. *Lancet*, **i**, 659–663.

McGoldrick, A., Macadam, A.J., Dunn, G. *et al.* (1995) Role of mutations G-480 and C-6203 in the attenuation phenotype of Sabin type 1 poliovirus. *J. Virol.*, **69**, 7601–7605.

McKnight, K.L. and Lemon, S.M. (1996) Capsid coding sequence is required for efficient replication of human rhinovirus 14 RNA. *J. Virol.*, **70**, 1941–1952.

Marvil, P., Knowles, N.J., Mockett, A.P.A., Britton, P., Brown, T.D.K. and Canavagh, D. (1999) Avian encephalomyelitis virus is a picornavirus and is most closely related to hepatitis. A virus. *J. Gen. Virol.*, **80**, 653–662.

Mason, P.W., Rieder, E. and Baxt, B. (1994) RGD sequence of foot-and-mouth disease virus is essential for infecting cells via the natural receptor but can be bypassed by an antibody dependent enhancement pathway. *Proc. Natl Acad. Sci. USA*, **91**, 1932–1936.

Matthews, D.A., Smith, W.W., Ferre, R.A. *et al.* (1994) Structure of human rhinovirus 3C protease reveals a trypsin-like polypeptide fold, RNA-binding site, and means for cleaving precursor polyprotein. *Cell*, **77**, 761–771.

Melchers, W.J., Hoenderop, J.G., Bruins Slot, H.J. *et al.* (1997) Kissing of the two predominant hairpin loops in the coxsackie B virus 3′ untranslated region is the essential structural feature of the origin of replication required for negative-strand RNA synthesis. *J. Virol.*, **71**, 686–696.

Melnick, J.R. (1996) Enteroviruses: polioviruses, coxsackieviruses, echoviruses, and newer enteroviruses. In: *Virology*, vol. 1. (eds Fields, B.N., Knipe, D.M. and Howley, P.M.), pp. 655–712. Lippincott-Raven, Philadelphia.

Melnick, J.L., Schmidt, N.J., Mirkovic, R.R., Chumakov, M.P., Lavrova, I.K. and Voroshilova, M.K. (1980) Identification of Bulgarian strain 258 of enterovirus 71. *Intervirology*, **12**, 297–302.

Melnick, J.L., Tagaya, I. and von Magnus, H. (1974) Enteroviruses 69, 70, and 71. *Intervirology*, **4**, 369–70.

Mendelsohn, C.L., Wimmer, E. and Racaniello, V.R. (1989) Cellular receptor for poliovirus: molecular cloning, nucleotide sequence, and expression of a new member of the immunoglobulin superfamily. *Cell*, **56**, 855–865.

Meyers, G., Wirblich, C. and Thiel, H.J. (1991) Rabbit hemorrhagic disease virus – molecular cloning and nucleotide sequencing of a calicivirus genome. *Virology*, **184**, 664–676.

Minor, P.D. (1990) Antigenic structure of picornaviruses. *Curr. Topics Immunol. Microbiol.*, **161**, 121–154.

Molla, A., Paul, A.V. and Wimmer, E. (1991) Cell-free, *de novo* synthesis of poliovirus. *Science*, **254**, 1647–1651.

Molla, A., Jang, S.K., Paul, A.V., Reuer, Q. and Wimmer, E. (1992) Cardioviral internal ribosomal entry site is functional in a genetically engineered dicistronic poliovirus. *Nature*, **356**, 255–257.

Molla, A., Hellen, C.U.T. and Wimmer, E. (1993a) Inhibition of proteolytic activity of poliovirus and rhinovirus 2A proteinases by elastase specific inhibitors. *J. Virol.*, **67**, 4688–4695.

Molla, A., Paul, A.V., Schmid, M., Jang, S.K. and Wimmer, E. (1993b) Studies on dicistronic polioviruses implicate viral proteinase 2Apro in RNA replication. *Virology*, **196**, 739–747.

Molla, A., Paul, A.V. and Wimmer, E. (1993c) Effects of temperature and lipophilic agents on poliovirus formation and RNA synthesis in a cell free system. *J. Virol.*, **67**, 5932–5938.

Molla, A., Harris, K.S., Paul, A.V., Shin, S.H., Mugavero, J. and Wimmer, E. (1994) Stimulation of poliovirus proteinase 3Cpro-related proteolysis by the genome-linked protein VPg and its precursor 3AB. *J. Biol. Chem.*, **269**, 27015–27020.

Moon, J.S., Domier, L.L., McCoppin, N.K., D'Arcy, C.J. and Jin, H. (1998) Nucleotide sequence analysis shows that *Rhopalosiphum padi* virus is a member of a novel group of insect-infecting RNA viruses. *Virology*, **243**, 54–65.

Morgan-Detjen, B., Lucas, J. and Wimmer, E. (1978) Poliovirus single-stranded and double-stranded RNA: differential infectivity in enucleated cells. *J. Virol.*, **27**, 582–586.

Mueller, S. and Wimmer, E. (1997) Expression of foreign proteins by poliovirus polyprotein fusion: analysis of genetic stability reveals rapid deletions and formation of cardiovirus-like open reading frames. *J. Virol.*, **72**, 20–31.

Mueller, S. and Wimmer, E. (1998) Expression of foreign proteins by poliovirus polyprotein fusion: Analysis of genetic stability reveals rapid deletions and formation, of cardiovirus-like open reading frames. *J. Virol.*, **72**, 20–31.

Murdin, A.D., Lu, H.H., Murray, M.G. and Wimmer, E. (1992) Poliovirus antigenic hybrids simultaneously expressing antigenic determinants from all three serotypes. *J. Gen. Virol.*, **73**, 607–611.

Murray, M.G., Bradley, J., Yang, X.-F., Wimmer, E., Moss, E.G. and Racaniello, V.R. (1988) Poliovirus host range is determined by a short amino acid sequence in neutralization antigenic site l. *Science*, **241**, 213–215.

Najarian, R., Caput, D., Gee, W. *et al.* (1985) Primary structure and gene organization of human hepatitis A virus. *Proc. Natl Acad. Sci. USA*, **82**, 2627–2631.

Nakashima, N., Sasaki, J., Tsuda, K., Yasunaga, C. and Noda, H. (1998) Properties of a new picorna-like virus of the brown-winged green bug, plautia stali. *J. Invertebr. Pathol.*, **71**, 151–158.

Neff, S., Sa-Carvalho, D., Rieder, E., Mason, P.W., Blystone, S.D. Brown, E.J. and Baxt, B. (1998) Foot-and-mouth disease virus virulent for cattle utilizes the integrin alpha(v)beta3 as its receptor. *J. Virol.*, **72**, 3587–3594.

Nicklin, M.J.H., Toyoda, H., Murray, M.G. and Wimmer, E. (1986) Proteolytic processing in the replication of polio and related viruses. *Biotechnology*, **4**, 33–42.

Nomoto, A. and Wimmer, E. (1987) In: *Genetic Studies of the Antigenicity and the Attenuation Phenotype of Poliovirus.* Cambridge University Press, Cambridge.

Nomoto, A., Lee, Y.F. and Wimmer, E. (1976) The 5′ end of poliovirus mRNA is not capped with m7G(5′)pppNp. *Proc. Natl Acad. Sci. USA*, **74**, 375–380.

Nomoto, A., Kitamura, N., Golini, F. and Wimmer, E. (1977a) The 5′ terminal structures of poliovirion RNA and poliovirus mRNA differ only in the genome-linked protein VPg. *Proc. Natl Acad. Sci. USA*, **74**, 5345–5349.

Nomoto, A., Morgan-Detjen, B., Pozzatti, R. and Wimmer, E. (1977b) The location of the polio genome protein in viral RNAs and its implication for RNA synthesis. *Nature*, **268**, 208–213.

Nomoto, A., Jacobson, A., Lee, Y., Dunn, J. and Wimmer, E. (1979) Defective interfering particles of poliovirus: mapping of the deletion and evidence that the deletions in the genome of DI (1), (2), (3) are located in the same region. *J. Mol. Biol.*, **128**, 179–196.

Novak, J.E. and Kirkegaard, K. (1994) Coupling between genome translation and replication in an RNA virus. *Genes Devel.*, **8**, 1726–1737.

Ohno, S. (1984) Repeats of base oligomers as the primordial coding sequences of the primeval earth and their vestiges in modern genes. *J. Mol. Evol.*, **20**, 313–321.

Orengo, C.A., Michie, A.D., Jones, S., Jones, D.T., Swindells, M.B. and Thornton, J.M. (1997) CATH – a hierarchic classification of protein domain structures. *Structure*, **5**, 1093–1108.

Palmenberg, A.C. (1989) Sequence alignments of picornaviral capsid proteins, In: *Molecular Aspects of Picornavirus Infection and Detection*, (eds Semler, B.L. and Ehrenfeld, E.). ASM Publications, Washington, DC.

Palmenberg, A.C. (1990) Proteolytic processing of picornaviral polyprotein. *Annu. Rev. Microbiol.*, **44**, 603–623.

Parsley, T.B., Towner, J.S., Blyn, L.B., Ehrenfeld, E. and Semler, B.L. (1997) Poly (rC) binding protein 2 forms a ternary complex with the 5′ terminal sequences of poliovirus RNA and the viral 3CD proteinase. *RNA*, **3**, 1124–1134.

Paul, A.V., Cao, X., Harris, K.S., Lama, J. and Wimmer, E. (1994) Studies with poliovirus polymerase 3D^{pol}: stimulation of poly (u) synthesis *in vitro* by purified poliovirus C protein 3AB. *J. Biol. Chem.*, **269**, 29173–29181.

Paul, A.V., Mugavero, J.A., Molla, A. and Wimmer, E. (1998a) Internal ribosomal entry site scanning of the poliovirus polyprotein: implications for proteolytic processing. *Virology*, **250**, 241–253.

Paul, A.V., van Boom, J.H., Filippov, D. and

Wimmer, E. (1998b) Protein-primed RNA synthesis by purified poliovirus RNA polymerase. *Nature*, **393**, 280–284.

Pelletier, J. and Sonenberg, N. (1988) Internal initiation of translation of eukaryotic mRNA directed by a sequence derived from poliovirus RNA. *Nature*, **334**, 320–325.

Pestova, T.V., Hellen, C.U.T. and Wimmer, E. (1994) A conserved AUG triplet in the 5′ non-translated region of poliovirus can function as an initiation codon *in vitro* and *in vivo*. *Virology*, **204**, 729–737.

Petterson, R., Flanegan, J., Rose, J. and Baltimore, D. (1977) 5′-terminal nucleotide sequences of polio-virus polyribosomal RNA and virion RNA are identical. *Nature*, **268**, 270–272.

Pfister, T. and Wimmer, E. (1999) Characterization of the nucleotide triphosphatase activity of poliovirus protein 2C reveals a mechanism by which guanidine inhibits replication of poliovirus. *J. Biol. Chem.*, **274**, 6992–7001.

Piccone, M.E., Rieder, E., Mason, P.W. and Grubman, M.J. (1995) The foot-and-mouth disease virus leader proteinase gene is not required for viral replication. *J. Virol.*, **69**, 5376–5382.

Pilipenko, E.V., Blinov, V.M., Romanova, L.I., Sinyakov, A.N., Maslova, S.V. and Agol, V.I. (1989a) Conserved structural domains in the 5′-untranslated region of picornaviral genomes: an analysis of the segment controlling translation and neurovirulence. *Virology*, **168**, 201–209.

Pilipenko, E.V., Blinov, V.M., Chernov, B.K., Dmitrieva, T.M. and Agol, V.I. (1989b) Conservation of the secondary structure elements of the 5′-untranslated region of cardio- and aphothovirus RNAs. *Nucleic Acids Res.*, **17**, 5701–5711.

Pilipenko, E.V., Gmyl, A.P., Maslova, S.V., Svitkin, Y.V., Sinyakov, A.N. and Agol, V.I. (1992) Prokaryotic-like *cis* elements in the cap-independent internal initiation of translation on picornavirus RNA. *Cell*, **68**, 119–131.

Pilipenko, E.V., Gmyl, A.P. and Agol, V.I. (1995) A model for rearrangements in RNA genomes. *Nucleic Acids Res*, **23**, 1870–1875.

Pilipenko, E.V., Poperechny, K.V., Maslova, S.V., Melchers, J.G., Bruins Slot, H.J. and Agol, V.I. (1996) *Cis*-element, oriR, involved in the initiation of (−) strand poliovirus RNA: a quasi-globular multi-domain RNA structure maintained by tertiary ("kissing") interactions. *EMBO J.*, **15**, 5428–5436.

Pincus, S.E., Diamond, D.C., Emini, E.A. and Wimmer, E. (1986) Guanidine-selected mutants of poliovirus: mapping of point mutations to polypeptide 2C. *J. Virol.*, **57**, 638–646.

Plotch, S.J., Palant, O. and Gluzman, Y. (1989) Purification and properties of poliovirus RNA polymerase expressed in *Escherichia coli*. *J. Virol.*, **63**, 216–225.

Porter, D.C., Ansardi, D.C., Choi, W.S. and Morrow, C.D. (1993) Encapsidation of genetically engineered poliovirus minireplicons which express human immunodeficiency virus type 1 Gag and Pol proteins upon infection. *J. Virol.*, **67**, 3712–3719.

Poyry, T., Kinnunen, L., Hyypia, T. *et al.* (1996) Genetics and phylogenetic clustering of enteroviruses. *J. Gen. Virol.*, **77**, 1699–1717.

Pringle, C.R. (1996) Virus taxonomy 1997. *Arch. Virol.*, **142**, 1727–1733.

Pulli, T., Koskimies, P. and Hyypia (1995) Molecular comparison of coxsackie A virus serotypes. *Virology*, **212**, 30–38.

Racaniello, V.R. and Baltimore, D. (1981) Cloned poliovirus complementary DNA is infectious in mammalian cells. *Science*, **214**, 916–919.

Raychaudhuri, G., Govindarajan, S., Shapiro, M., Purcell, R.H. and Emerson, S.U. (1998) Utilization of chimeras between human (HM-175) and simian (AGM-27) strains of hepatitis A virus to study the molecular basis of virulence. *J. Virol.*, **72**, 7467–7475.

Ren, R., Costantini, F., Gorgacz, E.J., Lee, J.J. and Racaniello, V.R. (1990) Transgenic mice expressing a human poliovirus receptor: a new model for poliomyelitis. *Cell*, **63**, 353–362.

Reuer, Q., Kuhn, R.J. and Wimmer, E. (1990) Characterization of poliovirus clones containing lethal and nonlethal mutations in the genome-linked protein VPg. *J. Virol.*, **64**, 2967–2975.

Richards, O.C. and Ehrenfeld, E. (1990) Poliovirus RNA replication. *Curr. Topics Microbiol. Immunol.*, **161**, 89–120.

Rivera, V.M., Welsh, J.D. and Maizel, J.V. (1988) Comparative sequence analysis of the 5′ noncoding region of the enteroviruses and rhinoviruses. *Virology*, **165**, 42–50.

Rodrigo, M.J. and Dopazo, J. (1995) Evolutionary analysis of the picornavirus family. *J. Mol. Evol.*, **40**, 362–371.

Rohll, J.B., Moon, D.H., Evans, D.J. and Almond, J.W. (1995) The 3′untranslated region of picornavirus RNA: features required for efficient genome replication. *J. Virol.*, **69**, 7835–7844.

Roivainen, M. and Hovi, T. (1987) Intestinal trypsin can significantly modify antigenic properties of polioviruses: implications for the use of inactivated poliovirus vaccine. *J. Virol.*, **61**, 3749–3753.

Romanova, L., Tolskaya, E., Kolesnikova, M. and Agol, V. (1980) Biochemical evidence for intertypic genetic recombination of polioviruses. *FEBS Lett.*, **118**, 109–112.

Romanova, L.I., Blonov, V.M., Tolskaya, E.A., Viktorova, E.G. and Kolenikova, M.S. (1986) The primary structure of intertypic poliovirus recombinants: a model of recombination between RNA genomes. *Virology*, **155**, 202–213.

Rossmann, M.G. (1987) The evolution of RNA viruses. *Bioessays*, **7**, 99–103.

Rossmann, M.G. and Johnson, J.E. (1989) Icosahedral RNA virus structure. *Annu. Rev. Biochem.*, **58**, 533–573.

Rossman, M.G., Arnold, E., Erikson, J.W. (1985) Structure of a human common cold virus and functional relationship to other picornaviruses. *Nature*, **317**, 145–153.

Rothberg, P., Harris, T., Nomoto, A. and Wimmer, E. (1978) The genome-linked protein of picornaviruses V. 04-(5′ Uridylyl)-tyrosine is the bond between the genome-linked protein and the RNA of poliovirus. *Proc. Natl Acad. Sci. USA*, **75**, 4868–4872.

Rueckert, R.R. (1996) *Picornaviridae*. In: *Virology* (eds Fields, B. *et al.*), pp. 609–654. Lippincott-Raven, Philadelphia.

Rux, J.J. and Burnett, R.M. (1998) Spherical viruses. *Curr. Opin. Struct. Biol.*, **8**, 142–149.

Ryan, M.D. and Flint, M. (1997) Virus-encoded proteinases of the picornavirus super-group. *J. Gen. Virol.*, **78**, 699–723.

Salas, M. (1991) Protein-priming of DNA replication. *Annu. Rev. Biochem.*, **60**, 39–71.

Sasaki, J., Nakashima, N., Saito, H. and Noda, H. (1998) An insect picorna-like virus, *Plautia stali* intestine virus, has genes of capsid proteins in the 3′ part of the genome. *Virology*, **244**, 50–58.

Schaffer, F.L. and Schwerdt, C.E. (1955) Crystallization of purified MEF-I poliomyelitis virus particles. *Proc. Natl Acad. Sci. USA*, **41**, 1020–1023.

Schlesinger, M.J. and Levinthal, C. (1963) Hybrid protein formation of *E. coli* alkaline phosphatase leading to *in vitro* complementation. *J. Mol. Biol.*, **7**, 1–12.

Semler, B., Hanecak, R. Anderson, C. and Wimmer, E. (1981a) Cleavage sites in the polypeptide precursors of poliovirus protein P2-X. *Virology*, **114**, 589–594.

Semler, B.L. Anderson, C.W., Kitamura, N., Rothberg, P.G., Wishart, W.L. and Wimmer, E. (1981b) Poliovirus replication proteins: RNA sequence encoding P3-lb and the site of pioteolytic processing. *Proc. Natl Acad. Sci. USA*, **78**, 3464–3468.

Semler, B.L., Dorner, A.J. and Wimmer, E. (1984) Production of infectious poliovirus from cloned cDNA is dramatically increased by SV40 transcription and replication signals. *Nucl. Acids Res.*, **12**, 5123–5141.

Shafren, D.R., Williams, D.T. and Barry, R.D. (1997) A decay-accelerating factor-binding strain of coxsackievirus B3 requires the coxsackievirus-adenovirus receptor protein to mediate lytic infection of rhabdomyosarcoma cells. *J. Virol.*, **71**, 9844–9848.

Shiroki, K., Ishii, T., Aoki, T., Kobashi, M., Ohta, S. and Nomoto, A. (1995) A new *cis*-acting element for RNA replication within the 5′ noncoding region of poliovirus type 1 RNA. *J. Virol.*, **69**, 6825–6832.

Skern, T. (1998) Picornain 3C. In *Handbook of Proteolytic Enzymes* (eds Barrett, A.J., Rawlings N.D. and Woessner, J.F.). Academic Press, London.

Solecki, D., Wimmer, E., Lipp, M. and Bernhardt, G. (1999) Identification and characterization of the *cis*-acting elements of the human CD155 gene core promoter. *J. Biol. Chem.*, in press.

Stanway, G. (1990) Structure, function and evolution of picornaviruses. *J. Gen. Virol.*, **71**, 2483–2501.

Staunton, D.E., Merluzzi, V.J., Rothlein, R., Barton, R. Marlin, S.D. and Springer, T.A. (1989) A cell adhesion molecule, ICAM-1, is the major surface receptor for rhinoviruses. *Cell*, **56**, 849–853.

Svitkin, Y.V., Hahn, H., Gingras, A.C., Palmenberg, A.C. and Sonenberg, N. (1998) Rapamycin and wortmannin enhance replication of a defective encephalomyocarditis virus. *J. Virol.*, **72**, 5811–5819.

Takata, H., Obuchi, M., Yamamoto, J. *et al.* (1998) L* protein of the DA strain of Theiler's murine encephalomyelitis virus is important for virus growth in a murine macrophage-like cell line. *J. Virol.*, **72**, 4950–4955.

Takeda, N., Kuhn, R.J., Yang, C.F., Takegami, T. and Wimmer, E. (1986) Initiation of poliovirus plus-strand RNA synthesis in a membrane complex of infected HeLa cells. *J. Virol.*, **60**, 43–53.

Takegami, T., Semler, B.L. Anderson, C.W. and Wimmer, E. (1983) Membrane fractions active in poliovirus RNA replication contain VPg precursor polypeptides. *Virology*, **128**, 33–47.

Tang, R.S., Barton, D.J., Flanegan, J.B. and Kirkegaard, K. (1997) Poliovirus RNA recombination in cell-free extracts. *RNA*, **3**, 624–633.

Tardy-Panit, M., Blondel, B., Martin, A., Tekaia, F., Horaud, F. and Delpeyroux, F. (1993) A mutation in the RNA polymerase of poliovirus type 1 contributes to attenuation in mice. *J. Virol.*, **67**, 4630–4638.

Tellier, R., Bukh, J., Emerson, S. and Purcell, R. (1996) Amplification of the full-length hepatitis A virus genome by long reverse transcription-PCR and transcription of infectious RNA directly from the amplicon. *Proc. Natl. Acad. Sci. USA*, **93**, 4370–4373.

Teterina, N.L., Gorbalenya, A.E., Egger, D., Bienz, K. and Ehrenfeld, E. (1997) Poliovirus 2C protein determinants of membrane binding and rearrangements in mammalian cells. *J. Virol.*, **71**, 8962–8972.

Todd, S., Towner, J.S. and Semler, B.L. (1997)

Translation and replication properties of the human rhinovirus genome *in vivo* and *in vitro*. *Virology*, **229**, 90–97.

Tolskaya, E., Romanova, L., Kolesnikova, M. and Agol, V. (1983) Intertypic recombination in poliovirus: genetic and biochemical studies. *Virology*, **124**, 121–132.

Tolskaya, E.A., Romanova, L.I., Kolesnikova, M.S., Gmyl, A.P., Gorbalenya, A.E. and Agol, V.I. (1994) Genetic studies on the poliovirus 2C protein, an NTPase. A plausible mechanism of guanidine effect on the 2C function and evidence for the importance of 2C oligomerization. *J. Mol. Biol.*, **236**, 1310–1323.

Tomko, R.P., Xu, R. and Philipson, L. (1997) HCAR and MCAR: the human and mouse cellular receptors for subgroup C adenoviruses and group B coxsackieviruses. *Proc. Natl Acad. Sci. USA*, **94**, 3352–3356.

Towner, J.S., Mazanet, M.M. and Semler, B.L. (1998) Rescue of defective poliovirus RNA replication by 3AB-containing precursor polyproteins. *J. Virol.*, **72**, 7191–7200.

Toyoda, H., Kohara, M., Kataoka, Y. *et al.* (1984) Complete nucleotide sequences of all three poliovirus serotype genomes: implication for genetic relationship, gene function and antigenic determinants. *J. Mol. Biol.*, **174**, 561–585.

Toyoda, H., Yang, C.F., Takeda, N., Nomoto, A. and Wimmer, E. (1987) Analysis of RNA synthesis of type 1 poliovirus by using an in vitro molecular genetic approach. *J. Virol.*, **61**, 2816–2822.

Tracy, S., Tu, Z., Chapman, N. and Hufnagel, G. (1995) Genetics of coxsackievirus B3 cardiovirulence. *Eur. Heart J.*, **16** (Suppl O), 15–17.

Tracy, S., Chapman, N.M., Romero, J. and Ramsingh, A.I. (1996) Genetics of coxsackievirus B cardiovirulence and inflammatory heart muscle disease. *Trends Microbiol.*, **4**, 175–179.

Turnbull-Ross, A.D., Mayo, M.A., Reavy, B. and Murant, A.F. (1993) Sequence analysis of the parsnip yellow fleck virus polyprotein: evidence of affinities with picornaviruses. *J. Gen. Virol.*, **74**, 555–561.

Van der Werf, S., Bradley, J., Wimmer, E., Studier, F.W. and Dunn, J.J. (1986) Synthesis of infectious poliovirus RNA by purified T7 RNA polymerase. *Proc. Natl Acad. Sci. USA*, **78**, 2330–2334.

Van der Wilk, F., Dullemans, A.M., Verbeek, M. and Van den Heuvel, J.F. (1997) Nucleotide sequence and genomic organization of *Acyrthosiphon pisum* virus. *Virology*, **238**, 353–362.

Ventoso, I., Barco, A. and Carrasco, L. (1998) Mutational analysis of poliovirus 2Apro. Distinct inhibitory functions of 2apro on translation and transcription. *J. Biol. Chem.*, **273**, 27960–27967.

Voroshilova, M. and Chumakov, M. (1959) Poliomyelitis-like properties of AB-IV Coxsackie A7 group of viruses. *Prog. Med. Virol.*, **2**, 106–107.

Ward, T., Powell, R.M., Pipkin, P.A., Evans, D.J., Minor, P.D. and Almond, J.W. (1998) Role for beta2-microglobulin in echovirus infection of rhabdomyosarcoma cells. *J. Virol.*, **72**, 5360–5365.

Waterson, A.P. and Wilkinsen, L. (1978) *An Introduction to the History of Virology.* Cambridge University Press, London.

WHO (1985) Expanded programme on immunization, global poliomyelitis eradication by the year 2000: manual for managers of immunization programmes on activities related to polio eradication. *WHO Bull.*, pp. 35–64.

Wimmer, E. (1982) Genome-linked proteins of viruses. *Cell*, **28**, 199–201.

Wimmer, E., Hellen, C.U.T. and Cao, X.M. (1993) Genetics of poliovirus. *Annu. Rev. Genet.*, **27**, 353–436.

Wimmer, E., Harber, J.J., Bibb, J.A., Gromeier, M., Lu, H.-H. and Bernhardt, G. (1994) Poliovirus receptors. In: *Cellular Receptors for Animal Viruses* (ed. Wimmer, E.), pp. 101–127. Cold Spring Harbor Laboratory Press, Cold Spring Harbor, NY.

Wu, M., Davidson, N. and Wimmer, E. (1978) An electron microscope study of proteins attached to poliovirus RNA and its replicative form (RF). *Nucl. Acids Res.*, **5**, 4711–4723.

Wutz, G., Auer, H., Nowotny, N., Grosse, B.,

Skern, T. and Kuechler, E. (1996) Equine rhinovirus serotypes 1 and 2: relationship to each other and to aphthoviruses and cardioviruses. *J. Gen. Virol.*, **77**, 1719–1730.

Xiang, W., Cuconati, A., Paul, A.V., Cao, X. and Wimmer, E. (1995a) Molecular dissection of the multifunctional poliovirus RNA-binding protein 3AB. *RNA*, **1**, 892–904.

Xiang, W., Harris, K.S., Alexander, L. and Wimmer, E. (1995b) Interaction between the 5′-terminal cloverleaf and 3AB/3CDpro of poliovirus is essential for RNA replication. *J. Virol.*, **69**, 3658–3667.

Xiang, W.K., Paul, A.V. and Wimmer, E. (1997) RNA signals in entero- and rhinovirus genome replication. In: *Recognition Signals on RNA Genomes*, Seminars in Virology (ed. Wimmer E.), pp. 256–273. Academic Press, New York.

Yamashita, T., Sakae, K., Tsuzuki, H. *et al.* (1998) Complete nucleotide sequence and genetic organization of Aichi virus, a distinct member of the Picornaviridae associated with acute gastroenteritis in humans. *J. Virol.*, **72**, 8408–8412.

Yang, W.-X., Terasaki, T., Shiroki, K. *et al.* (1997) Efficient delivery of circulating poliovirus to the central nervous system independently of poliovirus receptor. *Virology*, **229**, 421–428.

Yin-Murphy, M. (1973) Viruses of acute haemorrhagic conjunctivitis. *Lancet*, **i**, 545–546.

Yogo, Y. and Wimmer, E. (1972) Polyadenylic acid at the 3-terminus of poliovirus RNA. *Proc. Natl Acad. Sci. USA*, **69**, 1877–1882.

Zhao, W.D., Wimmer, E. and Lahser, F.C. (1999) Polivirus/Hepatitis C virus (Internal ribosomal entry site-core) chimeric viruses: Improved growth properties through modification of a proteolytic cleavage site and requirement for core RNA sequences but not core-related polypeptides. *J. Virol.*, **73**, 1546–1554.

Zoll, J., Galama, J.M., van Kuppeveld, F.J. and Melchers, W.J. (1996) Mengovirus leader is involved in the inhibition of host cell protein synthesis. *J. Virol.*, **70**, 4948–4952.

13

The Impact of Rapid Evolution of the Hepatitis Viruses

Juan I. Esteban, Maria Martell, William F. Carman and Jordi Gómez

There are five human hepatitis viruses known as hepatitis viruses A through E. However, apart from their liver tissue tropism and their capacity to cause hepatitis in the human host, these viruses show very little in common, especially in the molecular characteristics of their genomes. The recent application of molecular techniques for gene amplification and sequencing to the study of these viruses, especially those that cause persistent infections, has unravelled their significant genetic heterogeneity and their potential for rapid evolution. This chapter summarizes current knowledge on the genome heterogeneity of human hepatitis viruses and their biological implications, classified according to whether or not they cause chronic hepatitis.

HUMAN HEPATITIS VIRUSES THAT MAY CAUSE CHRONIC HEPATITIS

Hepatitis B Virus

Genome Organization

The genome is circular DNA of 3.2 kb, all of which codes for protein through four open reading frames that are partly overlapping and generate at least seven viral gene products (Figure 13.1). All mRNAs transcribed from the HBV genome use the same polyadenylation site. There are four promoters (preS1, S, C and X promoter) and two enhancers (I and II) embedded within the open reading frames. The C promoter controls transcription of the pregenomic and the precore mRNAs that encode the core protein, the hepatitis B e antigen (HBeAg) protein and DNA polymerase. The core promoter (nucleotides 1620–1785) also contains (between nucleotides 1687 and 1774) the enhancer II (Yee, 1989). This enhancer is also probably critical for the preferential expression of the S promoter in hepatocytes. HNF-3, HNF-4 and SP1 regulate core promoter and enhancer II activity (Guo *et al.*, 1993; Zhang *et al.*, 1993; Johnson *et al.*, 1995; Li *et al.*, 1995; Raney *et al.*, 1997). There are also three 5′ negative regulatory elements.

Replication Strategy

HBV replicates via an RNA to DNA step catalysed by reverse transcriptase. Four steps (reviewed in Nassal and Schaller, 1996) can be distinguished. First, the partially double-stranded DNA is converted into a 3200 bp covalently closed circular DNA (cccDNA) in the nucleus. Second, the cccDNA serves as a template for transcription of the pregenome and various messages. Next, translation allows packaging of the RNA pregenome and the viral polymerase

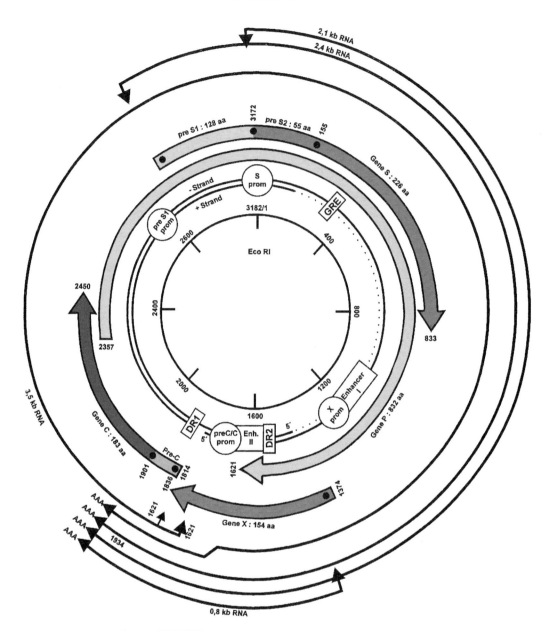

FIGURE 13.1 The genomic structure of HBV.

into the nucleocapsid. Synthesis of the DNA minus strand occurs after encapsidation. Finally, the polymerase synthesizes part of the plus DNA strand and nascent viruses are coated with envelope proteins and released.

The pregenome RNA is terminally redundant, so the pregenome is around 200 bp longer than the genome. Critical to the replicative strategy, direct repeat 2 (DR2) is near the polyadenylated

3′ end, but the DR1 sequence is at both ends. An encapsidation signal (epsilon) is found at the 5′ region of the RNA pregenome (Pollack and Ganem, 1993; Junker-Niepmann *et al.*, 1990). A primer protein is located at the 3′ end of the pregenome at DR1. After priming and during synthesis of the DNA minus strand, the pregenome is degraded through an RNase-H-like activity (as for retroviruses), leaving only

20 bp oligonucleotide at the 5′ end of the pre-genome (Radziwill *et al.*, 1990). This oligonucleotide, which contains the redundant DR1 sequence, hybridizes with DR2 on the minus-strand and synthesis of the plus-strand ensues.

Genotypes and Subtypes

Based on an 8% nucleotide divergence (Okamoto *et al.*, 1988; Norder *et al.*, 1992a,b), there are six genotypes, which tend to be distributed geographically. Four fundamental antigenic subtypes and five other sub-subtypes are antigenically defined, based on amino acid substitutions in the S protein. Residues 122 and 160 are critical, but variants such as at aa159 and 126/127 allow full antigenicity (Norder *et al.*, 1992a,b). Genotypes and subtypes probably evolved in ethnic backgrounds over centuries; the selection pressure is unknown. The clinical importance (see below) of a variant of HBsAg may be different depending upon the subtype

backbone. The genotypes do not appear to have any replication advantage over each other or different pathogenic potential, although there are hints that subtype adr may lead more often to long-term carriage (Shiina *et al.*, 1991) and genotype D may be associated with fulminant hepatitis more commonly than predicted from its prevalence in chronic carriers (Yasmin, 1997).

The Biological Significance of Variants of HBV

Precore/core gene HBeAg is a non-particulate secreted protein, whereas HBcAg forms the nucleocapsid. Nucleotide mutations, which are sometimes silent but usually translate as loss of the HBeAg start codon or an in-frame stop codon, regularly occur during seroconversion to anti-HBe (with loss of HBeAg) during chronic hepatitis (see below). The commonest of these is the stop codon at the end of the precore gene (Figure 13.2), within the HBeAg open reading

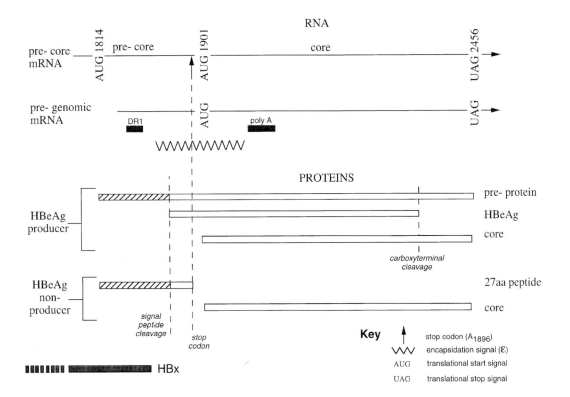

FIGURE 13.2   The precore/core region and A$_{1896}$

frame, termed A_{1896}. In our opinion, the following is now clear about A_{1896} (reviewed in Carman *et al.*, 1996; Miyakawa *et al.*, 1997). It occurs in patients with and without active hepatitis. In some people, it is found as a minority population during the HBeAg-positive phase and becomes dominant around or after appearance of anti-HBe. It is only seen in some genotypes (Lok *et al.*, 1994; Tong *et al.*, 1992), depending on stuctural constraints of epsilon. Although the widely held contention is that it is linked to severe disease, there are even some data that patients with A_{1896} have a better outcome than those without (Lindh *et al.*, 1996). However, this may be related to the infecting genotype. It does not *per se* influence response to interferon (Fattovich *et al.*, 1995; Lampertico *et al.*, 1995; Aikawa *et al.*, 1995); the fact that these people have been unable to clear the virus naturally indicates that they will have a poor response to immune-stimulating antivirals. Some anti-HBe-positive persons are infectious; A_{1896} may merely be a marker for the serological picture. Although HBeAg has been shown to inhibit replication, HBV DNA levels in A_{1896}-infected persons are not higher than in those who produce HBeAg. Finally, A_{1896}-containing strains are independently transmissible (Mphahlele *et al.*, 1997), even leading to acute hepatitis and clearance in an adult. Phylogenetic data show that these strains cluster together into multiple lineages within genotypes (Bollyky *et al.*, 1997; unpublished data). Transmission to infants usually leads to acute hepatitis (Raimondo *et al.*, 1993). Considering that anti-HBe-positive patients have become the commonest new presentation to hepatitis clinics, it may be that this is a strain that is becoming dominant. Thus, A_{1896} should be considered as a "wild-type" virus with biological properties that may or may not influence the clinical picture. Considering the number of studies on this single mutation, if we do not know by now if it does or not, we probably never will!

HBcAg contains B- and T-cell epitopes that appear critical for viral clearance. They are therefore likely to be under immune pressure. During the HBeAg-positive phase of chronic hepatitis, C-gene sequences from geographically/ethnically defined confines are relatively conserved, irrespective of disease activity (Carman *et al.*, 1995b). However, after seroconversion to anti-HBe and selection of A_{1896}, amino-acid substitutions collect within HBcAg. These occur rapidly soon after A_{1896} becomes dominant and are clustered within B-cell epitopes in those with ongoing disease (Carman *et al.*, 1997a). In cross-sectional studies of patients with active hepatitis, variants are also found within Th epitopes (Carman *et al.*, 1995b) but, in longitudinal studies, they tend to collect only in those who go on to remission. These substitutions are related to host factors (such as ethnicity or HLA) or only have significance in particular genotype backgrounds. The type of core variation seen in Japanese patients is quite different to Italians or Chinese. For example, a glycine to valine within a Th epitope was considered of interest by the authors, yet all mutations at this position seen in Chinese resulted in glycine. Similarly, in Italians, T67N was considered significant, whereas, in Japanese, N67T occurred. Perhaps this reflects the variable definition of a "standard" sequence in these studies as well as differences in ethnic background; in our studies, we have used a consensus of HBeAg-positive cases from the same region as a baseline, as these patients have very little variability. One particular variant, T12S, first described in Italians with chronic active hepatitis who had selected A_{1896}, has very recently been shown to inhibit T-cell proliferation of T-cells against T12. Other work (N. Naoumov and W.F. Carman, unpublished observations 1998) reveals that a number of variants found in patients with interferon non-responsiveness also lead to poor T-cell proliferation. Unfortunately, there is no formal evidence of B-cell epitope immune escape. If this is proved, then anti-HBc will take on a new role other than as a diagnostic marker. The evidence for HBeAg-negative variants being a direct cause of interferon non-responsiveness is not reproducible (Alexopoulou *et al.*, 1998).

The S gene   The envelope proteins of HBV are three: large, middle and small. The small protein is also known as HBsAg. There is a region between aa100 and aa160 termed the major hydrophilic region (MHR), to which most of the anti-HBs antibodies stimulated by vaccine bind. Variants of this region have been well described in three clinical situations (Figure 13.3), described below. Their relationship to the polymerase mutants that arise duing nucleoside analogue therapy is described later.

VACCINE-RELATED ESCAPE VARIANTS OF HBsAG   In studies of vaccinated children born to HBeAg-positive mothers, about 15% become anti-HBc-positive (evidence of infection, past or present). A proportion of these will be HBsAg-positive, i.e. they are carriers. About half of those are infected with variants of the MHR, the most consistently observed being G145R (glycine to arginine at aa145 of HBsAg; Carman *et al.*, 1990). The mothers often have 145G, but a significant proportion have a minority species of 145R (Figure 13.3). The four population-based studies from Singapore (Oon *et al.*, 1995), Indonesia, the UK (Ngui *et al.*, 1997) and the USA (Nainan *et al.*, 1997) have had remarkably similar results (Table 13.1). There are also a number of individual case reports describing variants in vaccinated infants, mostly from the

Far East (Fujii *et al.*, 1992; Okamoto *et al.*, 1992a; Hino *et al.*, 1995). The USA study showed, over-all, that 0.8% of vaccinated children born to HBeAg-positive mothers were infected with 145R; the other 4% with breakthrough were infected with 145G. Other variants also appear but, unlike 145R, often rapidly revert to the strain seen in the mother, indicating that 145R is not only replication-competent, as shown in chimpanzees (Ogata *et al.*, 1997), but its altered antigenicity gives it a significant survival advantage over 145G. Vaccination at birth is an ideal situation for selection of escape variants, being similar to liver transplantation, where high-titre anti-HBs preparations (HBIG; see below) are administered to prevent graft infection (Cariani *et al.*, 1995; Hawkins *et al.*, 1996; Carman *et al.*, 1996). So far, variants have not been seen in children exposed to carriers (family and other close contacts). Much of the anti-HBs response is directed towards the second loop of the MHR, so it is expected that variants in this region arise after such highly focused immune pressure. We are convinced that the

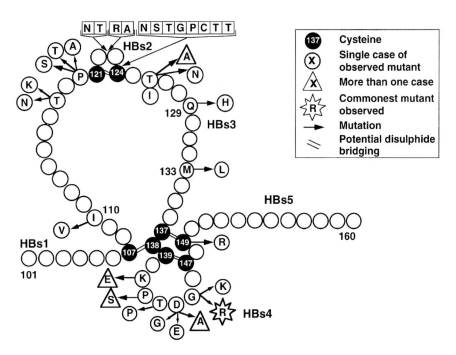

FIGURE 13.3 The major hydrophilic region of HBV surface antigen and some clinically important variants.

TABLE 13.1   The incidence of surface antigen variants in vaccinated children born to HBeAg positive mothers

Country	Vaccine regimen	Incidence of infection (anti-HBc/HBsAg) (%)	Proportion of variants (%)
UK	S vaccine + HBIG	6	12
USA	S vaccine + HBIG	9	23
Taiwan	S vaccine + HBIG	NG	33
Indonesia	Pre-S2 vaccine	8	40–75*
Indonesia	S vaccine	23	ND
Singapore	S vaccine + HBIG	12	39

NG = not given
*The lower figure represents the pairs where the sequence in the mother and child were different; the higher includes situations where the sequences were the same, but unusual

failure of the passive/active immunization protocol to control variant infection is due to altered antigenicity. Perhaps the emergence of this variant is the result of an imbalance between the humoral and cytotoxic T-cell immune responses. In acute infection, when variants do not become dominant, there is always a strong cytotoxic T-cell response. It is not known if there is any specific host factor that predisposes individuals to selection of G145R. Phylogenetic analyses (J. Wilson *et al.*, unpublished observations) indicate that R145 strains from around the world are not closely related, so they can arise from almost any background. However, selection of G145R in some strains leads to a translational stop codon in the overlapping polymerase protein.

Most cases have occurred in children with a low level of anti-HBs antibodies and the chimpanzee experiments indicate that high titres protect against infection with this variant (Ogata *et al.*, 1994). We do not know whether the lower levels of immunity often seen in immunocompromised and older vaccinated patients will be sufficient to prevent infection. As 145G infection does occur in vaccinees with low levels of immunity, poorly immunized people may have even greater susceptibility to HBV variants.

There are a number of crucial questions. First is the current prevalence. Naturally occurring 145R strains have been reported (Carman *et al.*, 1995a; Yamamoto *et al.*, 1994; Nainan *et al.*, 1997). On the other hand, large screening studies in Papua New Guinea, Sardinia and South Africa (Carman *et al.*, 1997b) failed to detect it. The second is whether there will be secondary spread in unvaccinated and vaccinated subjects. Unfortunately, this cannot be determined until the current generation of infected children reach sexual maturity. However, mathematical modelling (Wilson *et al.*, 1999) indicates that it will become the dominant viral strain. It will take a long time to emerge because, although highly contagious, HBV is not very infectious. This slow shift means it will also take several decades to eradicate, even with crossreactive vaccines, because of the chronic nature of the infection. Third, how often will it cause disease? 145R definitely causes chronic hepatitis; the first reported vaccinated case was HBeAg-positive and had chronic hepatitis. Whether novel vaccines such as those containing pre-S epitopes will resolve the issue is hard to predict; however, one study in Indonesia would indicate not, as the incidence of vaccine variants in the MHR was just as high, even in the absence of HBIG administration (Surya *et al.*, 1996).

HBsAg Variants and Diagnostic Assays   There are two possibilities to explain discrepancies in HBsAg reactivity between commercial assays. First, there may be very low levels of HBsAg. Second, they have variant sequences, which are not recognised by the utilized antibodies. There are some good examples of variants that are not recognized by some commercial HBsAg assays. The most dramatic are insertions in the region between cysteines at aa121 and 124 (Carman *et al.*, 1995a; Yamamoto *et al.*, 1994; Hou *et al.*, 1995). These appear also to have antigenic effects on distant epitopes within the MHR, explained by the interleaved nature of the linear and discontinuous epitopes in this antigenically complex region. It is not known whether these insertions affect neutralization, so they may not be vaccine-related. Studies are only beginning to address the antigenic importance and epidemiology of such variants. However, it is clear that, the harder one looks, the more will be found. For example (Carman *et al.*, 1997b), out of 2000 antenatal clinic attendees in Papua New Guinea, 5% of HBsAg-positive subjects were negative in a widely used monoclonal-based assay yet PCR-positive. Over 50% of these discordant samples had rare or unique variants of the MHR.

HBsAg Variants and HBIG Therapy   The combination of high-titre antibody (HBIG), low-titre virus and a large number of susceptible cells (the new liver) is an ideal breeding ground for mutants. One can consider four scenarios after transplantation for chronic HBV. First, HBIG therapy is successful and patients are PCR-negative. Second, patients receive HBIG but it is withdrawn because of side-effects: mutant selection seems to be uncommon. Third, patients become HBsAg-positive on therapy: such patients usually develop mutations in putative neutralizing epitopes. G145R is by far the commonest variant observed to arise in this situation. It appears to arise de novo, because it is not often seen in the pretransplant sample (Carman *et al.*, 1996; Protzer-Knolle *et al.*, 1998). Withdrawal of HBIG often leads to reversion to the pretransplant sequence (Ghany *et al.*, 1998), intimating that the new are not replicatively as efficient. It has been shown that HBIG does not bind to the selected viruses. Fourth, patients are PCR-positive but HBsAg-negative during therapy. Sometimes, a mutation that could affect secretion of HBsAg or has very significant effect on antigenicity is seen (Protzer-Knolle *et al.*, 1998; Carman *et al.*, 1999). Some approaches to diminish this selection pressure are to add nucleoside analogues such as lamivudine to HBIG or to add monoclonal antibodies that bind to the most important mutants along with HBIG. With the recent move to reduce use of human-plasma-derived products, monoclonal antibodies will probably become increasingly important.

Deletions and insertions in the pre-S regions during HBIG therapy can occur (Trautwein *et al.*, 1996). Studies are under way to examine their *in-vitro* effect and whether there is a significant amount of anti-pre-S in HBIG preparations.

The X protein/basal core promoter

Is Fulminant Hepatitis Due to Viral or Host Factors?   Consistent with a role for host factors is that an isolate from a case of fulminant hepatitis (FHB) had an identical sequence to that in the source patient with minimal hepatitis (Karayiannis *et al.*, 1995). Also, there have been outbreaks with only a minority of FHB cases, implying either that a host factor is involved or that the contact has a mixture of strains.

However, outbreaks of FHB disease (Oren *et al.*, 1989; Tanaka *et al.*, 1995) and the preferential transmission of FHB from anti-HBe carriers (Fagan *et al.*, 1986) imply the existence of a viral strain or strains. One implication is that A_{1896} is involved. Transmission of an A_{1896} strain from an FHB case to chimpanzees resulted in more severe hepatitis than would be expected (Ogata *et al.*, 1993). The finding of A_{1896}-containing strains in transmission pairs has lent support to the role of this variant. The pathogenesis of FHB in neonates, and the observation that transmission of A_{1896} strains to children seldom gives rise to chronic carriage (Raimondo *et al.*, 1993), may be partially explained by the absence of HBeAg as an immunomodulator. Thus, viral factors are probably also involved in FHB, rather than it being solely dependent on a strong immune response to a "normal" viral strain. A number of studies did not find a convincing association with A_{1896}, even in neonatal disease. Further, in acute infection with recovery, the HBeAg-negative variant may emerge (Carman *et al.*, 1991). Either there are multiple strains with A_{1896} (with different outcomes) or a host factor is crucial.

Unrelated FHB strains (Ogata *et al.*, 1993; Hasegawa *et al.*, 1994; Sato *et al.*, 1995) with A_{1896} had multiple substitutions concentrated in the basal core promoter (BCP)/enhancer II region, correlated with binding regions of transcription factors. Increased transcriptional efficiency may lead to very high levels of HBV DNA being produced, which may result, in certain individuals, in a massive immune response and therefore more severe disease. A genome from an FHB patient with A_{1896}, T_{1762} and A_{1764} (within the BCP) gave rise to very high levels of secreted HBV DNA, the major effect being on encapsidation (Baumert et al., 1996). Variability at nts 1766 and 1768, which are outside epsilon, seem to be important, implying that there is another region involved in encapsidation. We have completed a detailed phylogenetic analysis, which includes 20 new epidemiologically unrelated FHB sequences, revealing six clusters of fulminant strains each with distinct mutational patterns (Bollyky, 1997c). In keeping with the finding that contacts and index cases have highly related sequences is that contact sequences are found within the clusters. Variants were clustered in the promoters and enhancers and in HBx proteins. No unique variant was specifically associated with this disease in all cases, but the combination ("motifs") of nucleotide variants in the enhancer/promoter regions plus amino acid substitution in HBx was almost uniquely associated with this disease. Variants out of their usual genotype context were also found to be important. A_{1896} strains with such motifs were only found in FHB isolates, explaining why A_{1896} is common but FHB is rare. The transcription efficiency of the BCP from A_{1896}-associated fulminant strains was two to three times greater than non-A_{1896} fulminant strains and A_{1896} non-fulminant sequences (Yasmin 1997). Finally, oligonucleotides containing the variants failed to bind to some nuclear transcription factors, implying that a central pathogenetic mechanism in FHB is loss of inhibition of transcription.

CHRONIC HEPATITIS These variants were initially seen in anti-HBe-positive chronic carriers (Okamoto et al., 1994; Kidd-Ljunggren et al., 1997) at, or around, the RNA polymerase binding site. In 82% of those with mutations in the BCP, T_{1762} and/or A_{1764} was observed. In some chronic carriers, these two variants appeared as they seroconverted to anti-HBe without simultaneous precore mutations. It was proposed that these mutations are HBeAg-negative through regulation of transcription. In support (Kidd-Ljunggren et al., 1995),

90% of HBeAg-positive/anti-HBe-negative patients had the "usual" sequences at positions 1762 and 1764 but four of five HBeAg-negative/anti-HBe-positive patients had one or more mutations at these positions. However, subsequent studies have indicated that both types of variant commonly occur together and that T_{1762}/A_{1764} is genotype-related. As T_{1762}/A_{1764} is common in certain genotypes, this suggests unrecognized confounding factors in some studies. Some functional analyses do not support the claim that T_{1762} and A_{1764} down-regulate HBeAg production as an alternative mechanism to precore variants (Nishizono et al., 1995). However, the balance of data indicates that HBeAg and HBV DNA levels are slightly lower in cases with T_{1762} and A_{1764} (Sato et al., 1996). However, this area has far to run before it becomes clear what the role of variants is.

Anti-HBe-positive low-level viraemic carriers have been shown to have X-deleted strains, which, if confirmed, is an interesting mechanism of down-regulating replication and promoting viral persistence (Kurosaki et al., 1997). More than 15 different deletions of up to 20 base pairs around nucleotide 1770 have been identified (Kidd-Ljunggren et al., 1997). Interestingly, these deletions cluster around the binding site for transcription factors of the nuclear receptor family (Yu and Mertz, 1996, 1997), such as COUP-TF1 (a ubiquitous transcription factor of the steroid/thyroid nuclear receptor family) or liver enriched factor HNF4. Point mutations, deletions or insertions in this region also introduce new binding sites for the liver enriched transcription factors HNF1 and HNF3. COUP-TF1 is critical for strong activation of pregenomic and precore promoters (Günther et al., 1996), yet modification of this binding site can separate the regulation of these two messages (Buckwold et al., 1996; Yu and Mertz, 1996). Electrophoretic mobility shift assays showed that COUP-TF1 does indeed bind to nucleotides 1755–1768.

A further explanation, other than HBsAg variants, for the lack of serological markers is variants in the X gene (Laskus et al., 1994; Feitelson et al., 1995; Fukuda et al., 1995) that down-regulate the production of antigens and, consequently, antibodies.

The polymerase gene Finally, with patients being treated with lamivudine, famciclovir and other nucleoside analogues, variants of the polymerase gene have been arising. This has been well reviewed recently (Zoulim and Trepo, 1998), so will not be expanded on here. It is unlikely that these variants are "wild-type" viruses; they are often lost after withdrawal of the drug.

However, they do highlight a concept not yet discussed here: the overlapping nature of the S and P genes. A mutation in one can have deleterious effects in another. It is unlikely to be advantageous because otherwise it is likely to have happened already in Nature. As some of them are in the MHR of the overlapping S gene, there may be antigenic effects; in fact 145R can be selected in some patients on nucleoside analog therapy. The opposite may be even more important for therapeutic efficacy: MHR variants can lead to a reduced sensitivity of the polymerase gene for drug activity.

Is Chronic Carriage Due to Host or Viral Factors?

Host factors The likely reasons for chronicity are: (1) infection at birth with an immature (perhaps monoclonal-like) immune response; (2) acquired immunodeficiency; or (3) an HLA phenotype that can mount only a moderate or poor immune response. On balance, chronic carriage in adults appears to be primarily due to host factors. HLA class II alleles are associated with viral clearance (Thursz *et al.*, 1995). However, as only 5% of adults fail to clear HBV during acute infection, this can only be part of the story. Other genetically linked factors, such as mannose binding protein (Thomas *et al.*, 1996) or TNF-α alleles, are probably involved.

Viral variants Patients who are infected *ab initio* with HBeAg non-producer strains (precore variants) seldom go on to develop chronicity but usually get acute disease (Mphahlele *et al.*, 1997; Raimondo *et al.*, 1993), sometimes even fulminant hepatitis. This is probably because HBeAg induces tolerance of anti-HBc/e Th cells; if HBeAg is not present, this cannot occur.

However, this clearly does not explain the large number of infections due to HBeAg-producing strains in adults that lead to acute hepatitis and viral clearance. HBsAg is also believed to have an immunomodulatory effect, as yet formally unproven. There appears to be a sequential appearance of mutations in a number of genes. First, are those in the precore region (Brunetto *et al.*, 1989; Carman *et al.*, 1989; Okamoto *et al.*, 1990). As tolerance against HBcAg/HBeAg eventually diminishes, there ensues a battle between virus and host, with different outcomes depending on immune status and HLA phenotype. This process is associated with spikes in serum ALT (as a marker of immune mediated hepatocyte damage) with corresponding increases in anti-HBe immune complex and free IgM anti-HBc. Eventually the immune system wins the battle (and virus is cleared) or precore variants predominate (in a milieu of anti-HBe positivity). Next, (or simultaneously) comes the selection of mutants in the core protein, described above (Ehata *et al.*, 1992, 1993; Carman *et al.*, 1993; Chuang *et al.*, 1993; Boner *et al.*, 1995). It is not known when pre-SI and pre-S2 mutations appear, but they are associated most strongly with chronic hepatitis. Finally, variants seem to appear in the MHR (Kidd-Ljunggren *et al.*, 1995). This order will vary among individuals. Mutants which have lost the preS2 ATG (Gerken *et al.*, 1991; Santantonio *et al.*, 1992) probably have a similar role as precore mutants (switching off translation of an apparently non-essential protein), allowing immune escape. HBsAg and anti-HBs are in a state of flux during the seroconversion phase. The location, concentration and inherent antigenicity of antigens can affect the critical balance of immune response versus tolerance. This fluid situation remains until functional constraints on HBsAg do not allow further mutations and the anti-HBs responses are able to clear infected hepatocytes. This viral evolutionary cascade represents sequential attempts to survive the immune response; the downside is a gradual diminution of replication competence and infectability. It is likely that the majority of transmission occurs during acute infection or the HBeAg-positive phase of chronic carriage. Thus, although there is substantial micro-evolution (within individuals), little of this shows up in the macro-evolutionary picture (within populations) because many of these viruses are either not transmissible or are easily outcompeted by their accompanying "wild-type" viruses.

Hepatitis C Virus

Hepatitis C virus (HCV) is an enveloped virus classified in the family Flaviridae (Choo *et al.*, 1989; Houghton *et al.*, 1991; Miller and Purcell, 1990; Landford *et al.*, 1995). Its genome consists of a single unsegmented strand of RNA, with a plus polarity and 9600 nt in length. Viral particles have never been isolated and thus our knowledge of HCV genetic organization is derived from comparative analysis of its genetic sequence with that of other RNA viruses and from gene mapping both *in vitro* and in cell systems (Houghton, 1996). The genome comprises a 5′ untranslated region, which functions as an internal ribosome entry site, a long open reading frame, extending most of its genome length, which encodes a polyprotein precursor of about 3010 amino acids and a 3′ non-coding region containing an internal poly U/polypirimidine track (Houghton, 1996; Major and Feinstone, 1997; Figure 13.4). Two RNA species are found in infected hepatocytes (Fong *et al.*, 1991): a genomic RNA serving as mRNA for the synthesis of the viral polyprotein and a minus-stranded RNA form from which polymerization of the plus-stranded progeny is presumably initiated

(Fong *et al.*, 1991; Landford *et al.*, 1995). Recently, experiments of intrahepatic inoculation of *in-vitro* transcripts of full-length HCV cDNA have proved that HCV RNA alone is sufficient to cause disease (Kolykhalov *et al.*, 1997). HCV infection is associated with a variety of outcomes. Only 15% of exposed individuals recover completely from acute infection, which is usually asymptomatic, while 85% develop persistent infection. Of those with chronic infection, spontaneous histopathological improvement or even resolution may occur in less than 10% of cases. In an additional 25% the disease seems benign and non-progressive, while in the remaining 65% it tends to progress over time. The rate of progression among the latter is very variable and seems to be influenced by host and viral factors (NIH: Anonymous, 1997; Hoofnagle, 1997). It has been estimated that more than 200 million people worldwide are chronically infected with the hepatitis C virus.

Genomic Heterogeneity of HCV

Analysing significant numbers of cDNA clones of the hepatitis C virus from single isolates provided unequivocal evidence that the viral genome cannot be defined by a single sequence but rather by a population of variant sequences closely related to one another, i.e. a viral quasi-species (Martell *et al.*, 1992). In the lack of a suitable cell culture system, the mutation rate of HCV remains unknown, but the overall rate of fixation of mutations throughout the HCV

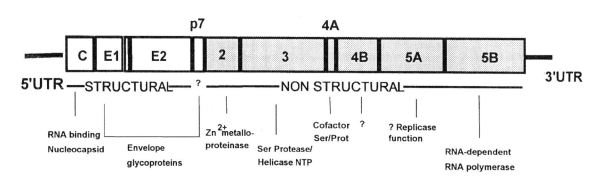

FIGURE 13.4 Organization of the hepatitis C virus genome and its encoded polyprotein. Relative location of the 5′NCR, structural genes (C = core; E1 and E2 = envelope genes 1 and 2), the p7 coding region (which product is also found in pestivirus-infected cells but not as a structural component of the mature virions), and non-structural genes (NS2 to NS5) and 3′NTR. Observed sizes of gene products (in kilodaltons), along with their known or putative function, are shown at the bottom.

genome has been estimated (by comparing the consensus sequences of isolates obtained at time intervals of 8–13 years from individual subjects) at approximately $1.44–1.92 \times 10^{-3}$ base substitutions per genome site per year (Ogata *et al.,* 1991; Okamoto *et al.,* 1992). This rate is unevenly distributed throughout the genome with that of the 5′ UTR, core and NS3 being significantly lower than those of other genes. The highest rate of fixation of mutations at the nucleotide and amino-acid levels corresponds to E1 and E2 regions with the greatest potential for variation corresponding to the hypervariable region-1 (HVR-1) of E2 (Ogata *et al.,* 1991; Okamoto *et al.,* 1992b). Interestingly, the consensus sequence of the HVR-1 changes at a higher rate during the acute phase of the infection (Yamaguchi *et al.,* 1994; Kao *et al.,* 1995) but the complexity (number of different HVR-1 sequences) of the viral population at this site is higher as the disease stage progresess (Honda *et al.,* 1994; Kurosaki *et al.,* 1994; Koizumi *et al.,* 1995). Similar results have been obtained with the core region (Kurosaki *et al.,* 1995). In contrast, it has been shown that the complexity of the viral quasispecies decreases in patients undergoing liver transplantation (Martell *et al.,* 1994). In summary, complexity of the HCV quasispecies is a dynamic parameter that changes sequentially during the course of infection as a result of random accumulation of mutations, and subsequent shaping by random events and by positive and negative selective pressures operating in the infected host. Analyses of HCV quasispecies have been reported from patients with acute and chronic infection (Honda *et al.,* 1994; Koizumi *et al.,* 1995; Kurosaki *et al.,* 1995; Yuki *et al.,* 1997; Farci *et al.,* 1998), from interferon-treated patients (Koizumi *et al.,* 1995; Le Guen *et al.,* 1997; Toyoda *et al.,* 1997b), and from donor–recipient (Kojima *et al.,* 1994; Hohne *et al.,* 1994; Esteban *et al.,* 1996) and mother–baby pairs (Weiner *et al.,* 1993; Ni *et al.,* 1997), and from a variety of biological samples, including serum, peripheral blood mononuclear cells and liver specimens. (Maggi *et al.,* 1997; Shimizu *et al.,* 1997) as well as different fragments of the same liver (positive- and negative-strand populations; Cabot *et al.,* 1997). On the other hand, the long-term evolutionary consequence of this heterogeneity among and within HCV isolates has been the generation of genetically distinct groups or genotypes, as a result of accumulation and fixation of group- or subgroup-specific mutations, a process that has probably taken centuries to evolve. Analysis of extensive sets of sequences from HCV isolates throughout the world has allowed identification of at least six major genotypes and a large number of subtypes (Tokita *et al.,* 1994, 1995; Bukh *et al.,* 1995; Simmonds, 1995). Genotypes 1, 2 and 3 show a broad geographical distribution, whereas others are mostly confined to specific geographical regions. Despite regional differences in genotype distribution, however, genotype 1b is not only the major variant in southern and eastern Europe and Japan but also in south-east Asia (Greene *et al.,* 1995). In many European countries, however, genotype distributions vary with age and risk group, reflecting rapid changes in genotype distribution (Nousbaum *et al.,* 1995). The recent construction of an infective cDNA clone of HCV, derived from genotype 1b strain for the coding region and genotype 1a strain for the 3′ and 5′ non-coding regions, has demonstrated that adequate collinear functionality within the genome is maintained, at least, between different subtypes (Yanagi *et al.,* 1998a, 1998b).

Implications of HCV Genome Heterogeneity

HCV Quasispecies and Viral Persistence   HCV does not integrate into the host genome yet persistence is the rule. Hence, other mechanisms might explain persistence. In addition, assuming some cytopathic potential, in order to persist, HCV must regulate its lytic potential and avoid elimination by the host immune system. Due to the quasispecies structure of its genome, HCV may use a variety of strategies to fulfil both requirements (Esteban *et al.,* 1997). HCV might down-regulate its cytopathic potential by the generation of defective genomes (Martell *et al.,* 1992; Higashi *et al.,* 1993), decreasing the overall viral yield, attenuating the cytopathic process and facilitating persistence (Barrett and Dimmock, 1986; Ahmed and Stevens, 1991; Kirkwood and Bangham, 1994). It has been proposed that, as for HIV (Meyerhans *et al.*, 1989), a great proportion of circulating HCV virions must contain defective genomes (Martell *et al.*, 1992; Higashi *et al.*, 1993). Also, decrease of replication by unknown mechanisms has been observed in most tissue culture systems developed for HCV (Shimizu and Yoshikura, 1995; Yoo *et al.*, 1995). Finally, reinoculation studies of chronically infected chimpanzees have shown that pre-existing, slower-replicating quasi-species outlive faster-replicating variants present in the second inoculum (Wyatt *et al.*, 1998). Cell variation and virus-cell coevolution might lead to the appearance of hepatocytes resistant to the cytopathic potential of HCV and capable of sustaining replication of cytopathic virions. Such a mechanism has been described in BHK cells infected with a cytopathic strain of foot and mouth disease virus (Martín Hernández *et al.*, 1994).

On the other hand, infected patients develop a broad and strong immune response against HCV which may limit viral production and contribute to liver damage. In most patients with chronic HCV infection, CTL responses to most HCV proteins have been demonstrated (Shirai *et al.*, 1994; Battegay *et al.*, 1995; Rehermann *et al.*, 1995), and all the necessary components for an effective CD8-mediated clearance of HCV-infected hepatocytes seem to be present in the infected liver. In addition, it seems that a strong HCV-specific CTL activity in the liver correlates with higher ALT levels (Ballardini *et al.*, 1995), lower viral titers and higher histological activity (Nelson *et al.*, 1995), suggesting that HCV-specific CTLs play an important role in host defence by regulating viral replication. In addition, there is evidence that humans and chimpanzees generate virus-neutralizing antibodies directed against B-cell epitopes present in the envelope glycoproteins within and outside the HVR-1 (Rosa *et al.*, 1996; Lechner *et al.*, 1998). Shortly after acute infection, virus-containing immune complexes can be detected in sera of most infected patients (Hijikata *et al.*, 1993; Kanto *et al.*, 1994) and their appearance usually coincides with a fall or a normalization of ALT levels (Kanto *et al.*, 1994) and a marked decrease in infectivity of the serum (Hijikata *et al.*, 1993; Shimizu and Yoshikura, 1995). Also, previous

neutralization of an HCV inoculum with positive chronic sera and immunization (Farci *et al.*, 1994) with recombinant E1–E2 heterodimers may prevent or attenuate infection upon rechallenge with the homologous acute-phase inoculum. With such a broad cellular and humoral immune response, genetic drift, i.e. mutations in neutralizing antibody or CTL epitopes, seems to be a contributing mechanism by which HCV avoids immune elimination and maintains persistence. Weiner *et al.* (1995) have shown that a single amino-acid change in a well-conserved CTL epitope in the helicase domain of NS3 abrogated recognition by intrahepatic CTLs, directed to the wild-type sequence, in an inoculated animal that developed persistent infection. Also, reduction of quasispecies complexity in immunosuppressed patients within the first post-liver transplantation period and in those co-infected with HIV and with reduced CD4 counts has been described (Toyoda *et al.*, 1997a). In both cases the reduction of the cellular immune pressure to the most fit (highly replicating) viral variants is in agreement with the presence of a homogeneous circulating quasi-species (Gretch *et al.*, 1994; Martell *et al.*, 1994; Lawal *et al.*, 1997; Yun *et al.*, 1997).

Escape from the humoral immune response is a mechanism potentially responsible for neutralization of circulating virions and prevention of reinfection, but has also been reported *in vivo* (Farci *et al.*, 1994). Rapid envelope sequence diversification has been repeatedly documented (Kumar *et al.*, 1994; van Doorn *et al.*, 1995) and appearance of escape variants to pre-existing antibodies, which may subsequently emerge as the predominant species during the course of the infection, may be another major role involved in chronicity (Okamoto *et al.*, 1992b; Scarcelli *et al.*, 1995). Also supporting the selective pressure exerted by the immune humoral response is the decreased rate of evolution shown by the quasispecies infecting agammaglobulinaemic patients (Kumar *et al.*, 1994) or the reduction in quasispecies diversity accompanying the decline in viral load commonly observed shortly after the acute phase of infection (Manzin *et al.*, 1998). Crossreactivity to pre-existing variants (Scarcelli *et al.*, 1995) and structural/functional constraints limiting variation in the primary sequence have been proposed as an explanation for the different rates of variation within the HVR-1 observed during the acute and chronic phases of the infection (Puntoriero *et al.*, 1998).

Quasispecies and liver damage   The damage caused during the infection ranges from minimal changes to cirrhosis of the liver and hepatocarcinoma, but little is known about the mechanism of hepatocyte injury in chronic HCV infection. There is evidence supporting both cytopathic effects of HCV and immune-mediated cell injury. Several observations suggest that HCV may be cytopathic. First, the histological picture of acute and chronic hepatitis suggests direct cytopathicity (Gerber, 1995). Second, among inoculated chimpanzees, detection of minus-strand HCV RNA by *in-situ* hybridization coincides and parallels aminotransferase elevations in the absence of inflammatory infiltrate for several weeks after inoculation (Negro *et al.*, 1998). Third, in some *in-vitro* cell culture systems, HCV replication is cytolitic, eventually leading to persistent low-level replication in surviving cells (Kato *et al.*, 1990; Yoo *et al.*, 1995). Also, the more severe course of infection in patients co-infected with HIV (Martin *et al.*, 1989; Pol *et al.*, 1995) or with agammaglobulinaemia (Bjoro *et al.*, 1994) would argue in favour of at least some direct viral damage. On the other hand, the broad and strong humoral and cellular immune response elicited by HCV infection suggests that the host immune system is involved in liver damage as well as in controlling viral replication (see below).

In relation to liver damage, detailed studies on the behaviour of infecting quasispecies have revealed a complex situation. While some authors (Hayashi *et al.*, 1997) have found a relationship between the degree of diversity of circulating viral populations and hepatic histology (Koizumi *et al.*, 1995), or with the average level of aminotransferases (Yuki *et al.*, 1997), such correlation has not been confirmed by others (Naito *et al.*, 1995). These discrepancies might be due to the differences in the methodology used or to differences in patient characteristics. Furthermore, it has been proved that the composition of the circulating viral population does not necessarily reflect the composition of the hepatic population (Cabot *et al.*, 1997) and that hepatic but not serum HCV RNA levels correlate with more severe liver inflammation (González-Peralta *et al.*, 1994), thus suggesting the need for a direct study of the intrahepatic viral dynamics in order to understand the mechanism of pathogenicity.

Quasispecies and response to interferon treatment   After Okada *et al.*, (1992) first suggested that the degree of variability of the HVR-1 domain correlated with interferon responsiveness, several studies have confirmed the correlation between quasispecies complexity and interferon responsiveness, so that patients with more complex populations respond less well or do not respond to treatment (Enomoto *et al.*, 1994; Kanazawa *et al.*, 1994; Koizumi *et al.*, 1995; Le Guen *et al.*, 1997). Recently, Polyak *et al.* (1997) have suggested that it is not the number of different variants but the degree of divergence between the variants making up the population that differentiates responders and non-responders. Discussion in this field is still open, as some other authors do not find differences between the complexity when correlated with viraemia levels (Hagiwara *et al.*, 1996; Shindo *et al.*, 1996). Alternatively, the quasispecies complexity has been related to the disappearance of HCV RNA only during the administration of the drug (Toyoda *et al.*, 1997b). Different genotypes, the viraemia level, the particular ability of each patient to eliminate infection, as well as the demographic and epidemiologic characteristics of the group studied and the methodology used to describe the complexity of the quasispecies, all probably complicate the interpretation of these contradictory results. What seems to be clear is that the number of different variants (Sakuma *et al.*, 1996) and the degree of sequence heterogenity decreases during IFN treatment in sustained or short-term responders (Shindo *et al.*, 1996; Yun *et al.*, 1996; González-Peralta *et al.*, 1997; Rossini *et al.*, 1997). Recently, there has been a description of a small region in the NS5-a locus of the 1b genotype of the virus, known as ISDR (interferon sensitivity determining region) where the presence of mutations in relation to a particular previously described sequence confers a higher sensitivity to interferon treatment (Enomoto *et al.*, 1995, 1996; Herion and Hoofnagle, 1997). NS5-a could be a potent inhibitor of double-strand RNA dependent protein kinase (PKR), one of the most important antiviral proteins induced by alfa-interferon, and mutation would inhibit NS5a–PKR interaction (Gale *et al.*, 1997). Nonetheless, the analysis of the viral population sequences before and after interferon treatment (Enomoto *et al.*, 1996) does not support a correlation between the appearance of specific mutant sequences and resistance to IFN. Neither has it been possible to correlate efficiency of treatment with differences in the quasispecies structure before treatment, nor with increased rates of fixation of mutations during treatment (Polyak *et al.*, 1998; Sáiz *et al.*, 1998). Instead it has been suggested that it is the balance between NS5-A sequences that could have a subtle regulatory influence on HCV replication (Pawlotsky *et al.*, 1998) and that mutation at a distance from the ISDR (Duverlie *et al.*, 1998) may be involved in interaction with PKR and, thus, sensitivity to interferon treatment.

Biological implications of HCV genotypes

Among the practical implications of HCV genotypes, those relating to serological or molecular diagnosis of HCV infection have been well established and can be summarized as follows. (1) Only the 5' NCR is well conserved among all known HCV isolates, but even this region contains variable domains – hence primers or probes deduced from these conserved domains must be used for universal detection of HCV-RNA (Bukh *et al.*, 1992; Smith *et al.*, 1995). (2) The NS4-based first-generation antibody assays detected anti-HCV in only one-third of genotype-2 or -3 infected blood donors, compared with 90% of those infected with type 1 (Simmonds, 1995). Currently available antibody-capture assays seem to work efficiently with all major genotypes, although weaker reactivity to the NS3-derived protein c33c, present in RIBA assay, may be seen among genotype-3 infected individuals (Simmonds *et al.*, 1995). Therefore, close attention to the influence of the genetic heterogeneity on the results of such tests is warranted.

Several studies have tried to correlate particular genotypes with a variety of clinical features and outcome of infection. Accumulated data can be summarized as follows: (1) There is no evidence for differences in pathogenicity among HCV variants. One possible exception is the observation that genotype-1b-infected liver transplant recipients develop acute or chronic hepatitis significantly more often and faster than those infected with other genotypes (Feray *et al.*, 1995). (2) A higher rate of long-term response to IFN treatment has been observed in patients infected with genotypes 2 and 3 as compared with those infected with type 1 (Bukh *et al.*, 1995; Simmonds, 1995). Genotypes 2 and 3 are each independent predictive factors of sustained response, along with viral load and absence of cirrhosis (Chemollo *et al.*, 1995). (3) Molecular epidemiology studies that can trace infectious sources have become an increasingly important application of HCV genotyping. Identification of the genotype and subtype of the presumed infectious source and its contacts is a necessary first step to selecting the appropriate primers for amplification and sequencing of variable (NS5 or E1/E2) regions and subsequent comparative analysis (Ohto *et al.*, 1994; Suzuki *et al.*, 1994; Esteban *et al.*, 1996).

Hepatitis Delta Virus

Hepatitis delta virus (HDV) is a unique subviral infectious agent that requires hepatitis B virus HBV for infection. HDV virion particles are aproximately 36 nm in diameter and have a coat made up of HbsAg, which encloses the RNA genome and the delta antigen (HDAg). HDV genome is a 1.68 kb highly structured single-stranded circular RNA with a high proportion (near 70%) of intramolecular complementary sequences (Kos *et al.*, 1986; Makino *et al.*, 1987; Wang *et al.*, 1992), resulting in a rod-like RNA structure under non-denaturing conditions. There are two domains in the delta RNA, each with features of two quite different RNA types (Figure 13.5). One quarter of the bases (left on figure) contain several characteristics resembling plant viroid or virusoid RNAs: a circular RNA, a rolling circle pathway replication and an

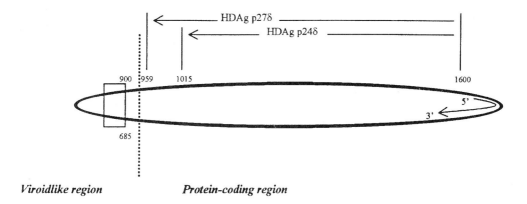

Viroidlike region   *Protein-coding region*

FIGURE 13.5 Genomic organization of the hepatitis D virus. The direction of the genomic sense is clockwise. Dotted line represents the boundary between the viroid-like region and the protein-coding region. Antigenomic strand RNA (residues 1600–959) codes for two forms of the delta antigen (HDAg). These two species of HDAg differ in having a termination codon (AUG) in one of the RNAs (residue 1015). Residue 685, site of rybozyme self-cleavage in the genomic strand; residue 900, site of ribozyme self-cleavage for the antigenomic strand; box, region of local tertiary structure and interaction.

autocatalytic cleavage and ligation activity (Sharmeen *et al.*, 1988, 1989; Wu and Lai, 1989; Wu *et al.*, 1989). Unlike viroid RNAs, however, the rest of the RNA genome (right on figure) encodes a virus-specific protein (HDAg), which constitutes the internal viral structural protein (Bonino *et al.*, 1984; Hoofnagle, 1989). The recent finding of a cellular mRNA in infected liver cells, encoding a protein strongly related to the HDAg, has suggested that delta RNA may have arisen when a free-living, self-replicating RNA "captured" a cellular mRNA encoding this protein (Brazas and Ganem, 1996; Robertson, 1996).

It is important to consider the obligatory dependence of this small RNA virus on the simultaneous presence of its helper virus. HDV may either co-infect or superinfect humans and chimpanzees in the presence of HBV (Rizzetto *et al.*, 1980a). Moreover, the data from the transplant model (Ottobrelli *et al.*, 1991) raise the question of whether, besides the HDV co-infection with HBV or HDV superinfection of HBV carriers, a third pathobiological mechanism of delta hepatitis may be HBV superinfection or reactivation of a clinically latent HDV infection. Thus, although the presence of HBV may be dispensable for HDV replication, HDV never-

theless needs foreign genetic information, such as is provided by a helper hepadnavirus, for its infection to be productive. Disease associated with HDV infection is frequently more severe than that due to the underlying HBV infection alone, but the severity of the disease can vary widely, from little evidence of chronic disease, which evolves into long-term stable cirrhosis, to severe chronic active hepatitis followed by rapidly degenerating active cirrhosis (Negro *et al.*, 1987; Bonino *et al.*, 1991; Rizzetto and Durazzo, 1991; Rizzetto *et al.*, 1992). The epidemiology of HDV resembles that of HBV but does not follow it as closely as one would expect. HDV infection is endemic in some geographical areas (countries in the Mediterranean basin, the Middle East, part of Africa and South America; Jacobson *et al.*, 1985). The prevalence of HDV infection, however, is relatively low among Orientals despite a high incidence of HBsAg-positive hepatitis (Rizzetto *et al.*, 1980b). Endemicity is low in the USA, where transmission is often related to intravenous drug use (Rizzetto *et al.*, 1991). Both geography and transmission route have been identified as possibly important determinants of disease severity.

Viral Heterogeneity

HDV differs from all other animal RNA viruses in that its genome replication does not utilize a viral polymerase but rather makes use of the host RNA polymerase II (MacNaughton *et al.*, 1991; Fu and Taylor, 1993). Although recent data suggest that an associated transcription factor, elongation factor SII, might have 3'→5' exonuclease activity (Izban and Luse, 1992), HDV RNA exibits significant genome heterogeneity in individual isolates (Chao *et al.*, 1990a, 1991; Wang *et al.*, 1992), a feature consistent with the quasispecies distribution of most RNA viruses. Studies of complete RNA sequences from sequential isolates (Weiner *et al.*, 1988; Lee *et al.*, 1992) have showed overall evolutionary rates of 3.18×10^{-3} and 5×10^{-3} substitutions/ nucleotide/year, respectively, similar to that of other RNA viruses. In these studies evolutionary rates have been shown to be different in different parts of the genome and appeared to correlate with changes in disease stage (Lee *et al.*, 1992).

A phenotypic consequence of the genome variability of HDV is the existence of two different forms of the delta antigen strictly dependent upon genome replication because of a specific site mutation (Weiner *et al.*, 1988; Luo *et al.*, 1990; Wang *et al.*, 1992). In every HDV-infected human or experimental animal, two HDAg species, which differ in structure by 19 aminoacids at the C terminus, are usually present (Bergmann and Gerin, 1986; Wang *et al.*, 1987). At the nucleotide level, the difference exists as a single base substitution at position 1015, changing the termination codon for the short form (195 aminoacids) from AUG to UGG, thus allowing the synthesis of a long 214 amino acid protein. These two forms also differ functionally: the short form (HDAg-p24) is required for RNA replication (Kuo *et al.*, 1989) while the long form (HDAg-p27) supresses viral RNA replication and is required for packaging of the HDV genome with HBsAg (Chao *et al.*, 1990b; Chang *et al.*, 1991; Glenn and White, 1991).

As expected, HDV isolates from different geographical areas are heterogeneous. Three genotypes of HDV have been defined based on genetic analysis of HDV isolates from different areas of the world: genotype I is the most widespread geographically; it has been found in Europe (Wang *et al.*, 1986, 1992; Saldanha *et al.*, 1990; Niro, *et al.*, 1997; Shakil *et al.*, 1997), North America (Makino *et al.*, 1987), north Africa (Zhang *et al.*, 1996), the Middle East (Lee *et al.*, 1992) the south Pacific (Chao *et al.*, 1991) and east Asia (Izban and Luse, 1992). The pathogensis of HDV infection in these areas varies from fulminant hepatitis to asymptomatic liver disease. Genotype II has been found only in east Asia (Taiwan and Japan; Imazeki *et al.*, 1991; Wu *et al.*, 1995; Lee *et al.*, 1996) and has been associated less frequently with fulminant hepatitis at the acute stage and a more favourable long-term clinical outcome at the chronic stage than type I infection, which also occurs in this region. Genotype III, the type more distantly related to the other genotypes, is exclusively associated with HDV infection in northern South America, where outbreaks of severe hepatitis with particularly high morbidity and mortality have been described (Imazeki *et al.*, 1991; Wu *et al.*, 1995; Lee *et al.*, 1996). The sequence heterogeneity of the HDV RNA might help to explain the variations observed in the degree of the severity of delta hepatitis and in the transmisibility of the virus.

HUMAN HEPATITIS VIRUSES THAT DO NOT CAUSE CHRONIC HEPATITIS

Hepatitis A Virus

The hepatitis A virus (HAV) is a single-stranded, positive sense RNA virus. Because of its unique features, it has been classified as the single species of a new genus, *Heparnavirus*, within the Picornaviridae family (Gust *et al.*, 1998). The HAV genome is a linear RNA molecule of approximately 7.5 kb and has a gene organization similar to that of picornaviruses, with which it shares little nucleotide sequence homology. As in other picornaviruses, the 5' end of the genome begins with a lengthy 5' untranslated region (5'-UTR), which contains a small protein (VPg) covalently bound to its 5' end

(Weitz *et al.*, 1986) and acts as an internal ribosome entry site (IRES; Cohen *et al.*, 1987). The coding region of the genome consists of a single long open reading frame of 6681 nucleotides. This has been arbitrarily divided into three regions, termed P1, which encodes the four capsid proteins (VP1 to VP4), and P2 and P3, which encode the non-structural proteins required for polyprotein processing and viral replication (Rueckert and Wimmer, 1984). The genome ends with a 3' non-coding region followed by a poly-A tail (Coulepis *et al.*, 1981).

Hepatitis A virus has been adapted to grow in cell culture (Provost and Hilleman, 1979) where the virus is not cytopathic, replicates slowly without interfering with host-cell macromolecular synthesis, and leads to the establishment of persistent infection. Cell-culture-adapted strains of HAV have led to the development of effective inactivated hepatitis A vaccines.

Infection with HAV, which is transmitted by the faecal–oral route, causes an acute, often asymptomatic, self-limiting disease of varying severity that never leads to chronic hepatitis (Battegay and Feinstone, 1997).

Genomic Heterogeneity

Early comparative studies of the nucleotide sequence of HAV isolates from different geographical regions suggested a high degree of conservation of the HAV genome (Ticehurst *et al.*, 1989). However, comparison of nucleic acid sequences in the region around the VP1/2A junction of simian and human isolates from different geographical regions has led to the classification of HAV strains into seven distinct genotypes (I–VII) differing in 15–25% of their nucleotide positions, and a number of subtypes (Robertson *et al.*, 1992). The majority of human isolates belong to genotypes I and III, while genotypes IV, V and VI have only been found in simian primates. Overall genetic diversity within this region of the genome, between individual strains of a given genotype, has been shown to be less than 3%, even among strains collected over many years, suggesting that the rate of accumulation of mutations in the HAV genome is very low.

In fact, despite the existence of different genotypes, there appears to be only one serotype of HAV causing infection in humans, which explains the cross-protection of human immune serum globulin and of inactivated HAV vaccines against infection with unrelated HAV strains (Lemon and Binn, 1983). The existence of a highly conserved single neutralization site, contributed for by the VP1 and VP3 structural proteins, which would be under strong structural and functional constraint, has been proposed and specific mutations associated with a neutralization escape phenotype have been mapped (Stapleton and Lemon, 1987; Nainan *et al.*, 1992; Ping and Lemon, 1992).

Implications of HAV Genome Heterogeneity

Despite the relative conservation of the HAV genome, the classification of HAV isolates into genotypes and subtypes has provided the basis for molecular epidemiological studies to track infectious sources in endemic and epidemic outbreaks (Normann *et al.*, 1995; Leino *et al.*, 1997; Taylor, 1997). Besides their utility in molecular epidemiology studies, no differences in the clinical course of the disease have been associated with infection with different HAV genotypes.

Although the presence of HAV particles containing defective genomes in clinical specimens has been reported (Siegl *et al.*, 1990), their role, if any, in the course of natural infections remains unknown. On the other hand, no specific genomic mutations in natural isolates of HAV have been associated with differences in clinical behaviour.

The situation is different in cell culture, in which rapidly replicating variants, attenuated strains and temperature-sensitive mutants have been selected and found to contain mutations in the 5' UTR and in the 2B and C coding regions (Cohen *et al.*, 1988; Day *et al.*, 1992; Emerson *et al.*, 1992; Zhang *et al.*, 1995). Some mutations in the 5'-UTR have been shown to promote viral replication in some cell-culture-adapted strains through an increase in viral translation (Schultz *et al.*, 1996) while others have been associated with an attenuated temperature-sensitive phenotype through an impairment in RNA synthesis (Shaffer *et al.*, 1995). Analysis of the molecular basis for viral cell culture and attenuation may provide the basis for the development of improved live-attenuated hepatitis A vaccines.

Hepatitis E Virus

Hepatitis E virus (HEV) is a non-enveloped virus 27–34 nm in diameter with a positive-strand RNA genome 7.6 kb in length (Bradley *et al.*, 1987; Reyes *et al.*, 1990). The RNA is polyadenylated at its 3′ end and comprises three positive-polarity open reading frames. Provisionally, HEV is classified as a new genus within the family Calciviridae. The largest open reading frame (ORF1) expands 5 kb from the 5′ end and encodes a putative RNA polimerase and other non-structural proteins. The second and the third open reading frames (ORF1 and ORF2) are made up of 2 kb and 0.37 kb encoding the capsid protein and an inmunogenic protein of unknown function, respectively. The third open reading frame overlaps the other two major ORFs (Tam *et al.*, 1991; Tsarev *et al.*, 1992). The virus provokes a mild-to-moderate disease and it never appears to cause a chronic disease although acute disease is associated with mortality among pregnant women (Khuroo *et al.*, 1981). HEV is transmitted by the faecal–oral route and is associated with low socio-economic status and periodic unimodal epidemics.

Heterogeneity

The sequence of the full-length genome has been determined for HEV strains from different geographical isolates of Asia and one from Mexico (Tam *et al.*, 1991; Aye *et al.*, 1992; Huang *et al.*, 1992; Tsarev *et al.*, 1992; Yin *et al.*, 1994). In addition, the genomes of several other strains from North America (Kwo *et al.*, 1997), Asia (Yin, *et al.*, 1993; Huang *et al.*, 1995) and Africa (Chatterjee *et al.*, 1997; Cuyck-Gandré *et al.*, 1997) have been partially sequenced. Comparison of these sequences has revealed conservation among the different isolates, with the most significant disparity between Asian strains and the Mexican strain. The ORF1 of the Mexican strain is over 25% different at the nucleotide level to other isolates, whereas strains from other geographical regions differ from each other by less than 10%. According to their genetic sequence, HEV isolates have been classified into three genotypes but one serotype appears to include all of these. Little variation (0.2–1.5%) (Yin, *et al.*, 1993; Panda and Jameel, 1997) has been found among isolates obtained at intervals of 3–4 years from the same region. This low rate of variation has led to the suggestion that HEVs from different geographical areas have diverged from each other in the past and have maintained a restricted circulation (Yin *et al.*, 1993). This geographical origin of HEV strain divergence is also supported by the higher degree of homology of the recently sequenced African strains to the Asian strains than to the Mexican one (Chatterjee *et al.*, 1997; Cuyck-Gandré *et al.*, 1997).

Implications

Epitope mapping has revealed immunodominant regions in the C-terminal regions of proteins encoded by ORF-2 and ORF-3 (Kaur *et al.*, 1992; Khudyakov *et al.*, 1994a). The primary structure of the antigenic region located at the C-terminus of ORF 3 has been shown to be highly conserved within the same strain but divergent among the most different strains. This variation has been demonstrated to change the antigenic properties of this region (Khudyakov *et al.*, 1994b). Concern that variation at this and other sites might yield false negatives in diagnostic assays based on detection of antibodies to synthetic peptides or recombinant antigens has encouraged the incorporation of multiple antigens from more than one strain in order to increase the sensitivity of current diagnostic assays. Regarding protective immunity, cross-challenge experiments in chimpanzees with most disease causing HCV strains have revealed cross-protection following infection (Arankalle *et al.*, 1995). Experimental immunoprophylaxis against HEV, based on recombinant antigens, appears to afford short-term protection and may be useful for pregnant women; nevertheless, sporadic and occasional epidemics are known to break out in a population tested to be IgG anti-HEV(ORF2)-positive (Panda *et al.*, 1995; Panda and Jameel, 1997). Whether the nucleic acid or amino-acid differences among strains may influence the transmissibility or infectivity of the virus in naive individuals or previously exposed individuals remains undetermined.

REFERENCES

Ahmed, R. and Stevens, J.G. (1991) *Fundamental Virology* (eds Fields, B.N. and Kniep, D.N.), pp. 241–265. Raven Press, New York.

Aikawa, T., Kanai, K., Kako, M. *et al.* (1995) Interferon-α2a for chronic hepatitis B with e antigen or antibody: comparable antiviral effects on wild-type virus and precore mutant. *J. Viral Hepat.*, **2**, 243–250.

Alexopolou, A., Owsianka, A.M., Kafiri, G., Dourakis, S., Carman, W.F. and Hadziyannis, S. (1998) Core variability does not affect response to interferon alpha in HBeAg negative chronic hepatitis B. *J. Hepatol.*, **29**, 345–351.

Arankalle, V.A., Chadha, M.S., Chobe, L.P., Nair, R. and Banerjee, K. (1995) Cross-challenge studies in rhesus monkeys employing different Indian isolates of hepatitis E virus. *J. Med. Virol.*, **46**, 358–636.

Aye, T.T., Uchida, T., Ma, X.Z. *et al.* (1992) Complete nucleotide sequence of a hepatitis E virus isolated from the Xinjiang epidemic (1986–1988) of China. *Nucl. Acids Res.*, **20**, 3512.

Ballardini, G., Groff, P., Pontisso, P. *et al.* (1995) Hepatitis C virus (HCV) genotype, tissue HCV antigens, hepatocellular expression of HLA-A,B,C and intercellular adhesion-1 molecules. Clues to pathogenesis of hepatocellular damage and response to interferon treatment in patients with chronic hepatitis C. *J. Clin. Invest.*, **95**, 2067–2075.

Barrett, A.D. and Dimmock, N.J. (1986) Defective interfering viruses and infections of animals. *Curr. Topics Microbiol. Immunol.*, **128**, 55–84.

Battegay, M. and Feinstone, S.M. (1997) Hepatitis A virus. In: *Viral Hepatitis* (ed. Wilson, R.A.), pp. 35–84. Marcel Dekker, New York.

Battegay, M., Fikes, J., Di Bisceglie, A.M. *et al.* (1995) Patients with chronic hepatitis C have circulating cytotoxic T cells which recognize hepatitis C virus-encoded peptides binding to HLA-A2.1 molecules. *J. Virol.*, **69**, 2462–2470.

Baumert, T.F., Rogers, S.A., Hasagawa, K. and Liang, T.J. (1996) Two core promotor mutations identified in a hepatitis B virus strain associated with fulminant hepatitis result in enhanced viral replication. *J. Clin. Invest.*, **98**, 2268–2276.

Bergmann, K.F. and Gerin, J.L. (1986) Antigens of hepatitis delta virus in the liver and serum of humans and animals. *J. Infect. Dis.*, **154**, 702–706.

Bjoro, K., Froland, S.S., Yun, Z., Samdal, H.H. and Haaland, T. (1994) Hepatitis C infection in patients with primary hypogammaglobulinemia after treatment with contaminated immune globulin. *N. Engl. J. Med.*, **331**, 1607–1611.

Bollyky, P., Yasmin, M., Holmes, E., Trautwein, C., Fagan, E.A. and Carman, W.F. (1997) Viruses from unrelated fulminant hepatitis B cases cluster in phylogenetically distinct lineages, each containing unique motifs of nucleotide and HBx variants which have increased transcriptional activity in vitro. *J. Hepatol.*, **26** (Suppl 1), 67.

Bonino, F., Hoyer, B., Shih, J., Rizzetto, M., Purcell, R.H. and Gerin, J.L. (1984) Delta hepatitis agent: structural and antigenic properties of the delta-associated particle. *Infect. Immun.*, **43**, 1000–1005.

Bonino, F., Brunetto, M.R. and Negro, F. (1991) Factors influencing the natural course of HDV hepatitis. *Prog. Clin. Biol. Res.*, **364**, 137–146.

Bradley, D.W., Krawczynski, K., Cook, E.H. Jr *et al.* (1987) Enterically transmitted non-A, non-B hepatitis: serial passage of disease in cynomolgus macaques and tamarins and recovery of disease-associated 27- to 34-nm viruslike particles. *Proc. Natl Acad. Sci. USA*, **84**, 6277–6281.

Brazas, R. and Ganem, D. (1996) A cellular homolog of hepatitis delta antigen: implications for viral replication and evolution. *Science*, **274**, 90–94.

Brunetto, M.R., Stemmler, M., Schodel, F. *et al.* (1989) Identification of HBV variants which cannot produce precore-derived HBeAg and may be responsible for severe hepatitis. *Ital. J. Gastroenterol.*, **21**, 151–154.

Buckwold, V.E., Xu, Z., Chen, M., Yen, T.S. and Ou, J.H. (1996) Effects of a naturally occurring mutation in the hepatitis B virus basal core promoter on precore gene expression and viral replication. *J. Virol.*, **70**, 5845–5851.

Bukh, J., Purcell, H. and Miller, R.H. (1992) Sequence analysis of the 5′ noncoding region of hepatitis C virus. *Proc. Natl Acad. Sci. USA*, **89**, 4942–4946.

Bukh, J., Miller, R.H. and Purcell, R.H. (1995) Genetic heterogeneity of hepatitis C virus: quasispecies and genotypes. *Sem. Liver Dis.*, **15**, 41–63.

Cabot, B., Esteban, J.I., Martell, M. *et al.* (1997) Structure of replicating hepatitis C virus (HCV) quasispecies in the liver may not be reflected by analysis of circulating HCV virions. *J. Virol.*, **71**, 1732–1734.

Cariani, E., Ravaggi, A., Tanzi, E. *et al.* (1995) Emergence of hepatitis B virus S gene mutant in a liver transplant recipient. *J. Med. Virol.*, **47**, 410–415.

Carman, W.F. (1996) Molecular variants of hepatitis B virus. *Clin. Lab. Med. (Hepatitis and Chronic Liver Disease)*, **16**, 407–428.

Carman, W.F., Jacyna, M.R., Hadziyannis, S. *et al.* (1989) Mutation preventing formation of e antigen in patients with chronic HBV infection. *Lancet*, **ii**, 588–591.

Carman, W.F., Zanetti, A.R., Karayiannis, P. *et al.* (1990) Vaccine-induced escape mutant of hepatitis B virus. *Lancet*, **336**, 325–329.

Carman, W.F., Hadziyannis, S., Karayiannis, P. *et al.* (1991) Association of the precore variant of HBV with acute and fulminant hepatitis B infection. In: *Viral Hepatitis and Liver Disease* (eds Hollinger, Lemon, S.M. and Margolis), pp. 216–219. Williams & Wilkins, Baltimore, MD.

Carman, W.F., Thomas, H.C. and Domingo, E. (1993) Viral genetic variation; hepatitis B virus as a clinical example. *Lancet*, **341**, 349–353.

Carman, W.F., Korula, J., Wallace, L., MacPhee, R., Mimms, L. and Decker, R. (1995a) Fulminant reactivation of hepatitis B due to envelope protein mutant of HBV that escaped detection by monoclonal HBsAg ELISA. *Lancet*, **345**, 1406–1407.

Carman, W.F., Thursz, M., Hadziyannis, S. *et al.* (1995b) HBeAg negative chronic active hepatitis: HBV core mutations occur predominantly in known antigenic determinants. *J. Viral Hepat.*, **2**, 77–84.

Carman, W.F., Trautwein, C., vanDeursen, F.J. *et al.* (1996) Hepatitis B virus envelope variation after transplantation with and without hepatitis B immune globulin prophylaxis. *Hepatology*, **24**, 489–493.

Carman, W.F., Boner, W.F., Fattovich, G. *et al.* (1997a) HBV core protein mutations are concentrated in B cell epitopes in progressive disease and T helper cell epitopes in clinical remission. *J. Infect. Dis.*, **175**, 1093–1100.

Carman, W.F., van Deursen, F.J., Mimms, L.T. *et al.* (1997b) The prevalence of surface antigen variants of hepatitis B virus in Papua New Guinea, South Africa and Sardinia. *Hepatology*, **26**, 1658–1666.

Carman, W.F., Owsianka, A., Wallace, L.A., Dow, B.C., Mutimer, D.J. (1999) Antigenic characterisation of pre- and post-liver transplant hepatitis B surface antigen sequences from patients treated with hepatitis B immune globulin. *J. Hepatol.* **30**.

Chang, F.L., Chen, P.J., Tu, S.J., Wang, C.J. and Chen, D.S. (1991) The large form of hepatitis delta antigen is crucial for assembly of hepatitis delta virus. *Proc. Natl Acad. Sci. USA*, **88**, 8490–8494.

Chao, Y.C., Chang, M.F., Gust, I. and Lai, M.M. (1990a) Sequence conservation and divergence of hepatitis delta virus RNA. *Virology*, **178**, 384–392.

Chao, M., Hsieh, S.Y. and Taylor, J (1990b) Role of two forms of hepatitis delta virus antigen: evidence for a mechanism of self-limiting genome replication. *J. Virol.*, **64**, 5066–5069.

Chao, Y.C., Lee, C.M., Tang, H.S., Govindarajan, S. and Lai, M.M. (1991) Molecular cloning and characterization of an isolate of hepatitis delta virus from Taiwan. *Hepatology*, **13**, 345–352.

Chatterjee, R., Tsarev, S., Pillot, J., Coursaget, P., Emerson, S.U. and Purcell, R.H. (1997)

African strains of hepatitis E virus that are distinct from Asian strains. *J. Med. Virol.*, **53**, 139–144.

Chemello, L., Bonetti, P., Cavalletto, L. *et al.* (1995) Randomized trial comparing three different regimens of alpha-2a-interferon in chronic hepatitis C. The TriVeneto Viral Hepatitis Group. *Hepatology*, **22**, 700–706.

Choo, Q.L., Kuo, G., Weiner, A.J., Overby, L.R., Bradley, D.W. and Houghton, M. (1989) Isolation of a cDNA clone derived from a blood-borne non-A, non-B viral hepatitis genome. *Science*, **244**, 359–362.

Chuang, W.-L., Omata, M., Ehata, T. *et al.* (1993) Precore mutations and core clustering mutations in chronic hepatitis B virus infection. *Gastroenterology*, **104**, 263–271.

Cohen, J.I., Ticehurst, J.R., Purcell, R.H., Buckler-White, A. and Baroudy, B.M. (1987) Complete nucleotide sequence of wild-type hepatitis A virus: comparison with different strains of hepatitis A virus and other picornaviruses. *J. Virol.*, **61**, 50–59.

Cohen, J.I., Rosenblum, B., Daemer, R.J. and Ticehurst, J.R. (1988) Attenuation of hepatitis A virus: molecular analysis. In: *Viral Hepatitis and Liver Disease* (ed. Zuckerman, A.J.), pp. 43–47. Alan R. Liss, New York.

Coulepis, A.G., Tannock, G.A., Locarnini, S.A. and Gust, I. (1981) Evidence that the genome of hepatitis A virus consists of a single-stranded RNA. *J. Virol.*, **37**, 473–477.

Cuyck-Gandré, H., Zhang, H.Y., Tsarev, S. *et al.* (1997) Characterization of hepatitis E virus (HEV) from Algeria and Chad by partial genome sequence. *J. Med. Virol.*, **53**, 340–347.

Day, S.P., Murphy, P., Brown, E.A. and Lemon, S.M. (1992) Mutations within the 5′ nontranslated region of hepatitis A virus RNA which enhance replication in BS-C-1 cells. *J. Virol.*, **66**, 6533–6540.

Duverlie, G., Khorsi, H., Castelain, S. *et al.* (1998) Sequence analysis of the NS5A protein of European hepatitis C virus 1b isolates and relation to interferon sensitivity. *J. Gen. Virol.*, **79**, 1373–1381.

Ehata, T., Omata, M., Yokosuka, O., Hosodo, K. and Ohto, M. (1992) Variations in codons 84–101 in the core nucleotide sequence correlate with hepatocellular injury in chronic hepatitis B virus infection. *J. Clin. Invest.*, **89**, 332–338.

Ehata, T., Omata, M., Chuang, W.L. *et al.* (1993) Mutations in core nucleotide sequence of hepatitis B virus correlate with fulminant and severe hepatitis. *J. Clin. Invest.*, **91**, 1206–1213.

Emerson, S.U., Huang, Y.K., McRill, C. and Lewis, M. (1992) Mutations in both the 2B and 2C genes of hepatitis A virus are involved in adaptation to growth in cell culture. *J. Virol.*, **66**, 650.

Enomoto, N., Kurosaki, M., Tanaka, Y., Marumo, F. and Sato, C. (1994) Fluctuation of hepatitis C virus quasispecies in persistent infection and interferon treatment revealed by single-strand conformation polymorphism analysis. *J. Gen. Virol.*, **75**, 1361–1369.

Enomoto, N., Sakuma, I., Asahina, Y. *et al.* (1995) Comparison of full-length sequences of interferon-sensitive and resistant hepatitis C virus 1b. Sensitivity to interferon is conferred by amino acid substitutions in the NS5A region. *J. Clin. Invest.*, **96**, 224–230.

Enomoto, E., Sakuma, I., Asahina, Y. *et al.* (1996) Mutations in the nonstructural protein 5A gene and response to interferon in patients with chronic hepatitis C virus 1b infection. *N. Engl. J. Med.*, **334**, 77–81.

Esteban, J.I., Gómez, J., Martell, M. *et al.* (1996) Transmission of hepatitis C virus by a cardiac surgeon. *N. Engl. J. Med.*, **334**, 555–560.

Esteban, J.I., Gómez, J., Martell, M. and Guardia, J. (1997) Hepatitis C virus. In: *Viral Hepatitis* (ed. Wilson, R.A.), pp. 147–216. Marcel Dekker, New York.

Fagan, E., Smith, P., Davison, F. and Williams, R. (1986) Fulminant hepatitis B in successive female sexual partners of two anti-HBe positive males. *Lancet*, **ii**, 538–540.

Farci, P., Alter, H.J., Wong, D.C. *et al.* (1994) Prevention of hepatitis C virus infection in chimpanzees after antibody-mediated *in vitro* neutralization. *Proc. Natl Acad. Sci. USA*, **91**, 7792–7796.

Farci, P., Shimoda, A., Melpolder, J. *et al.* (1998) The viral quasispecies in the natural history of HCV infection. International Meeting on hepatitis C virus and related viruses, Venice 170 (Abstract)

Fattovich, G., Colman, K., Thursz, M. *et al.*

(1995) Pre-core/core gene variation and interferon therapy. *Hepatology*, **22**, 1355–1362.

Feitelson MA, Duan LX, Guo J *et al.* (1995) Precore and X region mutants in hepatitis B virus infections among renal dialysis patients. *J. Viral Hepat.*, **2**, 19–31.

Feray, C., Gigou, M., Samuel, D. *et al.* (1995) Influence of the genotypes of hepatitis C virus on the severity of recurrent liver disease after liver transplantation. *Gastroenterology*, **108**, 1088–1096.

Fischer, S, Schüttler, CG, Repp, R *et al.* (1988) Complex regulation of enhancer II and pre-genomic promoter of HBV wildtype and variants. In: *Tagung der Gesellschaft für Virologie*. Hamburg.

Fong, T.L., Shindo, M., Feinstone, S.M., Hoofnagle, J.H. and Di Bisceglie, A.M. (1991) Detection of replicative intermediates of hepatitis C viral RNA in liver and serum of patients with chronic hepatitis C. *J. Clin. Invest.*, **88**, 1058–1060.

Fu, T.B. and Taylor, J. (1993) The RNAs of hepatitis delta virus are copied by RNA polymerase II in nuclear homogenates. *J. Virol.*, **67**, 6965–6972.

Fujii, H., Moriyama, K., Sakamoto, N. *et al.* (1992) Gly 145 to Arg substitution in HBs antigen of immune escape mutant of hepatitis B virus. *Biochem. Biophys. Res. Comm.*, **184**, 1152–1157.

Fukuda, R., Xuan-Thanh, N., Ishimura, N. *et al.* (1995) X gene and precore region mutations in the hepatitis B virus genome in persons positive for antibody to hepatitis B e antigen: comparison between asymptomatic "healthy" carriers and patients with severe chronic active hepatitis. *J. Infect. Dis.*, **172**, 1191–1197.

Gale, M.J.J., Korth, M.J., Tang, N.M. *et al.* (1997) Evidence that hepatitis C virus resistance to interferon is mediated through repression of the PKR protein kinase by the nonstructural 5A protein. *Virology*, **230**, 217–227.

Gerber, M.A. (1995) Pathobiologic effects of hepatitis C. *J. Hepatol.*, **22** (Suppl.), 83–86.

Gerken, G., Kremsdorf, D., Petit, M.A. *et al.* (1991) Hepatitis B defective virus with rearrangements in the preS gene during HBV chronic infection. *J. Hepatol.*, **13**, S93–S96.

Ghany, M.G., Ayola, B., Villamil, F.G. *et al.* (1998) Hepatitis B virus S mutants in liver transplant recipients who were reinfected despite hepatitis B immune globulin prophylaxis. *Hepatology*, **27**, 213–222.

Glenn, J.S. and White, J.M. (1991) Trans-dominant inhibition of human hepatitis delta virus genome replication. *J. Virol.*, **65**, 2357–2361.

González-Peralta, R.P., Davis, G.L. and Lau, J.Y. (1994) Pathogenetic mechanisms of hepatocellular damage in chronic hepatitis C virus infection. *J. Hepatol.*, **21**, 255–259.

González-Peralta, R.P., Liu, W.Z., Davis, G.L., Qian, K.P. and Lau, J.Y. (1997) Modulation of hepatitis C virus quasispecies heterogeneity by interferon-alpha and ribavirin therapy. *J. Viral Hepat.* **4**, 99–106.

Greene, W.K., Cheong, M.K., Ng, V. and Yap, K.W. (1995) Prevalence of hepatitis C virus sequence variants in South-East Asia. *J. Gen. Virol.*, **76**, 211–215.

Gretch, D., Corey, L., Wilson, J. *et al.* (1994) Assessment of hepatitis C virus RNA levels by quantitative competitive RNA polymerase chain reaction: high-titer viremia correlates with advanced stage of disease. *J. Infect. Dis.*, **169**, 1219–1225.

Günther, S., Piwon, N., Iwanska, A., Schilling, R., Meisel, H. and Will, H. (1996) Type, prevalence, and significance of core promoter/enhancer II mutations in hepatitis B viruses from immunosuppressed patients with severe liver disease. *J. Virol.*, **70**, 8318–8331.

Guo, W.M., Chen, M., Yen, T.S.B. and Ou, J.-H. (1993) Hepatocyte specific expression of the hepatitis B virus core promoter depends on both positive and negative regulation. *Mol. Cell. Biol.*, **13**, 443–448.

Gust, J., Coulepis, A.G., Feinstone, S.M. *et al.* (1998) Taxonomic classification of hepatitis A virus. *Intervirology*, **20**, 1–7.

Hagiwara, H., Hayashi, N., Kasahara, A. *et al.* (1996) Treatment with recombinant interferon-alpha 2a for patients with chronic hepatitis C: predictive factors for biochemical and virologic response. *Scand. J. Gastroenterol.*, **31**, 1021–1026.

Hasegawa, K., Huang, J., Rogers, S.A., Blum, H.E. and Liang, T.J. (1994) Enhanced replication of a hepatitis B virus mutant associated with an epidemic of fulminant hepatitis. *J. Virol.*, **68**, 1651–1659.

Hawkins, A., Gilson, R., Gilbert, N. *et al.* (1996) Hepatitis B virus surface mutations associated with infection after liver transplantation. *J. Hepatol.*, **11**, 8–14.

Hayashi, J., Kishihara, Y., Yamaji, K. *et al.* (1997) Hepatitis C viral quasispecies and liver damage in patients with chronic hepatitis C virus infection. *Hepatology*, **25**, 697–701.

Herion, D. and Hoofnagle, J.H. (1997) The interferon sensitivity determining region: all hepatitis C virus isolates are not the same. *Hepatology*, **25**, 769–771.

Higashi, Y., Kakumu, S., Yoshioka, K. *et al.* (1993) Dynamics of genome change in the E2/NS1 region of hepatitis C virus *in vivo*. *Virology*, **197**, 659–668.

Hijikata, M., Shimizu, Y.K., Kato, H. *et al.* (1993) Equilibrium centrifugation studies of hepatitis C virus: evidence for circulating immune complexes. *J. Virol.*, **67**, 1953–1958.

Hino, K., Okuda, M., Hashimoto, O., Ishiko, H. and Okita, K. (1995) Glycine-to-arginine substitution at codon 145 of HBsAg in two infants born to hepatitis B e antigen-positive carrier. *Dig.Dis. Sci.*, **40**, 566–570.

Hohne, M., Schreier, E. and Roggendorf, M. (1994) Sequence variability in the *env*-coding region of hepatitis C virus isolated from patients infected during a single source outbreak. *Arch. Virol.*, **137**, 25–34.

Honda, M., Kaneko, S., Sakai, A., Unoura, M., Murakami, S. and Kobayashi, K. (1994) Degree of diversity of hepatitis C virus quasispecies and progression of liver disease. *Hepatology*, **20**, 1144–1151.

Hoofnagle, J.H. (1989) Type D (delta) hepatitis (clinical conference; published erratum appears in J.A.M.A. (1989) **261**(24), 3552; see comments). *J.A.M.A.*, **261**, 1321–1325.

Hoofnagle, J.H. (1997) Hepatitis C: the clinical spectrum of disease. *Hepatology*, **26** (Suppl.1), 15S–20S.

Hou, J.L., Karayiannis, P., Thomas, H.C. *et al.* (1995) A unique insertion in the S gene of surface antigen negative HBV in Chinese carriers. *Hepatology*, **21**, 273–278.

Houghton, M. (1996) In: *Fields' Virology*, 3rd edn

(eds Fields, B.N., Knipe, D.N. and Howley, P.N.), pp. 1035–1057. Lippincott-Raven, Philadelphia, PA.

Houghton, M., Weiner, A., Han, J., Kuo, G. and Choo, Q.L. (1991) Molecular biology of the hepatitis C viruses: implications for diagnosis, development and control of viral disease. *Hepatology*, **14**, 381–388.

Huang, C.C., Nguyen, D., Fernandez, J. (1992) Molecular cloning and sequencing of the Mexico isolate of hepatitis E virus (HEV). *Virology*, **191**, 550–558.

Huang, R., Nakazono, N., Ishii, K., Kawamata, O., Kawaguchi, R. and Tsukada, Y. (1995) Existing variations on the gene structure of hepatitis E virus strains from some regions of China. *J. Med. Virol.*, **47**, 303–308.

Imazeki, F., Omata, M. and Ohto, M. (1991) Complete nucleotide sequence of hepatitis delta virus RNA in Japan. *Nucl. Acids. Res.*, **19**, 5439.

Izban, M.G. and Luse, D.S. (1992) The RNA polymerase II ternary complex cleaves the nascent transcript in a $3' \to 5'$ direction in the presence of elongation factor SII. *Genes Devel.*, **6**, 1342–1356.

Jacobson, I.M., Dienstag, J.L., Werner, B.G., Brettler, D.B., Levine, P.H. and Mushahwar, I.K. (1985) Epidemiology and clinical impact of hepatitis D virus (delta) infection. *Hepatology*, **5**, 188–191.

Johnson, J.L., Raney, A.K. and McLachlan, A. (1995) Characterization of a functional hepatocyte nuclear factor 3 binding site in the hepatitis B virus nucleocapsid promoter. *Virology*, **208**, 147–158.

Junker-Niepmann, M., Bartenschlager, R. and Schaller, H. (1990) A short *cis*-acting sequence is required for hepatitis B virus pregenome encapsidation and sufficient for packaging of foreign RNA. *EMBO J.*, **9**, 3389–3396.

Kanazawa, Y., Hayashi, N., Mita, E. *et al.* (1994) Influence of viral quasispecies on effectiveness of interferon therapy in chronic hepatitis C patients. *Hepatology*, **20**, 1121–1130.

Kanto, T., Hayashi, N., Takehara, T. *et al.* (1994) Buoyant density of hepatitis C virus recovered from infected hosts: two different features in sucrose equilibrium density-gradient centrifugation related to degree of liver inflammation. *Hepatology*, **19**, 296–302.

Kao, J.H., Chen, P.J., Lai, M.Y., Wang, T.H., and Chen, D.S. (1995) Quasispecies of hepatitis C virus and genetic drift of the hypervariable region in chronic type C hepatitis. *J. Infect. Dis.*, **172**, 261–264.

Karayiannis, P., Alexopoulou, A., Hadziyannis, S. *et al.* (1995) Fulminant hepatitis associated with hepatitis B virus antigen-negative infection: importance of host factors. *Hepatology*, **22**, 1628–1634.

Kato, N., Hijikata, M., Ootsuyama, Y. *et al.* (1990) Molecular cloning of the human hepatitis C virus genome from Japanese patients with non-A, non-B hepatitis. *Proc. Natl Acad. Sci. USA*, **87**, 9524–9528.

Kaur, M., Hyams, K.C., Purdy, M.A. *et al.* (1992) Human linear B-cell epitopes encoded by the hepatitis E virus include determinants in the RNA-dependent RNA polymerase. *Proc. Natl Acad. Sci. USA*, **89**, 3855–3858.

Khudyakov, Y., Favorov, M.O., Jue, D.L., Hine, T.K. and Fields, H.A. (1994a) Immunodominant antigenic regions in a structural protein of the hepatitis E virus. *Virology*, **198**, 390–393.

Khudyakov, Y., Khudyakova, N.S., Jue, D.L., Wells, T.W., Padhya, N. and Fields, H.A. (1994b) Comparative characterization of antigenic epitopes in the immunodominant region of the protein encoded by open reading frame 3 in Burmese and Mexican strains of hepatitis E virus. *J. Gen. Virol.*, **75**, 641–646.

Khuroo, M.S., Teli, M.R., Skidmore, S., Sofi, M.A. and Khuroo, M. (1981) Incidence and severity of viral hepatitis in pregnancy. *Am. J. Med.*, **70**, 252–255.

Kidd-Ljunggren, K., Ekdahl, K., Oberg, M., Kurratong, S. and Lolekha, S. (1995) Hepatitis B strains in Thailand: genomic variants in chronic carriers. *J. Med. Virol*, **47**, 454–461.

Kidd-Ljunggren, K., Oberg, M. and Kidd, A.H. (1997) Hepatitis B virus X gene 1751 to 1764 mutations: implications for HBeAg status and disease. *J. Gen. Virol.*, **78**, 1469–1478.

Kirkwood, T.B. and Bangham, C.R. (1994) Cycles, chaos, and evolution in virus cultures: a model of defective interfering particles. *Proc. Natl Acad. Sci. USA*, **91**, 8685–8689.

Koizumi, K., Enomoto, N., Kurosaki, M. *et al.* (1995) Diversity of quasispecies in various disease stages of chronic hepatitis C virus infection and its significance in interferon treatment. *Hepatology*, **22**, 30–35.

Kojima, M., Osuga, T., Tsuda, F., Tanaka, T. and Okamoto, H. (1994) Influence of antibodies to the hypervariable region of E2/NS1 glycoprotein on the selective replication of hepatitis C virus in chimpanzees. *Virology*, **204**, 665–721.

Kolykhalov, A.A., Agapov, E.V., Blight, K.J., Mihalik, K., Feinstone, S.M. and Rice, C.M. (1997) Transmission of hepatitis C by intrahepatic inoculation with transcribed RNA. *Science*, **277**, 570–574.

Kos, A., Dijkema, R., Arnberg, A.C., van der Meide, P.H. and Schellekens, H. (1986) The hepatitis delta (delta) virus possesses a circular RNA. *Nature*, **323**, 558–560.

Kumar, U., Monjardino, J. and Thomas, H.C. (1994) Hypervariable region of hepatitis C virus envelope glycoprotein (E2/NS1) in an agammaglobulinemic patient. *Gastroenterology*, **106**, 1072–1075.

Kuo, M.Y., Chao, M. and Taylor, J. (1989) Initiation of replication of the human hepatitis delta virus genome from cloned DNA: role of delta antigen. *J. Virol.*, **63**, 1945–1950.

Kurosaki, M., Enomoto, N., Marumo, F. and Sato, C. (1994) Evolution and selection of hepatitis C virus variants in patients with chronic hepatitis C. *Virology*, **205**, 161–169.

Kurosaki, M., Enomoto, N., Marumo, F. and Sato, C. (1995) Variations in the core region of hepatitis C virus genomes in patients with chronic hepatitis. *Arch. Virol.*, **140**, 1087–1094.

Kurosaki, M., Enomoto, N., Asahina, Y. *et al.* (1996) Mutations in the core promoter region of hepatitis B virus in patients with chronic hepatitis B. *J. Med. Virol.*, **49**, 115–123.

Kwo, P.Y., Schlauder, G.G., Carpenter, H.A. *et al.* (1997) Acute hepatitis E by a new isolate acquired in the United States. *Mayo Clin. Proc.*, **72**, 1133–1136.

Lampertico, P., Manzin, A., Rumi, M. G. *et al.* (1995) Hepatitis B virus precore mutants in

HBeAg carriers with chronic hepatitis treated with interferon. *J. Viral Hepat.*, **2**, 251–256.

Landford, R.E., Chavez, D., Chisari, F. and Sureau, C. (1995) Lack of detection of negative-extrahepatic tissues by the highly strand-specific rTth reverse transcriptase PCR. *J. Virol.*, **69**, 8079–8083.

Laskus, T., Rakela, J., Tong, M.J., Nowicki, M.J., Mosley, J.W. and Persing, D.H. (1994) Naturally occurring hepatitis B virus mutants with deletions in the core promoter region. *J. Hepatol.*, **20**, 837–841.

Lawal, Z., Petrik, J., Wong, V.S., Alexander, G.J. and Allain, J.P. (1997) Hepatitis C virus genomic variability in untreated and immunosuppressed patients. *Virology*, **228**, 107–111.

Lechner, S., Rispeter, K., Meisel, H. *et al.* (1998) Antibodies directed to envelope proteins of hepatitis C virus outside of hypervariable region 1. *Virology*, **243**, 313–321.

Lee, C.M., Bih, F.Y., Chao, Y.C., Govindarajan, S. and Lai, M.M. (1992) Evolution of hepatitis delta virus RNA during chronic infection. *Virology*, **188**, 265–273.

Lee, C.M., Changchien, C.S., Chung, J.C. and Liaw, Y.F. (1996) Characterization of a new genotype II hepatitis delta virus from Taiwan. *J. Med. Virol.*, **49**, 145–154.

Le Guen, B., Squadrito, G., Nalpas, B., Berthelot, P., Pol, S. and Brechot, C. (1997) Hepatitis C virus genome complexity correlates with response to interferon therapy: a study in French patients with chronic hepatitis C. *Hepatology*, **25**, 1250–1254.

Leino, T., Leinkki, P., Hyypia, T. *et al.* (1997) Hepatitis A outbreak amongst intravenous amphetamine abusers in Finland. *Scand. J. Infect. Dis.*, **29**, 213–216.

Lemon, S.M. and Binn, L.N. (1983) Antigenic relatedness of two strains of hepatitis A virus determined by cross-neutralization. *Infect. Immun.*, **42**, 418–420.

Li, M., Xie, Y., Wu, X., Kong, Y. and Wang, Y. (1995) HNF3 binds and activates the second enhancer, ENII, of hepatitis B virus. *Virology*, **214**, 371–378.

Lindh, M., Horal, P., Dhillon, A.P., Furuta, Y. and Norkrans, G. (1996) Hepatitis B virus carriers without precore mutations in hepatitis B e antigen-negative stage show more severe liver damage. *Hepatology*, **24**, 494–501.

Lok, A.S.F., Akarca, U. and Greene, S. (1994) Mutations in the pre-core region of the hepatitis B virus serve to enhance the stability of the secondary structure of the pre-genome encapsidation signal. *Proc. Natl Acad. Sci. USA*, **91**, 4077–4081.

Luo, G.X., Chao, M., Hsieh, S.Y., Sureau, C., Nishikura, K. and Taylor, J. (1990) A specific base transition occurs on replicating hepatitis delta virus RNA. *J. Virol.*, **64**, 1021–1027.

MacNaughton, T.B., Gowans, E.J., McNamara, S.P. and Burrell, C.J. (1991) Hepatitis delta antigen is necessary for access of hepatitis delta virus RNA to the cell transcriptional machinery but is not part of the transcriptional complex. *Virology*, **184**, 387–390.

Maggi, F., Fornai, C., Vatteroni, M.L. *et al.* (1997) Differences in hepatitis C virus quasispecies composition between liver, peripheral blood mononuclear cells and plasma *J. Gen. Virol.*, **78**, 1521–1525.

Major, M.E. and Feinstone, S.M. (1997) The molecular virology of hepatitis C. *Hepatology*, **25**, 1527–1538.

Makino, S., Chang, M.F., Shieh, C.K. *et al.* (1987) Molecular cloning and sequencing of a human hepatitis delta (delta) virus RNA. *Nature*, **329**, 343–346.

Manzin, A., Solforosi, L., Petrelli, E. *et al.* (1998) Evolution of hypervariable region 1 of hepatitis C virus in primary infection. *J. Virol.*, **72**, 6271–6276.

Martell, M., Esteban, J.I., Quer, J. *et al.* (1992) Hepatitis C virus (HCV) circulates as a population of different but closely related genomes: quasispecies nature of HCV genome distribution. *J. Virol.*, **66**, 3225–3229.

Martell, M., Esteban, J.I., Quer, J. *et al.* (1994) Dynamic behavior of hepatitis C virus quasispecies in patients undergoing orthotopic liver transplantation. *J. Virol.*, **68**, 3425–3436.

Martin, P., Di Bisceglie, A.M., Kassianides, C., Lisker-Melman, M. and Hoofnagle, J.H. (1989) Rapidly progressive non-A, non-B hepatitis in patients with human immunodeficiency virus infection. *Gastroenterology*, **97**, 1559–1561.

Martín Hernández, A.M., Carrillo, E.C., Sevilla,

N. and Domingo, E. (1994) Rapid cell variation can determine the establishment of a persistent viral infection. *Proc. Natl Acad. Sci. USA*, **91**, 3705–3709.

Meyerhans, A., Cheynier, R., Albert, J. *et al.* (1989) Temporal fluctuations in HIV quasispecies *in vivo* are not reflected by sequential HIV isolations. *Cell*, **58**, 901–910.

Miller, R.H. and Purcell, R.H. (1990) Hepatitis C virus shares amino acid sequence similarity with pestiviruses and flaviviruses as well as members of two plant virus supergroups. *Proc. Natl Acad. Sci. USA*, **87**, 2057–2061.

Miyakawa, Y., Okamoto, H. and Mayumi, M. (1997) The molecular basis of hepatitis B e antigen (HBeAg)-negative infections. *J. Viral Hepat.*, **4**, 1–8.

Mphahlele, M.J., Shattock, A.G., Boner, W., Quinn, J., McCormick, P.A. and Carman, W.F. (1997) Transmission of a homogeneous HBV population of A_{1896} containing strains leading to mild resolving acute hepatitis and seroconversion to anti-HBe in an adult. *Hepatology*, **26**, 743–746.

Nainan, O.V., Brinton, M.A. and Margolis, H.S. (1992) Indentification of amino acids located in the antibody binding sites of human hepatitis A virus. *Virology*, **191**, 984–987.

Nainan, O.V., Stevens, C.E., Taylor, R.E. and Margolis, H.S. (1997) Hepatitis B virus (HBV) antibody resistant mutants among mothers and infants with chronic HBV infection. In: *Viral Hepatitis and Liver Disease* (eds Rizzetto, M., Purcell, R.H., Gerin, J.L. and Verme, G.), pp. 132–134. Edizioni Minerva Medica, Turin.

Naito, M., Hayashi, N., Moribe, T. *et al.* (1995) Hepatitis C viral quasispecies in hepatitis C virus carriers with normal liver enzymes and patients with type C chronic liver disease. *Hepatology*, **22**, 407–412.

Nassal, M. and Schaller, H. (1996) Hepatitis B virus replication – an update. *J. Viral Hepat.*, **3**, 217–226.

Negro, F., Baldi, M., Bonino, F. *et al.* (1987) Chronic HDV (hepatitis delta virus) hepatitis. Intrahepatic expression of delta antigen, histologic activity and outcome of liver disease. *Hepatology*, **6**, 8–14.

Negro, F., Krawczynski, R., Quadri, R. *et al.*

(1998) Detection of intrahepatic (–) HCV RNA by strand specific, semi-quantitative RT-PCR in chronic hepatitis C. International Meeting on hepatitis C virus and related viruses, Venice, p. 168A (Abstract)

Nelson, D.R., Marousis, C.G., Davis, G.L. and Lau, J.Y.N. (1995) Defining the role of intrahepatic HCV-specific cytotoxic T-lymphocytes in chronic hepatitis C. *Hepatology*, **22** (Suppl.), 287A.

Ni, Y.H., Chang, M.H., Chen, P.J., Lin, H.H. and Hsu, H.Y. (1997) Evolution of hepatitis C virus quasispecies in mothers and infants infected through mother-to-infant transmission. *J. Hepatol.*, **26**, 967–974.

Ngui, S.L., O'Connell, S., Eglin, R.P., Heptonstall, J. and Teo, C.G. (1997) Low detection rate and maternal provenance of hepatitis B virus S gene mutants in cases of failed postnatal immunoprophylaxis in England and Wales. *J. Infect.Dis.*, **176**, 1360–1365.

NIH: anonymous (1997) National Institutes of Health consensus development conference panel statement: management of hepatitis C. *Hepatology*, **26** (Suppl. I), 2S–10S.

Niro, G.A., Smedile, A., Andriulli, A., Rizzetto, M., Gerin, J.L. and Casey, J.L. (1997) The predominance of hepatitis delta virus genotype I among chronically infected Italian patients. *Hepatology*, **25**, 728–734.

Nishizono, A., Hiraga, M., Kohno, K. *et al.* (1995) Mutations in the core promoter/enhancer II regions of naturally occurring hepatitis B virus variants and analysis of the effects on transcription activitities. *Intervirology*, **38**, 290–294.

Norder, H., Couroucé, A-M. and Magnius, L.O. (1992a) Molecular basis of hepatitis B virus serotype variations within the four major subtypes. *J. Gen. Virol.*, **73**, 3141–3145.

Norder, H., Hammas, B., Löfdahl, S., Couroucé A.M. and Magnius, L.O. (1992b) Comparison of the amino acid sequences of nine different serotypes of hepatitis B surface antigen and genomic classification of the corresponding hepatitis B virus strains. *J. Gen. Virol.*, **73**, 1201–1208.

Normann, A., Pfisterer-Hunt, M., Schade, S. *et al.* (1995) Molecular epidemiology of an outbreak of hepatitis A in Italy. *J. Med. Virol.*, **47**, 467–471.

Nousbaum, J.B., Pol, S., Nalpas, B., Landais, P., Berthelot, P. and Brechot, C. (1995) Hepatitis C virus type 1b (II) infection in France and Italy. Collaborative Study Group. *Ann. Intern. Med.*, **122**, 161–168.

Ogata, N., Alter, H.J., Miller, R.H. and Purcell, R.H. (1991) Nucleotide sequence and mutation rate of the H strain of hepatitis C virus. *Proc. Natl Acad. Sci. USA*, **88**, 3392–3396.

Ogata, N., Miller, R.H., Ishak, K.G. and Purcell. R.H. (1993) The complete nucleotide sequence of a pre-core mutant of hepatitis B virus implicated in fulminant hepatitis and its biological characterization in chimpanzees. *Virology*, **194**, 263–276.

Ogata, N., Zanetti, A.R., Yu, M., Miller, R.H. and Purcell, R.H. (1997) Infectivity and pathogenicity in chimpanzees of a surface gene mutant of hepatitis B virus that emerged in a vaccinated infant. *J. Infect. Dis.*, **175**, 511–523.

Ohto, H., Terazawa, S., Sasaki, N. *et al.* (1994) Transmission of hepatitis C virus from mothers to infants. The Vertical Transmission of Hepatitis C Virus Collaborative Study Group (see comments). *N. Engl. J. Med.*, **330**, 744–750.

Okada, S., Akahane, Y., Suzuki, H., Okamoto, H. and Mishiro, S. (1992) The degree of variability in the amino terminal region of the E2/NS1 protein of hepatitis C virus correlates with responsiveness to interferon therapy in viremic patients. *Hepatology*, **16**, 619–624.

Okamoto, H., Yano, K., Nozaki, Y. and Mishiro, S. (1992a) Mutations within the s gene of hepatitis-B virus transmitted from mothers to babies immunized with hepatitis-B immune globulin and vaccine. *Paediatr. Res.*, **32**, 264–268.

Okamoto, H., Kojima, M., Okada, S. *et al.* (1992b) Genetic drift of hepatitis C virus during an 8.2-year infection in a chimpanzee: variability and stability. *Virology*, **190**, 894–899.

Okamoto, H., Tsuda, F., Akahane, Y. *et al.* (1995) Molecular epidemiology of hepatitis B virus vaccine variants in Singapore. *Vaccine*, **13**, 699–702.

Okamoto, H., Tsuda, F., Sakugawa, H. *et al.* (1988) Typing hepatitis B virus by homology in nucleotide sequence: comparison of surface antigen subtypes. *J. Gen. Virol.*, **69**, 2575–2583.

Oon, C.-J., Lim, G.-K., Ye, Z. *et al.* (1995) Molecular epidemiology of hepatitis B virus vaccine variants in Singapore. *Vaccine*, **13**, 699–702.

Oren, I., Hershow, R.C., Ben-Porath, E. *et al.* (1989) A common-source outbreak of fulminant hepatitis B in a hospital. *Ann. Intern. Med.*, **110**, 691–698.

Ottobrelli, A., Marzano, A., Smedile, A. *et al.* (1991) Patterns of hepatitis delta virus reinfection and disease in liver transplantation. *Gastroenterology*, **101**, 1649–1655.

Panda, S.K. and Jameel, S. (1997) Hepatitis E virus: from epidemiology to molecular biology. *Viral Hepat.*, **3**, 227–251.

Panda, S.K., Nanda, S.K., Zafrullah, M., Ansari, I.H., Ozdener, M.H. and Jameel, S. (1995) An Indian strain of hepatitis E virus (HEV): cloning, sequence, and expression of structural region and antibody response in sera from individuals from an area of high-level HEV endemicity. *J. Clin. Microbiol.*, **33**, 2653–2659.

Pawlotsky, J.M., Germanidis, G., Neumann, A.U., Pellerin, M., Frainais, P.O. and Dhumeaux, D. (1998) Interferon resistance of hepatitis C virus genotype 1b: relationship to nonstructural 5A gene quasispecies mutations. *J. Virol.*, **72**, 2795–2805.

Ping, L.-H. and Lemon, S.M. (1992) Antigenic structure of human hepatitis A virus defined by analysis of escape mutants selected against murine monoclonal antibodies. *J. Virol.*, **66**, 2208–2216.

Pol, S., Trinh, T.N., Thiers, V. *et al.* (1995) Chronic hepatitis C of drug users: influence of HIV infection. *Hepatology*, **22** (Suppl), 340A.

Pollack J.R. and Ganem, D. (1993) An RNA stem-loop structure directs hepatitis B virus genomic RNA encapsidation. *J. Virol.*, **67**, 3254–3263.

Polyak, S.J., Faulkner, G., Carithers, R.L. Jr, Corey, L. and Gretch, D.R. (1997) Assessment of hepatitis C virus quasispecies heterogeneity by gel shift analysis: correlation with response to interferon therapy. *J. Infect. Dis.*, **175**, 1101–1107.

Polyak, S.J., McArdale, S., Liu, S.L. *et al.* (1998) Evolution of hepatitis C virus quasispecies in hypervariable region 1 and putative interferon sensitivity-determining region during interferon therapy and natural infection. *J. Virol.*, **72**, 4288–4296.

Protzer-Knolle, U., Naumann, U., Bartenschlager, R. *et al.* (1998) Hepatitis B virus with antigenically altered hepatitis B surface antigen is selected by high-dose hepatitis B immune globulin after liver transplantation. *Hepatology*, **27**, 254–263.

Provost, P.J. and Hilleman, M.R. (1979) Propagation of human hepatitis A virus in cell culture in vitro. *Proc. Soc. Exp. Biol. Med.*, **160**, 213–221.

Puntoriero, G., Meola, A., Lahm, A. *et al.* (1998) Towards a solution for hepatitis C virus hypervariability: mimotopes of hypervariable region 1 can induce antibodies cross-reacting with a large number of viral variants. *EMBO. J.*, **17**, 3521–3533.

Radziwill, G., Tucker, W. and Schaller, H. (1990) Mutational analysis of the hepatitis B virus P gene product: domain structure and RNase H activity. *J. Virol.*, **64**, 613–620.

Raimondo, G., Tanzi, E., Brancatelli, S. *et al.* (1993) Is the course of perinatal hepatitis B virus infection influenced by genetic heterogeneity of the virus? *J. Med. Virol.*, **40**, 87–90.

Raney, A.K., Johnson, J.L., Palmer, C.N. and McLachlan, A. (1997) Members of the nuclear receptor superfamily regulate transcription from the hepatitis B virus nucleocapsid promoter. *J. Virol.*, **71**, 1058–1071.

Rehermann, B., Chang, K.M., McHutchison, J.G. and Chisari, F.V. (1995) Comparative analysis of the cytotoxic T cell response in chronic HBV and HCV infection. *Hepatology*, **22** (Suppl.), 334A.

Repp, R., Keller, C., Borkhardt, A. *et al.* (1992) Detection of a hepatitis B virus variant with a truncated X gene and enhancer II. *Arch. Virol.*, **125**, 299–304.

Reyes, G.R., Purdy, M.A., Kim, J.P. *et al.* (1990) Isolation of a cDNA from the virus responsible for enterically transmitted non-A, non-B hepatitis. *Science*, **247**, 1335–1339.

Rizzetto, M. and Durazzo, M. (1991) Hepatitis delta virus (HDV) infections. Epidemiological and clinical heterogeneity. *J. Hepatol.*, **13** (Suppl. 4), S116–S118.

Rizzetto, M., Hoyer, B., Canese, M.G., Shih, J., Purcell, R.H. and Gerin, J.L. (1980a) Delta agent: association of delta antigen with hepatitis B surface antigen and RNA in serum of delta-infected chimpazees. *Proc. Natl Acad. Sci. USA*, **77**, 6124–6128.

Rizzetto, M., Purcell, R.H. and Gerin, J.L. (1980b) Epidemiology of HBV-associated delta agent: geographical distribution of anti-delta and prevalence in polytransfused HBsAg carrier. *Lancet*, **i**, 1215–1218.

Rizzetto, M., Ponzetto, A. and Forzani, I. (1991) Epidemiology of hepatitis delta virus: overview. *Prog. Clin. Biol. Res.*, **364**, 1–20.

Rizzetto, M., Hadziyannis, S., Hansson, B.G., Toukan, T. and Gust, I. (1992) Hepatitis delta virus infection in the world: epidemiological patterns and clinical expression. *Gastroenterol. Int.*, **5**, 18–32.

Robertson, H.D. (1996) How did replicating and coding RNAs first get together? *Science*, **274**, 66–67.

Robertson, B.H., Jansen, R.W., Khanna, B. *et al.* (1992) Genetic relatedness of hepatitis A virus strains recovered from different geographical regions. *J. Gen Virol.*, **73**, 1365–1377.

Rosa, D., Campagnoli, S., Moretto, C. *et al.* (1996) A quantitative test to estimate neutralizing antibodies to the hepatitis C virus: cytofluorimetric assessment of envelope glycoprotein 2 binding to target cells. *Proc. Natl Acad. Sci. USA*, **93**, 1759–1763.

Rossini, A., Ravaggi, A., Biasi, L. *et al.* (1997) Virological response to interferon treatment in hepatitis C virus carriers with normal aminotransferase levels and chronic hepatitis. *Hepatology*, **26**, 1012–1017.

Rueckert, R.R. and Wimmer, E. (1984) Systematic nomenclature of picornavirus proteins. *J. Virol.*, **50**, 957–959.

Sáiz, J.C., López-Labrador, F.X., Ampurdanés, S. *et al.* (1998) The prognostic relevance of the nonstructural 5A gene interferon sensitivity determining region is different in infections with genotype 1b and 3a isolates of hepatitis C virus. *J. Infect. Dis.*, **177**, 839–847.

Sakuma, I., Enomoto, N., Kurosaki, M.,

Marumo, F. and Sato, C. (1996) Selection of hepatitis C virus quasispecies during interferon treatment. *Arch. Virol.*, **141**, 1921–1932.

Saldanha, J.A., Thomas, H.C. and Monjardino, J.P. (1990) Cloning and sequencing of RNA of hepatitis delta virus isolated from human serum. *J. Gen. Virol.*, **71**, 1603–1606.

Santantonio, T., Jung, M.C., Schneider, R. *et al.* (1992) Hepatitis-B virus genomes that cannot synthesise pre-S2 proteins occur frequently and as dominant virus populations in chronic carriers in Italy. *Virology*, **188**, 948–952.

Sato, S., Suzuki, K., Akahane, Y. *et al.* (1995) Hepatitis B virus strains with mutations in the core promoter in patients with fulminant hepatitis. *Ann. Intern. Med.*, **122**, 241–248.

Scarcelli, E., Cerino, A., Esposito, C., Silini, E., Mondelli, M.U. and Traboni, C. (1995) Occurrence of antibodies reactive with more than one variant of the putative envelope glycoprotein (Gp70) hypervariable region 1 in viremic hepatitis C virus-infected patients. *J. Virol.*, **69**, 4407–4412.

Schultz, D.E., Honda, M., Whetter, L.E., McKnight, K.L. and Lemon, S.M. (1996) Mutations within the 5′ nontranslated RNA of cell culture-adapted hepatitis A virus which enhance cap-independent translation in cultured African green monkey kidney cells. *J. Virol.*, **70**, 1041–1049.

Shaffer, D.R., Emerson, S.U., Murphy, P.C., Govindarajan, S. and Lemon, S.M. (1995) A hepatitis A virus deletion mutant which lacks the first pyrimidine-rich tract of the 5′ nontranslated RNA remains virulent in primates after direct intrahepatic nucleic acid transfection. *J. Virol.*, **69**, 6600–6604.

Shakil, A.O., Hadziyannis, S., Hoofnagle, J.H., Di Bisceglie, A.M., Gerin, J.L. and Casey, J.L. (1997) Geographic distribution and genetic variability of hepatitis delta virus genotype I. *Virology*, **234**, 160–167.

Sharmeen, L., Kuo, M.Y., Dinter-Gottlieb, G. and Taylor, J. (1988) Antigenomic RNA of human hepatitis delta virus can undergo self-cleavage. *J. Virol.*, **62**, 2674–2679.

Sharmeen, L., Kuo, M.Y. and Taylor, J. (1989) Self-ligating RNA sequences on the antigenome of human hepatitis delta virus. *J. Virol.*, **63**, 1428–1430.

Shiina, S., Fujino, H., Uta, Y. *et al.* (1991) Relationship of HBsAg subtypes with HBeAg anti-HBe status and chronic liver disease. Part I: Analysis of 1744 HBsAg carriers. *Am. J. Gastroenterol.*, **86**, 866–871.

Shimizu, Y.K. and Yoshikura, H. (1995) In vitro systems for the detection of hepatitis C virus infection. *Viral Hepat. Rev.*, **1**, 59–65.

Shimizu, Y.K., Igarashi, H., Kanematu, T. *et al.* (1997) Sequence analysis of the hepatitis C virus genome recovered from serum, liver, and peripheral blood mononuclear cells of infected chimpanzees. *J. Virol.*, **71**, 5769–5773.

Shindo, M., Hamada, K., Koya, S., Arai, K., Sokawa, Y. and Okuno, T. (1996) The clinical significance of changes in genetic heterogeneity of the hypervariable region 1 in chronic hepatitis C with interferon therapy. *Hepatology*, **24**, 1018–1023.

Shirai, M., Okada, H., Nishioka, M. *et al.* (1994) An epitope in hepatitis C virus core region recognized by cytotoxic T cells in mice and humans. *J. Virol.*, **68**, 3334–3342.

Siegl, G., Nüesch, J.P.F. and Chastonay, J. (1990) Defective interfering particles of hepatitis A virus in cell cultures and clinical specimens. In: *New Aspects of Positive-strand RNA Viruses* (eds Brinton, M.A. and Heinz, F.X.), pp. 102–107. American Society for Microbiology, Washington, DC.

Simmonds, P. (1995) Variability of hepatitis C virus. *Hepatology*, **21**, 570–583.

Smith, D.B., Mellor, J., Harvis, L.M. *et al.* (1995) Variation of the hepatitis C virus 5′ non-coding region: implications for secondary structure, virus detection and typing. *J. Gen. Virol.*, **76**, 1749–1761.

Stapleton, J.T. and Lemon, S.M. (1987) Neutralization escape mutants define a dominant immunogenic neutralization site on hepatitis A virus. *J. Virol.*, **61**, 491–498.

Surya, L.G.P., Kishimoto, S., Sudaryat, S *et al.* (1996) Prevention of mother-to-infant transmission of hepatitis B virus with use of a pre-S2-containing vaccine in Bali, Indonesia. *Vacc. Res.* **5**.

Suzuki, K., Mizokami, M., Lau, J.Y. *et al.* (1994) Confirmation of hepatitis C virus transmission through needlestick accidents by molecular evolutionary analysis. *J. Infect. Dis.*, **170**, 1575–1578.

Tam, A.W., Smith, M.M., Guerra, M.E. *et al.* (1991) Hepatitis E virus (HEV): molecular cloning and sequencing of the full-length viral genome. *Virology*, **185**, 120–131.

Tanaka, S., Yoshiba, M., Iino, S. *et al.* (1995) A common-source outbreak of fulminant hepatitis B in hemodialysis patients induced by pre core mutant. *Kidney Int.*, **48**, 1972–1978.

Taylor, M.B. (1997) Molecular epidemiology of South African strains of hepatitis A virus: 1982–1996. *J. Med. Virol.*, **51**, 273–279.

Thomas H.C., Foster, G.R., Sumiya M. *et al.* (1996) Mutation of gene for mannose-binding protein associated with chronic hepatitis B viral infection. *Lancet*, **348**, 1417–1419.

Thursz, M., Kwiatkowski, D., Allsopp, C.E.M., Greenwood, B.M., Thomas, H.C. and Hill A.V..S. (1995) Association between an MHC class II allele and clearance of hepatitis B virus in the Gambia. *New. Engl. J. Med.*, **332**, 1065–1069.

Ticehurst, J.R., Cohen, J.I., Feinstone, S.M., Purcell, R.H., Jansen, R.W. and Lemon, S. M. (1989) In: *Molecular Aspects of Picornaviral Infection and Detection* (eds Semler, B.L. and Ehrenfeld, E.), pp. 27–50. American Society for Microbiology, Washington, DC.

Tokita, H., Okamoto, H., Tsuda, F. *et al.* (1994) Hepatitis C virus variants from Vietnam are classifiable into the seventh, eighth, and ninth major genetic groups. *Proc. Natl Acad. Sci. USA.*, **91**, 11022–11026.

Tokita, H., Okamoto, H., Luengrojanakul, P. *et al.* (1995) Hepatitis C virus variants from Thailand classifiable into five novel genotypes in the sixth (6b), seventh (7c, 7d) and ninth (9b, 9c) major genetic groups. *J. Gen. Virol.*, **76**, 2329–2335.

Tong, S.P., Li, J.S., Vitvitski, L. and Trèpo, C. (1992) Replication capacities of natural and artificial precore stop codon mutants of hepatitis B virus: relevance of pregenome encapsidation signal. *Virology*, **191**, 237–245.

Toyoda, H., Fukuda, Y., Koyama, Y., Takamatsu, J., Saito, H. and Hayakawa, T. (1997a) Effect of immunosuppression on composition of quasispecies population of hepatitis C virus in patients with chronic hepatitis C coinfected with human immunodeficiency virus. *J. Hepatol.*, **26**, 975–982.

Toyoda, H., Kumada, T., Nakano, S. *et al.* (1997b) Quasispecies nature of hepatitis C virus and response to alpha interferon: significance as a predictor of direct response to interferon. *J. Hepatol.*, **26**, 6–13.

Trautwein, C., Schrem, H., Tillmann, H.L. *et al.* (1996) Hepatitis B virus mutations in the preS genome before and after liver transplantation. *Hepatology*, **24**, 482–488.

Tsarev, S.A., Emerson, S.U., Reyes, G.R. *et al.* (1992) Characterization of a prototype strain of hepatitis E virus. *Proc. Natl Acad. Sci. USA*, **89**, 559–563.

Van Doorn, L.J., Capriles, I., Maertens, G. *et al.* (1995) Sequence evolution of the hypervariable region in the putative envelope region E2/NS1 of hepatitis C virus is correlated with specific humoral immune responses. *J. Virol.*, **69**, 773–778.

Wang, K.S., Choo, Q.L., Weiner, A.J. *et al.* (1986) Structure, sequence and expression of the hepatitis delta (delta) viral genome. *Nature*, **323**, 456.

Wang, K.S., Choo, Q.L., Weiner, A.J. *et al.* (1992) Structure, sequence and expression of the hepatitis delta (delta) viral genome genetic drift of hepatitis C virus during an 8.2-year infection in a chimpanzee: variability and stability. *Virology*, **190**, 894–899.

Weiner, A.J., Choo, Q.L., Wang, K.S. *et al.* (1988) A single antigenomic open reading frame of the hepatitis delta virus encodes the epitope(s) of both hepatitis delta antigen polypeptides p24 delta and p27 delta. *J. Virol.*, **62**, 594–599.

Weiner, A.J., Thaler, M.M., Crawford, K. *et al.* (1993) A unique, predominant hepatitis C virus variant found in an infant born to a mother with multiple variants. *J. Virol.*, **67**, 4365–4368.

Weiner, A., Erickson, A.L., Kansopon, J. *et al.* (1995) Persistent hepatitis C virus infection in a chimpanzee is associated with emergence of a cytotoxic T lymphocyte escape variant. *Proc. Natl Acad. Sci. USA*, **92**, 2755–2759.

Weitz, M., Baroudy, B.M., Maloy, W.L., Ticehurst, J.R. and Purcell, R.H. (1986) Detection of a genome-linked protein (VPg) of hepatitis A virus and its comparison with other picornaviral VPgs. *J. Virol.*, **60**, 124–130.

Wilson, J.N., Nokes, D.J. and Carman, W.F. (1999) Mathematical modelling of vaccine escape. *Vaccine*, in press.

Wu, H.N. and Lai, M.M. (1989) Reversible cleavage and ligation of hepatitis delta virus RNA. *Science*, **243**, 652–654.

Wu, H.N., Lin, Y.J., Lin, F.P., Makino, S., Chang, M.F. and Lai, M.M. (1989) Human hepatitis delta virus RNA subfragments contain an autocleavage activity. *Proc. Natl Acad. Sci. USA*, **86**, 1831–1835.

Wu, J.C., Choo, K.B., Chen, C.M., Chen, T.Z., Huo, T.I. and Lee, S.D. (1995) Genotyping of hepatitis D virus by restriction-fragment length polymorphism and relation to outcome of hepatitis D. *Lancet*, **346**, 939–941.

Wyatt, C.A., Andrus, L., Brotman, B., Huang, F., Lee, D.H. and Prince, A.M. (1998) Immunity in chimpanzees chronically infected with hepatitis C virus: role of minor quasispecies in reinfection. *J. Virol.*, **72**, 1725–1730.

Yamaguchi, K., Tanaka, E., Higashi, K. *et al.* (1994) Adaptation of hepatitis C virus for persistent infection in patients with acute hepatitis. *Gastroenterology*, **106**, 1344–1348.

Yamamoto, K., Horita, M., Tsuda, F. *et al.* (1994) Naturally occurring escape mutants of hepatitis B virus with various mutations in the S gene in carriers seropositive for antibody to hepatitis B surface antigen. *J. Virol.*, **68**, 2671–2676.

Yanagi, M., St. Claire, M., Shapiro, M., Emerson, S.U., Purcell, R.H. and Bukh, J. (1998a) A chimeric cDNA clone of hepatitis C virus (HCV) that encodes the complete open reading frame of a genotype 1b strain (HC-J4) is infectious *in vivo*. International Meeting on hepatitis C virus and related viruses, Venice, p. 025 (Abstract)

Yanagi, M., St. Claire, M., Shapiro, M., Emerson, S.U., Purcell, R.H. and Bukh, J. (1998b) The terminal structure of the 3′ untranslated region (UTR) of hepatitis C virus (HCV) is essential for infectivity *in vivo*. International

Meeting on hepatitis C virus and related viruses. Venice, p. 026 (Abstract)

Yasmin, M. (1997) Molecular biology of fulminant hepatitis B viruses. PhD thesis, University of Glasgow, Glasgow, UK.

Yee, J.-K. (1989) A liver-specific enhancer in the core promoter region of human hepatitis B virus. *Science*, **246**, 658–661.

Yin, S., Tsarev, S.A., Purcell, R.H. and Emerson, S.U. (1993) Partial sequence comparison of eight new Chinese strains of hepatitis E virus suggests the genome sequence is relatively stable. *J. Med. Virol.*, **41**, 230–241.

Yin, S., Purcell, R.H. and Emerson, S.U. (1994) A new Chinese isolate of hepatitis E virus: comparison with strains recovered from different geographical regions. *Virus Genes.*, **9**, 23–32.

Yoo, B.J., Selby, M.J., Choe, J. *et al.* (1995) Transfection of a differentiated human hepatoma cell line (Huh7) with in vitro-transcribed hepatitis C virus (HCV) RNA and establishment of a long-term culture persistently infected with HCV. *J. Virol.*, **69**, 32–38.

Yu, X. and Mertz, J.E. (1996) Promoters for synthesis of the pre-C and pregenomic mRNAs of human hepatitis B virus are genetically distinct and differentially regulated. *J. Virol.*, **70**, 8719–8726.

Yu, X. and Mertz, J. (1997) Differential regulation of the pre-C and pregenomic promoters of human hepatitis B virus by members of the nuclear receptor superfamily. *J. Virol.*, **71**, 9366–9374.

Yuki, N., Hayashi, N., Moribe, T. *et al.* (1997) Relation of disease activity during chronic hepatitis C infection to complexity of hypervariable region 1 quasispecies. *Hepatology*, **25**, 439–444.

Yun, Z.B., Odeberg, J., Lundeberg, J., Weiland, O., Uhlen, M. and Sonnerborg, A. (1996) Restriction of hepatitis C virus heterogeneity during prolonged interferon-alpha therapy in relation to changes in virus load. *J. Infect. Dis.*, **173**, 992–996.

Yun, Z., Barkholt, L. and Sonnerborg, A. (1997) Dynamic analysis of hepatitis C virus polymorphism in patients with orthotopic liver transplantation. *Transplantation*, **64**, 170–172.

Zhang, P., Raney, A.K. and McLachlan, A. (1993) Characterization of functional Sp1 transcription factor binding sites in the hepatitis B virus nucleocapsid promoter. *J. Virol.*, **67**, 1472–1481.

Zhang, H.-C., Ping, L.-H., Chao, S.F., Grace, K., Clarke, B. and Lemon, S.M. (1995) An infectious cDNA clone of cytopathic hepatitis A virus: genomic regions associated with rapid replication and cytopathic effect. *Virology*, **212**, 686–697.

Zhang, Y.Y., Tsega, E. and Hansson, B.G. (1996) Phylogenetic analysis of hepatitis D viruses indicating a new genotype I subgroup among African isolates. *J. Clin. Microbiol.*, **34**, 3023–3030.

Zoulim, F. and Trepo, C. (1998) Drug therapy for chronic hepatitis B: antiviral efficacy and influence of hepatitis B virus polymerase mutations on the outcome of therapy. *J. Hepatol.*, **29**, 151–168.

14

Antigenic Variation in Influenza Viruses

Robert G. Webster

INTRODUCTION

Influenza is the paradigm of a viral disease in which continued evolution of the virus is of paramount importance for annual epidemics and occasional pandemics of disease in humans. The gene pool of influenza A viruses in aquatic birds provides all the genetic diversity required for the emergence of pandemic influenza viruses for humans, lower animals and birds. In humans, pigs and horses, influenza A viruses show both antigenic drift and genetic shift. In contrast, there is evidence that avian influenza viruses in the aquatic bird reservoir are in evolutionary stasis. The purpose of this chapter is to establish that rapid evolution in influenza A viruses in humans and other mammals has continued since the beginning of recorded medical history and is dependent on periodic introductions of gene segments or entire influenza viruses into humans from the avian influenza virus gene pool. In aquatic wild birds, influenza virus appears to be fully adapted to its host and causes no disease signs.

Continuing evolution is most prominent in the surface glycoproteins of influenza viruses but also occurs in each of the eight gene segments of both type A and type B influenza viruses. This variability results from accumulation of molecular changes in the eight RNA segments, which can occur by a number of different mechanisms including: (1) point mutations (antigenic drift); (2) gene reassortment (genetic shift); (3) defective-interfering particles; and (4) RNA recombination. Each of these mechanisms may contribute to the evolution of influenza viruses.

Mutations, including substitutions, deletions and insertions, are one of the most important mechanisms for producing variation in influenza viruses. The lack of proofreading among RNA polymerases contributes to replication errors on the order of 1 in 10^4 bases (Holland *et al.*, 1982; Steinhauer and Holland, 1987). This contrasts with the much higher replication fidelity found among DNA polymerases, i.e. errors on the order of 1 in 10^9 bases per replication cycle. Each round of RNA virus replication results in a mixed population with many variants, most of which are not viable but some of which have potentially advantageous mutations that can become dominant under the right selective conditions. The surprise from recent studies on influenza viruses is not the extent of genetic variation but rather the apparent genetic stability of influenza viruses in aquatic avian reservoirs.

Since influenza viruses have segmented genomes, reassortment is an important mechanism for producing diversity very rapidly; it occurs among influenza A viruses in nature and is important in the appearance of pandemics in human populations. Reassortment has been demonstrated between influenza B virus strains and can occur between different co-circulating lineages in nature. There is no known influenza

B virus gene pool except in humans; consequently, genetic shift is not found in influenza B viruses.

Other mechanisms for producing genetic variation include defective-interfering particle-mediated interference and intermolecular recombination. Although defective-interfering particles can influence evolution by reducing the yields of non-defective particles and modifying pathogenicity (Steinhauer and Holland 1987), their role in influenza virus evolution has not received much attention. Although intermolecular recombination in negative-stranded viruses is rare, studies have demonstrated that it does occur (Khatchikian et al., 1989). This provides another mechanism for providing rapid evolutionary changes.

VARIATION: ANTIGENIC DRIFT AND GENETIC SHIFT

The influenza viruses are unique among the respiratory tract viruses in that they undergo significant antigenic variation. Both the surface antigens of the influenza A viruses undergo two types of antigenic variation: antigenic drift and genetic shift (Murphy and Webster, 1996). Antigenic drift involves minor antigenic changes in the HA and NA, whereas shift involves major antigenic changes in these molecules resulting from replacement of the gene segment.

Antigenic Drift

Antigenic drift occurs in both the HA and NA of influenza A and B viruses. The structure of both of these glycoproteins is established and the antigenic domains surround the conserved receptor binding site on the HA and the enzyme active center on the NA (Colman, 1989; Wiley et al., 1981). The detailed structure of one of these epitopes on the N9 NA was established by crystallization of the Fab fragment of monoclonal antibody (NC41) and comprises five peptide

loops located at the rim of the enzyme active site (Tulip et al., 1992).

Five antigenic domains (A–E) have been defined on the HA1 by comparative sequence analysis. Antigenic mapping with mouse monoclonal antibodies and sequence analysis of escape mutants confirm the number and location of the antigenic sites in H3 and H1 influenza viruses (Webster and Laver, 1980; Gerhard et al., 1981). Crystallographic studies of escape mutants have confirmed that antibodies bind to those regions of the molecule where amino acid substitutions permit escape from neutralization (Knossow et al., 1984).

After the appearance of a new subtype, antigenic differences between isolates can be detected using ferret antisera for analysis. Analysis with monoclonal antibodies indicates that major antigenic heterogeneity is detectable among different influenza virus isolates at any time (Stevens et al., 1987). Antigenic drift occurs by accumulation of a series of point mutations resulting in amino acid substitutions in antigenic sites A–E at the membrane distal region of the HA. These substitutions prevent binding of antibodies induced by the previous infection, and thus the virus can infect the host. In influenza A viruses of humans and other mammals, antibodies play a role in the selection of mutants, whereas in influenza A viruses in avian species, and in influenza B and C viruses, antibody selection does not appear to be involved in selection (Air et al., 1990).

Although antigenic drift can be mimicked in the laboratory by growth of the influenza viruses in the presence of monoclonal antibodies to a single antigenic site (e.g. to site A), it is not possible to select antigenic variants with mixtures of monoclonal antibodies to different antigenic sites (e.g. to sites A, B, C, etc.). The frequency of selection of variants by monoclonal antibodies to a single site *in vitro* is 10^{-5} (Yewdell et al., 1979); with a mixture of monoclonal antibodies to two different antigenic sites the frequency would be 10^{-10}. How then does selection of drift variants occur in humans who have polyclonal responses and a mixture of antibodies to each antigenic site? Analysis of postinfection human sera indicated a limited antibody repertoire (Wang et al., 1986). The restricted ability of human sera to neutralize monoclonal-antibody-selected escape mutants (Natali *et al.*, 1981) suggests that the selection of variants may be sequential, involving several different human hosts.

Cocirculation of Multiple Lineages

Sequence analyses of the HA of human H3N2 influenza A viruses have suggested that they evolve along a single lineage (Yamashita et al., 1988). However, separate sublineages of influenza A (H3N2) subtypes can cocirculate in humans for short periods. Two distinct groups of antigenic variants of the influenza A (H3N2) subtype cocirculated between 1989 and 1990. They were represented by the strains A/Beijing/353/89 and A/Shanghai/24/90, and related viruses cocirculated between 1990 and 1992 (Daly *et al.*, 1998).

Sequence analyses of the human H1N1 influenza viruses circulating in the human population reveal that separate lineages can cocirculate (Cox *et al.*, 1993). This causes added complications for the selection of human vaccine strains. Studies on avian, equine and swine influenza viruses indicate that, like human influenza A viruses, multiple lineages of HA can cocirculate. Multiple genetic variants of influenza B viruses have been shown to cocirculate during a single epidemic (Lu *et al.*, 1983), and two distinct lineages of influenza B viruses, represented by the strains B/Yamagata/16/88 and B/Victoria/2/87, have been maintained for at least 4 years in humans (Rota *et al.*, 1990).

Comparison of Evolutionary Patterns Among Influenza Viruses

Comparison of the sequence divergence among genes of type A, B and C viruses suggest that in humans influenza B and C viruses evolve more slowly than A viruses. Evolutionary rates calculated for influenza B are approximately twice as slow as for influenza A viruses, i.e. $1.1–1.96 \times 10^{-3}$ nucleotide substitutions per site per year for HA of influenza A H3N2 viruses (Cox *et al.*, 1993). For the HE gene of influenza C virus the rate is estimated to be 0.49×10^{-3} nucleotide substitutions per site per year, almost one-tenth of the value of the rates calculated for human influenza A HA genes (Muraki et al., 1996).

Among influenza A viruses there is extensive variation in the evolutionary rate and is lowest in avian influenza virus, highest in human influenza A viruses and intermediate in swine and equine influenza viruses. This is clearly seen in H3 influenza viruses: the H3 influenza viruses that appeared in 1968 have accumulated approximately 7.9 nucleotide and 3.4. amino substitutions per year; equine 2 (H3N8) influenza viruses that emerged earlier accumulated 3.1 nucleotides and 0.8 amino acids per year (Bean *et al.*, 1992), whereas avian influenza A viruses, in contrast, show even less variation in amino acids, with some viruses having not changed in 50 years. Consequently, avian influenza viruses remain antigenically conserved, and antigenic drift is slower in equine and swine influenza virus than in the human virus. Accumulation of mutations in the HA of avian influenza viruses is randomly distributed, indicating that immune selection does not play a role in antigenic drift (Shäfer *et al.*, 1993). Despite the slow rate of change, antigenic differences are found in influenza HA and NA subtypes of influenza viruses from avian species, and these probably reflect different phylogenetic lineages rather than recently accumulated changes in antigenic areas. Antigenic drift has been reported in swine (Kilbourne, 1978) and in equine influenza viruses (Mumford and Chambers, 1998), but the results can be confounded by comparisons of influenza viruses in different phylogenetic lineages.

The question of why influenza B and C viruses and influenza A viruses from different hosts have variable evolutionary rates is unresolved. The intrinsic error frequency of the RNA-replicating enzymes would be expected to be similar; however, the selective pressures imposed by the functional constraints of the viral proteins, by the immune status of the host and by adaptation in the host populations are all variables that influence the evolutionary rate.

There is controversy as to whether influenza evolution is a result of positive selection or random neutral change. Most protein sequences are more strongly conserved than nucleotide sequences. Random mutations among the 61 codons specifying amino acids will result in a coding change in 24% of the codons. In influenza A, about 50% of nucleotide changes in HA and NA sequences result in amino acid changes (Air *et al.*, 1990). In the HA1 of H3N2 viruses, there is strong support for positive darwinian selection (Fitch *et al.*, 1997). The most likely selection pressure is the immune system of the host. In influenza B viruses, there is less apparent direct selection to change the protein, with approximately 30% of nucleotide changes giving rise to amino acid substitutions in the surface glyoprotein molecules (Air *et al.*, 1990).

The very low rates of evolution of the HA and NA of avian influenza viruses in their aquatic bird reservoirs indicate that immune selection probably plays little role in selection; these viruses are optimally adapted to their host and randomly occurring mutations provide no selective advantage.

Variation in the Genes Encoding the Internal Proteins of Influenza Viruses

Each virus gene may evolve differently because of different selective pressure and evolutionary constraints. Genes that code for surface proteins (HA and NA) may be subject to strong selection pressure by neutralizing antibodies of host immune systems. Genes coding for internal proteins (e.g. NP) may not be subjected to strong host immune selection pressure but are thought to undergo significant host-specific adaptive evolution (Gammelin *et al.*, 1990; Gorman *et al.*, 1991). Internal protein genes such as those coding for polymerase (e.g. PB2) may have virus-specific functional constraints on evolution (Gorman *et al.*, 1990).

A comparison of coding and noncoding evolutionary rates for human virus genes shows that they are evolving at different rates. For example, the M1 and M2 genes of the M gene segment are evolving very differently from each other. M2, a minor influenza virus surface protein, is evolving much more rapidly than the internal M1 matrix protein, and a higher proportion of the nucleotide changes in the M2 gene are non-silent. M1 is evolving very slowly and shows almost no accumulation of coding

changes over a 55-year period (Ito *et al.*, 1991).

Homologous genes from host-specific virus strains also evolve at different rates. Evolutionary studies of influenza virus NP genes have shown that avian virus genes from aquatic birds are evolving more slowly than those in human viruses and that avian virus proteins are highly conserved, showing no net evolution over the past 60 years (Gorman *et al.*, 1991).

Genetic Shift

Serological and virological evidence suggests that since 1889 there have been six instances of the introduction of a virus bearing an HA subtype that had been absent from the human population for some time. For the HA there has been a cyclical appearance of the three human subtypes, with the sequential emergence of H2 viruses in 1889, H3 in 1900, H1 in 1918, H2 again in 1957, H3 again in 1968 and H1 again in 1977 (Figure 14.1).

Spanish Influenza-1918

Over the past two and half centuries, between 10 and 20 human influenza pandemics have swept the globe. The most devastating of these, the so-called Spanish flu of 1918, caused more than 20 million deaths and affected upward of 200 million people. The Spanish influenza pandemic has been extensively chronicled (Crosby *et al.*, 1976) and is described as "the greatest medical holocaust in history" (Waring, 1971), which "killed more people in a few months than all the armies of the 1914–18 war in four years" (Walters, 1978). The virus was remarkable in causing high mortality in the 15–34-year-old age group and depressed the life expectancy by more than 10 years (Walters, 1978). There seems to be general agreement that the American Expeditionary Force carried the virus to France in April 1918. Phylogenetic analysis indicates that the common ancestor of the human and classical swine influenza viruses originated between 1905 and 1914 (Gorman *et al.*, 1991). There were three waves of influenza during the final phase after the First World War: a mild

FIGURE 14.1   Recycling of influenza A viruses in humans? Seroarcheology suggests that H2N2 and H3N8 influenza viruses circulated in humans in 1918 and H1N1 influenza virus was introduced into humans in 1910. In 1957, the "Asian" H2N2 influenza virus re-emerged in Guiyang in southern China and in 1968 was replaced by "Hong Kong" H3N2, which was first detected in Hong Kong. These viruses were reassortants and contained avian gene segments (see Figure 14.2). In 1977, the "Russian" H1N1 influenza again appeared in humans and was probably reintroduced from a laboratory (see Figure 14.2).

summer wave and a lethal autumn wave in 1918 and, finally, in the winter of 1919 a nasty, but less lethal wave.

Seroarcheology, the detection of antibodies to earlier influenza virus infection in the sera of elderly people, has provided indirect evidence for the prior circulation of an H1 virus (Masurel and Heijtink, 1983). The partial nucleotide sequence of one Spanish influenza virus has been obtained from a formalin-fixed paraffin embedded lung sample from a soldier who died in 1918 (Taubenberger *et al.*, 1997). The partial sequence of the HA, NA, NP, M1, M2 genes indicates that the agent was a unique H1N1 influenza virus most closely related to classical swine influenza virus. It was noteworthy that HA did not possess the polybasic amino acid motif associated with high pathogenicity of avian H5 and H7 influenza viruses.

Genetic Shifts in the 20th Century

Since the first human influenza virus was isolated in 1933, antigenic shifts in type A influenza viruses have occurred in 1957 when the H2N2 subtype (Asian influenza) replaced the H1N1 subtype, in 1968 when the Hong Kong (H3N2) virus appeared, and in 1977 when the H1N1 virus reappeared (Figure 14.2). Each of these major genetic shifts have several characteristics in common: (1) their appearance was sudden; (2) they first occurred

in China; (3) they were antigenically distinct from the influenza viruses then circulating in humans; and (4) they were confined to the H1, H2, and H3 subtypes. On the basis of phylogenetic evidence, the most likely explanation for the appearance of new pandemic strains in humans is that they were derived from avian influenza viruses either after reassortment with the currently circulating human strain or by direct transfer. There is ample evidence for genetic reassortment between human and animal influenza A viruses in vivo (Webster *et al.*, 1971), and genetic reassortment has also been detected in humans (Cox *et al.*, 1983). Genetic and biochemical studies have concluded that the 1957 and 1968 strains arose by genetic reassortment. The 1957 Asian H2N2 strain obtained its HA, NA, and PB1 genes from an avian virus and the remaining five genes from the preceding human H1N1 strain (Scholtissek *et al.*, 1978; Kawaoka *et al.*, 1989; Figure 14.2).

The origin of the H1N1 Russian 1977 influenza virus is a mystery. This virus appeared in Anshan, northern China, in May 1977 and subsequently spread to the rest of the world; it is identical in all genes to the virus that caused a human influenza epidemic in 1950 (Nakajima *et al.*, 1978). Where was this virus for 27 years? The possible explanations include preservation in a frozen state, preservation in an animal reservoir or retention in an integrated, as yet undetected form in the genetic material of a human or lower

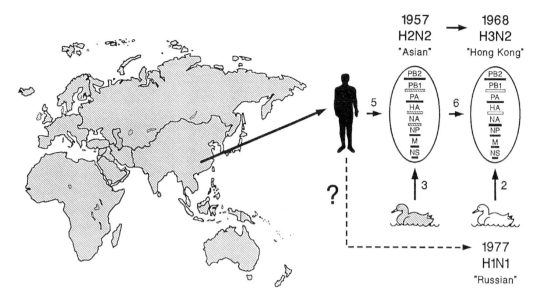

FIGURE 14.2 Genetic shift in human influenza pandemics in the 20th century. Phylogenetic evidence suggests that an influenza virus possessing eight gene segments from avian influenza reservoirs was transmitted to humans and pigs before 1918 and replaced earlier strains. The site of origin of this strain is unresolved, but historical evidence suggests that this virus was carried from North America to Europe by American troops and caused the catastrophic "Spanish" influenza pandemic of 1918. In 1957, the "Asian" pandemic virus acquired three genes (PB1, HA, and NA) from the avian influenza gene pool in aquatic birds by genetic reassortment and kept five other genes from the circulating human strain. After the Asian strain appeared, the H1N1 strains disappeared from humans. In 1968, the "Hong Kong" pandemic virus was circulating in humans. After the appearance of the Hong Kong strain, the H2N2 Asian strains were no longer detectable in humans. In 1977, the "Russian" H1N1 influenza virus, which had circulated in humans in 1950, reappeared and spread in children and young adults. This virus probably escaped from a laboratory and has continued to cocirculate with the H3N2 influenza viruses in the human population.

animal. The animal reservoir option is unlikely, for the accumulation of mutations would have continued. There is no evidence for integration of influenza genetic material into the host genome, leaving the most likely explanation that in 1977 the H1N1 virus was reintroduced to humans from a frozen source. It is possible that the re-emergence of this virus represents an escape from a laboratory.

The Hong Kong Incident

In 1997 an avian H5N1 influenza virus transmitted directly to humans in Hong Kong and six of the 18 infected persons died (de Jong *et al.*, 1997; Yuen *et al.*, 1997; Claas *et al.*, 1998; Subbarao *et al.*, 1998). A highly pathogenic avian influenza virus had been isolated several months before from domestic chickens in the New Territories of Hong Kong, where it had caused 70% mortality among 6800 chickens (Claas *et al.*, 1998). The available evidence is that the avian H5N1 virus was perpetuated in live poultry markets in Hong Kong and was the source of the viruses that infected the 18 humans (Shortridge *et al.*, 1998).

The H5N1 isolates from humans and the earlier chicken H5N1 viruses have been characterized extensively, and there is no doubt that the chicken H5N1 virus was the precursor of the human H5N1 strain. Both the chicken and human viruses belong to the H5 lineage of Eurasian avian influenza viruses. All gene segments were of Eurasian avian origin and were not derived by reassortment. The nucleotide sequence homology of the eight gene segments of the human and chicken H5N1 isolates varies from 98% for the polymerase A gene (PA) to 100% for the matrix (M) and non-structural (NS) genes. The hemagglutinin (HA) genes of the chicken and human isolates are 99.6% homologous, and the neuraminidase (NA) genes are 99.5% homologous.

Despite this high degree of homology, phylogenetic analysis showed that two subgroups cocirculated in Hong Kong, with differences in each gene. Figure 14.3 shows the phylogenetic subgroups based on HA1. In one group, most viruses possessed a carbohydrate side chain at residue 158 of HA1; the other group lacked the carbohydrate. The viruses of both subgroups possessed a series of basic amino acids (RERRRKKR) at the connecting peptide of HA1/HA2 and were highly pathogenic in chickens. The sequence encoding the receptor binding sites was identical in the two groups and indicated binding to terminal SAα2,3 galactose determinants, as was described for the index human virus (Claas *et al.*, 1998; Subbarao *et al.*, 1998). Viruses from both subgroups caused severe disease in humans (Bender *et al.*, 1998;

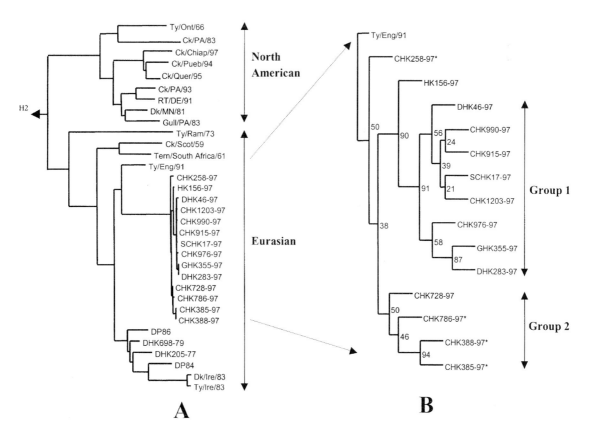

FIGURE 14.3   Phylogenetic trees of the HA1 gene of H5 influenza A viruses. **A.** The evolutionary relationship of the H5N1 Hong Kong viruses to various H5 viruses isolated previously in North America and Eurasia. The tree was generated by using the neighbor-joining algorithm in PHYLIP Version 3.57c software (Felsenstein, 1993), and is rooted to A/pintail/Praimorie/625/76 (H2N2). Horizontal lines are proportional to the numbers of nucleotide substitutions between branch points. **B.** A 100-replication bootstrap re-sampling of the H5 Hong Kong viruses from (**A**). which is rooted to A/Turkey/England/91. The number at each branch point indicates the percentage probability of the accuracy of the resulting topology. Asterisks indicate the presence of a potential glycosylation site at amino acid 158 of HA. The abbreviations for the viruses are given in Shortridge *et al.*, 1998.

TABLE 14.1 Comparison of nucleotide and amino-acid changes in genes and encoded proteins of H5 influenza viruses

Viruses	Genes	Analysed coding region	Mean no. of nucleotide changes (± SE)	Mean no. of amino acid changes(± SE)	Mean % of AAC/NC (± SE)
H5 Chicken Hong Kong*	HA1	77–1042	6.8 ± 3.6	0.6 ± 0.5	5.9 ± 5.2
H5 Human Hong Kong†		77–1042	9.5 ± 2.9	5.0 ± 2.4	50.5 ± 18.7
H5 Chicken Mexico‡		77–1042	40.8 ± 19.1	20.9 ± 10.1	57.3 ± 16.7
H5 Chicken Pennsylvania§		77–1042	5.0	3.0	60.0
H5 Duck¶		77–1042	41.5 ± 12.5	5.6 ± 0.8	15.1 ± 3.4
H5 Chicken Hong Kong*	NA	64–1381	11.6 ± 5.9	4.2 ± 2.1	36.6 ± 6.4
H5 Chicken Mexico‡		64–1381	12.0 ± 9.3	3.6 ± 2.7	37.9 ± 25.9
H5 Chicken Hong Kong*	PB2	34–2304	16.5 ± 9.6	7.0 ± 3.6	46.9 ± 14.8
H5 Chicken Mexico‡		34–2304	22.2 ± 18.1	5.4 ± 4.4	24.3 ± 0
Avian**		34–2304	294.7 ± 58.2	12.3 ± 4.1	4.9 ± 2.3
H5 Chicken Hong Kong*	PB1	34–2238	17.4 ± 11.9	4.2 ± 2.9	25.3 ± 10.9
H5 Chicken Mexico‡		34–2238	25.8 ± 21.1	3.6 ± 2.8	14.0 ± 0
Avian††		34–2238	96.3 ± 7.4	2.0 ± 0	2.1 ± 0.1
H5 Chicken Hong Kong*	PA	25–2172	13.1 ± 9.6	4.5 ± 2.7	51.3 ± 30.7
H5 Chicken Mexico‡		25–2172	24.4 ± 19.2	10.6 ± 8.5	41.3 ± 6.7
Avian‡‡		25–2172	71.0 ± 20.8	10.0 ± 2.5	10.6 ± 4.6
H5 Chicken Hong Kong*	NP	46–1549	9.7 ± 5.7	2.7 ± 1.2	35.5 ± 21.8
H5 Chicken Mexico‡		46–1549	12.2 ± 9.6	0.6 ± 0.4	5.0 ± 0
Avian§§		46–1549	72.0 ± 9.0	3.0 ± 1.2	4.3 ± 2.2

* CHK258–97, DHK46–97, CHK915–97, CHK283–97, CHK786–97, GHK355–97
† HK156, HK/481/97, HK/482/97, HK/483/97
‡ CMEX94, CHID94, CPUE1–94, CPUE2–94, CQUE94
§ Ck/PA/1/83, Ck/PA/1370/83
¶ DHK205–77, Dk/HK/342/78, DHK698–79, DPOTS84,
** Budg/Hok/77, RT/NJ/85, Ty/MN/80, Mal/NY/78
†† Ty/MN/80, Mal/NY/78, Gull/MD/77
‡‡ Pin/Alb/79, Ty/MN/80, RT/NJ/85, Dk/Hok/80,
§§ DHK205–77, DHK698–79, Dk/HK/7/75,Dk/Beij/78

Suarez *et al.*, 1998), but there was no evidence for person-to-person transmission. The NAs of both avian and human H5N1 strains have a unique 19-amino-acid deletion in the stalk region. Both the groups of H5N1 viruses are closely related antigenically, but monoclonal antibodies and ferret antiserum can distinguish the avian from the human viruses.

When an influenza virus is first introduced into a new host, it usually undergoes a period of rapid evolutionary change that is most apparent in antigenic drift in the HA and NA glycoproteins. Nucleotide changes that result in amino-acid changes are usually most numerous in the HA as a result of immunological selection, and less numerous in the internal genes. This was not the case with the H5N1 influenza viruses from poultry and humans in Hong Kong (Table 14.1).

In H5N1 viruses isolated from poultry in Hong Kong, the percentage of coding changes in the HA gene (5.9%) was an order of magnitude less than that found in H5N2 virus from chickens in Mexico (57.3%). This finding indicates a surprisingly low rate of evolution of the HA gene. Only the HA of influenza viruses from aquatic birds had a comparably low proportion of coding changes (H5 duck, 15.1%). After transmission to humans the H5 HA gene evolved rapidly, which is expected in a new host. The rates of amino-acid changes in the NA and other genes (PB2, PB1, PA, NP) in H5 viruses from Hong Kong and Mexico were higher than those found in influenza viruses from aquatic birds. The rapid rate of change is typical of a virus in a new host. Thus, the HA gene of the H5N1 influenza viruses isolated from poultry in Hong Kong is optimally adapted, while the other gene segments are evolving rapidly. After transmission to humans the HA genes showed the expected high rate of change. This reversal of the pattern of coding changes suggests that the HA is better adapted to chickens than are the other genes. The Hong Kong H5N1 viruses may be reassortants that acquired the HA gene from an H5 virus that is well adapted to domestic poultry, while the other seven genes may have come from another source, such as wild aquatic birds. The HA could possibly have come from A/goose/Guandong/1/96 (H5N1), a virus that

was highly lethal in geese (Xu *et al.*, 1998). Guangdong province lies adjacent to Hong Kong and provides much of Hong Kong's domestic poultry. The other seven genes are of Eurasian avian origin, but their source is unresolved. Additional sequence information from Eurasian influenza viruses should elucidate the origin of these gene segments.

No additional cases of H5N1 in humans or poultry in Hong Kong have been reported since the slaughter of poultry in the live-bird markets was initiated on 29 December 1997. The information that is now available underscores the wisdom of the slaughter of 1.6 million domestic fowl in Hong Kong in 1997. The H5N1 virus was evolving rapidly in chickens and was very likely to acquire mutations that would permit human-to-human spread.

DIFFERENCES IN THE EVOLUTION PATTERNS OF THE INFLUENZA VIRUS GENES IN BIRDS AND MAMMALS

As discussed previously (Webster *et al.*, 1992), waterfowl and other birds are believed to be the primary host for all of the influenza A virus strains that have been introduced into mammals. Multiple examples of each of the genome segments of influenza virus isolates from various waterfowl species, as well as those of virus isolated from mammals, have been sequenced and analyzed to determine their evolutionary relationships. For each of the genome segments, multiple coexisting lineages of viruses infecting waterfowl have been found. These lineages include those specific for geographical regions, those found in North America typically being distinct from those in Eurasia. The influenza viruses isolated from mammals are of two classes: (1) those that are closely related to avian virus and appear to have been only transiently established in the mammalian host – these include the transitory introduction into humans of classical swine influenza viruses, which can result in deaths, as occurred at Fort Dix (Gaydos *et al.*, 1997) and other isolated introductions

(Rota *et al.*, 1989; Wentworth *et al.*, 1994); and (2) those that have become established in a mammalian host and have diverged progressively from the avian virus ancestor. These include the avian H1N1 influenza viruses that have become established in pigs in Europe (Scholtissek *et al.*, 1998). These viruses have established a permanent lineage in their new host and are evolving rapidly. The established mammalian virus lineages include those isolated from horses, swine and humans. Each of the mammalian virus lineages shows a progressive accumulation of mutations and a good correlation of its date of isolation and its position on its respective branch of the phylogenetic tree. This consistent relationship is not seen among the avian viruses, so if there is a progressive accumulation of mutations in avian viruses with time, the time frame must be greater than the period covered by the available virus isolates.

ORIGIN OF INFLUENZA A, B, AND C VIRUSES

The general structural features and genome organization of influenza A, B, and C viruses suggest that they share a common ancestry distinct from other negative-strand RNA viruses (Desselberger *et al.*, 1980; Smith and Palese, 1989). Of the three virus types, A and B viruses are much more similar to each other in genome organization and protein homology than to C viruses (Lamb, 1983), which suggests that influenza C virus diverged well before the split between A and B viruses (Desselberger *et al.*, 1980). The more distant relationship of influenza C virus to the A and B viruses is demonstrated in phylogenetic analyses of NP genes (Gammelin *et al.*, 1990). Among the homologous genes that have been compared for influenza A and B viruses, the polymerase genes show the highest level of homology, and the organization and size of these genes have remained more similar than for other genes (Lamb, 1989). The 60% level of homology between the nucleotide sequences of PB1 genes of A and B viruses is

greater than the lowest homologies seen among nucleotide sequences of HA1 subunits of the 15 HA subtypes of influenza A viruses (31.3–60.6%) (Kawaoka *et al.*, 1990). By comparison, the nucleotide sequences of the HA1 subunits of influenza A and B viruses show 36% homology (Krystal *et al.*, 1982). Phylogenetic analysis of HA1 sequences shows that influenza B is more closely related to the HA lineages containing H8, H9 and H12 subtypes than to other HA lineages (Webster *et al.*, 1992).

CONCLUSION AND OUTLOOK

Antigenic variation in influenza viruses is dependent on the inherent variability of RNA and on the error-prone polymerase during replication. It provides a primitive strategy to avoid the immune mechanisms designed for virus control. Variation in the gene products permits the virus to stay one step ahead of the immune response.

Phylogenetic analyses of influenza A viruses have revealed that:

- All mammalian lineages originated from influenza viruses in aquatic birds and it is possible that influenza B viruses arose from the same source.
- There is geographical separation of avian influenza viruses into American and Eurasian lineages.
- Influenza viruses in their aquatic bird reservoir are in evolutionary stasis.
- After transfer to mammals there is rapid evolution.
- The human pandemics of this century have mainly occurred in south-east Asia and involved reassortment. This process probably occurs in pigs, for they possess receptors for both avian and human influenza viruses.
- The influenza viruses currently circulating in humans and pigs in North America originated by transmission of all genes from the avian reservoir prior to the 1918 Spanish influenza pandemic; some of the genes have

subsequently been replaced by others from the influenza gene pool in birds.

- The influenza virus gene pool in aquatic birds of the world is probably perpetuated by low-level transmission within that species throughout the year.

Given the existence in the aquatic bird reservoir of all known influenza A subtypes, we must accept the fact that influenza is not an eradicable disease. Prevention and control are the only realistic goals. If we assume that people, pigs and aquatic birds are the principal variables associated with the interspecies transfer of influenza virus and the emergence of new human pandemic strains, it is important to maintain surveillance in these species.

ACKNOWLEDGMENTS

This work was supported by US Public Health Service Research Grants AI-08831 from the National Institutes of Health, CORE Grant CA-21765 and American Lebanese Syrian Associated Charities (ALSAC). We thank Kimberly Hampton for manuscript preparation.

REFERENCES

Air, G.M., Gibbs, A.J., Laver, W.G. and Webster, R.G. (1990) Evolutionary changes in influenza B are not primarily governed by antibody selection. *Proc. Natl. Acad. Sci. USA*, **87**, 3884–3888.

Bean, W.J., Schell, M., Katz, J. *et al.* (1992) Evolution of the H3 influenza virus hemagglutinin from human and nonhuman hosts. *J. Virol.*, **66**, 1129–1138.

Bender, C., Huang, J., Hall, H., Klimov, A., Cox, N. and Subbarao, K. (1998) Molecular analysis of the surface protein genes of human (H5N1) influenza viruses. American Society of Virology, 17th Annual Meeting, University of British Columbia, Vancouver, BC, Canada, W29-7, p. 111.

Claas, E.C.J., Osterhaus, A.D., van Beek, R. *et al.* (1998) Human influenza A (H5N1) virus related to a highly pathogenic avian influenza virus. *Lancet*, **351**, 472–477.

Colman, P.M. (1989) Neuraminidase; enzyme and antigen. In: *The Influenza Viruses* (ed. Krug, R.M.), pp. 175–218. Plenum Press, New York.

Cox, N.J., Bai, Z.S. and Kendal, A.P. (1983) Laboratory-based surveillance of influenza A (HINI) and A (H3N2) viruses in 1980-81: antigenic and genomic analyses. *Bull. WHO*, **61**, 143–152.

Cox, N., Xu, X., Bender, C. *et al.* (1993) Evolution of hemagglutinin in epidemic variants and selection of vaccine viruses. In: *Options for the Control of Influenza II*, Proceedings of the International Conference on Options for the Control of Influenza, Courchevel, 27 September–2 October, 1992 (eds Hannoun, C., Kendal, A.P., Klenk, H.D. and Ruben, F.L.), pp. 223–230. Excerpta Medica, Amsterdam.

Crosby, A.W. (1976) *Epidemic and Peace, 1918*. Greenwood Press, Westport, CT.

Daly, J.M., Wood, J.M. and Roberton, J.S. (1998) Cocirculation and divergence of human influenza viruses. In: *Textbook of Influenza* (eds Nicholson, K.G., Webster, R.G. and Hay, A.J.), pp. 168–177. Blackwell Science, Oxford.

De Jong, J.C., Claas, E.C., Osterhaus, A.D., Webster, R.G. and Lim, W.L. (1997) A pandemic warning? (letter). *Nature*, **389**, 554.

Desselberger, U., Racaniello, V.R., Zazra, J.J. and Palese, P. (1980) The 3′ and 5′-terminal sequences of influenza A, B and C virus RNA segments are highly conserved and show partial inverted complementarity. *Gene*, **8**, 315–328.

Felsenstein, J.P. (1993) (Phylogenetic I reference package) version 3.5. Department of Genetics, University of Washington, Seattle, WA. Distributed by the author.

Fitch, W.M., Bush, R.M., Bender, C.A. and Cox, N.J. (1997) Long term trends in the evolution of H(3) HA1 human influenza type A. *Proc. Natl Acad. Sci. USA*, **94**, 7712–7718.

Gammelin, M., Altmuller, A., Reinhardt, U. *et al.* (1990) Phylogenetic analysis of nucleoproteins suggests that human influenza A viruses emerged from a 19th-century avian ancestor. *Mol. Biol. Evol.*, **7**, 194–200.

Gaydos, J.C., Hodder, R.A., Top, F.H. *et al.* (1977) Swine influenza A at Fort Dix, New Jersey (January – February 1976). II. Transmission and morbidity in units with cases. *J. Infect. Dis.*, **136** (Suppl.), S363–368.

Gerhard, W., Yewdell, J., Frankel, M.E. and Webster, R.G. (1981) Antigenic structure of influenza virus haemagglutinin defined by hybridoma antibodies. *Nature*, **290**, 713–717.

Gorman, O.T., Donis, R.O., Kawaoka, Y. and Webster, R.G. (1990) Evolution of influenza A virus PB2 genes: implications for evolution of the ribonucleoprotein complex and origin of human influenza A virus. *J. Virol.*, **64**, 4893–4902.

Gorman, O.T., Bean, W.J., Kawaoka, Y., Donatelli, I., Guo, Y.J. and Webster, R.G. (1991) Evolution of influenza A virus nucleoprotein genes: implications for the origin of H1N1 human and classical swine viruses. *J. Virol.*, **65**, 3704–3714.

Holland, J., Spindler, K., Horodyski, F., Grabau, E., Nichol, E. and VandePol, S. (1982) Rapid evolution of RNA genomes. *Science*, **215**, 1577–1585.

Ito, T., Gorman, O.T., Kawaoka, Y., Bean, W.J. and Webster, R.G. (1991) Evolutionary anaylsis of the influenza A virus M gene with comparison of the M1 and M2 proteins. *J. Virol.*, **65**, 5491–5498.

Kawaoka, Y., Krauss, S. and Webster, R.G. (1989) Avian-to-human transmission of the PB1 gene of influenza A viruses in the 1957 and 1968 pandemics. *J. Virol.*, **63**, 4603–4608.

Kawaoka, Y., Yamnikova, S., Chambers, T.M., Lvov, D.K. and Webster, R.G. (1990) Molecular characterization of a new hemagglutinin, subtype H14, of influenza A virus. *Virology*, **179**, 759–767.

Khatchikian, D., Orlich, M. and Rott, R. (1989) Increased viral pathogenicity after insertion of a 28S ribosomal RNA sequence into the haemagglutinin gene of an influenza virus. *Nature*, **340**, 156–157.

Kilbourne, E.D. (1978) Genetic dimorphism in influenza viruses: characterization of stably associated hemagglutinin mutants differing in antigenicity and biological properties. *Proc. Natl Acad. Sci. USA*, **75**, 6258–6262.

Knossow, M., Daniels, R.S., Douglas, A.R., Skehel, J.J. and Wiley, D.C. (1984) Three dimensional structure of an antigenic mutant of the influenza virus haemagglutinin. *Nature*, **311**, 678–680.

Krystal, M., Elliot, R.M., Benz, E.W. Jr, Young, J.F. and Palese, P. (1982) Evolution of influenza A and B viruses. Conservation of structural features in the hemagglutinin genes. *Proc. Natl Acad. Sci. USA*, **79**, 4800–4804.

Lamb, R.A. (1983) The influenza virus RNA segments and their encoded proteins. In: *Genetics of Influenza Viruses* (eds Palese, P. and Kingsbury D.W.), pp. 21–69. Springer-Verlag, New York.

Lamb, R.A. (1989) Genes and proteins of the influenza viruses. In: *The Influenza Viruses* (eds Krug, R.M., Fraenkel-Conrat, H. and Wagner, R.R.), pp. 1–88. Plenum Press, New York.

Lu, B.L., Webster, R., Brown, L. and Nerome, K. (1983) Heterogeneity of influenza B viruses. *Bull. WHO*, **61**, 681–687.

Masurel, N. and Heijtink, R.A. (1983) Recycling of H1N1 influenza A virus in man – a haemagglutinin antibody study. *J. Hyg.*, **90**, 397–402.

Mumford, J.A. and Chambers, T.M. (1998) Equine influenza. In: *Textbook of Influenza* (eds Nicholson, K.G., Webster, R.G. and Hay, R.G.), p. 146. Blackwell Science, Oxford.

Muraki, Y., Hongo, S., Sugawara, K., Kitame, F. and Nakamura, K. (1996) Evolution of the haemagglutinin-esterase gene of influenza C Virus. *J. Gen. Virol.*, **77**, 673–679.

Murphy, B.R. and Webster, R.G. (1996) Orthomyxoviruses. In: *Fields' Virology* (eds Fields, B.N., Knipe, D.M., Howley, P.M. *et al.*), ch. 46, pp. 1397–1445. Raven Press, New York.

Nakajima, K., Desselberger, U. and Palese, P. (1978) Recent human influenza A (H1N1) viruses are closely related genetically to strains isolated in 1950. *Nature*, **274**, 334–339.

Natali, A., Oxford, J.S. and Schild, G.C.U. (1981) Frequency of naturally occurring antibody to influenza virus antigenic variants selected with monoclonal antibody. *J. Hyg.*, **87**, 185–190.

Rota, P.A., Rocha, E.P., Harmon, M.W. *et al.* (1989) Laboratory characterization of a swine influenza virus isolated from a fatal case of human influenza. *J. Clin. Microbiol.*, **27**, 1413–1416.

Rota, P., Wallis, T., Harmon, M., Rota, J., Kendal, A. and Nerome, K. (1990) Co-circulation of two distinct evolutionary lineages of influenza type B virus since 1983. *Virology*, **175**, 59–68.

Schäfer, J.R., Kawaoka, Y., Bean, W.J., Suss, J., Senne, D. and Webster, R.G. (1993) Origin of the pandemic 1957 H2 influenza A virus and the persistence of its possible progenitors in the avian reservoir. *Virology*, **194**, 781–788.

Scholtissek, C., Hinshaw, V.S. and Olsen, C. (1998) Influenza in pigs and their role as the intermediate host. In: *Textbook of Influenza* (eds Nicholson, K.G., Webster, R.G. and Hay, A.J.), pp. 137–145. Blackwell Science, Oxford.

Scholtissek, C., Rohde, W., Harms, E. and Rott, R. (1978) Assignment of gene functions to RNA segments of influenza A2-Singapore and genetic relatedness to other influenza strains. In: *Negative Strand Viruses and the Host Cell* (eds Barry, R.D. and Mahy, B.W.J.), pp. 19–25. Academic Press, London.

Shortridge, K.F., Zhou, N.N., Guan, Y. *et al.* (1998) Characterization of avian H5N1 influenza viruses from poultry in Hong Kong. *Virology*, **252**, 333–342.

Smith, F.I. and Palese, P. (1989) Variation in influenza virus genes. In: *The Influenza Viruses* (ed. Krug, R.M.), pp. 319–359. Plenum Press, New York.

Steinhauer, D.A. and Holland, J.J. (1987) Rapid evolution of RNA viruses. *Annu. Rev. Microbiol.*, **41**, 409–433.

Stevens, D.J., Douglas, A.R., Skehel, J.J. and Wiley, D.C. (1987) Antigenic and amino acid sequence analysis of the variants of H1N1 influenza virus in 1986. *Bull. WHO*, **65**, 177–180.

Saurez, D.L., Perdue, M.L., Cox, N. *et al.* (1998) Comparisons of highly virulent H5N1 influenza A viruses isolated from humans and chickens from Hong Kong. *J. Virol.*, **72**, 6678–6688.

Subbarao, K., Klimov, A., Katz, J. *et al.* (1998) Characterization of an avian Influenza A (H5N1) virus isolated from a child with a fatal respiratory illness. *Science*, **279**, 393–396.

Taubenberger, J.K., Reid, A.H., Kraft, A.E., Bijwaard, K.E. and Fanning, T.G. (1997) Initial genetic characterization of the 1918 "Spanish" influenza virus. *Science*, **275**, 1793–1796.

Tulip, W.R., Varghese, J.N., Laver, W.G., Webster, R.G. and Coleman P.M. (1992) Refined crystal structure of the influenza virus N9 neuraminidase-NC 41 Fab complex. *J. Mol. Biol.*, **227**, 122–148.

Walters, J.H. (1978) Influenza 1918; the contemporary perspective. *Bull. NY Acad. Med.*, **54**, 855–864.

Wang, M.L., Skehel, J.J. and Wiley, D.C. (1986) Comparative analyses of the specificities of anti-influenza hemagglutinin antibodies in human sera. *J. Virol.*, **57**, 124–128.

Waring, J.I. (1971) *A History of Medicine in South Carolina, 1900–70*, p. 33. South Carolina Medical Association, Charleston, SC.

Webster, R.G., Bean, W.J., Gorman, O.T., Chambers, T.M. and Kawaoka, Y. (1992) Evolution and ecology of influenza A viruses. *Microbiol. Rev.*, **56**, 152–179.

Webster, R.G., Campbell, C.H. and Granoff, A. (1971) The *"in vivo"* production of "new" influenza A viruses. I. Genetic recombination between avian and mammalian influenza viruses. *Virology*, **44**, 317–328.

Webster, R.G. and Laver, W.G. (1980) Determination of the number of non-overlapping antigenic areas on Hong Kong (H3H2) influenza virus hemagglutinin with monoclonal antibodies and the selection of variants with potential epidemiological significance. *Virology*, **104**, 139–148.

Wentworth, D.E., Thompson, B.L., Xu, X. *et al.* (1994) An influenza A (H1N1) virus, closely related to swine influenza virus, responsible for a fatal case of human influenza. *J. Virol.*, **68**, 2051–2058.

Wiley, D.C., Wilson, I.A. and Skehel, J.J. (1981) Structural indentification of the antibody-binding sites of Hong Kong influenza haemagglutinin and their involvement in antigenic variation. *Nature*, **289**, 373–378.

Xu, X., Cox, N. and Guo, Y. (1998) Emergence of influenza A (H5N1) viruses in geese in China. American Society for Virology 17th Annual Meeting, University of British Columbia, Vancouver, BC, Canada, W-29–5, p.110.

Yamashita, M., Krystal, M., Fitch, W.M. and Palese, P. (1988) Influenza B virus evolution: co-circulating lineages and comparison of evolutionary pattern with those of influenza A and C viruses. *Virology*, **163**, 112–122.

Yewdell, J.W., Webster, R.G. and Gerhard, W.U. (1979) Antigenic variation in three distinct determinants of an influenza type A haemagglutinin molecule. *Nature*, **279**, 246–248.

Yuen, K.Y., Chan, P.K.S., Peiris, M. *et al.* (1998) Clinical features and rapid viral diagnosis of human disease associated with avian influenza A H5N1 virus. *Lancet*, **351**, 467–471.

DNA Virus Contribution to Host Evolution

Luis P. Villarreal

INTRODUCTION

Many viral infections have a profound impact on host fitness and survival and thus affect host evolution. For the most part, a virus-infected host appears to be under negative selection as virus replication will often result in morbidity and mortality. However, not all viral infections are pathogenic. DNA viruses in particular have a tendency to establish life-long and inapparent persistent infections. In bacteria, persistent infections are also known as temperate or proviral infections. Persistent infections by DNA viruses represent a distinctly different relationship from that of an acutely replicating and non-persisting infection and should result in distinct virus–host ecologies and biologies (for review see DeFilippis and Villarreal, 1998). However, the potential role that infections with DNA viruses have played in the evolution of their host has not been well examined. In this review I will consider both the theoretical and observed relationships between DNA virus and host and present evidence that persistent infections with DNA viruses can provide an important contribution to host evolution, such as being the likely source of host DNA replication genes.

A study of virus evolution presents many seemingly intractable practical problems. As there remains no obvious fossil record of past virus–host interactions, the best information comes from inferences based on viral and host genetic data. Those viral-like sequences remaining in their host now seem like remnants, whose original impact can only dimly be surmised. Furthermore, such data are unlikely to provide any information concerning what may have been major biological events mediated by virus. For example, the potential termination of a species by virus epidemic would leave no trace in the current genetic record. In addition, evolutionary science, whose conceptual synthesis in the 1930s (Dobzhansky, 1937) preceded the modern understanding of a virus as a molecular genetic parasite in the 1950s (Luria, 1959), has yet to fully appreciate the contributions of viral molecular genetic parasites and horizontal gene flow to host evolution, especially those of persistent or inapparent infections.

This chapter considers the evolution of DNA viruses as a separate topic. Most reviews on virus evolution have concentrated on RNA viruses with their correspondingly high mutation rates and genetically heterogeneous populations (Domingo *et al.*, 1996; Domingo and Holland, 1997; Garnett and Antia, 1998). The apparent genetic stability of DNA viruses makes them appear to evolve slowly and thus they have received much less attention. However, DNA viruses (and retroviruses) tend to have a very intimate molecular relationship with their host. As host genomes are DNA and as DNA replication is tightly regulated, DNA viruses must either parasitize this host DNA replication apparatus or override host regulation with viral replication proteins. DNA viruses therefore face common mechanistic problems that can require a very specific involvement with host molecular genetic function, which constitutes the viral habitat.

A common view of the origin of DNA viruses is that a simple DNA virus (lacking DNA polymerase genes, such as SV40) arose from a rogue cellular replicon that somehow became linked to a regulatory (initiating replication) protein and packaging genes, which allowed the replicon to become an autonomous DNA virus. It might be expected that such an event would be frequent in evolution and result in frequent and polyphyletic lineages for simple DNA viruses. More complex DNA viruses would also need to acquire additional replication and control proteins to become more independent of cellular regulation. Such a process would suggest a linkage between simple and more complex DNA viruses as they evolve to become more autonomous of host replication proteins. However, DNA virus evolution is not cumulative from lower order to higher order. A simple DNA virus does not evolve to a more complex virus nor does a satellite virus evolve to become an autonomous virus. Phylogenetic relationships between simple and more complex DNA viruses are not observed. Instead, distinct virus replication strategies appear to be conserved within viral classes and phylogenetic conservation can often be identified over long evolutionary time periods within such classes of virus. The escape of autonomous rogue cellular replicons appears therefore to be a very uncommon event at best. However, the incorporation of persistent viral replicons into cellular genome has been frequently observed, especially in lower organisms.

It is a fundamental requirement that genetic parasites should be able to differentiate self from host genome. How does a DNA virus distinguish its own genome from that of its host? The molecular basis of this differentiation will constitute the mechanism of genetic identity used by the virus and is closely linked to the

family of the virus. In many cases, this can be the interaction of a replication protein with a cognate DNA origin of replication, such as SV40 T-Ag interacting in a species-specific manner with cellular DNA polymerase-alpha-primase complex (Bhattacharyya *et al.*, 1995).

Alternatively, a different mode of DNA replication, such as protein-primed leading strand DNA or rolling circular replication (RCR replicons) might circumvent the regulated cellular DNA replication process. In other cases, viruses can specifically degrade cellular DNA (and other viral DNA) while protecting viral DNA using some virus-specific DNA modification, such as hydroxymethylcytosine replacing cytosine, as in T4 (and all T-even) bacteriophage, or by a sequestered virus-specific DNA-protein structure, as in herpes simplex virus. Viral DNA polymerases are also often origin-specific and show marked specificity for a DNA template complexed to a viral DNA binding (chromatin) protein (for references see Braithwaite and Ito, 1993, and Heringa and Argos, 1994b). Such identity systems and strategies tend to be highly conserved in individual DNA virus groups.

The Persistence of DNA Viruses

Persistent infections are common with many but not all DNA viruses. Persistence is the capacity of a viral infection to continue or re-emerge after an initial productive infection, followed at times by a period of non-production, from an individually infected host or host cell. As I use this term, persistence will include the condition known as latency. Latency is an infected state in which there is no detectable ongoing virus production, but in which production can be re-established. During this state, the molecular genetic distinction between host and viral genomes may be lost as the virus genome can act as a host genetic element. Yet even in this situation, virus in a persistently infected host cell must maintain a distinct molecular genetic identity if the resident virus is to retain the capacity to reactivate and spread to another host at a later time. For example, in lysogeny, the bacterial equivalent of a latent infection, bacteria harboring the prophage may undergo 100 virus-free generations and still be able to induce virus with the right stimulus. Similarly, a human infected at childhood with VZV does not re-express infectious virus until a zoster recurrence (shingles) 50 or more years later. In both these cases, an inapparently infected host has acquired a new genetic identity that will alter the way this merged virus–host complex responds to some future stimulus. Thus viral persistence is generally associated with specific episodic periods of virus production.

To ensure transmission, persistence must respond to a variety of specific host and environmental cues. Tissue damage, cell differentiation and death, DNA damage, superinfection with another virus, inflammation, immunosuppression, sexual reproduction and birth in an infected mother have all been identified as mechanisms associated with reactivation of persistent DNA viruses.

Virus/Host Fitness

Fitness is the genetic contribution by an individual's descendants to the future generations of a population. As fitness is a main determinant of evolutionary survival, the potential effect a DNA virus might have on host fitness must be considered as well as how viral fitness itself is achieved. Fitness, however, is a concept that is dependent on a transitory or instantaneous condition of selection and is therefore difficult to define and measure in specific experimental terms.

Infection with an acute virus is generally considered to impose a negative selection on the host population due to the disease induced as a consequence of virus replication. Thus viral fitness is positively associated with efficient viral replication and host fitness is negatively associated with viral replication. Such an acute virus–host relationship has been well developed from the perspective of host population biology and expressed as a mathematical model that resembles earlier predator–prey models in

which the virus is the predator and the host is the prey (Anderson and May, 1979; May and Anderson, 1979; May and Nowak 1995; Domingo et al., 1996; Domingo and Holland, 1997). Relative fitness is often used as a population-based measurement of viral fitness and is the number of reproductively successful progeny divided by the average number of progeny for the population. Such population-based modeling appears to be most suitable when applied to various human and animal viral epidemics, such as measles virus or the HIV epidemic. Stated simply, increased viral replication gives increased harm (virulence) to the host and will likewise increase the probability of virus transmission. With high viral replication rates it can be expected that most progeny virus will fail to be reproductively successful, although enough virus must find a suitable host to continue the infectious cycle. In these mathematical models, viral transmission rates are highly dependent on host population structure.

LONG-TERM AND SHORT-TERM LIFE STRATEGIES: ACUTE AND PERSISTENT MODELS FOR VIRUS BIOLOGY

Population-based mathematical models were originally used to measure allele distribution (Fisher, 1930). As noted above, these population growth parameters have also been used to measure viral fitness. Reproductive value (RV) and net reproductive output (R_o), which for a virus is viable offspring per individual per lifetime, are generally employed. These reproductive rates can also be considered as the fecundity of the virus. However, as noted by Giske et al. (1993), these population parameters are dimensionless ratios that take no account of variable generation times as they lack a temporal dimension. Variations in generation length would not matter. The only thing that would matter is the number of offspring produced during the lifetime of the parental virus. Yet the offspring survival is also crucial, especially if offspring must reach a specified "reproductive age".

If we consider the distinction between an

acute and persistent or latent viral infection, we can clearly see that there exist very large variations in viral generation times and total reproductive output, which appears to identify distinctly different life strategies between such viruses or states. In order to understand the potential impact of different viral life strategies on host evolution we should take into account the need for a time-dependent measurement of fitness. Qualitatively, Giske *et al.* argue that "fitness increases with an increase in both reproduction and juvenile survival" (Giske *et al.*, 1993). Thus the longevity of the offspring in addition to absolute reproduction rates can be crucial for success, especially if long-lived juveniles can reproduce at low numbers but with high reproductive probability.

If we apply this reasoning to DNA viral life strategies, we see that a virus can increase its fitness either by increasing its reproductive rate (offspring per parent generation; acute life strategy) or by increasing the survival time of the offspring, which can increase probability of transmission (persistent life strategy). One way to do this is for the virus to be maintained as a temporary host genetic element. In the case of a provirus, integrated into the host genome, we can now see that although the numbers of offspring generated by a provirus might be very small (equal to the numbers of host offspring), the probability of transmission could be near 1. If there is in addition some positive selection for the maintenance of a genomic virus, this could result in a host population in which essentially all members have the genomic virus.

These two states (acute and persistent) of viral life strategy can be considered analogous to the r verses K life strategies of MacArthur and Wilson (1967). Originally r and K represented coefficients for populations in island ecologies in which r was the intrinsic rate of population increase and K was the population equilibrium or ceiling. This analysis led to the identification of species that appear to have distinct life strategies, which were said to be either r- or K-selected. Here, high fecundity of relatively small offspring with low survival or reproductive probabilities (r/acute) are associated with harsh unpredictable environments, whereas predictable and constant environments select for a low number of larger offspring with high survival and reproductive probability in which the parental investment in offspring is large (K/persistent). In addition, the K life strategy is also associated with intense intraspecific competition between members of K-selected species. However, on closer inspection some basic K and r characteristics do not seem applicable to DNA viruses. Yet the concept of a specific "life strategy" does appear to apply broadly to many organisms. For a virus, the specific r coefficient would apply to the initial virus growth in a naive host population, such as the measles epidemics that depopulated Oceania in the 18th and 19th centuries, whereas the K coefficient would apply to the resulting steady-state or endemic infection, such as childhood measles infections in large populations. However, both of these parameters are in the context of an acute life-strategy virus. Persistent infections appear to have a distinct population biology. During persistence, the absolute numbers of persisting genomes or their reproductive rates is not what matters. One functional genome in the correct site (VZV in a ganglion, for example) may suffice to maintain the persistence. What matters is that persisting genomes survive to reactivate at the right time and place and make enough virus to infect new host or host offspring (i.e. attain the required reproductive age and have a high reproductive probability).

The two prominent life-strategies of DNA viruses can be called an acute life-strategy (a-virus) and a persistent life-strategy (p-virus). An a-virus will follow the population biology as previously described for epidemics and will have a high dependence on host population structure and density. A p-virus, however, will need to attain a high reproductive probability at a specific time and condition. A way to achieve this would be for the persisting virus to generally infect the offspring of its host. Thus a goal of persistence is to allow the virus to "survive" long enough to be present for the infection of new host, suggesting a minimal viral longevity at which propagation is most likely to be selected. In such a persistent life-strategy, virus generally moves from parent to offspring and viral transmission becomes much less dependent on the host population structure than with acute viral life-strategies. Therefore p-virus population biology will be directly reflected by host population biology. Three p-virus strategies for infecting new host can be considered. In one, the p-virus is very efficient, has a very high reproductive probability and simply keeps up with the generation of new host. In the second, the p-virus has a specific mechanism that compels the host to maintain it or negatively affects hosts that are encountered that are not infected. In the third, the p-virus provides a viral function that confers positive selection for the host to maintain the infection. Such infections will all generally appear to be commensal, as host and viral reproductive probabilities are linked. The above three persistent strategies would also be expected to have a significant effect on host evolution by altering the host response to the environment.

Nucleotide Word Bias Supports Acute and Persistent Phage Life-Strategies

With the accumulation of genomic sequence information, it has become clear that there exist simple sequence characteristics that identify specific types of organism. One such characteristic is the nucleotide word frequency. The frequency of di, tri and tetra nucleotide occurrence is far from random in genomes, as such word frequencies have a distinctive signature in all organisms (Karlin and Ladunga, 1994; Karlin and Burge, 1995; Blaisdell et al., 1996). Many such short nucleotide sequences also correspond to palindromic words that are recognized as restriction enzyme sites. As restriction enzymes are commonly found in both *Bacteria* and *Archaea* as antiviral defense systems (Wilson and Murray, 1991; Sowers, 1995), the occurrence of such sequence words can indicate how much selective pressure DNA viruses have exerted on their host genomes. Word bias and restriction-modification avoidance of palindromic restriction words in *Bacteria* and *Archaea*, suggests that during evolution bacterial DNA has been exposed to a wide spectrum of restric-

tion enzymes, most probably via mobile plasmids and prophages (Gelfand and Koonin, 1997). Interestingly, this nucleotide word bias is much less apparent in mycobacteria, which suggests that intracellular parasites might avoid virus-induced selective pressure by residing within a host. In keeping with this idea, mitochondrial genomes are essentially devoid of the palindrome word bias seen in free-living bacteria.

Another use of nucleotide word bias has been to examine viral DNA sequences of different organisms for similar word bias. In bacteria, DNA viruses can be considered to form two major classes; those that code for their own DNA replication proteins (such as various T-phages and PRD-1) and infect various species only in a lytic manner, and those viruses that do not encode DNA replication proteins (such as the lambdoid, P1, MU), which are able to establish lysogenic – proviral infections in specific host. In a report by Blaisdell *et al.* (1996), it was noted that the lytic phages have nucleotide word bias that is essentially random and does not reflect that of the host genome. In contrast, the temperate phages had nucleotide word bias that was essentially equivalent to that of the host genome. These oligonucleotide signatures (relative abundance) include low TA, high GC and normal CGCG, which are clearly seen in temperate (lambdoid, MU P1, P4, P22) phages, but not in lytic phages (T2, T4, T3, T7, PZA, PRD-1) or, to a lesser degree, in parasitic filamentous ssDNA phages (RCR, f1, 122), indicating a tight link between host genome and persisting lysogenic phage. Temperate phages have genomes that closely resemble those of their host whereas lytic phages do not. Given the variable nature of these enteric temperate phages, which includes members having distinctly different morphologies, a virus that can act as an independent plasmid (P1, single copy, with its own GACT methylase to allow plasmid maintenance), and also other temperate phages without low GATC (DAM recognition site) sequence and one phage that can act as a transposon (MU), there seems little otherwise in common among these phages except for their temperate lifestyle.

This appears to be strong support of the view that there exist two clear life-strategies (persistent and acute) for bacterial DNA viruses and that these strategies result in distinctly different evolutionary associations between viral and host genomes. Temperate phages were more similar to each other and their *E. coli* host than to any other phage, as if these viruses and their host were in a common genetic pool. In sharp contrast, the lytic phages were most distant from their host, and from each other and the temperate phages. The PRD-1 lytic phage (tectivirus) was most distant of all. PRD-1 is unique in several other regards as well (e.g. it has protein-primed linear dsDNA replication, contains a membrane internal to the capsid that remains out of the cell, and infects a very broad range of bacterial hosts that carry conjugative plasmids; see Blaisdell *et al.*, 1996).

BACTERIA–VIRUS EVOLUTION

The bacterial world can be considered as a specific habitat from the perspective of DNA viruses. As viruses numerically dominate the microbial and aquatic community (Cochlan *et al.*, 1993; Steward *et al.*, 1996), we expect that the bacterial world is highly influenced by virus evolutionary pressure, as supported above. The molecular habitat of both *Archaea* and *Bacteria* is similar in many respects. The genomes tend to be large circular molecules with single origins of replication (Krawiec and Riley, 1990). Cell-cycle control is distinctly different from eukaryotes and DNA synthesis can reinitiate during one round of cell division. Although chromatin-like proteins are sometimes found, they are not tightly bound or assembled into highly organized chromatin structures like eukaryotic chromosomes. Bacteria lack nuclei, allow coupled transcription and translation, have rigid cell walls and generally lack differentiated cell types. Their immune systems are non-adaptive and consist mainly of restriction modification enzymes. These are all basic characteristics of

bacteria to which a parasitic DNA virus must adapt and these characteristics distinguish the bacterial habitat for a virus from that of a eukaryotic host.

Phage Evolution

The genetic changes associated with bacterial virus evolution have probably been best characterized in studies of siphoveridae linear ds DNA phage isolated over a 30-year period from *Streptococcus thermophilis* used in dairy fermentation (as reviewed by Brussow *et al.*, 1998). Because of its economic importance, phage variation has been extensively studied. Both temperate and virulent phages were isolated (with extensive cross-hybridization between these two groups). Obligate virulent (lytic-only) phage that can outreplicate the temperate phage was observed to arise from deletions of the temperate phage lysogeny module, conserved in temperate phages. Genes such as integrase, cl repressor and immunity function were deleted (Figure 15.1). Mixed phage stocks (lytic/lysogenic) were commonly observed, although the majority of phage were not lytic (with some exceptions). Mitomycin C induction of lysogenic phage results in very similar phage (with few point changes in nucleotide sequence during a 2-year period) from similar starter cultures, but generally yields different phage isolates from ongoing industrial sites, which are open to wild new phage. Natural (raw milk) phage isolates have not been studied but would be expected to be genetically heterogeneous.

What is clear from these studies is that genetic relationships between different phages are distinctly modular and occurred via recombination of gene motifs between various different phages (Figure 15.1). The exchange of these motifs indicates a high rate of horizontal flow of various subgene domains into various other phage lineages, clearly suggesting that phage evolution is not following a simple lineage. Similar modular horizontal gene flow between different viruses has been observed with T4 phage and with P2 phage (Haggard-Ljungquist

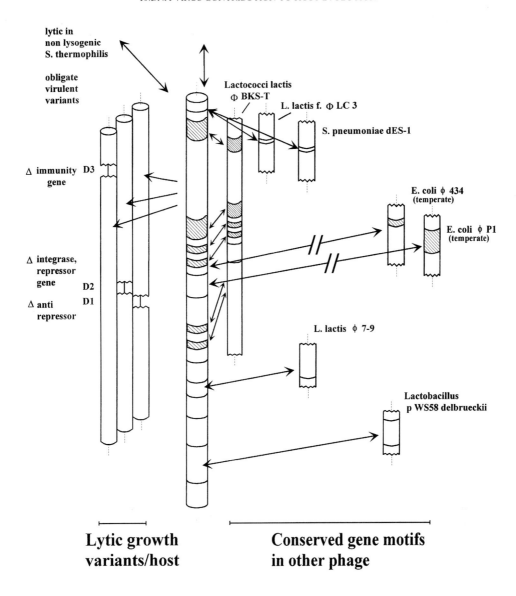

FIGURE 15.1 Phage evolution. Graphical representation of sequences similar to the phage Sfi21 of *S. thermophilis* found in other phage. Data from Brüssow *et al.*, 1998. Shaded areas represent conserved sub-gene motifs; gapped regions represent deletions.

et al., 1992), although in this case exchange was limited to the tail fiber region (Kutter *et al.*, 1995). Clearly, distance matrix and neighbor-joining methods for phylogenetic analysis would not usefully predict this modular phage evolutionary history.

Bacterial Genome Evolution and Phage Persistence

The study of bacterial pathogenicity serves as the most thoroughly examined model for the impressive adaptability of bacteria to their environment. DNA viruses play a major role in the evolution of host bacteria and their pathogenic adaption (Cheetham and Katz, 1995; Wain-Hobson, 1998). The best-established examples of this are the virulence toxin genes for diphtheria toxin, shiga-like toxin of *E. coli*, pyrogenic toxins of *Streptococcus*, and botulinum toxins (Holmes *et al.*, 1994; Wain-Hobson, 1998) all encoded by temperate bacteriophage. More recent GC fingerprint analysis of whole genomes confirms that virulence genes typically have a distinct make-up that confirms horizontal transmission (Krikau and Jahn, 1991; Wang *et al.*, 1995) and that the acquisition of this complex phenotype can occur in a single genetic event (McDaniel and Kaper, 1997). It is clear that horizontal transfer of genetic elements is also the major mode of genetic adaption to pathogenicity by bacteria. In addition, bacteriophage transduction is a well established way to move genetic material between bacteria, often integrating into tRNA genes. Also, plasmid-mediated resistance is commonly observed. The acquisition and the transfer of virulence phenotypes and genes occurs in clustered regions known as pathogenic islands (Finlay and Falkow, 1997). The common aspect of such pathogenic adaptability is that DNA with the clustered resistance genes has moved into bacteria horizontally by some transfer mechanism to become either a prophage, an episome or an integrated version of the genes flanked by some sequence that allows DNA mobilization. In the case of the *D. nodosus* virulence-associated region (Cheetham and Katz, 1995) these mobility genes are known to be similar to various bacteriophage and plasmid genes from other species in a radiated pattern reminicent of the evolution of the phage of *S. thermophilis*, shown in Figure 15.1.

In order to become mobilized, the pathogenic genes require the activity of an enzyme that can excise the DNA, such as an integrase gene. One well-studied integrase gene (which is also a bacterial virulence factor) is closely related to coliphage P4 primase replication gene (Cheetham and Katz, 1995). Such a genetic relationship suggests that phage may have contributed to the evolution of this DNA mobilization process. Given this inference, it is worth further considering P4 phage biology as a system that can

affect bacterial adaptability. P4 is highly prevalent, can be found in one-quarter of clinical *E. coli* isolates (Lindqvist *et al.*, 1993) and is known to be able to persist in various ways, such as an inapparent DNA (11 kb) episome. When P4 is persisting episomally, it replicates as an RCR or multicopy plasmid in which it expresses the *kil* gene for DNA replication. This gene is a member of the initiator protein family for rolling circle DNA replication and shows the consensus moieties found from Bacteria to the densoviruses and parvoviruses of eukaryotes (Table 15.1). Related genes include conjugal transfer proteins, which must nick DNA to initiate ssDNA transfer. This is one of the most conserved of all gene families (Gorbalenya *et al.*, 1990; Ilyina and Koonin, 1992).

P4 can also integrate into the bacterial chromosome and become lysogenic (Lindqvist *et al.*, 1993). P4 is parasitic to infection with a 35 kb P2 phage, and will initiate a lytic infection following P2 superinfection. When superinfected with P2, P4 will then amplify and replicate bidirectionally from a unique origin. Thus, with the P2/P4 phage system, we see both the persistent and acute life-strategy of a common bacterial virus with its host depending on superinfection with the helper virus (P4 is a p-virus, P2 is an a-virus). In addition, when the bacterial host is infected with a persistent and inapparent P4 phage, it has viral replication genes that could enhance the potential for evolution and adaption via DNA mobilization of the bacterial host.

Infections with temperate phage are a very common situation in all free-living bacteria studied to date. All sequenced free-living bacterial genomes in the Genbank database contain multiple temperate phage genomes as well as regions of phage. This means that temperate viral genomes must be an environmentally normal constituent of bacterial genomes. One clear consequence of these proviral infections is induction of host-cell immunity, at least to related lytic phages. In the case of P1 infection, cells are also immune to unrelated phage types, because of the virally encoded restriction modification enzymes (for references see Meyer, 1994), suggesting that strong competition exists between temperate viruses to occupy the host and exclude other viruses.

We are left with the clear view that, in bacteria and their DNA viruses, the major process of genetic adaption and evolution involves the horizontal transfer of genetic elements and that such a process can be either a direct result of virus infection or due to the presence of various parasitic genetic elements that use a virus-like replication/mobilization process. Considering how highly adaptable these bacterial species are, we are left to wonder why such an effective process does not appear to be used in eukaryotes.

Persistence and Complex Phenotypes: the Genesis of Bacterial Immunity

The main bacterial defense against DNA virus infection seems to be the restriction modification system (see Klieve et al., 1989; Wilson and Murray, 1991; Sowers, 1995). However, the evolutionary origin of the bacterial immune system against DNA viruses poses an interesting dilemma because restriction modification systems require the seemingly unlikely simultaneous creation of two distinct enzyme activities (a site-specific methylase and restriction endonuclease activity). Accounting for the sudden evolutionary origin of such "complex phenotypes" requiring multiple genes is a general and unresolved issue in evolutionary biology. The genesis of pathogenic forms of bacteria can also require the rapid acquisition of multiple genes. However, this dilemma can be resolved by considering the relationship of persisting genetic parasites to their host. Temperate phage P1 and P7 of *E. coli* encode type III restriction modification enzymes, most probably to exclude other phage. Such a lysogenic bacterium would thus have the genetic basis from which to generate a host antiviral immune system. Restriction of endonuclease activity resembles the endonuclease activity of integrase involved in transposition. The 5′ overhang is a common feature of many enzymes involved in transposition (transposase) as well as restriction (Mizuuchi, 1992; Klobutcher et al., 1993; Slater et al., 1995), such as the conjugative transposable element tn916

(Manganelli *et al.*, 1996). Restriction endonuclease could also have evolved from integrase activities (Heitman, 1993). Given the high rate of genetic recombination and adaption of phage relative to the cell, a more likely scenario is that a virus could provide for the evolutionary genesis of both these activities.

The presence of a persisting viral episome, such as P4, could therefore provide the raw genetic material to develop a host restriction enzyme. In addition, DNA methylation is involved in the control of the initiation timing of DNA replication (Slater *et al.*, 1995). A plasmid with a bidirectional origin of replication would require such an activity to independently control plasmid DNA replication. The report of Naito et al. (1995) provides further support for the idea that immunity and viral persistence are linked. They show that a parasitic plasmid that codes for both a restriction and modification enzyme, uses these two activities like a poison and antidote to compel infected bacteria to maintain the persisting episome, as its loss will result in the degradation of host DNA. Thus a single genetic event, becoming persistently infected by a genetic parasite, provides a new genomic identity system that can also provide immune systems for the host.

DNA Viruses of Archaea

As *Archaea* (archeaebacteria) are now thought to represent another entirely separate order (*Crebarchaeoata*) of life from Eubacteria (*Bacteria*), and also to be more related to eukaryotes (Woese *et al.*, 1990; Woese, 1998), it is worth considering if these organisms share many of the virus–host relationships that we have just considered for *E. coli*. Like *Bacteria*, *Archaea* also appear to have single circular genomes with circular ds DNA. These organisms are also known to posses restriction modification systems that protect against DNA virus infection (Klieve *et al.*, 1989; Sowers, 1995). The Archea exist in two kingdoms, the *Crenarchaeota* kingdom, which is more evolutionarily related to eukaryotic microalgae, and

TABLE 15.1   Conservation of functional motifs in the replication proteins of rolling circular replicons (RCRs); data reorganized from Ilyia and Koonin, 1992

Microviridae, Gram-negative and Archaebacteria plasmids								
1	phix174	A	(200–390)	FDTLTLAD	53	RLHFHAVHF	70	VGFYVAKYVN
2	G4	A	(241–431)	FDTLTLAD	53	RLHFHAVHF	70	VGFYVAKYVN
3	S13	A	(209–399)	FDTLTLAD	53	RLHFHAVHL	70	VGFYVAKYVN
4	SPV4	gp2	(76–218)	FVTLTYSD	50	RPHYHICFF	44	-ANYTARYTT
5	Chp1	p5	(140–284)	VLILTYDN	45	RMHWRMIVF	48	T-FYVARYVQ
6	BP186	A	(319–487)	FYTLTAPS	48	TPHWHMLMF	47	TG-YIAKYIS
7	EC67 2/3		(357–400; 1–116)	FYTITCPS	48	TVHWHLMCF	45	TS-YIAKYIS
8	PHASYL	ARP	(87–249)	FLTLTFRD	37	RIHYHLLVA	56	IGRYVGKYIS
9	pEPLX	ARP	(59–232)	FITLTLPP	50	ALHLHIVMV	92	ASAYMGKYLS
10	pHGN1	REP	(103–208)	MVTLTAST	54	YAHIHLGVF	58	LGAYLAAYMA
11	pGRBI	REP	(94–226)	MVTLTASS	48	YVHIHLGVF	57	LGAYLAAYMA
12	pEHSPN	REP	(50–182)	MVTLAASS	51	YVHISLGVF	47	LGAYLAAYMA
Gram-positive plasmids								
13	pBAA1	REP	(87–226)	FLTLTVRN	48	HPHFHVLIP	64	ISKYPVKDTD
14	pFTB14	REP	(119–258)	FLTLTVRN	48	HPHFHVLLP	63	ISKYPVKDTD
15	pLP1	REP	(108–244)	FLTLTVKN	49	NQHLHVLLF	56	TAKYEVKSAD
16	pUB110	REP	(124–255)	FLTLTVKN	49	NQHMHVLVC	55	TAKYPVYDTD
17	pC194	REP	(97–221)	FITLTTPN	48	NPHFHVLIA	49	MAKYSGKDSD
18	pLAB1000	REP	(106–236)	FLTLTAEN	47	HQHMHVLLF	56	TAKYQVKSKD
19	pBC1	REP	(31–168)	FLTLTVRN	53	HPHFHVLLC	67	VSKYPVKDTD
20	pKYM	REP	(116–252)	FLTLTVRN	46	HPHFHCLLM	59	TLKYSVKPED
21	pSK89	REP	(14–197)	FLTLTTPN	48	NPHFHVLMA	49	MRKYSGKDSD
22	pNost	REP	(135–262)	FVTLTVKN	50	HPHFHVLMM	49	VIKYSVKE$D
23	pTD1	REP	(114–240)	FITLTVKN	51	HPHYHILAA	50	VAKYSVKATD
Plant geminivirus								
24	ABMV	AC1	(15–110)	FLTYPQCS	32	EPHLHVLIQ	36	VKSYIDKDGD
25	PYMV	AC1	(15–110)	FLTYPQCS	32	EPBLHVLIQ	36	VKSYVEKDGD
26	BGMV	AC1	(15–110)	FLTYPRCT	32	EPHLHALIQ	36	VKEYIDKDGV
27	TGMV	AC1	(16–110)	FLTYPQCS	32	QPHLHVLIQ	36	VKTYIDKDGD
28	CLV	AC1	(14–109)	FLTYPKCS	32	EPHLHALIQ	36	VKSYLDKDGD
29	BCTV	C1	(15–110)	FLTYPQCS	32	QPHLHVLLQ	36	VKSYVDKDGD
30	TYLCV	C1	(13–108)	FLTYPNCS	32	EPHLHVLIQ	36	VKTYVEKDGN
31	MSV	C1	(18–107)	FLTYPKCP	32	SLHLHALLQ	30	VRDYILKEPL
32	DSV	C1	(15–104)	FLTYSKCD	32	SLHSHALVQ	30	VRTYTLKNPV
33	WDV	C1	(18–113)	FLTYPECT	32	SPHLHVLVQ	37	VRDYITKEVD
34	CSMV	C1	(42–131)	FLTYPRCP	32	EPHLHAFVQ	30	TLKYCMKHPE
35	MiSV	C1	(15–110)	FLTYPHCN	32	DPHLHVLIO	30	VFCYISKTNG
36	SLCV	C1	(15–110)	FLTYPRCD	33	SPHLHCLIQ	36	VKNYITKEGD
37	SSV	C1	(15–110)	FLTYSRCP	32	GYHIIIVLA	31	VRAYAMKNPV

Gram-positive plasmids

38	pADB201	REP	(12–118)	LLVYPDSA	34	KPHYHIVLA	30	MWRYMTH - - K
39	pMV158	REP	(11–128)	FLLYPESI	37	KAHYHVLYI	34	MYLYLTHESK
40	pE194	REP	(22–132)	FVLYPESA	32	KEHYHILVM	30	LVRYMIH - - M
41	PWV01	REP	(13–147)	FLLYPDSI	44	KPHYHVIYI	34	SYEYRTRPSK
42	pFX2	REP	(13–148)	FLLYPDSI	45	MPHYHVYIL	34	SYEYLTHESK
43	pLB4	REP	(32–137)	IVVYPESL	30	KSEYHLVLN	30	AVRYLTH - - M
44	pIJ101	REP	(62–234)	LVTFTARH	78	HPHIHAIVL	69	LAEYIAKTQD
45	pSB24	REP	(62–234)	LVTFTARH	76	HPHIHAIVL	71	LCEYIAKTQD
46	IS801	REP	(113–272)	HLVFTLPD	50	HPHVHLSVT	83	TGRYLKKPPI

Chamydial plasmids

47	?pCHL1	REP	(9–108)	FIKSPIHL	33	SSHYHALAA	42	LEAYGVKRYK
48	?pCpA1	REP	(8–112)	IIKSSLHL	36	PSHYliALA	12	IEAYGVKRYK
49	?ColE3 (E2)	REP	(33–131)	IAILARFI	40	NGHAHLLYA	32	DVNYSGLICK
50	?CAA	p52	(316–411)	FATLTALG	29	GQRWHTLVP	41	TATYALKEPV
51	?CFDV	p17	(25–148)	CFSSTESR	48	RSHFHITIC	49	ERTYCTSTSR

Parvovirus – animal

52	MVM	NS1	(175–243)			GWHCHVLIG	51	LLTYKHKQTK
53	CPV	NS1	(127–196)			GWHCHVLLH	51	ILTYRHKQTK
54	FPLV	NS1	(127–196)			GWIINIVLL	51	ILTYRHKQTK
55	MEV	NS1	(127–196)			GWHCHVLLR	51	ILTYRHKQTK
56	B19	NS1	(79–147)			GYHIHVVTG	50	IENYLMKKIP
57	ADV	NS1	(154–231)			QFHIHCCLG	60	PYKYFNKQTK
58	AAV	NS1	(88–162)			YPHMHVLVE	56	IPNYLLPKTQ
59	ADNV	NS1	(399–456)			GDHIHILFS	42	IL-YCIRYGI

		u a	h
Consensus		futltxxx	xpHuHuuux	uxxYuxxxx

At the bottom is the consensus sequence that represents each of these conserved domains. The first domain is not conserved in the animal parvoviral replication proteins.

the *Euryachaeota* kingdom, which includes the more distant halophilic species (Figure 15.2). Both these kingdoms have DNA viruses, but of very different types and with very different host relationships.

DNA viruses of thermophilic and hyperthermophilic *Archaea* (the Crenarchaeota kingdom) are well known and include virus types not found in *Bacteria*, such as the fuselloviridae, with an enveloped virion containing circular ds DNA, (Zillig *et al.*, 1996), and the lipothoviridae, which are flexible filaments containing linear ds DNA packaged into a core structure associated with DNA binding proteins (Schleper *et al.*, 1992). These viruses are highly prevalent, non-lytic or temperate and infected host survive infection. Curiously, lambdoid viruses (with head and tail) appear to be absent or rare in *Crenarchaeota Archaea*.

In halophilic and methanogenic *Archaea* (Euryachaeota kingdom), however, most characterized viruses are lytic and T-like (containing head with tails) with abroad host range and closely resembling T7 phage of *E. coli*, including having concatamer-based DNA replication (Nuttall and Dyall-Smith, 1993, 1995; Zillig *et al.*, 1996). This indicates that a common viral replication strategy exists between viruses of *Archaea* and *Bacteria*, in spite of very different cellular lifestyles, incompatable systems of transcription, and exteme variation in habitats, and suggest that these T-phage viral systems have existed prior to the divergence of *Archaea* from *Bacteria*. In addition, high-copy-level parasitic plasmids, such as pDL10, are widespread and, in the case of pDL10, will alleviate auxotrophic growth in its host or, like pTIK4, code for killer function (Zillig *et al.*, 1996).

Thus persistent infection with parasitic genetic elements is common in the Crenarchaeota archaebacteria and can in some cases provide a clear advantage to the host. It therefore seems that the two archaebacteria kingdoms have distinct relationships with their DNA viruses and that the Crenarchaeota kingdom, which is closer to eukaryotes, more commonly supports p-virus life strategy, and the Euryachaeota supports an a-virus life strategy as found in Bacteria.

The Dilemma of DNA Replication Genes: a Viral Role?

The amino acid sequences of replication proteins of *Bacteria* appear to be very distinct from those of archaebacteria (Edgell and Doolittle, 1997). None of the Archaea or eukaryotic proteins involved in DNA replication (origin recognition complex-DnaA/ORC, ss DNA binding protein, Rep A-primer, DnaG/polymerase-α, polymerase-Δ, helicase DnaB/Dna2, PCNA-sliding clamp polymerase cofactor, lagging fragment DNA ligase) shows significant primary sequence homology to those of *Bacteria* (Mushegian and Koonin, 1996; Edgell and Doolittle, 1997), yet they perform very similar functions. Such divergence between functional homologs has been called a non-orthologous substitution, leading some to propose that the common ancestor (cenancestor) of *Archaea* and *Bacteria* may have had an RNA genome.

This intrepretation suffers from the fact that *Archaea* and *Bacteria* share sequence similarity between many other proteins, including those that are involved in deoxynucleotide metabolism (Mushegian and Koonin, 1996). Genomic sequence analysis confirms that, despite the lack of shared replication genes, several hundred genes are orthologous between *Bacteria* and *Archaea* (Green *et al.*, 1993). Given the error catastrophe limit for RNA synthesis, is seems implausible that so many conserved genes could be viably maintained as an RNA genome.

Since these two orders also represent the deepest phylogenetic division, and the DNA replication process is functionally very similar between them, we are therefore left with the dilemma of needing to explain the origin of a distinct set of replication proteins in each order. In addition, as noted above, these two orders can support similar DNA viruses with related replication genes. It is, therefore, very interesting that in *Mathanococcus*, whose genome has no bacterial homolog to the origin recognition protein (ORC1), a clear homolog to ORC1 can be found present on persisting *Mathanococcus* plasmids. This ORC homolog may also be important for plasmid maintenance and replication (Gavin *et al.*, 1995; Loo *et al.*, 1995). In addition, it has long been established that the replication proteins (especially DNA polymerase) of eukaryotic DNA viruses and bacterial phage share considerable sequence similarity, thus linking bacterial and eukaryotic replication genes. This raises the question of whether DNA viruses might provide the evolutionary link between bacterial and archeal-eukaryotic replication genes.

Three families of DNA polymerase have been described that are denoted A (*E. coli* A-pol I), B (pol II) and C (pol III), in which the B family includes the replicative polymerase of eukaryotes (Wang, 1991; Diffley, 1994; Gavin *et al.*, 1995; Loo *et al.*, 1995). These sequence families cannot be aligned with each other, nor do eukaryotic cellular DNA polymerase genes show sequence similarity to either Bacteria or Archaea genes (Miller *et al.*, 1988; Braithwaite and Ito, 1993; Wang *et al.*, 1995). However, when initially sequenced, T4 DNA polymerase was observed to be very similar to eukaryotic and eukaryotic viral DNA polymerases (Spicer *et al.*, 1988). Furthermore, if the viral DNA polymerases are included, similarity can be seen between widely separated DNA viruses and their host DNA B polymerases, particularly with the DNA pols of herpes-like and T4 viruses (Bernstein and Bernstein, 1989; Braithwaite and Ito, 1993; Heringa and Argos, 1994a). Sufficient sequence similarity exists to allow the construction of phylogenetic trees, but the resulting trees do not appear to adhere to accepted biological relationships (Braithwaite and Ito, 1993). Curiously, in these trees viral DNA polymerase sequences often appear most similar to the root of the most closely related host DNA polymerases, and the non-orthologous character of bacterial and eukaryotic DNA polymerases is not seen when viral DNA polymerases are included. The most conserved three regions of these DNA polymerases have been noted to be important functional domains of the pol B family (Bernstein and Bernstein, 1989; Wang *et al.*, 1995). However, eukaryotic DNA pols differ from phage pols (T4, RB69) in that phage DNA pols have a "gap" relative to eukaryotic polymerases that corresponds to the most variable region between phage genes (Tommasino, 1991; Wang *et al.*, 1995). Thus the functional domains and their relative positions are maintained between

phage and eukaryotic DNA polymerases, which is consistent with a common lineage.

Various gene sequences have been used to make a universal tree of life. However, different trees result, depending on whether rRNA, EF or other genes are used, which presents a dilemma for understanding how these various genes could have been derived from the same lineage (for references see Forterre, 1997). The current strategy is to combine the results of multiple analysis to generate an average tree. However, such averaging does not address the discrepancies resulting from different genes. If the sequences from the replicative DNA polymerase family B are used to root the tree, the resulting tree does not show the three accepted domains (Bacteria, Archaea and Eucarya), and Eucarya is paraphylitic, which seems most incongruent with established patterns of evolution (Braithwaite and Ito, 1993). Why do these replication proteins, which would seem to be such a basic and conserved molecular process for cells, appear to be at odds with accepted patterns of evolution?

EUKARYOTES AS A DNA VIRUS HABITAT

Although the enzymes for DNA replication serve very similar functions in eukaryotes and bacteria, the eukaryotic orders present a rather distinct habitat for DNA viruses compared to bacteria. Eukaryotic genomes are segmented into several linear chromosomes. Chromosomes contain numerous origins (thousands) of DNA replication that are tightly controlled and initiate DNA replication only once in either the early or late S phase of the cell cycle (Taylor, 1984). The DNA is tightly packaged into a chromatin structure in a membrane-surrounded nucleus. Eukaryotic cells constitutively bind their ORI C protein to DNA-replication complex throughout the cell cycle, unlike bacterial systems, which transiently bind DNA A and prevent further binding by SeqA protein (reviews in Diffley, 1995, 1996, and Donovan and Diffley, 1996). It is very likely that this and other differences in molecular habitat limit the molecular strategies that eukaryotic DNA viruses can use. For example, all eukaryotic DNA viruses appear to enter host cells as DNA–protein complexes, and not as naked DNA, which is common in bacteria. Thermophile Archaea DNA phage may also enter cells in DNA-protein complexes. Many cells of eukaryotic organisms are terminally differentiated and, although some can amplify DNA, most such cells cannot continue to support DNA replication, thus limiting such cells as potential hosts for many DNA viruses. In addition, the host antiviral defense systems are very different in eukaryotes and bacteria. Except for the phycodnaviruses of microalgae, eukaryotes and their DNA viruses do not use restriction enzymes to degrade DNA.

Instead of restriction enzymes, eukaryotes have evolved both non-adaptive and adaptive immune responses to viruses that are not seen in bacteria. DNA methylation is seen in both bacterial (Barlow, 1993) and eukaryotic DNA. This DNA methylation may have originally been an antiviral response system (Doerfler, 1991, 1996) that later evolved to serve many other host functions. However, these activities are now distinct in bacteria and eukaryotes. In addition, apoptosis occurs in most eukaryotes and appears to be a common response to viral infection, and will exert a strong selective force on eukaryotic DNA viruses (for references see Teodoro and Branton, 1997). Curiously, although there are many examples of DNA viruses in eukaryotes, unlike bacterial temperate phage, there are few examples of viral genome integration being part of the normal life-strategy of eukaryotic DNA virus. The adaptive immune system of vertebrates introduces a major shift in the selective landscape for eukaryotic DNA viruses and requires that they either replicate and transmit very quickly, vary viral antigenicity or modulate host immune recognition in order to survive. Although there can be many antigenic types of eukaryotic DNA virus, they do not generally show high dynamic antigenic variability within one antigenic type. Therefore, host immune modulation would seem to be the crucial phenotype for these parasites. Thus, there seem to be many basic differences between the eukaryotic and bacterial habitat for DNA viral parasites.

EARLY EUKARYOTES AND THEIR DNA VIRUSES

The earliest eukaryotes are currently thought to be represented by the trichonomads, diplonomads and microsporidia, which include species such as *Giardia* and *Leishmania*. These organisms are aerotolerant anaerobes (Knoll, 1992) that are asexual, divide by binary fission, can have two nuclei (e.g. *Giardia lamblia*) and lack mitochondria, chloroplasts, Golgi bodies and endoplasmic reticulum. They live parasitically within the host (Vossbrinck *et al.*, 1987; Sogin and Silberman, 1998). As noted, universal trees of life have been proposed, as shown in Figure 15.2, in which the deepest division is between *Archaea* and *Bacteria*. Generally, eukaryotes are proposed to have branched from the thermophile Archaea lineage. Differences, however, exist in the placement of early eukaryotes depending on the sequence used to generate the tree. Shown in the figure are two trees based on the analysis of the small subunit of rRNA (Knoll, 1992) and the analysis of EF-G/2 sequence respectively. The two differ in that EF-G/2 places the unicellular *Chlorella* algae as representing the earliest eukaryote whereas rRNA small subunit places the microsporidia as representing the earliest eukaryote.

As both *Archaea* and *Bacteria* have a very high prevalence of infection by various DNA viruses, it is most curious, then, that few if any DNA viruses infect *Giardia* and *Microsporidium* species. The viruses that have been isolated from these organisms are all ds RNA viruses, such as GLV (Kasprzak and Majewska, 1995; Kohler and Wang, 1997). Unlike bacteria, these ds RNA viruses enter their host via endocytosis. In addition these ds RNA viruses show similarity to the ds RNA viruses of yeast and other fungi (Kasprzak and Majewska, 1995), which evolved later; this suggests that such ds RNA viruses are stably associated with their host on an evolutionary time scale. We currently have no explanation for the absence of DNA viruses (either lytic or persisting) from these organisms. Also, because these primitive eukaryotes are asexual and appear to lack any mechanism of DNA transduction, it is not clear how these

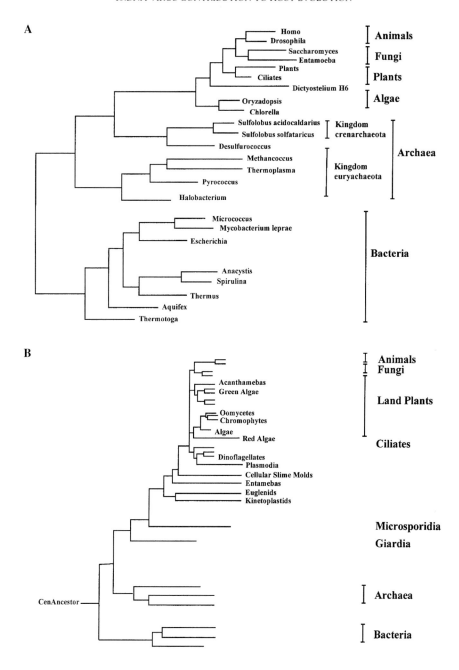

FIGURE 15.2 Proposed universal trees of life based on the analysis of the small subunit of RNA (**A**) and EF-G/2 sequence (**B**).

organisms genetically adapt to changes. Their parasitizing other organisms, however, raises another possibility. The bacterial ancestors of these eukaryotes were clearly under considerable selective pressure by DNA viruses, as demonstrated by nucleotide word bias indicated earlier. However, this nucleotide word bias is absent in parasitic *Mycoplasma genetialium* as well as being absent from mitochondrial DNA, suggesting that, once within its parasitized host, infection of these organisms by DNA viruses no longer poses a risk and is not a strong selective pressure (Gelfand and Koonin, 1997). If the early eukaryotes also escaped DNA viruses by becoming parasitic, this could account for the lack of such viruses now.

Microalgae/Eukaryotic Algae

The eukaryotic algae represent the oldest known eukaryote for which there exists clear geological data (Knoll, 1992) and, although some differences exist relative to microsporidia, algae may even predate *Giardia* according to EF-G2 phylogenetics (Figure 15.2B). Microalgae are single-cell or multicell plants (with chloroplast). These species include the unicellular asexual *Chlorella*-like green algae, which are normally endosymbionts in ciliate *P. bursaria* and hydra, as well as the multicellular, sexual brown algae (Van Etten *et al.*, 1991; Van Etten, 1994). *Chlorella* species are more related to terrestrial plants then to brown algae and it appears that *Chlorella* viral promoters are highly active in plant (and bacterial) cells (Mitra *et al.*, 1994). These algal species are highly prevalent in the oceans and fresh water (Reisser, 1993). It appears that all classes of eukaryotic algae have DNA viruses. These viruses are members of the phycodnavirus family, are water-stable and have rather large (140–190 nm) polyhedral capsids, with no external membrane, but containing a lipid component located inside the outer capsid. The viral genomes are large linear ds DNA (300 kb) with closed hairpin ends. In addition, these viruses resemble bacteriophage in that the capsid remains outside the cell wall after DNA penetration, and attached virus will bore a hole in the cell wall. Virus particles can be found at concentrations of 10^8–10^{11} per liter in fresh water and virus level correlates to spring blooms of phytoplankton (Van Etten *et al.*, 1991; Reisser, 1993).

Another unique aspect of phycodnaviruses is that the viral genome appears to be the only non-prokaryotic source of restriction enzymes which cuts di- and tri-nucleotides. *Chlorella* virus NY-2A encodes at least 12 DNA endonuclease/methyltransferase genes (Mitra and Higgins, 1994; Zhang *et al.*, 1998). That these are viral (not host) genes might suggest that phycodnaviruses are so highly prevalent that they have evolved restriction modification systems in order to preclude each other from their *Chlorella* host, and such occlusion is known to occur (Chase *et al.*, 1989).

The major classification of *Chlorella* virus is related to the New and Old World hosts that they infect. Phycodnaviruses are species-specific and will differentiate between strains and subspecies of algae by an unknown mechanism (Mueller *et al.*, 1996b). Interestingly, *Chlorella* will resist phycodnavirus virus infection when it is within its paramecium host (Van Etten, 1994). A well-studied member of this family is PBCV-1, which is unusual among DNA viruses in that it codes for hyaluronan synthase (DeAngelis *et al.*, 1997). This is an outer membrane bound enzyme linked to polysaccharide, closely associated with all higher animals and some pathogenic bacteria. That this DNA virus codes for these genes suggests one way in which such a complex phenotype might be able to move during evolution between organisms. PBCV-1 is generally lytic in its host, with no known lysogenic version, although it is possible that an unknown species of host could support a persistent infection. It is interesting that *Chlorella* species are haploid and asexual, although how this relates to phycodnavirus is not known. This virus appears therefore to be an a-virus in its natural host.

Evolutionarily, the phycodnaviruses are most similar to the cytoplasmic iridoviruses (via major capsid protein; Sonntag *et al.*, 1994), which are acute viruses of insects and aquatic vertebrates. In addition, the DNA polymerase sequences of the phycodnaviruses show clear similarity to the herpesvirus and other eukaryotic viral DNA polymerase genes (poxvirus, baculovirus, African swine fever; Van Etten *et al.*, 1991; Chen and Suttle, 1996). As PBCV-1 encodes its own DNA polymerase, it also resembles PDR and the T-even phages (PDR has a 15 kb linear DNA genome with capsid external to lipid membrane and only a lytic relationship with it host).

Unicellular and Filamentous Brown Algae

Brown-filamentous algae are distinct from *Chlorella*-like species in that they are diploid, sexually reproducing, multicellular and differentiating. Unlike PBCV-1, the phycodnaviruses that infect these brown algal host are also quite biologically distinct in that they generally establish persistent or even endogenous (genomic) infections (Mueller *et al.*, 1996b). The best studied of these algal viruses are *Ecotocarpus siliculosus* virus (EsV), *Feldmannia simplex* virus (FsV) and *Micromonas pusilla* virus (MPV). These viral genomes also differ from *Chlorella* virus as the DNA is circular, and, unlike CSV, only 1% of C and 3% of A nucleotides in the viral genomes are methylated. This type of methylated genome is a characteristic of prokaryotic organisms. In addition to and in contrast with the bacterial and *Chlorella* viruses, these viruses of the diploid algae are non-lytic yet code for a viral DNA polymerase (as well as other replication proteins). These viruses have circular DNA genomes that, in contrast to all the DNA viruses of higher eukaryotes, can integrate outside the yeast 2 μm circle (Meinhardt *et al.*, 1997). In some species of brown alga (such as *Feldmania* species – FsV), all members appear to harbor and express virus, probably due to integrated viral genomes, as transmission follows mendelian genetics (Sengco *et al.*, 1996).

FsV replication, however, is developmentally regulated and occurs in highly differentiated reproductive (zoospore-forming) structures of the plant (Mueller *et al.*, 1996a,b; Del Campo *et al.*, 1997). In vegetative algae cells, infection is inapparent, with about one copy per cell. The viral infection can interfere with zoospore development, but the degree of viral interference with zoospore formation varies widely, from mild to severe depending on the specific host species infected. Esv will infect *Feldmania* zoospores, without multiplication, and will cause malformations in host, while the viral DNA persists. EsV-infected host will express virus in reproductive structures, but there is no measurable effect on host growth and reproduction (Del Campo *et al.*, 1997). Thus phycodnavirus infections in these species are essential-

ly benign and inapparent in specific hosts and are also associated with reproductive systems of the host. This is consistent with a p-virus life-strategy.

Did Eukaryotic Polymerases Evolve from a DNA Virus?

The phycodnaviruses and their algal host are near the junction that separates Bacteria from Eukaryotes and infection with DNA viruses is highly prevalent in both these orders. If we examine the phylogenetic relationship of phycodnavirus DNA polymerase relative to cellular and viral polymerases, we can see links both to host polymerases and polymerases of *Archaea*, *Bacteria* and other related DNA viruses that were not seen with cellular DNA polymerase. If DNA viruses have played a role in eukaryotic host evolution, the viral progenitors should appear to be near the root of all eukaryotic DNA polymerases and provide the link that connects prokaryotic and eukaryotic genomes. Normally, viral DNA polymerases would be excluded from such an analysis of host evolution.

There are clear similarities within functionally crucial domains between bacterial phage plymerase (T4, PDR1), human pol-α and yeast pol 1, herpes and vaccinia viral polymerase, and yeast linear plasmids (Bernad *et al.*, 1987; Miller *et al.*, 1988; Spicer *et al.*, 1988; Braithwaite and Ito, 1993). All these DNA polymerase genes have a distinct modular nature with colinear and highly conserved functional motifs of decreasing similarity (denoted; I, II, III, IV, V and VI; Bernad *et al.*, 1987; Wang, 1991; Heringa and Argos, 1994b). The regions of greatest similarity between the phage DNA pols and CV1 (I, II, III) are the same regions that are most conserved between CV-1 and the various eukaryotic DNA pols and appear to be important for dNTP and PP_i binding (Wang, 1991). In addition, the T4 and φ29 phage polymerases are inhibited by the same compounds (aphidicolin, PAA, BuAdATP) that are characteristic of eukaryotic DNA polymerase alpha (Bernad *et*

(*Text continues on page 412*)

TABLE 15.2   Results of a Gapped-BLAST (Genbank) analysis from the DNA polymerase gene of *Feldmania* species phycodnavirus

				Score	E-value
gb	|AF013260|	AF013260	*Feldmania* sp. virus DNA-dependent DNA pol	1995	0.0
gb	|AF020193|	AF020193	Glycine max DNA polymerase delta (soy bean)	328	2e-88
emb	|X88928|	DMDNAPOLD	*D. melanogaster* mRNA for DNA polymerase-Δ	320	8e-86
gb	|L07734|	YSPDNAPOLD	*Schizosaccharomyces pombe* DNA polymerase-Δ	312	2e-83
emb	|X15477|	SCCDC2	Yeast *CDC2* gene for DNA polymerase III	307	5e-82
emb	|X61920|	SCPOL3	*S. cerevisiae POL3* gene for DNA polymerase II	305	2e-81
emb	|X88804|	CAPOL3GEN	*C. albicans POL3* gene	305	2e-81
gb	|M64715|	PFADNAPDB	*Plasmodium falciparum* DNA polymerase-Δ	302	2e-80
gbl	U83704|	MAU83704	*Mesocricetus auratus* DNA polymerase-Δ	296	1e-78
gb	|M80395|	BOVDPLDA	Bovine DNA polymerase-Δ	291	3e-77
gb	|M86836|	CHVDPOA	*Chlorella* virus DNA polymerase (*dpo*) gene	246	6e-77
gb	|M81735|	HUMDNAPOLD	Human DNA polymerase-Δ	290	9e-77
emb	|X59278|	SPPOL3	*S. pombe POL3* gene for DNA polymerase	289	2e-76
emb	|Z21848|	MMDPDCS	*M. musculus* DNA polymerase-Δ	289	2e-76
emb	|AJ222691|	RNDPDELTA	*Rattus norvegicus* DNA polymerase-Δ	288	3e-76
gb	|M86837|	CHVDPOB	*Chlorella* virus DNA polymerase (*dpo*) gene	247	7e-64
gb	|M31122|	HSVSPOLGBP	Herpesvirus saimiri major DNA binding protein	215	4e-54
gb	|U97553|	U97553	Murine herpesvirus 68 strain WUMS	214	5e-54
gb	|AF005370|	AF005370	Alcelaphine herpesvirus 1 L-DNA	210	9e-53
gb	|U20824|	EHVU20824	Equine herpesvirus 2	202	2e-50
gb	|U43400|	HHU43400	Human herpesvirus-7 (HHV7)	194	5e-48
gb	|AF033184|	AF033184	Rhesus cytomegalovirus	194	6e-48
gb	|U93872|	U93872	Kaposi's-sarcoma-associated herpesvirus	193	1e-47
gb	|L25706|	GPCDNPO	Guinea-pig cytomegalovirus	187	6e-46
gb	|U68299|	MCU68299	Mouse cytomegalovirus 1	180	8e-44
gb	|M63804|	HH6DNAPOL	Human herpesvirus 6 ORF R, 3' end, DNA poly	178	4e-43
gb	|L40431|	HSKUL30H	Gallid herpesvirus 2 DNA polymerase	163	1e-38
gb	|AF030027|	AF030027	Equine herpesvirus 4 strain	161	7e-38
emb	|AJ224971|	FHV224971	Feline herpesvirus 1 DNA polymerase gene	158	4e-37
emb	|X04771|	HEHSVIDP	Herpes simplex virus type 1 DNA polymerase	151	5e-35
gb	|U32975|	MPU32975	*Micromonas pusilla* virus MpV-SP1 DNA polymerase	150	8e-35
gb	|M29683|	YSCREV3	*S. cerevisiae* DNA polymerase (*rev3*) gene,	150	8e-35
dbj	|D12983|	PYWPOL	*Pyrococcus furiosus pol* gene for DNA polymerase	148	5e-34
gb	|U32976|	MPU32976	*Micromonas pusilla* virus MpV-SP2 DNA polymerase	145	4e-33
gb	|M16321|	HS2POL	HSV-2 DNA polymerase gene, complete cds.	143	1e-32
gb	|U47108|	TSU47108	*Thermococcus* sp. DNA polymerase A (PolA)	142	2e-32
emb	|Z86099|	HSV2HG52	Herpes simplex virus type 2 (strain HG52)	141	5e-32
emb	|Z54173|	PSP710POL	*Pyrococcus* sp. *pol* gene	140	9e-32
gb	|U32985|	PBU32985	*Paramecium bursaria Chlorella* virus CVA-1 DNA pol	140	1e-31
emb	|Z54174|	PAPABPOL	*P. abyssi pol* gene	139	3e-31
dbj	|D14980|	HS5TOWNE	Human cytomegalovirus gene for DNA polymerase	130	1e-28
gb	|M14709|	HS5POL	Human cytomegalovirus (HCMV) DNA polymerase	130	1e-28
gb	|AE001070|	AE001070	*Archaeoglobus fulgidus* section 37 of 172	127	1e-27
gb	|M86664|	HSECOMGEN	Equine herpesvirus 1	125	3e-27
gb	|U32983|	CSU32983	*Chrysochromulina* spp. virus CbV-PW1 DNA pol	125	5e-27
gb	|U50550|	RCU50550	Rat cytomegalovirus major DNA binding protein	122	3e-26
gb	|U59426|	OTU59426	*Oxytricha trifallax* DNA polymerase alpha	122	4e-26
emb	|X94677|	BHV1ULGNS	Bovine herpesvirus type 1 genes for UL 27	122	4e-26
gb	|U32984|	CSU32984	*Chrysochromulina* spp. virus CbV-PW3 DNA pol	120	1e-25
dbj	|D38574|	PYRDNAPB	*Pyrodictium occultum polB* gene for DNA pol	118	4e-25
gb	|U78172|	LDU78172	*Leishmania donovani* DNA polymerase-α	118	4e-25
gb	|U92269|	PHU92269	Phocine herpesvirus type 1 DNA polymerase	114	7e-24
emb	|X60951|	TBPOLG	*T. brucei* pol-associated gene 1 and *pol* gene	112	3e-23
gb	|AF035537|	AF035537	*Homo sapiens* DNA polymerase-ζ	112	3e-23
gb	|J03268|	YSCPOL1M	Yeast (*S. cerevisiae*) polymerase I gene	111	8e-23
gb	|L33366|	MVODNAPOL	*Methanococcus voltae* DNA polymerase gene	106	3e-21
gb	|U02001|	ONU02001	*Oxytricha nova* DNA polymerase-α	104	6e-21
gb	|U82173|	PAU82173	*Pichia angusta* DNA polymerase III gene	104	6e-21
gb	|L19042|	ASVDNAP	African swine fever virus DNA polymerase	104	1e-20
dbj	|AB004461|	AB004461	*Oryza sativa* mRNA for DNA polymerase-α	103	1e-20
gb	|AE000888|	AE000888	*Methanobacterium thermoautotrophicum* from . . .	103	2e-20
emb	|X73330|	ASFVDPOL	African swine fever virus gene for DNA polymerase	102	3e-20
emb	|X69673|	SPPOLAL	*S. pombe polA* gene for DNA polymerase-α	102	4e-20
gb	|U27575|	ASU27575	African swine fever virus PAAr-463 DNA pol	101	8e-20

TABLE 15.2   (*continued*)

				Score	E-value
gb	|U92875|	SSU92875	*Sulfolobus solfataricus* DNA polymerase gene	90	1e-16
gb	|M14793|	HS2POLD	Herpes simplex virus type 2 (strain 186) DNA pol	85	7e-15
gb	|M62646|	ECOPOLB	*E. coli* DNA polymerase II (polB)	82	6e-14
dbj	|D38573|	PYRDNAPA	*Pyrodictium occultum polA* gene for DNA pol	81	8e-14
dbj	|D90310|	DROPOLA	*D. melanogaster POLA* gene for DNA polymerase	81	8e-14
gb	|L18785|	PFAPOLA	*Plasmodium falciparum* DNA polymerase-α	80	2e-13
gb	|M38283|	ECODINA	*Escherichia coli* polymerase II (*dinA*) gene	80	2e-13
gb	|U00707|P	SU00707	*Pyrococcus* sp. Deep Vent DNA polymerase	79	4e-13
emb	|AL010220|	PFSC04011	*Plasmodium falciparum* DNA pol	79	4e-13
gb	|U89699|	U89699	*Halteria* sp. macronuclear DNA polymerase-α	78	9e-13
gb	|U89701|	U89701	*Paraurostyla viridis* macronuclear DNA polymerase	76	2e-12
emb	|X66049|	PPDNAPOLD	*P. polycephalum* mRNA (partial) for DNA pol	76	4e-12
gb	|U35732|	SAU35732	*Spodoptera ascovirus* DNA polymerase (DP)	76	4e-12
gb	|U33846|	SAU33846	*Sulfolobus acidocaldarius* DNA polymerase	73	2e-11
gb	|U89705|	U89705	*Uroleptus gallina* macronuclear DNA polymerase	73	2e-11
gb	|M74198|	THCVDPE	*Thermococcus litoralis* I-TliI endonuclease	68	6e-10
gb	|M14708|	HS5DNAPOL	Human cytomegalovirus DNA polymerase gene	68	6e-10
gb	|U89703|	U89703	*Stylonychia lemnae* macronuclear DNA polymerase	68	1e-09
gb	|AF031812|	AF031812	Ovine herpesvirus 2 DNA dependent DNA pol	66	4e-09
gb	|M87045|	HH6DNAPOL2	Human herpesvirus 6, strain AJ, DNA polymerase	64	2e-08
emb	|Y13030|	TSTYPOL	*Thermococcus* sp. *pol* gene	62	3e-08
gb	|U89706|	U89706	*Urostyla grandis* macronuclear DNA polymerase	62	4e-08
gb	|M31638|	FPVPOLDA	Fowl poxivirus DNA polymerase	62	6e-08
gb	|AF009136|	AF009136	*Leishmania amazonensis* DNA polymerase-α	62	6e-08
gb	|U67532|	U67532	*Methanococcus jannaschii* section 74 of 150	61	1e-07
gb	|AF009149|	AF009149	*Leishmania mexicana* DNA polymerase-α	60	2e-07
gb	|M63941|	PFADNAPDA	*Plasmodium falciparum* DNA polymerase-Δ	59	4e-07
gb	|AF009139|	AF009139	*Leishmania chagasi* DNA polymerase-α ·	59	5e-07
gb	|AF009141|	AF009141	*Leishmania donovani* DNA polymerase-α	59	5e-07
gb	|AF009147|	AF009147	*Leishmania infantum* DNA polymerase-α	59	5e-07
gb	|U89700|	U89700	*Holosticha* sp. macronuclear DNA polymerase-α	59	5e-07
gb	|AF009140|	AF009140	*Leishmania deanei* DNA polymerase-α	58	7e-07
emb	|X89500|	CHVPOLYM1	Canine herpesvirus DNA polymerase gene	57	1e-06
gb	|AF009138|	AF009138	*Leishmania braziliensis* DNA polymerase-α	57	1e-06
gb	|AF009148|	AF009148	*Leishmania major* DNA polymerase-α gene	57	1e-06
gb	|AF009135|	AF009135	*Leishmania aethiopica* DNA polymerase-α	56	3e-06
gb	|AF009142|	AF009142	*Endotrypanum monterogei* DNA polymerase-α	56	3e-06
gb	|AF009144|	AF009144	*Leishmania herreri* DNA polymerase-α	56	3e-06
gb	|AF031811|	AF031811	Bovine herpesvirus 4 DNA dependent DNA pol	55	4e-06
gb	|AF024570|	AF024570	*Mus musculus* DNA polymerase-Δ	55	4e-06
gb	|AF005477|	AF005477	Kaposi's-sarcoma-associated herpesvirus DNA pol	54	1e-05
gb	|AE000807|	AE000807	*Methanobacterium thermoautotrophicum* from . . .	50	2e-04
emb	|X89502|	CHVPOLYM3	Canine herpesvirus DNA polymerase gene	49	3e-04
gb	|U89704|	U89704	*Stylonychia mytilus* macronuclear DNA polymerase	48	6e-04
dbj	|D11476|	LPVDNAP	*Lymantria dispar* nuclear polyhedrosis virus	48	0.001
emb	|X57314|	CBVDNAPOL	*C. biennis* entomopoxvirus gene for DNA pol	45	0.005
gb	|U89702|	U89702	*Pleurotricha lanceolata* macronuclear DNA pol	45	0.006
gb	|U92874|	SSU92874	*Sulfolobus shibatae* N-terminal acetyl transf . . .	43	0.019
gb	|M75136|	IH1CG	Ictalurid herpesvirus 1 (channel catfish virus) DNA pol	42	0.056
gb	|U58368|	ONU58368	*Oxytricha nova* micronuclear DNA polymerase-α	42	0.056
gb	|U89697|	OTU89697	*Oxytricha trifallax* DNA polymerase-α	41	0.074
gb	|U63465|	HHU63465	Human herpesvirus 6B DNA polymerase gene	41	0.096
gb	|U00429|	PSU00429	*Pyrococcus* sp. (strain GI-J) DNA polymerase	40	0.22
gb	|M13213|	VACPOLD	Vaccinia virus DNA polymerase gene	38	0.64
gb	|M36339|	VACPOLRDA	Vaccinia virus RNA polymerase (rpo30)	38	0.64
emb	|Y08257|	SS56KBFR	*S. solfataricus* 56 kb DNA fragment	38	0.84
gb	|L22579|	VARCG	Variola major virus (strain Bangladesh-1975)	37	1.4
gb	|U63466|	HHU63466	Human herpesvirus 7 DNA polymerase gene	37	1.4
emb	|X87158|	PPDNAPOL1	*P. polycephalum* mRNA for DNA polymerase-α	36	3.2
gb	|AF034436|	AF034436	*Xestia c-nigrum* granulovirus DNA polymerase	36	3.2

The database used was the translated version of the non-redundant Genbank (June, 1998). The first 250 similar sequences were retrieved and the sequences were then filtered to eliminate subgene clones (less than 300 amino acids) and highly similar or equivalent sequences.

TABLE 15.3 Results of a Gapped-BLAST (Genbank) analysis from the DNA polymerase gene of the phycodnavirus *Chlorella* virus-1

				Score	E-value
gb	|M86836|	CHVDPOA	*Chlorella* virus DNA polymerase (*dpo*) gene	1327	0.0
emb	|X88804|	CAPOL3GEN	*C. albicans* POL3 gene	384	e-105
gb	|AF020193|	AF020193	Glycine max DNA polymerase-Δ	383	e-104
emb	|X61920|	SCPOL3	*S. cerevisiae* POL3 gene for DNA polymerase II	369	e-100
emb	|X59278|	SPPOL3	*S. pombe* POL3 gene for catalytic subunit DNA pol	369	e-100
emb	|X88928|	DMDNAPOLD	*D. melanogaster* mRNA for DNA polymerase-Δ	365	2e-99
emb	|X15477|	SCCDC2	Yeast *CDC2* gene for DNA polymerase III	359	1e-97
gb	|M80395|	BOVDPLDA	Bovine DNA polymerase-Δ catalytic	350	5e-95
gb	|M80397|	HUMDNAPOLC	Human DNA polymerase-Δ catalytic subunit	349	9e-95
gb	|U83704|	MAU83704	*Mesocricetus auratus* DNA polymerase-Δ	347	5e-94
emb	|Z21848|	MMDPDCS	*M. musculus* mRNA for DNA polymerase-Δ	346	1e-93
emb	|X62423|	PFPOLD	*P. falciparum* pol-Δ gene for DNA polymerase	343	9e-93
gb	|M64715|	PFADNAPDB	*Plasmodium falciparum* DNA polymerase-Δ	343	9e-93
emb	|AJ222691|	RNDPDELTA	*Rattus norvegicus* mRNA for DNA polymerase-Δ	341	3e-92
gb	|AF013260|	AF013260	*Feldmania* sp. virus DNA-dependent DNA pol	287	7e-76–86
gb	|M31122|	HSVSPOLGBP	Herpesvirus saimiri major DNA binding protein	264	4e-69
gb	|U75698|	KSU75698	Kaposi's-sarcoma-associated herpesvirus	262	2e-68
gb	|AF005370|	AF005370	Alcelaphine herpesvirus 1L-DNA	251	4e-65
gb	|AF037218|	AF037218	Human herpesvirus 7 (HHV7 strain RK)	244	5e-63
emb	|V01555|	EBV	Epstein–Barr virus (EBV) genome	232	2e-59
gb	|U20824|	EHVU20824	Equine herpesvirus 2, complete genome	230	7e-59
gb	|M63804|	HH6DNAPOL	Human herpesvirus 6 (HHV-6) ORF R, DNA pol	229	2e-58
gb	|U97553|	U97553	Murine herpesvirus 68 strain WUMS, complete	225	2e-57
gb	|AF029302|	AF029302	Rhesus monkey rhadinovirus H26-95	221	3e-56
gb	|U32975|	MPU32975	*Micromonas pusilla* virus MpV-SP1 DNA polymerase	218	3e-55
gb	|M73549|	HS5GLYPROB	Murine cytomegalovirus	217	9e-55
gb	|L25706|	GPCDNPO	Guinea-pig cytomegalovirus glycoprotein B and pol	212	2e-53
gb	|AF033184|	AF033184	Rhesus cytomegalovirus glycoprotein B gene and pol	210	9e-53
gb	|U32983|	CSU32983	*Chrysochromulina* spp. virus CbV-PW1 DNA polymerase	189	2e-46
emb	|AJ224971|	FHV224971	Feline herpesvirus 1 DNA polymerase gene	174	5e-42
gb	|M86664|	HSECOMGEN	Equine herpesvirus 1 complete genome	172	3e-41
gb	|AF035537|	AF035537	*Homo sapiens* DNA polymerase-Z (REV3) mRNA	164	4e-39
emb	|Z78205|	BHT1UL	Bovine herpesvirus type 1 UL22–35 genes	159	2e-37
gb	|U47108|	TSU47108	*Thermococcus* sp. DNA polymerase A (*PolA*) gene	156	2e-36
emb	|X06745|	HSPOLAR	Human mRNA for DNA polymerase-α subunit	154	7e-36
dbj	|D38574|	PYRDNAPB	*Pyrodictium occultum polB* gene for DNA pol	153	9e-36
gb	|U84155|	PWU84155	*Pyrococcus woesei* DNA-dependent DNA polymerase	153	9e-36
dbj	|D12983|	PYWPOL	*Pyrococcus furiosus pol* gene for DNA polymerase	153	9e-36
emb	|Z54174|	PAPABPOL	*P. abyssi pol* gene	152	3e-35
gb	|AE001070|	AE001070	*Archaeoglobus fulgidus* section 37 of 172 . . .	149	2e-34
emb	|X04771|	HEHSV1DP	Herpes simplex virus type 1 DNA polymerase	149	2e-34
emb	|X60951|	TBPOLG	*T. brucei* pol-associated gene 1 and *pol* gene	148	4e-34
gb	|M29683|	YSCREV3	*S. cerevisiae* DNA polymerase (*rev3*) gene	147	9e-34
gb	|U50550|	RCU50550	Rat cytomegalovirus major DNA binding protein	146	1e-33
emb	|X03181|	HEHSV1PO	Herpes simplex virus type 1 (HSV-1) dbp/pol	146	2e-33
gb	|M16321|	HS2POL	HSV-2 DNA polymerase gene	145	3e-33
gb	|L40431|	HSKUL30H	Gallid herpesvirus 2 DNA polymerase (UL30h)	143	1e-32
emb	|Z86099|	HSV2HG52	Herpes simplex virus type 2 (strain HG52)	143	1e-32
emb	|X69673|	SPPOLAL	*S. pombe polA* gene for DNA polymerase-α	141	5e-32
gb	|J03268|	YSCPOL1M	Yeast (*S. cerevisiae*) polymerase I gene	140	8e-32
gb	|L33366|	MVODNAPOL	*Methanococcus voltae* DNA polymerase gene	139	1e-31
gb	|AF030027|	AF030027	Equine herpesvirus 4 strain NS80567, complete . . .	139	2e-31
gb	|U78172|	LDU78172	*Leishmania donovani* DNA polymerase-α	136	2e-30
dbj	|AB004461|	AB004461	*Oryza sativa* mRNA for DNA polymerase-α	132	3e-29

TABLE 15.3 (*continued*)

				Score	E-value
gb	|U59426|	OTU59426	*Oxytricha trifallax* DNA polymerase-α	130	9e-29
gb	|U02001|	ONU02001	*Oxytricha nova* DNA polymerase-α	122	3e-26
gb	|U92269|	PHU92269	Phocine herpesvirus type 1 DNA polymerase	120	1e-25
dbj	|D14980|	HS5TOWNE	Human cytomegalovirus gene for DNA polymerase	119	2e-25
dbj	|D90310|	DROPOLA	*D. melanogaster* POLA gene for DNA polymerase	118	3e-25
gb	|L18785|	PFAPOLA	*Plasmodium falciparum* DNA polymerase-α	118	5e-25
gb	|U92875|	SSU92875	*Sulfolobus solfataricus* DNA polymerase gene	116	1e-24
emb	|X64466|	SSPOLS	*S. solfataricus* polS gene	116	2e-24
dbj	|D38573|	PYRDNAPA	*Pyrodictium occultum* polA gene for DNA pol	116	2e-24
gb	|S48157|	S48157 DNA	Polymerase-primase 180 kDa subunit (*Drosophila . . .*	91	9e-23
gb	|U33846|	SAU33846	*Sulfolobus acidocaldarius* DNA polymerase gene	109	2e-22
gb	|M13213|	VACPOLD	Vaccinia virus DNA polymerase gene	102	3e-20
gb	|L22579|	VARCG	Variola major virus (strain Bangladesh-1975) complete . . .	102	3e-20
gb	|AE000888|	AE000888	*Methanobacterium thermoautotrophicum* from . . .	101	6e-20
gb	|AF028831|	AF028831	*Cenarchaeum symbiosum* DNA polymerase gene	100	1e-19
gb	|U49979|	OVU49979	Orf virus E10R homolog gene	99	2e-19
gb	|AE000116|	ECAE000116	*Escherichia coli* K-12 MG1655 section 6 . . .	97	1e-18
gb	|M35371|	ECOPOLBDA	*E. coli* DNA polymerase II (*polB*) gene	97	1e-18
gb	|M38283|	ECODINA	*Escherichia coli* polymerase II (*dinA*) gene	96	3e-18
gb	|L19042|	ASVDNAP	African swine fever virus DNA polymerase	94	8e-18
gb	|U00707|	PSU00707	*Pyrococcus* sp. Deep Vent DNA polymerase	90	1e-16
gb	|U60315|	MCU60315	*Molluscum contagiosum* virus subtype 1	87	1e-15
gb	|M31638|	FPVPOLDA	Fowl poxvirus DNA polymerase	87	1e-15
gb	|U67532|	U67532	*Methanococcus jannaschii* section 74 of 150 of . . .	87	2e-15
gb	|M74198|	THCVDPE	*Thermococcus litoralis* I-TliI endonuclease (I . . .	77	1e-12
gb	|U35732|	SAU35732	*Spodoptera ascovirus* DNA polymerase (DP)	76	3e-12
gb	|AF034436|	AF034436	*Xestia c-nigrum* granulovirus DNA polymerase	75	4e-12
gb	|AF031812|	AF031812	Ovine herpesvirus 2 DNA dependent DNA pol	75	5e-12
gb	|U89701|	U89701	*Paraurostyla viridis* macronuclear DNA polymerase	75	7e-12
emb	|X66049|	PPDNAPOLD	*P. polycephalum* mRNA (partial) for DNA pol	73	2e-11
emb	|X57314|	CBVDNAPOL	*C. biennis* entomopoxvirus gene for DNA pol	72	4e-11
gb	|M63941|	PFADNAPDA	*Plasmodium falciparum* DNA polymerase-Δ	71	8e-11
gb	|U34036|	BRU34036	Bacteriophage RB69 DNA polymerase (43) gene	70	2e-10
gb	|AF031811|	AF031811	Bovine herpesvirus 4 DNA dependent DNA pol	69	3e-10
gb	|M14793|	HS2POLD	Herpes simplex virus type 2 (strain 186) DNA pol	68	5e-10
gb	|M14708|	HS5DNAPOL	Human cytomegalovirus DNA polymerase gene	68	9e-10
emb	|X89500|	CHVPOLYM1	Canine herpesvirus DNA polymerase gene	67	1e-09
gb	|U89703|	U89703	*Stylonychia lemnae* macronuclear DNA polymerase	66	2e-09
gb	|U89706|	U89706	*Urostyla grandis* macronuclear DNA polymerase-α	66	4e-09
gb	|U89699|	U89699	*Halteria* sp. macronuclear DNA polymerase-α	66	4e-09
gb	|U89705|	U89705	*Uroleptus gallina* macronuclear DNA polymerase . . .	66	4e-09
emb	|Y13030|	TSTYPOL	*Thermococcus* sp. pol gene	63	2e-08
gb	|AF005477|	AF005477	Kaposi's-sarcoma-associated herpesvirus DNA pol	63	2e-08
gb	|M87045|	HH6DNAPOL2	Human herpesvirus 6, strain AJ, DNA polymerase	63	2e-08
gb	|U11242|	NPVU11242	*Helicoverpa zea* nuclear polyhedrosis virus	61	7e-08
gb	|U89700|	U89700	*Holosticha* sp. macronuclear DNA polymerase-α	61	7e-08
dbj	|D16231|	D16231	*Bombyx mori* nuclear polyhedrosis virus DNA pol	57	1e-06
gb	|L33180|	NPHT3COMP	*Bombyx mori* nuclear polyhedrosis virus	57	1e-06
gb	|AE000807|	AE000807	*Methanobacterium thermoautotrophicum* from . . .	55	4e-06
gb	|U18677|	CFU18677	*Choristoneura fumiferana* nuclear polyhedrosis virus	55	5e-06
gb	|89704|	U89704	*Stylonychia mytilus* macronuclear DNA polymerase	54	9e-06
dbj	|D11476|	LPVDNAP	*Lymantria dispar* nuclear polyhedrosis virus	54	9e-06
gb	|L22858|	NPHCG	*Autographa californica* nuclear polyhedrosis virus	54	1e-05
gb	|U75930|	OPU75930	*Orgyia pseudotsugata* nuclear polyhedrosis virus	54	1e-05
gb	|U00429|	PSU00429	*Pyrococcus* sp. (strain GI-J) DNA polymerase	53	3e-05

TABLE 15.3   *(continued)*

				Score	E-value
gb	IAF024570I	AF024570	*Mus musculus* DNA polymerase-Δ	52	5e-05
emb	IX71597I	SSDNAPOL	*S. solfataricus* gene for DNA polymerase	50	2e-04
gb	IAF009136I	AF009136	*Leishmania amazonensis* DNA polymerase-α	49	3e-04
gb	IAF009138I	AF009138	*Leishmania braziliensis* DNA polymerase-α	49	4e-04
gb	IAF009149I	AF009149	*Leishmania mexicana* DNA polymerase-α	48	5e-04
gb	IU89702I	U89702	*Pleurotricha lanceolata* macronuclear DNA polymerase	48	5e-04
gb	IAF009148I	AF009148	*Leishmania major* DNA polymerase-α gene	48	9e-04
gb	IM37727I	ECODINA5	*E. coli* DNA polymerase II (dinA) gene	47	0.002
gb	IU67263I	HAU67263	*Helicoverpa armigera* nucleopolyhedrovirus DNA pol	46	0.002
gb	IAF009135I	AF009135	*Leishmania aethiopica* DNA polymerase-α	46	0.002
gb	IM75136I	IH1CG	Ictalurid herpesvirus 1 (channel catfish virus)	46	0.002
gb	IAE001057I	AE001057	*Archaeoglobus fulgidus* section 50 of 172 of . . .	46	0.003
gb	IAF009140I	AF009140	*Leishmania deanei* DNA polymerase-α gene	45	0.006
gb	IU58368I	ONU58368	*Oxytricha nova* micronuclear DNA polymerase-α	40	0.010
gb	IAF009142I	AF009142	*Endotrypanum monterogei* DNA polymerase-α	44	0.010
gb	IAF009144I	AF009144	*Leishmania herreri* DNA polymerase-α	44	0.010
gb	IAF009141I	AF009141	*Leishmania donovani* DNA polymerase-α	44	0.010
gb	IAF009147I	AF009147	*Leishmania infantum* DNA polymerase-α	44	0.010
emb	IX87158I	PPDNAPOL1	*P. polycephalum* mRNA for DNA polymerase-α	44	0.010
gb	IAF009139I	AF009139	*Leishmania chagasi* DNA polymerase-α	43	0.030
gb	IU89697I	OTU89697	*Oxytricha trifallax* DNA polymerase-α	41	0.067
emb	IX89502I	CHVPOLYM3	Canine herpesvirus DNA polymerase gene	41	0.087
gb	IU63456I	AHU63456	Aotine herpesvirus 1 DNA polymerase gene, pa . . .	36	2.2

The database used was the translated version of the non-redundant Genbank (June, 1998). The first 250 similar sequences were retrieved and the sequences were then filtered to eliminate subgene clones (less than 300 amino acids) and highly similar or equivalent sequences.

al., 1987), as is the herpesvirus DNA polymerase (Wang *et al.*, 1995). All this suggests a tighter link between bacterial DNA viruses and eukaryotic replication proteins than is seen with corresponding cellular DNA replication proteins. Furthermore, the T4 genome generally has more genes that are similar to eukaryotic genes or viruses (especially fungal introns) than *E. coli* genes, including the topoisomerase and ligase genes (for references see Kutter *et al.*, 1995).

A gapped BLAST analysis of all the predicted proteins related to the phycodnavirus DNA polymerases in the combined Genbank translated nucleotide database is shown in Tables 15.2 and 15.3. This list has been filtered to exclude subsequences and highly related or redundant genes and included sequences that are known or very likely to code for a gene. As can be seen, sequences related to phycodnaviral DNA pol (both FsV and CV-1) yield a large set of clearly similar proteins that belong to the B family of

DNA polymerases from numerous organisms, including most orders of Eukaryotes, many large eukaryotic DNA viruses (but not adenovirus), *Archaea*, *Bacteria* and several lytic bacterial phages. Interestingly, the FsV polymerase was not in the sequences most similar to CV-1 nor is the converse true. In both cases, it was unexpected that these viral polymerases were more similar to the host DNA polymerase-Δ from fungal, plant and other species. As oligonucleotide primers that amplify DNA pol sequences from *Chlorella* don't amplify sequences from brown algal viruses, it appears that these two viral polymerase genes differ significantly from each other and that algal viruses make up a phylogenetically diverse group (Chen and Suttle, 1996). In addition, both the alpha (primase interacting) and delta (extension synthesis) replicative DNA polymerases were similar to phycodnavirus polymerase, although the delta polymerases were clearly most similar.

Curiously, mouse and human DNA pol-Δ were more similar to both FsV and CV-1 DNA pol then were the closest herpes DNA pol; herpesvirus saimiri. The HSV 1 pol was one of the most dissimilar sequences.

Using the PAUP phylogenetic inference program to compare these polymerases to each other by neighbor joining and parsimony analysis, the dendogram in Figure 15.3 was generated. Although nine equally parsimonious trees actually resulted, these trees only differed in their tip placement and the overall node structures were conserved. As the tree shown in Figure 15.3 does not show tip detail, it accurately represents all these equally parsimonious trees. These results are consistent with other prior phylogenetic analyses, which included viral DNA polymerases (Braithwaite and Ito, 1993; Heringa and Argos, 1994b). Similar kinds of organism and polymerase (alpha, delta) are clearly clustered (higher eukaryotes, lower *Eukarya*, *Archaea*, *Bacteria*, pox viruses, herpes viruses, etc.), but the links between these clusters are at odds with accepted evolutionary relationships.

Overall, three distinct large clusters of polymerases are seen (labeled 1, 2 and 3 in Figure 15.3). The first cluster (1) contains many members of the beta and gamma herpesvirus family and contains HHV-6, cytomegalovirus and EBV. This cluster is at the root of the second cluster (2), which contains most of the phycodnaviruses (with the *Feldmania* viral sequence most resembling the root), which are at the base of the cluster containing the eukaryotic DNA polymerase-Δ genes. Joining this cellular pol-Δ cluster is another cluster that has the replicative DNA polymerases of various fungal and plasmodia species. The internodal distances within the overall cluster 1 are not large, suggesting that the parsimony algorithm has not artifactually forced the nodal connections. The remaining large cluster (3) is actually composed of three subclusters (a, b and c), which have the insect virus (spodopasco) off the common root for this set. The a-subcluster contains various lytic cytoplasmic DNA viruses, including vaccinia, molluscum contagosium and African swine fever virus families. Most curiously, lytic phage T4 is also in this cluster, as well as various

lower eukaryotic hypotrich species. Although T4 phage might have been considered a candidate percursor for the phycodnaviral DNA polymerase, its positioning on the dendogram away from the common roots of phycodnaviral genes make T4 seem more like a decendant than the ancestor of eukaryotic DNA viruses. Clearly, T4 DNA polymerase bears a striking similarity to eukaryotic DNA polymerases, as noted previously (Spicer *et al.*, 1988). The b-subcluster contains the replicative DNA pol-α from vertebrates and lower eukaryotes in addition to a set of sequences from Leishman species with fungal and *Plasmodium* genes at the base. The c-subcluster of group 3 contains both Archaea and Bacteria species and is hence labeled "prokaryotic". However, these are all related to the "nonessential" DNA polymerase II (pol *B*) gene of *E. coli* that lacks 5′ to 3′ exonuclease activity. The replicative DNA pol III of *E. coli* did not show significant similarity to the phycodnaviral DNA polymerase. In addition, various members of the herpesvirus family – canine, feline and including herpes simplex virus I/II and VZV (alphaherpesviruses) and some unclassified herpesvirus members – were outside these three main clusters and were connected at the root of all the other DNA polymerases.

It is most interesting to note that all the phycodnaviral DNA *pol* genes branch off from near the root of lineage that contains the replicative DNA pol-Δ cluster of all eukaryotes. Although this was previously observed with only the CV-1 sequence (Braithwaite and Ito, 1993), no significance was given to this observation. The analysis suggests that these viruses did not acquire their DNA polymerase from other eukaryotes, but acquired this gene prior to the evolution of the eukaryotic DNA *pol*-Δ gene or its ancestor. It therefore seems possible that these viral polymerases may be related to the original progenitor of host DNA polymerase delta.

A related, but more complex, situation may also apply to the DNA pol-α genes of all eukaryotes as well. Here, all the replicative DNA polymerases from *Archaea*, *Bacteria* and DNA pol-α from eukaryotes are predicted to have a common root that links these to the eukaryotic pol-Δ sequences. The sequence most similar to the base root, however, would be the DNA polymerase gene from insect DNA ascoviruses. In addition, it is observed that the connections between these DNA pol-α like and DNA pol-Δ like clusters are most similar to various phycodnaviral and insect DNA polymerases. In other words, if there is a common root for all these DNA polymerase genes, then it appears that viral DNA polymerases of early eukaryotes are most similar to that cenancestral gene. Given the non-orthologous nature of DNA replication proteins between bacteria and eukaryotes noted previously (Edgell and Doolittle, 1997) and the clear similarity that can be seen in phycodnaviral and cellular DNA polymerase, these results support the idea that a DNA virus with a distinct set of replication proteins could have been the ancestor of the eukaryotic replication system. Similar analysis (FAST A) of other DNA replication genes, two different PCNA genes and type II topoisomerase sequences of PBCV-1 phycodnavirus also identifies different species of plants or lower eukaryotes respectively as being most similar to these viral replication protein sequences (J. Van Etten, personal communication).

Clearly, additional analyses are needed in order to better evaluate these rather counterintuitive implications. In addition, if virus-derived subgene modules are also involved in this host evolution, as noted with *S. thermophilis* phage evolution earlier, only multiple module transfer to host would be clear from the analysis above. Still, we are left with the implication that the tree in Figure 15.3, which seems so incongruent with accepted biological relationships, is an accurate reflection of the horizontal transfer of replication genes that link otherwise separate lineages. This conclusion can also help explain why there appears to be some basic link between the life-strategies of entire orders of organisms and the viruses parasitizing them, because the host may sometimes be dependent on early viral parasites for subsequent host evolution. If this interpretation is correct, then our binary neighbor-joining method of phylogenetic analysis is unable to accurately predict evolutionary history, at least for transferred genes. Thus, new analytical methods that will consider and assess horizontal transfer of genes resulting in reticulated trees would be needed (Brower *et al.*, 1996; Page and Charleston, 1997). The horizontal (infectious) movement of p-elements in the world's population of *Drosophila melanogaster* that occurred during this century suggests that such processes are more common in eukaryotes than we have previously appreciated (Clark and Kidwell, 1997).

GENERAL CONSIDERATIONS

Could virus-mediated large-scale gene transmission allow the evolution of complex characters not attainable by point changes? Could viral infections aid in the genetic isolation and speciation of their host? Do persistent DNA viruses contribute in a different manner from acute DNA viruses to host evolution? Although we cannot clearly answer these questions, I have presented both theoretical arguments and many experimental observations that support the idea that persistent and acute viral life-strategies are distinct and can have major and different effects on their host populations. Many more possible examples from the eukaryotic DNA viruses could also have been presented that adhere to the idea that p-virus life-strategies are well conserved (including human herpes virus, polydnaviruses of wasps, mammalian polyomavirus, papillomavirus) and distinct from that of a-virus life-strategies (including insect baculoviruses, acquatic iridioviruses, carnivore parvoviruses, smallpox virus). The consequences of these agents for host evolution are clearly distinct.

What genetic elements can be traced back to the beginning of evolution? It would seem obvious that the replication system should be one such ancient element and we have seen

FIGURE 15.3 Phylogenetic analysis of sequences similar to the *Feldmania* species phycodnavirus DNA polymerase sequence. The filtered set of sequences from Table 15.2 were used for a PAUP analysis.

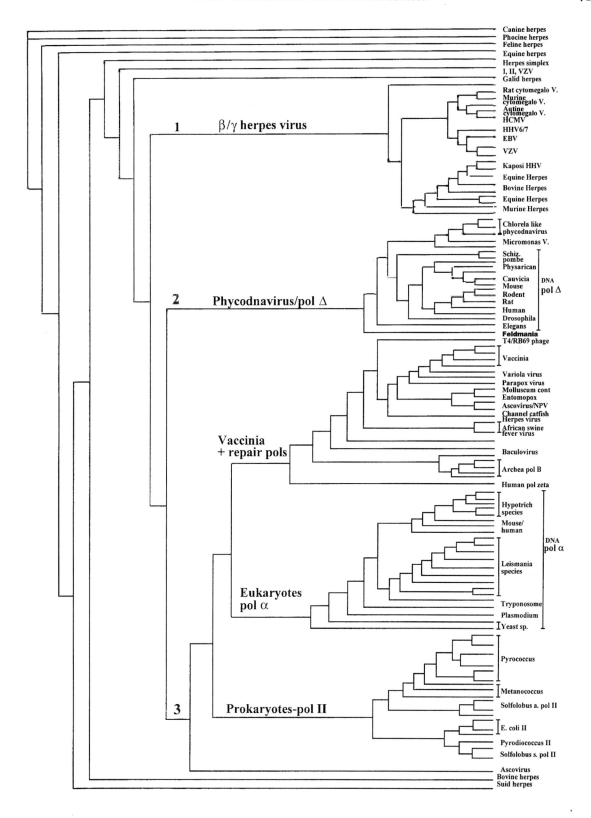

evidence that rolling circular replication proteins are indeed highly conserved throughout evolution. Yet we have considered the situation with the replicative cellular DNA polymerase and note that, unless viral DNA polymerases are included, a phylogenetic analysis does not link *Bacteria* with *Archaea/Eukarya* replication proteins. Could it be that viral replication systems should also be considered an evolutionarily conserved component in the lineage of host evolution? Recently, a sequence highly similar to *E. coli* DNA pol l (a "non-essential" family A DNA polymerase similar to T5/T7 DNA pol; Knopf, 1998), has been reported to be found widely dispersed in many eukaryotic sequences, yet there is no known homolog for pol l in eukaryotes (Sonnhammer and Wootton, 1998), leading the authors to consider that horizontal transmission might account for this observation.

T. Dobzhansky recognized mutation as the "raw material" on which natural selection acts. With B. McClintock's discovery of transposable elements in maize, mutagenic mechanisms became more horizontal in nature. The accumulating genomic sequence data show us many examples that suggest that horizontal transfer is more prevalent than has been accepted. It seems likely that the mutagenic process in evolution is generally enhanced by infectious agents. Perhaps Salvador Luria's (1959) speculation was more accurate than we later came to believe when he wrote, considering the role the temperate bacterial viruses might play in evolution of the host, that:

> May we not feel that in the virus, in their merging with the cellular genome and their re-emerging from them, we observe processes which, in the course of evolution, have created the successful genetic patterns that underlie all living things?

REFERENCES

Anderson, R.M. and May, R.M. (1979) Population biology of infectious diseases: Part I. *Nature*, **280**, 361–367.

Barlow, D.P. (1993) Methylation and imprinting: from host defense to gene regulation? *Science*, **260**, 309–310.

Bernad, A., Zaballos, A., Salas, M. and Blanco, L. (1987) Structural and functional relationships between prokaryotic and eukaryotic DNA polymerases. *EMBO J.*, **6**, 4219–4225.

Bernstein, H. and Bernstein, C. (1989) Bacteriophage T4 genetic homologies with bacteria and eucaryotes. *J. Bacteriol.*, **171**, 2265–2270.

Bhattacharyya, S., Lorimer, H.E. and Prives, C. (1995) Murine polyomavirus and simian virus 40 large T antigens produce different structural alterations in viral origin DNA. *J. Virol.*, **69**, 7579–7585.

Blaisdell, B.E., Campbell, A.M., and Karlin, S. (1996) Similarities and dissimilarities of phage genomes. *Proc. Natl Acad. Sci. USA*, **93**, 5854–5859.

Braithwaite, D.K. and Ito, J. (1993) Compilation, alignment, and phylogenetic relationships of DNA polymerases. *Nucl. Acids Res.*, **21**, 787–802.

Brower, A.V.Z., DeSalle, R. and Vogler, A. (1996) Gene trees, species trees, and systematics: a cladistic perspective. *Gene*, 423–450.

Brussow, H., Bruttin, A., Desiere, F., Lucchini, S. and Foley, S. (1998) Molecular ecology and evolution of *Stretococcus thermophilus* bacteriophages – a review. *Virus Genes*, **16**, 95–109.

Chase, T.E., Nelson, J.A., Burbank, D.E. and Van Etten, J.L. (1989) Mutual exclusion occurs in a *Chlorella*-like green alga inoculated with two viruses. *J. Gen. Virol.*, **70**, 1829–1836.

Cheetham, B.F. and Katz, M.E. (1995) A role for bacteriophages in the evolution and transfer of bacterial virulence determinants. *Mol. Microbiol.*, **18**, 201–208.

Chen, F. and Suttle, C.A. (1996) Evolutionary relationships among large double-stranded DNA viruses that infect microalgae and other organisms as inferred from DNA polymerase genes. *Virology*, **219**, 170–178.

Clark, J.B. and Kidwell, M.G. (1997) A phylogenetic perspective on P transposable element evolution in Drosophila. *Proc. Natl Acad. Sci. USA*, **94**, 11428–11433.

Cochlan, W.P., Wikner, J., Steward, G.F., Smith, D.C. and Azam, F. (1993) Spatial distribution of viruses, bacteria and chlorophyll a in neritic, oceanic and estuarine environments. *Marine Ecol. Prog. Ser.*, **92**, 77–87.

DeAngelis, P.L., Jing, W., Graves, M.V., Burbank, D.E. and Van Etten, J.L. (1997) Hyaluronan synthase of chlorella virus PBCV-1. *Science*, **278**, 1800–1803.

DeFilippis, V.R. and Villarreal, L.P. (1998) An introduction to the evolutionary ecology of viruses. In: *Viral Ecology* (ed. Hurst, C.). John Wiley, New York.

Del Campo, E., Ramazanov, Z., Garcia-Reina, G. and Mueller, D.G. (1997) Photosynthetic responses and growth performance of virus-infected and noninfected Ectocarpus siliculosus (Phaeophyceae). *Phycologia*, **36**, 186–189.

Diffley, J.F. (1994) Eukaryotic DNA replication. *Curr. Opin. Cell Biol.*, **6**, 368–372.

Diffley, J.F. (1995) The initiation of DNA replication in the budding yeast cell division cycle. *Yeast*, **11**, 1651–1670.

Diffley, J.F. (1996) Once and only once upon a time: specifying and regulating origins of DNA replication in eukaryotic cells. *Genes Devel.*, **10**, 2819–2830.

Dobzhansky, T. (1937) *Genetics and the Origin of Species*. Columbia University Press, New York.

Doerfler, W. (1991) Patterns of DNA methylation – evolutionary vestiges of foreign DNA inactivation as a host defense mechanism. A proposal. *Biol. Chem. Hoppe Seyler*, **372**, 557–564.

Doerfler, W. (1996) A new concept in (adenoviral) oncogenesis: integration of foreign DNA and its consequences. *Biochim. Biophys. Acta*, **1288**, F79–F99.

Domingo, E. and Holland, J.J. (1997) RNA virus mutations and fitness for survival. *Annu. Rev. Microbiol.*, **51**, 151–178.

Domingo, E., Escarmis, C., Sevilla, N. *et al.* (1996) Basic concepts in RNA virus evolution. *FASEB J.*, **10**, 859–864.

Donovan, S. and Diffley, J.F. (1996) Replication origins in eukaryotes. *Curr. Opin. Genet. Devel.*, **6**, 203–207.

Edgell, D.R. and Doolittle, W.F. (1997) Archaea and the origin(s) of DNA replication proteins. *Cell*, **89**, 995–998.

Finlay, B.B. and Falkow, S. (1997) Common themes in microbial pathogenicity revisited. *Microbiol. Mol. Biol. Rev.*, **61**, 136–169.

Fisher, R.A. (1930) *The Genetical Theory of Natural Selection*, Clarendon Press, Oxford.

Forterre, P. (1997) Protein versus rRNA: problems in rooting the Universal Tree of Life. *Asm. News*, **63**, 89–95.

Garnett, G.P. and Antia, R. (1998) *The Evolutionary Biology of Viruses* (ed. Morse, S.S.). Raven Press, New York.

Gavin, K.A., Hidaka, M. and Stillman, B. (1995) Conserved initiator proteins in eukaryotes. *Science*, **270**, 1667–1671.

Gelfand, M.S. and Koonin, E.V. (1997) Avoidance of palindromic words in bacterial and archaeal genomes: a close connection with restriction enzymes. *Nucl. Acids Res.*, **25**, 2430–2439.

Giske, J., Aksnes, D.L. and Forland, B. (1993) Variable generation times and Darwinian fitness measures. *Evol. Ecol.*, **7**, 233–239.

Gorbalenya, A.E., Koonin, E.V. and Wolf, Y.I. (1990) A new superfamily of putative NTP-binding domains encoded by genomes of small DNA and RNA viruses. *FEBS Lett.*, **262**, 145–148.

Green, P., Lipman, D., Hillier, L., *et al.* (1993). Ancient conserved regions in new gene sequences in the protein data base. *Science*, **259**, 1711–1716.

Haggard-Ljungquist, E., Halling, C. and Calendar, R. (1992) DNA sequences of the tail fiber genes of bacteriophage P2: evidence for horizontal transfer of tail fiber genes among unrelated bacteriophages. *J. Bacteriol.*, **174**, 1462–1477.

Heitman, J. (1993) On the origins, structures and functions of restriction-modification enzymes. *Genet. Eng. (NHY)*, **15**, 57–108.

Heringa, J. and Argos, P. (1994a) Evolution of viruses as recorded by their polymerase sequences. *Proc. Natl. Acad. Sci. USA*, 87–103.

Heringa, J. and Argos, P. (1994) The evolution of viruses as recorded by their polymerase sequences. In: *The Evolutionary Biology of Viruses* (eds Morse, S.S.), pp. 87–103. Raven Press, New York.

Holmes, R.K. and Schmitt, M.P. (1994) Bacteriophage toxins and disease. In: *Encyclopedia of Virology* (eds Webster, R.G. and

Granoff, A.), pp. 101–106. Academic Press, San Diego, CA.

Ilyina, T.V. and Koonin, E.V. (1992) Conserved sequence motifs in the initiator proteins for rolling circle DNA replication encoded by diverse replicons from eubacteria, eucaryotes and archaebacteria. *Nucl. Acids. Res.*, **20**, 3279–3285.

Karlin, S. and Burge, C. (1995) Dinucleotide relative abundance extremes: a genomic signature. *Trends Genet.*, **11**, 283–290.

Karlin, S. and Ladunga, I. (1994) Comparisons of eukaryotic genomic sequences. *Proc. Natl Acad. Sci. USA*, **91**, 12832–12836.

Kasprzak, W. and Majewska, A.C. (1995) Viruses of parasitic protozoa. *Wiadomosci Parazytologiczne*, **41**, 131–137.

Klieve, A.V., Hudman, J.F. and Bauchop, T. (1989) Inducible bacteriophages from ruminal bacteria. *Appl. Environ. Microbiol.*, **55**, 1630–1634.

Klobutcher, L.A., Turner, L.R. and LaPlante, J. (1993) Circular forms of developmentally excised DNA in *Euplotes crassus* have a heteroduplex junction. *Genes Dev.*, **7**, 84–94.

Knoll, A.H. (1992) The early evolution of eukaryotes: a geological perspective. *Science*, **256**, 622–627.

Knopf, C.W. (1998) Evolution of viral DNA-dependent DNA polymerases. *Virus Genes*, **16**, 47–58.

Kohler, S. and Wang, C.C. (1997) Site-specific binding of polymerase-containing articles of the *Giardia lamblia* double-stranded RNA virus to the viral plus-strand RNA. *Res. Virol.*, **148**, 311–321.

Krawiec, S. and Riley, M. (1990) Organization of the bacterial chromosome. *Microbiol. Rev.*, **54**, 502–539.

Krikau, M.F. and Jahn, C.L. (1991) Tec2, a second transposon-like element demonstrating developmentally programmed excision in *Euplotes crassus*. *Mol. Cell. Biol.*, **11**, 4751–4759.

Kutter, E., Gachechiladze, K., Poglazov, A. *et al.* (1995) Evolution of T4-related phages. *Virus Genes*, **11**, 285–297.

Lindqvist, B.H., Deho, G. and Calendar, R. (1993) Mechanisms of genome propagation and helper exploitation by satellite phage P4. *Microbiol. Rev.*, **57**, 683–702.

Loo, S., Fox, C.A., Rine, J., Kobayashi, R., Stillman, B. and Bell, S. (1995) The origin recognition complex in silencing, cell cycle progression, and DNA replication. *Mol. Cell. Biol.*, **6**, 741–756.

Luria, S.E. (1959) *Virus Growth and Variation* (eds Isaacs, A. and Lacey, B.W.), pp. 1–10. Cambridge University Press, London.

MacArthur, R.H. and Wilson, E.O. (1967) *The Theory of Island Biogeography*. Princeton University Press, Princeton, NJ.

McDaniel, T.K. and Kaper, J.B. (1997) A cloned pathogenicity island from enteropathogenic *Escherichia coli* confers the attaching and effacing phenotype on E. coli K-12. *Mol. Microbiol.*, **23**, 399–407.

Manganelli, R., Ricci, S. and Pozzi, G. (1996) Conjugative transposon Tn916: evidence for excision with formation of 5′-protruding termini. *J. Bacteriol.*, **178**, 5813–5816.

May, R.M. and Anderson, R.M. (1979) Population biology of infectious disease: Part II. *Nature*, **280**, 455–461.

May, R.M. and Nowak, M.A. (1995) Coinfection and the evolution of parasite virulence. *Proc. Roy. Soc. Lond. B Biol. Sci.*, **261**, 209–215.

Meinhardt, F., Schaffrath, R. and Larsen, M. (1997) Microbial linear plasmids. *Appl. Microbiol. Biotechnol.*, **47**, 329–336.

Meyer, J. (1994) P1 bacteriophage. In: *Encyclopedia of Virology* (eds Webster, R.G. and Granoff, A.), Academic Press, San Diego, CA.

Miller, M.A., Korn, D. and Wang, T.S. (1988) The evolutionary conservation of DNA polymerase alpha. *Nucl. Acids Res.*, **16**, 7961–7973.

Mitra, A. and Higgins, D.W. (1994) The *Chlorella* virus adenine methyltransferase gene promoter is a strong promoter in plants. *Plant Mol. Biol.*, **26**, 85–93.

Mitra, A., Higgins, D.W. and Rohe, N.J. (1994) A *Chlorella* virus gene promoter functions as a strong promoter both in plants and bacteria. *Biochem. Biophys. Res. Commun.*, **204**, 187–194.

Mizuuchi, K. (1992) Transpositional recombination: mechanistic insights from studies of mu and other elements. *Annu. Rev. Biochem.*, **61**, 1011–1051.

Mueller, D.G., Braeutigam, M. and Knippers, R.

(1996a) Virus infection and persistence of foreign DNA in the marine brown alga Feldmannia simplex (Ectocarpales, Phaeophyceae). *Phycologia*, **35**, 61–63.

Mueller, D.G., Sengco, M., Wolf, S. *et al.* (1996b) Comparison of two DNA viruses infecting the marine brown algae *Ectocarpus siliculosus* and *E. fasciculatus*. *J. Gen. Virol.*, **77**, 2329–2333.

Mushegian, A.R. and Koonin, E.V. (1996) A minimal gene set for cellular life derived by comparison of complete bacterial genomes. *Proc. Natl Acad. Sci. USA*, **93**, 10268–10273.

Naito, T., Kusano, K. and Kobayashi, I. (1995) Selfish behavior of restriction-modification systems. *Science*, **267**, 897–899.

Nuttall, S.D. and Dyall-Smith, M.L. (1993) HF1 and HF2: novel bacteriophages of halophilic archaea. *Virology*, **197**, 678–684.

Nuttall, S.D. and Dyall-Smith, M.L. (1995) Halophage HF2: genome organization and replication strategy. *J. Virol.*, **69**, 2322–2327.

Page, R.D. and Charleston, M.A. (1997) From gene to organismal phylogeny: reconciled trees and the gene tree/species tree problem. *Mol. Phylogenet. Evol.*, **7**, 231–240.

Reissér, W. (1993) Viruses and virus-like particles of freshwater and marine eukaryotic algae-a review. *Arch. Protistenkunde*, **143**, 257–265.

Schleper, C., Kubo, K. and Zillig, W. (1992) The particle SSV1 from the extremely thermophilic archaeon Sulfolobus is a virus: demonstration of infectivity and of transfection with viral DNA. *Proc. Natl Acad. Sci. USA*, **89**, 7645–7649.

Sengco, M.R., Brautigam, M., Kaap, M. and Mueller, D.G. (1996) Detection of virus DNA in *Ectocarpus siliculosus* and *E. fasciculatus* (Phaeophyceae) from various geographic areas. *Eur. J. Phycol.*, **31**, 73–78.

Slater, S., Wold, S., Lu, M., Boye, E., Skarstad, K. and Kleckner, N. (1995) *E. coli* SeqA protein binds oriC in two different methyl-modulated reactions appropriate to its roles in DNA replication initiation and origin sequestration. *Cell*, **82**, 927–936.

Sogin, M.L. and Silberman, J.D. (1998) Evolution of the protists and protistan parasites from the perspective of molecular systematics. *Int. J. Parasitol.*, **28**, 11–20.

Sonnhammer, E.L.L. and Wootton, J.C. (1998) Widespread eukaryotic sequences, highly similar to bacterial DNA polymerase I, looking for functions. *Curr. Biol.*, **7**, R463–R465.

Sonntag, K.C., Schnitzler, P., Koonin, E.V. and Darai, G. (1994) Chilo iridescent virus encodes a putative helicase belonging to a distinct family within the "DEAD/H" superfamily: implications for the evolution of large DNA viruses. *Virus Genes*, **8**, 151–158.

Sowers, K.R. (1995) *Restriction-modification Systems of Methanogenic Archaea*. Cold Spring Harbor Laboratory Press, Cold Spring Harbor, NY.

Spicer, E.K., Rush, J., Fung, C., Reha-Krantz, L.J., Karam, J.D. and Konigsberg, W.H. (1988) Primary structure of T4 DNA polymerase. Evolutionary relatedness to eucaryotic and other procaryotic DNA polymerases. *J. Biol. Chem.*, **263**, 7478–7486.

Steward, G.F., Smith, D.C. and Azam, F. (1996) Abundance and production of bacteria and viruses in the Bering and Chukchi Seas. *Marine Ecol. Prog. Ser.*, **131**, 287–300.

Taylor, J.H. (1984) Origins of replication and gene regulation. *Mol. Cell Biochem.*, **61**, 99–109.

Teodoro, J.G. and Branton, P.E. (1997) Regulation of apoptosis by viral gene products. *J. Virol.*, **71**, 1739–1746.

Tommasino, M. (1991) Killer system of *Kluyveromyces lactis*: the open reading frame 10 of the pGK12 plasmid encodes a putative DNA binding protein. *Yeast*, **7**, 245–252.

Van Etten, J.L. (1994) Algal viruses In: *Encyclopedia of Virology* (eds Webster, R.G. and Granoff, A.), pp. 35–40. Academic Press, San Diego, CA.

Van Etten, J.L., Lane, L.C. and Meints, R.H. (1991) Viruses and viruslike particles of eukaryotic algae. *Microbiol. Rev.*, **55**, 586–620.

Vossbrinck, C.R., Maddox, J.V., Friedman, S., Debrunner-Vossbrinck, B.A. and Woese, C.R. (1987) Ribosomal RNA sequence suggests microsporidia are extremely ancient eukaryotes. *Nature*, **326**, 411–414.

Wain-Hobson, Simon (1998) Is antigenic variation of HIV important for AIDS and what might be expected in the future? In: *The Evolutionary Biology of Viruses*. (ed Morse, S.S.). Raven Press, New York.

Wang, C.C., Yeh, L.S. and Karam, J.D. (1995) Modular organization of T4 DNA polymerase. Evidence from phylogenetics. *J. Biol. Chem.*, **270**, 26558–26564.

Wang, T.S. (1991) Eukaryotic DNA polymerases. *Annu. Rev. Biochem.*, **60**, 513–552.

Wilson, G.G. and Murray, N.E. (1991) Restriction and modification systems. *Annu. Rev. Genet.*, **25**, 585–627.

Woese, C. (1998) The universal ancestor. *Proc. Natl Acad. Sci. USA*, **95**, 6854–6859.

Woese, C.R., Kandler, O. and Wheelis, M.L. (1990) Towards a natural system of organisms: proposal for the domains Archaea, Bacteria, and Eucarya. *Proc. Natl Acad. Sci. USA*, **87**, 4576–4579.

Zhang, Y., Nelson, M., Nietfeldt, J. *et al.* (1998) *Chlorella* virus NY-2A encodes at least 12 DNA endonuclease/methyltransferase genes. *Virology*, **240**, 366–375.

Zillig, W., Prangishvilli, D., Schleper, C. *et al.* (1996) Viruses, plasmids and other genetic elements of thermophilic and hyperthermophilic Archaea. *FEMS Microbiol. Rev.*, **18**, 225–236.

16

Parvovirus Variation and Evolution

Colin R. Parrish and Uwe Truyen

INTRODUCTION TO PARVOVIRUSES AND THEIR PROPERTIES

Parvoviruses comprise a family of small viruses that have a non-enveloped capsid that contains a linear single-stranded DNA genome of between 4500 and 5250 nt. The viruses are very widely distributed in nature and they infect many different vertebrate and invertebrate hosts (Cotmore and Tattersall, 1987; Murphy *et al.*, 1995). The two subfamilies within the family Parvoviridae are the Parvovirinae, which infect vertebrate hosts, and the Densovirinae, which infect invertebrates. Within the Parvovirinae the three recognized genera are the parvoviruses (autonomous parvoviruses including various rodent parvoviruses, the parvoviruses of carnivores including canine parvovirus (CPV), feline panleukopenia virus (FPV) and Aleutian mink disease virus (ADV)); the erythroviruses (human B19 virus and related viruses of primates; and the dependoviruses (adeno-associated viruses (AAV) of humans and other hosts, which primarily replicate in cells that are co-infected with an adenovirus or herpesvirus). In the Densovirinae the three genera are *Densovirus*, *Iteravirus* and *Brevidensovirus*. Densovirinae infect many different invertebrate hosts from the class Insecta, including members of the orders Lepidoptera, Diptera and Orthoptera. Poorly characterized parvoviruses appear to infect members of the order Decapoda

(shrimps and prawns) from the class Crustacea.

The parvoviruses are genetically simple and have between one and three transcriptional promoters depending on the particular virus. Through a variety of strategies those give rise to messages for between one and four non-structural proteins, and between two and four capsid proteins. Although there has been work at a variety of levels on many different parvoviruses, most detailed studies of DNA replication have been of the genomes of the minute virus of mouse (MVM) parvovirus and the human AAV type-2 dependovirus (reviewed by: Berns and Linden, 1995; Cotmore and Tattersall, 1995; Hong *et al.*, 1994). Gene regulation is also best understood for those two viruses, as well as for LuIII, the B19 human parvovirus and for ADV. Studies of the viral capsids and capsid protein structures and functions have been conducted on a number of different viruses, with the most complete model being that of CPV and the closely related FPV (reviewed by Agbandje *et al.*, 1995).

Viral Gene Functions

NS1 and Rep 78/68

The large non-structural protein encoded within the left side of the genome encodes the NS1 and Rep proteins in the parvoviruses and dependoviruses respectively (Cotmore and Tattersall,

1987, 1995; Berns, 1990). These multifunctional proteins are required for DNA replication and for regulation of viral gene expression; they have site-specific nickase activity, ATPase and helicase activities, and they are also covalently attached to the 5′ end of the viral DNA.

NS2 and Rep 52/40

The NS2 proteins of MVM or the rat virus LuIII appear to be dispensable for viral replication in host cells other than mouse or rat cells respectively, but they are required for efficient translation or assembly of the virus capsid proteins in the natural host cells (Li and Rhode, 1991; Naeger *et al.*, 1993; Cotmore *et al.*, 1997). The mechanism involved in this function is not known, but as NS2 from MVM specifically interacts with members of the 14-3-3 family of proteins (Brockhaus *et al.*, 1996), it may be involved in phosphorylation of the capsid proteins. The NS2 protein in CPV does not appear to be required for virus replication in dog and cat cells in tissue culture or in dogs (Wang *et al.*, 1998).

The specific functions of the smaller Rep proteins of AAV are poorly understood. They lack the DNA-binding functions and replication functions of the larger Rep proteins, but do retain the helicase function and stimulate the production of ssDNA (Chejanovski and Carter, 1989; Im and Muzyczka, 1992; Smith and Kotin, 1998).

Capsid Protein Genes

In the MVM- and CPV-related viruses, VP1 and VP2 are produced by alternative splicing of the viral mRNA (Cotmore and Tattersall, 1987). The proteins overlap in sequence so that the entire 65–67 kDa sequence of VP2 is found within the VP1 sequence, with VP1 having a unique N-terminal extension of between 120 and 150 amino acids, depending on the virus. A total of 60 copies of the proteins make up the 25 nm diameter capsid, with about 90% of the protein being VP2 and 10% VP1. In full (DNA-containing) capsids VP2 may be cleaved to VP3 by cellular proteases.

The capsid is the primary target of neutraliz-

ing antibody, and for most of the mammalian viruses humoral immunity, including maternal antibody or passively administered IgG, protects animals against infection (Parrish, 1995). Antibodies also appear to be important in recovery from infection, as antibody treatments can arrest CPV replication in dogs and can terminate chronic infections with the human B19 parvovirus (Brown *et al.*, 1994; Brown and Young, 1995; Young, 1995).

In the case of ADV, antibodies produced in most virus-infected mink reduce viral titers, but in many animals the virus is not efficiently neutralized by the antibody and a persistent chronic infection is established (Porter, 1986; Alexandersen *et al.*, 1994; Bloom *et al.*, 1994). The level of virus replication and intracellular gene expression is reduced by the presence of circulating antibodies, although the mechanism by which that occurs is not understood (Alexandersen *et al.*, 1988, 1994).

The antigenic structure of the capsid has been examined in detail for several parvoviruses, and there are most probably significant differences between the various viruses. Conformation-dependent neutralizing epitopes are important targets of neutralizing antibodies on the exposed surface of CPV and B19 capsids (Yoshimoto *et al.*, 1991; Brown *et al.*, 1992; Saikawa *et al.*, 1993; Strassheim *et al.*, 1994; Kawase *et al.*, 1995). Linear epitopes have also been defined on the capsid proteins, including the N-terminus of VP2, which is exposed to the exterior of CPV (Langeveld *et al.*, 1993). There appear to be differences in the topologies of the capsid proteins of the different parvoviruses, since, although the VP1-unique sequence is inside the CPV and MVM capsids, that sequence is exposed on the outside of virus-like particles assembled after coexpression of VP1 and VP2 of B19 in insect cells, and anti-VP1 antibodies neutralize infectious B19 (Anderson *et al.*, 1995; Weichert *et al.*, 1998).

The role of antigenic variation and immune selection in the epidemiology of the parvoviruses is not well understood, and the analysis is complicated by the fact that some antigenic variation is closely associated with changes in the virus capsids that alter host range or other properties (Chang, *et al.*, 1992; Strassheim *et al.*,

1994; Parker and Parrish, 1997). Although genetically widely separated viruses can be distinguished by neutralization or hemagglutination inhibition with polyclonal sera, within a particular virus group there appears to be only minor antigenic variation even among viruses that have been separated by decades. Variation of capsid epitopes can be detected using monoclonal antibody analysis, but the epidemiological significance of any antigenic differences is not well understood. Most animals that have recovered from infection appear to be resistant to re-infection by antigenically related viruses, even if those are variant at one or more epitope.

EPIDEMIOLOGY OF THE PARVOVIRUSES

Most autonomous parvoviruses cause acute infections of their host animals that last for less than 10 days, by which time the virus is cleared by the developing host immunity, and infectious virus or infected cells cannot be readily demonstrated. Prolonged replication occurs frequently in ADV-infected mink, where the virus apparently persists and continues to replicate in a number of tissues for the life of the animal (Porter, 1986; Bloom *et al.*, 1994). For B19 in humans a small number of chronic and persistent infections have been described in individuals who are immune-suppressed, or who for other reasons do not develop effective immunity (Brown *et al.*, 1994; Young, 1995).

Human AAV integrates its DNA into the host cell genome, and the integrated genome remains latent without any detectable virus gene expression or DNA replication. The virus becomes reactivated and replicates to high levels when the cell is infected by a helper virus such as adenovirus or herpesvirus (Berns, 1990; Berns and Linden, 1995). AAV is widespread in humans and generally causes non-clinical infections.

The mechanisms of transmission among the parvoviruses vary. For the carnivore parvoviruses related to CPV the virus uses fecal–oral transmission, replicating in the intestine and shedding in the faeces (Parrish, 1995).

Some of the rodent parvoviruses appear to also replicate in the intestine, and they may also be transmitted through urine after replication in the kidney. The human B19 virus is thought to be transmitted by respiratory routes (Brown and Young, 1995; Young, 1995).

HOST OR HELPER VIRUS GENE PRODUCTS INTERACTING WITH THE VIRUS

Because of the small size of their genome and the small number of gene products encoded, the parvoviruses require a great many functions supplied by the host cell. The autonomous parvoviruses replicate only in cells that are in mitotic S-phase as they require many different functions for their DNA replication that are only expressed by the cell during mitosis (Cotmore and Tattersall, 1995). AAV requires a number of cellular functions, as well as products supplied by the adenovirus or herpes helper viruses (reviewed by Berns, 1990). Adenovirus functions required for replication and gene expression include many early gene products, including E1A, E1B and E4, some of which may act indirectly by their effects on the host cell functions. Adenovirus late gene functions appear to be required for AAV replication.

PHYLOGENETIC RELATIONSHIPS AMONG THE MEMBERS OF THE FAMILY PARVOVIRUS

Little analysis has been done of the overall genetic relationships between the parvoviruses, although a comparison of the capsid protein sequences has been reported (Chapman and Rossmann, 1993). A comparison of the sequences of conserved regions of the genome, in particular of the NS1 gene, shows that all the parvoviruses have certain sequences in common, and they are most probably related through a distant common ancestor. As shown in Figure 16.1, the viruses were readily subdivided into several distinct

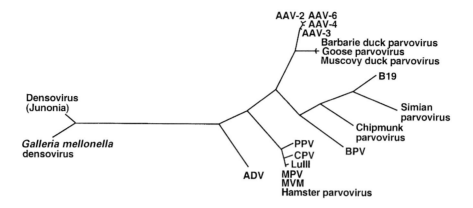

FIGURE 16.1 The phylogenetic relationships between the parvoviruses determined from about 220 residues from a conserved region of the non-structural protein 1 (NS1) or Rep proteins. Sequences from Genbank were translated, and the region equivalent to the sequence between residues 354 and 562 of MVMi was aligned using the Clustal algorithm of DNAStar. The most parsimonious relationships were determined using the Branch and Bound algorithm of the program PAUP version 3.1.1. Thirty minimal phylogenies were derived, which had the same overall topology but which differed in the arrangements of closely related sequences within some clades. Branch lengths are proportional to the number of differences between the sequences.

clades, which in some cases appeared to be associated with the hosts of origin. The insect viruses (densovirus) were genetically distinct from the vertebrate viruses (Figure 16.1). The AAVs from humans were all closely related, and are most closely related to parvoviruses from geese and ducks. Although the erythroviruses from primates (B19 and simian parvoviruses) were most closely related to each other, it appears that B19 is a distinct virus and was not recently introduced into humans from the primate viruses that have been analysed. Most viruses from rats, mice and hamsters were found to be within the same clade as CPV and several related viruses of carnivores and porcine parvovirus (PPV), while ADV was found to be quite distantly related to all the other vertebrate viruses. It is not possible to determine the times of divergence of the various viruses, as no molecular clock of the parvovirus sequences is known. However, it is likely that most of the viruses are evolving independently from their hosts, and that in the past mutations in the viruses have caused them to alter their host ranges and enter new host populations.

RECENT EVOLUTION AND TEMPORAL VARIATION

The variation of only a few parvoviruses has been studied in detail. Sequence variation of the B19 viruses in humans have been examined in several studies, where viruses from chronic and acute infections have also been compared. CPV and the closely related viruses of other carnivores appear to cause exclusively acute infections, while ADV in mink and some other carnivores causes mostly chronic and persistent infections.

Human B19 and Other Erythroviruses

The variation of the B19 human parvovirus has been studied using restriction enzyme analysis, single-strand polymorphism comparison and/or sequencing of various regions of the viral genome. Because a variety of approaches have been used to compare sequences or markers from different regions of the genomes, it is often difficult to correlate the results obtained, or to determine the significance of the different virus "types" defined in the various studies. General conclusions that have been reported by a number of studies are that between 1% and 4% sequence variation is found between isolates. There appears to be global spread of the viruses, as there is often close similarity between isolates collected from various regions of the world and at various times over the past two decades,

although viruses collected from one geographical area are generally more similar to each other. There is a tendency for viruses from patients with persistent infections to have a higher variation compared to viruses from patients with acute infections.

Comparing seven isolates collected in Italy between 1989 and 1994 from one geographical area, where there was no obvious connection between the isolates, showed a maximum variation of 0.61% within the 2400–3400 nt sequence within the amino terminal end of the VP genes (Gallinella *et al.*, 1995). These viruses were 0.7% and 0.77% different on average from the prototype Wi and Au sequences of viruses collected in the UK and USA, respectively. Of the 22 nucleotide sequence difference detected, nine resulted in amino acid sequence differences, and of those seven were between residues 4 and 114

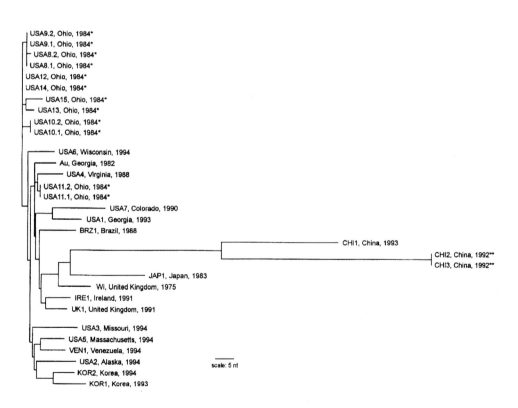

FIGURE 16.2 The phylogenetic relationships between 31 B19 human parvovirus sequences based on the VP1/VP2 coding region, with the country and year of isolation of each virus indicated. *, isolates from a single community-wide outbreak of B19; **, viruses from a mother and child with epidemiologically related B19 infection. Reproduced from Erdman *et al.*, 1996, with permission.

in VP1 sequence, while two were within the VP1/VP2 common region (Gallinella *et al.*, 1995).

Fifty B19 isolates collected throughout the world were examined by the single-stranded conformational polymorphism (SSCP) assay using sequences from a 283 nt region of the non-structural gene (1399–1682), and that showed five different SSCP types among the isolates (Kerr *et al.*, 1995). All the variation was due to single or double silent nucleotide sequence differences within the region analysed. There was found to be variation in the distribution of the different types both geographically and temporally, as type 3 strains predominated in Japan and the UK while type 4 predominated in the USA. In Japan a number of SSCP types were detected among the strains collected between 1981 and 1987, while viruses collected between 1990 and 1994 were mostly type 3 (Kerr *et al.*, 1995).

The sequences of the complete VP1 and VP2 gene region (2343 nt) of 29 isolates from 25 infected patients in various regions of the world were compared to each other, and to the two published Wi and Au sequences. Those viruses included 10 from an outbreak in Ohio, USA, one of which was a mother–child pair, and other isolates from throughout the USA, UK, Brazil, Ireland, Venezuela, Korea, Japan and China (Erdman *et al.*, 1996; Figure 16.2). The sequences differed by as few as 2 nucleotides and up to as many as 99 (4.2%), and by between 0 and 13 (1.7%) amino-acid sequence differences. No insertions or deletions were observed. Variation of nucleotides was found throughout the VP1 and VP2 genes, but the amino acid sequences clustered into three regions of the gene – in the VP1-unique region, around the junction of the VP1 and VP2 coding regions and within the VP1 and VP2 overlapping region (Figure 16.3).

In that study isolates from the outbreak in Ohio could be divided into two classes, represented by seven and two samples each (Erdman *et al.*, 1996). Within each group the sequences differed by only a few nucleotides (Figure 16.2), and where multiple isolates were collected from four individuals only a single difference was found in the sequences from one of the people. It appeared that the virus strains had a world-wide distribution, as genetically closely related viruses were obtained from various regions of the world. However, there was some geographical clustering of the strains – viruses from China formed a distinct clade, while isolates from the USA were generally clustered, as were the isolates from Korea (Erdman *et al.*, 1996; Figure 16.2).

Viruses recovered from chronic and acute infections were examined in several studies, and in most cases no specific correlation was seen between any particular disease syndrome and any virus type (Mori *et al.*, 1987; Gallinella *et al.*, 1995; Kerr *et al.*, 1995). In one study a twofold higher genomic variability was seen in the NS1, VP1 and VP2 genomic regions of B19 sequences recovered from cases of persistent infection (arthritis or other chronic diseases) compared to those from acute infections (Hemauer *et al.*, 1996). A distinct genome type was detected by restriction enzyme analysis in a case of B19-induced encephalopathy, although the specific relationship of that variation to the disease seen has not been established (Umene and Nunoue, 1995)

CPV and Related Viruses

CPV, FPV and the related viruses of other carnivores represent more than one type of evolutionary scenario. Because CPV emerged as an apparently new disease, there has been a particular interest in examining the evolution of the CPV strains compared to the long-existing FPV and related viruses. Besides sequence analysis, studies of these viruses have also included examining and comparing antigenic relationships, host range properties, capsid structures and also genetic mapping of the properties that differ between the various viruses.

Sequence Analysis

Analysis of the DNA sequences of viruses from dogs, cats, raccoons, mink, and arctic foxes gives a general understanding of how these viruses evolve under varying circumstances. Because host range and antigenic variation

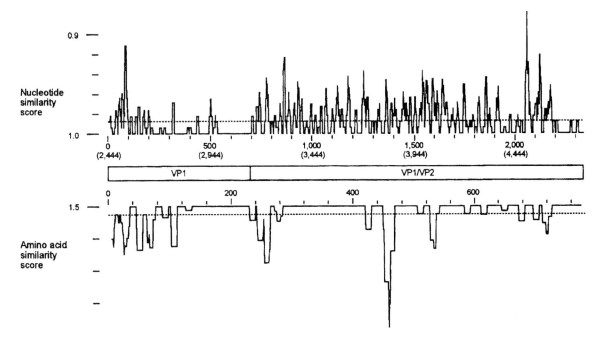

FIGURE 16.3 Variation of the nucleotide and deduced amino acid sequences of the 31 aligned B19 *VP1/VP2* genes. The extent of deflection of the solid line represents the degree of heterogeneity between sequences within a 10-nucleotide or -amino-acid window. The average similarity score for the entire alignment is shown as a dotted line. Reproduced from Erdman *et al.*, 1996, with permission.

appears to be primarily determined by the capsid protein gene, most analysis has focused on that gene, with less analysis of the *NS1* gene. The capsid protein gene sequences showed that all canine isolates formed a single clade, and they most probably derived from a single common ancestral virus sequence that arose only a single time (Figure 16.4; Truyen *et al.*, 1995). The sequences of viruses isolated from cats, mink, raccoons or foxes could not be clearly distinguished from each other, suggesting that there was interspecies transmission of those latter viruses.

Although restriction analysis of CPV and related virus sequences suggested a relationship between CPV and vaccine strains of FPV (Tratschin *et al.*, 1982), more detailed analysis of the virus DNA sequences indicated that the CPV isolates were not more closely related to those vaccine strains, and in fact may have been more closely related to a virus of red foxes (Truyen *et al.*, 1998).

Evolution of the VP1 and VP2 genes can be divided into two types. The CPV isolates appeared to be under strong selection, as there was a predominance of non-synonymous over synonymous nucleotide substitutions, and a progressive accumulation of sequence changes compared to the ancestral CPV sequence. Genetic analysis showed that much of the capsid protein gene differences between CPV and the feline, mink, raccoon and fox viruses was associated with phenotypic properties of the virus – including changes in host range, antigenicity, and sialic acid binding (Figure 16.5; Parrish, 1991; Chang *et al.*, 1992; Horiuchi *et al.*, 1994; Parker and Parrish, 1997). The common ancestor of the CPVs most probably existed in Europe between 1972 and 1976, and the rate of sequence substitution of the *VP1/VP2* gene of the CPV isolates was calculated to be 2×10^{-4} per nt per year for the first 19 years after 1978 (Parrish *et al.*, 1991; Truyen *et al.*, 1995). However, it is almost certain that this does not

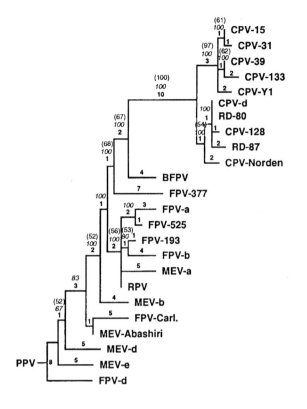

FIGURE 16.4 Phylogenetic relationships of capsid protein gene sequences of viruses from dogs (CPV), Asiatic raccoon dogs (RD), cats (FPV), mink (MEV), raccoons (RPV) and foxes (BFPV). The phylogeny was rooted using the sequence of the PPV capsid protein gene as an outgroup. Numbers indicate the number of nucleotides differing within each branch, numbers in italics indicate the number of trees that had that particular branch, and numbers in brackets indicate the bootstrap support for that particular branch in the phylogeny. Reproduced from Truyen *et al.*, 1995, with permission.

represent the rate of a general "molecular clock" for parvovirus sequences, as most of the changes within CPV are clearly under strong selection. In contrast, FPV and other virus sequences were less easy to analyse as there are no clear ancestral sequences and the closest outgroup, that of pig parvovirus, was only 74% identical with FPV, and hence was not ideal. For the FPV-related viruses, there was a lower proportion of non-synonymous changes, and it appears that those viruses are not under as strong selection as the viruses from dogs (Truyen *et al.*, 1995).

The *NS1* sequences showed similar overall phylogenetic relationships. There were about half as many nucleotide substitutions in the *NS1* gene compared to the capsid protein gene of the same viruses, and there was a lower proportion of non-synonymous sequence differences (Truyen *et al.*, 1995).

Antigenic Variation

Monoclonal antibody (MAb) analysis showed that the virus capsids from dogs all differed from those from cats, mink or raccoons in at least two specific neutralizing epitopes – one present only on CPV and the other present on FPV isolates (Mochizuki, 1989; Parrish and Carmichael, 1983; Parrish *et al.*, 1984; Strassheim *et al.*, 1994). Further analysis showed that there was further antigenic variation among the CPV isolates collected during different years after 1978 (Parrish *et al.*, 1985, 1991). CPV isolates col-

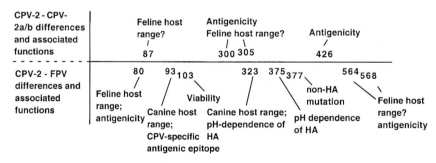

FIGURE 16.5 Locations of specific functions within the *VP2* gene of CPV and FPV, defined using genetic mapping or site-directed mutagenesis of naturally occurring differences between viruses. Above the line are *VP2* residues that differ between CPV type-2 and CPV type-2a or CPV type-2b, and below the line are residues that differ between CPV type-2 and FPV. Numbers indicate the location of the residue in the sequence of the *VP2* gene.

lected during 1978 and early 1979, after the emergence of the virus, were designated CPV type-2 (CPV-2) to distinguish them from the previously existing parvovirus, minute virus of canine. Those viruses were all antigenically identical worldwide (Parrish and Carmichael, 1983; Parrish et al., 1985, 1991). However, between 1979 and 1980 an antigenically variant strain of CPV was identified in many different countries of the world, which was termed CPV type-2a (CPV-2a; Parrish et al., 1985, 1991). CPV-2a isolates only differed in the *VP1/VP2* gene in only five or six amino acids compared to CPV type-2 isolates, but they had lost two epitopes and gained two different neutralizing epitopes on the capsid (Figure 16.6). Around 1984 a further antigenically variant virus strain was detected, which differed in one amino acid sequence within a neutralizing epitope in the capsid, and that virus was designated CPV type-2b (Parrish et al., 1991; Figure 16.6). Despite the small numbers of differences between those viruses, they each became globally distributed within a year or two of first being detected

(Parrish et al., 1988a; Senda et al., 1988; de Ybanez et al., 1995; Truyen et al., 1996b). As described below, those viruses differed in host range as well as antigenicity (Truyen et al., 1996a).

Natural antigenic variation of FPV and MEV isolates has also been described (Parrish and Carmichael, 1983; Parrish et al., 1984; Mochizuki et al., 1989). The variation in many cases was due to small numbers of changes in the sequence of the capsid protein gene – in particular due to differences of residue 300, which was in the center of one of the major neutralizing antigenic sites of the virus (Strassheim et al., 1994; Wikoff et al., 1994).

Hemagglutination and Sialic Acid Binding

CPV and FPV isolates from animals all appear to be naturally hemagglutinating, although non-hemagglutinating mutants have been derived in tissue culture (Barbis et al., 1992). Hemagglutination (HA) is a reflection of viral binding to sialic acid on erythrocyte

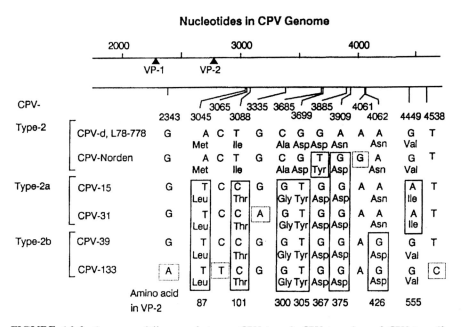

FIGURE 16.6 Sequence differences between CPV type-2, CPV type-2a and CPV type-2b virus isolates, showing the sequences within the *VP1/VP2* gene that vary between isolates of the different strains of virus, their location within the gene, and coding differences. Reproduced from Parrish et al., 1991, with permission.

glycoproteins and perhaps glycolipids. FPV or MEV isolates HA only in buffers with pHs below 6.6, while the CPV isolates HA up to at least pH 7.5 (Carmichael *et al.*, 1980; Parrish *et al.*, 1988b; Parrish, 1991; Chang *et al.*, 1992). Those viruses specifically bind N-glycolylneuraminic acid (NGNA), which is the sialic acid on feline erythrocytes of the A blood group (about 70% of most breeds of domestic cat), but which is not present on the erythrocytes of most breeds of dog (Yasue *et al.*, 1978; Andrews *et al.*, 1992; C.R. Parrish, unpublished results). The viruses do not bind N-acetylneuraminic acid, which is present on dog erythrocytes or on feline blood group B cells. As the same sialic acids are generally found on host tissues, it is likely that this binding can influence the virus–host interaction *in vivo*.

Host Range

The host range differences between the various viruses are complex, and the viruses differ in their ability to infect cells in culture and animals. All viruses replicated in feline cells tested in tissue culture, but only the isolates from dogs replicated in cultured canine cells, indicating that CPV had a property enabling it to replicate in dogs and that this was due to a change in the ability of the virus to infect individual canine cells (Truyen and Parrish, 1992).

However, when the viruses were inoculated into animals some different host ranges were observed (Truyen and Parrish, 1992). FPV isolates replicated efficiently in cats but CPV isolates showed varying feline replication, depending on the strain of CPV being tested. CPV-2 isolates did not replicate in cats but CPV-2a and CPV-2b isolates both replicated efficiently in cats after experimental inoculation (Truyen *et al.*, 1996a). That this was a natural host range for the viruses was confirmed by the finding that CPV-2a and CPV-2b were also isolated from between 10% and 20% of cats that had natural parvovirus disease in Japan, Germany and the USA (Mochizuki *et al.*, 1993; Truyen *et al.*, 1995, 1996b).

The host range for dogs also differed from that seen in tissue culture. CPV isolates replicated in a number of lymphoid tissues and the intestinal epithelium of dogs. FPV did not replicate detectably in dog cells in tissue culture but replicated in canine thymus and bone marrow cells *in vivo* but not in the peripheral lymphoid tissues or the intestine (Truyen and Parrish, 1992).

Genetic Mapping of Host Range and Other Specific Properties of CPV and FPV

Infectious plasmid clones of CPV and FPV isolates were analysed by recombination mapping and site-directed mutagenesis (Parrish *et al.*, 1988; Chang *et al.*, 1992; Horiuchi *et al.*, 1994). The specific differences between CPV and FPV – canine host range, antigenic structure and the pH dependence of HA – were all due to two sequence changes between CPV and FPV within the capsid protein gene. Residue 93 was Lys in FPV and Asn in CPV, and residue 323 was Asp in FPV and Asn in CPV (Figure 16.7). Changing either residue 93 or 323 in the CPV sequence to the FPV residue resulted in a virus that did not infect dogs or dog cells efficiently, while if both residues in FPV were changed to Asn, as in CPV, that mutant infected dog cells and dogs. The CPV-specific antigenic epitope recognized by monoclonal antibodies could be introduced by having an Asn at residue 93 in either virus background (Figure 16.7; Chang *et al.*, 1992, Agbandje *et al.*, 1993).

The pH dependence of HA which differed between CPV and FPV was also controlled primarily by residue 323, and both CPV and FPV showed a CPV-like pH dependence of HA (i.e. HA at pH up to pH 7.5) when residue 323 was an Asn, and an FPV-like HA (HA only below pH 6.6) when 323 was an Asp. An Asp–Asn difference of VP2 residue 375 also affected the pH dependence of HA, although that depended on the sequence at position 323, which showed a dominant effect over the residue at position 375 (Chang *et al.*, 1992).

Other Capsid Sequences Affecting Canine Host Range

Although VP2 residues 93 and 323 were the main determinants of canine host range of CPV, other sequences in the capsid also affect the abil-

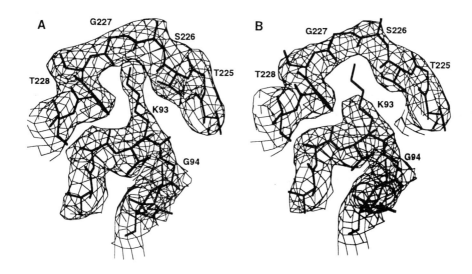

FIGURE 16.7 Structure of the capsid proteins of the CPV and FPV in the region of the sequence 93, which differed between an Asn (CPV) and Lys (FPV), and which in part controlled canine host range. The mesh indicates the electron density as determined for (**A**) FPV and (**B**) CPV. The atomic model for FPV is shown superimposed on the density in each case, showing the difference in the position of lysine 93 (K93) in FPV compared to the asparagine in CPV at that position. This difference also caused specific antigenic differences between the CPV and FPV. Reproduced from Agbandje *et al.*, 1993, with permission.

ity of the virus to infect canine cells. This was seen in a mutant of CPV-2 derived after extended passage in feline cells, where the virus lost the ability to infect dogs or dog cells in culture, and also lost reactivity to several neutralizing monoclonal antibodies (Parrish and Carmichael, 1986). The host range change was due to sequence differences of VP2 residues 300 (Ala to Asp) and 301 (Thr to Ile). A series of site-directed mutants prepared at or near those positions in the capsid structure showed that most changes reduced the efficiency of virus infection of dog cells, indicating that the overall structure of the region was important for maintenance of canine host range (Parker and Parrish, 1997). The atomic structure of one mutant of CPV (with the change of VP2 residue 300 from Ala to Asp) showed that the new Asp side chain formed a hydrogen bond with the Lys side chain of a threefold-related VP2 molecule, and perhaps stabilized the structure of the capsid (Llamas-Saiz *et al.*, 1996). Although that region was not directly associated with the canine host

range differences between FPV and CPV-2, in CPV-2a and CPV-2b a series of differences (87 Met–Leu; 300 Ala–Gly; 305 Asp–Tyr) were present in the vicinity of this region of the structure (Figure 16.6; Parrish *et al.*, 1991). These differences may enhance the ability of the virus to infect dogs and perhaps contributed to the rapid spread of the newer strains of CPV in dogs throughout the world.

Feline Host Range

CPV-2 isolates did not replicate detectably in cats but CPV-2a and CPV-2b both replicated efficiently. Recombinants between CPV-2 and FPV could only replicate in cats efficiently if FPV sequences from the two ends of the capsid protein gene were included in the recombinant (Truyen *et al.*, 1994). If only one or the other region was included the virus would replicate to a higher titer in the cat but not to the level of FPV (Figure 16.8).

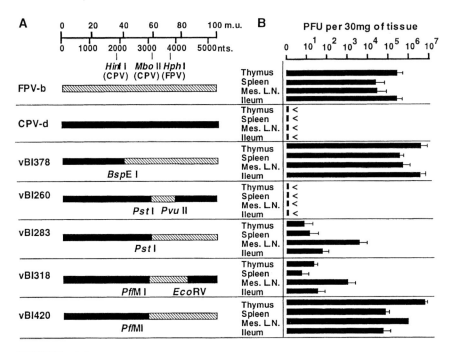

FIGURE 16.8 Genetic mapping of the feline host range difference between CPV type-2 and FPV. The figure shows the recombinants that were prepared between CPV type-2 (solid bars) and FPV (cross-hatched bars) and the virus titers in the thymus, spleen, mesenteric lymph node and ileum of cats 4 days after inoculation. Reproduced from Truyen *et al.*, 1994, with permission.

Aleutian Disease Virus

A number of studies have shown that there is significantly greater variation in the genomes of ADV isolates than is seen for most of the other parvoviruses. ADV isolates normally grow only in animals, but certain strains have been adapted to grow in feline cells in tissue culture. The genomic sequences showed a 2.5% sequence difference between a wild-type pathogenic strain of ADV and a tissue-culture-adapted virus, and there was a hypervariable sequence within the capsid protein gene at a position that is most likely within the first loop of the VP2 protein structure (Figure 16.9A, B; Bloom *et al.*, 1988; Oie *et al.*, 1996). ADV has also been recovered from ferrets, skunks and raccoons, and these viruses appear to be very similar to each other, although whether there is natural transmission between these hosts is not known. The

FIGURE 16.9 A. Comparison of the sequences of a 161 nt region of the ADV genome (ADV-G sequence 3036–3196) that includes a hypervariable region of the capsid protein gene. Viruses were obtained from mink (ADVG, DF, TR, RE, ZK8, Utah, Ontario, Pullman), skunk (Skunk) and raccoons (RAC, RE). **B**. The translation of the same sequence shown in **A**, showing VP2 amino acids 211–263. **C**. Location within the NS protein gene of variation between the four different classes (types) of ADV NS1 sequence that were present in an ADV isolate (Type 1) or found in three inocula prepared from infected mink (Types 2, 3 and 4). Nucleotide sequence differences in the longer open reading frame (LORF) are shown by the upward lines, while the amino-acid sequence differences are shown by the downward lines. Within the short ORF or middle ORF (MORF), the amino-acid substitutions are shown by lines that are 1.5 times longer. Reproduced from Gottschalck *et al.*, 1994, with permission. **D**. Dendrogram showing the relationships between the amino acid sequences of the four types of ADV described in **C**. The horizontal lines are proportional to the distance between the sequences, and the percentage differences are indicated. Reproduced from Gottsohalck *et al.*, 1994, with permission.

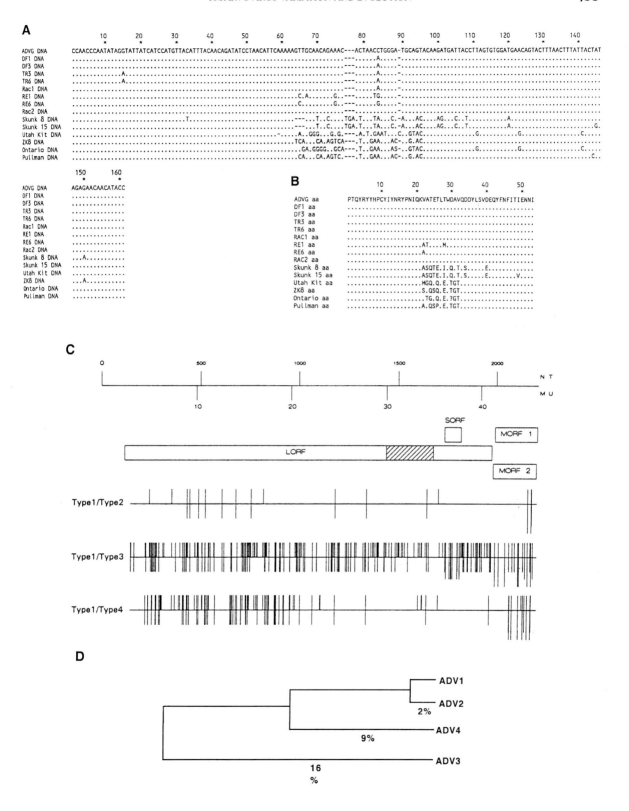

raccoon and skunk viruses appear very similar to viruses isolated from farmed mink, although there are differences between the isolates in the hypervariable region of the capsid protein gene (Oie *et al.*, 1996).

The spectrum of ADV sequences of the non-structural and structural protein genes of the genome was examined for experimentally infected Danish mink. Many different sequences were found in mink inoculated with a single inoculum. The sequences differed by up to 5% and they also differed markedly in the highly variable region (Gottschalck *et al.*, 1991). Since ADV can establish long-term persistent infections with continuously circulating virus, it is likely that mixed infections occur and that experimental inocula prepared in mink may contain more than one virus strain.

Comparing virus non-structural gene sequences (nt 123–2208) from different ADV isolates or virus stocks prepared from infected mink showed that there was extensive variation between the viruses in the different preparations. Some of those viruses had been originally prepared from pools of tissues from infected mink, and most had been repeatedly passaged through mink to prepare the working stocks (Gottschalck *et al.*, 1994). Variation of up to 11.4% was seen between the nucleotide sequences and up to 16% between the amino acid sequences (Figure 16.9C), and that variation was found distributed throughout the sequences compared (Figure 16.9D; Gottschalck *et al.*, 1994).

The host range of the ADV for mink or tissue cultured cells was mapped using recombination between infectious plasmid clones, and that was a property of a 530 base region within the capsid protein gene (Bloom *et al.*, 1993). That region controlled replication in cultured cells, as well as the alternative replication in mink.

Rodent Parvoviruses

Although the rodent parvoviruses MVM, LuIII and H1 have been intensively studied for their genetic and biochemical properties, information about the variation and evolution of the rodent viruses in nature is still being obtained. Many of the earliest viral isolates studied appear to be rodent viruses that were isolated as possible contaminating agents in tissue cultures or transplantable tumors, and in some cases their true origin and the degree of tissue culture adaptation are not fully understood. Those viruses include: the minute virus of mice prototype strain (MVMp) isolated from a murine adenovirus stock; H1 virus, which was isolated from the HEP-1 human tumor transplanted into rats and is most probably a rat virus; LuIII isolated from human cells (reviewed in Tattersall and Ward, 1978); and a tissue culture isolate of MVM named MVM-Cutter (MVM-(c); Besselsen *et al.*, 1996). A number of viruses have been recently detected in mice (mouse parvovirus (MPV)), hamsters (hamster parvovirus (HPV)) and rats (rat parvovirus (RPV)), which appear to be widespread in rodent colonies (Ball-Goodrich and Johnson, 1994; Besselsen *et al.*, 1996; Jacoby *et al.*, 1996). Many of those viruses are difficult or impossible to grow in tissue culture. The known and suspected rodent parvoviruses are more than 79% identical in DNA sequence, while viruses that were considered to be of the same type and given the same name are generally more than 95% identical (Figure 16.10). Apart from the relationships between the viruses at the sequence level, little is known about the details of the evolutionary or host range similarities or differences between these viruses, or their variation over time or between hosts.

The host range or tissue tropisms of some of the rodent viruses are controlled through changes in the capsid protein gene, and in the structures of the capsids. Two strains of MVM differ in their ability to infect fibroblast-like or lymphoid mouse cells, and the critical determinant of the infection appears to be as few as two specific sequence differences in the VP2 protein, residues 317 and 321 (Ball-Goodrich and Tattersall, 1992). From the sequence and structure of MVMp, the changes involved VP2 residues 317 Ala to Thr and 321 Glu to Gly, which appear to alter the hydrogen bonding of the VP2 protein (Llamas-Saiz *et al.*, 1997).

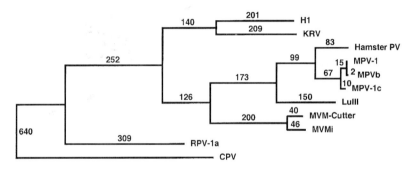

FIGURE 16.10 The phylogenetic relationships between various rodent parvoviruses, determined from the complete sequence of the *VP1/VP2* gene nucleic acid sequence, with the sequence of the CPV *VP1/VP2* gene used as an outgroup. The phylogeny was determined using the Branch-and-Bound algorithm of the program PAUP version 3.1.1, and the single minimal phylogeny is shown. Viruses are H1 and Kilham rat virus (KRV), hamster parvovirus, mouse parvoviruses (MPV), LuIII, MVM, rat parvovirus (RPV) and CPV.

CONCLUSIONS

Our understanding of the evolution of the parvoviruses is still being developed, but there are some interesting conclusions that can be drawn at this time.

- In general the viruses are genetically stable in nature – in most cases there is between 1% and about 6% sequence differences between viruses from the same or closely related hosts. There is some evidence that viruses from chronic infections have a higher sequence variation than those from acute infections, and this is particularly seen in studies of the ADV.

- The evolution of CPV provides an interesting example of the process of virus emergence through host range variation and evolution. Although that type of change is not common, such a genetic change only has to happen once under the right circumstances for the variant virus to become established in its new host. Once emerged, CPV has been undergoing a series of changes leading to further adaptation to its new host, and also perhaps due to immune pressure of the host.

- In most cases the temporal rates of variation cannot be estimated, as the viruses are endemic and appear to be in equilibrium with their hosts. The rate determined for CPV after it emerged was 2×10^{-4}/nt per year. This rapid rate may be due to modification of the ssDNA in the capsid, to a high number of replication cycles, to the process of host adaptation, or to some combination of these pressures.

REFERENCES

Agbandje, M., Parrish, C.R. and Rossmann, M.G. (1995) The structure of parvoviruses. *Sem. Virol.*, **6**, 299–309.

Alexandersen, S., Bloom, M.E. and Wolfinbarger, J. (1988) Evidence of restricted viral replication in adult mink infected with Aleutian disease of mink parvovirus. *J. Virol.*, **62**, 1495–1507.

Alexandersen, S., Storgaard, T., Kamstrup, N., Aasted, B. and Porter, D.D. (1994) Pathogenesis of Aleutian mink disease parvovirus infection: effects of suppression of antibody response on viral mRNA levels and on development of acute disease. *J. Virol.*, **68**, 738–749.

Anderson, S., Momoeda, M., Kawase, M., Kajigaya, S. and Young, N.S. (1995) Peptides derived from the unique region of B19 parvovirus minor capsid protein elicit neutralizing antibodies in rabbits. *Virology*, **206**, 626–632.

Andrews, G.A., Chavey, P.S., Smith, J.E. and Rich, L. (1992) *N*-glycolylneuraminic acid and *N*-acetylneuraminic acid define feline blood group A and B antigens. *Blood*, **79**, 2485–2491.

Ball-Goodrich, L.J. and Johnson, E. (1994) Molecular characterization of a newly recognized mouse parvovirus. *J. Virol.*, **68**, 6476–6486.

Ball-Goodrich, L.J. and Tattersall, P. (1992) Two amino acid substitutions within the capsid are coordinately required for acquisition of fibrotropism by the lymphotropic strain of minute virus of mice. *J. Virol.*, **66**, 3415–3423.

Barbis, D.P., Chang, S.-F. and Parrish, C.R. (1992) Mutations adjacent to the dimple of canine parvovirus capsid structure affect sialic acid binding. *Virology*, **191**, 301–308.

Berns, K.I. (1990) Parvovirus replication. *Microbiol. Rev.*, **54**, 316–329.

Berns, K.I. and Linden, R.M. (1995) The cryptic life style of adeno-associated virus. *Bioessays*, **17**, 237–245.

Besselsen, D.G., Pintel, D.J., Purdy, G.A. *et al.* (1996) Molecular characterization of newly recognized rodent parvoviruses. *J. Gen. Virol.*, **77**, 899–911.

Bloom, M.E., Alexandersen, S., Perryman, S., Lechner, D. and Wolfinbarger, J.B. (1988) Nucleotide sequence and genomic organization of Aleutian mink disease parvovirus (ADV): sequence comparisons between a nonpathogenic and a pathogenic strain of ADV. *J. Virol.*, **62**, 2903–2915.

Bloom, M.E., Berry, B.D., Wei, W., Perryman, S. and Wolfinbarger, J.B. (1993) Characterization of chimeric full-length molecular clones of Aleutian mink disease parvovirus (ADV): identification of a determinant governing replication of ADV in cell culture. *J. Virol.*, **67**, 5976–5988.

Bloom, M.E., Kanno, H., Mori, S. and Wolfinbarger, J.B. (1994) Aleutian mink disease: puzzles and paradigms. *Infect. Agents Dis.*, **3**, 1–23.

Brockhaus, K., Plaza, S., Pintel, D.J., Rommelaere, J. and Salome, N. (1996) Nonstructural proteins NS2 of minute virus of mice associate in vivo with 14–3–3 protein family members. *J. Virol.*, **70**, 7527–7534.

Brown, K.E. and Young, N.S. (1995) Parvovirus B19 infection and hematopoiesis. *Blood Rev.*, **9**, 176–182.

Brown, K.E., Young, N.S. and Liu, J.M. (1994) Molecular, cellular and clinical aspects of parvovirus B19 infection. *Crit. Rev. Oncol. Hematol.*, **16**, 1–31.

Brown, C.S., Jensen, T., Meloen, R.H. *et al.* (1992) Localization of an immunodominant domain on baculovirus-produced parvovirus B19 capsids: correlation to a major surface region on the native virus particle. *J. Virol.*, **66**, 6989–6996.

Carmichael, L.E., Joubert, J.C., Pollock, R.V.H. (1980) Hemagglutination by canine parvovirus: serologic studies and diagnostic applications. *Am. J. Vet. Res.*, **40**, 784–791.

Chang, S.F., Sgro, J.Y. and Parrish, C.R. (1992) Multiple amino acids in the capsid structure of canine parvovirus coordinately determine the canine host range and specific antigenic and hemagglutination properties. *J. Virol.*, **66**, 6858–6567.

Chapman, M.S. and Rossmann, M.G. (1993) Structure, sequence, and function correlations among parvoviruses. *Virology*, **194**, 491–508.

Chejanovsky, N. and Carter, B.J. (1989) Mutagenesis of an AUG codon in the adeno-associated virus rep gene: effects on viral DNA replication. *Virology*, **173**, 120–128.

Cotmore, S.F. and Tattersall, P. (1987) The autonomously replicating parvoviruses of vertebrates. *Adv. Virus Res.*, **33**, 91–174.

Cotmore, S.F. and Tattersall, P. (1995) DNA replication in the autonomous parvoviruses. *Sem. Virol.*, **6**, 271–281.

Cotmore, S.F., D'Abramo, A.M. Jr, Carbonell, L.F., Bratton, J. and Tattersall, P. (1997) The NS2 polypeptide of parvovirus MVM is required for capsid assembly in murine cells. *Virology*, **231**, 267–280.

De Ybanez, R.R., Vela, C., Cortes, E., Simarro, I. and Casal, J.I. (1995) Identification of types of canine parvovirus circulating in Spain. *Vet. Rec.*, **136**, 174–175.

Erdman, D.D., Durigon, E.L., Wang, Q.-Y. and Anderson, L.J. (1996) Genetic diversity of human parvovirus B19: sequence analysis of the VP1/VP2 gene from multiple isolates. *J. Gen. Virol.*, **77**, 2767–2774.

Gallinella, G., Venturoli, S., Gentilomi, G., Musiani, M. and Zerbini, M. (1995) Extent of sequence variability in a genomic region coding for capsid proteins of B19 parvovirus. *Arch. Virol.*, **140**, 1119–1125.

Gottschalck, E., Alexandersen, S., Cohn, A., Poulsen, L.A., Bloom, M.E. and Aasted, B. (1991) Nucleotide sequence analysis of Aleutian mink disease parvovirus shows that multiple virus types are present in infected mink. *J. Virol.*, **65**, 4378–4386.

Gottschalck, E., Alexandersen, S., Storgaard, T., Bloom, M.E. and Aasted, B. (1994) Sequence comparison of the non-structural genes of four different types of Aleutian mink disease parvovirus indicates an unusual degree of variability. *Arch. Virol.*, **138**, 213–231.

Hemauer, A., von Pobotzki, A., Gigler, A. *et al.* (1996) Sequence variability among different parvovirus B19 isolates. *J. Gen. Virol.*, **77**, 1781–1785.

Hong, G., Ward, P. and Berns, K.I. (1994) Intermediates of adeno-associated virus DNA replication *in vitro*. *J. Virol.*, **68**, 2011–2015.

Horiuchi, M., Goto, H., Ishiguro, N. and Shinagawa, M. (1994) Mapping of determinants of the host range for canine cells in the genome of canine parvovirus using canine parvovirus/mink enteritis virus chimeric viruses. *J. Gen. Virol.*, **75**, 1319–1328.

Im, D.S. and Muzyczka, N. (1992) Partial purification of adeno-associated virus Rep78, Rep52, and Rep40 and their biochemical characterization. *J. Virol.*, **66**, 1119–1128.

Jacoby, R.O., Ball-Goodrich, L.J., Besselsen, D.G., McKisic, M.D., Riley, L.K. and Smith, A.L. (1996) Rodent parvovirus infections. *Lab. Anim. Sci.*, **46**, 370–380.

Kawase, M., Momoeda, M., Young, N.S. and Kajigaya, S. (1995) Most of the VP1 unique region of B19 parvovirus is on the capsid surface. *Virology*, **211**, 359–366.

Kerr, J.R., Curran, M.D., Moore, J.E. *et al.* (1995) Genetic diversity in the non-structural gene of parvovirus B19 detected by single-stranded conformational polymorphism assay (SSCP) and partial nucleotide sequencing. *J. Virol. Meth.*, **53**, 213–222.

Langeveld, J.P., Casal, J.I., Vela, C. *et al.* (1993) B-cell epitopes of canine parvovirus: distribution on the primary structure and exposure on the viral surface. *J. Virol.*, **67**, 765–772.

Li, X. and Rhode, S.L. (1991) Nonstructural protein NS2 of parvovirus H-1 is required for efficient viral protein synthesis and virus production in rat cells *in vivo* and *in vitro*. *Virology*, **184**, 117–130.

Llamas-Saiz, A.L., Agbandje-McKenna, M., Parker, J.S.L., Wahid, A.T.M., Parrish, C.R. and Rossmann, M.G. (1996) Structural analysis of a mutation in canine parvovirus which controls antigenicity and host range. *Virology*, **225**, 65–71.

Llamas-Saiz, A.L., Agbandje-McKenna, M., Wikoff, W.R., Bratton, J., Tattersall, P. and Rossman, M.G. (1997) Structure determination of minute virus of mice. *Acta Cryst.*, **D53**, 93–102.

Mochizuki, M., Konishi, S., Ajiki, M. and Akaboshi, T. (1989) Comparison of feline parvovirus subspecific strains using monoclonal antibodies against a feline panleukopenia virus. *Nippon Juigaku Zasshi*, **51**, 264–272.

Mochizuki, M., Harasawa, R. and Nakatani, H. (1993) Antigenic and genomic variabilities among recently prevalent parvoviruses of canine and feline origin in Japan. *Vet. Microbiol.*, **38**, 1–10.

Mori, J., Beattie, P., Melton, D.W., Cohen, B.J. and Clewley, J.P. (1987) Structure and mapping of the DNA of human parvovirus B19. *J. Gen. Virol.*, **68**, 2797–2807.

Murphy, F.A., Fauquet, C.M., Mayo, M.A. *et al.* (1995) International Committee on Taxonomy of Viruses, International Union of Microbiological Societies. Virology Division Virus taxonomy/G: classification and nomenclature of viruses: Sixth report of the

International Committee on Taxonomy of Viruses. *Arch. Virol. (Suppl.)*, **viii**.

Naeger, L.K., Salome, N. and Pintel, D.J. (1993) NS2 is required for efficient translation of viral mRNA in minute virus of mice-infected murine cells. *J. Virol.*, **67**, 1034–1043.

Oie, K.L., Durrant, G., Wolfinbarger, J.B. *et al.* (1996) The relationship between capsid protein (VP2) sequence and pathogenicity of Aleutian mink disease parvovirus (ADV): a possible role for raccoons in the transmission of ADV infections. *J. Virol.*, **70**, 852–861.

Parker, J.S.L. and Parrish, C.R. (1997) Canine parvovirus host range is determined by the specific conformation of an additional region of the capsid. *J. Virol.*, **71**, 9214–9222.

Parrish, C.R. (1991) Mapping specific functions in the capsid structure of canine parvovirus and feline panleukopenia virus using infectious plasmid clones. *Virology*, **183**, 195–205.

Parrish, C.R. (1995) Pathogenesis of feline panleukopenia virus and canine parvovirus. *Baillière's Clin. Haematol.*, **8**, 57–71.

Parrish, C.R. and Carmichael, L.E. (1983) Antigenic structure and variation of canine parvovirus, feline panleukopenia virus, and mink enteritis virus. *Virology*, **129**, 401–414.

Parrish, C.R. and Carmichael, L.E. (1986) Characterization and recombination mapping of an antigenic and hostrange mutation of canine parvovirus. *Virology*, **148**, 121–132.

Parrish, C.R., Gorham, J.R., Schwartz, T.M. and Carmichael, L.E. (1984) Characterisation of antigenic variation among mink enteritis virus isolates. *Am. J. Vet. Res.*, **45**, 2591–2599.

Parrish, C.R., O'Connell, P. H., Evermann, J. F. and Carmichael, L. E. (1985) Natural variation of cannine parvovirus. *Science*, **230**, 1046–1048.

Parrish, C.R., Have, P., Foreyt, W.J., Evermann, J.F., Senda, M. and Carmichael, L.E. (1988a) The global spread and replacement of canine parvovirus. *J. Gen. Virol.*, **69**, 1111–1116.

Parrish, C.R., Aquadro, C.F. and Carmichael, L.E. (1988b) Canine host range and a specific epitope map along with variant sequences in the capsid protein gene of canine parvovirus and related feline, mink and raccoon parvoviruses. *Virology*, **166**, 293–307.

Parrish, C.R., Aquadro, C., Strassheim, M.L.,

Evermann, J.F., Sgro, J.-Y. and Mohammed, H. (1991) Rapid antigenic-type replacement and DNA sequence evolution of canine parvovirus. *J. Virol.*, **65**, 6544–6552.

Porter, D.D. (1986) Aleutian disease – persistent parvovirus infection in mink with a maximal but ineffective host immune response. *Prog. Med. Virol.*, **33**, 42–60.

Saikawa, T., Anderson, S., Momoeda, M., Kajigaya, S. and Young, N.S. (1993) Neutralizing linear epitopes of B19 parvovirus cluster in the VP1 unique and VP1–VP2 junction regions. *J. Virol.*, **67**, 3004–3009.

Senda, M., Hirayama, N., Itoh, O. and Yamamoto, H. (1988) Canine parvovirus: strain difference in haemagglutination activity and antigenicity. *J. Gen. Virol.*, **69**, 349–354.

Smith, R.H. and Kotin, R.M. (1988) The Rep52 gene product of adeno-associated virus is a DNA helicase with 3′-to-5′ polarity. *J. Virol.*, **72**, 4874–4881.

Strassheim, L.S., Gruenberg, A., Veijalainen, P., Sgro, J.-Y. and Parrish, C.R. (1994) Two dominant neutralizing antigenic determinants of canine parvovirus are found on the threefold spike of the virus capsid. *Virology*, **198**, 175–184.

Tattersall, P. and Ward, D.C. (1978) The parvoviruses – an introduction. In: *Replication of Mammalian Parvoviruses* (eds Ward, D.C. and Tattersall, P.), pp. 3–12. Cold Spring Harbor Laboratory Press, Cold Spring Harbor, NY.

Tratschin, J.-D., McMaster, G.K., Kronauer, G. and Siegl, G. (1982) Canine parvovirus: relationship to wild-type and vaccine strains of feline panleukopenia virus and mink enteritis virus. *J. Gen. Virol.*, **61**, 33–41.

Truyen, U. and Parrish, C.R. (1992) Canine and feline host ranges of canine parvovirus and feline panleukopenia virus: distinct host cell tropisms of each virus *in vitro* and *in vivo.*, *J. Virol.*, **66**, 5399–5408.

Truyen, U., Agbandje, M. and Parrish, C.R. (1994) Characterization of the feline host range and a specific epitope of feline panleukopenia virus. *Virology*, **200**, 494–503.

Truyen, U., Gruenberg, A., Chang, S.F., Obermaier, B., Veijalainen, P. and Parrish, C.R. (1995) Evolution of the feline-subgroup parvoviruses and the control of canine host range *in vivo. J. Virol.*, **69**, 4702–4710.

Truyen, U., Evermann, J.F., Vieler, E. and Parrish, C.R. (1996a) Evolution of canine parvovirus involved loss and gain of feline host range. *Virology*, **215**, 186–189.

Truyen, U., Platzer, G. and Parrish, C.R. (1996b) Antigenic type distribution among canine parvoviruses in dogs and cats in Germany. *Vet. Rec.*, **138**, 365–366.

Truyen, U., Geissler, K., Parrish, C.R., Hermanns, W. and Siegl, G. (1998) No evidence for a role of modified live virus vaccines in the emergence of canine parvovirus. *J. Gen. Virol.*, **79**, 1153–1158.

Umene, K. and Nunoue, T. (1995) A new genome type of human parvovirus B19 present in sera of patients with encephalopathy. *J. Gen. Virol.*, **76**, 2645–2651.

Wang, D., Yuan, W., Davis, I. and Parrish, C.R. (1998) Nonstructural protein-2 and the replication of canine parvovirus. *Virology*, **240**, 273–281.

Weichert, W.S., Parker, J.S.L., Wahid, A.T.M., Chang, S.-F., Meier, E. and Parrish, C.R. (1998) Assaying for structural variation in the capsid of parvovirus and its role in infection. *Virology*, **250**, 106–117.

Wikoff, W.R., Wang, G., Parrish, C.R., Cheng, R.H., Strassheim, M.L., Baker, T.S. and Rossmann, M.G. (1994) The structure of a neutralized virus: canine parvovirus complexed with neutralizing antibody fragment. *Structure*, **2**, 595–607.

Yasue, S., Handa, S., Miyagawa, S., Inoue, J., Hasegawa, A., Yamakawa, T. (1978) Difference in form of sialic acid in red blood cell glycolipids of different breeds of dogs. *J. Biochem. (Tokyo)*, **83**, 1101–1107.

Yoshimoto, K., Rosenfeld, S., Frickhofen, N. *et al.* (1991) A second neutralizing epitope of B19 parvovirus implicates the spike region in the immune response. *J. Virol.*, **65**, 7056–7060.

Young, N.S. (1995) B19 parvovirus. *Baillière's Clin. Haematol.*, **8**, 25–56.

The Molecular Evolutionary History of the Herpesviruses

Duncan J. McGeoch and Andrew J. Davison

INTRODUCTION

The herpesviruses are a numerous group of large DNA viruses widely distributed in nature. The range of disease associations and biological characteristics found across the group is substantial. The DNA genomes are heterogeneous in terms of size, base composition, gene content and organization. Work on DNA sequence determination, ongoing since early in the 1980s, has now accumulated an extensive set of herpesvirus sequences, including (as of middle 1998) 17 complete, published genome sequences (Table 17.1). The various data in these several categories should all contribute to considering the evolutionary history of the group. This field has been both challenging and fruitful: our overall understanding of the descent of herpesviruses and of processes in their evolution remains incomplete, but progress has certainly been made and illuminating insights achieved – some quite unexpected.

Historically, herpesviruses were defined by their characteristic virion architecture: an icosahedral capsid, of diameter near 125 nm and with 162 capsomeric spikes, embedded in an amorphous proteinaceous matrix ("tegument") and enveloped with a lipid membrane carrying external glycoproteins to give an overall diameter of near 215 nm (Schrag *et al.*, 1989). In recent years, DNA sequences for genes of many herpesviruses have been used to demonstrate that the viruses isolated from mammalian and avian hosts do indeed form an evolutionarily related group, albeit highly diverged (this finding has effectively changed the route for assignment of viruses as herpesviruses: common practice now is to use limited DNA sequences from a novel virus as first proof of herpesvirus membership). Although little is known of herpesviruses of reptiles, limited sequence information has very recently become available for certain herpesviruses isolated from turtles, and these also belong to the mammalian/avian group (Quackenbush *et al.*, 1998). On the other hand, it is now clear that herpesviruses isolated from fish and amphibian hosts form a second grouping that is only tenuously related to the mammalian and avian viruses. For our purposes in this chapter we deal with this dichotomy by restricting the bulk of the text to treatment of herpesviruses of mammals and birds, and postponing topics involving the viruses of fish and amphibians to a late part of the text.

Classification of herpesviruses primarily on the basis of biological characteristics produced a scheme with three subfamilies, the *Alpha-*, *Beta-* and *Gammaherpesvirinae* (Roizman *et al.*, 1992, 1995). This has now been overtaken by criteria of relatedness based on molecular sequences; overall the previous scheme has proved to be

Origin and Evolution of Viruses
ISBN 0–12–220360-7

TABLE 17.1   Herpesvirus lineages and sequenced genomes

Subfamily Alphaherpesvirinae		
α1 lineage	Herpes simplex virus type 1	X14112
	Herpes simplex virus type 2	Z86099
α2 lineage	Varicella-zoster virus	X04370
	Equine herpesvirus 1	M86664
	Equine herpesvirus 4	AF030027
	Bovine herpesvirus 1	AJ004801
α3 lineage	Marek's disease virus (avian)	No complete sequence
α4 lineage	Infectious laryngotracheitis virus (avian)	No complete sequence
Subfamily Betaherpesvirinae		
β1 lineage	Human cytomegalovirus	X17403, Y13735, U33331
	Murine cytomegalovirus	U68299
β2 lineage	Human herpesvirus 6	X83413
	Human herpesvirus 7	U43400, AF037218
Subfamily Gammaherpesvirinae		
γ1 lineage	Epstein–Barr virus	V01555, M35547
γ2 lineage	Human herpesvirus 8	U75698
	Herpesvirus saimiri	X64346
	Equine herpesvirus 2	U20824
	Alcelaphine herpesvirus 1	AF005370
	Murine herpesvirus 68	U97553
Fish herpesviruses		
	Channel catfish virus	M75136

Viruses whose genomes are completely sequenced are listed (with their sequence library accession numbers), except for the α3 and α4 lineages, where no complete sequences are available.

impressively robust, with only occasional corrections necessary. A summary outline is provided in Table 17.1 of the herpesvirus subfamilies and major lineages within these according to our present understanding. This corresponds in most respects with the taxonomy presented in the forthcoming (1999) Seventh Report of the International Committee on Taxonomy of Viruses. However, for discussing evolution in this chapter we wish to avoid use of the still imperfect taxonomic nomenclature, in particular regarding assignments of genera. In order to illustrate the diversity of diseases caused by herpesviruses and the corresponding diversity in their underlying biological properties, we employ the device of enumerating the herpesviruses presently known to have the human species as their natural host, and in effect allow-

ing the list of their properties, in Table 17.2, to speak for itself. At present eight human herpesviruses are known, which represent all the major lineages populated by mammalian viruses (α1, α2, β1, β2, γ1, γ2). Genome sizes range from 125 to 245 kb, and genome base compositions from 36% to 70% G+C. Most of these viruses occur at high prevalence in human populations, although one (HSV2) is characteristically sexually transmitted and the prevalence of the most recently discovered (HHV8) remains uncertain. All establish a lifetime presence in a host, involving a range of mechanisms of latency. In their latent aspects, three are neurotropic (HSV1, HSV2, VZV), three are lymphotropic (HHV6, HHV7, EBV), one is involved with the monocyte lineage (HCMV), and again the situation with HHV8 is still to be clarified. Two have

TABLE 17.2   Human herpesviruses

Virus	Abbrev.	Lineage	Genome Size (kb)	G+C Content (%)	Genome Type*	Encoded Proteins†	Pathology
Herpes simplex virus type 1	HSV1	α1	152	68	E	74	Recurrent epithelial lesions; rarer, serious neural disease
Herpes simplex virus type 2	HSV2	α1	155	70	E	74	As for HSV1
Varicella-zoster virus	VZV	α2	125	46	D	69	Primary infection chickenpox; recurrence as shingles
Human cytomegalovirus	HCMV	β1	245‡	57‡	E	227‡	Cause of congenital abnormalities; severe infections in immunocompromised
Human herpesvirus 6	HHV6	β2	159	43	A	85§	Primary infection exanthem subitum
Human herpesvirus 7	HHV7	β2	153	36	A	84§	No definitive disease association
Epstein–Barr virus	EBV	γ1	184**	60**	C	83**	Primary infection mononucleosis; associated with Burkitt's lymphoma and other neoplasias
Human herpesvirus 8	HHV8	γ2	141	53††	B	81	Associated with Kaposi's sarcoma and other neoplasias

* See Figure 17.1.

† Estimates of number of encoded proteins are not precise, as discussed in the text.

‡ Genomic data for HCMV are inclusive of regions absent from the sequence for strain AD169 (Chee *et al.*, 1990; Cha *et al.*, 1996; Dargan *et al.*, 1997).

§ Estimates for numbers of proteins encoded by HHV6 and HHV7 are from Megaw *et al.*, 1998.

** Genomic data for EBV are inclusive of a region absent from the sequence for strain B95-8 (Parker *et al.*, 1990).

†† G+C content for HHV8 excludes terminal reiterations.

an involvement with human cancers (EBV and HHV8). Infections can range from inapparent through disabling to lethal. All this for one host species, albeit the most urgently studied.

We aim in this chapter to provide an overview of the evolutionary history of the herpesviruses. Our treatment is in essence comparative: we use genome sequences and encoded protein sequences to discern processes that have occurred during evolution and to construct a phylogenetic description of the mammalian and avian viruses. We next move to a finer level, treating aspects of specific genes and gene sub-systems. The topics in this part of the chapter were selected on the basis of the varied contri-butions they could make to illuminating the overall picture; they do not constitute a general tour of herpesvirus genes. We then turn to her-pesviruses of fish and amphibians to recapitu-late our descriptions in terms of those very dif-ferent entities. Finally, we attempt to draw the threads together to summarize the current situ-ation in our understanding of herpesvirus evo-lution.

COMPARATIVE DESCRIPTIONS OF HERPESVIRUS GENOMES

Structures of Herpesvirus DNAs

Herpesvirus genomes consist of single mole-cules of double-stranded DNA. They cover a substantial range of sizes: among the fully sequenced genomes, the smallest is VZV at 125 kb and the HCMV at 245 kb. In every case most of the genome consists of so-called "unique" sequences, and the majority of genomes also contain large repeated elements (of the order of 10^3–10^4 bp in size). Distinct copies of these major repeats are considered to possess indistinguish-able sequences, with homogeneity being main-tained by recombination. Different patterns of placement of unique and repeat sequences define at least six genome types, as shown in Figure 17.1. In most cases, viruses in a sublin-eage are of the same genome type, but this clas-sification does not necessarily correlate with evolutionary relatedness. For instance, viruses

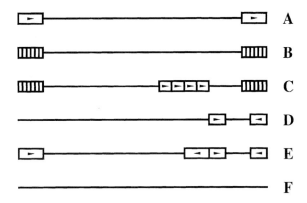

FIGURE 17.1 Sequence arrangements of herpesvirus DNAs. A schematic representation is given of the layouts in herpesvirus DNAs of unique sequences (shown as single lines) and repeat elements (shown as open boxes, with rela-tive orientations indicated for large repeats by arrowheads). In types B and C the number of terminal reiterations, and in type C the number of internal repeats, are considered vari-able. In type E, the sequences of the large and small repeat elements are distinct. Type F is intended to encompass genomes with very small or absent terminal ·repeat sequences. Adapted from Roizman et al., 1995.

in the α1 group have Type E genomes, and so does HCMV in the β1 group, but the R_L, R_S and U_S elements of HCMV are evolutionarily dis-tinct from the corresponding elements of the α1 viruses. Major repeat elements are clearly dynamic entities on an evolutionary timescale, as evidenced by comparisons of the S regions of α1 and α2 genomes. These all consist of an unique sequence (U_S) flanked by a pair of repeats (IR_S and TR_S) in opposing orientations. Comparison of the gene contents of S regions shows that gene orders are variable although clearly related (Davison and McGeoch, 1986). Certain genes that in some genomes are in the U_S segment are in others duplicated, by location in the flanking repeats; and in other compar-isons, certain genes are seen to be located in U_S alternatively near one extremity or the other, rel-ative to the overall layout of genes. These move-ments can be accounted for as the result of recombination events between two copies of the S region, with one end of a double crossover located legitimately in a flanking repeat and the other illegitimately between heterologous points in the two copies of U_S.

In addition to these large features of genomic structure, other finer scale aspects bear describing in an evolutionary context.

The most prominent concerns base compositions of herpesvirus DNAs, which at the whole genome level cover an impressively wide range, from 32–75% G+C content (Roizman *et al.*, 1992). There is no general correlation between virus lineage and base composition – the α2 group, for instance, contains canine herpesvirus (32% G+C) and pseudorabies virus (74% G+C). Wide differences in base composition between regions in one genome are also commonplace. Generally, large repeat elements exhibit a higher G+C content than the unique sequences in the same genome. The unique sequence DNA of herpesvirus saimiri (HVS) is 34% G+C, while the flanking repeats are 71% G+C (Albrecht *et al.*, 1992). In HSV1 and HSV2, with overall G+C contents of 68% and 70% respectively, the region with the highest G+C content is the 6600 bp R_s with close to 80% G+C, and this is inclusive of 3900 bp of coding sequence for one large protein (McGeoch *et al.*, 1986; Dolan *et al.*, 1998). At such extreme base compositions, changes relative to homologous genes with lesser compositional bias cannot be accommodated only through synonymous changes in the coding sequence, so that overall amino acid composition of encoded protein also changes.

Although compositionally biased DNA sequences have been widely observed in other contexts, including genomes of eukaryotes and prokaryotes, the effects seen in herpesviruses are very striking by any standards. The extremity of base composition to which even protein coding sequences have evolved argues for the action of powerful selective forces, but the nature of these remains obscure. We can discern three functional classes of components, that would (1) introduce base substitutions, (2) impart directionality to the substitutions fixed and (3) impose a differential effect between unique and repeat sequences (McGeoch *et al.*, 1986). The first of these roles and also the second could be ascribed to the herpesvirus DNA polymerase and to imbalances in dNTP pools in infected cells. In order to account for the differential effect on repeat sequences we have to invoke recombination processes between repeat copies in an intracellular pool of genome sequences, to accelerate changes relative to those in unique sequences. A recombinational mechanism could also be responsible for imposing directionality, in a form of biased gene conversion.

Examination of patterns of dinucleotide frequencies has shown marked differences among herpesviruses (Honess *et al.*, 1989). In the α1 and α2 groups the dinucleotide frequencies exhibit only small deviations from random expectation. In γ genomes there is a markedly low occurrence of 5'-CG. In β1 and β2 genomes this restriction of 5'-CG occurs only in the major immediate early gene regions. This phenomenon, the "CpG shortage", has been known for many years as a global feature of vertebrate nuclear DNA, and is considered to result from methylation of cytosine residues in DNA, in the sequence 5'-CG, to give 5-methylcytosine; any deamination of the 5-methylcytosine then transforms the sequence to 5'-TG, which is presumably not subject to any repair process. A 5'-CG deficit in the herpesvirus context thus appears to indicate that over an evolutionary period of time the virus DNA has existed in a cellular environment where methylation of DNA cytosine was active, and in a form that made it susceptible to this modification.

The final fine structure feature of herpesvirus genomes to mention is the occurrence of families of short, tandemly reiterated sequences. Such tandem repeats are found in all herpesvirus groups, with single copy sizes in the range, very roughly, of 5–100 bp. Simple arrays of identical elements and also more complicated mixtures of distinct repeat motifs occur. Copy numbers of repeat units range up to many tens. Some arrays clearly occur within protein coding sequences. Where comparative data are available for different strains or clones of the same virus species, it is seen that at least some repeat families can exhibit variation among isolates both in copy number and in sequence of the repeat unit. These effects are presumed to be mediated by recombination or by slippage during DNA replication. Finally, it is our personal impression that length variations in homopolymeric tracts in herpesvirus DNAs arise with some facility, as evidenced by their occurrence in virus isolates, molecular clones and mutants, and that these could be regarded as the lower unit-length limit case of tandem reiteration.

Gene Complements of Herpesviruses

Unique regions in herpesvirus genomes are generally closely packed with protein coding genes, but there are exceptions – the genome of equine herpesvirus 2, for instance, contains substantial "voids" of heteropolymeric but evidently non-coding DNA (Telford *et al.*, 1995). In major repeat elements protein coding sequences also occur but are typically arranged sparsely. As summarized in Table 17.2 for the human viruses, estimates of the number of protein coding genes range from 69 for VZV to over 220 for HCMV. Some comment on the nature of these estimates is required. At the upper end of the range, with the genomes of cytomegaloviruses (β1), the numbers are imprecise because of the relatively small amount of experimental analysis of gene structure and expression that has been carried out for these viruses, because of suspected occurrence of extensive splicing and because some listed open reading frames may not be genuinely protein-coding. Our view is that for HCMV and murine cytomegalovirus the current estimates of gene number may be on the high side. These kinds of limitation are much less severe among the smaller genomes of the α subfamily. Particularly in the case of the extensively studied HSV1, however, another kind of uncertainty exists. Various novel transcripts and encoded polypeptides have been described over the past decade, of which we consider that not all may correspond to genuine functional entities. Thus, our best estimate of the number of HSV genes is 74, but inclusion of additional possibilities would bring the list to over 80 (McGeoch *et al.*, 1988; Dolan *et al.*, 1998); this topic is discussed further in the section on Evolution. We cannot attempt here a comprehensive account of gene functions of herpesviruses. Instead, we have summarized functions known for HSV1 (the best studied herpesvirus) into several categories, and in Table 17.3 have listed numbers of genes in each category. In addition to mRNAs from protein

coding genes, the mammalian herpesviruses also express a heterogeneous collection of other RNA species that do not encode proteins, and these are in effect ignored in this text.

Herpesvirus genome sequences are widely diverged, so that for comparing genes between viruses it is generally most useful to proceed by way of their encoded amino-acid sequences. Systematic comparisons of genes from pairs of herpesviruses have shown unambiguously that some 30 genes have homologues in all fully sequenced mammalian herpesviruses (and, on the basis of incomplete data, this looks likely to be applicable also to the avian herpesviruses). To these can be added another 10 genes, which present less definitive cases (in terms of marginal amino acid sequence similarity, or patterns of hydrophilic and hydrophobic residues, or equivalent genomic locations) but whose ubiquity we regard as overall reasonably convincing. Present understanding is thus that there are 40 genes that have detectable counterparts across the α, β and γ subfamilies; these conserved genes have been termed the "core" set (Davison and Taylor, 1987; McGeoch, 1989). Examination of the genomic locations of the core set of genes has shown that within each virus subfamily their layout is nearly universally constant, with the same orders and relative orientations. One exception concerns the S region of α genomes, as outlined above. Among the three subfamilies, short-range ordering of core genes is mostly maintained, but on the larger scale there is substantial rearrangement of several blocks of genes. In all cases the core genes occupy roughly the central part of the long unique region (U_L). Genes that do not belong to the core set are found mainly towards the extremities of U_L and in other genomic regions (repeat elements and U_S), and to a lesser extent interspersed among core genes. We emphasize that this description of comparative gene layout presents only a digest of global features from a complex picture.

We can examine the roles of proteins encoded in the core set using the functional and structural categories already employed for HSV1 genes, as shown in Table 17.3. This analysis demonstrates that genes for DNA replication and processing systems are (with one exception) universal, while those for

TABLE 17.3   Functional and structural classes of HSV1 proteins

Class of protein	Number of Proteins	Core	Non-core
1. Concerned with genomic DNA	20		
(a) Central DNA synthetic machinery	7	6	1
(b) Peripheral enzymes	5	3*	2
(c) Processing and packaging of DNA	8	8	0
2. Control and modulation	10		8
3. Virion	30		
(a) Capsid assembly and structure	6	6	0
(b) Tegument	10	3	7
(c) Surface and membrane	14	5	9
4. Other and unknown	14	7	7
Totals	**74**	**40**	**34**

Proteins were assigned to one category only, that judged most characteristic of the function. Thus, certain proteins that are found in the tegument are listed under "Control and modulation".

* Two apparently conserved enzymes are not enzymatically active in the β lineage versions; see text.

enzymes of nucleotide metabolism and repair are not all so. The non-conserved HSV1 DNA replication protein is concerned with initiation of DNA synthesis from origins of replication; this part of the DNA replicative machinery turns out to differ substantially between different lineages of herpesviruses, while downstream parts of the machinery are common to all lineages; this topic is raised again in the section on Evolution. Genes for control functions are almost all not universal, and indeed the two registered in the core set were assigned as "control" only on relaxed criteria – they specify a protein kinase of unknown function and a post-transcriptional mRNA-processing protein. Genes for capsid assembly and structural proteins are, unsurprisingly, universal, unlike many of those for tegument proteins and virion surface proteins. The numbers for the set of tegument proteins lack precision because this group is still incompletely defined experimentally and because certain tegument proteins were pre-emptively assigned to the "control" class. The universal surface proteins present an interesting list: three (glycoprotein B and the glycoprotein-H:glycoprotein-L complex) are well studied, are known to be essential in HSV1 and several other viruses, and are involved in cell entry and cell-to-cell spread (Spear, 1993), while in contrast the remaining two (glycoproteins M and N) are non-essential in HSV1 and are of unknown functional significance (MacLean et al., 1991; Pyles et al., 1992).

In summary, most of the core gene complement of herpesviruses is concerned with specifying the icosahedral capsid structure, with synthesizing and packaging progeny DNA and with cell entry and exit mechanisms, while genes supplying all well-characterized control functions and also those for most virion tegument and surface components are specific to subfamilies or to lineages within subfamilies. Non-core gene sets also encompass other novel capabilities, some of which are touched on later (Section 4). For β and γ herpesviruses the numbers of non-core genes assigned as of unknown function are higher than for HSV1, and for the β1 viruses in particular the number of genes that would be listed as putatively encoding virion glycoproteins and control proteins is large.

PHYLOGENY OF HERPESVIRUSES

As already described, herpesvirus genomes differ substantially in terms of large scale features, including gene complements and organization of unique and repeat elements. Putting those aspects aside, it is clear from comparisons of homologous gene sequences between viruses that the genomes have also undergone very extensive processes of base substitution and of local addition/deletion events. Differences in overall base composition represent one result of the substitution process, and indeed illustrate cogently the extent of base substitution. As a specific instance of magnitude of divergences, HSV1 and HSV2, which are very closely related by herpesvirus standards, show overall in their protein coding regions a synonymous (silent) divergence of near 0.5 substitutions per synonymous site, with values for individual gene pairs forming an apparently homogeneous distribution (Dolan *et al.*, 1998). Non-synonymous divergences are all lower, and of course vary more from gene to gene. For most other pairwise comparisons of gene sequences across the range of herpesviruses, synonymous divergences are too high to estimate (i.e. the synonymous sites are in effect saturated with mutations that have occurred since divergence of the lineages of such viruses), and indeed only for the genes encoding the most highly conserved proteins are non-synonymous divergences low enough to allow good estimates.

We have employed substitutional differences between herpesviruses, using members of the core set of genes, to give quantitative estimates of relationships across the herpesvirus family through building phylogenetic trees (McGeoch and Cook, 1994; McGeoch *et al.*, 1995). The high levels of divergence, both of substitutions and addition/deletion changes, constitute an important limitation, which restricted our analysis to using encoded amino-acid sequences in preference to DNA sequences, and to concentrating on eight particularly suitable genes (as judged primarily by the quality of alignment achievable for a set of homologous sequences). In our first analyses, phylogenetic trees were constructed mainly via the neighbour-joining method.

Branching patterns obtained were consistent from gene to gene except for minor local discrepancies where the data did not provide adequate resolution to discriminate closely spaced branchpoints with confidence. Relative proportions of branches were maintained from gene to gene within each subfamily, subject to stochastic noise, but differences were observed for the deep branches separating subfamilies. We have recently been conducting an extensive re-examination of herpesvirus phylogeny with currently available sequences, employing the superior (but computationally demanding) approach of maximum-likelihood modelling and including allowance for differences in rates of change among sites. These investigations give strong support to the previously obtained herpesvirus tree, although they further emphasize that length estimates of the deepest branches are subject to large uncertainties.

Our analyses have yielded a consistent picture of herpesvirus phylogeny, in terms of unambiguous division of the mammalian and avian herpesvirus species into the three subfamilies and establishment of most details of branching patterns within subfamilies and their sublineages. In addition, it became clear that many elements in the branching patterns within each of the major sublineages were congruent with branching patterns for the corresponding lineages of mammalian host species. This observation was initially made with the α1 and α2 groups using glycoprotein B gene sequences (McGeoch and Cook, 1994), and has since been extended to examples in the β1 and γ2 sublineages and to other gene sets (D. J. McGeoch, unpublished work). Such correspondence between virus and host lines of descent argues strongly that these virus lineages arose by way of cospeciation with the host. In turn, accepting this hypothesis means that dates from palaeontology for host evolutionary events can be applied to divergence events between pairs of herpesviruses, so that absolute rates of sequence change can then be estimated for this virus family. Our present understanding of the herpesvirus phylogenetic tree is shown in Figure 17.2A, which aims to include as many virus species as possible by combining data from several genes, and Figure 17.2B then summarizes the situation by indicating only the major virus lineages.

The palaeontological datings invoked to interpret the tree's structure range from 18 million years before the present (Myears BP) for divergence of certain New World monkeys through to 60 Myears BP for divergence of the perissodactyl and artiodactyl lineages in the *Ungulata*, and with an effective upper bound of interpretability corresponding to the mammalian radiation of around 60–80 million years ago (Benton, 1990). We are thus invoking a timescale on a grand order, and in line with this the estimated rates of sequence change are low. We estimated for glycoprotein B genes that non-synonymous substitutions have occurred at a rate near 2.7×10^{-8} substitutions per site per year in each lineage, and synonymous substitutions at 10^{-7} substitutions per site per year (this latter is an order of magnitude estimate only; McGeoch and Cook, 1994). Analysis of restriction site variation for isolates of HSV1 from different human populations and use of prehistoric human migration datings to provide a time base gave a compatible estimate of substitution rate (Sakaoka *et al.*, 1994). By these and other measures, we consider that herpesvirus DNA sequences have mutated approximately 10–100 times faster than equivalent classes of sequences in the host genome (and around five orders of magnitude slower than, for instance, influenza virus and HIV).

Features of the tree that support cospeciation of mammalian viruses include: (1) ordering of the Old World and New World primate viruses in group $\alpha 1$; (2) ordering of primate, perissodactyl and artiodactyl ungulate viruses in group $\alpha 2$; (3) ordering of rodent and primate cytomegaloviruses (group $\beta 1$); and ordering of Old and New World primate viruses in group $\gamma 2$. Interestingly, certain regions of the herpesvirus tree that presented closely spaced or incompletely resolved branchpoints mirror corresponding problem areas in palaeontological studies of the host lineages. These are: first, the $\alpha 2$ grouping of carnivore, perissodactyl and artiodactyl ungulate viruses; and second, the unresolved branching details in the $\beta 1$ lineage of rodent and primate cytomegaloviruses. Within the sublineages there are only two loci

that do not conform readily to a cospeciational scenario. The first is the placement of bovine herpesvirus 2 (BHV2) in the $\alpha 1$ lineage, and this is interpreted as resulting from a host-switching event that occurred at some point on the lineage leading to BHV2. The second concerns the $\gamma 2$ grouping, where we have not been able to obtain unambiguous results for branching patterns for five lineages (Figure 17.2A). These problems appeared to be associated particularly with inclusion of murine herpesvirus 68 sequences in datasets; this situation represents an unresolved aspect in our analytical efforts (unpublished work of D. J. McGeoch).

Regarding deeper branchings in the tree, the two avian lineages are interpreted as indicating ancient changes of host species; most economically, from mammalian to avian hosts. The other branchings that generate the set of sublineages might correspond to expansion of ancient viruses into new niches in terms of host cell tropisms or other strategic aspects. Our tree depicts divergence of the α lineages from the progenitor of the β and γ lineages as the earliest event. This is based, first, on observations that for most genes the α lineage amino-acid sequences are scored as more distant from each of the other two lineages than they are from each other, and secondly on analysis of herpesvirus uracil glycosylase genes using eukaryotic and prokaryotic homologues as outgroups (McGeoch *et al.*, 1995). For the timescale shown, estimates deeper than 60 million years before the present are increasingly sensitive to values of input parameters in the modelling process, and are therefore tentative. The analyses thus indicate that at a remote period, of the order of 200 Myears BP, there existed a herpesvirus species that was the last common ancestor from which all present lineages of mammalian and avian herpesviruses evolved. Early mammals were extant at this date, but we can have no direct knowledge of the nature of the host species harbouring the ancestral virus. We presume that the core genes as defined above were all present in the ancestral virus genome. To the core set we can add three more genes that are not present in all contemporary viruses, whose distribution indicates that they are likely to be ancestral but have been lost in certain lineages. These are the genes

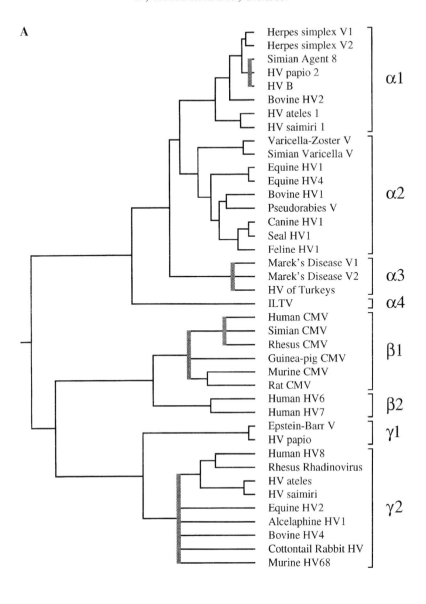

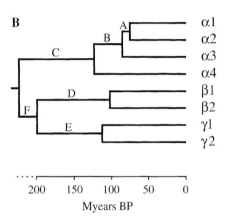

encoding thymidine kinase, the small subunit of ribonucleotide reductase and a protein involved in initiation of DNA replication; we return to them in the next section. We can thus identify 43 genes as the minimal ancestral set. The ancestral virus thus defined was already a well-developed herpesvirus in terms of its capsid structure, its possession of at least some tegument and surface proteins, and its DNA replication arrangements.

We consider that phylogeny of mammalian and avian herpesviruses has now reached a rather developed state as regards its major features, and indeed it occupies a leading position among DNA virus families in terms of breadth, detail and robustness, and especially in possessing a convincing chronology; this achievement has evidently been favoured both by availability of large sequence datasets and by the particular characteristics of herpesvirus evolution. Nonetheless, two limiting aspects of the analysis should not be forgotten. The first of these derives from limitations of present analytical methods, particularly as applied to the widely diverged sequences of herpesviruses. Analysis of herpesvirus phylogeny thus remains an active research subject in terms of applying new and superior methods. The second concerns the unavoidable fact that only a subset of genomic characteristics could be employed in the analysis, namely substitutional differences in encoded amino-acid sequences of a relatively small set of genes. There is no apparent reason to suspect that this is not a generally valid measure; at least within subfamilies, sequences have evolved at consistent relative rates for the set of core genes employed. However, instances of marked non-uniformity in levels of substitution have begun to emerge, involving certain non-core genes in EBV and HHV8, emphasizing that caution is required.

EVOLUTION OF HERPESVIRUS GENE SYSTEMS

Origins of Herpesvirus Genes

Among the 43 genes inferred to have been present in the genome of the common ancestor of present day mammalian herpesviruses, nine are clearly related by similarities in encoded amino sequences to cellular genes, to the extent that they and the cellular genes must have common evolutionary origins (Table 17.4). Notably, these all encode enzymes, of which all but one are involved in DNA metabolism and replication. We hold as almost axiomatic that these genes must have been acquired from some other genetic element, with the primary source a cellular genome. In contrast, for none of the proteins assigned as capsid, tegument or virion surface components is there any substantive indication of relatedness with a cellular protein. Gene capture remains, nonetheless, an obviously attractive origin for members of this class of herpesvirus genes. In all of the completely sequenced herpesvirus genomes there are

TABLE 17.4   Mammalian herpesvirus ancestral genes with cellular homologues

HSV gene	Function of encoded protein
UL2	Uracil-DNA glycosylase
UL5	Component of DNA helicase
UL9	DNA helicase active at origin of DNA replication
UL13	Protein kinase
UL23	Deoxynucleoside kinase (thymidine kinase)
UL30	Catalytic subunit of DNA polymerase
UL39	Large subunit of ribonucleotide reductase
UL40	Small subunit of ribonucleotide reductase
UL50	Deoxyuridine triphosphatase

FIGURE 17.2   Composite phylogenetic tree for mammalian and avian herpesviruses. **A**. A phylogenetic tree is shown for 40 herpesvirus species, based on amino-acid sequences from eight genes in 16 species, and mainly on glycoprotein B genes only for the remainder. This tree represents an interim stage in a continuing project, and will change in fine details with further work. Five regions where closely spaced branchpoints or insufficient data gave uncertainties are shown as multifurcations and marked by shaded lines. V, virus; HV, herpesvirus; CMV, cytomegalovirus; ILTV, infectious laryngotracheitis virus. **B**. The tree in **A** is reduced to the major sublineages only. Internal branches are labelled for identification in the text. A timescale is indicated, after McGeoch et al., 1995.

among non-core genes further examples of genes that have clear non-herpesviral homologues. Some of these are restricted to a subfamily, some to a sublineage, some to a virus species; evidently gene capture has proceeded throughout herpesvirus evolution. The encoded products are diverse, including nucleotide anabolic enzymes, various immune system proteins, transcriptional regulators, cell-state regulatory proteins and a parvovirus DNA replication protein. All of these are presumed to have been captured from another genome (in most cases the host cell's), and it seems likely that the ongoing characterization of cellular genomes will throw up further examples of herpesvirus/cell homologues.

In their forms in herpesvirus genomes, these genes are generally devoid of introns, suggesting that a mechanism involving reverse transcription of mRNA species may have been commonly involved in capture. Experimental studies have been made of interactions between herpesvirus and retrovirus genomes in mixed infections of the α3 avian herpesvirus Marek's disease virus (MDV) and a chicken retrovirus, and MDV isolates with retroviral sequences inserted into their genome obtained; in most cases the stably inserted retrovirus sequences were single LTRs (long terminal repeats; reviewed by Brunovskis and Kung, 1996). One serotype of MDV was found to contain sequences resembling LTRs, which may well represent the mutated remains of natural retroviral insertion events (Isfort et al., 1992). In principle, then, retroviruses or other retrotransposing entities may have been involved in gene capture by herpesviruses or, by LTR insertion, in disrupting coding sequences or altering patterns of transcription. Such processes could thus have been of widespread significance in herpesvirus evolution, although direct evidence is likely to remain unobtainable.

Examples are visible in all three herpesvirus subfamilies of multigene families, presumed to have arisen through successive gene duplication events. These are most developed in the β1 group, and account for a substantial part of the increased size of β1 genomes relative to those of other herpesviruses (Chee et al., 1990; Rawlinson et al., 1996). In HCMV, for example,

there are three pairs of related genes and six larger families (Chee et al., 1990). Members of families occur both in tandem arrays and distributed across the genome. The products of several families are membrane glycoproteins, there is a group of homologues of G-protein coupled receptors, and one family of proteins with possible regulatory roles (US22 family). Some of these families occur also in the β2 HHV6 and HHV7 (Gompels et al., 1995; Nicholas, 1996), emphasizing that their presences has had very long-term stability, and thus that the members of the family must be presumed to have distinct functional significances. The precise natures of such distinct roles remain rather obscure. Another form of duplication is observed with the dUTPase gene of herpesviruses: comparisons of amino acid sequences of herpesvirus dUTPases with those from all other sources indicate that the herpesvirus gene has experienced an intragenic duplication, which has resulted in a longer polypeptide chain with conserved motifs differentially preserved in one or the other of the two portions (McGeoch, 1990). As touched on below, this development may have been associated with gain of some distinct, presently unknown, functional role for this herpesvirus protein.

It is interesting to consider whether genes may have arisen de novo in herpesvirus evolution (McGeoch and Davison, 1995). Our position, considering the large size of the genomes and the extent and complexity of genomic changes during the evolution of the family, is that gene development in this mode might well have been not uncommon; the challenge, however, is to identify convincing candidates for such genes. One such class would be authenticated gene pairs with extensively overlapping reading frames, where we presume that one of the pair developed in situ, and some such cases occur in herpesvirus genomes. As an example of a gene whose coding region does not overlap another and that may have arisen de novo, we discuss gene US12 of HSV1 and HSV2. US12, expressed in the immediate early class, appears unique to the α1 lineage, and is dispensable for virus growth. Its product is active in interdicting antigen presentation in infected cells, by bind-

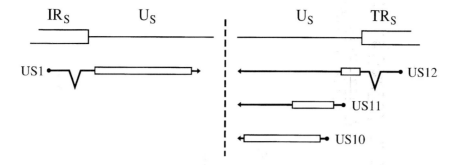

FIGURE 17.3 Genomic organization around the HSV1 *US12* gene. The layouts of genes at each end of the HSV1 U$_S$ region are shown. At the top of the figure, the left and right extremities of U$_S$ are indicated by solid lines and the abutting parts of repeats IR$_S$ and TR$_S$ by open boxes. In the lower part of the figure, transcripts for genes *US1*, *US10*, *US11* and *US12* are shown as heavy lines with coding regions as open boxes. 5′ termini of mRNAs are marked by solid circles and 3′ termini by arrowheads. After McGeoch *et al.*, 1985, and Rixon and McGeoch, 1985.

ing to the TAP peptide transporter (Früh *et al.*, 1995; Hill *et al.*, 1995) – i.e. it should provide a selectable advantage for the virus but is nonetheless essentially a "luxury" function. The encoded HSV1 polypeptide is only 88 amino acids in length (86 in HSV2), and the active part is contained in the *N*-terminal 35 residues (Galocha *et al.*, 1997). When purified and in solution it has no detectable stable secondary structure, so that it is more appropriately viewed as a TAP-binding oligopeptide than as a globular protein (P.N. Barlow, H. W. M. Moss and D. J. McGeoch, unpublished data). The *US12* gene lies at one end of the U$_S$ region, with its coding region overlying the transcription initiation sequences of the downstream *US11* gene. *US12* is transcribed from an immediate early promoter located in the immediately adjacent TR$_S$ element (Figure 17.3), and the twin copy of this promoter in the IR$_S$ element drives expression of gene *US1* at the opposite extremity of U$_S$ (McGeoch *et al.*, 1985; Rixon and McGeoch, 1985). Since *US1* encodes a substantial protein species (420 amino acid residues), it is attractive to regard the *US1/US12* promoter sequence as having evolved primarily for expressing gene *US1*. In this view the association with the *US12* region then occurred as a by-product, perhaps in a

genomic rearrangement that moved the R$_S$/U$_S$ boundaries (as discussed in an earlier section), so that the resulting immediate early transcription of the undeveloped *US12* locus then provided fertile conditions for genesis of a new function. These aspects, of lineage specificity, a function that is both non-essential and selectable, small size and lack of ordered structure of the product, and characteristics of the genomic location together build a persuasive case for a history of generation of *US12 de novo*.

As mentioned earlier, additional putative protein coding genes of HSV1 have been described in the last several years of whose functionality we remain sceptical. These typically overlap substantially or completely with recognized genes, and thus at face value could represent functions that have arisen *de novo*. These entities fall in two classes. In the first, an mRNA species is found that is 3′-coterminal with the mRNA for an established gene but whose 5′ terminus is downstream of that of the standard mRNA. Translation of the novel mRNA initiates within the reading frame of the established gene and is translated in the same phase to give an *N*-terminally truncated version of the complete gene product. It is beyond dispute that two core genes (HSV1 *UL26* and *UL26.5*) are arranged thus and are essential for capsid formation (see Rixon, 1993). As yet, however, there is no evidence that the truncated protein products in other such systems (*UL8.5*, *UL12.5*, *UL15.5* and *US1.5*) exert distinct phenotypes (Dolan *et al.*, 1998).

The second class of putative additional genes, which we wish to term *quasigenes*, consists of a proposed protein coding region overlapping that of a recognized gene in the opposing orientation (Bruni and Roizman, 1996; Randall and Roizman, 1997; Randall *et al.*, 1997; Chang *et al.*, 1998). Candidates in HSV1 are *UL27.5*, *UL43.5*, *ORF P* and *ORF O* (the latter two overlapping gene *RL1*). All would require unusual transcriptional properties (such as frameshifting, editing or splicing) in order to be translated similarly in HSV2 and none is conserved in other herpesviruses. None is essential for viral replication (but *UL27.5* has not yet been tested). Moreover, none appears to have undergone independent selection in any region: all apparent amino-acid sequence conservation between HSV1 and HSV2 can be accounted for as an indirect consequence of conservation of the recognized gene on the antiparallel strand. It is possible that quasigenes are evolutionary newcomers that have arisen in HSV1 and that confer niche-specific advantages, but that their selection has had insufficient time to be reflected in conservation of encoded amino-acid sequences independent of that of the antiparallel gene. This interpretation finds support in the conclusion that the *ORF P* protein inhibits splicing (Bruni and Roizman, 1996). Clearly, the presence of at least four such genes in HSV1 would challenge prevailing views of molecular evolution in herpesviruses and potentially in the wider range of organisms. We should not deny their reality on such grounds, but at present we consider it more likely that quasigenes are an artefact resulting from high G+C content (which imparts a general low frequency of stop-codon sequences in potential reading frames), and that the expressed polypeptides are irrelevant to infection. If quasigenes have any function, exploration of an RNA antisense mechanism, such as that reported for *ORF P* (Randall and Roizman, 1997), may prove a more promising avenue.

Initiation of DNA Replication

This sub-section and the next both address aspects of the genome replication machinery of herpesviruses; this clearly constitutes an ancient set of functions in the virus family, and the theme of our treatment concerns how such a central system has been modulated and changed during the evolutionary development of the virus subfamilies. Across the family of mammalian herpesviruses, certain aspects of arrangements for replication of the genome are universal and others are variable. All viruses have an equivalent set of six genes that encode the proteins needed for replicative elongation of DNA chains during lytic infection, and the set of genes for subsequent processing and packaging of nascent DNA appears to be similarly invariant (Challberg, 1996). However, there are distinct ways in which initiation of DNA synthesis is organized, and there are differences between virus subfamilies in the manner in which supply of dNTP precursors is ensured. In addition, in at least one sublineage (γ1; Epstein–Barr virus, EBV) there is a separate DNA replication system for maintenance of genome copies when the virus is in a latent state in dividing cells (Yates, 1996). In this condition the genome exists as a circular nuclear episome and its replication requires only one virus gene product (EBNA-1), which acts on an origin of replication distinct from the lytic origins to induce DNA copying by host cell machinery.

In the α subfamily, there is a gene (HSV1 *UL9*) whose product is a DNA helicase that binds to origins of DNA replication, and in association with other proteins acts to initiate synthesis of progeny DNA (Challberg, 1996; Boehmer and Lehman, 1997). The origins of replication (three in the HSV1 genome) are short, around 140 bp, with dyad symmetry and a central A+T rich segment. An equivalent situation exists in β2 viruses, with a *UL9* homologue and a similar origin structure (Gomples *et al.*, 1995; Nicholas, 1996). The *UL9* homologues occur in an equivalent location in α and β2 genomes, in terms of the identities and relative orientations of neighbouring genes. In β1, γ1 and γ2 viruses, however, there appears to be no *UL9* homologue, and the mechanism of DNA chain initiation is more complex (at least in β1 and γ1 viruses: in γ2 viruses the situation is not well characterized). In the γ1 and β1 viruses, origins of lytic replication are larger, bipartite elements containing complex sets of repeats (see Challberg, 1996; Yates, 1996). Initiation of replication is thought to involve both a virus-encoded transcriptional regulator acting on one of the segments of the origin and a host protein or proteins interacting with the other. Examination of the summary phylogenetic tree (Figure 17.2B) shows that the mechanism involving *UL9* is likely to represent the ancestral state, with adoption of a new mode of chain initiation occurring independently in the branch leading to the β1 viruses and in internal branch E. The significance of this change remains uninvestigated: we suggest that the more elaborate mechanism may allow superior control over the onset of DNA synthesis, by criteria of whether both virus and cellular gene expression are in appropriate states.

We close this topic by noting two other observations which may also indicate that herpesvirus systems for initiation of DNA replication are particularly susceptible to evolutionary tinkering. The first is that comparison of genome sequences for the α subfamily shows that replication origins differ from virus to virus in both number and location. The second point concerns the presence in the genome of HHV6 of an extra gene homologous to the *rep* gene of the helper dependent parvovirus adeno-associated virus (AAV; Thomson *et al.*, 1991). This has very probably been acquired from AAV via the process of non-homologous recombination by which AAV replicative forms integrate into host cell DNA. AAV *rep* encodes a protein, Rep 78/68, that possesses DNA-binding and helicase activities and acts in initiation of AAV DNA replication, and that also has a role in regulation of gene expression in AAV infected cells. The HHV6 version of the gene has been shown to be competent both in supporting parvovirus DNA replication and as a modulator of gene activity (Thomson *et al.*, 1994). While its significance for HHV6 remains unknown, its presence in HHV6 certainly imparts a flavour of novel possibilities for initiation of DNA synthesis.

Enzymes of Nucleotide Metabolism

All α and γ herpesviruses encode three enzymes active in pathways of synthesis of dNTPs for DNA replication, namely thymidine kinase (TK), deoxyuridine triphosphatase (dUTPase) and ribonucleotide reductase (RR). The last of these has two subunits (RR1 and RR2) encoded by separate viral genes. Some viruses encode additional enzymes, namely thymidylate synthase (VZV and certain γ2 viruses) and dihydrofolate reductase (certain γ2 viruses; Davison and Scott, 1986; Albrecht *et al.*, 1992; Russo *et al.*, 1996). The standard rationale for occurrence of these herpesvirus genes is that in some aspect of its life cycle the virus requires to replicate DNA in a cell type that has inadequate pools of DNA precursors, and that direct supply of nucleotide synthesizing capability alleviates this situation. In contrast, viruses of the β subfamily do not have genes for TK and RR2, and parsimony criteria indicate that these genes are ancestral but were lost in the common lineage leading to the β1 and β2 viruses (internal branch D in Figure 17.2B). Additionally, inspection of the putative RR1 and dUTPase sequences of β viruses indicates that these proteins are actually unlikely to possess the relevant enzymatic activity, since they both lack sequence features otherwise generally conserved. This argument is clearest in the case of dUTPase: the β virus candidates do not have five local conserved motifs which are thought to be directly involved in dUTPase function (Cedergren-Zeppezauer *et al.*, 1992), and their assignment as dUTPase homologues is based on the presence of an additional motif unique to herpesvirus dUTPases (and considered not to be involved in the enzymatic mechanism) and on the genomic location of the gene. We therefore come to the conclusion that the strategy of the virus genome supplying enzymes of nucleotide synthesis was completely abandoned near the start of the β herpesvirus lineage. This correlates neatly with observations that cytomegaloviruses switch on systems of host DNA synthesis at an early stage in the infectious cycle (St Jeor *et al.*, 1974; Mocarski, 1996); presumably host enzymes then provide adequate dNTP synthesis capacity to support virus DNA replication.

The underlying logic of these changes as evolutionary developments remains obscure. An intriguing aspect of this account is that in β genomes the reading frames for the RR1 and dUTPase homologues have remained open through the extensive evolutionary changes that have occurred in the subfamily's existence, rather than being lost as with the TK and RR2 homologues. This strongly suggests that each encoded protein continues to have some function, distinct from the known enzymatic activity. As noted above, the β subfamily dUTPase homologues still have a herpesvirus-specific motif, and we suggest that this relates to an unknown extra role that all herpesvirus dUTPase proteins possess and that correlates with intragenic duplication of the herpesvirus dUTPase gene (see earlier). Finally, this account provides a cautionary footnote to the appearance of the RR2 and dUTPase genes in the core conserved set as defined earlier.

Cell Transformation Systems

The subject of possible oncogenicity of herpesviruses has a long history, but at present the only mammalian herpesviruses that are usually regarded as tumorigenic are in the γ1 and γ2 groups, and these are treated here. The avian α3 lineage also contains oncogenic species. In the γ1 lineage, the genes of EBV that are involved in the so-called latent cycle in transforming B lymphocytes to an indefinite growth phenotype have been studied extensively (Henderson *et al.*, 1994; Ring, 1994). The gene products comprise six nuclear proteins (EBNA-1, EBNA-2, EBNA-3A, EBNA-3B, EBNA-3C and EBNA-LP), three membrane proteins (LMP-1, LMP-2A and LMP-2B) and also two small non-translated RNAs. EBNA-1 has already been mentioned for its involvement in the maintenance replication of episomal EBV DNA. The EBNA genes are scattered across about 100 kb of the EBV genome and are expressed by way of a transcription unit that runs across most of the genome and is then differentially spliced to give an assortment of mRNA species from which the individual EBNA proteins are translated (Henderson *et al.*,

1994). The three EBNA-3 species are encoded in a set of tandemly arranged genes and are distantly related: evidently the well-diverged products of duplication events. The LMP-1 and LMP-2 proteins are encoded in separate, highly spliced transcription units. LMP-2A and -2B proteins are the closely similar results of differential splicing of one gene, notable in that it runs across the genomic termini and so is only assembled by circularization of the genome. There are positive and negative interactions among these latent proteins (and with cellular proteins) in control of their expression and in cell transformation activities. Present understanding is that the most important species for transformation are EBNA-2 and EBNA-3C (both modulators of gene expression) and LMP-1. The latent genes thus comprise an elaborate subsystem, which we presume to have evolved solely within the γ1 lineage, and there are no clear homologues known outside this lineage. EBNA-1, -2 and -LP all contain extended repeated sections, so may have developed at least partially *in situ*. The LMP proteins are reminiscent of cellular species with multiple transmembrane segments, and the EBNA-3 proteins are large and mostly heteropolymeric; from these characteristics this set are likely to have originated by gene capture events.

There are two variants of EBV, known as EBV1 and EBV2. These differ in their distribution: EBV2 is common in Africa and New Guinea, less so in Caucasian populations (Zimber *et al.*, 1986). There is no clear distinction between the two types in pathology, but there are differences in biology of cell transformation – EBV1 appears more efficient in transforming B cells, and transformed lines grow more vigorously (Rickinson *et al.*, 1987). The complete sequence is known for EBV1 (B95-8 isolate; Baer *et al.*, 1984) but unfortunately only localized sequences have so far been determined for EBV2 (Dambaugh *et al.*, 1984; Sample *et al.*, 1990; Lees *et al.*, 1993). At the level of restriction nuclease profiles of whole genomic DNAs, the two are very similar. The available information gives the following picture of the relationship between the two genomes. First, corresponding sequences are likely to be very closely similar over most of the DNAs, diverged by around

0.01 or 0.02 substitutions per site (Lees *et al.,* 1993) – that is, by about the distance found between two isolates of, for instance, HSV1. Secondly, however, the divergences between the sequences for the *EBNA-2* gene and for the whole of the *EBNA-3* region are very much larger (Dambaugh *et al.,* 1984; Sample *et al.,* 1990). For the *EBNA-2* coding sequences, the synonymous divergence is 0.42 substitutions per site and the non-synonymous divergence 0.29 substitutions per site. For *EBNA-3A, EBNA-3B* and *EBNA-3C* the values are closely grouped: 0.13, 0.15 and 0.13 for synonymous divergences, and 0.08, 0.10 and 0.11 for non-synonymous divergences. (These divergence values were produced by the present authors.) Thus, for the *EBNA-2* gene and the *EBNA-3* genes, the divergences are of the order of five- to 30-fold greater than seen elsewhere in the genomes. For *EBNA-2,* the two EBV types are diverged to roughly the extent seen in comparison of HSV1 and HSV2.

We regard these observations as highly interesting and challenging. This phenomenon, of substantial differences that are localized to a subset of the latent genes, hints at a mode of herpesvirus evolution distinct from and altogether more dynamic than the slow, cospeciational divergence described above, while its evolutionary and mechanistic basis remains obscure. The closeness of the major parts of the genomes argues that the events generating the two current types took place quite recently in evolutionary terms, perhaps of the order of 10^5–10^6 years ago. The divergent genes strikingly include known key players in cell transformation. The sharpness of the boundaries between diverged and closely similar genome segments is also notable: all of the sequences of the *EBNA-2* gene and the three contiguous *EBNA-3* genes are diverged, and nothing besides. In principle, an underlying selective pressure could be supplied by an immunological means (immune evasion by loss of a particular set of T-cell epitopes) or by an advantage conferred through a difference in functionality, for instance in cell transformation; we personally find the latter more easily credible. An obvious means of generating the two virus types would be by recombination between a proto-EBV and an unknown primate γ1 virus or viruses. An alternative

might be by recombination within an intracellular pool of virus genomes acting to concentrate substitutions that were in some manner positively selectable. Both these possible mechanisms have unattractive aspects, and the topic remains an intriguing mystery.

Research on cell transformation by the γ2 herpesviruses is less developed. In HVS, one gene (*STP*) has been identified as sufficient for transformation of the growth phenotype of tissue culture cells; there are no counterparts of the EBNA genes. Different subtypes of HVS display large differences in the size and sequence characteristics of their STP proteins (Jung *et al.,* 1991). Study of the recently discovered human γ2 virus HHV8 is adding new depth to our view of cell transformation and oncogenesis by γ2 viruses. First, HHV8 has a particularly extensive set of homologues of cell genes, some of which probably contribute to oncogenesis at least in auxiliary roles (Russo *et al.,* 1996; Neipel *et al.,* 1997). These number at least 12, and include genes encoding immunomodulatory proteins such as a viral IL-6 and CC chemokines. It has been pointed out that many of the effects that these HHV8 genes are considered to exert on infected cells, probably as part of tumorigenesis, are achieved also by EBV, but characteristically in its case via induction of cell genes rather than by homologous virus genes (Neipel *et al.,* 1997; Moore and Chang, 1998).

Two other facets of the HHV8 genome that may bear on transformation are currently emerging. The first concerns a HHV8-specific gene, *K1,* adjacent to one extremity of the genome, that apparently encodes a Type 1 membrane glycoprotein. *K1* is probably not to be taken as homologous to the *STP* gene of HVS, although there are similarities between the two, in genomic location and in the encoded product having a hydrophobic C-terminal section. Sequencing studies of HHV8 isolates are revealing that two loci within *K1,* each of about 20 codons, are hypervariable and fall into at least three sublineages (Lee *et al.,* 1998; Nicholas *et al.,* 1998; P. M. Cook, D. Whitby, D. Nalwanga-Kakoola *et al.,* unpublished data). There is no exact equivalent of this phenomenon known for any other herpesvirus, although in a general way it resembles the large subtype differences seen in HVS *STP*. The K1 protein has been shown to have transforming potential when transferred to a non-transforming strain of HVS (Lee *et al.*, 1998). The second aspect concerns another HHV8-specific gene, *K15*, at the opposite end of the genome, whose eight exons encode a putative membrane protein with 12 transmembrane segments (thus similar to EBV LMP-2) plus a cytoplasmic C-terminal domain (M. A. Glenn, L. Rainbow, A. J. Davison and T.F. Schulz, unpublished data). Two forms of *K15* gene are recognized that are highly diverged throughout their lengths, exons and introns alike. The *K15* variability thus resembles the situation described above for the *EBNA-2* and *EBNA-3* regions of EBV1 and EBV2. Available information on sequences of HHV8 isolates indicates that, except for *K1* and *K15*, the genome is highly conserved.

HERPESVIRUSES OF FISH AND AMPHIBIANS

Herpesviruses of vertebrates other than mammals and birds have been known for many years. We should emphasize that such viruses have been assigned as herpesviruses almost entirely on the basis of their virion structure. Compared with the wealth of information available on the genomes of the mammalian/avian herpesviruses, genomic data for herpesviruses of other vertebrates are sparse: the complete sequence for channel catfish virus (CCV; Davison, 1992) and limited sequences for two salmonid herpesviruses (Bernard and Mercier, 1993; Davison, 1998), an amphibian herpesvirus (A.J. Davison, unpublished data) and three closely related turtle herpesviruses (Quackenbush *et al.*, 1998). While it has not been possible to carry out phylogenetic comparisons to anywhere near the depth achieved for mammalian herpesviruses, several interesting evolutionary features for the fish and amphibian herpesviruses are emerging.

The genomes of CCV, salmonid herpesvirus 1 (SalHV1) and ranid herpesvirus 1 (RaHV1; also termed Lucké tumour herpesvirus of frogs) are 134, 174 and 217 kb in size, and thus fall within the size range populated by mammalian herpesvirus genomes. The CCV structure is type A, like that of β2 viruses, and the SalHV1 structure is type D, like that of α2 viruses (Figure 17.1). The predicted amino-acid sequences of CCV's complement of encoded proteins, however, show little relationship to those of mammalian herpesviruses. The only significant similarities exist in proteins (mostly enzymes) that also have counterparts in genomes other than those of herpesviruses and therefore could have been acquired by gene capture, in most cases from the host cell, independently of acquisition by the mammalian/avian herpesviruses. Examples include DNA polymerase and dUTPase, where conservation appears to be limited to local sequence motifs. By criteria of amino-acid sequence characteristics, CCV DNA polymerase is assigned to a subset of DNA polymerases distinct from the group related to cellular δ-polymerase to which α, β and γ herpesviruses belong (Knopf, 1998). CCV dUTPase resembles non-herpesviral dUTPase rather than the herpesviral form with longer chain length described above. No gene specific to mammalian herpesviruses, such as those encoding capsid, tegument or envelope proteins, has any homologue in CCV detectable by amino-acid sequence similarity.

The single candidate for a conserved gene that might indicate common ancestry with the mammalian/avian herpesviruses is the spliced CCV gene comprising reading frames 62, 69 and 71. This is clearly homologous by sequence similarity to a core gene of the mammalian viruses (HSV1 *UL15*), which contains two exons and encodes a putative ATPase involved in packaging DNA into the capsid. This gene also has a distant relative in bacteriophage T4 but no cellular equivalent has been described. Analysis of CCV capsid structure by cryoelectron microscopic imaging and computer reconstruction has shown that fine details of the capsid's architecture are closely similar for CCV and HSV1 (Booy *et al.*, 1996). In addition, the CCV capsid may assemble by a mechanism similar to that for α herpesviruses (Davison and Davison, 1995). Although not overwhelming, the evidence is consistent with CCV and the mammalian/avian herpesviruses sharing a common evolutionary origin for their capsid structures but having diverged so extensively that similarities at the protein sequence level have all but disappeared. However, there is no hint of common ancestry for other virion components or in other gene systems. Given the distinct phylogenetic association of CCV DNA polymerase and the lack of detectable counterparts of other proteins comprising the DNA replicative machinery of mammalian herpesviruses, it is possible that genes for DNA replication were acquired independently in the lineages of CCV and of the mammalian herpesviruses, either by way of introduction of a new capability to the virus or by substitution. This interpretation then raises the possibility that the last common ancestor of CCV and the mammalian herpesviruses was a simpler entity that possessed genes encoding the capsid but may have lacked other parts of the genetic capabilities of today's herpesviruses. It is not possible to estimate with any precision when the two lineages were established. The event is likely to have been very ancient, speculatively having occurred when bony fish separated from other vertebrates – of the order of 440 Myears BP (Benton, 1990). There is a potential fallacy, or at least a narrowness of vision, to register concerning this long view: given that the assignments as herpesviruses were defined by possession of a characteristic capsid, it ought not to be surprising that at the greatest distances this structure should be all that is visible. Other lineages of genes than those specifying the capsid are also essential elements in the modern states of herpesviruses, but our account has not been aimed primarily at tracing their antecedents.

Comparisons of the complete encoded protein sequences for CCV and the limited sequences available for salmonid herpesviruses (SalHV1 and SalHV2) show that the latter two viruses are related most closely to each other, and then more distantly to CCV. Qualitatively CCV and SalHV1 thus appear as representing two distinct divisions of the fish herpesviruses that are equivalent to the subfamilies of the mammalian/avian herpesviruses. The existence of at least 18 CCV genes with SalHV1 counterparts, located in U_L in rearranged blocks, is also reminiscent of the situation with mammalian herpesviruses in different subfamilies (Davison,

1998). Analysis of a 40 kb region of the RaHV1 genome indicates that this virus is a distant relative of CCV and SalHV1 rather than of mammalian herpesviruses. The CCV gene counterparts in this region are also rearranged. Thus, this amphibian herpesvirus appears to define a third division of the CCV-related viruses. These evaluations demonstrate conclusively that the similarities in gross genome structures described above, of CCV with the β2 viruses and SalHV1 with the α2 viruses, have arisen independently in the two virus groups.

Sequence data for a part of the DNA polymerase gene of turtle herpesviruses indicate that these viruses belong with the mammalian/avian branch, perhaps as a novel offshoot of the α subfamily. The limited extent of presently available sequences and also the narrow base in terms of ranges of virus and host species prompt caution, but if confirmed and generalized this finding would place a gulf between amphibian and reptilian herpesviruses. Herpesvirus phylogeny may be yet further expanded as a result of the characterization of an oyster herpesvirus (Le Deuff et al., 1994). Early indications are that this virus is not significantly related to either of the two major branches and may instead define a third branch, potentially populated by invertebrate herpesviruses (A. J. Davison, unpublished data). The possible large-scale relationships in what we can now view as the herpesvirus superfamily are summarized in Figure 17.4.

CONCLUSIONS

The main concern of this chapter has been with aspects of the macroevolution of herpesviruses. By comparing genomes, gene sequences and gene sets we have been able to outline various processes that have taken place in the evolution of these viruses and to discern major events in their evolutionary history. The overall scheme of herpesvirus evolution that we have elaborated places the development of herpesviruses in much the same timeframe as that of the vertebrates. The basis for this assignment is the extensive congruence of branching patterns of

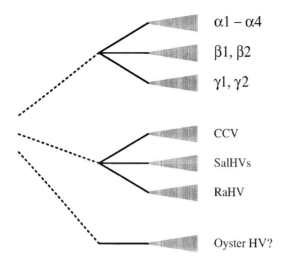

α1 – α4

β1, β2

γ1, γ2

CCV

SalHVs

RaHV

Oyster HV?

FIGURE 17.4 Deep phylogeny of herpesviruses. A schematic view is given of the organization of the three major lineages of herpesviruses and their possible relationships, as discussed in the text. Subfamily groupings of the mammalian/avian viruses, the counterparts of these for fish and amphibian viruses, and the proposed but still tentative grouping of invertebrate (oyster) viruses, are shown as shaded triangles. Evolution of such groupings from common ancestors is indicated by solid lines and a hypothetical overall common origin (for capsid genes at least) by dashed, incompletely joined lines.

the viruses within each sublineage of the mammalian herpesviruses and the branching patterns of their host lineages, which strongly supports a basis of cospeciation. Extrapolation of the resulting timescale then suggests that a common ancestor of today's mammalian herpesviruses existed around the time of emergence of the first mammals. This ancestral species was already a developed herpesvirus and thus must also have had an extensive history. A possible glimpse of that earlier history is afforded by comparison of present day mammalian and fish herpesviruses: the common capsid architecture argues for a common descent, while the gene complements appear deeply distinct. A resolution of this puzzle envisages descent from a prototypic entity that was in effect a preherpesvirus, possessing genes that specified capsid structure but lacking significant other parts of the modern gene complement. A well-grounded understanding of the origins

and evolutionary rationale of this hypothetical creature may be beyond our grasp.

The apparent widespread occurrence of very long-term associations of herpesvirus and host species, to the extent of cospeciation, implies directly that extensive coevolution must have occurred, moulding the genomes of both host and virus. We presume that we see the outcome of this process in the present day phenotypes of the viruses – these are very successful parasites, many infecting a high proportion of the host population and maintaining that infection for the host's lifetime, while in terms of effects on host population they are not severe pathogens. Conversely, the high virulence seen in certain herpesviral infections of a host of a species close to but distinct from the natural host (for instance, the lethal disease caused by simian herpesvirus B in humans) also serves as an indicator of the moderating outcome of coevolution. However, we suspect that the extended nature of the coevolutionary association may well have acted to incorporate adaptations of both host and virus so fundamentally into the genetic organization and capabilities of the contemporary species that it could be very difficult to recognize and dissect out specific adaptations. We should also note here that the emphasis we have placed on cospeciation should not obscure or diminish the fact that other modes of evolutionary development of herpesviruses certainly exist. Our phylogenetic trees showed only infrequent examples of changes in host species that persisted as long-term lineages, but cross-species infections occur at observable levels, say with humans and other primates or between captive animals. Alteration of niche in terms of host tissue tropism represents another evolutionary mode, which we invoked in connection with ancient branchings in the herpesvirus tree, and this can also be seen in more recent evolution, for instance with HSV1 and HSV2. Regarding novel aspects of herpesvirus evolution, we look forward to resolution of the mechanisms generating the locus-specific divergences described for EBV and HHV8 with great anticipation and sense of the radically unknown.

We have not addressed any microevolutionary aspects of herpesviruses – the changes occurring in contemporary populations of viruses, or events in their recent history. Studies have in fact been carried out on variation in herpesvirus isolates, for instance with HSV and EBV. There are two aspects that make this line of research hard in comparison to equivalent work on RNA viruses and retroviruses, namely the large sizes of herpesvirus genomes and their slow rates of change. Nonetheless, certain advances have been made. Restriction nuclease profiles of HSV isolates have been used to demonstrate that variation among isolates is consistent with a close association between virus strains and human ethnic groups over the last 10^5 years (Sakaoka et al., 1994). Several studies have examined regions in the genomes of EBV isolates for possible effects of immune pressure exerted by cytotoxic T lymphocytes in selected human populations; interestingly, in one case such effects were detected (de Campos-Lima et al., 1993) and in others not (Lee et al., 1995; Khanna et al., 1997). Other EBV studies have sought correlations between EBNA gene sequence variants and disease associations of EBV isolates (Snudden et al., 1995; Bhatia et al., 1996; Gutiérrez et al., 1997). Our evaluation is that such work is in its very early stages, and will advance much further with modern capabilities. A future goal for study of herpesvirus evolution will be to unify results from such study of contemporary variation with the long timescale account we have sketched in this chapter. Parenthetically, such unification may not be wholly simple: in our text above we registered caution in extrapolating distance and rate estimations to very deep events in the herpesvirus tree, and there are also technical complications that apply to extrapolation in the opposite direction to a short timeframe.

Finally, we are all too aware that our account has been heavily phenomenological – we have outlined the effects of evolutionary processes on herpesvirus genomes and given a timescale to certain aspects, but we have not achieved any very detailed insight into the evolutionary forces that have operated on these genomes, or the opportunities that were associated with new genomic elaborations. Given that strategic mechanisms in herpesvirus infections are highly complicated and still only partially under-stood, and given the complex histories sketched above, it is rather likely that such a functional account of the evolution of these viruses will only ever be fragmentary. Conversely, we also believe that evolutionary analysis must comprise an essential part of the effort toward understanding herpesvirus strategies.

REFERENCES

Albrecht, J.-C., Nicholas, J., Biller, D. et al. (1992) Primary structure of the herpesvirus saimiri genome. J. Virol., **66**, 5047–5058.

Baer, R., Bankier, A.T., Biggin, M.D. et al. (1984) DNA sequence and expression of the B95–8 Epstein–Barr virus genome. Nature (Lond.), **310**, 207–211.

Benton, M.J. (1990) Vertebrate Palaeontology. HarperCollins Academic, London.

Bernard, J. and Mercier, A. (1993) Sequence of two Eco RI fragments from salmonis herpesvirus 2 and comparison with ictalurid herpesvirus 1. Arch. Virol., **132**, 437–442.

Bhatia, K., Raj, A., Gutiérrez, M.I. et al. (1996) Variation in the sequence of Epstein Barr virus nuclear antigen 1 in normal peripheral blood lymphocytes and in Burkitt's lymphomas. Oncogene, **13**, 177–181.

Boehmer, P.E. and Lehman, I.R. (1997) Herpes simplex virus DNA replication. Annu. Rev. Biochem., **66**, 347–384.

Booy, F.P., Trus, B.L., Davison, A.J. and Steven, A.C. (1996) The capsid architecture of channel catfish virus, an evolutionarily distant herpesvirus, is largely conserved in the absence of discernible sequence homology with herpes simplex virus. Virology, **215**, 134–141.

Bruni, R. and Roizman, B. (1996) Open reading frame P – a herpes simplex virus gene repressed during productive infection encodes a protein that binds a splicing factor and reduces synthesis of viral proteins made from spliced mRNA. Proc. Natl Acad. Sci. USA, **93**, 10423–10427.

Brunovskis, P. and Kung, H.-J. (1996) Retrotransposition and herpesvirus evolution. Virus Genes, **11**, 259–270.

Cedergren-Zeppezauer, E.S., Larsson, G., Nyman, P.O. *et al.* (1992) Crystal structure of a dUTPase. *Nature (Lond.),* **355**, 740–743.

Cha, T.-A., Tom, E., Kemble, G.W., Duke, G.M., Mocarski, E.S. and Spaete, R.R. (1996) Human cytomegalovirus clinical isolates carry at least 19 genes not found in laboratory strains. *J. Virol.,* **70**, 78–83.

Challberg, M. (1996) Herpesvirus DNA replication. In: *DNA Replication in Eukaryotic Cells* (ed. De Pamphilis, M.L.), pp. 721–750. Cold Spring Harbor Laboratory Press, Cold Spring Harbor, NY.

Chang, Y.E., Menotti, L., Filatov, F., Campadelli-Fiume, G. and Roizman, B. (1998) U$_L$ 27.5 is a novel γ$_2$ gene antisense to the herpes simplex virus 1 gene encoding glycoprotein B. *J. Virol.,* **72**, 6056–6064.

Chee, M.S., Bankier, A.T., Beck, S. *et al.* (1990) Analysis of the protein coding content of the sequence of human cytomegalovirus strain AD169. *Curr. Topics Microbiol. Immunol.,* **154**, 125–169.

Dambaugh, T., Hennessy, K., Chamnankit, L. and Kieff, E. (1984) U2 region of Epstein–Barr virus DNA may encode Epstein–Barr nuclear antigen 2. *Proc. Natl Acad. Sci. USA,* **81**, 7632–7636.

Dargan, D.J., Jamieson, F.E., Maclean, J., Dolan, A., Addison, C. and McGeoch, D.J. (1997) The published DNA sequence of human cytomegalovirus AD169 lacks 929 base pairs affecting genes UL42 and UL43. *J. Virol.,* **71**, 9833–9836.

Davison, A.J. (1992) Channel catfish virus: a new type of herpesvirus. *Virology,* **186**, 9–14.

Davison, A.J. (1998) The genome of salmonid herpesvirus 1. *J. Virol.,* **72**, 1974–1982.

Davison, A.J. and Davison, M.D. (1995) Identification of structural proteins of channel catfish virus by mass spectrometry. *Virology,* **206**, 1035–1043.

Davison, A.J. and McGeoch, D.J. (1986) Evolutionary comparisons of the S segments in the genomes of herpes simplex virus type 1 and varicella-zoster virus. *J. Gen. Virol.,* **67**, 597–611.

Davison, A.J. and Scott, J.E. (1986) The complete DNA sequence of varicella-zoster virus. *J. Gen. Virol.,* **67**, 1759–1816.

Davison, A.J. and Taylor, P. (1987) Genetic relations between varicella-zoster virus and Epstein–Barr virus. *J. Gen. Virol.,* **68**, 1067–1079.

De Campos-Lima, P.-O., Gavioli, P., Zhang, Q.-J. *et al.* (1993) HLA-A11 epitope loss isolates of Epstein–Barr virus from a highly A11+ population. *Science,* **260**, 98–100.

Dolan, A., Jamieson, F.E., Cunningham, C., Barnett, B.C. and McGeoch, D.J. (1998) The genome sequence of herpes simplex virus type 2. *J. Virol.,* **72**, 2010–2021.

Früh, K., Ahn, K, Djaballah, H. *et al.* (1995) A viral inhibitor of peptide transporters for antigen presentation. *Nature (Lond.),* **375**, 415–418.

Galocha, B., Hill, A., Barnett, B.C. *et al.* (1997) The active site of ICP47, a herpes simplex virus encoded inhibitor of the MHC encoded peptide transporter TAP, maps to the N-terminal 35 residues. *J. Exp. Med.,* **185**, 1565–1572.

Gompels, U.A., Nicholas, J., Lawrence, G. *et al.* (1995) The DNA sequence of human herpesvirus-6: structure, coding content, and genome evolution. *Virology,* **209**, 29–51.

Gutiérrez, M.I., Raj, A., Spangler, G. *et al.* (1997) Sequence variations in EBNA-1 may dictate restriction of tissue distribution of Epstein–Barr virus in normal and tumour cells. *J. Gen. Virol.,* **78**, 1663–1670.

Henderson, S.A., Huen, D. and Rowe, M. (1994) Epstein–Barr virus transforming proteins. *Sem. Virol.,* **5**, 391–399.

Hill, A., Jugovic, P., York, I. *et al.* (1995) Herpes simplex virus turns off the TAP to evade host immunity. *Nature (Lond.),* **375**, 411–415.

Honess, R.W., Gompels, U.A., Barrell, B.G. *et al.* (1989) Deviations from expected frequencies of CpG dinucleotides in herpesvirus DNAs may be diagnostic of differences in the states of their latent genomes. *J. Gen. Virol.,* **70**, 837–855.

Isfort, R., Jones, D., Kost, R., Witter, R. and Kung, H.-J. (1992) Retrovirus insertion into herpesvirus *in vitro* and *in vivo. Proc. Natl Acad. Sci. USA,* **89**, 991–995.

Jung, J.U., Trimble, J.J., King, N.W., Biesinger, B., Fleckenstein, B.W. and Desrosiers, R.C. (1991) Identification of transforming genes of sub-

group A and C strains of *Herpesvirus saimiri.*
Proc. Natl Acad. Sci. USA, **88**, 7051–7055.

Khanna, R., Slade, R.W., Poulsen, L. *et al.* (1997) Evolutionary dynamics of genetic variation in Epstein–Barr virus isolates of diverse geographical origins: evidence for immune pressure-independent genetic drift. *J. Virol.,* **71**, 8340–8346.

Knopf, C.W. (1998) Evolution of viral DNA-dependent DNA polymerases. *Virus Genes,* **16**, 47–58.

Le Deuff, R.M., Nicolas, J.L., Renault, T. and Cochennec, N. (1994) Experimental transmission of a herpes-like virus to axenic larvae of Pacific oyster, *Crassostrea gigas. Bull. Eur. Assoc. Fish Pathol.,* **14**, 69–72.

Lee, S.P., Morgan, S., Skinner, J. *et al.* (1995) Epstein–Barr virus isolates with the major HLA B35.01-restricted cytotoxic T-lymphocyte epitope are prevalent in a highly B35.01-positive African population. *Eur. J. Immunol.,* **25**, 102–110.

Lee, H., Veazey, R., Williams, K. *et al.* (1998) Deregulation of cell growth by the K1 gene of Kaposi's sarcoma-associated herpesvirus. *Nat. Med.,* **4**, 435–440.

Lees, J.F., Arrand, J.E., de V. Pepper, S. *et al.* (1993) The Epstein–Barr virus candidate vaccine antigen gp340/220 is highly conserved between virus types A and B. *Virology,* **195**, 578–586.

McGeoch, D.J. (1989) The genomes of the human herpesviruses: contents, relationships and evolution. *Annu. Rev. Microbiol.,* **43**, 235–265.

McGeoch, D.J. (1990) Protein sequence comparisons show that the "pseudoproteases" encoded by poxviruses and certain retroviruses belong to the deoxyuridine triphosphatase family. *Nucl. Acids Res.,* **18**, 4105–4110.

McGeoch, D.J. and Cook, S. (1994) Molecular phylogeny of the Alphaherpesvirinae subfamily and a proposed evolutionary timescale. *J. Mol. Biol.,* **238**, 9–22.

McGeoch, D.J. and Davison, A.J. (1995) Origins of DNA viruses. In: *Molecular Basis of Virus Evolution* (eds Gibbs, A.J., Calisher, C.H. and Garcia-Arenal, F.), pp. 67–75. Cambridge University Press, Cambridge.

McGeoch, D.J., Dolan, A. Donald, S. and Rixon,

F.J. (1985) Sequence determination and genetic content of the short unique region in the genome of herpes simplex virus type 1. *J. Mol. Biol.,* **181**, 1–13.

McGeoch, D.J., Dolan, A., Donald, S. and Brauer, D.T.K. (1986) Complete DNA sequence of the short repeat region in the genome of herpes simplex virus type 1. *Nucl. Acids Res.,* **14**, 1727–1745.

McGeoch, D.J., Dalrymple, M.A., Davison, A.J. *et al.* (1988) The complete DNA sequence of the long unique region in the genome of herpes simplex virus type 1. *J. Gen. Virol.,* **69**, 1531–1574.

McGeoch, D.J., Cook, S., Dolan, A., Jamieson, F.E. and Telford, E.A.R. (1995) Molecular phylogeny and evolutionary timescale for the family of mammalian herpesviruses. *J. Mol. Biol.,* **247**, 443–458.

MacLean, C.A., Efstathiou, S., Elliott, M.L., Jamieson, F.E. and McGeoch, D.J. (1991) Investigation of herpes simplex virus type 1 genes encoding multiply inserted membrane proteins. *J. Gen. Virol.,* **72**, 897–906.

Megaw, A.M., Rapaport, D., Avidor, B., Frenkel, N. and Davison, A.J. (1998) The DNA sequence of the RK strain of human herpesvirus 7. *Virology,* **244**, 119–132.

Mocarski, E.S. Jr (1996) Cytomegaloviruses and their replication. In: *Fields' Virology,* 3rd edn (eds Fields, B.N., Knipe, D.M. and Howley, P.M.), pp. 2447–2492. Lippencott-Raven, Philadelphia, PA.

Moore, P.S. and Chang, Y. (1998) Antiviral activity of tumor-suppressor pathway: clues from molecular piracy by KSHV. *Trends Genet.,* **14**, 144–150.

Neipel, F., Albrecht, J.-C. and Fleckenstein, B. (1997) Cell-homologous genes in the Kaposi's sarcoma-associated rhadinovirus human herpesvirus 8: determinants of its pathogenicity? *J. Virol.,* **71**, 4187–4192.

Nicholas, J. (1996) Determination and analysis of the complete nucleotide sequence of human herpesvirus 7. *J. Virol.,* **70**, 5975–5989.

Nicholas, J., Zong, J.-C., Alcendor, D.J. *et al.* (1998) Novel organizational features, captured cellular genes, and strain variability within the genome of KSHV/HHV8. *Monogr. Natl Cancer Inst.,* **23**, 79–88.

Parker, B.D., Bankier, A., Satchwell, S., Barrell, B. and Farrell, P.J. (1990) Sequence and transcription of Raji Epstein–Barr virus DNA spanning the B95–8 deletion. *Virology*, **179**, 339–346.

Pyles, B.R., Sawtell, N.M. and Thompson, R.L. (1992) Herpes simplex virus type 1 dUTPase mutants are attenuated for neurovirulence, neuroinvasiveness, and reactivation from latency. *J. Virol.*, **66**, 6706–6713.

Quackenbush, S.L., Work, T.M., Balazs, G.H. *et al.* (1998) Three closely related herpesviruses are associated with fibropapillomatosis in marine turtles. *Virology*, **246**, 392–399.

Randall, G. and Roizman, B. (1997) Transcription of the derepressed open reading frame P of herpes simplex virus 1 precludes the expression of the antisense $\gamma_1 34.5$ gene and may account for the attenuation of the mutant virus. *J. Virol.*, **71**, 7750–7757.

Randall, G., Lagunoff, M. and Roizman, B. (1997) The product of ORF O located within the domain of herpes simplex virus 1 genome transcribed during latent infection binds to and inhibits *in vitro* binding of infected cell protein 4 to its cognate DNA site. *Proc. Natl Acad. Sci. USA*, **94**, 10379–10384.

Rawlinson, W.D., Farrell, H.E. and Barrell, B.G. (1996) Analysis of the complete DNA sequence of murine cytomegalovirus. *J. Virol.*, **70**, 8833–8849.

Rickinson, A.B., Young, L.S. and Rowe, M. (1987) Influence of the Epstein–Barr virus nuclear antigen EBNA2 on the growth phenotype of virus-transformed B cells. *J. Virol.*, **61**, 1310–1317.

Ring, C.J.A. (1994) The B cell-immortalizing functions of Epstein–Barr virus. *J. Gen. Virol.*, **75**, 1–13.

Rixon, F.J. (1993) Structure and assembly of herpesviruses. *Sem. Virol.*, **4**, 135–144.

Rixon, F.J. and McGeoch, D.J. (1985) Detailed analysis of the mRNAs mapping in the short unique region of herpes simplex type 1. *Nucl. Acids Res.*, **13**, 953–973.

Roizman, B., Desrosiers, R.C., Fleckenstein, B., Lopez, C., Minson, A.C. and Studdert, M.J. (1992) The family Herpesviridae: an update. *Arch. Virol.*, **123**, 425–449.

Roizman, B., Desrosiers, R.C., Fleckenstein, B.,

Lopez, C., Minson, A.C. and Studdert, M.J. (1995) *Herpesviridae*. In: *Virus Taxonomy*, pp. 114–127. *Arch. Virol. Suppl.*, **10**,

Russo, J.J., Bohenzky, R.A., Chien, M.-C. *et al.* (1996) Nucleotide sequence of the Kaposi sarcoma-associated herpesvirus (HHV8). *Proc. Natl Acad. Sci. USA*, **93**, 14862–14867.

Sakaoka, H., Kurita, K., Iida, Y. *et al.* (1994) Quantitative analysis of genomic polymorphism of herpes simplex virus type 1 strains from six countries: studies of molecular evolution and molecular epidemiology of the virus. *J. Gen. Virol.*, **75**, 513–527.

Sample, J., Young, L., Martin, B. *et al.* (1990) Epstein–Barr virus types 1 and 2 differ in their EBNA-3A, EBNA-3B, and EBNA-3C genes. *J. Virol.*, **64**, 4084–4092.

Schrag, J.D., Prasad, B.V.V., Rixon, F.J. and Chiu, W. (1989) Three-dimensional structure of the HSV1 nucleocapsid. *Cell*, **56**, 651–660.

Snudden, D.K., Smith, P.R., Lai, D., Ng, M.-H. and Griffin, B.E. (1995) Alterations in the structure of the EBV nuclear antigen, EBNA1, in epithelial cell tumours. *Oncogene*, **10**, 1545–1552.

Spear, P.G. (1993) Entry of alphaherpesviruses into cells. *Sem. Virol.*, **4**, 167–180.

St Jeor, S.C., Albrecht, T.B., Funk, F.D. and Rapp, F. (1974) Stimulation of cellular DNA synthesis by human cytomegalovirus. *J. Virol.*, **13**, 353–362.

Telford, E.A.R., Watson, M.S., Aird, H.C., Perry, J. and Davison, A.J. (1995) The DNA sequence of equine herpesvirus 2. *J. Mol. Biol.*, **249**, 520–528.

Thomson, B.J., Efstathiou, S. and Honess, R.W. (1991) Acquisition of the human adeno-associated virus type-2 *rep* gene by human herpesvirus type-6. *Nature (Lond.)*, **351**, 78–80.

Thomson, B.J., Weindler, F.W., Gray, D., Schwaab, V. and Heilbronn, R. (1994) Human herpesvirus 6 (HHV-6) is a helper virus for adeno-associated virus type 2 (AAV-2) and the AAV-2 *rep* gene homologue in HHV-6 can mediate AAV-2 DNA replication and regulate gene expression. *Virology*, **204**, 304–311.

Yates, J.L. (1996) Epstein–Barr virus DNA replication. In: *DNA Replication in Eukaryotic Cells* (ed. De Pamphilis, M.L.), pp. 751–773. Cold

Spring Harbor Laboratory Press, Cold Spring Harbor, NY.

Zimber, U., Adldinger, H.K., Lenoir, G.M. *et al.* (1986) Geographical prevalence of two Epstein–Barr virus types. *Virology*, **154**, 56–66.

African Swine Fever Virus: A Missing Link Between Poxviruses and Iridoviruses?

José Salas, María L. Salas and Eladio Viñuela

THE THREE FAMILIES OF CYTOPLASMIC DNA VIRUSES INFECTING ANIMALS

Cytoplasmic DNA viruses infecting animals are classified into three families: the Poxviridae, the Iridoviridae, and the Asfarviridae family where African swine fever virus (ASFV) is presently included. The family Poxviridae comprises a large group of enveloped, double-stranded DNA viruses that infect vertebrates (Chordopoxvirinae subfamily) or insects (Entomopoxvirinae subfamily; Moss, 1996). These subfamilies are further subdivided into different genera. The genus *Orthopoxvirus* of the Chordopoxvirinae subfamily includes vaccinia virus, the best known and prototypal member of the poxviruses, as well as variola virus, which caused smallpox until its eradication through immunization with the attenuated vaccinia virus (Fenner, 1996).

The poxvirus particles are brick-shaped and consist of an external envelope, a surface layer and an internal biconcave core that contains the virus DNA in a nucleoprotein complex. The genome is a linear double-stranded DNA of about 130 to about 300 kb, with terminal inverted repetitions (TIR) of variable lengths (approximately 10 kb in the vaccinia virus genome) and

with the ends covalently closed by A+T-rich hairpin loops, which form incompletely base-paired structures and exist in two forms that are inverted and complementary in sequence (Moss, 1996). The complete DNA sequence of vaccinia virus (Goebel *et al.*, 1990), variola virus (Massung *et al.*, 1994) and molluscum contagiosum virus (Senkevich *et al.*, 1996) have been determined, allowing the identification of 163–198 open reading frames (ORF). Poxvirus DNA replication occurs exclusively in the cytoplasm of the infected cell and involves the formation of concatemeric intermediates, which are resolved to produce the mature unit-length molecules (Moyer and Graves, 1981; Baroudy *et al.*, 1982).

Poxvirus particles contain all the enzymes and factors required for the synthesis and modification by polyadenylation and capping of early mRNA (Moss, 1990). Two other classes of mRNA, intermediate and late, are synthesized during the virus infectious cycle. The intermediate class is synthesized after the initiation of viral DNA replication but before the expression of late genes (Vos and Stunnenberg, 1988). Early viral proteins are required for transcription of intermediate genes, and intermediate genes encode transcription factors for late gene expression. Late genes, in turn, encode early

Origin and Evolution of Viruses
ISBN 0–12–220360–7

transcription factors and enzymes, such as an RNA polymerase, poly A polymerase and the capping enzyme, which are packaged into the virions for the subsequent round of infection, indicating that viral gene expression is regulated in a cascade fashion (Baldick and Moss, 1993).

The Iridoviridae family comprises four genera of viruses infecting vertebrate and invertebrate hosts. The genera *Iridovirus* and *Chloriridovirus* include the iridescent insect viruses of small (120–140 nm) and large size (>180 nm) respectively. Frog virus 3 (FV3), one of the most widely studied members of this family, is the type species of the genus *Ranavirus*, and lymphocystis disease virus, which causes the formation of benign tumors in several fish species, belongs to the genus *Lymphocystivirus*. All the members of this family are deoxyviruses with icosahedral morphology (Willis, 1990).

The iridovirus genome is a double-stranded DNA molecule of 150–350 kb, which is circularly permuted and terminally redundant (Goorha and Murti, 1982; Darai *et al.*, 1983; Delius *et al.*, 1984; Ward and Kalmakoff, 1987). This genome structure, unique among eukaryotic viruses, resembles that of some prokaryotic viruses, such as bacteriophage T4. A property common to the iridoviruses infecting vertebrates is the high degree of methylation of their DNA at cytosine residues in CpG sequences (Willis and Granoff, 1980; Darai *et al.*, 1983; Wagner *et al.*, 1985). This characteristic is not found in the insect iridoviruses.

Although the morphogenesis of the iridoviruses takes place in specialized areas of the cytoplasm, designated virus assembly sites, it has been shown in the case of FV3 that the replication of the virus DNA is a two-stage process, with the synthesis of unit-length genomes, which occurs in the nucleus, followed by a second stage of replication in the cytoplasm with the formation of large concatemers, which are the precursors of the mature viral DNA molecules (Goorha, 1982). This mechanism of DNA replication resembles in some aspects that described for bacteriophages with circularly permuted and terminally redundant genomes.

Iridovirus early mRNA synthesis also occurs in the nucleus, with the participation of the host cell RNA polymerase II modified by a virion factor. Late mRNAs are synthesized in the cytoplasm, but the enzymes involved in late transcription remain to be identified. The transcripts lack poly A tails at their 3'-end, but they are capped and methylated at the 5'-end (Willis *et al.*, 1990).

ASFV is an enveloped deoxyvirus that infects different species of suids and causes a severe disease in domestic pigs. ASFV also propagates in ticks of the genus *Ornithodoros*, being the only known arbovirus that contains DNA (Viñuela, 1985; Costa, 1990). Because of its icosahedral morphology and cytoplasmic site of multiplication, ASFV has been classified for many years as an iridovirus. However, with increasing knowledge of the molecular biology of ASFV, it became evident that the virus differed in many aspects from the iridoviruses and, at the same time, some striking similarities with the poxviruses in genome structure and the strategies used for gene expression were revealed. For these reasons, ASFV has been excluded from the Iridoviridae family and is presently the only member of a separate virus family, the Asfarviridae.

ASFV, because of the properties that it shares with both the poxviruses and the iridoviruses, could be considered, in a way, as a poxvirus with the morphology of an iridovirus and might thus represent an evolutive link between these viruses. The properties of ASFV, poxviruses and iridoviruses, which lend support to this view, are compared below.

COMPARATIVE PROPERTIES OF ASFV, POXVIRUSES, AND IRIDOVIRUSES

Genome Structure

The ASFV genome is a linear, double-stranded DNA molecule of 170–190 kb. The ends are covalently closed by hairpin loops that, like those of poxvirus DNA, are composed of incompletely paired A and T residues and are present in two equimolar forms that, when compared in opposite polarities, are inverted and

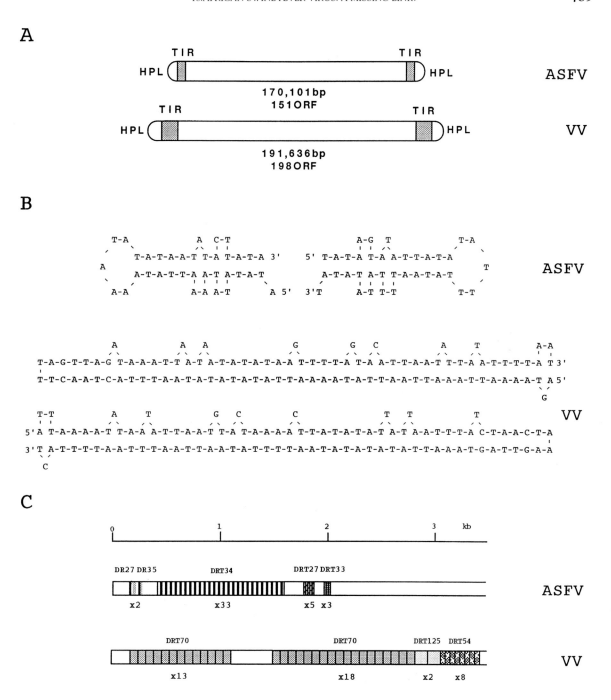

FIGURE 18.1 Structure of the ASFV and vaccinia virus (VV) genome. **A.** Schematic representation of ASFV and VV DNA. The number of base pairs (bp) and ORFs is indicated. HPL, hairpin loop; TIR, terminal inverted repetition. **B.** Nucleotide sequences of the two forms of the hairpin loops of ASFV and VV. **C.** Structure of the TIRs of ASFV and VV. The ASFV TIR is 2134 nucleotides long and consists of unique sequences (in white) interspersed with two types of direct repeats (DR) and three sets of direct repeats in tandem (DRT). The VV TIR is 10 000 nucleotides long and contains four sets of direct repeats in tandem separated by a unique sequence (in white) at the terminal 3.5 kb of the TIR, followed by a unique sequence of 6.5 kb not represented in the figure. The number of nucleotides of each repeat and the number of repeats in each set are indicated.

complementary (González *et al.*, 1986; Figure 18.1). Another characteristic of the ASFV genome is the presence of TIRs that resemble in structure those of the poxviruses and are identical at both genome ends (de la Vega *et al.*, 1994; Meireles and Costa, 1994). The ASFV TIRs are about 2.1 kb long and consist of unique sequences interspersed with five different sets of repeated sequences (Figure 18.1). The central region of the TIR is formed by 33 copies in tandem of 34 bp-long repeated sequence.

Therefore, the structure of the ASFV genome is similar in many aspects to that of the DNA of poxviruses, but is unrelated to the structure of the iridovirus genome, which, as mentioned above, is circularly permuted and terminally redundant.

DNA Replication

For many years, ASFV has been thought to multiply exclusively in the cytoplasm of the infected cell, as the poxviruses do. However, a role for the host-cell nucleus in ASFV multiplication was suggested by the finding that the virus does not grow in enucleated Vero cells nor is viral DNA synthesized (Ortín and Viñuela, 1977), raising the possibility that the host-cell nucleus might be a site of viral DNA replication. This possibility is supported by recent evidence that indicates that the replication of the viral DNA is initiated in the nucleus of the infected cell (García-Beato *et al.*, 1992b). This nuclear stage is followed by a second phase of replication in the cytoplasm. The replication of ASFV DNA is therefore similar to that of FV3 in the existence of nuclear and cytoplasmic stages of DNA replication, although the molecular mechanisms of DNA synthesis are probably different, since both viruses have distinct DNA structures. On the other hand, it is striking that poxviruses and ASFV, which have similar DNA structures that suggest a common mechanism of DNA synthesis, differ in the requirement of the nucleus for DNA replication. The precise role of the host nucleus in ASFV DNA synthesis is not well established at present. The need for a nuclear protein, possibly involved in the initiation of DNA replication and not codified by the virus genome, might account for this nuclear dependence.

Little is known at present about the molecular mechanism of ASFV DNA synthesis. In the case of the poxviruses, the replication of the DNA involves the synthesis of concatemeric intermediates, which are then resolved to generate mature genomes. Analysis of poxvirus replicating DNA with restriction enzymes has shown that these concatemers are head-to-head and tail-to-tail-linked molecules. The sequence obtained for the genomic junction within these concatemers is also consistent with this structure. The presence of head-to-head and tail-to-tail junctions in replicating poxvirus DNA suggests a self-priming replication model (Moyer and Graves, 1981; Baroudy *et al.*, 1982). In this model, a nick is proposed to occur at one end of the genome, which provides a free 3'-end for priming. Head-to-head and tail-to-tail-linked molecules that may be replicative intermediates have also been detected in ASFV-infected cells (González *et al.*, 1986; Caeiro *et al.*, 1990). It remains to be ascertained whether the formation of these concatemeric intermediates occurs during the nuclear or cytoplasmic stages of ASFV DNA replication. In the case of FV3, large concatemers are synthesized only during the cytoplasmic phase of replication. These concatemers, however, have a different structure from those generated in the replication of ASFV and poxvirus DNA.

Several proteins of viral origin may be involved in the replication of the DNA of these viruses. Both vaccinia virus and ASFV encode a replicative DNA polymerase belonging to the alpha family of polymerases, as well as a DNA ligase (Earl *et al.*, 1986; Colinas *et al.*, 1990; Rodríguez *et al.*, 1993b; Yáñez and Viñuela, 1993; Hammond *et al.*, 1992). The iridovirus lymphocystis disease virus has also been shown to contain a gene coding for an alpha-type DNA polymerase (Tidona and Darai, 1997). In addition, ASFV and lymphocystis disease virus encode a protein that has sequence similarity to the proliferating cell nuclear antigen (PCNA) and that could act as a processivity factor for the viral DNA polymerase (Yáñez *et al.*, 1995); Tidona and Darai, 1997). A type I topoisomerase that may be involved in the replication of the viral

DNA is encoded by vaccinia virus (Shuman and Moss, 1987), while ASFV codifies for a type II topoisomerase whose possible role in DNA replication remains to be determined (García-Beato et al., 1992a). In addition, these viruses code for enzymes involved in the synthesis of deoxyribonucleotides, such as thymidine kinase (Hruby and Ball, 1982; Weir et al., 1982; Blasco et al., 1990; Martín-Hernández and Tabarés, 1991; Schnitzler et al., 1991), thymidylate kinase (Smith et al., 1989; Yáñez et al., 1993b) and ribonucleotide reductase (Slabaugh et al., 1988; Tengelsen et al., 1988; Boursnell et al., 1991; Tidona and Darai, 1997).

The analysis of the ASFV genome has revealed that it also encodes the essential enzymes of a DNA base excision repair (BER) system (Yáñez et al., 1995; Oliveros et al., 1997). These enzymes are a glycosylase, an apurinic/apyridinic endonuclease, a DNA polymerase of 20 kDa belonging to the pol X family, and the above-mentioned DNA ligase. This BER system may be a crucial defense mechanism to repair modified bases in the virus DNA, since the macrophages, which are the target cells in natural ASFV infections, generate reactive oxygen species for their microbicidal functions that may cause oxidative damage to the viral DNA. So far, no such system has been described in the poxviruses and the iridoviruses.

As already mentioned, the DNA of vertebrate iridoviruses is heavily methylated. In this connection, a putative restriction-modification system similar to that found in bacteria has been identified in the iridoviruses. This system, consisting of a DNA methyltransferase and a protein homologous to type II restriction endonuclease specific for a CCGG target sequence, might be able to degrade the host cell DNA, while the viral DNA would be protected from cleavage by specific methylation (Kaur et al., 1995; Tidona and Darai, 1997).

Transcription

Early RNA

ASFV particles contain, as those of poxviruses, all the enzymatic machinery required for the synthesis and modification of early mRNA. These activities include an RNA polymerase, a poly A polymerase and a capping enzyme (Kuznar et al., 1980; Salas et al., 1981). In addition, the virions contain protein kinase (Polatnick et al., 1974; Salas et al., 1988a; Baylis et al., 1993) and nucleic-acid-dependent ATPase activities (Kuznar et al., 1981), although the role of these enzymes in transcription remains to be determined. The possible involvement of the viral topoisomerase II in early transcription is suggested by the finding that the specific inhibitor of this enzyme, coumermycin A1, inhibits in-vitro transcription by ASF virions (Salas et al., 1983). The RNAs synthesized by the virion-associated RNA polymerase have methylated CAP structures at their 5′ ends and poly A tails at their 3′ ends, as the eukaryotic and poxvirus RNAs (Salas et al., 1981).

Early genes are transcribed in ASFV-infected cells in the presence of protein or DNA synthesis inhibitors by the virion-associated RNA polymerase (Salas et al., 1986). Transcription of these genes starts at a short distance from the translation initiation ATG of the corresponding ORF. Runs of seven or more thymidylate residues (7T motifs) are signals for 3′ end formation of the early RNAs (Almazán et al., 1992). The 7T motifs are similar to the T5NT sequences that act as termination signal for vaccinia virus early RNAs (Yuen and Moss, 1986).

Intermediate RNA

An intermediate class of ASFV RNAs, with the characteristics of poxvirus intermediate transcripts, has been identified (Rodríguez et al., 1996). The synthesis of these RNAs is initiated after the onset of viral DNA replication, but before the transcription of late genes. Two ASFV genes give rise to intermediate RNAs. One of them, gene I243L, encodes a protein that is similar to the members of the cellular TFIIS transcription factors and to the RPO30 subunit of vaccinia virus RNA polymerase (Rodríguez et al., 1992), which is responsible for VITF-1 (intermediate transcription factor) activity and is also related to the TFIIS family of transcription factors (Rosales et al., 1994). The protein encoded by the other intermediate gene of ASFV, I226R, shows no significant similarity to other proteins in the databases. However, a possible role for *I226R* in the regulation of late transcription has been proposed, taking into account that several of the vaccinia virus intermediate genes encode late transcription factors.

Late RNA

ASFV late genes are transcribed after the initiation of viral DNA replication and at later times than intermediate genes. Like the early RNAs, the ASFV late RNAs are of defined sizes, have short leader sequences before the translational start site, and their 3' ends map within the conserved 7T motif (Almazán *et al.*, 1993; Rodríguez *et al.*, 1996). In contrast, poxvirus late RNAs are heterogeneous in size and lack defined 3' ends (Cooper *et al.*, 1981; Mahr and Roberts, 1984). The late transcription machinery of the poxviruses does not recognize the early termination T5NT signal.

A cascade model for the regulation of poxvirus gene expression has been proposed, since the factors involved in the transcription of each class of gene are synthesized during the expression of the previous temporal classes (Baldick and Moss, 1993). The existence of the same temporal classes of genes in ASFV-infected cells supports a similar mode of regulation for ASFV gene expression.

A very different strategy is used for the transcription of FV3 DNA. FV3 particles do not contain RNA polymerase, and RNA synthesis depends on the host cell RNA polymerase II (Goorha, 1981). By contrast, the cellular enzyme is not involved in the transcription of ASFV genes, since viral RNA synthesis in ASFV-infected cells is not inhibited by α-amanitin or 5,6-dichloro-1-β-D-ribofuranosyl benzimidazole, two specific inhibitors of that enzyme (Salas *et al.*, 1988b). The role of RNA polymerase II during vaccinia virus multiplication is not clear at present. Although results from experiments using α-amanitin or enucleated or UV-irradiated cells may suggest a role of the cellular polymerase during the infection, there is no direct evidence that this enzyme is involved in the transcription of late poxviral genes (Moss, 1996).

Genes Involved in Transcription

Both vaccinia virus and ASFV encode a multi-subunit RNA polymerase (Yáñez *et al.*, 1993a, 1995), as well as a guanylyltransferase for the synthesis of the 5' CAP structure (Pena *et al.*, 1993). A vaccinia virus poly A polymerase gene has been identified and characterized (Gershon *et al.*, 1991; Gershon and Moss, 1992, 1994), but a homologous gene has not been found in the ASFV genome, although the viral RNAs are polyadenylated. As already mentioned, both viruses encode proteins homologous to the cellular TFSII transcription factor and, in the case of vaccinia virus, genes coding for additional factors specific for the different stages of transcription have been identified (Moss, 1996). The iridoviruses encode polypeptides related to the largest and second largest subunits of eukaryotic RNA polymerase II, as well as to one of the small subunits of this enzyme, but there is no evidence for the existence of enzymes involved in RNA modification (Schnitzler *et al.*, 1994; Tidona and Darai, 1997). In addition, lymphocystis disease virus codes for homologs of early and late transcription factors of poxviruses, although the role of these proteins in transcription has not been addressed (Tidona and Darai, 1997).

Virion Structure and Morphogenesis

The ASFV particle, with an icosahedral morphology and a diameter of 200 nm, consists of a DNA containing nucleoid, a thick protein layer designated the core shell, a lipid envelope surrounding this shell and the capsid (Andrés *et al.*, 1997). The extracellular viruses possess an outer envelope, which is acquired by budding through the plasma membrane (Breese and DeBoer, 1966). This morphology is very similar to that of the iridoviruses that infect vertebrates (Carrascosa *et al.*, 1984; Goorha and Granoff, 1994), which also acquire an envelope by budding at the cell membrane (Goorha and Granoff, 1994). ASFV morphogenesis is a complex process that occurs in cytoplasmic areas termed viral factories. The virus particles assemble from membranous structures derived from the endoplasmic reticulum (Rouiller *et al.*, 1998; Andrés *et al.*, 1998), which become polyhedral immature virions by the progressive formation of the capsid on their convex surface (Andrés *et al.*, 1997; García-Escudero *et al.*, 1998). Underneath the lipid envelope, the core shell and the DNA-containing nucleoid are consecutively formed. Several structural proteins that compose the core shell are produced by proteolytic processing of polyprotein precursors (Simón-Mateo *et al.*, 1993, 1997; Andrés *et al.*, 1997). Interestingly, ASFV protein p72, which is the major component of the capsid, has homology with the capsid protein of iridoviruses (Schnitzler and Darai, 1993). On the other hand, neither the ASFV nor the iridovirus particles contain glycoproteins (Elliot and Kelly, 1980; del Val *et al.*, 1986; Goorha and Granoff, 1994). Although few data are available at present on the morphogenesis of the iridoviruses, it is likely that their assembly pathway might be similar to that of ASFV, given the structural resemblance between them.

The morphology of the poxviruses, which has been described above, is quite different to that of ASFV and iridoviruses. It is interesting, however, to mention that the precursor viral membranes are also derived from the endoplasmic reticulum (Sodeik *et al.*, 1993), and that several major structural proteins of the core are originated by proteolytic processing of precursors (Moss and Rosenblum, 1973), although these precursors are polypepides and not polyproteins, as in the case of ASFV. On the other hand, externalization of the poxvirus particles, in contrast to ASFV and iridoviruses, does not occur by a budding process but through fusion of the enveloped particles with the cell membrane.

Virus–Host Interactions

Many viruses have developed strategies to evade the inflammatory and immunological responses of the host to the infection. Cytokines, such as interleukin-1, TNF and interferon, regulate these cellular responses and may be antiviral or destroy the virus-infected cell. The poxviruses and ASFV have evolved mechanisms to avoid cytokine action,

although the strategies used by these viruses appear to be different. Thus, the poxviruses codify for homologs of individual cytokine receptors (Alcamí and Smith, 1992, 1995; Spriggs *et al.*, 1992; Palumbo *et al.*, 1994), while ASFV contains an IκB-like gene to inhibit the activity of the nuclear factor NFκB, which controls the expression of cytokine genes involved in the inflammatory and immune responses (Powell *et al.*, 1996; Revilla *et al.*, 1998). In the case of the iridoviruses, the only gene identified so far that might be involved in the modulation of the host immune response is the lymphocystis disease virus gene homologous to the TNF receptor (Tidona and Darai, 1997). An alternative strategy used by poxviruses to block interleukin-1 action is based on the acquisition of a gene coding for a protein called CrmA, which is homologous to members of the serine protease inhibitor (SERPIN) superfamily and that prevents the intracellular conversion of the inactive precursor of interleukin-1 to the active form (Pickup *et al.*, 1986; Palumbo *et al.*, 1989; Ray *et al.*, 1992).

ASFV has also developed a mechanism to block the action of cytotoxic T lymphocytes on virus-infected cells. The virus codifies for a homolog of the T-cell adhesion molecule CD2, a plasma membrane protein that mediates the interaction between the lymphocytes and the infected cells. The viral homolog could block this interaction by sequestering the CD2 ligand in the infected cell (Rodríguez *et al.*, 1993a).

Programmed cell death or apoptosis is a host-cell response against virus infection that would prevent the production and spread of viral progeny. Many viruses have developed mechanisms to inhibit the induction of apoptosis during the infection cycle. Thus, ASFV encodes homologs of the cellular proto-oncogen Bcl-2 (Neilan *et al.*, 1993; Afonso *et al.*, 1996; Brun *et al.*, 1996; Revilla *et al.*, 1997) and of the inhibitor of apoptosis (IAP) protein (Chacón *et al.*, 1995), while the CrmA protein of poxviruses is also able to prevent the apoptosis of the infected cells. As in the case of ASFV, the iridovirus *Chilo* iridescent virus has been shown to encode a protein homologous to IAPs (Birnbaum *et al.*, 1994).

EVOLUTIONARY RELATIONSHIP OF ASFV WITH POXVIRUSES AND IRIDOVIRUSES

The classification of viruses into families is based mainly on certain properties such as virion morphology, genome structure, strategies of gene expression and site of replication. It is thought that viruses that share the same properties, therefore belonging to the same family, are evolutionarily related, a suggestion that is strengthened by the finding of sequence homology of their genomes. However, in some cases, viruses belonging to different families show certain similarities in the strategy of replication and even in genes encoding proteins with the same function, which may suggest a phylogenetic relationship between these families.

As just mentioned, one of the criteria used in virus taxonomy is the site of virus multiplication and assembly. Most animal DNA viruses replicate either in the nucleus or in the cytoplasm of the infected cell. The iridoviruses are an exception to this and replicate in both the nucleus and the cytoplasm. Because of this characteristic, the iridoviruses have been considered as intermediate forms in the evolution of nuclear and cytoplasmic DNA viruses (Goorha and Granoff, 1994). ASFV, on the other hand, also requires the cell nucleus to initiate the replication of DNA but, in contrast to the iridoviruses, is not dependent on the nuclear transcriptional machinery, since the virion contains all the enzymes necessary for early mRNA synthesis. In this feature, ASFV is similar to the poxviruses, which are the only known DNA viruses that multiply exclusively in the cytoplasm. In relation to the evolution of the nuclear and cytoplasmic DNA viruses, it is also interesting to mention that the iridoviruses have a genome structure and replication mechanisms similar to certain DNA phages, and also possess, as do bacteria, a restriction-modification system. Thus, the iridoviruses may have diverged before ASFV and the poxviruses, while ASFV would occupy, in the evolutionary tree of cytoplasmic DNA viruses, an intermediate stage between the iridoviruses and the poxviruses. On the other hand, and in relation to the evolutionary history of ASFV, it should be mentioned that the phylogenetic analysis of the viral topoisomerase II gene indicates that the ASFV gene was diverged from an ancestral cellular gene before the divergence of protozoa, yeasts and metazoa, implying the existence of an ancestor of ASFV at least as old as the appearance of lower eukaryotes (García-Beato et al., 1992a).

Is ASFV, therefore, a missing link between the poxviruses and the iridoviruses? When comparing the properties of these viruses shown in Figure 18.2, it can be noticed that, while poxviruses and iridoviruses have unique properties not shared by the other families (continuous line boxes), ASFV shares its properties with either the iridoviruses or the poxviruses (dashed line boxes). The morphological resemblance between ASFV and the iridoviruses and the fact that the capsid proteins of both viruses are homologous strongly suggests the existence of an evolutionary relationship between these viruses. On the other hand, the striking similarities between ASFV and the poxviruses in genome structure and in the strategies used for gene expression also point to a relationship of ASFV with the poxvirus family. Altogether, ASFV could be viewed as a poxvirus "disguised" as an iridovirus, a view that underlines the likely position of ASFV as a link between the poxvirus and iridovirus families.

In relation to the evolutive history of ASFV and the poxviruses, an interesting finding is that the hairpin loop termini of ASFV are very similar in size, sequence and structure to those of the linear plasmids of spirochetes of the genus Borrelia, the 37-nucleotide sequence of the ASFV hairpin being 67% identical to the 38-nucleotide loop of Borrelia (Hinnebusch and Barbour, 1991). Likewise, the hairpins of the poxviruses show some sequence similarity with those of the Borrelia plasmids. This telomeric similarity raises the possibility of a common ancestry of ASFV, the poxviruses and the Borrelia plasmids. Alternatively – and since ASFV and one species of Borrelia, B. duttoni, share the same arthropod vector in Africa, Ornithodoros moubata – it is also possible that the plasmid of B. duttoni originated through a horizontal genetic transfer from ASFV. Such a genetic transfer could also have occurred between ASFV and poxvirus ancestors.

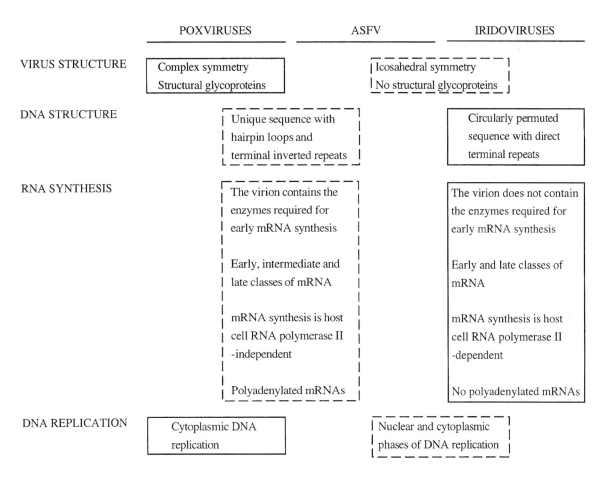

FIGURE 18.2 Comparative properties of poxviruses, ASFV and iridoviruses. Properties unique to the poxviruses or the iridoviruses are enclosed in continuous-line boxes and properties shared by ASFV with either the poxviruses or the iridoviruses are enclosed in dashed-line boxes.

A comprehensive comparison of the DNA sequences now available for ASFV, several poxviruses and the iridovirus lymphocystis disease virus may allow us to establish more precisely the evolutionary relationships between these viruses. So far, these comparisons have been limited to a few genes. The analysis of one of them, coding for the thymidine kinase, reveals that the lymphocystis disease virus gene diverged before those of ASFV and the poxviruses, while the ASFV enzyme is more distantly related than the poxvirus thymidine kinase to their cellular homologs (Blasco *et al.*, 1990; Schnitzler *et al.*,

1991). These data are in accordance with the proposed placing of the iridoviruses at the bottom of the evolutionary tree of cytoplasmic DNA viruses as well as with the intermediate position of ASFV in this tree. It should be noted, however, that, although these viruses may have a common ancestor and share certain properties related to the replication cycle, they have further diverged to acquire different sets of genes for the modulation of the host response to infection, developing different mechanisms to avoid the host immune response and to prevent the induction of apoptosis during infection.

REFERENCES

Afonso, C.L., Neilan, J.G., Kutish, G.F. and Rock, D.L. (1996) An African swine fever virus Bcl-2 homolog 5HL suppresses apoptotic cell death. *J. Virol.*, **70**, 4858–4863.

Alcamí, A. and Smith, G.L. (1992) A soluble receptor for interleukin-1β encoded by vaccinia virus – a novel mechanism of virus modulation of the host response to infection. *Cell*, **71**, 153–167.

Alcamí, A. and Smith, G.L. (1995) Cytokine receptors encoded by poxviruses: a lesson in cytokine biology. *Immunol. Today*, **16**, 474–478.

Almazán, F., Rodríguez, J.M., Andrés, G., Pérez, R., Viñuela, E. and Rodríguez, J.F. (1992) Transcriptional analysis of multigene family 110 of African swine fever virus. *J. Virol.*, **66**, 6655–6667.

Almazán, F., Rodríguez, J.M., Angulo, A., Viñuela, E. and Rodríguez, J.F. (1993) Transcriptional mapping of a late gene coding for the p12 attachment protein of African swine fever virus. *J. Virol.*, **67**, 553–556.

Andrés, G., Simón-Mateo, C., and Viñuela, E. (1997) Assembly of African swine fever virus: role of polyprotein pp220. *J. Virol.*, **71**, 2331–2341.

Andrés, G., García-Escudero, R., Simón-Mateo, C. and Viñuela, E. (1998) African swine fever virus is enveloped by a two-membraned collapsed cisterna derived from the endoplasmic reticulum. *J. Virol.*, **72**, 8988–9001.

Baldick, C.J. Jr and Moss, B. (1993) Characterization and temporal regulation of mRNAs encoded by vaccinia virus intermediate-stage genes. *J. Virol.*, **67**, 3515–3527.

Baroudy, B.M., Venkatesam, S., and Moss, B. (1982) Structure and replication of vaccinia virus telomeres. Cold Spring Harbor Symp. *Quant. Biol.*, **47**, 723–729.

Baylis, S.A., Banham, A.H., Vydelingum, S., Dixon, L.K. and Smith, G.L. (1993) African swine fever virus encodes a serine protein kinase which is packaged into virions. *J. Virol.*, **67**, 4549–4556.

Birnbaum, M.J., Clem, R.J. and Miller, L.K. (1994) An apoptosis-inhibiting gene from a nuclear polyhedrosis virus encoding a polypeptide with Cys/His sequence motifs. *J. Virol.*, **68**, 2521–2528.

Blasco, R., López-Otín, C., Muñoz, M., Bockamp, E.-O., Simón-Matero, C. and Viñuela, E. (1990) Sequence and evolutionary relationships of African swine fever virus thymidine kinase. *Virology*, **178**, 301–304.

Boursnell, M., Shaw, K., Yáñez, R.J., Viñuela, E. and Dixon, L. (1991) The sequences of ribonucleotide reductase genes from African swine fever virus show considerable homology with those of the Orthopoxvirus, vaccinia virus. *Virology*, **184**, 411–416.

Breese, S.S. Jr and DeBoer, C.J. (1966) Electron microscope observation of African swine fever virus in tissue culture cells. *Virology*, **28**, 420–428.

Brun, A., Rivas, C., Esteban, M., Escribano, J.M. and Alonso, C. (1996) African swine fever virus gene *A179L*, a viral homolog of *bcl2*, protects cells from programmed cell death. *Virology*, **225**, 227–230.

Caeiro, F., Meireles, M., Ribeiro, G. and Costa, J.V. (1990) *In vitro* DNA replication by cytoplasmic extracts from cells infected with African swine fever virus. *Virology*, **179**, 87–94.

Carrascosa, J.L., Carazo, J.M., Carrascosa, A.L., García, N., Santisteban, A. and Viñuela, E. (1984) General morphology and capsid fine structure of African swine fever virus. *Virology*, **132**, 160–172.

Chacón, M.R., Almazán, F., Nogal, M.L., Viñuela, E. and Rodriguez, J.F. (1995) The African swine fever virus IAP homolog is a late structural polypeptide. *Virology*, **214**, 670–674.

Colinas, R.J., Goebel, S.J., Davis, S.W., Johnson, G.P., Norton, E.K. and Paoletti, E. (1990) A DNA ligase gene in the Copenhagen strain of vaccinia virus is nonessential for viral replication and recombination. *Virology*, **179**, 267–275.

Cooper, J.A., Wittek, R., and Moss, B. (1981) Extension of the transcriptional and translational map of the left end of the vaccinia virus genome to 21 kilobase pairs. *J. Virol.*, **39**, 733–745.

Costa, J.V. (1990) African swine fever virus. In:

Molecular Biology of Iridoviruses (ed. Darai, G.), pp. 247–270. Kluwer Academic, Dordrecht.

Darai, G., Anders, K., Delius, H.G., Gelderblom, H., Samalecos, C. and Flügel, R.M. (1983) Analysis of the genome of fish lymphocystis disease virus isolated directly from epidermal tumours of pleuronectes. *Virology*, **126**, 466–479.

Del Val, M., Carrascosa, J.L. and Viñuela, E. (1986) Glycosylated components of African swine fever virus particles. *Virology*, **152**, 39–49.

De la Vega, I., González, A., Blasco, R., Calvo, V. and Viñuela, E. (1994) Nucleotide sequence and variability of the inverted terminal repetitions of African swine fever virus. *Virology*, **201**, 152–156.

Delius, H., Darai, G., and Flügel, R.M. (1984) DNA analysis of insect iridescent virus 6: evidence for circular permutation and terminal redundancy. *J. Virol.*, **49**, 609–614.

Earl, P.L., Jones, E.V., and Moss, B. (1986) Homology between DNA polymerases of poxviruses, herpesviruses, and adenoviruses: nucleotide sequence of the vaccinia virus DNA polymerase gene. *Proc. Natl Acad. Sci. USA*, **83**, 3659–3663.

Elliot, R.M., and Kelly, D.C. (1980) Frog virus 3 replication: induction and intracellular distribution of polypeptides in infected cells. *J. Virol.*, **33**, 28–51.

Fenner, F. (1996) Poxviruses. In: *Fields' Virology* (eds Fields, B., Knipe, D.M. and Howley, P.M.) pp. 2673–2702. Lippincott-Raven, Philadelphia, PA.

García-Beato, R., Freije, J.M.P., López-Otín, C., Blasco, R., Viñuela, E. and Salas, M. L. (1992a) A gene homologous to topoisomerase II in African swine fever virus. *Virology*, **188**, 938–947.

García-Beato, R., Salas, M.L., Viñuela, E. and Salas, J. (1992b) Role of the host cell nucleus in the replication of African swine fever virus. *Virology*, **188**, 637–649.

García-Escudero, R., Andrés, G., Almazán, F. and Viñuela, E. (1998) Inducible gene expression from African swine fever virus recombinants: analysis of the major capsid protein p 72. *J. Virol.*, **72**, 3185–3195.

Gershon, P.D., and Moss, B. (1992) Transition from rapid processive to slow non-processive polyadenylation by vaccinia virus poly A polymerase catalytic subunit is regulated by the net length of the poly A tail. *Genes Devel.*, **6**, 1568–1575.

Gershon, P.D., and Moss, B. (1994) Uridylate-containing RNA sequences determine the specificity for binding and polyadenylation by the catalytic subunit of vaccinia virus poly A polymerase. *EMBO J.*, **12**, 4705–4714.

Gershon, P.D., Ahn, B.Y., Garfield, M. and Moss, B. (1991) Poly A polymerase and a dissociable polyadenylation stimulatory factor encoded by vaccinia virus. *Cell*, **66**, 1269–1278.

Goebel, S.J., Johnson, G.P., Perkus, M.E., Davis, S.W., Winslow, J.P. and Paoletti, E. (1990) The complete DNA sequence of vaccinia virus. *Virology*, **179**, 247–266.

González, A., Talavera, A., Almendral, J.M. and Viñuela, E. (1986) Hairpin loop structure of African swine fever virus DNA. *Nucl. Acids Res.*, **14**, 6835–6844.

Goorha, R. (1981) Frog virus 3 requires RNA polymerase II for its replication. *J. Virol.*, **37**, 496–499.

Goorha, R. (1982) Frog virus 3 DNA replication occurs in two stages. *J. Virol.*, **43**, 519–528.

Goorha, R. and Granoff, A. (1994) Frog virus 3. In: *Encyclopedia of Virology* (eds Webster, R.G. and Granoff A.), pp. 503–508. Academic Press, London.

Goorha, R. and Murti, K.G. (1982) The genome of FV3, an animal DNA virus is circularly permuted and terminally redundant. *Proc. Natl Acad. Sci. USA*, **79**, 248–252.

Hammond, J.M., Kerr, S.M., Smith, G.L. and Dixon, L.K. (1992) An African swine fever virus gene with homology to DNA ligases. *Nucl. Acids Res.*, **20**, 2667–2671.

Hinnebusch, J. and Barbour, G. (1991) Linear plasmids of *Borrelia burgdorferi* have a telomeric structure and sequence similar to those of eukaryotic virus. *J. Bacteriol.*, **173**, 7233–7239.

Hruby, D.E. and Ball, L.A. (1982) Mapping and identification of the vaccinia virus thymidine kinase gene. *J. Virol.*, **43**, 403–409.

Kaur, K., Rohozinski, J. and Goorha, R. (1995) Identification and characterization of frog virus 3 DNA methyltransferase gene. *J. Gen. Virol.*, **76**, 1937–1943.

Kuznar, J., Salas, M.L. and Viñuela, E. (1980) DNA-dependent RNA polymerase in African swine fever virus. *Virology*, **101**, 169–175.

Kuznar, J., Salas, M.L. and Viñuela, E. (1981) Nucleoside triphosphate phosphohydrolase activities in African swine fever virus. *Arch. Virol.*, **96**, 307–310.

Mahr, A. and Roberts, B.E. (1984) Arrangement of late RNAs transcribed from a 7.1 kilobase *Eco* RI vaccinia virus DNA fragment. *J. Virol.*, **49**, 510–520.

Martín-Hernández, A.M. and Tabarés, E. (1991) Expression and characterization of the thymidine kinase gene of African swine fever virus. *J. Virol.*, **65**, 1046–1052.

Massung, R.F., Liu, L.-I., Qi, J. *et al.* (1994) Analysis of the complete genome of smallpox variola major virus strain Bangladesh-1975. *Virology*, **201**, 215–240.

Meireles, M., and Costa, J.V. (1994) Nucleotide sequence of the telomeric region of the African swine fever virus genome. *Virology*, **203**, 193–196.

Moss, B. (1990) Regulation of vaccinia virus transcription. *Annu. Rev. Biochem.*, **54**, 661–688.

Moss, B. (1996) Poxviridae. The viruses and their replication. In: *Fields' Virology* (eds Fields, B., Knipe, D.M. and Howley, P.M.), pp. 2637–2672. Lippincott-Raven, Philadelphia, PA.

Moss, B., and Rosenblum, E.N. (1973) Protein cleavage and poxvirus morphogenesis: tryptic peptide analysis of core precursors accumulated by blocking assembly with rifampicin. *J. Mol. Biol.*, **81**, 267–269.

Moyer, R.W. and Graves, R.L. (1981) The mechanism of cytoplasmic orthopoxvirus DNA replication. *Cell*, **27**, 391–401.

Neilan, J.G., Lu, Z., Afonso, C.L., Kutish, G.F., Sussman, M.D. and Rock, D.L. (1993) An African swine fever virus gene with similarity to the proto-oncogen *bcl-2* and the Epstein–Barr virus gene *BHFR1*. *J. Virol.*, **67**, 4391–4394.

Oliveros, M., Yáñez, R.J., Salas, M.L., Salas, J., Viñuela, E., and Blanco, L. (1997) Characterization of an African swine fever virus 20-kDa DNA polymerase involved in DNA repair. *J. Biol. Chem.*, **272**, 30899–30910.

Ortín, J. and Viñuela, E. (1977) Requirement of cell nucleus for African swine fever virus replication in VERO cells. *J. Virol.*, **21**, 902–905.

Palumbo, G., Pickup, D.J., Frederickson, T.N., McIntyre, L.J. and Buller, R.M.L. (1989) Inhibition of an inflammatory response is mediated by 38-kDa protein of cowpox virus. *Virology*, **172**, 262–273.

Palumbo, G.J., Buller, R., and Glasgow, W.C. (1994) Multigenic evasion of inflammation by poxviruses. *J. Virol.*, **68**, 1737–1749.

Pena, L., Yáñez, R.J., Viñuela, E., and Salas, M.L. (1993) African swine fever virus guanylyl-transferase. *Virology*, **193**, 319–328.

Pickup, D.J., Ink, B., Hu, W., Ray, C.A. and Joklik, W.K. (1986) Hemorrhage in lesions caused by cowpox virus is induced by a viral protein that is related to plasma protein inhibitors of serine proteases. *Proc. Natl Acad. Sci. USA*, **83**, 7698–7702.

Polatnick, J., Pan, I.C. and Gravell, M. (1974) Protein kinase activity in African swine fever virus. *Arch. Gesamte Virusforsch.*, **44**, 156–159.

Powell, P.P., Dixon, L.K. and Parkhouse, R.M. (1996) An IκB homolog encoded by African swine fever virus provides a novel mechanism for downregulation of proinflammatory cytokine response in host macrophages. *J. Virol.*, **70**, 8527–8533.

Ray, C.A., Black, R.A., Kronheim, S.R. *et al.* (1992) Viral inhibition of inflammation-cowpox virus encodes an inhibitor of interleukin-1β converting enzyme. *Cell*, **69**, 597–604.

Revilla, Y., Cebrián, A., Baixerás, E., Martínez-A., C., Viñuela, E. and Salas, M.L. (1997) Inhibition of apoptosis by the African swine fever virus *Bcl-2* homologue: role of the BH1 domain. *Virology*, **228**, 400–404.

Revilla, Y., Callejo, M., Rodríguez, J.M. *et al.* (1998) Inhibition of nuclear factor κB activation by a virus-encoded IκB-like protein. *J. Biol. Chem.*, **273**, 5405–5411.

Rodríguez, J.M., Salas, M.L. and Viñuela, E. (1992) Genes homologous to ubiquitin-conjugating proteins and eukaryotic transcription factor SII in African swine fever virus. *Virology*, **186**, 40–52.

Rodríguez, J.M., Yáñez, R.J., Almazán, F., Viñuela, E. and Rodriguez, J.F. (1993a)

African swine fever virus encodes a CD2 homolog responsible for the adhesion of erythrocytes to infected cells. *J. Virol.*, **67**, 5312–5320.

Rodríguez, J.M., Yáñez, R.J., Rodriguez, J.F., Viñuela, E. and Salas, M.L. (1993b) The DNA polymerase-encoding gene of African swine fever virus: sequence and transcriptional analysis. *Gene*, **136**, 103–110.

Rodríguez, J.M., Salas, M.L. and Viñuela, E. (1996) Intermediate class of mRNAs in African swine fever virus. *J. Virol.*, **70**, 8584–8589.

Rosales, R., Harris, N., Ahn, B.-Y. and Moss, B. (1994) Purification and identification of a vaccinia virus-encoded intermediate stage promoter-specific transcription factor that has homology to eukaryotic transcription factor SII (TFIIS) and an additional role as a viral RNA polymerase subunit. *J. Biol. Chem.*, **269**, 14260–14267.

Rouiller, I., Brookes, S.M., Hyatt, A.D., Windsor, M. and Wileman, T. (1998) African swine fever virus is wrapped by the endoplasmic reticulum. *J. Virol.*, **72**, 2373–2387.

Salas, M.L., Kuznar, J. and Viñuela, E. (1981) Polyadenylation, methylation and capping of the RNA synthesized *in vitro* by African swine fever virus. *Virology*, **113**, 484–491.

Salas, M.L., Kuznar, J. and Viñuela, E. (1983) Effect of rifamycin derivatives and coumermycin Al on *in vitro* RNA synthesis by African swine fever virus. *Arch. Virol.*, **77**, 77–80.

Salas, M.L., Rey-Campos, J., Almendral, J.M., Talavera, A. and Viñuela, E. (1986) Transcription and translation maps of African swine fever virus. *Virology*, **152**, 228–240.

Salas, M.L., Salas, J. and Viñuela, E. (1988a) Phosphorylation of African swine fever virus proteins *in vitro* and *in vivo*. *Biochimie*, **70**, 627–635.

Salas, J., Salas, M.L. and Viñuela, E. (1988b) Effect of inhibitors of the host cell RNA polymerase II on African swine fever virus multiplication. *Virology*, **164**, 280–283.

Schnitzler, P. and Darai, G. (1993) Identification of the gene encoding the major capsid protein of fish lymphocystis disease virus. *J. Gen. Virol.*, **74**, 2143–2150.

Schnitzler, P., Handermann, M., Szépe, O. and Darai, G. (1991) The primary structure of the thymidine kinase gene of fish lymphocystis disease virus. *Virology*, **182**, 835–840.

Schnitzler, P., Sonntag, K.-C., Müller, M. *et al.* (1994) Insect iridescent virus type 6 encodes a polypeptide related to the largest subunit of eukaryotic RNA polymerase II. *J. Gen. Virol.*, **75**, 1557–1567.

Senkevich, T., Bugert, J.J., Sisler, J.R., Koonin, E.V., Darai, G. and Moss, B. (1996) Genome sequence of a human tumorigenic poxvirus: prediction of specific host response-evasion genes. *Science*, **273**, 813–816.

Shuman, S. and Moss, B. (1987) Identification of a vaccinia virus gene encoding a type I DNA topoisomerase. *Proc. Natl Acad. Sci. USA*, **84**, 7478–7482.

Simón-Mateo, C., Andrés, G. and Viñela, E. (1993) Polyprotein processing in African swine fever virus: a novel gene expression strategy for a DNA virus. *EMBO J.*, **12**, 2977–2987.

Simón-Mateo, C., Andrés, G., Almazán, F. and Viñuela, E. (1997) Proteolytic processing in African swine fever virus: evidence for a new structural polyprotein, pp62. *J. Virol.*, **71**, 5799–5804.

Slabaugh, M.B., Roseman, N., Davis, R. and Mathews, C. (1988) Vaccinia virus-encoded ribonucleotide reductase: sequence conservation of the gene for the small subunit and its amplification in hydroxyurea-resistant mutants. *J. Virol.*, **62**, 519–527.

Smith, G.L., de Carlos, A. and Chan, Y.S. (1989) Vaccinia virus encodes a thymidylate kinase gene: sequence and transcriptional mapping. *Nucl. Acids Res.*, **17**, 7581–7590.

Sodeik, B., Doms, R.W., Ericsson, M. *et al.* (1993) Assembly of vaccinia virus: role of the intermediate compartment between the endoplasmic reticulum and the Golgi stacks. *J. Cell Biol.*, **121**, 521–541.

Spriggs, M.K., Hruby, D.E., Maliszewski, C.R. *et al.* (1992) Vaccinia and cowpox viruses encode a novel secreted interleukin-1-binding protein. *Cell*, **71**, 145–152.

Tengelsen, L.A., Slabaugh, M.B., Bibler, J.K. and Hruby, D.E. (1988) Nucleotide sequence and molecular genetic analysis of the large subunit of ribonucleotide reductase encoded by vaccinia virus. *Virology*, **164**, 121–131.

Tidona, C.A. and Darai, G. (1997) The complete DNA sequence of lymphocystis disease virus. *Virology*, **230**, 207–216.

Viñuela, E. (1985) African swine fever virus. *Curr. Topics Microbiol. Immunol.*, **116**, 151–170.

Vos, J.C. and Stunnenberg, H.G. (1988) Derepression of a novel class of vaccinia virus genes upon DNA replication. *EMBO J.*, **7**, 3487–3492.

Wagner, H., Simon, D., Werner, E., Gelderblom, H., Darai, G. and Flügel, R.M. (1985) Methylation pattern of DNA of fish lymphocystis disease virus. *J. Virol.*, **53**, 1005–1007.

Ward, V.K. and Kalmakoff, J. (1987) Physical mapping of the DNA genome of insect iridescent virus type 9 from *Wiseana* spp. larvae. *Virology*, **160**, 507–510.

Weir, J.P., Bajszar, G. and Moss, B. (1982) Mapping of the vaccinia virus thymidine kinase gene by marker rescue and by cell-free translation of selected mRNA. *Proc. Natl Acad. Sci. USA*, **79**, 1210–1214.

Willis, D.B. (1990) Taxonomy of iridoviruses. In: *Molecular Biology of Iridoviruses* (ed. Darai, G.), pp. 1–12. Kluwer Academic, Dordrecht.

Willis, D.B. and Granoff, A. (1980) FV3 DNA is heavily methylated at CpG sequences. *Virology*, **107**, 250–257.

Willis, D.B., Thompson, J.P. and Beckman, W. (1990) Transcription of frog virus 3. In: *Molecular Biology of Iridoviruses* (ed. Darai, G.), pp. 173–186. Kluwer Academic, Dordrecht.

Yáñez, R.J. and Viñuela, E. (1993) African swine fever virus encodes a DNA ligase. *Virology*, **193**, 531–536.

Yáñez, R.J., Boursnell, M., Nogal, M.L., Yuste, L. and Viñuela, E. (1993a) African swine fever virus encodes two genes which share significant homology with the two largest subunits of DNA-dependent RNA polymerase. *Nucl. Acids Res.*, **21**, 2423–2427.

Yáñez, R.J., Rodríguez, J.M., Rodríguez, J.F., Salas, M.L. and Viñuela, E. (1993b) African swine fever virus thymidylate kinase gene: sequence and transcriptional mapping. *J. Gen. Virol.*, **74**, 1633–1638.

Yáñez, R.J., Rodríguez, J.M., Nogal, M.L. *et al.* (1995) Analysis of the complete sequence of African swine fever virus. *Virology*, **208**, 249–278.

Yuen, L., and Moss, B. (1986) Multiple 3′ ends of mRNA encoding vaccinia virus growth factor occur within a series of repeated sequences downstream of T clusters. *J. Virol.*, **60**, 320–323.

Index